Lecture Notes in Computer Science 2142

Edited by G. Goos, J. Hartmanis and J. van Leeuwen

Springer
Berlin
Heidelberg
New York
Barcelona
Hong Kong
London
Milan
Paris
Tokyo

Laurent Fribourg (Ed.)

Computer Science Logic

15th International Workshop, CSL 2001
10th Annual Conference of the EACSL
Paris, France, September 10-13, 2001, Proceedings

Volume Editor

Laurent Fribourg
LSV (Ecole Normale Supérieure de Cachan and CNRS)
61, Avenue du Président Wilson, 94230 Cachan, France
E-mail: fribourg@lsv.ens-cachan.fr

Cataloging-in-Publication Data applied for

Die Deutsche Bibliothek - CIP-Einheitsaufnahme

Computer science logic : 15th international workshop ; proceedings /
CSL 2001, Paris, France, September 10 - 13, 2001. Laurent Fribourg (ed.). -
Berlin ; Heidelberg ; New York ; Barcelona ; Hong Kong ; London ; Milan ;
Paris ; Tokyo : Springer, 2001
(Annual Conference of the EACSL ... ; 2001)
(Lecture notes in computer science ; Vol. 2142)
ISBN 3-540-42554-3

CR Subject Classification (1998): F.4, I.2.3-4, F.3

ISSN 0302-9743
ISBN 3-540-42554-3 Springer-Verlag Berlin Heidelberg New York

Springer-Verlag Berlin Heidelberg New York
a member of BertelsmannSpringer Science+Business Media GmbH

http://www.springer.de

Typesetting: Camera-ready by author, data conversion by Olgun Computergrafik
Printed on acid-free paper SPIN 10840185 06/3142 5 4 3 2 1 0

Preface

The Annual Conference of the European Association for Computer Science Logic, CSL 2001, was held in Paris, Palais de la Mutualité, on September 10–13, 2001. This was the 15th in a series of annual meetings, originally intended as International Workshops on Computer Science Logic, and the 10th to be held as the Annual Conference of the EACSL. The conference was organized by Laboratoire Spécification et Vérification (CNRS & ENS Cachan).

The CSL 2001 program committee selected 39 of 91 submitted papers for presentation at the conference and publication in this proceedings volume. The submitted papers originated from 26 different countries. Each paper was refereed by at least three reviewers.

In addition to the contributed papers, the scientific program of CSL 2001 included three invited talks (Jean-Yves Girard, Peter O'Hearn, and Jan Van den Bussche). This volume includes the papers provided by the invited speakers as well as the selected contributed papers. The topics of the papers include: linear logic, descriptive complexity, semantics, higher-order programs, modal logics, verification, automata, λ-calculus, induction, equational calculus, and constructive theory of types.

I am most grateful to the members of the program committee and all the referees for their thorough work. I am also particularly indebted to François Laroussinie, helped by Patricia Bouyer, Nicolas Markey, and Philippe Schnoebelen, for the successful organization of this event. Special thanks to Emmanuel Fleury for the design of the beautiful "Notre-Dame de Paris" poster.

June 2001 Laurent Fribourg

Program Commitee

Andrea Asperti (U. Bologna)
Julian Bradfield (U. Edinburgh)
René David (U. Savoie)
Gilles Dowek (INRIA)
Laurent Fribourg (ENS Cachan) (**chair**)
Daniel Leivant (Indiana U.)
David McAllester (AT&T)
Johann Makowsky (Technion Haifa)
Aart Middeldorp (U. Tsukuba)
Catuscia Palamidessi (Penn. State)
Frank Pfenning (CMU)
Philippe Schnoebelen (ENS Cachan)
Iain Stewart (U. Leicester)
Moshe Vardi (Rice U.)
Philip Wadler (Avaya Labs)
Thomas Wilke (U. Kiel)

Referees

Andreas Abel
Thorsten Altenkirch
Arnon Avron
Steve Awodey
Philippe Balbiani
David Basin
Peter Baumgartner
Michael Benedikt
Ulrich Berger
Rudolf Berghammer
Didier Bert
Leopoldo Bertossi
Egon Boerger
Guillaume Bonfante
Olivier Bournez
Hubert Comon
Paolo Coppola
Tristan Crolard
Anuj Dawar
Jean-Paul Delahaye
Stéphane Demri
Joëlle Despeyroux
Mariangiola Dezani
Roberto Di Cosmo
Roy Dyckhoff
Zoltan Esik
Maribel Fernandez
Nissim Francez
Didier Galmiche
Ian Gent
Alfons Geser
Herman Geuvers
Neil Ghani
Eduardo Gimenez
Jean Goubault-Larrecq
Erich Graedel
Etienne Grandjean
Martin Grohe
Orna Grumberg
Stefano Guerrini
Joe Halpern
Michael Hanus
Hugo Herbelin
Tin Kam Ho
Josh Hodas
Ian Hodkinson
Doug Howe
Rick Hull
Florent Jacquemard
David Janin
Marcin Jurdzinski
Shmuel Katz
Claude Kirchner
Ralf Kuesters
Dietrich Kuske
Juliana Küster Filipe
Yves Lafont
Yassine Lakhnech
Arnaud Lallouet
Richard Lassaigne
Giacomo Lenzi
Jordi Levy
Leonid Libkin
Patrick Lincoln
John Longley
Denis Lugiez
Christoph Lüth
Christopher Lynch
Jim Lynch
Ian Mackie
Grant Malcolm
Julian Marino
Jean-Yves Marion
Simone Martini
Marteen Marx
Dale Miller
Alexandre Miquel
Cesar Munoz
Paliath Narendran
Joachim Niehren
Robert Nieuwenhuis
Ulf Nilsson
K. Nour
Luke Ong
Vincent van Oostrom
Fiedrich Otto
Martin Otto
Luca Padovani
Nicolas Peltier
Laure Petrucci
Claudine Picaronny
Brigitte Pientka
Andreas Podelski
Axel Poigné
Jeff Polakow
Randy Pollack

John Power
Christophe Raffalli
Femke van Raamsdonk
Alexander Razborov
Laurent Régnier
Mark Reynolds
Christophe Ringeissen
Giuseppe Rosolini
Paul Rozière
Claudio Sacerdoti Coen
Ildiko Sain
Irene Schena
Thomas Schwentick
Alex Simpson
Robert Staerk
Ludwig Staiger
Ian Stark
Colin Stirling
Howard Straubing
Aaron D. Stump
Taro Suzuki
Armando Tacchella
Jean-Marc Talbot
Sophie Tison
Konstantinos Tourlas
Ralf Treinen
Alasdair Urquhart
Pawel Urzyczyn
Tarmo Uustalu
Jan Van den Bussche
Kumar N. Verma
René Vestergaard
Luca Viganò
Laurent Vigneron
Igor Walukiewicz
Pascal Weil
Benjamin Werner

Local Organizing Committee

Patricia Bouyer
Laurent Fribourg
François Laroussinie
Nicolas Markey
Philippe Schnoebelen

Sponsors

CNRS
ENS Cachan
France Télécom
INRIA
LSV
Ministère de la Recherche

Table of Contents

Invited Papers

Contributed Papers

Descriptive Complexity

Semantics

Higher-Order Programs

Modal Logics

Verification

Automata

Lambda-Calculus

Induction

Equational Calculus

Constructive Theory of Types

Local Reasoning about Programs that Alter Data Structures

Peter O'Hearn[1], John Reynolds[2], and Hongseok Yang[3]

[1] Queen Mary, University of London
[2] Carnegie Mellon University
[3] University of Birmingham and University of Illinois at Urbana-Champaign

Abstract. We describe an extension of Hoare's logic for reasoning about programs that alter data structures. We consider a low-level storage model based on a heap with associated lookup, update, allocation and deallocation operations, and unrestricted address arithmetic. The assertion language is based on a possible worlds model of the logic of bunched implications, and includes spatial conjunction and implication connectives alongside those of classical logic. Heap operations are axiomatized using what we call the "small axioms", each of which mentions only those cells accessed by a particular command. Through these and a number of examples we show that the formalism supports local reasoning: A specification and proof can concentrate on only those cells in memory that a program accesses.
This paper builds on earlier work by Burstall, Reynolds, Ishtiaq and O'Hearn on reasoning about data structures.

1 Introduction

Pointers have been a persistent trouble area in program proving. The main difficulty is not one of finding an in-principle adequate axiomatization of pointer operations; rather there is a mismatch between simple intuitions about the way that pointer operations work and the complexity of their axiomatic treatments. For example, pointer assignment is operationally simple, but when there is aliasing, arising from several pointers to a given cell, then an alteration to that cell may affect the values of many syntactically unrelated expressions. (See [20, 2, 4, 6] for discussion and references to the literature on reasoning about pointers.)

We suggest that the source of this mismatch is the global view of state taken in most formalisms for reasoning about pointers. In contrast, programmers reason informally in a local way. Data structure algorithms typically work by applying local surgeries that rearrange small parts of a data structure, such as rotating a small part of a tree or inserting a node into a list. Informal reasoning usually concentrates on the effects of these surgeries, without picturing the entire memory of a system. We summarize this local reasoning viewpoint as follows.

> To understand how a program works, it should be possible for reasoning and specification to be confined to the cells that the program actually accesses. The value of any other cell will automatically remain unchanged.

L. Fribourg (Ed.): CSL 2001, LNCS 2142, pp. 1–19, 2001.

Local reasoning is intimately tied to the complexity of specifications. Often, a program works with a circumscribed collection of resources, and it stands to reason that a specification should concentrate on just those resources that a program accesses. For example, a program that inserts an element into a linked list need know only about the cells in that list; there is no need (intuitively) to keep track of all other cells in memory when reasoning about the program.

The central idea of the approach studied in this paper is of a "spatial conjunction" $P * Q$, that asserts that P and Q hold for separate parts of a data structure. The conjunction provides a way to compose assertions that refer to different areas of memory, while retaining disjointness information for each of the conjuncts. The locality that this provides can be seen both on the level of atomic heap assignments and the level of compound operations or procedures. When an alteration to a single heap cell affects P in $P * Q$, then we know that it will not affect Q; this gives us a way to short-circuit the need to check for potential aliases in Q. On a larger scale, a specification $\{P\}C\{Q\}$ of a heap surgery can be extended using a rule that lets us infer $\{P * R\}C\{Q * R\}$, which expresses that additional heap cells remain unaltered. This enables the initial specification $\{P\}C\{Q\}$ to concentrate on only the cells in the program's footprint.

The basic idea of the spatial conjunction is implicit in early work of Burstall [3]. It was explicitly described by Reynolds in lectures in the fall of 1999; then an intuitionistic logic based on this idea was discovered independently by Reynolds [20] and by Ishtiaq and O'Hearn [7] (who also introduced a spatial implication $P -\!\!* Q$, based on the logic BI of bunched implications [11, 17]). In addition, Ishtiaq and O'Hearn devised a classical version of the logic that is more expressive than the intuitionistic version. In particular, it can express storage deallocation.

Subsequently, Reynolds extended the classical version by adding pointer arithmetic. This extension results in a model that is simpler and more general than our previous models, and opens up the possibility of verifying a wider range of low-level programs, including many whose properties are difficult to capture using type systems. Meanwhile, O'Hearn fleshed out the theme of local reasoning sketched in [7], and he and Yang developed a streamlined presentation of the logic based on what we call the "small axioms".

In this joint paper we present the pointer arithmetic model and assertion language, with the streamlined Hoare logic. We illustrate the formalism using programs that work with a space-saving representation of doubly-linked lists, and a program that copies a tree.

Two points are worth stressing before continuing. First, by *local* we do not merely mean *compositional* reasoning: It is perfectly possible to be compositional and global (in the state) at the same time, as was the case in early denotational models of imperative languages. Second, some aspects of this work bear a strong similarity to semantic models of local state [19, 15, 16, 13, 12]. In particular, the conjunction $*$ is related to interpretations of syntactic control of interference [18, 10, 12], and the Frame Rule described in Section 3 was inspired by the idea of the expansion of a command from [19, 15]. Nevertheless, local reasoning about state is not the same thing as reasoning about local state: We are proposing here that specifications and reasoning themselves be kept confined, and this is an issue whether or not we consider programming facilities for *hiding* state.

2 The Model and Assertion Language

The model has two components, the store and the heap. The store is a finite partial function mapping from variables to integers. The heap is indexed by a subset `Locations` of the integers, and is accessed using indirect addressing $[E]$ where E is an arithmetic expression.

$$\mathtt{Ints} \triangleq \{..., -1, 0, 1, ...\} \qquad \mathtt{Variables} \triangleq \{x, y, ...\}$$

$$\mathtt{Atoms}, \mathtt{Locations} \subseteq \mathtt{Ints} \qquad \mathtt{Locations} \cap \mathtt{Atoms} = \{\},\ \mathtt{nil} \in \mathtt{Atoms}$$

$$\mathtt{Stores} \triangleq \mathtt{Variables} \rightharpoonup_{fin} \mathtt{Ints} \qquad \mathtt{Heaps} \triangleq \mathtt{Locations} \rightharpoonup_{fin} \mathtt{Ints}$$

$$\mathtt{States} \triangleq \mathtt{Stores} \times \mathtt{Heaps}$$

In order for allocation to always succeed, we place a requirement on the set `Locations`: For any positive integer n, there are infinitely many sequences of length n of consecutive integers in `Locations`. This requirement is satisfied if we take `Locations` to be the non-negative integers. (In several example formulae, we will implicitly rely on this choice.) Then we could take `Atoms` to be the negative integers, and `nil` to be -1.

Integer and boolean expressions are determined by valuations

$$[\![E]\!]s \in \mathtt{Ints} \qquad [\![B]\!]s \in \{true, false\}$$

where the domain of $s \in$ `Stores` includes the free variables of E or B. The grammars for expressions are as follows.

$$E, F, G ::= x, y, ... \mid 0 \mid 1 \mid E + F \mid E \times F \mid E - F$$

$$B ::= \mathtt{false} \mid B \Rightarrow B \mid E = F \mid E < F \mid \mathtt{isatom?}(E) \mid \mathtt{isloc?}(E)$$

The expressions $\mathtt{isatom?}(E)$ and $\mathtt{isloc?}(E)$ test whether E is an atom or location.

The assertions include all of the boolean expressions, the points-to relation $E \mapsto F$, all of classical logic, and the spatial connectives `emp`, $*$ and $-\!\!*$.

$$\begin{array}{rll} P, Q, R ::= & B \mid E \mapsto F & \text{Atomic Formulae} \\ \mid & \mathtt{false} \mid P \Rightarrow Q \mid \forall x. P & \text{Classical Logic} \\ \mid & \mathtt{emp} \mid P * Q \mid P -\!\!* Q & \text{Spatial Connectives} \end{array}$$

Various other connectives are defined as usual: $\neg P = P \Rightarrow \mathtt{false}$; $\mathtt{true} = \neg(\mathtt{false})$; $P \vee Q = (\neg P) \Rightarrow Q$; $P \wedge Q = \neg(\neg P \vee \neg Q)$; $\exists x.\ P = \neg \forall x. \neg P$.

We use the following notations in the semantics of assertions.

1. $dom(h)$ denotes the domain of definition of a heap $h \in$ `Heaps`, and $dom(s)$ is the domain of $s \in$ `Stores`;

2. $h\#h'$ indicates that the domains of h and h' are disjoint;
3. $h * h'$ denotes the union of disjoint heaps (i.e., the union of functions with disjoint domains);
4. $(f \mid i \mapsto j)$ is the partial function like f except that i goes to j. This notation is used both when i is and is not in the domain of f.

We define a satisfaction judgement $s, h \models P$ which says that an assertion holds for a given store and heap. (This assumes that $\mathrm{Free}(P) \subseteq dom(s)$, where $\mathrm{Free}(P)$ is the set of variables occurring freely in P.)

$$s, h \models B \quad \text{iff } [\![B]\!]s = \mathit{true}$$

$$s, h \models E \mapsto F \text{ iff } \{[\![E]\!]s\} = dom(h) \text{ and } h([\![E]\!]s) = [\![F]\!]s$$

$$s, h \models \mathtt{false} \quad \text{never}$$

$$s, h \models P \Rightarrow Q \text{ iff if } s, h \models P \text{ then } s, h \models Q$$

$$s, h \models \forall x.P \quad \text{iff } \forall v \in \mathtt{Ints}.\ [s \mid x \mapsto v], h \models P$$

$$s, h \models \mathtt{emp} \quad \text{iff } h = [\,] \text{ is the empty heap}$$

$$s, h \models P * Q \text{ iff } \exists h_0, h_1.\ h_0\#h_1,\ h_0 * h_1 = h,\ s, h_0 \models P \text{ and } s, h_1 \models Q$$

$$s, h \models P \mathrel{-\!\!*} Q \text{ iff } \forall h'.\text{ if } h'\#h \text{ and } s, h' \models P \text{ then } s, h * h' \models Q$$

Notice that the semantics of $E \mapsto F$ is "exact", where it is required that E is the only active address in the current heap. Using $*$ we can build up descriptions of larger heaps. For example, $(10 \mapsto 3) * (11 \mapsto 10)$ describes two adjacent cells whose contents are 3 and 10.

On the other hand, $E = F$ is completely heap independent (like all boolean and integer expressions). As a consequence, a conjunction $(E = F) * P$ is true just when $E = F$ holds in the current store and when P holds for the same store and some heap contained in the current one.

It will be convenient to have syntactic sugar for describing adjacent cells, and for an exact form of equality. We also have sugar for when E is an active address.

$$\begin{array}{lll} E \mapsto F_0, \ldots, F_n & \stackrel{\Delta}{=} & (E \mapsto F_0) * \cdots * (E + n \mapsto F_n) \\ E \doteq F & \stackrel{\Delta}{=} & (E = F) \wedge \mathtt{emp} \\ E \mapsto - & \stackrel{\Delta}{=} & \exists y. E \mapsto y \qquad\qquad (y \notin \mathrm{Free}(E)) \end{array}$$

A characteristic property of $\doteq$ is the way it interacts with $*$:

$$(E \doteq F) * P \quad \Leftrightarrow \quad (E = F) \wedge P.$$

As an example of adjacency, consider an "offset list", where the next node in a linked list is obtained by adding an offset to the position of the current node. Then the formula

$$(x \mapsto a, o) * (x + o \mapsto b, -o)$$

describes a two-element, circular, offset list that contains a and b in its head fields and offsets in its link fields. For example, in a store where $x = 17$ and $o = 25$, the formula is true of a heap

17	a		42	b
18	25		43	-25

The semantics in this section is a model of (the Boolean version of) the logic of bunched implications [11, 17]. This means that the model validates all the laws of classical logic, commutative monoid laws for `emp` and $*$, and the "parallel rule" for $*$ and "adjunction rules" for $-\!\!*$.

$$\frac{P \Rightarrow Q \quad R \Rightarrow S}{P * R \Rightarrow Q * S}$$

$$\frac{P * R \Rightarrow S}{P \Rightarrow R -\!\!* S} \qquad \frac{P \Rightarrow R -\!\!* S \quad Q \Rightarrow R}{P * Q \Rightarrow S}$$

Other facts, true in the specific model, include

$$\big((E \mapsto F) * (E' \mapsto F') * \mathtt{true}\big) \Rightarrow E \neq E' \qquad \mathtt{emp} \Leftrightarrow \forall x. \neg(x \mapsto - * \mathtt{true})$$

See [21] for a fuller list.

3 The Core System

In this section we present the core system, which consists of axioms for commands that alter the state as well as a number of inference rules. We will describe the meanings for the various commands informally, as each axiom is discussed.

There is one axiom for each of four atomic commands. We emphasize that the right-hand side of := is *not* an expression occurring in the forms $x := [E]$ and $x := \mathtt{cons}(E_1, ..., E_k)$; $[\cdot]$ and `cons` do not appear within expressions. Only $x := E$ is a traditional assignment, and it is the only atomic command that can be described by Hoare's assignment axiom. In the axioms x, m, n are assumed to be distinct variables.

THE SMALL AXIOMS

$$\{E \mapsto -\}\,[E] := F\,\{E \mapsto F\}$$

$$\{E \mapsto -\}\,\mathtt{dispose}(E)\,\{\mathtt{emp}\}$$

$$\{x \doteq m\}x := \mathtt{cons}(E_1, ..., E_k)\{x \mapsto E_1[m/x], ..., E_k[m/x]\}$$

$$\{x \doteq n\}\,x := E\,\{x \doteq (E[n/x])\}$$

$$\{E \mapsto n \wedge x = m\}\,x := [E]\,\{x = n \wedge E[m/x] \mapsto n\}$$

THE STRUCTURAL RULES

Frame Rule

$$\frac{\{P\}C\{Q\}}{\{P * R\}C\{Q * R\}} \; \text{Modifies}(C) \cap \text{Free}(R) = \{\}$$

Auxiliary Variable Elimination

$$\frac{\{P\}\, C\, \{Q\}}{\{\exists x.P\}\, C\, \{\exists x.Q\}} \; x \notin \text{Free}(C)$$

Variable Substitution

$$\frac{\{P\}\, C\, \{Q\}}{(\{P\}\, C\, \{Q\})[E_1/x_1, ..., E_k/x_k]} \; \begin{array}{l} \{x_1, ..., x_k\} \supseteq \text{Free}(P, C, Q), \text{ and} \\ x_i \in \text{Modifies}(C) \text{ implies} \\ E_i \text{ is a variable not free in any other } E_j \end{array}$$

Rule of Consequence

$$\frac{P' \Rightarrow P \quad \{P\}\, C\, \{Q\} \quad Q \Rightarrow Q'}{\{P'\}\, C\, \{Q'\}}$$

The first small axiom just says that if E points to something beforehand (so it is active), then it points to F afterwards, and it says this for a small portion of the state in which E is the only active cell. This corresponds to the operational idea of $[E] := F$ as a command that stores the value of F at address E in the heap. The axiom also implicitly says that the command does not alter any variables; this is covered by our definition of its Modifies set below.

The `dispose`(E) instruction deallocates the cell at address E. In the post-condition for the `dispose` axiom `emp` is a formula which says that the heap is empty (no addresses are active). So, the axiom states that if E is the sole active address and it is disposed, then in the resulting state there will be no active addresses. Here, the exact points-to relation is necessary, in order to be able to conclude `emp` on termination.

The $x :=$ `cons`$(E_1, ..., E_k)$ command allocates a contiguous segment of k cells, initialized to the values of $E_1, ..., E_k$, and places in x the address of the first cell from the segment. The precondition of the axiom uses the exact equality, which implies that the heap is empty. The axiom says that if we begin with the empty heap and a store where $x = m$, we will obtain k contiguous cells with appropriate values. The variable m in this axiom is used to record the value of x before the command is executed.

We only get fixed-length allocation from $x :=$ `cons`$(E_1, ..., E_k)$. It is also possible to formulate an axiom for a command $x :=$ `alloc`(E) that allocates a segment of length E; see [21].

We have also included small axioms for the other two commands, but they are less important. These commands are not traditionally as problematic, because they do not involve heap alteration.

The small axioms are so named because each mentions only the area of heap accessed by the corresponding command. For $[E] := F$ and $x := [E]$ this is one cell, in the axioms for `dispose` or `cons` precisely those cells allocated or deallocated are mentioned, and in $x := E$ no heap cells are accessed.

The notion of free variable referred to in the structural rules is the standard one. Modifies(C) is the set of variables that are assigned to within C. The Modifies set of each of $x := \mathtt{cons}(E_1, ..., E_k)$, $x := E$ and $x := [E]$ is $\{x\}$, while for $\mathtt{dispose}(E)$ and $[E] := F$ it is empty. Note that the Modifies set only tracks potential alterations to the store, and says nothing about the heap cells that might be modified.

In this paper we treat the Rule of Consequence semantically. That is, when the premisses $P' \Rightarrow P$ and $Q \Rightarrow Q'$ are true in the model for arbitrary store/heap pairs, we will use the rule without formally proving the premisses.

The Frame Rule codifies a notion of local behaviour. The idea is that the precondition in $\{P\}C\{Q\}$ specifies an area of storage, as well as a logical property, that is sufficient for C to run and (if it terminates) establish postcondition Q. If we start execution with a state that has additional heap cells, beyond those described by P, then the values of the additional cells will remain unaltered. We use $*$ to separate out these additional cells. The invariant assertion R is what McCarthy and Hayes called a "frame axiom" [9]. It describes cells that are not accessed, and hence not changed, by C.

As a warming-up example, using the Frame Rule we can prove that assigning to the first component of a binary cons cell does not affect the second component.

$$\cfrac{\cfrac{\{x \mapsto a\}\, [x] := b\, \{x \mapsto b\}}{\{(x \mapsto a) * (x+1 \mapsto c)\}\, [x] := b\, \{(x \mapsto b) * (x+1 \mapsto c)\}}\ \text{Frame}}{\{x \mapsto a, c\}\, [x] := b\, \{x \mapsto b, c\}}\ \text{Syntactic Sugar}$$

The overlap of free variables between $x + 1 \mapsto c$ and $[x] := b$ is allowed here because Modifies($[x] := b$) $= \{\}$.

4 Derived Laws

The small axioms are simple but not practical. Rather, they represent a kind of thought experiment, an extreme take on the idea that a specification can concentrate on just those cells that a program accesses.

In this section we show how the structural rules can be used to obtain a number of more convenient derived laws (most of which were taken as primitive in [20, 7]). Although we will not explicitly state a completeness result, along the way we will observe that weakest preconditions or strongest postconditions are derivable for each of the individual commands. This shows a sense in which nothing is missing in the core system, and justifies the claim that each small axiom gives enough information to understand how its command works.

We begin with $[E] := F$. If we consider an arbitrary invariant R then we obtain the following derived axiom using the Frame Rule with the small axiom as its premise.

$$\{(E \mapsto -) * R\}\,[E] := F\,\{(E \mapsto F) * R\}$$

This axiom expresses a kind of locality: Assignment to $[E]$ affects the heap cell at position E only, and so cannot affect the assertion R. In particular, there is no need to generate alias checks within R. With several more steps of Auxiliary Variable Elimination we can obtain an axiom that is essentially the one from [20]:

$$\{\exists x_1, \cdots, x_n. (E \mapsto -) * R\}\,[E] := F\,\{\exists x_1, \cdots, x_n. (E \mapsto F) * R\}$$

where $x_1, ..., x_n \notin \text{Free}(E, F)$.

For allocation, suppose $x \notin \text{Free}(E_1, ..., E_k)$. Then a simpler version of the small axiom is

$$\{\texttt{emp}\} x := \texttt{cons}(E_1, ..., E_k) \{x \mapsto E_1, ..., E_k\}$$

This can be derived using rules for auxiliary variables and Consequence. If, further, R is an assertion where $x \notin \text{Free}(R)$ then

$$\dfrac{\dfrac{\{\texttt{emp}\}\, x := \texttt{cons}(E_1, ..., E_k)\, \{x \mapsto E_1, ..., E_k\}}{\{\texttt{emp} * R\}\, x := \texttt{cons}(E_1, ..., E_k)\, \{(x \mapsto E_1, ..., E_k) * R\}}\ \text{Frame}}{\{R\}\, x := \texttt{cons}(E_1, ..., E_k)\, \{(x \mapsto E_1, ..., E_k) * R\}}\ \text{Consequence}$$

The conclusion is the strongest postcondition, and a variant involving auxiliary variables handles the case when $x \in \text{Free}(R, E_1, ..., E_k)$.

As an example of the use of these laws, recall the assertion $(x \mapsto a, o) * (x + o \mapsto b, -o)$ that describes a circular offset-list. Here is a proof outline for a sequence of commands that creates such a structure.

$$\begin{array}{l}
\{\texttt{emp}\} \\
\quad x := \texttt{cons}(a, a) \\
\{x \mapsto a, a\} \\
\quad t := \texttt{cons}(b, b) \\
\{(x \mapsto a, a) * (t \mapsto b, b)\} \\
\quad [x + 1] := t - x \\
\{(x \mapsto a, t - x) * (t \mapsto b, b)\} \\
\quad [t + 1] := x - t \\
\{(x \mapsto a, t - x) * (t \mapsto b, x - t)\} \\
\{\exists o. (x \mapsto a, o) * (x + o \mapsto b, -o)\}
\end{array}$$

The last step, which is an instance of the Rule of Consequence, uses $t - x$ as the witness for o. Notice how the alterations in the last two commands are done locally. For example, because of the placement of $*$ we know that $x + 1$ must be different from t and $t + 1$, so the assignment $[x + 1] := t - x$ cannot affect the $t \mapsto b, b$ conjunct.

If we wish to reason backwards, then $-\!\!*$ can be used to express weakest preconditions. Given an arbitrary postcondition Q, choosing $(E \mapsto F) -\!\!* Q$ as the invariant gives a valid precondition for $[E] := F$

$$\dfrac{\dfrac{\{E \mapsto -\}\,[E] := F\,\{E \mapsto F\}}{\{(E \mapsto -) * ((E \mapsto F) -\!\!* Q)\}\,[E] := F\,\{(E \mapsto F) * ((E \mapsto F) -\!\!* Q)\}}\text{ Frame}}{\{(E \mapsto -) * ((E \mapsto F) -\!\!* Q)\}\,[E] := F\,\{Q\}}\text{ Consequence}$$

The Consequence step uses an adjunction rule for $*$ and $-\!\!*$. The precondition obtained is in fact the weakest: it expresses the "update as deletion followed by extension" idea explained in [7]. The weakest precondition for allocation can also be expressed with $-\!\!*$.

The weakest precondition for `dispose` can be computed directly, because the Modifies set of $\texttt{dispose}(E)$ is empty.

$$\dfrac{\dfrac{\{E \mapsto -\}\,\texttt{dispose}(E)\,\{\texttt{emp}\}}{\{(E \mapsto -) * R\}\,\texttt{dispose}(E)\,\{\texttt{emp} * R\}}\text{ Frame}}{\{(E \mapsto -) * R\}\,\texttt{dispose}(E)\,\{R\}}\text{ Consequence}$$

The conclusion is (a unary version of) the axiom for `dispose` from [7].

The weakest precondition axiom for $x := E$ is the usual one of Hoare. For $x := [E]$ is it similar, using $\exists$ to form a "let binder" (where $n \notin \mathrm{Free}(E, P, x)$.

$$\{P[E/x]\}x := E\{P\}$$
$$\{\exists n.\,(\texttt{true} * E \mapsto n) \wedge P[n/x]\}x := [E]\{P\}$$

The formal derivations of these laws from the small axioms make heavy use of Variable Substitution and Auxiliary Variable Elimination; the details are contained in Yang's thesis [24].

Another useful derived law for $x := [E]$ is for the case when $x \notin \mathrm{Free}(E, R)$, $y \notin \mathrm{Free}(E)$, and when the precondition is of the form $(E \mapsto y) * R$. Then,

$$\{(E \mapsto y) * R\}\,x := [E]\,\{(E \mapsto x) * R[x/y]\}.$$

5 Beyond the Core

In the next few sections we give some examples of the formalism at work. In these examples we use sequencing, `if-then-else`, and a construct `newvar` for declaring a local variable. We can extend the core system with their usual Hoare logic rules.

$$\frac{\{P \wedge B\}\,C\,\{Q\} \quad \{P \wedge \neg B\}\,C'\,\{Q\}}{\{P\}\,\texttt{if}\,B\,\texttt{then}\,C\,\texttt{else}\,C'\{Q\}} \qquad \frac{\{P\}\,C_1\,\{Q\} \quad \{Q\}\,C_2\,\{R\}}{\{P\}\,C_1; C_2\,\{R\}}$$

$$\frac{\{P\}\,C\,\{Q\}}{\{P\}\,\texttt{newvar}\,x.\,C\,\{Q\}}\; x \notin \mathrm{Free}(P, Q)$$

We will also use simple first-order procedures. The procedure definitions we need will have the form

$\texttt{procedure}\; p(x_1, ..., x_n; y)$
B

where $x_1, ..., x_n$ are variables not changed in the body B and y is a variable that is assigned to. Procedure headers will always contain all of the variables occurring freely in a procedure body. Accordingly, we define

$$\begin{aligned} \text{Modifies}(p(x_1, ..., x_n; y)) &= \{y\} \\ \text{Free}(p(x_1, ..., x_n; y)) &= \{x_1, ..., x_n, y\}. \end{aligned}$$

We will need these clauses when applying the structural rules. In the examples the calling mechanism can be taken to be either by-name for all the parameters, or by-value on the x_i's and by-reference on y.

Procedures are used in Section 7 mainly to help structure the presentation, but in Section 6 we also use recursive calls. There we appeal to the standard partial correctness rule which allows us to use the specification we are trying to prove as an assumption when reasoning about the body [5].

Our treatment in what follows will not be completely formal. We will continue to use the Rule of Consequence in a semantic way, and we will make inductive definitions without formally defining their semantics. Also, as is common, we will present program specifications annotated with intermediate assertions, rather than give step-by-step proofs.

6 Tree Copy

In this section we consider a procedure for copying a tree. The purpose of the example is to show the Frame Rule in action.

For our purposes a tree will either be an atom a or a pair (τ_1, τ_2) of trees. Here is an inductive definition of a predicate $\texttt{tree}\,\tau\,i$ which says when a number i represents a tree τ.

$$\begin{aligned} \texttt{tree}\,a\,i &\stackrel{\Delta}{\iff} i = a \land \texttt{isatom?}(a) \land \texttt{emp} \\ \texttt{tree}\,(\tau_1, \tau_2)\,i &\stackrel{\Delta}{\iff} \exists x, y.\,(i \mapsto x, y) * (\texttt{tree}\,\tau_1\,x * \texttt{tree}\,\tau_2\,y) \end{aligned}$$

These two cases are exclusive. For the first to be true i must be an atom, where in the second it must be a location.

The $\texttt{tree}\,\tau\,i$ predicate is "exact", in the sense that when it is true the current heap must have all and only those heap cells used to represent the tree. If τ has n pairs in it and $s, h \models \texttt{tree}\,\tau\,i$ then the domain of h has size $2n$.

The specification of the CopyTree procedure is

$$\{\texttt{tree}\,\tau\,p\}\ \texttt{CopyTree}(p; q)\ \{(\texttt{tree}\,\tau\,p) * (\texttt{tree}\,\tau\,q)\}.$$

and here is the code.

```
procedure CopyTree(p; q)
newvar i, j, i', j'.
{tree τ p}
if isatom?(p) then
    {τ = p ∧ isatom?(p) ∧ emp}
```

$\{(\mathtt{tree}\,\tau\,p) * (\mathtt{tree}\,\tau\,p)\}$
  $q := p$
$\{(\mathtt{tree}\,\tau\,p) * (\mathtt{tree}\,\tau\,q)\}$
else
$\{\exists \tau_1, \tau_2, x, y.\, \tau \doteq (\tau_1, \tau_2) * (p \mapsto x, y) * (\mathtt{tree}\,\tau_1\,x) * (\mathtt{tree}\,\tau_2\,y)\}$
  $i := [p];\ j := [p+1];$
$\{\exists \tau_1, \tau_2.\, \tau \doteq (\tau_1, \tau_2) * (p \mapsto i, j) * (\mathtt{tree}\,\tau_1\,i) * (\mathtt{tree}\,\tau_2\,j)\}$
  $\mathtt{CopyTree}(i; i');$
$\{\exists \tau_1, \tau_2.\, \tau \doteq (\tau_1, \tau_2) * (p \mapsto i, j) * (\mathtt{tree}\,\tau_1\,i) * (\mathtt{tree}\,\tau_2\,j) * (\mathtt{tree}\,\tau_1\,i')\}$
  $\mathtt{CopyTree}(j; j');$
$\{\exists \tau_1, \tau_2.\, \tau \doteq (\tau_1, \tau_2) * (p \mapsto i, j) * (\mathtt{tree}\,\tau_1\,i) * (\mathtt{tree}\,\tau_2\,j) * (\mathtt{tree}\,\tau_1\,i')$
  $*(\mathtt{tree}\,\tau_2\,j')\}$
  $q := \mathsf{cons}(i', j')$
$\{\exists \tau_1, \tau_2.\, \tau \doteq (\tau_1, \tau_2) * (p \mapsto i, j) * (\mathtt{tree}\,\tau_1\,i) * (\mathtt{tree}\,\tau_2\,j) * (\mathtt{tree}\,\tau_1\,i')$
  $*(\mathtt{tree}\,\tau_2\,j') * (q \mapsto i', j')\}$
$\{(\mathtt{tree}\,\tau\,p) * (\mathtt{tree}\,\tau\,q)\}$

Most of the steps are straightforward, but the two recursive calls deserve special comment. In proving the body of the procedure we get to use the specification of `CopyTree` as an assumption. But at first sight the specification does not appear to be strong enough, since we need to be sure that $\mathtt{CopyTree}(i; i')$ does not affect the assertions $p \mapsto i, j$ and $\mathtt{tree}\,\tau_2\,j$. Similarly, we need that $\mathtt{CopyTree}(j; j')$ does not affect $\mathtt{tree}\,\tau_1\,i'$.

These "does not affect" properties are obtained from two instances of the Frame Rule:

$$\frac{\{\mathtt{tree}\,\tau_1\,i\}\ \mathtt{CopyTree}(i; i')\ \{(\mathtt{tree}\,\tau_1\,i) * (\mathtt{tree}\,\tau_1\,i')\}}{\begin{array}{l}\{\tau \doteq (\tau_1, \tau_2) * (p \mapsto i, j) * (\mathtt{tree}\,\tau_1\,i) * (\mathtt{tree}\,\tau_2\,j)\} \\ \mathtt{CopyTree}(i; i') \\ \{\tau \doteq (\tau_1, \tau_2) * (p \mapsto i, j) * (\mathtt{tree}\,\tau_1\,i) * (\mathtt{tree}\,\tau_2\,j) * (\mathtt{tree}\,\tau_1\,i')\}\end{array}}$$

and

$$\frac{\{\mathtt{tree}\,\tau_2\,j\}\ \mathtt{CopyTree}(j; j')\ \{(\mathtt{tree}\,\tau_2\,j) * (\mathtt{tree}\,\tau_2\,j')\}}{\begin{array}{l}\{\tau \doteq (\tau_1, \tau_2) * (p \mapsto i, j) * (\mathtt{tree}\,\tau_1\,i) * (\mathtt{tree}\,\tau_2\,j) * (\mathtt{tree}\,\tau_1\,i')\} \\ \mathtt{CopyTree}(j; j') \\ \{\tau \doteq (\tau_1, \tau_2) * (p \mapsto i, j) * (\mathtt{tree}\,\tau_1\,i) * (\mathtt{tree}\,\tau_2\,j) * (\mathtt{tree}\,\tau_1\,i') * (\mathtt{tree}\,\tau_2\,j')\}.\end{array}}$$

Then, the required triples for the calls are obtained using Auxiliary Variable Elimination to introduce $\exists \tau_1, \tau_2$. (It would also have been possible to strip the existential at the beginning of the proof of the `else` part, and then reintroduce it after finishing instead of carrying it through the proof.)

This section illustrates two main points. First, if one does not have some way of representing or inferring frame axioms, then the proofs of even simple programs with procedure calls will not go through. In particular, for recursive programs attention to framing is essential if one is to obtain strong enough induction hypotheses. The `CopyTree` procedure could not be verified without the

Frame Rule, unless we were to complicate the initial specification by including some explicit representation of frame axioms.

Second, the specification of `CopyTree` illustrates the idea of a specification that concentrates only on those cells that a program accesses. And of course these two points are linked; we need some way to infer frame axioms, or else such a specification would be too weak.

7 Difference-Linked Lists

The purpose of this section is to illustrate the treatment of address arithmetic, and also disposal. We do this by considering a space-saving representation of doubly-linked lists.

Conventionally, a node in a doubly-linked list contains a data field, together with a field storing a pointer n to the next node and another storing a pointer p to the previous node. In the difference representation we store $n - p$ in a single field rather than have separate fields for n and p. In a conventional doubly-linked list it is possible to move either forwards or backwards from a given node. In a difference-linked list given the current node c we can lookup the difference $d = n - p$ between next and previous pointers. This difference does not, by itself, give us enough information to determine either n or p. However, if we also know p we can calculate n as $d + p$, and similarly given n we can obtain p as $n - d$. So, using the difference representation, it is possible to traverse the list in either direction as long as we keep track of the previous or next node as we go along.

A similar, more time-efficient, representation is sometimes given using the xor of pointers rather than their difference.

We now give a definition of a predicate $\mathtt{dl}$. If we were working with conventional doubly-linked lists then $\mathtt{dl}\, a_1 \cdots a_n\, (i, i', j, j')$ would correspond to

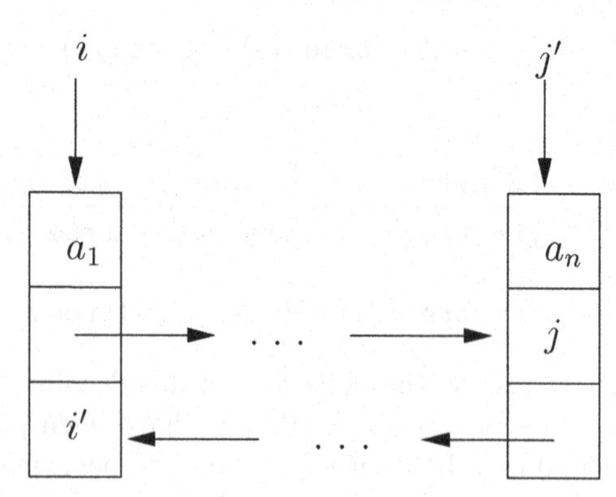

Typically, a doubly-linked list with front i and back j' would satisfy the predicate $\mathtt{dl}\, \alpha\, (i, \mathtt{nil}, \mathtt{nil}, j')$. The reason for the internal nodes i' and j is to allow us to consider partial lists, not terminated by $\mathtt{nil}$.

A definition of $\mathtt{dl}$ for conventional doubly-linked lists was given in [20]. The main alteration we must make is to use

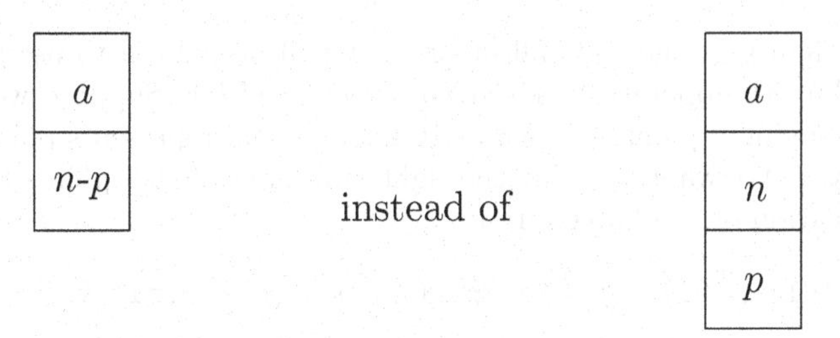

to represent a node.

Here is the definition.

$$\begin{aligned}\texttt{dl}\,\epsilon\,(i,i',j,j') &\stackrel{\Delta}{\iff} \texttt{emp} \wedge i=j \wedge i'=j' \\ \texttt{dl}\,a\alpha\,(i,i',k,k') &\stackrel{\Delta}{\iff} \exists j.(i \mapsto a, j-i') * \texttt{dl}\,\alpha\,(j,i,k,k')\end{aligned}$$

We are using juxtaposition to represent the consing of an element a onto the front of a sequence α, and ϵ to represent the empty sequence. As a small example, $\texttt{dl}\,ab\,(5,1,3,8)$ is true of

5	a	8	b
6	8-1	9	3-5

It is instructive to look at how this definition works for a sequence consisting of a single element, a. For $\texttt{dl}\,a\,(i,i',j,j')$ to hold we must have $\exists x.(i \mapsto a, x - i') * \texttt{dl}\,\epsilon\,(x,i,j,j')$; we can pick x to be j, as suggested by the $i=j$ part of the case for ϵ. We are still left, however, with the requirement that $i=j'$, and this in fact leads us to the characterization $i \mapsto a, j-i' \wedge i=j'$ of $\texttt{dl}\,a\,(i,i',j,j')$.

Thus, a single-lement list exemplifies how the ϵ case is arranged to be compatible with the operation of consing an element onto the front of a sequence. The roles of the $i=j$ and $i'=j'$ requirements are essentially reversed for the dual operation, of adding a single element onto the end of a sequence. This operation is characterized as follows.

$$\texttt{dl}\,\alpha a\,(i,i',k,k') \Leftrightarrow \exists j'.\,\texttt{dl}\,\alpha\,(i,i',k',j') \;*\; k' \mapsto a, k-j'$$

In the examples to come we will also use the following properties.

$$\begin{aligned} j' \neq \texttt{nil} \wedge \texttt{dl}\,\alpha\,(i,\texttt{nil},j,j') &\Rightarrow \exists \beta, a, k.\, \alpha \doteq \beta a * \\ &\qquad \texttt{dl}\,\beta\,(i,i',j',k) \;*\; j' \mapsto a, j-k \\ \texttt{dl}\,\alpha\,(i,i',j,\texttt{nil}) &\Rightarrow \texttt{emp} \wedge \alpha=\epsilon \wedge i'=\texttt{nil} \wedge i=j \\ \texttt{dl}\,\alpha\,(\texttt{nil},i',j,j') &\Rightarrow \texttt{emp} \wedge \alpha=\epsilon \wedge j=\texttt{nil} \wedge i'=j' \end{aligned}$$

Doubly-linked lists are often used to implement queues, because they make it easy to work at either end. We axiomatize an enqueue operation.

Rather than give the code all at once, it will be helpful to use a procedure to encapsulate the operation of *setting a right pointer*. Suppose we are in the position of having a pointer j', whose difference field represents pointing on the right to, say, j. We want to swing the right pointer so that it points to k instead. The specification of the procedure is

$$\{\mathtt{dl}\,\alpha\,(i,\mathtt{nil},j,j')\}\,\mathtt{setrptr}(j,j',k;i)\{\mathtt{dl}\,\alpha\,(i,\mathtt{nil},k,j')\}.$$

Notice that this specification handles the $\alpha = \epsilon$ case, when j' does not point to an active cell.

Postponing the definition and proof of `setrptr` for a moment, we can use it to verify a code fragment for putting an value a on the end of a queue.

$\{\mathtt{dl}\,\alpha\,(front,\mathtt{nil},\mathtt{nil},back)\}$
$\quad t := back;$
$\{\mathtt{dl}\,\alpha\,(front,\mathtt{nil},\mathtt{nil},t)\}$
$\quad back := \mathtt{cons}(a,\mathtt{nil}-t);$
$\{\mathtt{dl}\,\alpha\,(front,\mathtt{nil},\mathtt{nil},t) * back \mapsto a,\mathtt{nil}-t\}$
$\quad \mathtt{setrptr}(\mathtt{nil},t,back;front)$
$\{\mathtt{dl}\,\alpha\,(front,\mathtt{nil},back,t) * back \mapsto a,\mathtt{nil}-t\}$
$\{\mathtt{dl}\,\alpha a\,(front,\mathtt{nil},\mathtt{nil},back)\}$

The code creates a new node containing the value a and the difference $\mathtt{nil}-t$. Then, the procedure call $\mathtt{setrptr}(\mathtt{nil},t,back;front)$ swings the right pointer associated with t so that the next node becomes $back$. In the assertions, the effect of $back := \mathtt{cons}(a,\mathtt{nil}-t)$ is axiomatized by tacking $*(back \mapsto a,\mathtt{nil}-t)$ onto its precondition. This sets us up for the call to `setrptr`; because of the placement of $*$ we know that the call will not affect $(back \mapsto a,\mathtt{nil}-t)$. More precisely, the triple for the call is obtained using Variable Substitution to instantiate the specification, and the Frame Rule with $(back \mapsto a,\mathtt{nil}-t)$ as the invariant.

Finally, here is an implementation of $\mathtt{setrptr}(j,j',k;i)$.

$\{\mathtt{dl}\,\alpha\,(i,\mathtt{nil},j,j')\}$
$\quad \mathtt{if}\ j' = \mathtt{nil}\ \mathtt{then}$
$\quad \{\alpha = \epsilon \wedge \mathtt{emp} \wedge j' = \mathtt{nil}\}$
$\qquad i := k$
$\quad \{\alpha = \epsilon \wedge \mathtt{emp} \wedge j' = \mathtt{nil} \wedge i = k\}$
$\quad \mathtt{else}$
$\quad \{\exists \alpha', b, p.\, (\alpha \doteq \alpha' b) * \mathtt{dl}\,\alpha'\,(i,\mathtt{nil},j',p) * (j' \mapsto b, j-p)\}$
$\qquad \mathtt{newvar}\ d.\ d := [j'+1]; [j'+1] := k+d-j$
$\quad \{\exists \alpha', b, p.\, (\alpha \doteq \alpha' b) * \mathtt{dl}\,\alpha'\,(i,\mathtt{nil},j',p) * (j' \mapsto b, k-p)\}$
$\{\mathtt{dl}\,\alpha\,(i,\mathtt{nil},k,j')\}$

The tricky part in the verification is the `else` branch of the conditional, where the code has to update the difference field of j' appropriately so that k becomes the next node of j'. It updates the field by adding k and subtracting j; since the field initially stores $j-p$, where p is the address of the previous node, such calculation results in the value $k-p$.

The use of the temporary variable d in the `else` branch is a minor irritation. We could more simply write $[j'+1] := k + [j'+1] - j$ if we were to allow nesting of $[\cdot]$. An unresolved question is whether, in our formalism, such nesting could be dealt with in a way simpler than compiling it out using temporary variables.

Now we sketch a similar development for code that implements a dequeue operation. In this case, we use a procedure $\texttt{setlptr}(i, i', k; j')$, which is similar to `setrptr` except that it swings a pointer to the left instead of to the right.

$$\{\texttt{dl}\,\alpha\,(i, i', \texttt{nil}, j')\}\ \texttt{setlptr}(i, i', k; j')\ \{\texttt{dl}\,\alpha\,(i, k, \texttt{nil}, j')\}$$

The dequeue operation removes the first element of a queue and places its data in x.

$$
\begin{array}{l}
\{\texttt{dl}\,a\alpha\,(\mathit{front}, \texttt{nil}, \texttt{nil}, \mathit{back})\}\\
\{\exists n'.\, \mathit{front} \mapsto a, n' - \texttt{nil} * \texttt{dl}\,\alpha\,(n', \mathit{front}, \texttt{nil}, \mathit{back})\}\\
\ \ x := [\mathit{front}];\ d := [\mathit{front}+1];\ n := d + \texttt{nil};\\
\{x \doteq a * \mathit{front} \mapsto a, n - \texttt{nil} * \texttt{dl}\,\alpha\,(n, \mathit{front}, \texttt{nil}, \mathit{back})\}\\
\ \ \texttt{dispose}(\mathit{front}); \texttt{dispose}(\mathit{front}+1);\\
\{x \doteq a * \texttt{dl}\,\alpha\,(n, \mathit{front}, \texttt{nil}, \mathit{back})\}\\
\ \ \texttt{setlptr}(n, \mathit{front}, \texttt{nil}; \mathit{back})\\
\{x \doteq a * \texttt{dl}\,\alpha\,(n, \texttt{nil}, \texttt{nil}, \mathit{back})\}
\end{array}
$$

This code stores the data of the first node in the variable x and obtains the next pointer n using arithmetic with the difference field. The placement of $*$ sets us up for disposing front and $\mathit{front}+1$: The precondition to these two commands is equivalent to an assertion of the form $(\mathit{front} \mapsto a) * (\mathit{front}+1 \mapsto n' - \texttt{nil}) * R$, which is compatible with what is given by two applications of the weakest precondition rule for `dispose`. After the disposals have been done, the procedure call $\texttt{setlptr}(n, \mathit{front}, \texttt{nil}; \mathit{back})$ resets the difference field of the node n so that its previous node becomes `nil`.

The code for $\texttt{setlptr}(i, i', k; j')$ is as follows.

$$
\begin{array}{l}
\{\texttt{dl}\,\alpha\,(i, i', \texttt{nil}, j')\}\\
\ \ \texttt{if}\ i = \texttt{nil}\ \texttt{then}\\
\ \ \{\alpha = \epsilon \wedge \textbf{emp} \wedge i = \texttt{nil}\}\\
\ \ \ \ \ j' := k\\
\ \ \{\alpha = \epsilon \wedge \textbf{emp} \wedge i = \texttt{nil} \wedge k = j'\}\\
\ \ \texttt{else}\\
\ \ \{\exists \alpha', a, n.\, (\alpha \doteq a\alpha') * \texttt{dl}\,\alpha'\,(n, i, \texttt{nil}, j') * (i \mapsto a, n - i')\}\\
\ \ \ \ \ [i+1] := [i+1] + i' - k\\
\ \ \{\exists \alpha', a, n.\, (\alpha \doteq a\alpha') * \texttt{dl}\,\alpha'\,(n, i, \texttt{nil}, j') * (i \mapsto a, n - k)\}\\
\{\texttt{dl}\,\alpha\,(i, k, \texttt{nil}, j')\}
\end{array}
$$

8 Memory Faults and Tight Specifications

In this paper we will not include a semantics of commands or precise interpretation of triples, but in this section we give an informal discussion of the semantic properties of triples that the axiom system relies on.

Usually, the specification form $\{P\}C\{Q\}$ is interpreted "loosely", in the sense that C might cause state changes not described by the pre and postcondition. This leads to the need for explicit frame axioms. An old idea is to instead consider a "tight" interpretation of $\{P\}C\{Q\}$, which should guarantee that C only alters those resources mentioned in P and Q; unfortunately, a precise definition of the meaning of tight specifications has proven elusive [1]. However, the description of local reasoning from the Introduction, where a specification and proof concentrate on a circumscribed area of memory, requires something like tightness. The need for a tight interpretation is also clear from the small axioms, or the specifications of `setlptr`, `setrptr` and `CopyTree`.

To begin, the model here calls for a notion of memory fault. This can be pictured by imagining that there is an "access bit" associated with each location, which is on iff the location is in the domain of the heap. Any attempt to read or write a location whose access bit is off causes a memory fault, so if E is not an active address then $[E] := E'$ or $x := [E]$ results in a fault. A simple way to interpret $\mathtt{dispose}(E)$ is so that it faults if E is not an active address, and otherwise turns the access bit off.

Then, a specification $\{P\}C\{Q\}$ holds iff, whenever C is run in a state satisfying P: (i) it does not generate a fault; and (ii) if it terminates then the final state satisfies Q. (This is a partial correctness interpretation; the total correctness variant alters (ii) by requiring that there are no infinite reductions.) For example, according to the fault-avoiding interpretation, $\{17 \mapsto -\}\,[17] := 4\,\{17 \mapsto 4\}$ holds but $\{\mathtt{true}\}\,[17] := 4\,\{17 \mapsto 4\}$ does not. The latter triple fails because the empty heap satisfies $\mathtt{true}$ but $[17] := 4$ generates a memory fault when executed in the empty heap.

In the logic, faults are precluded by the assumptions $E \mapsto -$ and $E \mapsto n$ in the preconditions of the small axioms for $[E] := E'$, $x := [E]$ and $\mathtt{dispose}(E)$.

The main point of this section is that this fault-avoiding interpretation of $\{P\}C\{Q\}$ gives us a precise formulation of the intuitive notion of tightness. (We emphasize that this requires faults, or a notion of enabled action, and we do not claim that it constitutes a general analysis of the notion of tight specification.)

> The avoidance of memory faults in specifications ensures that a well-specified program can only dereference (or dispose) those heap cells guaranteed to exist by the precondition, or those which are allocated during execution.

Concretely, if one executes a program proved to satisfy $\{P\}C\{Q\}$, starting in a state satisfying P, then memory access bits are unnecessary. A consequence is that it is not necessary to explicitly describe all the heap cells that don't change, because those not mentioned automatically stay the same.

Fault avoidance in $\{P\}C\{Q\}$ ensures that if C is run in a state strictly larger than one satisfying P, then any additional cells must stay unchanged; an attempt to write any of the additional cells would falsify the specification, because it would generate a fault when applied to a smaller heap satisfying P. For example, if $\{17 \mapsto -\}\,C\,\{17 \mapsto 4\}$ holds then $\{(17 \mapsto -) * (19 \mapsto 3)\}\,C\,\{(17 \mapsto 4) * (19 \mapsto 3)\}$ should as well, as mandated by the Frame Rule, because any attempt to dereference address 19 would falsify $\{17 \mapsto -\}\, C\, \{17 \mapsto 4\}$ if we give C a state where the access bit for 19 is turned off. (This last step is delicate, in that one could entertain operations, such as to test whether an access bit is on, which contradict it; what is generally needed for it is a notion which can be detected in the logic but not the programming language.)

9 Conclusion

We began the paper by suggesting that the main challenge facing verification formalisms for pointer programs is to capture the informal local reasoning used by programmers, or in textbook-style arguments about data structures. Part of the difficulty is that pointers exacerbate the *frame problem* [9, 1]. (It is only part of the difficulty because the frame problem does not, by itself, say anything about aliasing.) For imperative programs the problem is to find a way, preferably succinct and intuitive, to describe or imply the frame axioms, which say what memory cells are not altered by a program or procedure. Standard methods, such as listing the variables that might be modified, do not work easily for pointer programs, because there are often many cells not directly named by variables in a program or program fragment. These cells might be accessed by a program by following pointer chains in memory, or they might not be accessed even when they are reachable.

The approach taken here is based on two ideas. The first, described in Section 8, to use a fault-avoiding interpretation of triples to ensure that additional cells, active but not described by a precondition, are not altered during execution. The second is to use the $*$ connective to infer invariant properties implied by these tight specifications.

The frame problem for programs is perhaps more approachable than the general frame problem. Programs come with a clear operational semantics, and one can appeal to concrete notions such as a program's footprint. But the methods here also appear to be more generally applicable. It would be interesting to give a precise comparison with ideas from the AI literature [22], as well as with variations on Modifies clauses [1, 8]. We hope to report further on these matters – in particular on the ideas outlined in Section 8 – in the future. (Several relevant developments can be found in Yang's thesis [24].)

There are several immediate directions for further work. First, the interaction between local and global reasoning is in general difficult, and we do not mean to imply that things always go as smoothly as in the example programs we chose. They fit our formalism nicely because their data structures break naturally into disjoint parts, and data structures that use more sharing are more difficult to handle. This includes tree representations that allow sharing of subtrees, and graph structures. Yang has treated a nontrivial example, the Shorr-Waite graph marking algorithm, using the spatial implication $-\!\!*$ is used to deal with the sharing found there [23]. More experience is needed in this direction. Again, the challenging problem is not to find a system that is adequate in principle, but rather is to find rules or reasoning idioms that cover common cases simply and naturally.

Second, the reasoning done in examples in this paper is only semi-formal, because we have worked semantically when applying the Rule of Consequence. We know of enough axioms to support a number of examples, but a comprehensive study of the proof theory of the assertion language is needed. Pym has worked out a proof theory of the underlying logic BI [17] that we can draw on. But here we use a specific model of BI and thus require an analysis of properties special to that model. Also needed is a thorough treatment of recursive definitions of predicates.

Finally, the examples involving address arithmetic with difference-linked lists are simplistic. It would be interesting to try to verify more substantial programs that rely essentially on address arithmetic, such as memory allocators or garbage collectors.

Acknowledgements

O'Hearn would like to thank Richard Bornat, Cristiano Calcagno and David Pym for discussions about local reasoning and bunched logic. He was supported by the EPSRC, under the "Verified Bytecode" and "Local Reasoning about State" grants. Reynolds was supported by NSF grant CCR-9804014. Yang was supported by the NSF under grant INT-9813854.

References

1. A. Borgida, J. Mylopoulos, and R. Reiter. On the frame problem in procedure specifications. *IEEE Transactions of Software Engineering*, 21:809–838, 1995.
2. R. Bornat. Proving pointer programs in Hoare logic. *Mathematics of Program Construction*, 2000.
3. R.M. Burstall. Some techniques for proving correctness of programs which alter data structures. *Machine Intelligence*, 7:23–50, 1972.
4. C. Calcagno, S. Isthiaq, and P. W. O'Hearn. Semantic analysis of pointer aliasing, allocation and disposal in Hoare logic. Proceedings of the Second International ACM SIGPLAN Conference on Principles and Practice of Declarative Programming, 2000.
5. P. Cousot. Methods and logics for proving programs. In J. van Leeuwen, editor, *Handbook of Theoretical Computer Science*, volume B, pages 843–993. Elsevier, Amsterdam, and The MIT Press, Cambridge, Mass., 1990.
6. C. A. R. Hoare and J. He. A trace model for pointers and objects. In Rachid Guerraoui, editor, *ECCOP'99 - Object-Oriented Programming, 13th European Conference*, pages 1–17, 1999. Lecture Notes in Computer Science, Vol. 1628, Springer.
7. S. Isthiaq and P.W. O'Hearn. BI as an assertion language for mutable data structures. In *Conference Record of the Twenty-Eighth Annual ACM Symposium on Principles of Programming Languages*, pages 39–46, London, January 2001.
8. K. R. M. Leino and G. Nelson. Data abstraction and information hiding. Technical Report Reearch Report 160, Compaq Systems Research Center, Palo Alto, CA, November 2000.

9. J. McCarthy and P. Hayes. Some philosophical problems from the standpoint of artificial intelligence. *Machine Intelligence*, 4:463–502, 1969.
10. P. W. O'Hearn. Resource interpretations, bunched implications and the $\alpha\lambda$-calculus. In *Typed λ-calculus and Applications*, J-Y Girard editor, L'Aquila, Italy, April 1999. Lecture Notes in Computer Science 1581.
11. P. W. O'Hearn and D. J. Pym. The logic of bunched implications. *Bulletin of Symbolic Logic*, 5(2):215–244, June 99.
12. P. W. O'Hearn and J. C. Reynolds. From Algol to polymorphic linear lambda-calculus. *J. ACM*, 47(1):267–223, January 2000.
13. P. W. O'Hearn and R. D. Tennent. Parametricity and local variables. *J. ACM*, 42(3):658–709, May 1995. Also in [14], vol 2, pages 109–164.
14. P. W. O'Hearn and R. D. Tennent, editors. *Algol-like Languages.* Two volumes, Birkhauser, Boston, 1997.
15. F. J. Oles. *A Category-Theoretic Approach to the Semantics of Programming Languages.* Ph.D. thesis, Syracuse University, Syracuse, N.Y., 1982.
16. F. J. Oles. Functor categories and store shapes. In O'Hearn and Tennent [14], pages 3–12. Vol. 2.
17. D. J. Pym. *The Semantics and Proof Theory of the Logic of Bunched Implications.* Monograph to appear, 2001.
18. J. C. Reynolds. Syntactic control of interference. In *Conference Record of the Fifth Annual ACM Symposium on Principles of Programming Languages*, pages 39–46, Tucson, Arizona, January 1978. ACM, New York. Also in [14], vol 1.
19. J. C. Reynolds. The essence of Algol. In J. W. de Bakker and J. C. van Vliet, editors, *Algorithmic Languages*, pages 345–372, Amsterdam, October 1981. North-Holland, Amsterdam. Also in [14], vol 1, pages 67-88.
20. J. C. Reynolds. Intuitionistic reasoning about shared mutable data structure. In Jim Davies, Bill Roscoe, and Jim Woodcock, editors, *Millennial Perspectives in Computer Science*, pages 303–321, Houndsmill, Hampshire, 2000. Palgrave.
21. J. C. Reynolds. Lectures on reasoning about shared mutable data structure. *IFIP Working Group 2.3 School/Seminar on State-of-the-Art Program Design Using Logic.* Tandil, Argentina, September 2000.
22. M. Shanahan. *Solving the Frame Problem: A Mathematical Investigation of the Common Sense Law of Inertia.* MIT Press, 1997.
23. H. Yang. An example of local reasoning in BI pointer logic: the Schorr-Waite graph marking algorithm. Manuscript, October 2000.
24. H. Yang. *Local Reasoning for Stateful Programs.* Ph.D. thesis, University of Illinois, Urbana-Champaign, Illinois, USA, 2001 (expected).

Applications of Alfred Tarski's Ideas in Database Theory*

Jan Van den Bussche

University of Limburg (LUC)
B-3590 Diepenbeek, Belgium

Abstract. Many ideas of Alfred Tarski – one of the founders of modern logic – find application in database theory. We survey some of them with no attempt at comprehensiveness. Topics discussed include the genericity of database queries; the relational algebra, the Tarskian definition of truth for the relational calculus, and cylindric algebras; relation algebras and computationally complete query languages; real polynomial constraint databases; and geometrical query languages.

Alfred Tarski, 1901–1983

To Dirk

1 Introduction

Alfred Tarski was one of the founders of modern logic, and a philosopher and mathematician of extraordinary breadth and depth. It is therefore not surprising that many of his ideas find application also in database theory, a field of

* I thank Janos Makowsky for having proposed me to write and present this paper. I owe a lot to Dirk Van Gucht, database theorist and Tarski fan, for having taught me so much during the past ten years, about database theory as well as about Tarski.

L. Fribourg (Ed.): CSL 2001, LNCS 2142, pp. 20–37, 2001.

theoretical computer science where logic plays an important role. In this year of Tarski's hundredth anniversary, it seems desirable to survey some of these applications. We will not attempt to be comprehensive, however.

2 Relational Database Queries and Logical Notions

To begin our discussion, we fix some infinite universe $\mathbb{U}$ of atomic data elements. In a set-theoretic formalization they would play the role of "urelemente".

In the relational approach to database management, introduced by Codd [19], we define a *database schema* $\mathcal{S}$ as a finite set of relation names, each with an associated arity. A *relational database* $\mathbf{D}$ *with schema* $\mathcal{S}$ then assigns to each $R \in \mathcal{S}$ a *finite* n-ary relation $R^{\mathbf{D}} \subseteq \mathbb{U}^n$, where n is the arity of R.

We store information in a database so that we can retrieve it later. The answer of a *query* to a relational database is again a relation: this is very convenient as it allows us to compose queries, or to store answers of queries as additional relations in the database. The answers 'yes' or 'no' are represented by the nonempty and the empty relation of arity 0, respectively. For example, let $\mathcal{S} = \{R\}$ where the arity of R is 2. So, databases over $\mathcal{S}$ can be identified with finite binary relations on $\mathbb{U}$. Some examples of queries we might want to ask such databases are:

1. Is there an identical pair in R? (Answer: nullary relation.)
2. What are the elements occurring in the left column of R, but not in the right one? (Answer: unary relation.)
3. What are the 5-tuples $(x_1, x_2, x_3, x_4, x_5)$ such that (x_1, x_2), (x_2, x_3), (x_3, x_4), and (x_4, x_5) are all in R? (Answer: five-ary relation.)
4. What is the transitive closure of R? (Answer: binary relation.)
5. Which pairs of elements (x_1, x_2) are such that the sets $\{y \mid (x_1, y) \in R\}$ and $\{y \mid (x_2, y) \in R\}$ are nonempty and have the same cardinality? (Answer: binary relation.)
6. Is the cardinality of R a prime number? (Answer: nullary relation.)

At the most general level, we could formally define an *n-ary query on $\mathcal{S}$* as a function q from databases $\mathbf{D}$ with schema $\mathcal{S}$ to finite relations $q(\mathbf{D}) \subseteq \mathbb{U}^n$. However, this definition is much too liberal. To illustrate this, let us take the same example schema $\mathcal{S}$ as above, and a, b and c three different elements of $\mathbb{U}$. Now consider the database $\mathbf{D}_0$ where $R^{\mathbf{D}_0} = \{(a,b), (a,c)\}$, and a unary query q_0 on $\mathcal{S}$ that maps $\mathbf{D}_0$ to the singleton $\{b\}$. This query does not seem "logical:" given the information provided by $\mathbf{D}_0$, there is no reason to favor b above c, as b and c are completely symmetric in $\mathbf{D}_0$. Note that none of the example queries given above has this "unlogical" nature: each of them can be answered purely on the basis of the information present in the database, and this is how it should be.

How can we formalize this intuitive notion of a "logical" query? Tarski has shown us how [60]. Consider the following cumulative hierarchy of universes $\mathbb{U}_0$, $\mathbb{U}_1$, $\mathbb{U}_2$, and so on, and their union U^*:

$$\mathbb{U}_0 := \mathbb{U}, \quad \mathbb{U}_{n+1} := \mathbb{U} \cup \mathcal{P}(\mathbb{U}_n), \quad \mathbb{U}^* := \bigcup_n \mathbb{U}_n$$

Here $\mathcal{P}$ denotes the powerset operation. Most mathematical objects we want to construct on top of $\mathbb{U}$ can be formalized as elements of $\mathbb{U}^*$. For example, by the ordered pair construction $(x, y) := \{\{x\}, \{x, y\}\}$, ordered pairs of elements of $\mathbb{U}$ live in $\mathbb{U}_2$, and thus binary relations on $\mathbb{U}$ live in $\mathbb{U}_3$. Database queries also live in $\mathbb{U}^*$. For example, a unary query on binary relations, being itself a binary relation from binary relations to unary relations, lives in $\mathbb{U}_6$. More generally, any notion involving objects living in $\mathbb{U}^*$, such as a property of such objects, or a relation among such objects, can itself be formalized as an object living in $\mathbb{U}^*$.

Tarski now calls such a notion *logical* if it is left invariant by all possible permutations of $\mathbb{U}$. So, $P \in \mathbb{U}^*$ is logical if $f(P) = P$ for every permutation f of $\mathbb{U}$, where permutations of $\mathbb{U}$ are extended to $\mathbb{U}^*$ in the canonical manner. For example, no singleton $\{a\}$ with $a \in \mathbb{U}$ is logical: there is no purely logical reason to single out any particular atomic data element. The whole set $\mathbb{U}$ is logical, and so is the empty set. The identity relation $\{(x, x) \mid x \in \mathbb{U}\}$ is logical, and so is the diversity relation $\{(x, y) \mid x, y \in \mathbb{U},\ x \neq y\}$. The higher we go up in the cumulative hierarchy, the more complex logical notions we find. In particular, queries may or may not be logical. For example, the "unlogical" query q_0 from above is indeed not logical in the sense of Tarski. For if it were, it would have to be invariant under the transposition $t = (b\ c)$ and thus would have to contain not only the pair $(\mathbf{D}_0, \{b\})$ but also the pair $(t(\mathbf{D}_0), \{t(b)\}) = (\mathbf{D}_0, \{c\})$, which is impossible as q_0 is a function. On the other hand, all the earlier example queries 1–6 are readily seen to be logical.

Unaware of this[1], Chandra and Harel [16], and independently Aho and Ullman [6], based on practical considerations, pointed out the following "universality property" (as A&U called it), or "consistency criterion" (as C&H called it) for database queries. It is now generally known as the *genericity* of database queries[2], and says that for any query q, databases $\mathbf{D}_1$ and $\mathbf{D}_2$, and permutation f of $\mathbb{U}$, if $f(\mathbf{D}_1) = \mathbf{D}_2$, then also $f(q(\mathbf{D}_1)) = q(\mathbf{D}_2)$. Clearly, a query is generic in this sense if and only if it is logical. So, interestingly, Tarski's definition of logical notion somehow inevitably turned out to hold for database queries.

We note that Tarski saw his definition in the context of Klein's Erlanger Programm [66] in which different geometries are identified with the groups of transformations under which the fundamental notions of the geometry in question are left invariant. For example, topology could be defined as the geometry whose notions are left invariant by continuous transformations (homeomorphisms). According to Tarski then, logic is the "geometry" whose notions are left invariant by *all* transformations.

3 The Relational Algebra and First-Order Queries

A fundamental insight of Codd was that many complex operations performed on data files can be expressed as combinations of five basic operators on relations:

---

[1] The paper cited [60] was only published in 1986, but is based on a talk delivered by Tarski twenty years earlier.

[2] The specific term 'genericity' was first used for this purpose by Hull and Yap [36] and caught on.

1. *union* of two relations of the same arity;
2. *difference* between two relations of the same arity;
3. *cartesian product:* if r is of arity n and s is of arity m, then $r \times s$ equals $\{(x_1, \ldots, x_n, y_1, \ldots, y_m) \mid (x_1, \ldots, x_n) \in r \text{ and } (y_1, \ldots, y_m) \in s\}$;
4. *projection:* if $\bar{x} = (x_1, \ldots, x_n)$ is an n-tuple and $i_1, \ldots, i_p \in \{1, \ldots, n\}$, then $\pi_{i_1,\ldots,i_p}(\bar{x})$ equals $(x_{i_1}, \ldots, x_{i_p})$; if r is an n-ary relation then $\pi_{i_1,\ldots,i_p}(r)$ equals $\{\pi_{i_1,\ldots,i_p}(\bar{x}) \mid \bar{x} \in r\}$;
5. *selection:* if r is of arity n and $i, j \in \{1, \ldots, n\}$, then $\sigma_{i=j}(r)$ equals $\{(x_1, \ldots, x_n) \in r \mid x_i = x_j\}$.

A query on a schema $\mathcal{S}$ is said to be *expressible in the relational algebra* if it can be defined by an expression built from the relation names of $\mathcal{S}$ using the above five operators. For example, example query no. 2 from the previous section is easily expressed as $\pi_1(R) - \pi_2(R)$, and example query no. 3 as $\pi_{1,2,4,6,8}\sigma_{2=3}\sigma_{4=5}\sigma_{6=7}(R \times R \times R \times R)$.

A classical theorem of Codd [20] identifies the queries expressible in the relational algebra with the first-order queries. An n-ary query q on $\mathcal{S}$ is called *first-order* if there is a first-order formula $\varphi(x_1, \ldots, x_n)$ over the relational vocabulary $\mathcal{S}$, such that for every database $\mathbf{D}$,

$$q(\mathbf{D}) = \{(a_1, \ldots, a_n) \in |\mathbf{D}|^n \mid \mathbf{D} \models \varphi[a_1, \ldots, a_n]\}.$$

Here, $|\mathbf{D}|$ denotes the *active domain* of $\mathbf{D}$, consisting of all elements of $\mathbb{U}$ actually occurring in one of the relations of $\mathbf{D}$. In evaluating $\mathbf{D} \models \varphi[a_1, \ldots, a_n]$, we view $\mathbf{D}$ as an $\mathcal{S}$-structure (in the sense of model theory) with (finite) domain $|\mathbf{D}|$. Codd referred to first-order logic used to express queries in this way as the *relational calculus.*

Tarski being one of the founders of modern logic, it is not surprising that the first-order queries owe a lot to him. We mention just two things:

1. The now-standard definition of satisfaction of a formula in a structure was originally conceived by Tarski [55]. Thanks to Codd, every student of computer science gets in touch with the syntax and the Tarskian semantics of first-order logic, in the form of the relational calculus seen in the databases course.
2. In view of the previous section, we should also ask ourselves whether first-order queries actually satisfy the genericity criterion, or, equivalently, are they logical in the sense of Tarski? They sure are: in fact, already in 1936 Tarski and Lindenbaum [44] noted that not just first-order, but full typed higher-order logic can define only logical notions. Nowadays this sounds like a tautology, but back in the days when modern logic was still in the process of being defined, this was a fundamental observation.

In connection with Codd's theorem two more of Tarski's ideas find an application in database theory. We discuss them separately in the following subsections.

3.1 Relational Completeness

Codd thought of his theorem as a completeness result for the relational algebra: the class of first-order queries was the reference level of expressiveness query languages should aim for. Codd called a query language *relationally complete* if it could express all first-order queries. Later, people started to realize that a lot of interesting queries are *not* first-order [6,16,17]. For example, of the list of example queries given in the previous section, queries no. 4, 5 and 6 are not first-order.

So, relational completeness is not everything. However, there still is a sense in which the relational algebra (or equivalently, first-order logic) can be considered a "complete" database query language, as was independently discovered by Bancilhon [8] and Paredaens [47]. They showed that for any database $\mathbf{D}$, and any relation $r \subseteq |\mathbf{D}|^n$ such that every automorphism of $\mathbf{D}$ is also an automorphism of r, there exists a first-order query q such that $q(\mathbf{D}) = r$. Here, an automorphism of $\mathbf{D}$ (r) is a permutation f of $\mathbb{U}$ such that $f(\mathbf{D}) = \mathbf{D}$ ($f(r) = r$). Note that the conditions of the theorem are necessary: for *any* generic n-ary query q, $q(\mathbf{D}) \subseteq |\mathbf{D}|^n$ and has at least the automorphisms of $\mathbf{D}$.

This "BP-completeness" of first-order logic, as it came to be called, actually follows from an early model-theoretic insight of Tarski, and another early model-theoretic theorem known as Beth's theorem. When he introduced the notion of elementary equivalence of structures [53,54], Tarski noted that two *finite* structures are elementary equivalent only if they are isomorphic. Actually, given a finite structure $\mathbf{D}$ one can always write a single first-order sentence that is satisfied by any structure $\mathbf{D}'$ if and only if $\mathbf{D}'$ is isomorphic to $\mathbf{D}$.

Now let $\mathbf{D}$, over $\mathcal{S}$, and r be as in the BP-completeness theorem. Let $\mathcal{S}'$ be the expansion of $\mathcal{S}$ with an extra n-ary relation name R, and let $\mathbf{D}'$ be the expansion of $\mathbf{D}$ to $\mathcal{S}'$ by putting $R^{\mathbf{D}'} := r$. As $\mathbf{D}'$ is a finite structure, we can, by the above, write a first-order sentence φ such that any database $\mathbf{B}$ over $\mathcal{S}'$ satisfies φ iff $\mathbf{B}$ is isomorphic to $\mathbf{D}'$. Take any two $\mathbf{B}_1$ and $\mathbf{B}_2$ with $\mathbf{B}_1 \models \varphi$ and $\mathbf{B}_2 \models \varphi$; so there are permutations f_1 and f_2 of $\mathbb{U}$ so that $f_1(\mathbf{B}_1) = \mathbf{D}'$ and $f_2(\mathbf{B}_2) = \mathbf{D}'$. But then $f_2 f_1^{-1}$ is an automorphism of $\mathbf{D}$ and hence, by assumption, also of r. So, $f_2 f_1^{-1}(r) = r$, whence $R^{\mathbf{B}_1} = f_1^{-1}(r) = f_2^{-1}(r) = R^{\mathbf{B}_2}$.

We thus observe that φ implicitly defines R in terms of $\mathcal{S}$. By Beth's theorem, there is a first-order formula $\psi(\bar{x})$ over $\mathcal{S}$ such that in any model of φ the equivalence $\forall \bar{x}(R(\bar{x}) \leftrightarrow \psi)$ holds. This holds in particular in $\mathbf{D}'$ itself, by which we conclude that $\psi(\mathbf{D}) = r$.

We have thus easily derived the BP-completeness of first-order logic from some of the earliest model-theoretic results that were established. Still we recommend Paredaens's direct proof, which uses the relational algebra and is very elegant[3].

[3] Recently, Cohen, Gyssens and Jeavons [21] showed that even the relational algebra without the union and difference operators, but with nonequality selections $\sigma_{i \neq j}$, is already BP-complete, on condition the active domain is directly available as one of the relations (or a projection of them), and has at least 3 elements.

3.2 Cylindric Set Algebras

Take a first-order formula φ, and let the different variables occurring in it, free or bound, be $x_1, \ldots, x_n$. When we follow the Tarskian semantics of φ on a structure A and determine inductively for every subformula ψ the set ψ^A of n-tuples $(a_1, \ldots, a_n) \in A^n$ under which ψ is true in A, we notice that at every inductive step we perform one of the following three operations on these n-ary relations:

1. *union*, to evaluate $\vee$;
2. *complementation* with respect to A^n, to evaluate $\neg$; and
3. *cylindrification* along dimension i, to evaluate $\exists x_i$.

By the cylindrification along dimension i of a relation $r \subseteq A^n$, with $i \in \{1, \ldots, n\}$, we mean the operation

$$\gamma_i(r) := \{(a_1, \ldots, a_n) \in A^n \mid \exists a \in A : (a_1, \ldots, a_{i-1}, a, a_{i+1}, \ldots, a_n) \in r\}.$$

These three operations, together with the constant relations $\delta_{ij} = \{(a_1, \ldots, a_n) \in A^n \mid a_i = a_j\}$, called the *diagonals* and needed to evaluate equality atoms $x_i = x_j$, constitute the *full n-dimensional cylindric set algebra with base A*. Cylindric set algebras are canonical examples of the general class of abstract *cylindric algebras,* which has an equational definition in the usual style abstract algebraic structures are defined. Cylindric algebras are the same to first-order logic as Boolean algebras are to propositional logic, and they were introduced by Tarski and his collaborators [31,32,33,34].

We thus see that a relational algebra in much the same spirit as Codd's was already considered by Tarski[4]. Concretely, let $\mathcal{S}$ be a schema, and take n strictly larger than the arity of every relation name in $\mathcal{S}$. We can build up *n-CSA expressions over* $\mathcal{S}$ from the relation names in $\mathcal{S}$ and the constants δ_{ij} using the operators $e_1 \cup e_2$, $\neg e$, and $\gamma_i(e)$. When evaluating an n-CSA expression e on a database $\mathbf{D}$ with schema $\mathcal{S}$, the operators and constants are interpreted as in the full n-dimensional cylindric set algebra with base $|\mathbf{D}|$. Each relation name R is interpreted as the n-ary relation $R^{\mathbf{D}} \times |\mathbf{D}|^{n-m}$, if the arity of R is m. We then have:

Theorem 1. *An n-ary query is expressible in n-CSA in the sense just defined, if and only if it is expressible by a first-order formula using at most n different variables.*

Proof. Let the n variables be $x_1, \ldots, x_n$. The only thing we have to show is that atomic first-order formulas of the form $R(\ldots)$ can be expressed in n-CSA. Only for formulas of the form $R(x_1, \ldots, x_m)$ this is immediate by the expression R. The following example will suffice to explain the argument. Let $n = 3$, $m = 2$, and consider the formula $R(x_2, x_3)$. Note that the expression R expresses the

[4] Imielinski and Lipski [37] were the first to point out the connection between Codd's relational algebra and cylindric set algebras.

formula $R(x_1, x_2)$. We first copy the second column of R into its third column (which is "free:" recall that R stands for $R^{\mathbf{D}} \times |\mathbf{D}|$). Then we cylindrify the second column, after which we copy the first column to the now free second column. We finally cylindrify the first column. Formally, the following n-CSA expression expresses $R(x_2, x_3)$:

$$\gamma_1(\gamma_2(R \cap \delta_{2,3}) \cap \delta_{1,2})$$

In general, the existence of a "free column" guarantees us the room to perform the necessary transpositions on the columns to go from $R(x_1, \ldots, x_m)$ to $R(x_{\rho(1)}, \ldots, x_{\rho(m)})$ for an arbitrary permutation ρ of $\{1, \ldots, n\}$.

The trick used in the above proof is again an idea of Tarski [59]: he used it to give a substitution-free axiom system for first-order logic. Note how crucial it is in this respect that n is strictly larger than the arity of each relation name. Without this condition, the theorem does not seem to hold, although we have not proven this formally. The idea is that, with R binary for example, there are only a finite number of non-equivalent 2-CSA expressions over the schema $\{R\}$. However, there are infinitely many non-equivalent formulas with 2 variables over this schema.

The above theorem gives us a relational algebra for n-variable first-order logic. Bounded-variable fragments of first-order and infinitary logic were vigorously investigated in finite-model theory over the last decade [25,46], for a large part motivated by database theory, in particular the seminal paper by Abiteboul and Vianu [5]. (Another major motivation is descriptive complexity [38].)

4 Relation Algebras

In parallel with his work on cylindric algebras, Tarski also promoted *relation algebras* [18,51,61]. Like cylindric algebras, these are again a generalization of Boolean algebras, but in another direction. They have again an abstract equational definition, but we will only be concerned here with the operations of the *proper relation algebra with base A*, where A is a set of atomic data elements. These operations are defined on *binary* relations over A only and comprise the following:

1. *union*;
2. *complementation* with respect to A^2;
3. *composition:* $r \odot s := \{(x, y) \mid \exists z : (x, z) \in r \text{ and } (z, y) \in s\}$;
4. *conversion:* $\breve{r} := \{(x, y) \mid (y, x) \in r\}$.

For the remainder of this section, we fix a database schema $\mathcal{S}$ with all relation names binary. This is not a heavy restriction: an n-ary relation can easily and naturally be represented by n binary relations, and, as advocates of the *Decomposed Storage Model* argue quite convincingly [22,41], it can even be beneficial to do so from a systems engineering point of view.

Again we can build expressions starting from the relation names in $\mathcal{S}$, and the constant *Id*, using the four operators above. We will call these *RA-expressions*. On a database $\mathbf{D}$ with schema $\mathcal{S}$, an RA-expression e can be evaluated by interpreting the relation names as given by $\mathbf{D}$, interpreting the constant *Id* as the identity relation on $|\mathbf{D}|$, and interpreting the operations relative to base set $|\mathbf{D}|$. The result is a binary relation $e(\mathbf{D})$. So, RA-expressions always express binary queries.

Queries expressible in RA are clearly first-order, and actually a substantial number of first-order queries are expressible in RA. For example:

1. Recalling example query no. 2 from Section 2,
$$(Id \cap R \odot (Id \cup \neg Id)) - (Id \cap (Id \cup \neg Id) \odot R)$$
expresses $\{(x,x) \mid \exists y\, R(x,y) \land \neg\exists z\, R(z,x)\}$.
2. Recalling example query no. 3, $R \odot R \odot R \odot R$ expresses $\{(x_1, x_5) \mid \exists x_2, x_3, x_4 (R(x_1,x_2) \land R(x_2,x_3) \land R(x_3,x_4) \land R(x_4,x_5)\}$.
3. $R \odot (\neg Id \cap \breve{R} \odot R)$ expresses $\{(x,y) \mid R(x,y) \land \exists z \neq y : R(x,z)\}$.

Note that the above three example RA queries can actually already be expressed with a first-order formula using only 3 distinct variables. Indeed, in the first formula we could have reused the bound variable y instead of introducing a new bound variable z. The second query can be equivalently expressed as

$$\{(x,y) \mid \exists z \Big(\exists y(\exists z(R(x,z) \land R(z,y)) \land R(y,z)) \land R(z,y)\Big)\}$$

using just 3 variables. This is no coincidence; it is readily verified that any RA query is in FO^3 (first-order queries expressible using only 3 variables). Tarski and Givant [61] showed the converse: a binary query on binary relations is expressible in RA precisely when it is in FO^3.

4.1 From RA to FO by Pairing

So, RA seems rather limited in expressive power. However, Tarski and Givant showed also that in the presence of a "pairing axiom," RA becomes equally powerful as full first-order logic. We give a nice concretization of this general idea, due to Gyssens, Saxton and Van Gucht [28].

Consider the following two "pairing" operations on a binary relation r on some base set A:

- *left tagging:* $r^{\lhd} := \{(x,(x,y)) \mid (x,y) \in r\}$;
- *right tagging:* $r^{\rhd} := \{((x,y),y) \mid (x,y) \in r\}$.

Note that the resulting relations are not on A anymore, but on $A \cup A^2$. This suggests to build up a universe $\mathbb{U}^+$ on top of $\mathbb{U}$, similar to the universe $\mathbb{U}^*$ we considered in Section 2:

$$\mathbb{U}_0^+ := \mathbb{U}, \quad \mathbb{U}_{n+1}^+ := \mathbb{U}_n^+ \cup (\mathbb{U}_n^+)^2, \quad \mathbb{U}^+ := \bigcup_n \mathbb{U}_n^+.$$

Left tagging and right tagging can now be naturally viewed as operations on binary relations on $\mathbb{U}^+$.

Our objective, of course, is to add the pairing operations to RA. A problem with this, however, is that when we want to evaluate an expression containing these operations on a database $\mathbf{D}$, it no longer makes sense to perform the complementation operation with respect to $|\mathbf{D}|$, as we are really working in $|\mathbf{D}|^+$. We cannot complement with respect to the full $\mathbb{U}^+$ either, as this could yield infinite relations, and we have been working with finite databases from the outset. A simple way out is to redefine complementation relative to the active domain of a relation. So, for any binary relation r, we define

$$\neg r := \{(x,y) \in |r|^2 \mid (x,y) \notin r\},$$

where $|r| = \{x \mid \exists y : (x,y) \in r \text{ or } (y,x) \in r\}$. A second modification we make to RA is that we throw away Id, since we don't need it anymore: for example, $R \cap Id$ is expressible as $R^{\lhd} \odot (R^{\lhd})^{\smile} \cup (R^{\rhd})^{\smile} \odot R^{\rhd}$.

We denote RA, modified as just described, and enriched with the pairing operations, by RA$^+$. Evaluating an RA$^+$-expression e on a database $\mathbf{D}$ in general yields a binary relation $e(\mathbf{D})$ on $|\mathbf{D}|^+$. Such binary relations can represent n-ary relations on $|\mathbf{D}|$. For example, we can represent the ternary relation $\{(a,b,c),(d,e,f)\}$ as $\{(a,(b,c)),(d,(e,f))\}$. Using such representations, we leave it as an exercise to the reader to simulate Codd's relational algebra in RA$^+$. So RA$^+$ has the full power of the first-order queries.

We conclude that Tarski produced two alternatives for Codd's relational algebra: cylindric set algebra, and relation algebra with pairing. From a systems engineering perspective, Codd's algebra remains of course very attractive [48].

4.2 A Computationally Complete Query Language Based on RA$^+$

Recall that on the most general level, given a schema $\mathcal{S}$, we defined a *generic n-ary query on $\mathcal{S}$* to be any (possibly partial) function from databases with schema $\mathcal{S}$ to finite n-ary relations on $\mathbb{U}$ that is invariant under every permutation of $\mathbb{U}$. Genericity allows us to give a standard definition of when a query is "computable." Note that this is not immediate, because standard computability works with concrete data objects, like natural numbers, or strings over a finite alphabet, while our atomic data elements in $\mathbb{U}$ remain abstract.

The computability definition, given in a seminal paper by Chandra and Harel [16], goes as follows. Let $\mathbf{D}$ be a database and suppose the cardinality of $|\mathbf{D}|$ is m. Any bijection from $|\mathbf{D}| \to \{1,\dots,m\}$ is called an *enumeration* of $\mathbf{D}$. The image of $\mathbf{D}$ under an enumeration yields a concrete object with which we can deal using the standard computational models. We now say that query q is *computable* if there is a computable function C in the standard sense, such that for every database $\mathbf{D}$, $q(\mathbf{D})$ is defined if and only if C is defined on every enumeration X of $\mathbf{D}$, and in this case $C(X)$ always equals an enumeration of $q(\mathbf{D})$.

We can capture the class of computable generic queries by making RA$^+$ into a programming language. It suffices to add variables holding finite binary relations on $\mathbb{U}^+$, assignment statements of the form $X := e$, where X is a variable and e is an RA$^+$-expression over the relation names in $\mathcal{S}$ and the variables, and to build up programs from these statements using composition and while-loops of the form '**while** $e \neq \varnothing$ **do** P'. We give programs the obvious operational semantics. For example, the following program computes the transitive closure of R:

$X := R$;
while $(X \odot R) - X \neq \varnothing$ **do**
  $X := X \cup X \odot R$.

Every program expresses a query by designating one of the variables as the output variable. This query is evidently generic and computable.

Since the programming language just described is quite similar to the original query language 'QL' first proved to be computationally complete by Chandra and Harel [16], it does not harm to refer to it by the same name. Using as before a representation of n-ary relations on $\mathbb{U}$ by binary relations on $\mathbb{U}^+$, we then have:

Theorem 2 ([16], see also [3,4,1]). *Every computable generic query is expressible by a QL-program.*

We feel this result is a nice a-posteriori confirmation of Tarski's conviction that relation algebra (with pairing) is a formalism with all the expressive power one needs.

We conclude this section with three remarks. First, given that RA$^+$ expressions evaluate to relations on $\mathbb{U}^+$ rather than on $\mathbb{U}$, one could generalize the notion of query somewhat to yield relations on $\mathbb{U}^+$, rather than on $\mathbb{U}$, as output. Let us refer to this generalized notion of query as *+-query*. The notions of genericity and computable readily generalize to +-queries. Then under these generalizations, the language QL just defined is still computationally complete: every computable generic +-query on binary relations is expressible by a program in QL.

Second, in view of the universe $\mathbb{U}^*$ considered in Section 2, of which $\mathbb{U}^+$ is only a subuniverse, we can generalize +-queries further to *$*$-queries* which now yield relations on $\mathbb{U}^*$ as output. QL is then no longer complete [63], but can be easily reanimated by adding a third tagging operation:

- *set tagging:* $r^{\triangle} := \{(x, \{y \mid (x,y) \in r\}) \mid \exists y : (x,y) \in r\}$.

This is an operation on binary relations on $\mathbb{U}^*$ rather than on $\mathbb{U}^+$. We then again have that QL enriched with set tagging is computationally complete for the generic $*$-queries [23,35].

Finally, note that although the tagging operations introduce pairs and sets of elements, these pairs and sets are still treated by the RA operations as atomic abstract elements. So it is natural to replace every element of $\mathbb{U}^* - \mathbb{U}$ occurring in the output of a $*$-query applied to a database $\mathbf{D}$ by a fresh new element in $\mathbb{U}$ not in $|\mathbf{D}|$. The class of abstract database transformations that can be obtained from computable generic $*$-queries in this way has a purely algebraic characterization [2,64].

5 Constraint Databases

Until now we have worked with relational databases over an unstructured universe $\mathbb{U}$ of atomic data elements. That the atomic data elements remain abstract and uninterpreted is one of the identifying features of classical database theory, and corresponds in practice to the generic bulk-processing nature of database operations. However, in reality $\mathbb{U}$ usually does have a structure of its own, in the form of predicates and functions defined on it, and there are applications where we want to take this structure into account. An important case, on which we will focus in this and the next section, is that of *spatial* databases containing information with a geometrical interpretation.

Suppose we want to store points in the plane in our database. In this case $\mathbb{U}$ is $\mathbb{R}$, the set of real numbers. We use a schema $\mathcal{S}$ with a binary relation name S. In a database $\mathbf{D}$ with schema $\mathcal{S}$, $S^{\mathbf{D}}$ then is a finite set of pairs of real numbers, i.e., a finite set of points in the plane. In the presence of an "interpreted" universe such as $\mathbb{R}$ we need to make a crucial but natural extension to the notion of first-order query: we make the interpreted predicates and functions of our universe available in our first-order formulas. Concretely, in the case of $\mathbb{R}$, we now use formulas over the vocabulary $\mathcal{S}$ expanded with the vocabulary $(<, +, \cdot, 0, 1)$ of ordered fields.

For example, suppose we want to ask whether all points in the database lie on a common circle around the origin. It is tempting to write the following formula for this purpose:

$$\exists r \forall x, y (S(x, y) \rightarrow x^2 + y^2 = r^2)$$

However, we should remember from Section 3 that we agreed to evaluate first-order formulas over the active domain of the database. In contrast, the above formula, and in particular the quantifier $\exists r$, is intended to be evaluated over the whole of $\mathbb{R}$. Since the radius of the circle does not need to be a coordinate of a point in the database, the formula is therefore incorrect as an expression of our query. A correct formula under the active-domain semantics is the following:

$$\exists x_1, y_1 \forall x, y (S(x, y) \rightarrow x^2 + y^2 = x_1^2 + y_1^2))$$

This example shows that the active-domain semantics for first-order formulas is not very natural in the case of spatial databases. It was fine in the case of an uninterpreted universe $\mathbb{U}$, because there, all elements of $\mathbb{U}$ not in the active domain of a database look alike with respect to that database [7]. In contrast, no two reals look alike in first-order logic over the reals with signature $(<, +, \cdot, 0, 1)$.

We thus have two ways of evaluating a first-order formula $\varphi(\bar{x})$ on a real database $\mathbf{D}$. We view $\mathbf{D}$ as an expansion of the structure $(\mathbb{R}, <, +, \cdot, 0, 1)$ with the relations of $\mathbf{D}$. When we then write $\mathbf{D} \models_{\text{adom}} \varphi[\bar{a}]$ for some tuple $\bar{a}$ of reals, we mean that $\varphi[\bar{a}]$ becomes true in $\mathbf{D}$ when we let each quantifier in φ range over $|\mathbf{D}|$ only. When we write $\mathbf{D} \models_{\text{natural}} \varphi[\bar{a}]$, we mean that $\varphi[\bar{a}]$ becomes true when we let each quantifier range over the whole of $\mathbb{R}$. The natural semantics is definitely more natural than the active-domain semantics, but is it really more powerful? The answer is no: Benedikt and Libkin [11] gave an algorithm that turns any formula φ into another formula ψ such that for every real database $\mathbf{D}$ we have $\mathbf{D} \models_{\text{natural}} \varphi$ iff $\mathbf{D} \models_{\text{active}} \psi$. From now on we will stick to the natural semantics.

Given the natural semantics, the new issue arises that the result of a first-order query to a real database can easily be infinite, even though the database is finite. For example, the following first-order query returns all points in the convex closure of S:

$$\{(x,y) \mid \exists x_1, y_1, x_2, y_2, \lambda : S(x_1,y_1) \wedge S(x_2,y_2) \wedge 0 \leqslant \lambda \leqslant 1 \\ \wedge (x,y) = \lambda(x_1,y_1) + (1-\lambda)(x_2,y_2)\}$$

How do we represent these infinite sets?

A solution to this issue was proposed by Kanellakis, Kuper and Revesz in their novel framework of *constraint databases* [40]. Call the above formula φ, and take for example the simple database $\mathbf{D}_0$ with just two points in it: $S^{\mathbf{D}_0} = \{(0,0),(1,1)\}$. To evaluate φ on $\mathbf{D}_0$ we can try the following simple idea called *plug-in evaluation:* replace in φ every atomic subformula of the form $S(u,v)$ by a corresponding formula defining $S^{\mathbf{D}_0}$, i.e., the formula $(u = 0 \wedge v = 0) \vee (u = 1 \wedge v = 1)$. We get the following formula which is purely over the reals only; it no longer mentions any database relations:

$$\{(x,y) \mid \exists x_1, y_1, x_2, y_2, \lambda : \\ ((x_1,y_1) = (0,0) \vee (x_1,y_1) = (1,1)) \\ \wedge ((x_2,y_2) = (0,0) \vee (x_2,y_2) = (1,1)) \\ \wedge 0 \leqslant \lambda \leqslant 1 \wedge (x,y) = \lambda(x_1,y_1) + (1-\lambda)(x_2,y_2)\}$$

This formula defines the infinite set of points in $\mathbb{R}^2$ on the closed line segment between $(0,0)$ and $(1,1)$ and symbolically represents the answer of our query on our database.

We can always perform plug-in evaluation of a first-order query on a real database, provided the numbers occurring in the database are rational so that we can effectively write down the formula defining the answer. In general, the subsets of $\mathbb{R}^n$ that are definable (as n-ary relations) by first-order formulas over $\mathbb{R}$ are known as the *semi-algebraic sets* [10,14]. They have quite nice properties.

Is this representation of semi-algebraic sets by real formulas workable? Can we, e.g., effectively decide whether the set defined by a given real formula is nonempty? Thanks to Tarski's decision procedure for the first-order theory of the reals [52], the representation is indeed quite workable. The computational complexity of Tarski's original procedure is very high, but over the years there has been steady progress in algorithms for real algebraic geometry [9,15,30,45]. A crucial parameter in the computational complexity of current algorithms is the number of quantifiers in the formula. In the case of plug-in evaluation this number depends only on the query formula and not on the database, which implies a polynomial time complexity.

Tarski actually gave a complete quantifier elimination procedure for real formulas, so we can define every semi-algebraic set already by a quantifier-free formula. To arrive at the concept of constraint database, we make one final but logical step: we allow semi-algebraic sets not just as outputs of queries, but also as inputs. Specifically, we remove the restriction that each relation in the database must be finite, and require instead that each relation must be a semi-algebraic set. In a *constraint database* we thus no longer store any actual tuples, but we store a collection of quantifier-free formulas, one for each relation name in the schema of the database. Plug-in evaluation is still possible: given a first-order formula φ and a constraint database $\mathbf{D}$, we replace in φ every atomic subformula of the form $S(u,v)$, with S a relation name from the database schema, by the real formula $\gamma(u,v)$ defining $S^{\mathbf{D}}$. The result of all these replacements is a real formula defining $\varphi(\mathbf{D})$. Since thanks to Tarski we can work with quantifier-free formulas, the computational complexity of deciding nonemptiness and many other algorithms remains polynomial-time like we had with finite databases.

Constraint databases has been an active research area over the past decade [39,43,49,62,43], and the constraint database concept is not at all limited to the reals. In principle it works for any universe $\mathbb{U}$ with a certain structure (consisting of predicates and functions) that admits effective quantifier elimination (and for which truth of atomic sentences is decidable)[5]. In this respect, recall that, as Tarski himself noted early on [57], his quantifier elimination for the reals implies that *every subset of* $\mathbb{R}$ *definable by a first-order formula with parameters over the reals is the union of a finite number of intervals.* This property alone is already responsible for a lot of the nice properties (referred to as "tame topology" [65]) of semi-algebraic sets. Any universe that has this property is called *o-minimal.* The "collapse" of the natural semantics for first-order logic on finite database to the active-domain semantics, which we pointed out earlier, does not just hold under the reals but under *any* universe that is o-minimal and admits quantifier elimination [11].

6 Geometric Queries

We conclude our survey by making the circle complete and returning to the first topic we discussed: the notion of genericity for database queries, now reconsidered in the new setting of spatial databases.

We begin by noting that, when thinking about spatial data, the real numbers as "urelemente" are not the right level of abstraction. They show up merely as a convenient representation, via coordinates, for the real urelemente which are the points in our geometric space. Let us work as before in the real plane $\mathbb{R}^2$ (everything we will say in this section generalizes to arbitrary $\mathbb{R}^d$). This means in general that we only work with database schemas $\mathcal{S}$ where the arity of every relation name is a multiple of 2. In a *geometric database* with schema $\mathcal{S}$, each relation, of arity $2n$, say, is interpreted as an n-ary relation on $\mathbb{R}^2$. Following

[5] Interesting recent work even considers universes that are non-numeric, such as strings or terms [12,13,24].

the constraint database approach outlined in the previous section, each relation is semi-algebraic. We call such databases *geometric.* An *n-ary geometric query* over $\mathcal{S}$ then is a function mapping geometric databases with schema $\mathcal{S}$ to n-ary relations on $\mathbb{R}^2$.

Recalling our discussion in Section 2, we can now again call a geometric query *generic* if it is invariant under all permutations of $\mathbb{R}^2$ (note: $\mathbb{R}^2$, not $\mathbb{R}$). This does make sense: these are the queries that treat the points as uninterpreted abstract data elements. They are exactly the classical generic queries under the unstructured universe $\mathbb{U}$ where this $\mathbb{U}$ happens to be $\mathbb{R}^2$. Many interesting geometric queries are not so "absolutely" generic, however, and this is as it should be: as we mentioned at the end of Section 2, Tarski viewed logic as an "extreme" kind of geometry. Thus, by considering various other groups of transformations of $\mathbb{R}^2$, corresponding to various geometrical interpretations of our spatial data, we can reach the geometric queries that fit the particular interpretation.

Let us illustrate this using the group of affinities, i.e., the permutations of $\mathbb{R}^2$ that preserve betweenness. We call a geometric query *affine-generic* if it is invariant under all affinities of $\mathbb{R}^2$. For example, consider the following geometric queries over a set of points S (i.e., a binary relation S):

1. Is S nonempty? In first-order: $\exists x, y\, S(x,y)$
2. Is S convex? In first-order: $\forall x_1, y_1, x_2, y_2, \lambda : (S(x_1,y_1) \wedge S(x_2,y_2) \wedge 0 \leqslant \lambda \leqslant 1) \rightarrow S(\lambda(x_1,y_1) + (1-\lambda)(x_2,y_2))$
3. Is S a circle? In first-order: $\exists r, x_0, y_0 \forall x, y : S(x,y) \leftrightarrow (x-x_0)^2+(y-y_0)^2 = r^2$

Query 1 is absolutely generic; query 2 is only affine-generic; and query 3 is not affine-generic (the notion of "circle" does not exist in affine geometry).

We now come to a natural question: how can we logically characterize the affine-generic first-order geometric queries? Tarski brings us inspiration. In his work on the axiomatization of elementary geometry [58,50], Tarski considered first-order logical formalisms with variables ranging over points in the geometric space, and elementary predicates on these points as dictated by the geometry to be formalized. For example, elementary affine geometry in the real plane corresponds to doing first-order logic in the structure $(\mathbb{R}^2, \beta)$, where β denotes the ternary betweenness predicate on $\mathbb{R}^2$: $\beta(p,q,r)$ holds if point p lies on the closed line segment between points q and r.

Inspired by this, we can view a geometric database $\mathbf{D}$ with schema $\mathcal{S}$ as a constraint database over the interpreted universe $(\mathbb{R}^2, \beta)$ rather than over the universe $(\mathbb{R}, <, +, \cdot, 0, 1)$. Note that under this alternative view, the schema changes: the arity of each relation name S goes from $2n$ to n. We denote this "halved" schema by $\mathcal{S}'$. First-order queries under the alternative view now are expressed by first-order formulas over the vocabulary $(\mathcal{S}', \beta)$ rather than $(\mathcal{S}, <, +, \cdot, 0, 1)$. Let us refer to the original class of first-order queries as FO[$\mathbb{R}$] and to the new one as FO[β]. Example queries 1 and 2 from above are expressed in FO[β] as follows:

1. $\exists p\, S(p)$
2. $\forall p, q, r : (S(p) \wedge S(q) \wedge \beta(r,p,q)) \rightarrow S(r)$

Example query 3 is not in FO[β]: indeed, query 3 is not affine-generic, and FO[β] clearly contains only affine-generic queries. It is also clear that FO[β] is a subclass of FO[$\mathbb{R}$]: we represent every point variable p by a pair of real variables (p_1, p_2), and $\beta(p, q, r)$ is easily expressible as $(q_1 - p_1) \cdot (p_2 - r_2) = (q_2 - p_2) \cdot (p_1 - r_1) \wedge 0 \leqslant (q_1 - p_1) \cdot (p_1 - r_1) \wedge 0 \leqslant (q_2 - p_2) \cdot (p_2 - r_2)$.

Interestingly, the converse holds as well:

Theorem 3 ([29]). *Every affine-generic geometric query in* FO[$\mathbb{R}$] *is in* FO[β].

The proof uses the original observation by Tarski that the geometrical constructions of $<$, $+$ and $\cdot$ of points on a line can be defined in first-order logic over β.

Other geometric interpretations can be captured in a similar way: for example, if we use instead of the betweenness predicate, the (4-ary) equal-distance predicate, we obtain euclidean rather than affine geometry. Capturing topological queries (invariant under all homeomorphisms) is much more difficult [42,27].

7 Conclusion

We have surveyed some ideas and results from database theory related to Tarski's ideas. We have neither been comprehensive with respect to Tarski's work, nor with respect to database theory, and probably not even with respect to the applications of the former in the latter. A great source to learn more about Tarski are his Collected Papers [26]. A great source to learn more about database theory are the proceedings of the annual ACM Symposium on Principles of Database Systems published by ACM Press, and the biannual International Conference on Database Theory published in Springer's Lecture Notes in Computer Science.

References

1. S. Abiteboul, R. Hull, and V. Vianu. *Foundations of Databases.* Addison-Wesley, 1995.
2. S. Abiteboul and P.C. Kanellakis. Object identity as a query language primitive. *Journal of the ACM*, 45(5):798–842, 1998.
3. S. Abiteboul and V. Vianu. Procedural languages for database queries and updates. *Journal of Computer and System Sciences*, 41(2):181–229, 1990.
4. S. Abiteboul and V. Vianu. Datalog extensions for database queries and updates. *Journal of Computer and System Sciences*, 43(1):62–124, 1991.
5. S. Abiteboul and V. Vianu. Computing with first-order logic. *Journal of Computer and System Sciences*, 50(2):309–335, 1995.
6. A.V. Aho and J.D. Ullman. Universality of data retrieval languages. In *Conference Record, 6th ACM Symposium on Principles of Programming Languages*, pages 110–120, 1979.
7. A.K. Aylamazyan, M.M. Gilula, A.P. Stolboushkin, and G.F. Schwartz. Reduction of the relational model with infinite domains to the case of finite domains. *Doklady Akademii Nauk SSSR*, 286(2):308–311, 1986. In Russian.

8. F. Bancilhon. On the completeness of query languages for relational data bases. In *Proceedings 7th Symposium on Mathematical Foundations of Computer Science*, volume 64 of *Lecture Notes in Computer Science*, pages 112–123. Springer-Verlag, 1978.
9. S. Basu. *Algorithms in Semi-Algebraic Geometry.* PhD thesis, New York University, 1996.
10. R. Benedetti and J.-J. Risler. *Real Algebraic and Semi-Algebraic Sets.* Hermann, 1990.
11. M. Benedikt and L. Libkin. Relational queries over interpreted structures. *Journal of the ACM*, 47(4):644–680, 2000.
12. M. Benedikt, L. Libkin, T. Schwentick, and L. Segoufin. String operations in query languages. In *Proceedings 20th ACM Symposium on Principles of Database Systems*, 2001.
13. A. Blumensath and E. Grädel. Automatic structures. In *Proceedings 15th IEEE Symposium on Logic in Computer Science*, pages 51–62, 2000.
14. J. Bochnak, M. Coste, and M.-F. Roy. *Real Algebraic Geometry.* Springer-Verlag, 1998.
15. B.F. Caviness and J.R. Johnson, editors. *Quantifier elimination and cylindrical algebraic decomposition.* Springer, 1998.
16. A. Chandra and D. Harel. Computable queries for relational data bases. *Journal of Computer and System Sciences*, 21(2):156–178, 1980.
17. A. Chandra and D. Harel. Structure and complexity of relational queries. *Journal of Computer and System Sciences*, 25:99–128, 1982.
18. L.H. Chin and A. Tarski. Distributive and modular laws in the arithmetic of relation algebras. *University of California Publications in Mathematics – New Series*, 1(9):341–384, 1951.
19. E. Codd. A relational model for large shared databanks. *Communications of the ACM*, 13(6):377–387, 1970.
20. E. Codd. Relational completeness of data base sublanguages. In R. Rustin, editor, *Data Base Systems*, pages 65–98. Prentice-Hall, 1972.
21. D. Cohen, M. Gyssens, and P. Jeavons. Derivation of constraints and database relations. In E.C. Freuder, editor, *Principles and Practice of Constraint Programming*, volume 1118 of *Lecture Notes in Computer Science*, pages 468–481, 1996.
22. G.P. Copeland and S. Khoshafian. A decomposition storage model. In *Proceedings of ACM-SIGMOD International Conference on Management of Data*, volume 14:4 of *SIGMOD Record*, pages 268–279. ACM Press, 1985.
23. E. Dahlhaus and J.A. Makowsky. Query languages for hierarchic databases. *Information and Computation*, 101(1):1–32, 1992.
24. E. Dantsin and A. Voronkov. Expressive power and data complexity of query languages for trees and lists. In *Proceedings 19th ACM Symposium on Principles of Database Systems*, pages 157–165, 2000.
25. H.-D. Ebbinghaus and J. Flum. *Finite Model Theory.* Springer, second edition, 1999.
26. S.R. Givant and R.N. McKenzie, editors. *Alfred Tarski, Collected Papers.* Birkhäuser, 1986.
27. M. Grohe and L. Segoufin. On first-order topological queries. In *Proceedings 15th IEEE Symposium on Logic in Computer Science*, pages 349–360, 2000.
28. M. Gyssens, L.V. Saxton, and D. Van Gucht. Tagging as an alternative to object creation. In J.C. Freytag, D. Maier, and G. Vossen, editors, *Query Processing For Advanced Database Systems*, chapter 8. Morgan Kaufmann, 1994.

29. M. Gyssens, J. Van den Bussche, and D. Van Gucht. Complete geometric query languages. *Journal of Computer and System Sciences*, 58(3):483–511, 1999.
30. J. Heintz, T. Recio, and M.-F. Roy. Algorithms in real algebraic geometry and applications to computational geometry. In J. Goodman, R. Pollack, and W. Steiger, editors, *Discrete and Computational Geometry*, volume 6 of *DIMACS Series in Discrete Mathematics and Theoretical Computer Science*. AMS-ACM, 1991.
31. L. Henkin, J.D. Monk, and A. Tarski. *Cylindric Algebras. Part I.* North-Holland, 1971.
32. L. Henkin, J.D. Monk, and A. Tarski. *Cylindric Algebras. Part II.* North-Holland, 1985.
33. L. Henkin, J.D. Monk, A. Tarski, H. Andréka, and I. Németi. *Cylindric Set Algebras*, volume 883 of *Lecture Notes in Mathematics*. Springer-Verlag, 1981.
34. L. Henkin and A. Tarski. Cylindric algebras. In R.P. Dilworth, editor, *Lattice Theory*, volume 2 of *Proceedings of Symposia in Pure Mathematics*, pages 83–113. American Mathematical Society, 1961.
35. R. Hull and J. Su. Algebraic and calculus query languages for recursively typed complex objects. *Journal of Computer and System Sciences*, 47(1):121–156, 1993.
36. R. Hull and C.K. Yap. The format model, a theory of database organization. *Journal of the ACM*, 31(3):518–537, 1984.
37. T. Imielinski and W. Lipski. The relational model of data and cylindric algebras. *Journal of Computer and System Sciences*, 28:80–102, 1984.
38. N. Immerman. *Descriptive Complexity*. Springer, 1999.
39. P.C. Kanellakis. Constraint programming and database langauges: a tutorial. In *Proceedings 14th ACM Symposium on Principles of Database Systems*, pages 46–53, 1995.
40. P.C. Kanellakis, G.M. Kuper, and P.Z. Revesz. Constraint query languages. *Journal of Computer and System Sciences*, 51(1):26–52, August 1995.
41. S. Khoshafian, G.P. Copeland, T. Jagodi, H. Boral, and P. Valduriez. A query processing strategy for the decomposed storage model. In *Proceedings of the Third International Conference on Data Engineering*, pages 636–643. IEEE Computer Society, 1987.
42. B. Kuijpers and V. Vianu. Topological queries. In Libkin et al. [43], chapter 10.
43. L. Libkin, G. Kuper, and J. Paredaens, editors. *Constraint Databases*. Springer, 2000.
44. A. Lindenbaum and A. Tarski. On the limitations of the means of expression of deductive theories. In *Logic, Semantics, Metamathematics. Papers from 1923–1938* [56], pages 384–392.
45. B. Mishra. Computational real algebraic geometry. In J.E. Goodman and J. O'Rourke, editors, *Handbook of Discrete and Computational Geometry*. CRC Press, 1997.
46. M. Otto. *Bounded variable logics and counting: a study in finite models*, volume 9 of *Lecture Notes in Logic*. Springer, 1997.
47. J. Paredaens. On the expressive power of the relational algebra. *Information Processing Letters*, 7(2):107–111, 1978.
48. R. Ramakrishnan and J. Gehrke. *Database Management Systems*. McGraw-Hill, second edition, 2000.
49. P.Z. Revesz. Constraint databases: a survey. In L. Libkin and B. Thalheim, editors, *Semantics in Databases*, volume 1358 of *Lecture Notes in Computer Science*, pages 209–246. Springer, 1998.
50. W. Schwabhäuser, W. Szmielew, and A. Tarski. *Metamathematische Methoden in der Geometrie*. Springer-Verlag, 1983.

51. A. Tarski. On the calculus of relations. *Journal of Symbolic Logic*, 6:73–89, 1941.
52. A. Tarski. *A Decision Method for Elementary Algebra and Geometry*. University of California Press, 1951.
53. A. Tarski. Some notions and methods on the borderline of algebra and metamathematics. In *Proceedings of the International Congress of Mathematicians, Cambridge, Mass, 1950*, volume 1, pages 705–720. American Mathematical Society, 1952.
54. A. Tarski. Contributions to the theory of models, I and II. *Indagationes Mathematicae*, 16:572–581 and 582–588, 1954. Volume III, which contains the list of references, is in volume 17 of the same journal.
55. A. Tarski. The concept of truth in formalized languages. In *Logic, Semantics, Metamathematics. Papers from 1923–1938* [56], pages 152–278.
56. A. Tarski. *Logic, Semantics, Metamathematics. Papers from 1923–1938*. Clarendon Press, Oxford, 1956.
57. A. Tarski. On definable sets of real numbers. In *Logic, Semantics, Metamathematics. Papers from 1923–1938* [56], pages 110–142.
58. A. Tarski. What is elementary geometry? In L. Henkin, P. Suppes, and A. Tarski, editors, *The Axiomatic Method, with Special Reference to Geometry and Physics*, pages 16–29. North-Holland, 1959.
59. A. Tarski. A simplified formalization of predicate logic with identity. *Archiv für Mathematische Logik und Grundlagenforschung*, 7:61–79, 1965.
60. A. Tarski. What are logical notions? *History and Philosophy of Logic*, 7:143–154, 1986. Edited by J. Corcoran.
61. A. Tarski and S. Givant. *A Formalization of Set Theory Without Variables*, volume 41 of *AMS Colloquium Publications*. American Mathematical Society, 1987.
62. J. Van den Bussche. Constraint databases: a tutorial introduction. *SIGMOD Record*, 29(3):44–51, 2000.
63. J. Van den Bussche and J. Paredaens. The expressive power of complex values in object-based data models. *Information and Computation*, 120:220–236, 1995.
64. J. Van den Bussche, D. Van Gucht, M. Andries, and M. Gyssens. On the completeness of object-creating database transformation languages. *Journal of the ACM*, 44(2):272–319, 1997.
65. L. van den Dries. *Tame Topology and O-Minimal Structures*. Cambridge University Press, 1998.
66. I.M. Yaglom. *Felix Klein and Sophus Lie: evolution of the idea of symmetry in the nineteenth century*. Birkhäuser, Boston, 1988.

Locus Solum: From the Rules of Logic to the Logic of Rules

Jean-Yves Girard

CNRS, Institut de Mathematiques de Luminy
163 Avenue de Luminy, case 907
13288 Marseille cedex 9
girard@iml.univ-mrs.fr

Abstract. Logic is no longer about a preexisting external reality, but about its own protocols, its own geometry. Typically the negation is not about saying "NOT", but about the mirror, the duality "I" vs. "the world" ...
The new approach encompasses the old one, typically if "I" win, "the world" loses, i.e., wins "NOT". When logical artifacts are identified with their own rules of production, LOCATIVE phenomenons arise. In particular, one realises that usual logic (including linear logic) is SPIRITUAL, i.e., up to isomorphism. But there is a deeper locative level, with indeed a more regular structure. Typically the usual (additive) conjunction has the value of categorical product in usual logic, and enjoys commutativity, associativity, etc. up to isomorphism.
In ludics, what corresponds is a plain intersection $G \cap H$, which is really associative, commutative, etc. (no isomorphisms); it contains the usual conjunction as a delocalised case $\varphi(G) \cap \psi(H)$. Incidentally this shows that the categorical view of logic – if very useful – is wrong... Nature abhors an isomorphism!
LUDICS is a monist approach to logic–without this nonsense distinction syntax/semantics/meta – just plain logical artifacts, period.

Biblio: Locus Solum, available at:
http://iml.univ-mrs.fr/ girard/Articles.html

L. Fribourg (Ed.): CSL 2001, LNCS 2142, p. 38, 2001.

The Expressive Power of Horn Monadic Linear Logic

Max Kanovich[1,2]

[1] Russian State University for the Humanities, Moscow, Russia
[2] Computer and Information Science Dept., Univ. of Pennsylvania,
200 S. 33rd Str. Philadelphia, PA 19104
kanovich@saul.cis.upenn.edu

Abstract. The most fundamental results of monadic second-order decidability, beyond the decidability of just pure monadic second-order logic, deal with the decidability of the monadic second-order theories of one and two successors and the decidability of the monadic second-order theory of linear order (Büchi, Rabin). Having moved from *sets* to *multisets*, we refine the underlying logic as linear logic. In contrast to the classical results, we prove the undecidability of *just pure monadic linear logic*, even if we use nothing but Horn formulas built up of unary predicates, in which no functional symbols are present. As for affine logic (linear logic plus weakening), we prove the undecidability of the Horn fragment of affine logic, which involves only one binary predicate ("linear order") and a fixed finite number of unary predicates, and which contains no functional symbols at all. We also show the undecidability of the $\exists$-free Horn fragment of monadic affine logic in the presence of only one constant symbol ("zero") and only one unary functional symbol ("successor"), and a fixed finite number of unary predicate symbols.
Along these lines, we obtain the undecidability of the *optimistic protocol completion* even for the class of communication protocols with two participants such that either of them is a finite automaton provided with one register capable of storing one atomic message, all the predicates used are at most unary, and no compound messages are in the use.

1 Motivations and Summary

The most fundamental results of monadic second-order decidability, beyond the decidability of just pure monadic second-order logic, deal with the decidability of the monadic second-order theories of one and two successors and the decidability of the monadic second-order theory of linear order [1,26,27,11]. Since the traditional monadic theories are dealing with *sets*, it involves the traditional Boolean logic over atomic formulas ($x \in y$). Having moved from *sets* to *multisets*, we refine the underlying logic as linear logic [9][1]. In particular, within the linear logic

[1] Here we confine ourselves to the so-called multiplicative-exponential fragment of linear logic without "additives". The *Horn* fragments under consideration are the simplest fragments of the multiplicative-exponential linear logic. As for the full propositional multiplicative-additive-exponential linear logic, it is undecidable [21].

L. Fribourg (Ed.): CSL 2001, LNCS 2142, pp. 39–53, 2001.

framework, the fact that "two copies of x are in y" can be expressed as a simple formula: $((x \in y) \otimes (x \in y))$. Contrary to what might have been expected, we prove the undecidability of *just pure monadic linear logic*, even if we use nothing but Horn formulas built up of unary predicates, which involve no functional symbols at all. This undecidability result is formulated in terms of communication protocols [25,6,2,22,8]. As for verifying formal specifications of protocols, prior to analysis of protocol properties related to intruders, we have to show that our *formal specification* of a given protocol meets the protocol rules and conventions at least under ideal conditions in the absence of an intruder. We prove that the problem of the *optimistic protocol completion: "Whether the protocol participants can finish playing in accordance with a protocol (when no one interferes with network transmission)"*, is undecidable, even if each of the participants is a finite automaton provided with one register capable of storing one atomic message, all the predicates in question are at most unary, and no compound messages are in the use. As compared to [3,7], the undecidability of the secrecy problem there is essentially based on the Cook encoding of Turing machines in terms of classical *binary* predicates. Since the classical monadic cases are decidable, their results are not directly translated into the pure monadic case.

From the Horn point of view, we accomplish the full picture. The propositional pure Horn linear logic is decidable but its complexity is of that of the reachability problem for Petri nets [10,13]. (The decision problem for the full propositional multiplicative-exponential fragment of linear logic is still open.) The next step - that is toward the pure monadic Horn fragment of linear logic, is shown here to yield undecidability.

In the second part of the paper we consider 'monadic' *affine logic*, that is linear logic with the Weakening rule. The full *propositional* affine logic is known to be decidable [17]. As for *pure monadic affine logic*, the problem is open. For affine logic, the *faithfulness* of our encoding in Theorem 3 fails exactly in the most subtle case of the zero-test. Nevertheless, in contrast to [26,1], we prove the undecidability of the Horn fragment of pure monadic affine logic, which, in addition, has one binary predicate symbol ("linear order"), as well as the undecidability of the Horn fragment of monadic affine logic, which involves only one constant ("zero") and one unary functional symbol ("successor").

The undecidability proofs are established by the following scheme. The program of a machine M is specified in terms of linear logic formulas taken from a certain Horn fragment in such a way that any M's computation can be "easily" transformed into a derivation for a certain 'target' sequent. The most complicated direction is the opposite one - that of extraction of an M's computation from a given derivation for the 'target' sequent. This *faithfulness* problem can be resolved by analysis of the particular derivations specified in the case of each of the encoding techniques. We sort out the *faithfulness* problem with the help of the comprehensive computational interpretation developed for the Horn fragments of linear and affine logics [13,14,15] and its generalizations.

2 Horn Linear Logic

Within the monadic paradigm, *atomic* formulas are of the form $(x \in y)$. On the other hand, any formula of the form $(x \in \mathcal{P})$, with $\mathcal{P}$ being a constant, can be thought of as a *unary* predicate, say $P(x)$. Therefore, the first-order Horn formulas, which invoke only unary predicates, can be treated as being inside the monadic language.

Definition 1. *A signature* Σ *consists of predicate symbols* Q_1, Q_2,... *with their arity, functional symbols* f_1, f_2,..., *and constants* c_1, c_2,.... *An* atomic formula *is* **1** *or a formula of the form:* $Q_j(t_1, t_2, .., t_k)$, *with* t_i *being* functional terms. *An* elementary product X *is defined as an expression of the form*

$$X := (P_1(t_{1,1}, .., t_{1,k_1}) \otimes P_2(t_{2,1}, .., t_{2,k_2}) \otimes \cdots \otimes P_m(t_{m,1}, .., t_{m,k_m}))$$

where P_1,P_2,..,P_m *are predicate symbols,* $t_{1,1}$,..,t_{1,k_1},...,$t_{m,1}$,..,t_{m,k_m} *are terms.*

We say that elementary products X *and* X' *are* equivalent, *and write* $X \simeq X'$, *if* X *does not differ from* X' *modulo commutativity and associativity of* $\otimes$.

A Horn formula *is defined as a closed formula of the form*

$$\forall x_1..x_n \, (X(x_1, .., x_n, a_1, .., a_p) \multimap \exists y_1..y_m \, Y(x_1, .., x_n, y_1, .., y_m, b_1, .., b_q))$$

where $X(x_1, .., x_n, a_1, .., a_p)$ *and* $Y(x_1, .., x_n, y_1, .., y_m, b_1, .., b_q)$ *are elementary products, and the list* a_1,..,a_p, b_1,..,b_q *contains all the constants occurring there.*

There is a clear isomorphism between elementary products (modulo commutativity) and finite multisets consisting of atomic formulas. E. g., the collection of facts $\{R(f(c)), S(d), S(d)\}$ is represented by a product of the form: $(R(f(c)) \otimes S(d) \otimes S(d))$, and vice versa. Thus any closed elementary product $W(a_1, .., a_n)$ is conceived of as a description of a certain *configuration*, or *situation*, in the system. A Horn formula $\forall\overline{x}(X(\overline{x}) \multimap \exists\overline{y}\, Y(\overline{x}, \overline{y}))$ is conceived of as an *instruction* to transform configurations in the *multiset rewriting* manner. Namely, a "part" of the form $X(\overline{t})$ within the current configuration can be replaced with a "part" of the form $Y(\overline{t}, \overline{d})$, where $\overline{d}$ is the vector of *fresh* constants generated in case. E. g., a typical programming assignment: $r := f(r)$, can be axiomatized as: $\forall x\, (R(x) \multimap R(f(x)))$, where the intended meaning of $R(v)$ is that: "The register r contains v".

Definition 2. *Let* Γ *be a set of Horn formulas. A* scenario $\mathcal{S}$ *is a course of events* in accordance with Γ, *which starts in a certain* initial configuration W. *Formally,* scenario $\mathcal{S}$ *is a rooted chain of vertices* v_0, v_1, v_2,... *such that*

(a) Every vertex v *is labelled by a 'configuration', that is a closed elementary product* $\mathtt{Config}(v)$. *For root* v_0, $\mathtt{Config}(v_0) := W$.

(b) Every edge (v, w) *is labelled by a closed* Horn sequent *of the form*

$$X(t_1, .., t_n, a_1, .., a_p) \vdash Y(t_1, .., t_n, d_1, .., d_m, b_1, .., b_q), \qquad (1)$$

such that d_1,..,d_m *are distinct* fresh *constants: each of these* d_j *neither occurs in* Γ, *nor occurs in* W, *nor occurs in the Horn sequents the previous edges are labelled by, and for some* $V(e_1, .., e_k)$:

$$\begin{cases} \mathtt{Config}(v) \simeq (X(t_1, .., t_n, a_1, .., a_p) \otimes V(e_1, .., e_k)), \\ \mathtt{Config}(w) \simeq (Y(t_1, .., t_n, d_1, .., d_m, b_1, .., b_q) \otimes V(e_1, .., e_k)), \end{cases}$$

and Γ *contains a formula, a 'general pattern' of (1), of the form:*

$$\forall x_1..x_n \, (X(x_1, .., x_n, a_1, .., a_p) \multimap \exists y_1..y_m \, Y(x_1, .., x_n, y_1, .., y_m, b_1, .., b_q)).$$

What we need here is to extend the *comprehensive* computational interpretation for Horn linear logic introduced in [13,14,15] to the first-order case.

Theorem 1 (Scenarios $\Longrightarrow$ Proofs). *Let Γ be a set consisting of Horn formulas. Let $W(a_1,..,a_p)$ and $Z(d_1,..,d_k,b_1,..,b_q)$ be closed elementary products, where $a_1,..,a_p$, $d_1,..,d_k$, $b_1,..,b_q$ are constants, and each of the constants $b_1,..,b_q$ occurs either in Γ or in $W(a_1,..,a_p)$. Then any scenario $\mathcal{S}$ in accordance with Γ, which leads from $W(a_1,..,a_p)$ to $Z(d_1,..,d_k,b_1,..,b_q)$, can be transformed into a linear logic derivation of a sequent of the form:*

$$!\Gamma,\, W(a_1,..,a_p) \vdash \exists z_1,..,z_k\, Z(z_1,..,z_k,b_1,..,b_q). \quad (2)$$

Here $!\Gamma$ stands for the set resulting from putting ! before each formula in Γ.

Proof. The desired proof is assembled by induction on $\mathcal{S}$. □

Theorem 2 (Proofs $\Longrightarrow$ Scenarios). *Let Γ consist of Horn formulas. Given a linear logic derivation for a sequent of the form (2), we can form a scenario $\mathcal{S}$ in accordance with Γ, which leads from $W(a_1,..,a_p)$ to $Z(s_1,..,s_k,b_1,..,b_q)$, for some closed terms $s_1,..,s_k$. Furthermore, these 'programs' $s_1,..,s_k$ can be taken in such a way that each of their functional symbols occurs in Γ or in $W(a_1,..,a_p)$.*

Proof. We assemble the corresponding chain-like scenarios by induction on the cut-free derivations running from their leaves to their roots (see [13,14,15]). □

3 Just Pure Monadic Case: No Functional Symbols

The undecidability of *pure monadic linear logic* is established in terms of protocols, which define the communication framework between two or more agents [25,6,2,22,8]. We consider here a *finite state message passing model*, wherein a *protocol*, as a theater director, is dealing with a cast of actors, each of them is a *finite* automaton, supplied with a fixed *finite* memory. Every actor is keeping up its part stipulated by the protocol. (*Intruders* may follow their own policy.) Besides their pure finite automaton actions, the actors may give/take the cues to/from the others. Since the actors cannot communicate directly on a local stage, they send *messages* on the *network*, ("From Alice to Bob: below is my public key"), or, in turn, receive them as *network* messages. As stipulated by the protocol, the actors have a chance to come out with their *nonce-word* messages, ("From Bob to Alice: below is my secret encrypted with your public key").

A prior property of a protocol is that of the *protocol completion* under 'optimistic' conditions: *Whether the actors can finish playing (when no one interferes with network transmission).*

According to the Dolev-Yao paradigm [6], *messages* are composed of indivisible abstract values by means of certain functions, like *encryption, decryption, signing, pairing*, etc. In this section we confine ourselves to the simplest case where *no functions are present*, and thereby all messages are *indivisible*.

We show undecidability of the *optimistic protocol completion* for such a restricted class of protocols, where, in addition, there are only two participants:

A ("Alice") and B ("Bob"), each provided with a *unique* register capable of storing one atomic message.

We will encode a *deterministic* 2-counter Minsky machine M [24], whose program, a list of *instructions* $I_1,\ldots,I_e$, is re-organized in such a way that M jumps from one counter to the other "by turns", namely, for certain disjoint finite sets of 'states' a_0, a_1,.., a_i,.. (associated with r_1) and b_0, b_1,.., b_j,.., (associated with r_2) each of M's instructions is of one of the following labelled forms $(i, j \geq 1)$ [2]:

"jump"	(O) a_i : **goto** b_j;	(O) b_j : **goto** a_k;
"increment"	(I) a_i : $r_1 := r_1+1$; **goto** b_j;	(I) b_j : $r_2 := r_2+1$; **goto** a_k;
"decrement"	(II) a_i : $r_1 := r_1-1$; **goto** b_j;	(II) b_j : $r_2 := r_2-1$; **goto** a_k;
"zero-test"	(III) a_i : **if** $(r_1 = 0)$ **goto** b_j;	(III) b_j : **if** $(r_2 = 0)$ **goto** a_k;

No different instructions are labelled by the same label. States a_1 and a_0 are the *initial* and *final* states of M, respectively. The *M's configuration* where M is in state m, and k_1 and k_2 are the current values of counters r_1 and r_2, respectively, is denoted by $(m; k_1, k_2)$. A *computation* performed by M is a sequence of M's configurations such that each step is made by one of the above instructions.

Within the language of our encoding, labels a_0, a_1,.., a_i,.. and b_0, b_1,.., b_j,.., will work as *constants*. We invoke also the following predicates:

(a) $N(x)$ is intended to show that: "Message x is on the network".
For m of the form a_i or b_j, $N(m)$ is also conceived of as: "M is in state m".
(b) $A_1(x)$ means that: "The unique register in the memory of A contains x", which is also conceived of as: "x is active with respect to M's counter r_1".
(c) $B_1(x)$ tells us that: "The unique register in the memory of B contains x", which is also conceived of as: "x is active with respect to M's counter r_2".
(d) The states of automaton A are named by propositions:
p_1^A, q_0^A, r^A, s_0^A, s_1^A,.., s_i^A,..., respectively.
(e) The states of automaton B are named by propositions:
p_1^B, q_0^B, r^B, s_0^B, s_1^B,.., s_j^B,..., respectively.

Given a number n, the role played by A ("Alice") is defined as follows (each of A's actions is attached by a Horn formula, its formal axiomatization):

(1) Being in her *initial* state p_1^A, A generates a *nonce*, say n_A, (a random number, which differs from all a_i and b_j, and the previous nonces, if any), stores it, sends n copies of n_A on the network, and goes to s_1^A:

$$\alpha_n := (p_1^A \multimap \exists y\,(s_1^A \otimes A_1(y) \otimes \underbrace{N(y) \otimes N(y) \otimes \cdots \otimes N(y)}_{n \texttt{ times}})) \tag{3}$$

(2) Being in s_i^A $(i \geq 1)$, A "performs" the instruction I of the form: a_i :... **goto** b_j, and signals that it is B's turn next to "perform" the instruction labelled by b_j:

(2a) Case O: I is of the form: a_i : **goto** b_j.
A sends message b_j on the network, and goes to r^A:

$$\gamma_I := (s_i^A \multimap (r^A \otimes N(b_j))) \tag{4}$$

[2] Suppose that two "old" instructions I and I' are consecutively dealing with the same counter r_1. Then we provide the desired "jumps" from r_1 to r_2 and back to r_1 by 'wedging' an instruction (O) between modified I and I'.

(2b) Case I: I is of the form: $a_i : r_1 := r_1{+}1$; **goto** b_j.
A sends two messages: b_j and the currently stored n_A, on the network, and goes to r^A:

$$\gamma_I := \forall x\,((s_i^A \otimes A_1(x)) \multimap (r^A \otimes A_1(x) \otimes N(x) \otimes N(b_j))) \tag{5}$$

(2c) Case II: I is of the form: $a_i : r_1 := r_1{-}1$; **goto** b_j.
A is waiting for a network message of the same form n_A as the nonce currently stored in the memory of A. Having received such a message, A intercepts it, sends b_j on the network, and goes to r^A:

$$\gamma_I := \forall x\,((s_i^A \otimes A_1(x) \otimes N(x)) \multimap (r^A \otimes A_1(x) \otimes N(b_j))) \tag{6}$$

(2d) Case III: I is of the form: a_i : **if** $(r_1 = 0)$ **goto** b_j.
A generates a new nonce n'_A, substitutes it for the old one, sends b_j on the network, and goes to r^A:

$$\gamma_I := \forall x\,((s_i^A \otimes A_1(x)) \multimap \exists y\,(r^A \otimes A_1(y) \otimes N(b_j))) \tag{7}$$

(3) Being in s_0^A, A cleanses up her memory, sends b_0 on the network, and goes to her *final* state q_0^A:

$$\sigma^A := \forall x\,((s_0^A \otimes A_1(x)) \multimap (q_0^A \otimes N(b_0))) \tag{8}$$

(4) Being in r^A, A is waiting for a network message of the form a_i. Having received such a message, A intercepts it, and goes to s_i^A:

$$\rho_i^A := ((r^A \otimes N(a_i)) \multimap s_i^A) \tag{9}$$

In turn, the role of B ("Bob") is to perform the following "mirror" actions:
(1) Being in his *initial* state p_1^B, B generates a random number, *nonce* n_B, stores it, and goes to r^B:

$$\beta_0 := (p_1^B \multimap \exists y\,(r^B \otimes B_1(y))) \tag{10}$$

(2) Being in r^B, B is waiting for a network message of the form b_j. Having received such a message, B intercepts it, and goes to s_j^B:

$$\rho_j^B := ((r^B \otimes N(b_j)) \multimap s_j^B) \tag{11}$$

(3) Being in s_j^B $(j \geq 1)$, B "performs" the instruction I of the form: b_j : ...**goto** a_k, and signals that it is A's turn next to "perform" the instruction labelled by a_k:
(3a) Case O: I is of the form: b_j : **goto** a_k.
B sends message a_k on the network, and goes to r^B:

$$\gamma_I := (s_j^B \multimap (r^B \otimes N(a_k))) \tag{12}$$

(3b) Case I: I is of the form: $b_j : r_2 := r_2{+}1$; **goto** a_k.
B sends two messages: a_k and the stored n_B, on the network, and goes to r^B:

$$\gamma_I := \forall x\,((s_j^B \otimes B_1(x)) \multimap (r^B \otimes B_1(x) \otimes N(x) \otimes N(a_k))) \tag{13}$$

(3c) Case II: I is of the form: $b_j : r_2 := r_2 - 1$; **goto** a_k.
B is waiting for a network message of the same form n_B as the nonce currently stored in the memory of B. Having received such a message, B intercepts it, sends a_k on the network, and goes to r^B:

$$\gamma_I := \forall x\,((s_j^B \otimes B_1(x) \otimes N(x)) \multimap (r^B \otimes B_1(x) \otimes N(a_k))) \tag{14}$$

(3d) Case III: I is of the form: b_j : **if** $(r_2 = 0)$ **goto** a_k.
B generates a new nonce n'_B, substitutes it for the old one, sends a_k on the network, and goes to r^B:

$$\gamma_I := \forall x\,((s_j^B \otimes B_1(x)) \multimap \exists y\,(r^B \otimes B_1(y) \otimes N(a_k))) \tag{15}$$

(4) Being in s_0^B, B cleanses up his memory, and goes to his *final* state q_0^B:

$$\sigma^B := \forall x\,((s_0^B \otimes B_1(x)) \multimap q_0^B) \tag{16}$$

We abbreviate as: $(N(m))^0 := \mathbf{1}$, and $(N(m))^k := \underbrace{N(m) \otimes N(m) \otimes \cdots \otimes N(m)}_{k \text{ times}}$.

Lemma 1. *Let a total situation of the whole system* "participants+network" *be of the form* $(i \geq 1)$*:*

$$(s_i^A \otimes A_1(n_A) \otimes r^B \otimes B_1(n_B) \otimes (N(n_A))^{k_1} \otimes (N(n_B))^{k_2}), \tag{17}$$

"A is in state s_i^A and keeps n_A in her memory, B is in state r^B and keeps n_B in his memory, there are k_1 copies of n_A and k_2 copies of n_B on the network." (Such a (17) is said to represent an M's configuration of the form $(a_i; k_1, k_2)$.)

Then the only action that can be performed is that of A with some γ_I, invoking an M's instruction I of the form: $a_i : \ldots$ **goto** *b_j. The effect of the action is a total situation of the form:*

$$(r^A \otimes A_1(n'_A) \otimes r^B \otimes B_1(n_B) \otimes (N(n_A))^{k'_1} \otimes (N(n_B))^{k_2} \otimes N(b_j)). \tag{18}$$

The next move can be made only by B with the help of ρ_j^B, resulting in a total situation of the form:

$$(r^A \otimes A_1(n'_A) \otimes s_j^B \otimes B_1(n_B) \otimes (N(n_A))^{k'_1} \otimes (N(n_B))^{k_2}), \tag{19}$$

which intends to represent the next M's configuration $(b_j; k'_1, k_2)$, the yield of I.

Similarly, with A and B interchanging their roles, any continuation from (19) involves the simulation of the M's instruction labelled by b_j, which leads to a total situation of the form:

$$(s_k^A \otimes A_1(n'_A) \otimes r^B \otimes B_1(n''_B) \otimes (N(n_A))^{k'_1} \otimes (N(n_B))^{k''_2}). \tag{20}$$

Proof. (11) is not applicable on (17), since any *nonce* differs from b_j. □

Lemma 2. *If I is not a zero-test, then $n'_A = n_A$, and (19) is a correct representation of $(b_j; k'_1, k_2)$:*

$$(r^A \otimes A_1(n'_A) \otimes s_j^B \otimes B_1(n_B) \otimes (N(n'_A))^{k'_1} \otimes (N(n_B))^{k_2}). \tag{21}$$

The case of a zero-test I of the form: a_i : **if** *$(r_1 = 0)$* **goto** *b_j, is more subtle because of $n'_A \neq n_A$.*

(a) *We can guarantee the 'correct' form (21) of (19) whenever* $(a_i; k_1, k_2)$ *and* $(b_j; k'_1, k_2)$ *are consecutive* M*'s configurations within* a legal M*'s computation, and thereby* $k'_1 = k_1 = 0$.

(b) *For* $k_1 \geq 1$, *(19) has an occurrence of* $N(n_A)$, *where* n_A *differs from the new* n'_A *in the memory of* A. *By construction, we could never have removed this occurrence of the obsolete* $N(n_A)$.

Definition 3. *Let* Γ_n *consist of Horn formulas (3)-(16):*

$$\Gamma_n := \{\alpha_n, \sigma^A, \rho_0^A, \rho_1^A, ..., \rho_i^A, ..., \beta_0, \sigma^B, \rho_0^B, \rho_1^B, ..., \rho_j^B, ...,\} \cup \bigcup_{k=1}^{e} \{\gamma_{I_k}\}.$$

Starting from the initial *total situation* $(p_1^A \otimes p_1^B)$ *of the whole system:* participants+network, *the protocol is said to be* completed *whenever it gets into the* final *total situation:* $(q_0^A \otimes q_0^B)$, *that is, in formal terms, there is a scenario* $\mathcal{S}$ *in accordance with* Γ_n, *which leads from* $(p_1^A \otimes p_1^B)$ *to* $(q_0^A \otimes q_0^B)$.

Theorem 3. *For any* n, *there is an exact correspondence between the following three sets:*

(i) *the computations performed by* M, *which lead to its final configuration* $(a_0; 0, 0)$, *starting from its initial configuration* $(a_1; n, 0)$,
(ii) *the linear logic derivations for a sequent of the form:* $!\Gamma_n, (p_1^A \otimes p_1^B) \vdash (q_0^A \otimes q_0^B)$,
(iii) *and scenarios* $\mathcal{S}$ *in accordance with* Γ_n, *which lead from* $(p_1^A \otimes p_1^B)$ *to* $(q_0^A \otimes q_0^B)$.

Proof. Let us sketch the main ideas.

(A) [Protocol Scenarios $\Longleftrightarrow$ LL Proofs] is provided by Theorems 1 and 2.

(B) [M's Computations $\Longrightarrow$ Protocol Scenarios]. By means of actions α_n and β_0 performed by A and B, respectively, the initial $(p_1^A \otimes p_1^B)$ is changed into a total situation of the form:

$$(s_1^A \otimes A_1(n_A) \otimes r^B \otimes B_1(n_B) \otimes (N(n_A))^n \otimes (N(n_B))^0), \tag{22}$$

which represents the initial M's configuration $(a_1; n, 0)$.
According to Lemmas 1 and 2, any M's computation can be simulated step-by-step by the actions of A and B, getting into a "counterpart" of the final $(a_0; 0, 0)$:

$$(s_0^A \otimes A_1(n'_A) \otimes r^B \otimes B_1(n'_B) \otimes (N(n'_A))^0 \otimes (N(n'_B))^0). \tag{23}$$

By the "cleansing" σ^A and σ^B, our simulating scenario is *completed* in $(q_0^A \otimes q_0^B)$.

(C) [Scenarios $\Longrightarrow$ M's Computations]. Without loss of generality, suppose that the 'initial' M's instruction, which is labelled by a_1, is the trivial "jump": a_1 : **goto** b_1.

On the way from $(p_1^A \otimes p_1^B)$ to $(q_0^A \otimes q_0^B)$ any scenario $\mathcal{S}$ is to lead, first, from $(p_1^A \otimes p_1^B)$ to the total situation representing $(b_1; n, 0)$:

$$(r^A \otimes A_1(n_A) \otimes s_1^B \otimes B_1(n_B) \otimes (N(n_A))^n \otimes (N(n_B))^0), \tag{24}$$

and then, eventually, to a total situation of the form:

$$(s_0^A \otimes A_1(n'_A) \otimes r^B \otimes B_1(n'_B) \otimes (N(n'_A))^0 \otimes (N(n'_B))^0), \quad (25)$$

just before finishing in $(q_0^A \otimes q_0^B)$. Furthermore, all these steps from (24) to (25) are inductively controlled by Lemma 1.

Let us consider an intermediate step, say from (17), representing $(a_i; k_1, k_2)$, to (19), which invokes γ_I, with I being of the form a_i :... **goto** b_j. (See Lemma 1)

For I, not being a zero-test, the enabling conditions of γ_I provide that (19) represents $(b_j; k'_1, k_2)$, and M can perform a *legal* move from $(a_i; k_1, k_2)$ to $(b_j; k'_1, k_2)$ by means of its instruction I.

For I, being a zero-test, had k_1 happened positive, it would have produced an "untouchable" occurrence of some $N(n_A)$ (see Lemma 2), contrary to the fact that $\mathcal{S}$ ends in the $N(n_A)$-free $(q_0^A \otimes q_0^B)$. Thus $k_1 = 0$, and thereby M can *legally* move from $(a_i; k_1, k_2)$ to $(b_j; k'_1, k_2)$, as well.

Bringing all together, we can construct the desired M's computation from $(a_1; n, 0)$ to $(a_0; 0, 0)$. □

Corollary 1. (a) *The Horn fragment of pure monadic linear logic, which contains only three unary predicate symbols, and a fixed, but large enough, finite number of 0-ary predicate symbols, and which contains no functional symbols at all, is undecidable.*

(b) *The protocol completion is undecidable even for the class of protocols with two participants such that each of the two is a finite automaton provided with one register capable of storing one atomic message, all the predicates used are at most unary, and no compound messages are in the use.*

Proof. Take a fixed M whose halting problem is undecidable. □

4 Monadic Affine Logic

Affine logic is linear logic with the weakening rule: $\frac{\Gamma, A, \Theta \vdash \Phi}{\Gamma, \Theta \vdash \Phi}$ and $\frac{\Gamma \vdash \Theta, A, \Phi}{\Gamma \vdash \Theta, \Phi}$.

Section 2 is extended to Horn affine logic as follows.

We will say that a scenario $\mathcal{S}$ in accordance with Γ *weakly leads to* $Z(c_1, .., c_k)$, starting from $W(a_1, .., a_p)$, if for some $\widetilde{Z}(c_1, .., c_k, e_1, .., e_q)$ and $V(c_1, .., c_k, e_1, .., e_q)$ such that $\widetilde{Z}(c_1, .., c_k, e_1, .., e_q) \simeq (Z(c_1, .., c_k) \otimes V(c_1, .., c_k, e_1, .., e_q)$, scenario $\mathcal{S}$ leads from $W(a_1, .., a_p)$ to $\widetilde{Z}(c_1, .., c_k, e_1, .., e_q)$ in the 'strict' sense of Section 2.

Theorem 4 (Scenarios $\Longleftrightarrow$ Proofs). *Let Γ consist of Horn formulas. Then both Theorems 1 and 2 remain valid if "leads" is replaced there with "weakly leads", and "linear logic" is replaced there with "affine logic".*

Proof. Similar to the proofs of Theorems 1 and 2. □

4.1 One Binary Relation

We encode a Turing machine M that has one tape, which is one-way infinite to the right. The program of M is: $I_1, \ldots, I_s$. The initial and final states of M are l_1 and l_0, respectively. In our encoding we invoke the following predicates:

(a) Propositions L_i are intended to show that: "M is in state l_i".
(b) $E(x, y)$ means that: "Cell y is just the next to the right to cell x".
(c) The intended meaning of $E_0(x)$ is that: "x is the last cell".
(d) $S(x)$ is intended to indicate that: "x is scanned by M".
(e) For any tape symbols ξ, $C_\xi(x)$ means that: "Cell x contains symbol ξ".

The key point of our encoding is that, given a finite tape, any number of new blank cells can be consecutively 'added' by repeatedly applying the following axiom TAPE ("0" in $C_0(y)$ serves for the blank symbol):

$$\mathtt{TAPE} := \forall x\,(E_0(x) \multimap \exists y\,(E(x,y) \otimes C_0(y) \otimes E_0(y))) \tag{26}$$

Lemma 3. *According to Theorem 4, the provable sequent:*

$$!\mathtt{TAPE},\ E_0(c_5) \vdash \exists y_1 y_2\,(E(c_5, y_1) \otimes E(y_1, y_2) \otimes C_0(y_1) \otimes C_0(y_2) \otimes E_0(y_2)),$$

correlates with a scenario leading from $E_0(c_5)$ *to an elementary product of the form:* $(E(c_5, c_6) \otimes E(c_6, c_7) \otimes C_0(c_6) \otimes C_0(c_7) \otimes E_0(c_7))$, *which represents the 2-cell extension of the tape:* $\boxed{c_5}$, *filled up with 0s:* $\boxed{c_5}\rightarrow\boxed{c_6}\rightarrow\boxed{c_7}$.

An M's configuration - that "M scans j-th cell in state l_k, when a string $\xi_1\xi_2..\xi_i..\xi_t$ is written left-justified on the tape consisting of t cells", is represented as a product of the form: $(L_k \otimes S(c_j) \otimes \bigotimes_{i=1}^{t} C_{\xi_i}(c_i) \otimes \bigotimes_{i=1}^{t-1} E(c_i, c_{i+1}) \otimes E_0(c_t))$. Each of the M's instructions is axiomatized as follows:

(a) The instruction I: *"if in state l_i looking at symbol ξ, replace it by η, move the tape head one cell to the right along the tape, and go into state l_j"*, is specified by the Horn axiom: $\beta_I := \forall x, y\,((L_i \otimes S(x) \otimes C_\xi(x) \otimes E(x,y)) \multimap (L_j \otimes S(y) \otimes C_\eta(x) \otimes E(x,y)))$.
(b) The instruction I: *"if in state l_i looking at symbol ξ, replace it by η, move the tape head one cell to the left along the tape, and go into state l_j"*, is specified by the Horn axiom: $\beta_I := \forall x, y\,((L_i \otimes S(y) \otimes C_\xi(y) \otimes E(x,y)) \multimap (L_j \otimes S(x) \otimes C_\eta(y) \otimes E(x,y)))$.
(c) The instruction I: *"if in state l_i looking at symbol ξ, replace it by η, and go into state l_j (without move)"*, is specified by the Horn axiom: $\beta_I := \forall x\,((L_i \otimes S(x) \otimes C_\xi(x)) \multimap (L_j \otimes S(x) \otimes C_\eta(x)))$.

Theorem 5. *For any input string $\zeta_1\zeta_2..\zeta_n$, there is an exact correspondence between the terminated computations performed by M, starting with the given input $\zeta_1\zeta_2..\zeta_n$, and the affine logic derivations for the following sequent, where $\mathcal{B}_M := \beta_{I_1}, \beta_{I_2}, .., \beta_{I_s}$, and c_1, c_2, .., c_n are distinct constants:*

$$!\mathtt{TAPE},\ !\mathcal{B}_M,\ L_1,\ S(c_1),\ \bigotimes_{i=1}^{n} C_{\zeta_i}(c_i),\ \bigotimes_{i=1}^{n-1} E(c_i, c_{i+1}),\ E_0(c_n) \vdash L_0. \tag{27}$$

Proof. Let us sketch the main points.

(A) [M's Computations $\Longrightarrow$ Proofs]. The initial M's configuration is represented by:

$$(L_1 \otimes S(c_1) \otimes \bigotimes_{i=1}^{n} C_{\zeta_i}(c_i) \otimes \bigotimes_{i=1}^{n-1} E(c_i, c_{i+1}) \otimes E_0(c_n)). \tag{28}$$

A given terminated M's computation is performed within the finite tape of some length t, so that the length of each of the M's configurations in question is assumed to be t. We expand the input $\zeta_1..\zeta_n$ to $\zeta_1..\zeta_n\zeta_{n+1}..\zeta_t$, by setting: $\zeta_k = 0$ for $k > n$. By our construction, this M's computation can be redefined as a scenario, leading from a closed elementary product of the expanded 'initial' form:

$$(L_1 \otimes S(c_1) \otimes \bigotimes_{i=1}^{t} C_{\zeta_i}(c_i) \otimes \bigotimes_{i=1}^{t-1} E(c_i, c_{i+1}) \otimes E_0(c_t)), \tag{29}$$

(representing the initial M's configuration with the input $\zeta_1..\zeta_t$), to:

$$(L_0 \otimes S(c_j) \otimes \bigotimes_{i=1}^{t} C_{\kappa_i}(c_i) \otimes \bigotimes_{i=1}^{t-1} E(c_i, c_{i+1}) \otimes E_0(c_t)), \tag{30}$$

representing the final M's configuration with the string $\kappa_1..\kappa_t$ on the tape. Our TAPE provides the way from (28) to (29) (see Lemma 3).
Having summarized, we obtain a scenario, which weakly leads from (28) to L_0 in accordance with $\mathtt{TAPE}, \mathcal{B}_M$. Theorem 4 yields a derivation for (27).

(B) [Proofs $\Longrightarrow$ M's Computations]. By Theorem 4, any proof for (27) can be transformed into a scenario $\mathcal{S}$ that weakly leads from (28) to L_0 in accordance with $\mathtt{TAPE}, \mathcal{B}_M$. Since each of the steps of $\mathcal{S}$ is controlled either by TAPE or by some β_I, any intermediate $\mathtt{Config}(v)$ is to be of the form:

$$(L_k \otimes S(c_j) \otimes \bigotimes_{i=1}^{m} C_{\xi_i}(c_i) \otimes \bigotimes_{i=1}^{m-1} E(c_i, c_{i+1}) \otimes E_0(c_m)), \tag{31}$$

representing thereby a certain M's configuration. From M's point of view, TAPE just expands the tape with one blank cell. The enabling conditions of β_I provide that M can get into the configuration represented by (31) by means of I applied to the previous one.
Bringing together all the steps of $\mathcal{S}$, we construct an M's computation that, starting from the initial M's configuration with $\zeta_1..\zeta_n$ represented by (28), leads to an M's configuration with state l_0, corresponding to the last step of $\mathcal{S}$. □

Corollary 2. *In contrast to the decidability of the monadic second-order theory of linear order [26], the Horn fragment of pure first-order affine logic, which contains only one binary predicate symbol, four unary predicate symbols, and a fixed, but large enough, finite number of 0-ary predicate symbols, and which contains no functional symbols at all, is undecidable.*

Proof. Take an M with 2 tape symbols, whose halting problem is undecidable. □

4.2 One Zero c, One Successor f, no $\exists$-Quantifier

We encode here an old-fashioned 2-counter Minsky machine M, whose program is $I_1,\ldots, I_s$, each I is of one of the following types:
(I) $l_i : r_m := r_m{+}1;$ **goto** l_j; (II) $l_i : r_m := r_m{-}1;$ **goto** l_j;
(III) $l_i :$ **if** $(r_m = 0)$ **goto** l_j;
Here l_i and l_j are *states*, $i \geq 1$, and r_m represents the m-th counter.
In our encoding we invoke the following predicates:

(a) Propositions L_i are intended to tell us that: "M is in state l_i".
(b) The intended meaning of $R_m(x)$ is that: "counter r_m contains x".
(c) A configuration $(l_i; k_1, k_2)$ is represented by: $(L_i \otimes R_1(f^{k_1}(c)) \otimes R_2(f^{k_2}(c)))$.

Each of the instructions is axiomatized in the following way:

(I) An instruction I of the form (I) is axiomatized by
$\psi_I := \forall x\,((L_i \otimes R_m(x)) \multimap (L_j \otimes R_m(f(x))))$.
(II) An instruction I of the form (II) is axiomatized by
$\psi_I := \forall x\,((L_i \otimes R_m(f(x))) \multimap (L_j \otimes R_m(x)))$.
(III) The zero-test I of the form (III) is axiomatized by
$\psi_I := ((L_i \otimes R_m(c)) \multimap (L_j \otimes R_m(c)))$.

Theorem 6. *For any k_1 and k_2, there is an exact correspondence between the computations, performed by M, which lead from $(l_1; k_1, k_2)$ to $(l_0; 0, 0)$, and the affine logic derivations (linear logic derivations) for the following sequent, where $\Psi_M := \psi_{I_1}, \psi_{I_2}, .., \psi_{I_s}$, and $f^k(c) := \underbrace{f(f(\ldots f}_{k \text{ times}}(c)))$:*

$$!\Psi_M,\ L_1,\ R_1(f^{k_1}(c)),\ R_2(f^{k_2}(c)) \vdash (L_0 \otimes R_1(c) \otimes R_2(c)).$$

Proof. There is a direct correlation between the M's computations, leading from $(l_1; k_1, k_2)$ to $(l_0; 0, 0)$, and the scenarios, which lead from $(L_1 \otimes R_1(f^{k_1}(c)) \otimes R_2(f^{k_2}(c)))$ to $(L_0 \otimes R_1(c) \otimes R_2(c))$ in accordance with Ψ_M. It remains to apply Theorem 4 (or Theorems 1 and 2 in the LL case). □

Corollary 3. *In contrast to decidability of the monadic second-order theories of one and two successors [1,26], the $\exists$-free Horn fragment of monadic affine logic, as well as the $\exists$-free Horn fragment of monadic linear logic, containing only one "zero" and one "successor", two unary predicate symbols, and a fixed, but large enough, finite number of propositions, are undecidable.*

5 Concluding Remarks

One can read Girard's original 1986 translations of quantified classical logic into linear logic as providing an undecidability proof of the general quantified case. Since the classical monadic cases are decidable, we need new techniques to prove the results that contrast with results for classical logic and complement previous results on linear logic.

The main result of the paper is that pure monadic linear logic is undecidable, even in the Horn function-symbol-free fragment. This result uses a *novel* encoding of 2-counter machines. Given what we know of related results, and of the computational complexity of linear logic, it is not surprising to encode such commands as *goto, increment*, and *decrement*. The main new idea relates to *testing for zero*, and also the structure of the machines used is organized to emphasize connections with communication protocols. Although the decision problem for the full propositional multiplicative-exponential fragment of linear logic remains open, from the Horn point of view, we have accomplished the full picture. The propositional pure Horn linear logic is decidable but its complexity is of that of the reachability problem for Petri nets [10,13]. The next step - that is toward the pure monadic Horn fragment of linear logic, is shown here to yield undecidability.

From this undecidability result, an argument for the undecidability of protocol analysis of the Dolev-Yao variety is put forward. We have shown that the *optimistic protocol completion* is undecidable even for the class of protocols with two participants such that either of them is a finite automaton provided with one register to store one atomic message, all the predicates used are at most unary, and no compound messages are in the use. In the light of this result the role of decomposition and composition of messages seems to be overestimated: even on the monadic stage without any composition rules we run into difficulties with the overwhelming power of *nonces*.

One could claim that the undecidability of the problem of *optimistic protocol completion* in the absence of an intruder is not so interesting, since the standard security protocol settings include some model of the intruder, who, for instance, can control the network, and can duplicate and delete messages (but not the state of protocol participants), and such an intruder model would clearly disrupt our encoding of 2-counter machines. But, many of the approaches for analyzing and reasoning about protocols based on certain formal specification languages, and, prior to analysis of protocol properties related to unreliable broadcast, we have to show that our *formal specification* of a given protocol meets the protocol rules and conventions at least under ideal conditions when no one interferes with network transmission.

As compared to [3,7], the undecidability of the secrecy problem there is essentially based on the Cook encoding of Turing machines in terms of classical *binary* predicates. The use of *nonces* and *compound messages* is critical there (in [7] there is an a priori bound on the depth of messages, so encryption and decryption are less important). Since the classical monadic cases are decidable, their results are not directly translated into the pure monadic case. On the other hand, our technique of the undecidability of the *optimistic protocol completion* can provide undecidability of the secrecy problem even in the case where *no compound messages are present.* (It might have shed some light on the fact why the *replay attack* is so popular).

We have also investigated the decision problems of *monadic affine logic.* In contrast to the classical results of the decidability of the monadic second-order theories of one and two successors and the decidability of the monadic second-order theory of linear order [1,26], we have shown:

(a) the undecidability of the Horn fragment of affine logic, which involves only one binary predicate symbol, four unary predicate symbols, and a fixed finite number of propositions, and which contains no functional symbols at all,

(b) and the undecidability of the $\exists$-free Horn fragment of monadic affine logic, which involves only one "zero" and one "successor", two unary predicate symbols, and a fixed finite number of propositions.

References

1. J.R.Büchi. On a decision method in restricted second-order arithmetic. In: *Logic, Methodology and Philosophy of Science,* Proc. of the 1960 Congress, Stanford, CA, 1962, p.1–11.
2. Burrows M., Abadi M., Needham R., A Logic of Authentication, *ACM Transactions on Computer Systems*, 8(1), (1990) 18-36.
3. Cervesato,I., Durgin, N.A., Lincoln, P.D., Mitchell, J.C., and Scedrov, A., A meta-notation for protocol analysis, 12-th IEEE Computer Security Foundations Workshop, Mordano, Italy, June 28-30, 1999.
4. Cervesato, I., Durgin, N.A., Mitchell, J.C., Lincoln, P.D., and Scedrov, A., Relating strands and multiset rewriting for security protocol analysis, *13-th IEEE Computer Security Foundations Workshop*, Cambridge, U.K., July 3-5, 2000, pp.35-51.
5. Iliano Cervesato, Nancy Durgin, Max Kanovich. and Andre Scedrov. Interpreting Strands in Linear Logic. In: "2000 Workshop on Formal Methods and Computer Security", 12th Int'l Conference on Computer Aided Verification (CAV 2000) Satellite Workshop, July, 2000, Chicago, USA.
6. Dolev D., Yao A., On the Security of Public Key Protocols, *IEEE Transactions on Information Theory*, 29(2), (1983) 198-208.
7. Durgin, N.A., Lincoln, P.D., Mitchell, J.C., and Scedrov, A., Undecidability of bounded security protocols, Workshop on Formal Methods and Security Protocols (FMSP'99), Trento, Italy, July 5, 1999. Electronic proceedings
8. Durgin, N.A. and Mitchell, J.C., Analysis of Security Protocols. In Calculational System Design, ed. M. Broy and R. Steinbruggen, IOS Press, 1999, pp.369–395.
9. J.-Y.Girard. Linear logic. *Theoretical Computer Science,* 50:1, 1987, pp.1-102.
10. V.Gehlot and C.A.Gunter. Normal process representatives. In *Proc. 5-th Annual IEEE Symposium on Logic in Computer Science, Philadelphia,* June 1990.
11. Yuri Gurevich. Monadic second-order theories. In "Model-Theoretical Logics" (ed. J. Barwise and S. Feferman) Springer-Verlag, 1985, pp.479-506.
12. J.S.Hodas and D.Miller. Logic programming in a fragment of intuitionistic linear logic. *Information and Computation,* 110(2), 1994, pp.327-365.
13. Max Kanovich. Horn Programming in Linear Logic is NP-complete. In *Proc. 7-th Annual IEEE Symposium on Logic in Computer Science, Santa Cruz,* June 1992, pp.200-210
14. Max Kanovich. Linear logic as a logic of computations, *Annals of Pure and Applied Logic*, 67 (1994) pp.183-212
15. Max Kanovich. The complexity of Horn fragments of linear logic. *Annals of Pure and Applied Logic*, 69:195–241, 1994.
16. Max Kanovich. The Direct Simulation of Minsky machines in Linear logic, In J.-Y. Girard, Y. Lafont, and L. Regnier, editors, *Advances in Linear Logic*, London Mathematical Society Lecture Notes, Vol. 222, pp.123-145. Cambridge University Press, 1995.
17. A.P.Kopylov. Decidability of Linear Affine Logic, In *Proc. 10-th Annual IEEE Symposium on Logic in Computer Science, San Diego, California,* June 1995, pp.496-504

18. Yves Lafont. The undecidability of second order linear logic without exponentials. *Journal of Symbolic Logic*, 61:541–548, 1996.
19. Yves Lafont and Andre Scedrov. The undecidability of second order multiplicative linear logic. *Information and Computation*, 125:46–51, 1996.
20. Yves Lafont. The finite model property for various fragments of linear logic, Journal of Symbolic Logic 62 (4), p.1202-1208.
21. P.Lincoln, J.Mitchell, A.Scedrov, and N.Shankar. Decision Problems for Propositional Linear Logic. *Annals of Pure and Applied Logic*, 56 (1992) pp.239-311.
22. Meadows C., The NRL Protocol Analyzer: An overview, *Journal of Logic Programming*, Vol. 26, No. 2, (1996) 113-131.
23. D.Miller. A Multiple-Conclusion Meta-Logic. *Theoretical Computer Science* 165(1), 1996, pp.201-232.
24. M.Minsky. Recursive unsolvability of Post's problem of 'tag' and other topics in the theory of Turing machines. *Annals of Mathematics,* 74:3:437-455, 1961.
25. Needham R., Schroeder M., Using Encryption for Authentication in large networks of computers, *Communications of the ACM*, 21(12), (1978) 993-999.
26. Michael Rabin. Decidability of second order theories and automata on infinite trees. *Trans. Am. Math. Soc.*, 141, 1969, pp.1-35.
27. Michael Rabin. Decidable theories. In *Handbook of mathematical logic* (ed. Barwise), 1977, pp.595-629.
28. Michael Sipser. *Introduction to the Theory of Computation*, PWS Publishing, 1997.

Non-commutativity and MELL in the Calculus of Structures

Alessio Guglielmi and Lutz Straßburger
Technische Universität Dresden
Fakultät Informatik - 01062 Dresden - Germany
Alessio.Guglielmi@Inf.TU-Dresden.DE and Lutz.Strassburger@Inf.TU-Dresden.DE

Abstract *We introduce the* calculus of structures*: it is more general than the sequent calculus and it allows for cut elimination and the subformula property. We show a simple extension of multiplicative linear logic, by a self-dual non-commutative operator inspired by* CCS*, that seems not to be expressible in the sequent calculus. Then we show that multiplicative exponential linear logic benefits from its presentation in the calculus of structures, especially because we can replace the ordinary, global promotion rule by a local version. These formal systems, for which we prove cut elimination, outline a range of techniques and properties that were not previously available. Contrarily to what happens in the sequent calculus, the cut elimination proof is modular.*

1 Introduction

The sequent calculus [5] is very appropriate for classical logic, but it has some problems in dealing with more refined logics like linear logic [6]. Observing certain logical relations in the sequent calculus might be impossible. In this paper we show a calculus, called the *calculus of structures*, which is able to overcome those difficulties.

We call *calculus* a framework, like natural deduction or the sequent calculus, for specifying logical systems. We say *formal system* to indicate a collection of inference rules in a given calculus. A *derivation* is a composition of instances of inference rules, a *proof* is a derivation free from hypotheses.

A proof in the sequent calculus is a tree, and branching occurs when two-premise rules are used. The two branches are statements that proofs exist for both premises. At the meta level, we say that the left branch is a proof *and* the right branch is a proof. In classical logic, this 'and' corresponds to the 'and' at the object level. This is not the case in other logics, like in linear logic.

Another founding property of the sequent calculus is the pivotal rôle of main connectives. Given a main connective in the conclusion, a rule gives meaning to it by saying that the conclusion is provable if subformulae obtained by removing the connective are in turn provable.

These two properties together have remarkable success in making the study of systems independent of their semantics, which is important if a semantics is incomplete, missing or still under development, as often happens in computer science. The problem is that the sequent calculus is unnecessarily rigid for some logics. We can relax the 'and' branching between premise trees, and abandon the decomposing of the conclusion around the main connective of one of its formulae. The question is whether we can do so while keeping the good properties, cut elimination especially.

The calculus of structures draws from a very simple principle, which is very dangerous if not realised with care. The inference rules are of the kind $\rho \dfrac{S\{T\}}{S\{R\}}$, where premise and conclusion are *structures*, i.e., formulae subject to certain equivalences

L. Fribourg (Ed.): CSL 2001, LNCS 2142, pp. 54–68, 2001.

(associativity, commutativity, units, ...). A structure $S\{R\}$ is a structure context $S\{\ \}$, whose hole is filled by the structure R. The rule scheme ρ above specifies that if a structure matches R, in a context $S\{\ \}$, it can be rewritten as specified by T, in the same context $S\{\ \}$ (or vice versa if one reasons top-down). A rule corresponds to implementing in the formal system *any axiom* $T \Rightarrow R$, where $\Rightarrow$ stands for the implication we model in the system. The danger lies in the words 'any axiom'.

In fact, rules could be used as axioms of a generic Hilbert system, where there is no special, structural relation between T and R. But then all the good proof theoretical properties would be lost. Our challenge is to design inference rules in a way that is conservative enough to allow us to prove cut elimination, and such that they possess the subformula property. Still we have to be liberal enough to overcome the problems of rigidity mentioned above.

It is important to note that the calculus of structures is *more general* than the sequent calculus, for logics with De Morgan rules. Any system that admits a one-sided presentation can be ported, trivially, to the calculus of structures. But, since we can do more, we want to use the new expressive capabilities to get new logics, or to make old logics better. We will do both things in this paper (without paying a big price).

Rules come in pairs, $\rho{\downarrow}\,\dfrac{S\{T\}}{S\{R\}}$ (down version) and $\rho{\uparrow}\,\dfrac{S\{\bar{R}\}}{S\{\bar{T}\}}$ (up version), where $\bar{U}$ is the negation of U and S stands for any context. This duality derives from the duality between $T \Rightarrow R$ and $\bar{R} \Rightarrow \bar{T}$. We would like to dispose of the up rules without affecting provability—after all, $T \Rightarrow R$ and $\bar{R} \Rightarrow \bar{T}$ are equivalent statements in many logics. The cut rule splits into several up rules, and this makes for a modular decomposition of the cut elimination argument, since we can get rid of up rules one after the other. This is one the main achievements of our paper (in [7], p. 15, Girard deems as 'rather shocking' this lack of modularity in the sequent calculus).

Derivations in the calculus of structures are chains of instances of rules. Contrarily to what happens in the sequent calculus, whose derivations are trees, our derivations have a *top-down* symmetry. This allows for new manipulations of derivations. For example, permuting down certain rules, like the cut, is easier than in the sequent calculus; entire derivations may be flipped upside down and negated and they still are valid derivations; and so on. The most important consequence of the new symmetry is that the cut rule $\mathsf{i}{\uparrow}\,\dfrac{S\{(R,\bar{R})\}}{S\{\bot\}}$ becomes top-down symmetric to the identity rule $\mathsf{i}{\downarrow}\,\dfrac{S\{1\}}{S\{[R,\bar{R}]\}}$ (here, (R,T) and $[R,T]$ denote the conjunction and the disjunction of R and T, and 1 and $\bot$ are the conjunctive and disjunctive units). It is then possible to reduce the cut rule to its atomic variant $\mathsf{a}{\uparrow}\,\dfrac{S\{(a,\bar{a})\}}{S\{\bot\}}$, the same way as identity can be just required for atoms in most systems in the sequent calculus. The reduction of cut to its atomic form simplifies the cut elimination argument, since there is no more interaction between a cut's principal formula and the structure of the proof.

We believe that the development of a calculus must be driven by its systems. Here we develop two systems inside the calculus of structures. The first one, in Sect. 2, is system BV (Basic system V) [8]. It is equivalent to multiplicative linear logic plus mix, extended by a non-commutative self-dual operator. System BV is motivated by the desire to grasp a sequential operator, like that of CCS [12], in a logical system, especially from a proof-search perspective. The logic obtained seems not to be expressible in the sequent calculus, certainly not in a simple way, while in our calculus it is straightforward. System BV is just a first, but crucial step toward a logical system encompassing languages of distributed computation. The methodology for designing systems, induced by the calculus of structures, is outlined in that section.

We start from a very simple observation. A basic reaction in CCS is $a\,|\,\bar{a} \to 0$: the two parallel processes a and $\bar{a}$ communicate and rewrite to the empty process 0. This naturally corresponds to the identity axiom in logic, if we express complementation in CCS by negation; the parallel composition '|' corresponds to disjunction (linear logic's multiplicative disjunction corresponds remarkably well, see for example [10]). Consider now sequential composition, as in the process $a.b$: the dual of this process must be $\overline{a.b} = \bar{a}.\bar{b}$, since $a.b \mid \bar{a}.\bar{b} \to^* 0$. Then, we need a self-dual non-commutative logical operator for modelling sequential composition. We are not committing to CCS: we just observe that, as witnessed by CCS, there is a natural way of seeing parallel and sequential compositions in a logical system.

In Sect. 3 the system ELS (multiplicative Exponential Linear logic in the calculus of Structures) is shown [16]. A first reason to study this system, which is equivalent to sequent calculus's MELL, is to see how our calculus performs on a system that is studied already elsewhere. We get a surprising result: the promotion rule can be made local, what is unlikely in the sequent calculus.

There is another reason for studying MELL in our calculus: we plan to enrich BV with contraction, in the hope of making it Turing equivalent. To this purpose, we need exponentials to control contraction, because we do not want to destroy the good behaviour of multiplicative disjunction with respect to parallel composition (what is known as 'resource sensitivity').

For both systems BV and ELS we state *decomposition theorems*: rules in derivations can be rearranged in a highly structured way (impossible in the sequent calculus) where subsystems of a given system are applied in sequence. Decomposition results allow us greatly to simplify the cut elimination proofs and are (still mysteriously) linked to other features of the systems under study. These theorems are welcome because proving cut elimination in the calculus of structures can be harder than in the sequent calculus, due to the more liberal applicability of inference rules.

We also prove cut elimination for both systems, and, overall, the argument is quite different than the usual one in the sequent calculus. Exploring the new methodology is by itself interesting, because there is the possibility of characterising the property of cut elimination in a more systematic way than before.

This paper only deals with syntax: our sole purpose is to present the calculus of structures and its properties. MELL is, of course, semantically well-known, and then so is ELS. System BV has been discovered by *trace semantics* [8].

2 Non-commutativity

A system in our calculus requires a language of structures. These are sort of intermediate expressions between formulae and sequents. Here we define the language for systems BV and SBV, and we call it BV. Intuitively, $[S_1, \ldots, S_h]$ corresponds to a sequent in linear logic, whose formulae are connected by pars, and associativity and commutativity are taken into account. The structure $(S_1, \ldots, S_h)$ corresponds to the times connection of $S_1, \ldots, S_h$; it is associative and commutative. The structure

$\langle S_1; \dots; S_h \rangle$ is associative and *non-commutative*: this corresponds to the new logical relation we introduce. All the details for this section can be found in [8].

2.1 Definition There are infinitely many *positive literals* and *negative literals*. Literals, positive or negative, are denoted by a, b, ... *Structures* are denoted by S, P, Q, R, T, U and V. The structures of the language BV are generated by

$$S ::= a \mid \circ \mid [\underbrace{S, \dots, S}_{>0}] \mid (\underbrace{S, \dots, S}_{>0}) \mid \langle \underbrace{S; \dots; S}_{>0} \rangle \mid \bar{S} \quad ,$$

where $\circ$, the *unit*, is not a literal; $[S_1, \dots, S_h]$ is a *par structure*, $(S_1, \dots, S_h)$ is a *times structure* and $\langle S_1; \dots; S_h \rangle$ is a *seq structure*; $\bar{S}$ is the *negation* of the structure S. Structures with a hole that does not appear in the scope of a negation are denoted by $S\{\ \}$. The structure R is a *substructure* of $S\{R\}$, and $S\{\ \}$ is its *context*. We simplify the indication of context in cases where structural parentheses fill the hole exactly: for example, $S[R,T]$ stands for $S\{[R,T]\}$.

Structures come with equational theories establishing some basic, decidable algebraic laws by which structures are indistinguishable. There is an analogue in the laws of associativity, commutativity, idempotency, and so on, usually imposed on sequents. We will see these laws together with the inference rules. It would be possible, of course, to introduce the equational laws by inference rules. But, having dropped connectives, our choice makes matters much clearer.

The next step in defining a system is giving its inference rules. The following definition is general, i.e., it holds for any system, not just BV.

2.2 Definition An (*inference*) *rule* is any scheme $\rho \frac{T}{R}$, where ρ is the *name* of the rule, T is its *premise* and R is its *conclusion*. Rule names are denoted by ρ and π. A (*formal*) *system*, denoted by $\mathscr{S}$, is a set of rules. A *derivation* in a system $\mathscr{S}$ is a finite or infinite chain of instances of rules of $\mathscr{S}$, and is denoted by Δ. A derivation can consist of just one structure. The topmost structure in a derivation, if present, is called its *premise*; if present, the lowest structure is called *conclusion*. A derivation Δ whose premise is T, conclusion is R, and whose rules are in $\mathscr{S}$ is denoted by $\begin{array}{c} T \\ \Delta \| \mathscr{S} \\ R \end{array}$.

A typical rule has shape $\rho \frac{S\{T\}}{S\{R\}}$ and specifies a step of rewriting, by the implication $T \Rightarrow R$, inside a generic context $S\{\ \}$. Rules with empty contexts correspond to the case of the sequent calculus. It is important to note that the notion of derivation is top-down symmetric. Logical axioms for the given systems will be given separately from the rules. They will induce the concept of *proof*, and their introduction is our way of breaking the symmetry and observing the usual proof theoretical properties, like cut elimination. We will be dealing with proofs only later in the section.

Let us see a system that deals with the new non-commutative logical relation. It is made by two sub-systems: one for *interaction* and the other for *structure*. The interaction fragment deals with negation, i.e., duality. It corresponds to identity and cut in the sequent calculus. In our calculus these rules become mutually top-down symmetric and both admit decompositions into their atomic counterparts.

The structure fragment corresponds, mainly, to logical rules in the sequent calculus; it defines the logical relations. Differently from the sequent calculus, the logical relations need not be defined in isolation, rather complex contexts can be taken into consideration. In the following system, as well as in the system in the next section, we consider *pairs* of logical relations, one inside the other.

Associativity

$$[\vec{R},[\vec{T}]] = [\vec{R},\vec{T}]$$
$$(\vec{R},(\vec{T})) = (\vec{R},\vec{T})$$
$$\langle\vec{R};\langle\vec{T}\rangle;\vec{U}\rangle = \langle\vec{R};\vec{T};\vec{U}\rangle$$

Commutativity

$$[\vec{R},\vec{T}] = [\vec{T},\vec{R}]$$
$$(\vec{R},\vec{T}) = (\vec{T},\vec{R})$$

Unit

$$[\circ,\vec{R}] = [\vec{R}]$$
$$(\circ,\vec{R}) = (\vec{R})$$
$$\langle\circ;\vec{R}\rangle = \langle\vec{R};\circ\rangle = \langle\vec{R}\rangle$$

Negation

$$\bar{\circ} = \circ$$
$$\overline{[R_1,\dots,R_h]} = (\bar{R}_1,\dots,\bar{R}_h)$$
$$\overline{(R_1,\dots,R_h)} = [\bar{R}_1,\dots,\bar{R}_h]$$
$$\overline{\langle R_1;\dots;R_h\rangle} = \langle\bar{R}_1;\dots;\bar{R}_h\rangle$$
$$\bar{\bar{R}} = R$$

Singleton

$$[R] = (R) = \langle R\rangle = R$$

Contextual Closure

if $R = T$ then $S\{R\} = S\{T\}$

$$\mathsf{a}{\downarrow}\,\frac{S\{\circ\}}{S[a,\bar{a}]} \qquad \mathsf{a}{\uparrow}\,\frac{S(a,\bar{a})}{S\{\circ\}}$$

Interaction

Structure (core)

$$\mathsf{s}\,\frac{S([R,T],U)}{S[(R,U),T]}$$

$$\mathsf{q}{\downarrow}\,\frac{S\langle[R,T];[R',T']\rangle}{S[\langle R;R'\rangle,\langle T;T'\rangle]}$$

$$\mathsf{q}{\uparrow}\,\frac{S(\langle R;T\rangle,\langle R';T'\rangle)}{S\langle(R,R');(T,T')\rangle}$$

Fig. 1 **Left:** *Syntactic equivalence* $=$ *for* BV **Right:** *System* SBV

2.3 Definition The structures of the language BV are equivalent modulo the relation $=$, defined at the left of Fig. 1. There, $\vec{R}$, $\vec{T}$ and $\vec{U}$ stand for finite, non-empty sequences of structures (sequences may contain ',' or ';' separators as appropriate in the context). At the right of the figure, system SBV is shown (Symmetric, or Self-dual, Basic system V). The rules $\mathsf{a}{\downarrow}$, $\mathsf{a}{\uparrow}$, s, $\mathsf{q}{\downarrow}$ and $\mathsf{q}{\uparrow}$ are called respectively *atomic interaction*, *atomic cut* (or *atomic cointeraction*), *switch*, *seq* and *coseq*. The *down fragment* of SBV is $\{\mathsf{a}{\downarrow},\mathsf{s},\mathsf{q}{\downarrow}\}$, the *up fragment* is $\{\mathsf{a}{\uparrow},\mathsf{s},\mathsf{q}{\uparrow}\}$.

Negation is involutive and can be pushed directly over atoms. The unit $\circ$ is self-dual and common to the three logical relations. One may think of it as a convenient way of expressing the empty sequence. Of course, rules become very flexible in the presence of such a unit. For example, the following notable derivation is valid:

$$\mathsf{q}{\downarrow}\,\frac{\mathsf{q}{\uparrow}\,\dfrac{(a,b)}{\langle a;b\rangle}}{[a,b]} \quad = \quad \mathsf{q}{\downarrow}\,\frac{\mathsf{q}{\uparrow}\,\dfrac{(\langle a;\circ\rangle,\langle\circ;b\rangle)}{\langle[a,\circ];[\circ,b]\rangle = \langle(a,\circ);(\circ,b)\rangle}}{[\langle a;\circ\rangle,\langle\circ;b\rangle]} \quad .$$

Here is a derivation for the CCS reaction $a.b \mid \bar{a}.\bar{b} \to^* 0$:

$$\mathsf{q}{\downarrow}\,\frac{\mathsf{a}{\downarrow}\,\dfrac{\mathsf{a}{\downarrow}\,\dfrac{\circ}{[b,\bar{b}]}}{\langle[a,\bar{a}];[b,\bar{b}]\rangle}}{[\langle a;b\rangle,\langle\bar{a};\bar{b}\rangle]} \quad .$$

Please note that $[\langle a;b\rangle,\langle\bar{b};\bar{a}\rangle]$ admits no derivation where both $[a,\bar{a}]$ and $[b,\bar{b}]$ interact. As the reader may notice, the correspondence with CCS is truly straightforward. The instance of the rule $\mathsf{q}{\downarrow}$ above can not be expressed in the sequent calculus, because

1. there should be two premises $\vdash a,\bar{a}$ 'and' $\vdash b,\bar{b}$, but we would have big problems with cut elimination, essentially because 'and' is too strong;
2. there is no principal connective in the conclusion, rather there are *two* of them to be considered *together*, namely, the two seq relations between a and b and between $\bar{a}$ and $\bar{b}$.

We do not mean that similar logics cannot be expressed in any other calculus. For example, Retoré does it in [13, 14], in proof nets. His logic is very close to ours, possibly the same, but the exact correspondence is at present unknown. None has been able to define in the sequent calculus a self-dual non-commutative relation that lives with commutative ones. We should mention the work [2, 15] by Abrusci and Ruet: they mix commutative and non-commutative relations in a sequent system, but instead of one self-dual sequential connective, they have two mutually dual ones.

A way of understanding the rule s is by considering linear logic's times rule $\otimes\dfrac{\vdash A,\Phi \quad \vdash B,\Psi}{\vdash A\otimes B,\Phi,\Psi}$. This rule is mimicked by

$$\mathsf{s}\dfrac{\mathsf{s}\dfrac{([R_A,T_\Phi],[U_B,V_\Psi])}{[([R_A,T_\Phi],U_B),V_\Psi]}}{[(R_A,U_B),T_\Phi,V_\Psi]} \quad ,$$

where R_A, U_B, T_Φ and V_Ψ correspond to the formulae A, B and the multisets of formulae Φ and Ψ. The two s instances could be swapped: the substructures in the par context can be brought inside the times structure independently. We have no combinatorial explosion in the splitting of a times context [9, 11], which depends on the impossibility, in the sequent calculus, of representing the middle structure in the derivation above. In fact, the lazy splitting algorithm of [9] is here represented naturally and simply.

System SBV is designed to ensure the subformula property: all the rule premises are made of substructures of the conclusions, except for the cut rule. This is of course a key ingredient in consistency arguments, and a basis for proof search.

2.4 Definition The following rules are called *interaction* and *cut* (or *cointeraction*):

$$\mathsf{i}{\downarrow}\dfrac{S\{\circ\}}{S[R,\bar{R}]} \quad \text{and} \quad \mathsf{i}{\uparrow}\dfrac{S(R,\bar{R})}{S\{\circ\}} \quad ;$$

R and $\bar{R}$ are called *principal structures*.

The sequent calculus rule $\mathsf{cut}\dfrac{\vdash A,\Phi \quad \vdash A^\perp,\Psi}{\vdash \Phi,\Psi}$ is realised as

$$\mathsf{i}{\uparrow}\dfrac{\mathsf{s}\dfrac{\mathsf{s}\dfrac{([R_A,T_\Phi],[\overline{R_A},V_\Psi])}{[([R_A,T_\Phi],\overline{R_A}),V_\Psi]}}{[(R_A,\overline{R_A}),T_\Phi,V_\Psi]}}{[T_\Phi,V_\Psi]} \quad .$$

The next theorem states the reduction of the interaction rules to atomic form.

2.5 Definition A rule ρ is *strongly admissible* for the system $\mathscr{S}$ if $\rho \notin \mathscr{S}$ and for every instance $\rho\dfrac{T}{R}$ there exists a derivation $\begin{array}{c}T\\ \|\mathscr{S}\\ R\end{array}$. The systems $\mathscr{S}$ and $\mathscr{S}'$ are *strongly equivalent* if for every derivation $\begin{array}{c}T\\ \|\mathscr{S}\\ R\end{array}$ there exists a derivation $\begin{array}{c}T\\ \|\mathscr{S}'\\ R\end{array}$, and vice versa.

2.6 Theorem *The rules* $\mathsf{i}{\downarrow}$ *and* $\mathsf{i}{\uparrow}$ *are strongly admissible for the systems* $\{\mathsf{a}{\downarrow},\mathsf{s},\mathsf{q}{\downarrow}\}$ *and* $\{\mathsf{a}{\uparrow},\mathsf{s},\mathsf{q}{\uparrow}\}$, *respectively.*

Proof Structural induction on the principal structure. We show the inductive cases of $\mathsf{i}{\uparrow}$:

$$\mathsf{i}{\uparrow}\dfrac{\mathsf{i}{\uparrow}\dfrac{\mathsf{q}{\uparrow}\dfrac{S(\langle P;Q\rangle,\langle\bar{P};\bar{Q}\rangle)}{S\langle(P,\bar{P});(Q,\bar{Q})\rangle}}{S(Q,\bar{Q})}}{S\{\circ\}} \quad \text{and} \quad \mathsf{i}{\uparrow}\dfrac{\mathsf{i}{\uparrow}\dfrac{\mathsf{s}\dfrac{\mathsf{s}\dfrac{S(P,Q,[\bar{P},\bar{Q}])}{S(Q,[(P,\bar{P}),\bar{Q}])}}{S[(P,\bar{P}),(Q,\bar{Q})]}}{S(Q,\bar{Q})}}{S\{\circ\}} \quad .$$

□

2.7 Definition We call *core* the set of rules, different than atomic (co)interaction ones, that appear in the reduction of interaction and cut to atomic form. Rules, other than (co)interactions, that are not in the core are called *non-core*. The core of SBV is $\{\mathsf{s}, \mathsf{q}\!\downarrow, \mathsf{q}\!\uparrow\}$, called SBVc; there are no non-core rules in SBV.

2.8 Remark Let ρ be a rule and π be its *corule*, i.e., π is obtained by swapping and negating premise and conclusion in ρ. The rule π is then strongly admissible for the system $\{\mathsf{i}\!\downarrow, \mathsf{i}\!\uparrow, \mathsf{s}, \rho\}$, because each instance $\pi\,\dfrac{S\{T\}}{S\{R\}}$ can be replaced by

$$\mathsf{i}\!\downarrow\dfrac{S\{T\}}{\mathsf{s}\,\dfrac{S(T,[R,\bar{R}])}{\rho\,\dfrac{S[R,(T,\bar{R})]}{\mathsf{i}\!\uparrow\dfrac{S[R,(T,\bar{T})]}{S\{R\}}}}} \quad .$$

The main idea for getting decomposition and cut elimination theorems is studying the permutability of rules. To get a decomposition theorem, instances are moved up or down along the derivation until a certain scheme is obtained. To get cut elimination, 'evil' rules, corresponding to cuts to be eliminated, are permuted up a proof until they reach the logical axiom and disappear.

2.9 Definition A rule ρ *permutes over* π if $\rho \neq \pi$ and for all $\pi\,\dfrac{Q}{\rho\,\dfrac{U}{P}}$ there is $\rho\,\dfrac{Q}{\pi\,\dfrac{V}{P}}$, for some V; if $\begin{array}{c}Q\\ \rho\,\overline{V}\\ \Vert \mathscr{S}\cup\{\pi\}\\ P\end{array}$ exists, for some system $\mathscr{S}$, we say that ρ *permutes by* $\mathscr{S}$ *over* π.

In the sequent calculus, identity rules are leaves of the derivation trees, of course. They can be put at the top in our calculus, too, but the dual is also true of cuts: they can be driven down with no effort. Here is the decomposition theorem.

2.10 Theorem *For every derivation* $\begin{array}{c}T\\ \Vert \mathsf{SBV}\\ R\end{array}$ *there is a derivation* $\begin{array}{c}T\\ \Vert \{\mathsf{a}\!\downarrow\}\\ Q\\ \Vert \mathsf{SBVc}\\ P\\ \Vert \{\mathsf{a}\!\uparrow\}\\ R\end{array}$, *for some structures* P *and* Q.

Proof The rule $\mathsf{a}\!\downarrow$ permutes over $\mathsf{a}\!\uparrow$ and permutes by SBVc over s, $\mathsf{q}\!\downarrow$ and $\mathsf{q}\!\uparrow$. Take the topmost instance of $\mathsf{a}\!\downarrow$ and move it upward until it reaches the top. Proceed inductively downward by moving up each $\mathsf{a}\!\downarrow$ instance until only $\mathsf{a}\!\downarrow$ instances are above it. Perform dually for $\mathsf{a}\!\uparrow$. □

Derivations are reduced to three-phase ones: a 'creation' phase, a middle phase where atoms are shuffled by rules in the core, and a 'destruction' phase.

It is time to break the top-down symmetry by making asymmetric *observations*: we want to detect *proofs*. To do so, we admit inference rules with no premise, called *logical axioms*. For SBV we have:

2.11 Definition The following (logical axiom) rule is called *unit*: $\circ\!\downarrow\,\dfrac{}{\circ}$. The system in Fig. 2 is called BV (Basic system V).

$$\circ\!\downarrow \frac{}{\circ} \qquad \mathsf{a}\!\downarrow \frac{S\{\circ\}}{S[a,\bar{a}]} \qquad \mathsf{s}\frac{S([R,T],U)}{S[(R,U),T]} \qquad \mathsf{q}\!\downarrow \frac{S\langle[R,T];[R',T']\rangle}{S[\langle R;R'\rangle,\langle T;T'\rangle]}$$

Fig. 2 *System* BV

2.12 Definition A *proof*, denoted by Π, is a finite derivation whose top is an instance of a logical axiom. A system $\mathscr{S}$ *proves* R if there is in $\mathscr{S}$ a proof Π whose conclusion is R, written $\begin{array}{c}\Pi \| \mathscr{S} \\ R\end{array}$. A rule ρ is *admissible* for the system $\mathscr{S}$ if $\rho \notin \mathscr{S}$ and for every proof $\begin{array}{c}\| \mathscr{S}\cup\{\rho\} \\ R\end{array}$ there exists a proof $\begin{array}{c}\| \mathscr{S} \\ R\end{array}$. Two systems are *equivalent* if they prove the same structures.

To get cut elimination, so as to have a system whose rules all enjoy the subformula property, we could just get rid of a↑, by proving its admissibility for the other rules. But we can do more than that: the whole up fragment of SBV, except for s (which also belongs to the down fragment), is admissible. This suggests a *modular* scheme for proving cut elimination, which, as a matter of fact, scales up to the much more complex case of MELL, in Sect. 3:

1 rules in the non-core up fragment of the system are trivially admissible for the core, plus interaction and their (down) corules (see 2.8);

2 prove admissibility for the up rules in the core;

3 show admissibility of a↑.

The decomposition into several up rules is very beneficial when systems are extended: the cut elimination proof of the smaller system can be largely reused for the bigger one, since it relies on mutual permutability of rules. (There are no non-core rules in SBV, we will see the general case in Sect. 3.)

We have to prove the equivalence of SBV ∪ {o↓} and BV. The first step is to show the admissibility of q↑. The proof of the theorem outlines our typical technique, which uses *super rules* to keep track of the context while permuting up a rule to be eliminated.

2.13 Theorem *The rule* q↑ *is admissible for* BV ∪ {a↑}.

Proof The rule q↑ can be generalised by a certain rule m↑ (called *comerge* and derived from semantics); m↑ permutes by {s, q↓} over a↓, s and q↓. By 2.10 a given proof can be transformed into

$$\begin{array}{c} \| \mathsf{BV}\cup\{\mathsf{q}\uparrow\} \\ P \\ \| \{\mathsf{a}\uparrow\} \\ R \end{array} \quad = \quad \begin{array}{c} \Pi' \| \mathsf{BV} \\ \mathsf{m}\uparrow \dfrac{T}{Q} \\ \| \mathsf{SBVc} \\ P \\ \| \{\mathsf{a}\uparrow\} \\ R \end{array} ,$$

where the top instance of q↑ has been called m↑. The m↑ instance can be permuted up until it disappears against o↓. Repeat inductively downward for all q↑ instances. □

The last step is getting rid of the a↑ instances.

2.14 Theorem *The rule* $\mathsf{a}\uparrow$ *is admissible for* BV.

Proof Similar to the previous one. We need the following fact: In BV, replace s by the rule $\mathsf{ds}\downarrow \frac{S([R,T],U)}{S[(R,U),T]}$ (*deep switch*), where R is not a proper times structure (i.e., there are no non-unit P and Q such that $R=(P,Q)$); the resulting system, called BVd, is equivalent to BV (the argument is not trivial). Transform the upper BV portion of the given proof into a BVd one. Then drive up the topmost $\mathsf{a}\uparrow$ instance by using the super rule $\mathsf{sa}\uparrow \frac{S(R\{a\},T\{\bar{a}\})}{S[R\{\circ\},T\{\circ\}]}$, which permutes by $\{\mathsf{s},\mathsf{q}\downarrow\}$ over $\mathsf{a}\downarrow$, $\mathsf{ds}\downarrow$ and $\mathsf{q}\downarrow$. The two $\mathsf{a}\downarrow$ instances that apply to the principal literals created by the $\mathsf{a}\uparrow$ instance must be permuted up preliminarily, until they reach the top of the proof. Proceed inductively downward. □

This completes the proof of cut elimination. The strategy we followed is completely deterministic, so the procedure is confluent.

Here comes consistency; a similar argument, exploiting the top-down symmetry, becomes hard in the sequent calculus, due to the difficulty in flipping derivations.

2.15 Theorem *If* R *is provable in* BV *then* $\bar{R}$ *is not provable, provided* $R \neq \circ$.

Proof A proof of R is like $\begin{array}{c} \circ\downarrow \frac{}{\circ} \\ \mathsf{a}\downarrow \frac{\circ}{[a,\bar{a}]} \\ \Vert \mathsf{BV} \\ R \end{array}$. Get $\begin{array}{c} \bar{R} \\ \Vert \mathsf{SBV} \\ (a,\bar{a}) \end{array}$ by flipping the given proof. If $\bar{R}$ is provable, then $(a,\bar{a})$ is provable in $\mathsf{SBV} \cup \{\circ\downarrow\}$ and, by 2.13 and 2.14, in BV: impossible. □

2.16 Remark If we restrict BV by disallowing seq structures, we get a system equivalent to MLL (Multiplicative Linear Logic) plus mix and nullary mix [1]. The proof of this is very similar to the proof of 3.12.

Systems equivalent to MLL with constants and without mix can be easily designed in our calculus, but they are not extensible to seq. Other reasons for collapsing the constants into $\circ$ come from external semantic arguments (see [8]).

3 Multiplicative Exponential Linear Logic

All general notions from Sect. 2 apply here. In the following, only what changes in the systems for MELL is defined. The main differences between our presentation and the sequent calculus one are: rules apply anywhere deep into structures, the switch rule replaces times, the promotion rule is decomposed into a local variant. Details can be found in [16].

3.1 Definition We denote by MELL (Multiplicative Exponential Linear Logic) the system in the sequent calculus whose formulae are generated by

$$A ::= a \mid \bot \mid 1 \mid A \parr A \mid A \otimes A \mid ?A \mid !A \mid A^{\bot} \quad ,$$

whose sequents are expressions of the kind

$$\vdash A_1, \ldots, A_h \quad , \qquad \text{for } h \geqslant 0 \quad ,$$

where the commas between formulae stand for multiset union, and whose rules are shown in Fig. 3. Formulae are denoted by A and B, multisets of formulae by Φ and Ψ. Negation obeys De Morgan rules.

Let us define the language of structures ELS (multiplicative Exponential Linear logic in the calculus of Structures). The multiplicatives are denoted as in Sect. 2; for the exponentials we use ? and !. Structures of ELS and formulae of MELL are in a trivial, mutual correspondence.

$$\mathsf{id}\,\frac{}{\vdash A, A^{\perp}} \qquad \mathsf{cut}\,\frac{\vdash A, \Phi \quad \vdash A^{\perp}, \Psi}{\vdash \Phi, \Psi} \qquad \mathbin{⅋}\,\frac{\vdash A, B, \Phi}{\vdash A \mathbin{⅋} B, \Phi} \qquad \otimes\,\frac{\vdash A, \Phi \quad \vdash B, \Psi}{\vdash A \otimes B, \Phi, \Psi} \qquad \perp\,\frac{\vdash \Phi}{\vdash \perp, \Phi}$$

$$\mathsf{wk}\,\frac{\vdash \Phi}{\vdash ?A, \Phi} \qquad \mathsf{ct}\,\frac{\vdash ?A, ?A, \Phi}{\vdash ?A, \Phi} \qquad \mathsf{dr}\,\frac{\vdash A, \Phi}{\vdash ?A, \Phi} \qquad \mathsf{pr}\,\frac{\vdash A, ?B_1, \ldots, ?B_h}{\vdash !A, ?B_1, \ldots, ?B_h}\; h \geqslant 0 \qquad \mathsf{1}\,\frac{}{\vdash 1}$$

Fig. 3 *System* MELL

3.2 Definition The structures of ELS are generated by

$$S ::= a \mid \perp \mid 1 \mid [\underbrace{S, \ldots, S}_{>0}] \mid (\underbrace{S, \ldots, S}_{>0}) \mid ?S \mid !S \mid \bar{S} \quad ,$$

where $\perp$ and 1 are *units*; $[S, \ldots, S]$ is a *par structure*, $(S, \ldots, S)$ is a *times structure*; $?S$ is a *why-not structure* and $!S$ is an *of-course structure*; $\bar{S}$ is the *negation* of S.

3.3 Definition The functions $\underline{\cdot}_{\mathsf{S}}$ and $\underline{\cdot}_{\mathsf{L}}$, from formulae to structures and vice versa, are as follows:

$$\begin{aligned}
\underline{a}_{\mathsf{S}} &= a \;, & \underline{a}_{\mathsf{L}} &= a \;, \\
\underline{\perp}_{\mathsf{S}} &= \perp \;, & \underline{\perp}_{\mathsf{L}} &= \perp \;, \\
\underline{1}_{\mathsf{S}} &= 1 \;, & \underline{1}_{\mathsf{L}} &= 1 \;, \\
\underline{A \mathbin{⅋} B}_{\mathsf{S}} &= [\underline{A}_{\mathsf{S}}, \underline{B}_{\mathsf{S}}] \;, & \underline{[R_1, \ldots, R_h]}_{\mathsf{L}} &= \underline{R_1}_{\mathsf{L}} \mathbin{⅋} \cdots \mathbin{⅋} \underline{R_h}_{\mathsf{L}} \;, \\
\underline{A \otimes B}_{\mathsf{S}} &= (\underline{A}_{\mathsf{S}}, \underline{B}_{\mathsf{S}}) \;, & \underline{(R_1, \ldots, R_h)}_{\mathsf{L}} &= \underline{R_1}_{\mathsf{L}} \otimes \cdots \otimes \underline{R_h}_{\mathsf{L}} \;, \\
\underline{?A}_{\mathsf{S}} &= ?\underline{A}_{\mathsf{S}} \;, & \underline{?R}_{\mathsf{L}} &= ?\underline{R}_{\mathsf{L}} \;, \\
\underline{!A}_{\mathsf{S}} &= !\underline{A}_{\mathsf{S}} \;, & \underline{!R}_{\mathsf{L}} &= !\underline{R}_{\mathsf{L}} \;, \\
\underline{A^{\perp}}_{\mathsf{S}} &= \overline{\underline{A}_{\mathsf{S}}} \;, & \underline{\bar{R}}_{\mathsf{L}} &= (\underline{R}_{\mathsf{L}})^{\perp} \;,
\end{aligned}$$

where $h > 0$. The function $\underline{\cdot}_{\mathsf{S}}$ extends to sequents by $\underline{\vdash}_{\mathsf{S}} = \perp$ and

$$\underline{\vdash A_1, \ldots, A_h}_{\mathsf{S}} = [\underline{A_1}_{\mathsf{S}}, \ldots, \underline{A_h}_{\mathsf{S}}] \quad , \qquad \text{for } h > 0 \quad .$$

It would be entirely possible to take MELL as presented above and transport it trivially into the calculus of structures. At that point, all of the proof theory possible in the sequent calculus would still be possible in our calculus. Instead, we collapse dereliction (dr) and contraction (ct) into absorption (which is a known, easy trick) and use the peculiarities of the calculus of structures to deal differently with times ($\otimes$) and promotion (pr). This way we get new properties.

3.4 Definition The structures of ELS are considered equivalent modulo the relation $=$, defined at the left of Fig. 4. There, $\vec{R}$ and $\vec{T}$ stand for finite, non-empty sequences of structures. At the right of the figure, system SELS is shown (Symmetric, or Self-dual, multiplicative Exponential Linear logic in the calculus of Structures). The rules $\mathsf{a}{\downarrow}$, $\mathsf{a}{\uparrow}$ and s are called, as in system SBV, *atomic interaction*, *atomic cut* (or *atomic cointeraction*) and *switch*. The rules $\mathsf{p}{\downarrow}$, $\mathsf{w}{\downarrow}$ and $\mathsf{b}{\downarrow}$ are called, respectively, *promotion*, *weakening* and *absorption*, and their corules get a prefix co- before their name. The *down fragment* of SELS is $\{\mathsf{a}{\downarrow}, \mathsf{s}, \mathsf{p}{\downarrow}, \mathsf{w}{\downarrow}, \mathsf{b}{\downarrow}\}$, the *up fragment* is $\{\mathsf{a}{\uparrow}, \mathsf{s}, \mathsf{p}{\uparrow}, \mathsf{w}{\uparrow}, \mathsf{b}{\uparrow}\}$.

The reader can check that the equations in Fig. 4 are equivalences in MELL.

3.5 Definition The following rules are *interaction* and *cut* (or *cointeraction*):

$$\mathsf{i}{\downarrow}\,\frac{S\{1\}}{S[R, \bar{R}]} \qquad \text{and} \qquad \mathsf{i}{\uparrow}\,\frac{S(R, \bar{R})}{S\{\perp\}} \quad .$$

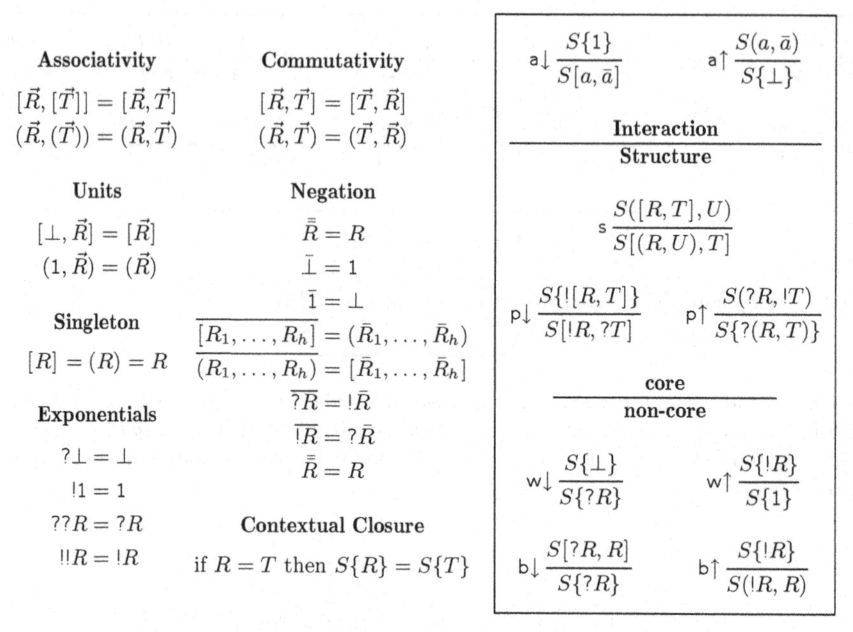

Fig. 4 **Left:** *Syntactic equivalence* $=$ *for* ELS **Right:** *System* SELS

$$
1\!\downarrow \frac{}{1} \qquad \mathsf{a}\!\downarrow \frac{S\{1\}}{S[a,\bar{a}]} \qquad \mathsf{s}\, \frac{S([R,T],U)}{S[(R,U),T]} \qquad \mathsf{p}\!\downarrow \frac{S\{![R,T]\}}{S[!R,?T]} \qquad \mathsf{w}\!\downarrow \frac{S\{\bot\}}{S\{?R\}} \qquad \mathsf{b}\!\downarrow \frac{S[?R,R]}{S\{?R\}}
$$

Fig. 5 *System* ELS

Like for system SBV, we have the following two propositions, which say: 1) the general interaction and cut rules can be decomposed into their atomic forms; 2) the cut rule is as powerful as the whole up fragment of the system, and vice versa (and the same holds for the interaction rule with respect to the down fragment).

3.6 Proposition *The rules* $\mathsf{i}\!\downarrow$ *and* $\mathsf{i}\!\uparrow$ *are strongly admissible for systems* $\{\mathsf{a}\!\downarrow, \mathsf{s}, \mathsf{p}\!\downarrow\}$ *and* $\{\mathsf{a}\!\uparrow, \mathsf{s}, \mathsf{p}\!\uparrow\}$, *respectively.*

Proof Similar to the proof of 2.6. □

3.7 Proposition *Every rule* $\rho\!\uparrow$ *in system* SELS *is strongly admissible for the system* $\{\mathsf{i}\!\downarrow, \mathsf{i}\!\uparrow, \mathsf{s}, \rho\!\downarrow\}$.

Proof See 2.8. □

3.8 Definition The *core* of SELS is the system $\{\mathsf{s}, \mathsf{p}\!\downarrow, \mathsf{p}\!\uparrow\}$, denoted by SELSc.

3.9 Definition The following (logical axiom) rule is called *one*: $1\!\downarrow \frac{}{1}$.

As we did in Sect. 2, we put our logical axiom into the down fragment of SELS.

3.10 Definition System ELS is shown in Fig. 5.

As a quick consequence of 3.6 and 3.7 we get:

3.11 Theorem $\mathsf{ELS} \cup \{\mathsf{i}\!\uparrow\}$ *and* $\mathsf{SELS} \cup \{1\!\downarrow\}$ *are strongly equivalent.*

The system $\mathsf{SELS} \cup \{1\downarrow\}$ is equivalent to MELL:

3.12 Theorem *If R is provable in $\mathsf{SELS} \cup \{1\downarrow\}$ then $\vdash \underline{R}_{\mathsf{L}}$ is provable in MELL, and if $\vdash \Phi$ is provable in MELL then $\underline{\vdash \Phi}_{\mathsf{S}}$ is provable in $\overline{\mathsf{SELS}}^{\mathsf{L}} \cup \{1\downarrow\}$.*

Proof For every rule $\rho \dfrac{S\{T\}}{S\{R\}}$ in SELS the sequent $\vdash (\underline{T}_{\mathsf{L}})^{\perp}, \underline{R}_{\mathsf{L}}$ is provable in MELL. Then the sequent $\vdash (\underline{S\{T\}}_{\mathsf{L}})^{\perp}, \underline{S\{R\}}_{\mathsf{L}}$ is provable. Use this and $\mathsf{cut} \dfrac{\vdash \underline{S\{T\}}_{\mathsf{L}} \quad \vdash (\underline{S\{T\}}_{\mathsf{L}})^{\perp}, \underline{S\{R\}}_{\mathsf{L}}}{\vdash \underline{S\{R\}}_{\mathsf{L}}}$ inductively over a given proof $\rho \dfrac{\overset{\| \mathsf{SELS} \cup \{1\downarrow\}}{S\{T\}}}{S\{R\}}$. Conversely, given a proof in MELL, transform it by an easy induction, proceeding from its root, into a proof in $\mathsf{SELS} \cup \{1\downarrow\}$. We only show the case of promotion, where the derivation Δ exists by induction hypothesis:

$$
\mathsf{p}\downarrow \dfrac{\mathsf{p}\downarrow \dfrac{\mathsf{p}\downarrow \dfrac{\overset{\Delta \| \mathsf{SELS}}{1\downarrow \dfrac{}{!1}}}{![\underline{A}_{\mathsf{S}}, ?\underline{B_1}_{\mathsf{S}}, \dots, ?\underline{B_h}_{\mathsf{S}}]} \quad \vdots}{[![\underline{A}_{\mathsf{S}}, ?\underline{B_1}_{\mathsf{S}}], ??\underline{B_2}_{\mathsf{S}}, \dots, ??\underline{B_h}_{\mathsf{S}}]}}{[!\underline{A}_{\mathsf{S}}, ??\underline{B_1}_{\mathsf{S}}, \dots, ??\underline{B_h}_{\mathsf{S}}]} \quad .
$$

□

An argument along these lines shows that for every cut free proof in MELL we can obtain a proof in ELS. Therefore, $\mathsf{i}\uparrow$ is admissible for ELS, by the cut elimination theorem for MELL [6]. In other words, the whole up fragment of SELS is admissible for ELS. However, we obtain this result for the calculus of structures by using the sequent calculus. Since we want to use our calculus for logics that cannot be captured by the sequent calculus, we must be able to prove cut elimination within our calculus, with no detour. The first step is a decomposition theorem.

3.13 Theorem *For every derivation* $\begin{matrix} T \\ \| \mathsf{SELS} \\ R \end{matrix}$ *there is a derivation* $\begin{matrix} T \\ \| \{\mathsf{b}\uparrow\} \\ T_1 \\ \| \{\mathsf{w}\downarrow\} \\ T_2 \\ \| \{\mathsf{a}\downarrow\} \\ T_3 \\ \| \mathsf{SELSc} \\ R_3 \\ \| \{\mathsf{a}\uparrow\} \\ R_2 \\ \| \{\mathsf{w}\uparrow\} \\ R_1 \\ \| \{\mathsf{b}\downarrow\} \\ R \end{matrix}$, *for some structures R_1, R_2, R_3, T_1, T_2, T_3.*

Proof The decomposition is done in three steps: $\mathsf{b}\uparrow$ and $\mathsf{b}\downarrow$ instances are separated, then $\mathsf{w}\downarrow$ and $\mathsf{w}\uparrow$, and then $\mathsf{a}\downarrow$ and $\mathsf{a}\uparrow$. The first step is very difficult (see [16]), the other two are rather trivial. □

If we just consider proofs instead of derivations, all top instances of $\mathsf{b}\uparrow$ become trivial: their premises and conclusions are equal to 1. Moreover, all $\mathsf{w}\uparrow$ instances can be removed by using 3.7 and 3.6.

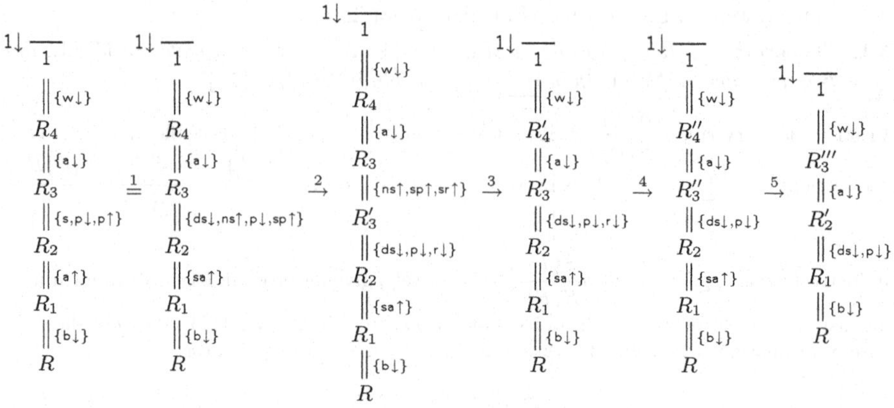

$$
\begin{array}{c}
1\!\downarrow\frac{}{1}\\
\Vert\{\mathsf{w}\!\downarrow\}\\
R_4\\
\Vert\{\mathsf{a}\!\downarrow\}\\
R_3\\
\Vert\{\mathsf{s},\mathsf{p}\!\downarrow,\mathsf{p}\!\uparrow\}\\
R_2\\
\Vert\{\mathsf{a}\!\uparrow\}\\
R_1\\
\Vert\{\mathsf{b}\!\downarrow\}\\
R
\end{array}
\;\overset{1}{=}\;
\begin{array}{c}
1\!\downarrow\frac{}{1}\\
\Vert\{\mathsf{w}\!\downarrow\}\\
R_4\\
\Vert\{\mathsf{a}\!\downarrow\}\\
R_3\\
\Vert\{\mathsf{ds}\!\downarrow,\mathsf{ns}\!\uparrow,\mathsf{p}\!\downarrow,\mathsf{sp}\!\uparrow\}\\
R_2\\
\Vert\{\mathsf{sa}\!\uparrow\}\\
R_1\\
\Vert\{\mathsf{b}\!\downarrow\}\\
R
\end{array}
\;\overset{2}{\rightarrow}\;
\begin{array}{c}
1\!\downarrow\frac{}{1}\\
\Vert\{\mathsf{w}\!\downarrow\}\\
R_4\\
\Vert\{\mathsf{a}\!\downarrow\}\\
R_3\\
\Vert\{\mathsf{ns}\!\uparrow,\mathsf{sp}\!\uparrow,\mathsf{sr}\!\uparrow\}\\
R'_3\\
\Vert\{\mathsf{ds}\!\downarrow,\mathsf{p}\!\downarrow,\mathsf{r}\!\downarrow\}\\
R_2\\
\Vert\{\mathsf{sa}\!\uparrow\}\\
R_1\\
\Vert\{\mathsf{b}\!\downarrow\}\\
R
\end{array}
\;\overset{3}{\rightarrow}\;
\begin{array}{c}
1\!\downarrow\frac{}{1}\\
\Vert\{\mathsf{w}\!\downarrow\}\\
R'_4\\
\Vert\{\mathsf{a}\!\downarrow\}\\
R'_3\\
\Vert\{\mathsf{ds}\!\downarrow,\mathsf{p}\!\downarrow,\mathsf{r}\!\downarrow\}\\
R_2\\
\Vert\{\mathsf{sa}\!\uparrow\}\\
R_1\\
\Vert\{\mathsf{b}\!\downarrow\}\\
R
\end{array}
\;\overset{4}{\rightarrow}\;
\begin{array}{c}
1\!\downarrow\frac{}{1}\\
\Vert\{\mathsf{w}\!\downarrow\}\\
R''_4\\
\Vert\{\mathsf{a}\!\downarrow\}\\
R''_3\\
\Vert\{\mathsf{ds}\!\downarrow,\mathsf{p}\!\downarrow\}\\
R_2\\
\Vert\{\mathsf{sa}\!\uparrow\}\\
R_1\\
\Vert\{\mathsf{b}\!\downarrow\}\\
R
\end{array}
\;\overset{5}{\rightarrow}\;
\begin{array}{c}
1\!\downarrow\frac{}{1}\\
\Vert\{\mathsf{w}\!\downarrow\}\\
R'''_3\\
\Vert\{\mathsf{a}\!\downarrow\}\\
R'_2\\
\Vert\{\mathsf{ds}\!\downarrow,\mathsf{p}\!\downarrow\}\\
R_1\\
\Vert\{\mathsf{b}\!\downarrow\}\\
R
\end{array}
$$

Fig. 6 *Cut elimination for* $\mathsf{SELS}\cup\{1\!\downarrow\}$

3.14 Theorem *For every proof* $\begin{array}{c}\overline{\Vert}\,\mathsf{SELS}\cup\{1\downarrow\}\\ R\end{array}$ *there is a proof* $\begin{array}{c}1\!\downarrow\frac{}{1}\\ \Vert\{\mathsf{w}\!\downarrow\}\\ R_4\\ \Vert\{\mathsf{a}\!\downarrow\}\\ R_3\\ \Vert\,\mathsf{SELSc}\\ R_2\\ \Vert\{\mathsf{a}\!\uparrow\}\\ R_1\\ \Vert\{\mathsf{b}\!\downarrow\}\\ R\end{array}$, *for some structures* R_1, R_2, R_3, R_4.

Proof It is a trivial variation of 3.13. □

The decomposition theorem is of great value for the cut elimination proof, because all instances of b↓ are already below the instances of p↑ and a↑ that have to be eliminated. This means that we do not have to deal with absorption (nor contraction), which are known to be most problematic in a cut elimination proof.

3.15 Theorem *The systems* $\mathsf{SELS}\cup\{1\!\downarrow\}$ *and* ELS *are equivalent.*

Proof The proof is similar to that for BV: we eliminate in order w↑, b↑, p↑ and a↑. For w↑ and b↑ we use 3.14. For a↑ and p↑ we use the super rules:

$$
\mathsf{sa}\!\uparrow\frac{S([a,P],[\bar{a},Q])}{S[P,Q]} \qquad\text{and}\qquad \mathsf{sp}\!\uparrow\frac{S([?R,P],[!T,Q])}{S[?(R,T),P,Q]} \quad .
$$

We also need the rule r↓ and its super corule sr↑:

$$
\mathsf{r}\!\downarrow\frac{S\{?[R,T]\}}{S[?R,?T]} \qquad\text{and}\qquad \mathsf{sr}\!\uparrow\frac{S([!R,P],[!T,Q])}{S[!(R,T),P,Q]} \quad .
$$

We then use the rule ns↑ (*non-deep switch*) which defines all instances of s that are not instances of ds↓ (see 2.14). Fig. 6 shows the steps of the transformation. We start from a decomposed proof produced by 3.14. Then we replace all instances of s either by ds↓ or ns↑, and all instances of p↑ and a↑ by sp↑ and sa↑, respectively. While permuting up the rules ns↑ and sp↑ over ds↓ and p↓ in Step 2, the rules sr↑ and r↓ are introduced. In Steps 3 and 4, the rules ns↑, sp↑ and sr↑, and then the rule r↓ are eliminated. In the last step the rule sa↑ is eliminated. □

4 Conclusions and Future Work

We have shown, in the calculus of structures, the system BV, which is an extension of MLL (Multiplicative Linear Logic) and which is not expressible in the sequent calculus in any known way. Research is currently going on finally to prove that it is impossible to capture BV in the sequent calculus. System BV is interesting for computer science because it models a typical notion of sequentialisation. We then extended MLL to MELL in our calculus, and we got a system whose promotion rule is local, as opposed to what is possible in the sequent calculus, where promotion is global. The new system does not present unnecessary non-determinism in dealing with the times connective.

The question is whether a new calculus is justified, given that the competition is the venerable sequent calculus. We answer yes for the following reasons:

1 *Simplicity*: The calculus of structures is more general than the sequent calculus (for logics with involutive negation), but is not more complicated. The case of multiplicative exponential linear logic shows that a simple system, deeply different than MELL, can be designed. System BV yields with very simple means a logic that defeats sequent calculus.

2 *Power*: The calculus of structures unveils properties and possibilities of analyses, like decomposition, that are not available in the sequent calculus.

3 *Modularity*: Proving cut-elimination is modular: if one enlarges a system, the work done for the smaller system can be used for the new. Moreover, the cut elimination argument for any given system is decomposed into separate pieces. This stems from the possibility of dealing with cut the same way we could with identity in the sequent calculus: our calculus makes use of a new symmetry.

One reason for these achievements is the applicability of rules deeply into structures, which allows for a lazy bookkeeping of the context. For example, the times rule in the sequent calculus must make an early choice of the splitting of its context, which is not the case in our calculus. The same happens with promotion: pieces of context can be brought inside the scope of an of-course one by one.

Another reason behind our results is the dropping of the idea of connective. In the calculus of structures, instead of defining connectives, rules define mutual relations of logical relations. Typical rules in the up fragment of a system are not definable in the sequent calculus, yet they are just simple duals of ordinary sequent calculus rules. Without much complication, we can then decompose the cut rule into its atomic form, which is the key to modularity.

One possible problem with our calculus is that, since rules apply anywhere deep into structures, proof search can be very non-deterministic. Research is in progress in our group to focus proofs not only along lines induced by the logical relations [3, 11], but also based on the depth of structures.

Classical logic is also studied. One can easily port 'additive' rules to our calculus, but the question, again, is whether we can get decomposition and a modular cut elimination proof. Recent work, in preparation, by Brünnler and Tiu, shows that classical logic enjoys a presentation whose rules are all local, and cut is admissible [4].

The next step will be to bring exponentials (and contraction) to system BV. The experiment performed in this paper shows that the operation is entirely practical in our calculus, and it would yield better results than proof nets [13, 14], which have notorious difficulties with exponentials. The resulting calculus will be Turing equivalent. Our hope is that MELL will be proved decidable (the question is still open): if this happened, it would mean that the edge is crossed by our self-dual non-commutative logical relation (the tape of a Turing machine?).

Finally, we have a further prototype system, inspired by traces [8], in which also the contraction rule is atomic. We are not able yet to prove cut elimination for it. If we were successful, we would obtain a totally distributed formalism, in the sense of computer science, which would also be a first class proof theoretical system.

References

[1] Samson Abramsky and Radha Jagadeesan. Games and full completeness for multiplicative linear logic. *Journal of Symbolic Logic*, 59(2):543–574, June 1994.

[2] V. Michele Abrusci and Paul Ruet. Non-commutative logic I: The multiplicative fragment. *Annals of Pure and Applied Logic*, 101(1):29–64, 2000.

[3] Jean-Marc Andreoli. Logic programming with focusing proofs in linear logic. *Journal of Logic and Computation*, 2(3):297–347, 1992.

[4] Kai Brünnler and Alwen Tiu. A local system for classical logic. In preparation.

[5] Gerhard Gentzen. Investigations into logical deduction. In M. E. Szabo, editor, *The Collected Papers of Gerhard Gentzen*, pages 68–131. North-Holland, Amsterdam, 1969.

[6] Jean-Yves Girard. Linear logic. *Theoretical Computer Science*, 50:1–102, 1987.

[7] Jean-Yves Girard. *Proof Theory and Logical Complexity, Volume I*, volume 1 of *Studies in Proof Theory*. Bibliopolis, Napoli, 1987. Distributed by Elsevier.

[8] Alessio Guglielmi. A calculus of order and interaction. Technical Report WV-99-04, Dresden University of Technology, 1999. On the web at: http://www.ki.inf.tu-dresden.de/~guglielm/Research/Gug/Gug.pdf.

[9] Joshua S. Hodas and Dale Miller. Logic programming in a fragment of intuitionistic linear logic. *Information and Computation*, 110(2):327–365, May 1994.

[10] Dale Miller. The π-calculus as a theory in linear logic: Preliminary results. In E. Lamma and P. Mello, editors, *1992 Workshop on Extensions to Logic Programming*, volume 660 of *Lecture Notes in Computer Science*, pages 242–265. Springer-Verlag, 1993.

[11] Dale Miller. Forum: A multiple-conclusion specification logic. *Theoretical Computer Science*, 165:201–232, 1996.

[12] Robin Milner. *Communication and Concurrency*. International Series in Computer Science. Prentice Hall, 1989.

[13] Christian Retoré. Pomset logic: A non-commutative extension of classical linear logic. In Ph. de Groote and J. R. Hindley, editors, *TLCA'97*, volume 1210 of *Lecture Notes in Computer Science*, pages 300–318, 1997.

[14] Christian Retoré. Pomset logic as a calculus of directed cographs. In V. M. Abrusci and C. Casadio, editors, *Dynamic Perspectives in Logic and Linguistics*, pages 221–247. Bulzoni, Roma, 1999. Also available as INRIA Rapport de Recherche RR-3714.

[15] Paul Ruet. Non-commutative logic II: Sequent calculus and phase semantics. *Mathematical Structures in Computer Science*, 10:277–312, 2000.

[16] Lutz Straßburger. MELL in the calculus of structures. Technical Report WV-2001-03, Dresden University of Technology, 2001. On the web at: http://www.ki.inf.tu-dresden.de/~lutz/els.pdf.

Quadratic Correctness Criterion for Non-commutative Logic

Virgile Mogbil

Institut de Mathématiques de Luminy, UPR 9016 - CNRS
163 Av. Luminy Case 907, 13288 Marseille Cedex 09, France
mogbil@iml.univ-mrs.fr

Abstract. The multiplicative fragment of Non commutative Logic (called MNL) has a proof nets theory [AR00] with a correctness criterion based on long trips for cut-free proof nets. Recently, R.Maieli has developed another criterion in the Danos-Regnier style [Mai00]. Both are in exponential time. We give a quadratic criterion in the Danos contractibility criterion style.

1 Introduction

Non commutative Logic (NL) is a unification of linear logic [Gir87] and cyclic linear logic [Gir89,Yet90,Abr91] (a classical conservative extension of the Lambek calculus [Lam58]). It includes all linear connectives: multiplicatives, additives, exponentials and constants. Recents results [AR00,Rue00,MR00] introduce proof nets, sequent calculus, phase semantics and all the importants theorems like cut elimination and sequentialisation. The central notion is the structure of order varieties. Let α be an order variety on a base set $X \cup \{x\}$, provided a point of view (the element x) α can be seen as a partial order on X. Order varieties can be presented in different ways by changing the point of view and are invariant under the change of presentation: one uses rootless planar trees called seaweeds. Thus this structure allows focusing on any formula to apply a rule.

Proof nets are graph representations of NL derivations. Then a proof net with conclusion A is obtained as an interpretation of a sequent calculus proof of A: we say that it can be sequentialized. But the corresponding cut-free derivation of the formula A is not unique in general. It introduces some irrelevant order on the sequent rules. For instance, a derivation Π ending with $\vdash A \parr B, C \parr D$ implies an order on the two rules introducing the principal connectives of $A \parr B$ and $C \parr D$, but the proof net corresponding to Π does not depend on such order.

A contracting proof structure is a hypergraph built in accordance with the syntax of proof nets and seaweeds. A proof structure is a particular contracting one. To know if a such structure is a proof net or not, we use a correctness criterion. The Maieli one is in the Danos-Regnier criterion style: at first it uses a switching condition and tests if we obtain an acyclic connected graph. Then for each ∇ link, we check the associated order varieties.

It is known that the proof nets of multiplicative linear logic have a linear time correctness criterion [Gue99]. The first step towards a linear algorithm is

L. Fribourg (Ed.): CSL 2001, LNCS 2142, pp. 69–83, 2001.

to have a contractibility criterion (the Danos one [Dan90]) which can be seen as a parsing algorithm. One can reformulate it in terms of a sort of unification. Then a direct implementation leads a quasi-linear algorithm, and sharp study give the exact complexity. Up to now, there was no polynomial criterion for MNL.

Here we present a set of shrinking rules for MNL proof structures characterising MNL proof nets as the only structures that contract to a seaweed. We show that this contractibility criterion is quadratic. This idea is extended by a presentation as a parsing algorithm. So this work criterion.

Notations. One writes $X \uplus Y$ for the disjoint union of the sets X and Y. Let ω and τ be orders respectively on the sets X and Y. Let x be in X. One writes $\omega[\tau/x]$ the order on $(X\backslash\{x\}) \cup Y$ defined by $\omega[\tau/x](y,z)$ iff $\omega(y,z)$ or $\tau(y,z)$ or $\omega(y,x)$ if $z \in Y$ or $\omega(x,z)$ if $y \in Y$. Let f and g be positive functions. One writes $g(n) = \Theta(f(n))$ to denote that $f = \mathcal{O}(g)$ and $g = \mathcal{O}(f)$.

2 Order Varieties

2.1 Order Varieties and Orders

Definition 1 (order varieties). *Let X be a set. An* order variety *on X is a ternary relation α which is:*

$$\begin{cases} \textit{cyclic:} & \forall x,y,z \in X, \alpha(x,y,z) \Rightarrow \alpha(y,z,x), \\ \textit{anti-reflexive:} & \forall x,y \in X, \neg\alpha(x,x,y), \\ \textit{transitive:} & \forall x,y,z,t \in X, \alpha(x,y,z) \textit{ and } \alpha(z,t,x) \Rightarrow \alpha(y,z,t), \\ \textit{spreading:} & \forall x,y,z,t \in X, \alpha(x,y,z) \Rightarrow \alpha(t,y,z) \textit{ or } \alpha(x,t,z) \textit{ or } \alpha(x,y,t). \end{cases}$$

Definition 2 (series-parallel orders). *Let ω and τ be two partial orders on disjoint sets X and Y respectively. Their* serial sum *(resp.* parallel sum*) $\omega < \tau$ (resp. $\omega \parallel \tau$) is a partial order on $X \cup Y$ defined respectively by:*

$$\begin{array}{ll} (\omega < \tau)(x,y) & \textit{iff } x <_\omega y \textit{ or } x <_\tau y \textit{ or } (x \in X \textit{ and } y \in Y), \\ (\omega \parallel \tau)(x,y) & \textit{iff } x <_\omega y \textit{ or } x <_\tau y. \end{array}$$

Definition 3 (closure). *Let $\omega = (X,<)$ be a partial order on X and $z \in X$. Let $\overset{z}{<}$ denote the binary relation: $x \overset{z}{<} y$ iff $x < y$ and z is comparable neither with x nor y. The* closure *of ω is the ternary relation $\overline{\omega}$ on X defined by:*

$$\begin{array}{lllll} \overline{\omega}(x,y,z) & \textit{iff } x<y<z & \textit{or } y<z<x & \textit{or } z<x<y & \textit{or} \\ & x \overset{z}{<} y & \textit{or } y \overset{x}{<} z & \textit{or } z \overset{y}{<} x. & \end{array}$$

Facts 1. i) If ω is a partial order on X then $\overline{\omega}$ is an order variety on X,

ii) The closure identifies serial and parallel sums of partial orders on disjoint sets.

Definition 4 (gluing). *Let ω and τ be two partial orders on disjoint sets X and Y respectively. The* gluing $\omega * \tau$ *of ω and τ is the following order variety on $X \cup Y$:*

$$\omega * \tau = \overline{\omega < \tau} = \overline{\omega \parallel \tau} = \overline{\tau < \omega}$$

Definition 5. *Let α be an order variety on a set X and $x \in X$. The order α_x* induced by α and x *is the partial order on $X \backslash \{x\}$ defined by:*

$$\alpha_x(y,z) \textit{ iff } \alpha(x,y,z)$$

One writes x for the unique partial order on $\{x\}$.

Proposition 1. *Let α be an order variety on a set X, $x \in X$ and ω be a partial order on $X \backslash \{x\}$. Then*

$$\alpha_x * x = \alpha \quad \textit{and} \quad (\omega * x)_x = \omega$$

Fact 2. Let α be an order variety on a non-empty set. α is series-parallel iff there exists a series-parallel order ω such that $\alpha = \overline{\omega}$. In other words, series-parallel order varieties are exactly those can be represented by series-parallel orders.

Definition 6 (seaweed). *Let $\alpha = \overline{\omega}$ be a series-parallel order variety on X ($\sharp X \geq 2$) such that ω is written as a (non-unique) binary tree T with leaves labelled by elements of X, and root and nodes labelled by $\bullet$ (serial composition) or $\circ$ (parallel composition).*

A seaweed *S representing α is a rootless planar tree with leaves labelled by elements of X and ternary nodes labelled by $\bullet$ or $\circ$, defined by removing the root of T:*

$$\alpha = \overline{\omega < \tau} = \omega * \tau = \overline{\omega \parallel \tau}$$

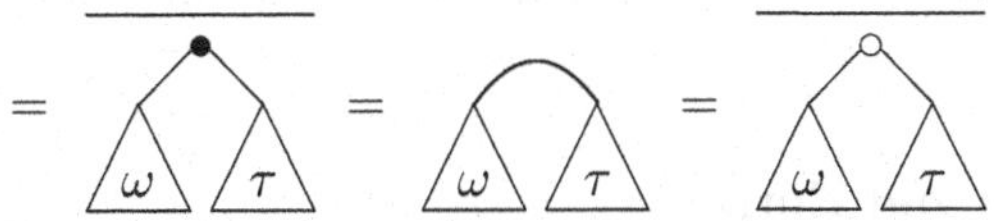

By convention orders are represented with top root and then seaweeds are oriented anti-clockwise:

$$\overline{a < b < c} =$$

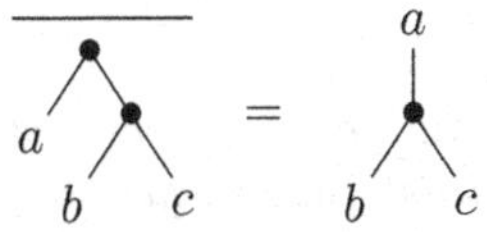

One extends the definition of seaweeds to the rootless planar trees on n-ary-nodes ($n \geq 3$).

Definition 7 (normal form). *Let α be a series-parallel order variety. Let the seaweeds representing α be considered modulo assocativity of $\circ$ and $\bullet$: there is not two nodes linked with a same label, and there is not binary or unary nodes. The equivalence class of such seaweeds modulo commutativity of $\circ$ has a unique representative which is said in* normal form.

The uniqueness comes from the next proposition.

Remark 1. A seaweed is in normal form if it has n-ary nodes and verifies that all paths between two leaves are a sequence of alternate $\bullet$ and $\circ$ nodes. Afterwards for a seaweed (not specially in normal form) we denote such alternate paths between arbitrary leaves x and y by the following figure:

x ···· y

This notation does not presuppose that this alternate path starts by a $\bullet$-node and finishes by a $\circ$-node.

Example 1. Let α be the closure of $[((a < b) < c) \parallel d] < [e \parallel (f < (g < h) < i)]$. Then the path between d and g is

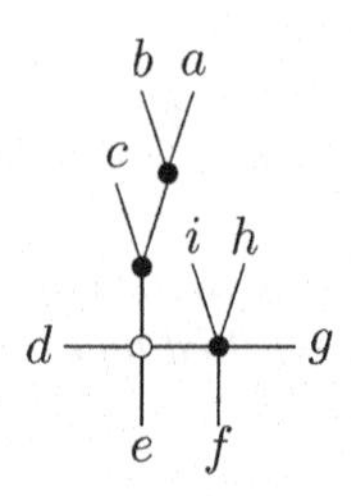

To be convenient we only use seaweeds in normal form. So $\circ$-nodes are commutative. When it is not ambiguous, we use an order variety instead of its representation.

2.2 Seesaw and Entropy

Definitions 8. *Let ω and τ be series-parallel orders on a same given set. The equivalence relation* seesaw *is defined by $\overline{\omega} = \overline{\tau}$. The relation* entropy $\trianglelefteq$ *is defined by $\omega \trianglelefteq \tau$ iff $\omega \subseteq \tau$ and $\overline{\omega} \subseteq \overline{\tau}$.*

Proposition 2. *In the case of series-parallel orders, seesaw (resp. entropy) turns out to be the least equivalence $\backsim$ (resp. the least reflexive transitive relation) given by:*

$$(\omega_1 \parallel \omega_2) \backsim (\omega_1 < \omega_2) \qquad \textit{(resp. } \omega[\omega_1 \parallel \omega_2] \trianglelefteq \omega[\omega_1 < \omega_2] \textit{)}$$

Facts 3. i) Entropy is a partial order, compatible with restriction and the serial and parallel sums of orders,

ii) entropy between orders corresponds to inclusion of order varieties: let α and β be order varieties on X, and $x \in X$, we have

$$\alpha \subseteq \beta \text{ iff } \alpha_x \trianglelefteq \beta_x.$$

This is independent from the choice of x,

iii) entropy is performed on seaweeds by changing some •-nodes into ○-nodes.

2.3 Wedge and Identification

Definitions 9 (wedge). *Let $(\omega_i)_{i\in I}$ be a non empty family of partial orders on a same set. The* wedge $\bigwedge_{i\in I}\omega_i$ *is the largest partial order (w.r.t. $\trianglelefteq$) such that*

$$(\bigwedge_{i\in I}\omega_i) \trianglelefteq \omega_i \textit{ for all } i \in I.$$

Let $(\alpha_i)_{i\in I}$ be a non empty family of order varieties on a set X. The wedge $\bigwedge_{i\in I}\alpha_i$ *is*

$$(\bigwedge_{i\in I}(\alpha_i)_x) * x$$

for an arbritrary $x \in X$.

Facts 4. i) Partial orders on a given set form a complete inf-semi-lattice for entropy and wedge,

ii) the wedge is not intersection in general,

iii) the wedge is not series-parallel in general, even if all ω_i are series-parallel,

iv) the wedge (partially) commutes with restriction:

$$\text{if } Y \subseteq |\omega_i| \text{ then } (\bigwedge_{i\in I}\omega_i) \upharpoonright Y \trianglelefteq (\bigwedge_{i\in I}\omega_i \upharpoonright Y),$$

v) the two notions of wedge are related by:

$$(\bigwedge_{i\in I}\alpha_i)_x = \bigwedge_{i\in I}(\alpha_i)_x \quad \text{and} \quad (\bigwedge_{i\in I}\omega_i) * x = \bigwedge_{i\in I}(\omega_i * x)$$

Definition 10 (identification). *Let α be an order variety on a set $X \uplus \{x\} \uplus \{y\}$, and let $z \notin X \cup \{x, y\}$. The* identification $\alpha[z/x, y]$ *of x and y into z in α is the order variety defined by:*

$$\alpha[z/x, y] = \alpha \upharpoonright_{X\cup\{x\}} [z/x] \ \wedge \ \alpha \upharpoonright_{X\cup\{y\}} [z/y]$$

Lemma 1. *i) $\alpha[z/x, y]_z * (x \parallel y) \subseteq \alpha$,*

ii) Let α be an order variety on $X \uplus \{x\} \uplus \{y\}$ and ω be a partial order on X such that $\omega * (x \parallel y) \subseteq \alpha$. Then $\omega * (x \parallel y) \subseteq \alpha[z/x,y]_z * (x \parallel y)$, or equivalently $\omega \trianglelefteq \alpha[z/x,y]_z$.

Proof. See the proof of lemma 3.35 in [Rue00]. □

Definition 11. *Let α be a series-parallel order variety represented by a seaweed S. We define the seaweed $S\langle z/x,y\rangle$ by the following sequence on the alternate path between x and y in S:*

1. *fis_{xy}: transform every $\circ$-node belong the path between x and y. This is called "fission":*

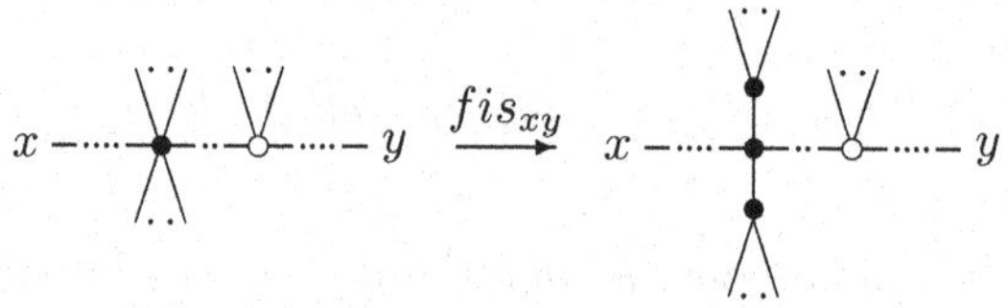

x ⋯ y $\xrightarrow{fis_{xy}}$ x ⋯ y

2. *ent_{xy}: apply entropy belong the path between x and y:*

x ⋯ y $\xrightarrow{ent_{xy}}$ x ⋯ y

3. *ass_{xy}: apply associativity belong the path between x and y:*

x ⋯ y $\xrightarrow{ass_{xy}}$ x ⋯ y

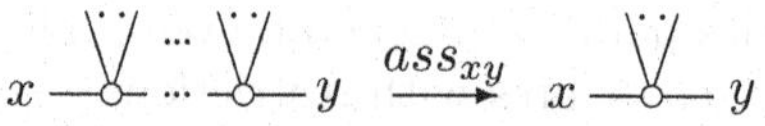

4. *substitute z for $x \parallel y$.*

Lemma 2. *i) Identification in order varieties is monotonic (for the inclusion),*
ii) If v denotes a map such that $v(S)$ is the order variety corresponding to the seaweed S then, for S and T seaweeds,

$$v(S) \subseteq v(T) \implies v(S\langle z/x,y\rangle) \subseteq v(T\langle z/x,y\rangle)$$

Proof. Let α and β be order varieties on a set X such that $\alpha \subseteq \beta$. We have $\alpha[z/x,y] \subseteq \beta[z/x,y]$ i.e. identification is monotonic because the wedge is clearly monotonic. On the seaweeds, the only nodes which are different in the representation of α and β are the $\circ$-nodes in the representation of α which correspond to $\bullet$-nodes in the representation of β. If so,

- by definition, for all $x, y \in X$, $fis_{xy}(\alpha)$ and $fis_{xy}(\beta)$ represent always the same included order varieties,

- all differents nodes on the path between x and y in $ent_{xy}(fis_{xy}(\alpha))$ become $\circ$-nodes and stay $\circ$-nodes in $ent_{xy}(fis_{xy}(\beta))$,
- all others are unchanged.

Hence the order variety represented by $ent_{xy}(fis_{xy}(\alpha))$ is included in the one which is represented by $ent_{xy}(fis_{xy}(\beta))$ □

Proposition 3. *Let α be a series-parallel order variety on a set $X \uplus \{x\} \uplus \{y\}$, and let $z \notin X \cup \{x, y\}$. If the seaweed S represents α then the seaweed $S\langle z/x, y\rangle$ represents the identification $\alpha[z/x, y]$.*

Proof. Using the notations of lemma 2,

$\supseteq$) With the hypothesis, we have that $\alpha[z/x, y]_z * (x \parallel y) \subseteq \alpha$. Then by the previous lemma,

$$v((\alpha[z/x, y]_z * (x \parallel y))\langle z/x, y\rangle) \subseteq v(\alpha\langle z/x, y\rangle)$$

So by definition of $S\langle z/x, y\rangle$, we obtain that

$$\alpha[z/x, y]_z * z \subseteq v(\alpha\langle z/x, y\rangle)$$

For all $u \in |\alpha|$ $\alpha_u * u = \alpha$, thus

$$\alpha[z/x, y] \subseteq v(\alpha\langle z/x, y\rangle)$$

$\subseteq$) By definition, $fis_{xy}(\alpha)$ represents the same order variety as α and for all order variety β, $v(ent_{xy}(\beta)) \subseteq \beta$. Thus $v(ent_{xy}(fis_{xy}(\alpha))) \subseteq \alpha$. Then we again have that $v(S\langle z/x, y\rangle)_z * (x \parallel y) \subseteq \alpha$. Then by definition and as identification is monotonic we have

$$v(S\langle z/x, y\rangle)_z * z \subseteq \alpha[z/x, y] \qquad i.e. \qquad v(S\langle z/x, y\rangle) \subseteq \alpha[z/x, y]$$

□

3 MNL Proof Nets

We restrict us to the multiplicative fragment of NL i.e. to the formulae built from atoms $a, a^{\perp}, \ldots$, the commutative conjonction and disjonction (resp. $\otimes$ and $⅋$) and the non commutative conjonction and disjonction (resp. $\odot$ and ∇).

Definitions 12 (links and proof structures). *A* link *is an object for which the* premises *(input edges) and the* conclusions *(output edges) are two disjoint sets of vertices:*

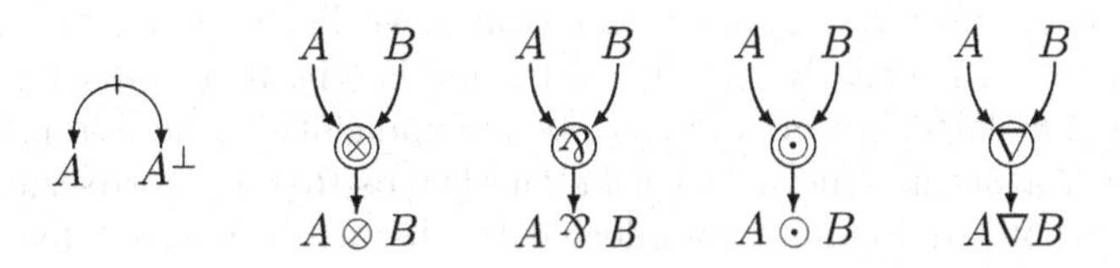

A proof structure *G over the vertices $V(G)$ is a set of links such that:*

- *every vertex in $V(G)$ is a conclusion of (only) one link,*
- *every vertex in $V(G)$ either is a* conclusion of G *(i.e. is not a premise of any link of G) or is a premise of (only) one link,*
- *the set γ of the conclusions of G (writen $G \vdash \gamma$) is not empty.*

3.1 Maieli Correctness Criterion

Definitions 13 (Switchings). *Let G a proof structure. A* switching s *for G is given by mutilating one premise-edge for each ∇-link and $⅋$-link. Any ∇-link (resp. $⅋$-link) admits a left/right mutilation wich is called the* left/right switch *of ∇ (resp. $⅋$). Any switching s for a proof structure G induces a graph on $V(G)$ which is called the* switched proof structure $s(G)$.

Fact 5. If a switched proof structure S induced by a proof structure $G \vdash \gamma$ is acyclic and connected then (viewing $\otimes$-nodes as $\circ$-nodes and $\odot$-nodes as $\bullet$-nodes, and effacing binary nodes implie that) S is a seaweed which represents a series-parallel order variety on γ.

Definition 14 (Suitable conclusion). *Let $G \vdash \gamma$ be a proof structure and s be a switching for G. Let a vertex of $s(G)$ labelled $A\nabla B$. A* conclusion suited to $A\nabla B$ *is a vertex $C \in \gamma$ such that there is no paths from $A\nabla B$ to C in $s(G)$ which is oriented in G.*

Definition 15 (M-correctness). *A proof structure G is* M-correct *iff for any switching s:*

1. *the switched proof structure $s(G)$ is acyclic and connected,*
2. *for any ∇-link labelled $A\nabla B$, for any suitable conclusion C, the intersection of the paths AB, AC and BC in the seaweed $s(G)$ is a $\odot$-node in G with the following anti-clockwise order:*

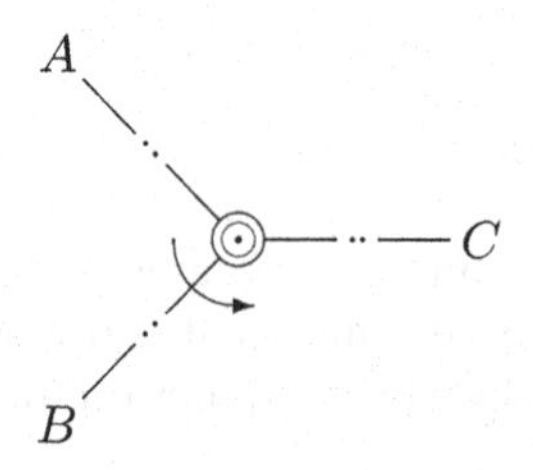

Theorem 1 ([Mai00]). *A proof structure G is M-correct iff G is sequentialisable.*

In the commutative fragment (multiplicative linear logic) the Maieli correctness criterion is exactly the Danos-Regnier's (the first step in the previous definition). The latter is well known to be in exponential time: if n is the number of $⅋$-links in a proof structure G then the Danos-Regnier correctness criterion checks 2^n graphs and cannot be inferred by the inspection of a fixed subset of the switches of G. So the Maieli correctness criterion is at least in exponential time.

3.2 The Size of a Proof Structure

If we call *size* of a proof structure G the number of registers $size(G)$ required for the memorisation of G on some ramdom access machine (RAM) then in any non redundant coding, $size(G)$ is linear in the number of vertices of G i.e. $size(G) = \Theta(|V(G)|)$. Moreover, since the number of links in G is linear in the number of vertices of G, $size(G) = \Theta(|G|)$ also. In the following, one shall analyse the worst case asymptotic complexity of correctness in terms of $size(G)$.

Remark 2. It is usual to describe a proof net with only one conclusion: built a tree of ⅋-links of the conclusions. This description does not improved the worst case asymptotic complexity.

4 Sequent Calculus

Definition 16. *A* sequent $\vdash \alpha$ *consists of a series-parallel order variety* α *of formula occurences.*

Identity group

$$\frac{}{\vdash A * A^{\perp}}\ \text{(identity)} \qquad \frac{\vdash \omega * A \quad \vdash \omega' * A^{\perp}}{\vdash \omega * \omega'}\ \text{(cut)}$$

Structural group

$$\frac{\vdash \beta}{\vdash \alpha}\ \text{(entropy)},\ \alpha \subseteq \beta$$

Logic group

$$\frac{\vdash \omega * A \quad \vdash \omega' * B}{\vdash (\omega' < \omega) * A \odot B} \qquad \frac{\vdash \omega * (A < B)}{\vdash \omega * A \nabla B}$$

$$\frac{\vdash \omega * A \quad \vdash \omega' * B}{\vdash (\omega \parallel \omega') * A \otimes B} \qquad \frac{\vdash \omega * (A \parallel B)}{\vdash \omega * A ⅋ B}$$

We can have a sequent calculus without an explicit rule for entropy: only the ⅋-rule need this rule. So we can substitute the entropy rule and the ⅋-rule by the following one given in [AR00]:

$$\frac{\vdash \alpha[A, B]}{\vdash \alpha[A⅋B/A, B]}\ (⅋\star\text{-rule})$$

where $\alpha[A⅋B/A, B]$ is the identification of definition 10. Indeed in the multiplicative fragment the two versions are equivalent: by lemma 1, we have

- $\alpha[A⅋B/A, B]_{A⅋B} * (A \parallel B) \subseteq \alpha$, so entropy and ⅋-rule can mimic the ⅋⋆-rule,
- $\omega*(A \parallel B) \subseteq \alpha$ implies $\omega*(A \parallel B) \subseteq \alpha[A⅋B/A, B]_{A⅋B}*(A \parallel B)$, so ⅋⋆-rule is an optimized version of ⅋-rule where entropy has been minimized.

See [Rue00] for a detailled explaination and consequences of removing the entropy rule in the full NL.

5 Contractibility Criterion

Definition 17 (Contracting proof structure). *A* contracting proof structure *G over the vertices $V(G)$ is a set of links and seaweeds such that:*

- *every vertex in $V(G)$ either is a conclusion of (only) one link or is an extremity of (only) one seaweed,*
- *every vertex in $V(G)$ either is a* conclusion of *G (i.e. is not a premise of any link of G) or is a premise of (only) one link or is an extremity of (only) one seaweed,*
- *the set γ of the conclusions of G (writen $G \vdash \gamma$) is not empty.*

We consider the following system of rewriting rules called *contraction rules* which is applied from contracting proof sub-structures to seaweeds:

- no rules for axiom-link, $\otimes$-link, $\odot$-link: an axiom-link is already a seaweed, a $\otimes$-link is viewed as a $\circ$-node and a $\odot$-link as $\bullet$-node,
- associativity rules, sequential rules and par rule:

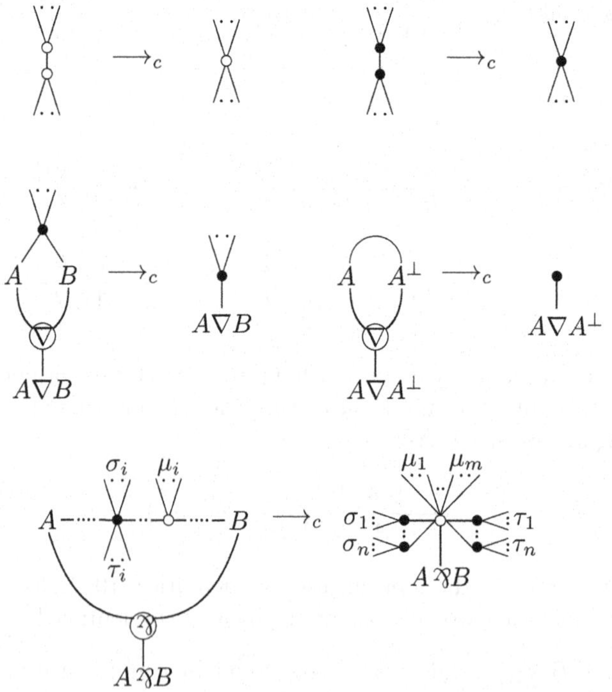

$\longrightarrow_c$ $\longrightarrow_c$

A B $\longrightarrow_c$ $A \nabla B$ A $A^\perp$ $\longrightarrow_c$ $A \nabla A^\perp$

$A \nabla B$ $A \nabla A^\perp$

σ_i μ_i μ_1 μ_m

A B $\longrightarrow_c$ σ_1 τ_1 σ_n τ_n

τ_i $A \parr B$

$A \parr B$

The par contraction rule corresponds to the transformation of a seaweed S and a $\parr$-link in $S\langle A \parr B / A, B\rangle$. We have $| \, n - m \, | \leq 1$ due to the alternate path between A and B.

Note that proof structures are particular contracting proof structures.

Definition 18 (Contractibility criterion). *A contracting proof structure G is c-correct if $\rightarrow_c^*$ reduces G to a seaweed.*

Theorem 2 (Confluence). *The system of contraction rules is confluent.*

Proof. There is no problems to do interactions with local rules like ∇-rule and associativity rules. The cases $⅋$-rule vs $⅋$-rule and $⅋$-rule vs ∇-rule are treated in [Mog01]. The $⅋$-rule vs associativity rules are exactly the same as vs ∇-rule. □

Theorem 3 (Sequentialisation). *A proof structure G is c-correct iff G is sequentialisable.*

The proof can be deduced from the sequentialisation theorem from next section by using proposition 3.

Corollary 1 (Correctness). *A proof structure G is c-correct iff G is M-correct.*

This correctness criterion acts on an initial contracting proof structure G with $size(G)$ links and nodes of seaweeds (recall that axiom-links are seaweeds). Let $n = \Theta(size(G))$ be the sum of weighted number of links and the number of nodes. The analysis of each step of reduction shows that the number of links always decreases and that:

- the associativity decreases the number of nodes of the seaweed,
- the ∇-rule decreases the number of links without changing the number of nodes. In the degenerated case, to assign a weight of 2 to ∇-links allows to decrease n.
- the $⅋$-rule acts on an alternate path. Let r and s be respectively the number of $\bullet$-nodes and $\circ$-nodes on this path. The contraction rule reduces the $r+s$ nodes to $2r+1$ nodes with $\mid r-s \mid \leq 1$ due to the alternate. Then in the worst case, the difference is of 2. So to assign a weight of 3 to $⅋$-links allows to decrease n.

So in the worst case (when G is c-correct), the number of steps of reduction in this criterion is linear in $size(G)$. Each step of reduction is a choice of a rule and the application of this rule. This decreases n down to 0.

Expect in the case of $⅋$-rule, the complexity of choosing a rule is linear in $size(G)$. In order to enable the choice of $⅋$-rule to have the same complexity mark each seaweed with an integer.

Applying a reduction rule is linear in $size(G)$ in the worst case: the associativity rules and the ∇-rules are in constant time, the $⅋$-rule is linear in the length of the path. Indeed this latter rule consists of an $S\langle z/x, y\rangle$ operation of some A and B into $A⅋B$: this requires a linear time for fis_{AB} as well as for ent_{AB} and for ass_{AB}.

Therefore this correctness criterion is in quadratic time.

6 Parsing

In the previous section, we are dealing with contracting proof structure i.e. with seaweeds. Here is the same quadratic time parsing algorithm that checks the correctness of a proof structure but the objects are directly order varieties. From the sequent calculus one can find a non determinist algorithm for the sequentialisation of proof structures. We present here a determinist reformulation. In order to show this, we introduce the parsing box which contains an order variety: let α be an order variety on a set X,

α

is called the *parsing box* α. This a kind of link without premises which has one conclusion for each element of X. We use the following set of parsing rules $\rightarrow_p$:

$A \quad A^\perp \quad \longrightarrow_p \quad A * A^\perp$

$\omega * A \quad B * \omega' \quad A \otimes B \quad \longrightarrow_p \quad (\omega \parallel \omega') * A \otimes B$

$\alpha[\ A\ ,\ B\ ] \quad A \wp B \quad \longrightarrow_p \quad \omega * A \wp B$

where $\omega * A \wp B = \alpha[A \wp B / A, B]$

$\omega * A \quad B * \omega' \quad A \odot B \quad \longrightarrow_p \quad (\omega' < \omega) * A \odot B$

$\omega * (A < B) \quad A \nabla B \quad \longrightarrow_p \quad \omega * A \nabla B$

By the properties of $\rightarrow_c$ and proposition 3, we obtain the confluence of $\rightarrow_p$.

Lemma 3. *If Π is a proof in cut-free MNL of $\vdash \alpha$ then we can naturally associate with Π a proof net Π^- which reduces to the parsing box $\beta \supseteq \alpha$.*

Proof. The proof net Π^- is defined by induction on Π as follow:

Case 1: Π is an axiom $\vdash A * A^\perp$; one must define Π^- as the axiom link: it is reduced to the parsing box $A * A^\perp$.

Case 2: Π is obtained by a $\otimes$-rule from λ_1 and λ_2 which are respectively the proofs of $\vdash \omega * A$ et $\vdash B * \omega'$; By induction hypothesis, λ_1^- and λ_2^- are respectively reduced to the parsing boxes $\beta * A \supseteq \omega * A$ and $B * \beta' \supseteq B * \omega'$. Then we define Π^- as the tensor on A and B of λ_1 and λ_2: it is reduced to the parsing box $(\beta \parallel \beta') * A \otimes B \supseteq (\omega \parallel \omega') * A \otimes B$.

Case 3: Π is obtained by a $\parr$-rule from λ which is a proof of $\vdash \alpha[A, B]$; By induction hypothesis, λ is reduced to the parsing box $\beta[A, B] \supseteq \alpha[A, B]$. Then we define Π^- as the par on A and B of λ: it is reduced to the parsing box $\beta[A \parr B/A, B] \supseteq \alpha[A \parr B/A, B]$ by lemma 2.

Case 4: Π is obtain by an entropy rule from λ which is a proof of $\vdash \beta$ with $\beta \supseteq \alpha$. Then we define Π^- as λ.

Case 5: Π is obtained by a $\odot$-rule or a ∇-rule; one can build Π^- like respectively in cases 2 and 3 if we recall that $\beta \subseteq \alpha[\omega < \omega']$ implies $\beta[\omega < \omega']$. □

Lemma 4. *If a proof net λ is reduced to the parsing box α then we can find a proof Π in sequent calculus of $\vdash \alpha$ such that $\Pi^- = \lambda$.*

Proof. By induction on the length of the reduction:

i) one step of reduction: λ is an axiom link which is reduced in the parsing box $A * A^\perp$. The claim is proved by taking as Π the axiom $\vdash A * A^\perp$.

ii) several steps of reduction: the system of parsing rules is confluent, so the last rule applied to λ is one of the followings:

- Tensor parsing rule: we have a proof net λ reduced in a parsing box $\beta = (\omega \parallel \omega') * A \otimes B$. So by the last step, there are the proof nets λ_1 and λ_2 reduced respectively in the parsing boxes $\omega * A$ and $B * \omega'$. By induction hypothesis, there is the proofs Π_1 and Π_2 in sequent calculus resp. of $\vdash \omega * A$ and $\vdash B * \omega'$ such that $\Pi_1^- = \lambda_1$ and $\Pi_2^- = \lambda_2$.
  So by taking as Π the tensor of $\vdash \omega * A$ and $\vdash B * \omega'$ we obtain a proof of $\vdash \beta$ such that $\Pi^- = \lambda$.
- Par parsing rule: we have a proof net λ reduced in a parsing box $\beta = \alpha[A \parr B/A, B]$. So by the last step, there is a proof net λ_1 reduced in a parsing box $\alpha[A, B]$. By induction hypothesis, there is a proof Π_1 in sequent calculus of $\vdash \alpha[A, B]$ such that $\Pi_1^- = \lambda_1$. One can take as Π the $\parr\star$-rule of $\alpha[A, B]$ then $\Pi^- = \lambda$.
- The others parsing rules can be treated as in previous cases. □

Theorem 4 (Sequentialisation). *Let us say that the proof structure G is p-correct when $\to_p$ reduces G to a parsing box. Then, G is p-correct iff G is sequentialisable.*

Proof. Deduce from lemma 3 and 4. □

Corollary 2. *A proof structure G is p-correct iff G is M-correct.*

7 Conclusion

These criteria are like the others one from the cut management point of view. Given a sequent calculus proof of NL with cuts P, there is an associate proof net with cuts. The standard cut elimination gives a cut-free proof net which can be sequentialised in a cut-free sequent calculus proof. Then this proof can be obtained from P by cut elimination. The question is to know what happens during the intermediate steps of cut elimination: is there a correctness criterion? i.e. is there a sequentialisation theorem extended to proof nets with cuts? In the commutative part of NL, the sequentialisation of proof nets with cuts can be solved by seeing a cut like a tensor for correctness. Detailed explanations can be found in [Laf95]. But these cannot be done here[1]. So how to deal with a contractibility correctness criterion for proof nets with cuts?

The obtained correctness criterion is quadratic but there is a linear alternative in the case of linear logic [Gue99]. This result comes from a reformulation of Danos contractibility criterion which is essentially based on unification. This gives a quasi-linear time algorithm. Guerrini's approach does not trivially generalize to this case. One can derive a trivial unification algorithm for NL from the parsing one but without improving the complexity. In fact, the needed information to make the unification is exactly that which is contained in the structure of order varieties. This new Danos contractibility style criterion for NL is a first step to obtain a linear correctness criterion.

Acknowledgements

I would like to thank R. Maieli and P. Ruet for theirs fruitful discussions. Thanks also to the anonymous referees for theirs comments and criticism.

References

Abr91. V. M. Abrusci. Phase semantics and sequent calculus for pure non-commutative classical linear propositional logic. *Journal of Symbolic Logic*, 56(4):1403–1451, 1991.

AR00. V. M. Abrusci and P. Ruet. Non commutative logic I: the multiplicative fragment. *Annals of Pure and Applied Logic*, 101:29–64, 2000.

Dan90. V. Danos. *Une application de la logique linéaire à l' étude des processus de normalisation (principalement de λ-calcul)*. PhD thesis, Université Denis Diderot, Paris 7, 1990.

Gir87. J.-Y. Girard. Linear logic. *Theoretical Computer Science*, 50, 1987.

Gir89. J.-Y. Girard. Towards a geometry of interaction. In John W. Gray and Andre Scedrov, editors, *Categories in Computer Science and Logic*, volume 92 of *Contemporary Mathematics*, pages 69–108, Providence, Rhode Island, 1989. American Mathematical Society.

[1] Solutions are given in [AR00] but they are not so elegant.

Gue99. S. Guerrini. Correctness of multiplicative proof nets is linear. In *Fourteenth Annual IEEE Symposium on Logic in Computer Science*, pages 454–463. IEEE Computer Science, 1999.

Laf95. Y. Lafont. From proof nets to interaction nets. In J.-Y. Girard, Y. Lafont, and L. Regnier, editors, *Advances in Linear Logic*, pages 225–247. Cambridge University Press, 1995. Proceedings of the Workshop on Linear Logic, Ithaca, New York, June 1993.

Lam58. J. Lambek. The mathematics of sentence structure. *Amer. Math. Mon.*, 65(3):154–170, 1958.

Mai00. R. Maieli. A new correctness criterion for multiplicative non-commutative proof-nets. Technical report, Institut de mathématiques de Luminy, 2000.

Mog01. V. Mogbil. *Sémantique des phases, réseaux de preuve et divers problèmes de décision en logique linéaire*. PhD thesis, Université de la méditerranée - Aix-Marseille II, 2001.

MR00. R. Maieli and P. Ruet. Non commutative logic III: focusing proofs. Technical Report 2000-14, insitut de Mathématiques de Luminy, 2000.

Rue00. P. Ruet. Non commutative logic II: sequent calculus and phase semantics. *Math. Struct. in Comp. Sci*, 10(2), 2000.

Yet90. D. N. Yetter. Quantales and (noncommutative) linear logic. *Journal of Symbolic Logic*, 55(1):41–64, 1990.

Capture Complexity by Partition

Yijia Chen[1,*] and Enshao Shen[2,**]

[1] Université de Paris-Sud, Laboratoire de Recherche en Informatique, Bâtiment 490,
F-91405 Orsay Cedex, France
Yijia.Chen@lri.fr

[2] Shanghai Jiao Tong University, Department of Computer Science,
Shanghai 200030, China
shen-es@cs.sjtu.edu.cn

Abstract. We show in this paper a special extended logic, partition logic based on so called partition quantifiers, is able to capture some important complexity classes **NP**, **P** and **NL** by its natural fragments. The Fagin's Theorem and Immerman-Vardi's Theorem are rephrased and strengthened into a uniform partition logic setting. Also the dual operators for the partition quantifiers are introduced to expose some of their important model-theoretic properties. In particular they enable us to show a 0-1 law for the partition logic, even when finite variable infinitary logic is adjunct to it. As a consequence, partition logic cannot count without built-in ordering on structures. Considering its better theoretical properties and tools than those of second order logic, partition logic may provide us with an alternative, yet uniform insight for descriptive complexity.

1 Introduction

From finite model theory, or more precisely, the theory of descriptive complexity, we know that all important complexity classes have their own natural logic counterparts. In other words, for each of these complexity classes, there exists a logical language capable for defining exactly those problems effectively checkable in this complexity class. The first of such correspondence is due to Fagin[4], which equates nondeterministic polynomial time with existential second order logic Σ_1^1. Some of the major results are summarized by the following table:

Complexity Class	Logic
NP	Existential Second Order Logic
P	Least Fixed Point Logic
NL	Transitive Closure Logic

Aside from the assurance of the machine-independent description of complexity classes, these results pave the way for logical approach to complexity

* Supported by Calife, a project of "Réseau National de Recherche en Télécommunications".

** Supported in part by National Foundation of Natural Science of China.

L. Fribourg (Ed.): CSL 2001, LNCS 2142, pp. 84–98, 2001.

issues, culminating in Immerman's famous proof of **NL** = co-**NL**[10]. Nevertheless compared with the Turing Machine, the unified machine model behind complexity classes, those logics seem more or less incoherent. For instance, we have different ways to reach them from first order logic: by adding higher order quantifiers (e.g. second order logic) or recursive operators (e.g. least fixed-point and transitive closure logic), and the latter enjoys an inductive flavor explicitly. Some effort has already been made to unify the logic theories, one is by Grädel[6], he identified some fragments of second order logic, say second order Horn and Krom logics, to capture **P** and **NL** respectively. Other approach is to augment the first order logic by a series of Lindström quantifiers which are based on some particular complete problems[1], such as using *Hamiltonian Path Operators* to capture **NP**[18].

Partition Logic arises from the ubiquitous mathematical operation as: partition a set into several disjoint union, over each partition subset certain property is satisfied homogeneously. One typical example is the congruent relation. H.-D. Ebbinghaus first in 1990's introduced the *Partition Quantifiers* to mimic such phenomena logically. This idea may go further back to Maltiz, yet with some extra infinite cardinality constraint[13], which is well beyond finite model theory. As it can define the connectivity of graph and some other non-first-order properties, partition logic is surely second order in nature. However it possesses some nice model-theoretic properties which are not shared by second order logic, such as the downward Löwenheim-Skolen-Tarski theorem and the Tarski Chain theorem[16]. In the meantime partition quantifiers can be looked as a special kind of monotone Lindström ones, so their Ehrenfeucht-Fraissé game is more elegant and tractable than that of second order logic[15]. Therefore in a sense, partition logic is locates in the lower level (near first order logic) of the fragment-spectrum of second order logic.

The attempt to apply partition logic in computer science started at a series of papers [14,15,17], in which it was proved that on word and tree structures, monadic fragment of partition logic is equivalent to monadic second order logic by the second author. He also found a natural fragment of partition logic equivalent to transitive closure logic, while Imhof showed another sublogic of it corresponds to bounded fixed point logic[8]. All these facts demonstrate that partition logic incorporates the recursive mechanism in a succinct form. In this paper, we show that partition logic may serve as a uniform platform to accommodate the most important part of complexity spectrum, i.e. **NP**, **P** and **NL**. Our main theorem provides a unified characterization of **NP** and **P** in partition logic on finite ordered structures, in which some key parameters of the machine is explicitly related to those of the partition quantifier, thereby giving the Turing machine a clearer logical reflection. Meanwhile we will also prove a 0-1 law for partition logic, thus without ordering, it even cannot define some very simple counting problem like Parity over arbitrary finite structures.

The paper is organized as follows: we give the definition of partition logic in Section 2, some examples are also provided. Section 3 reviews its relation with transitive closure logic and least fixed point logic by means of the dual operators of partition quantifiers, while the capturing of **NP** and **P** by partition logic is demonstrated in section 4. Section 5 is devoted to its 0-1 law.

We assume the reader has some basic knowledge of finite model theory, especially those results compiled in the preceding table, comprehensive details and references can be found in [3,11].

2 Preliminary

In this paper, only finite relational vocabularies are considered. Unless otherwise declared, structures are not necessarily finite but with two elements at least. We use FO, SO to denote first order and second order logics respectively. $\mathsf{FV}(\psi)$ is the set of free variables in ψ of a certain logic. $|A|$ is the cardinality of the set A, also we overload $|\overline{a}|$ for the length of a vector or a sequence of elements $\overline{a}$.

The language of partition logic is the enlargement of FO by a new formation rule: for any $k, m, n > 0$, if $\varphi(\overline{x}, \overline{y})$ is a well-formed formula with $|\overline{x}| = mk$ and $|\overline{y}| = nk$, then $\overset{k}{\mathsf{P}}{}^{m,n}_{\overline{x};\overline{y}}\varphi(\overline{x}, \overline{y})$ is also well-formed, and in which $\overline{x}, \overline{y}$ are bound.

Definition 1. *Given a structure $\mathcal{A}$ and $\overline{e} \in A^{|\overline{z}|}$, $\mathcal{A} \models \overset{k}{\mathsf{P}}{}^{m,n}_{\overline{x};\overline{y}}\varphi(\overline{x}, \overline{y}, \overline{z})[\overline{e}]$ where $\mathsf{FV}(\varphi) \subseteq \{\overline{x}, \overline{y}, \overline{z}\}$, iff there is a partition of A^k: $A^k = U \dot{\cup} V$, $U \neq \emptyset \neq V$, such that $\mathcal{A} \models \varphi[\overline{a}_0 \ldots \overline{a}_{m-1}, \overline{b}_0 \ldots \overline{b}_{n-1}, \overline{e}]$ for arbitrary $\overline{a}_0, \ldots, \overline{a}_{m-1} \in U$ and $\overline{b}_0, \ldots, \overline{b}_{n-1} \in V$.*

Obviously partition logic is a fragment of SO and also a monotone Lindström logic. For convenience, we will write $\mathsf{P}^{m,n}_{\overline{x};\overline{y}}\varphi$ in lieu of $\overset{1}{\mathsf{P}}{}^{m,n}_{\overline{x};\overline{y}}\varphi$, and these quantifiers are particularly named monadic partition quantifiers[17]. For any $k, m, n > 0$, let $\mathrm{FO}(\overset{k}{\mathsf{P}}{}^{m,n})$ be the extension of FO with $\overset{k}{\mathsf{P}}{}^{m,n}$ only. We abbreviate

$$\mathrm{FO}(\overset{\omega}{\mathsf{P}}{}^{1,1}) = \bigcup_{k<\omega} \mathrm{FO}(\overset{k}{\mathsf{P}}{}^{1,1}),$$

$$\mathrm{FO}(\overset{\omega}{\mathsf{P}}{}^{\omega,1}) = \bigcup_{k,m<\omega} \mathrm{FO}(\overset{k}{\mathsf{P}}{}^{m,1}),$$

$$\mathrm{FO}(\overset{\omega}{\mathsf{P}}{}^{\omega,\omega}) = \bigcup_{k,m,n<\omega} \mathrm{FO}(\overset{k}{\mathsf{P}}{}^{m,n}).$$

It is not very hard to show whenever $k \geq k', m \geq m'$ and $n \geq n'$, $\mathrm{FO}(\overset{k}{\mathsf{P}}{}^{m,n}) \geq \mathrm{FO}(\overset{k'}{\mathsf{P}}{}^{m',n'})$, so for example, $(\overset{3}{\mathsf{P}}{}^{1,5}_{\overline{x};\overline{y}}R\overline{x}\overline{y}) \wedge (\overset{2}{\mathsf{P}}{}^{4,2}_{\overline{x}';\overline{y}'}R'\overline{x}'\overline{y}') \in \mathrm{FO}(\overset{3}{\mathsf{P}}{}^{4,5}) \leq \mathrm{FO}(\overset{\omega}{\mathsf{P}}{}^{\omega,\omega})$. Meanwhile let $\mathrm{FO}(\mathrm{pos}\ \overset{\omega}{\mathsf{P}}{}^{\omega,\omega})$ denote the sublogic of $\mathrm{FO}(\overset{\omega}{\mathsf{P}}{}^{\omega,\omega})$ consisting of the formulae in which all partition quantifiers occur positively, i.e. within the scope of an even number of negation signs.

Example 1. One of the simplest properties that partition logic can deal with is the connectivity of (directed) graphs which is undefinable in FO,

$$\underline{\mathbf{Conn}} := \neg\mathsf{P}^{1,1}_{x;y}\neg Exy, \quad \text{where } E \text{ is a binary relation symbol.}$$

Namely, the graph can not be divided into two parts between which there is no cross edge. So $\mathcal{A} = (A, E^{\mathcal{A}})$ is strongly connected iff $(A, E^{\mathcal{A}}) \models \underline{\mathbf{Conn}}$. Meanwhile the reachability of two vertices can be characterized by:

$$\underline{\mathbf{Path}}(u,v) := \neg \mathsf{P}^{1,1}_{x;y}[\neg Exy \wedge y \neq u \wedge x \neq v],$$

which states that we can not divide the graph into two parts without cross edge to separate u and v. Then e is reachable to f in $(A, E^{\mathcal{A}})$ iff $(A, E^{\mathcal{A}}) \models \underline{\mathbf{Path}}[e, f]$. Surely we have

$$\models \underline{\mathbf{Conn}} \leftrightarrow \forall uv[u \neq v \rightarrow \underline{\mathbf{Path}}(u,v)].$$

Example 2. Though the definition of partition quantifiers concerns only bipartitions, $\mathrm{FO}(\overset{\omega}{\mathsf{P}}{}^{\omega,\omega})$ can also deal with properties built upon multi-partitions.

$$\begin{aligned}
\underline{\mathbf{4\text{-}Color}} &:= \exists u_0u_1u_2u_3\ \overset{2}{\mathsf{P}}{}^{4,4}_{\overline{x}_0\overline{x}_1\overline{x}_2\overline{x}_3;\overline{y}_0\overline{y}_1\overline{y}_2\overline{y}_3}[\psi_1 \wedge \psi_2],\\
\psi_1 :=&\quad (x_{00} = 0 \rightarrow (x_{01} \neq u_2 \wedge x_{01} \neq u_3))\\
&\wedge(y_{00} = 0 \rightarrow (y_{01} \neq u_0 \wedge y_{01} \neq u_1))\\
&\wedge(x_{00} = 1 \rightarrow (x_{01} \neq u_1 \wedge y_{01} \neq u_3))\\
&\wedge(y_{00} = 1 \rightarrow (y_{01} \neq u_0 \wedge y_{01} \neq u_2)),\\
\psi_2 :=&\quad (x_{00} = 0 \wedge x_{10} = 1 \wedge x_{01} = x_{11}\\
&\quad \wedge x_{20} = 0 \wedge x_{30} = 1 \wedge x_{21} = x_{31}) \rightarrow \neg Ex_{01}x_{21}\\
&\wedge(x_{00} = 0 \wedge y_{00} = 1 \wedge x_{01} = y_{01}\\
&\quad \wedge x_{10} = 0 \wedge y_{10} = 1 \wedge x_{11} = y_{11}) \rightarrow \neg Ex_{01}x_{11}\\
&\wedge(y_{00} = 0 \wedge x_{00} = 1 \wedge y_{01} = x_{01}\\
&\quad \wedge y_{10} = 0 \wedge x_{10} = 1 \wedge y_{11} = x_{11}) \rightarrow \neg Ey_{01}y_{11}\\
&\wedge(y_{00} = 0 \wedge y_{10} = 1 \wedge y_{01} = y_{11}\\
&\quad \wedge y_{20} = 0 \wedge y_{30} = 1 \wedge y_{21} = y_{31}) \rightarrow \neg Ey_{01}y_{21},
\end{aligned}$$

where each $\overline{x}_i = x_{i0}x_{i1}$ and similar for $\overline{y}_i$. Note $0, 1$ are the boolean constants which can be easily eliminated by first order existential quantification. A graph $\mathcal{A} \models \underline{\mathbf{4\text{-}color}}$ iff $\mathcal{A}$ can be 4 colored in a way such that each color is used at least once, which in case $|A| \geq 4$, is equivalent to 4-colorability problem. Assume $U|V$ is the partition that makes $\underline{\mathbf{4\text{-}color}}$ satisfied, let $U_i = \{e \mid \langle i, e\rangle \in U\}$ and $V_i = \{e \mid \langle i, e\rangle \in V\}$ for $i = 0, 1$, then it induces 4 disjoint partition subsets of A, $U_0 \cap V_0$, $U_0 \cap V_1$, $U_1 \cap V_0$ and $U_1 \cap V_1$. While ψ_1 ensures that each of them has a non-empty witness u_i, and any two points in the same subset being not adjacent is expressed by ψ_2. Clearly 3-colorability can be defined likewise.

3 Dual Operators of Partition Quantifiers

The classical extended logic capturing the graph reachability is the transitive closure logic, i.e. FO(TC)[9]. Example 1 invokes the following relation between $\mathrm{FO}(\overset{\omega}{\mathsf{P}}{}^{1,1})$ and FO(TC).

Theorem 1. [14] *The "duality" between* $\overset{k}{\mathsf{P}}{}^{1,1}$ *and* TC,

$$\models [\mathrm{TC}^k_{\overline{x},\overline{y}}\psi(\overline{x},\overline{y})](\overline{u},\overline{v}) \leftrightarrow \neg \overset{k}{\mathsf{P}}{}^{1,1}_{\overline{x};\overline{y}}[\overline{x} \neq \overline{v} \wedge \overline{y} \neq \overline{u} \wedge \neg\psi(\overline{x},\overline{y})],$$

$$\models \overset{k}{\mathsf{P}}{}^{1,1}_{\overline{x};\overline{y}}\psi(\overline{x},\overline{y}) \leftrightarrow \exists\overline{u}\,\overline{v}[\neg(\mathrm{TC}^k_{\overline{x},\overline{y}}\neg\psi(\overline{x},\overline{y}))(\overline{u},\overline{v})].$$

Hence $\mathrm{FO}(\overset{\omega}{\mathsf{P}}{}^{1,1}) = \mathrm{FO}(\mathrm{TC})$, and as a result, $\mathrm{FO}(\overset{\omega}{\mathsf{P}}{}^{1,1})$ captures **NL** on finite ordered structures. The above "duality" also inspires the next modification of partition quantifiers to so called *pseudo transitive closure operators* TP.

Definition 2. *Let* $\psi(\overline{u},\overline{v},\overline{z}) = [\overset{k}{\mathsf{TP}}{}^{m,n}_{\overline{x};\overline{y}}\varphi(\overline{x},\overline{y},\overline{z})](\overline{u},\overline{v})$, *where* $|\overline{u}| = |\overline{v}| = k$, $|\overline{x}| = mk$, $|\overline{y}| = nk$ *and* $\mathsf{FV}(\varphi) \subseteq \{\overline{x},\overline{y},\overline{z}\}$. *Given* $\overline{e},\overline{f} \in A^k$ *and* $\overline{g} \in A^{|\overline{z}|}$, *if for any partition* $U|V$ *of* A^k *with* $\overline{e} \in U$ *and* $\overline{f} \in V$, *there exist* $\overline{a}_0,\ldots,\overline{a}_{m-1} \in U$ *and* $\overline{b}_0,\ldots,\overline{b}_{n-1} \in V$ *such that* $\mathcal{A} \models \varphi[\overline{a}_0\ldots\overline{a}_{m-1},\overline{b}_0\ldots\overline{b}_{n-1},\overline{g}]$, *then we say* $\mathcal{A} \models \psi[\overline{e},\overline{f},\overline{g}]$.

The correspondence between $\overset{k}{\mathsf{P}}{}^{m,n}$ and $\overset{k}{\mathsf{TP}}{}^{m,n}$ is superficial, and we can see $\overset{k}{\mathsf{TP}}{}^{1,1}$ is exactly k-dimensional TC by Theorem 1:

$$\models [\overset{k}{\mathsf{TP}}{}^{m,n}_{\overline{x};\overline{y}}\psi(\overline{x},\overline{y})](\overline{u},\overline{v}) \leftrightarrow \neg \overset{k}{\mathsf{P}}{}^{m,n}_{\overline{x};\overline{y}}[\overline{x} \neq \overline{v}^1 \wedge \overline{y} \neq \overline{u} \wedge \neg\psi(\overline{x},\overline{y})],$$

$$\models \overset{k}{\mathsf{P}}{}^{m,n}_{\overline{x};\overline{y}}\psi(\overline{x},\overline{y}) \leftrightarrow \exists\overline{u}\,\overline{v}[\neg(\overset{k}{\mathsf{TP}}{}^{m,n}_{\overline{x};\overline{y}}\neg\psi(\overline{x},\overline{y}))(\overline{u},\overline{v})].$$

The following lemma shows some essential similarities between TP and TC, which was proved in [16], yet without introducing TP.

Lemma 1.
(1) *For any* $\mathcal{A}$ *and* $\overline{e},\overline{f},\overline{g} \in A^k$, *if both* $\mathcal{A} \models \overset{k}{\mathsf{TP}}{}^{m,n}_{\overline{x};\overline{y}}\psi[\overline{e},\overline{f}]$ *and* $\mathcal{A} \models \overset{k}{\mathsf{TP}}{}^{m,n}_{\overline{x};\overline{y}}\psi[\overline{f},\overline{g}]$, *then we have* $\mathcal{A} \models \overset{k}{\mathsf{TP}}{}^{m,n}_{\overline{x};\overline{y}}\psi[\overline{e},\overline{g}]$.
(2) *Given* $\mathcal{A} \subseteq \mathcal{B}$ *and* $\overline{e},\overline{f} \in A^k$ *with* $\mathcal{A} \models \overset{k}{\mathsf{TP}}{}^{m,n}_{\overline{x};\overline{y}}\psi[\overline{e},\overline{f}]$, *for any* $\overline{u} \in A^{mk}$ *and* $\overline{v} \in A^{nk}$, $\mathcal{A} \models \psi[\overline{u},\overline{v}]$ *implies* $\mathcal{B} \models \psi[\overline{u},\overline{v}]$, *then* $\mathcal{B} \models \overset{k}{\mathsf{TP}}{}^{m,n}_{\overline{x};\overline{y}}\psi[\overline{e},\overline{f}]$.
(3) *Let* $\tau = \{R\}$, *where* R *is a* $k(m+n)$*-ary relation symbol, for a* τ*-structure* $\mathcal{A}$, $\mathcal{A} \models [\overset{k}{\mathsf{TP}}{}^{m,n}_{\overline{x};\overline{y}}\mathrm{R}\overline{x}\,\overline{y}][\overline{e},\overline{f}]$ *where* $\overline{e},\overline{f} \in A^k$, *and* D *is the diagram of* $\mathcal{A}$, *there exists a finite subset* $D_{\overline{e},\overline{f}}$ *of* D *satisfying* $D_{\overline{e},\overline{f}} \models [\overset{k}{\mathsf{TP}}{}^{m,n}_{\overline{x};\overline{y}}\mathrm{R}\overline{x}\,\overline{y}](\underline{\overline{e}},\underline{\overline{f}})$, *where* $\underline{\overline{e}}$, $\underline{\overline{f}}$ *are new constant symbol sequences interpreted by* $\overline{e}$ *and* $\overline{f}$ *respectively.*

Intuitively, (1) guarantees the transitivity of TP, (2) means TP is closed under extensions, and in (3), $D_{\overline{e},\overline{f}}$ bears witness to the satisfaction of TP on $\overline{e},\overline{f}$, which may be imagined as the finite "path" in $\mathcal{A}$ that connects $\overline{e}$ and $\overline{f}$. The last property plays a crucial role in the proof of certain model-theoretic theorems of

[1] Here $\overline{x} \neq \overline{v}$ stands for $x_{00} \neq v_0 \vee x_{01} \neq v_1 \vee \cdots \vee x_{0k-1} \neq v_{k-1}$.

$\mathrm{FO}(\overset{\omega}{\mathsf{P}}^{\omega,\omega})$[16], also it can be used to embed partition logic into $\mathcal{L}_{\omega_1\omega}$. But as Example 2 shows that 3-colorability is axiomatizable in it, $\mathrm{FO}(\overset{\omega}{\mathsf{P}}^{\omega,\omega})$ is not inside finite variable infinitary logic $\mathcal{L}^{\omega}_{\infty\omega}$ by a result of Dawar[7].

The naive model-checking algorithm derived directly from the definition of $\overset{k}{\mathsf{P}}^{m,n}$ is unavoidably of exponential time, but Theorem 1 has already implied some **NL** algorithm for $\overset{k}{\mathsf{P}}^{1,1}$. To step forward more, we can design a **P** algorithm for any $\overset{k}{\mathsf{P}}^{m,1}$ by the following embedding result of $\overset{k}{\mathsf{P}}^{m,1}$ in the least fixed point logic, FO(LFP).

Proposition 1. *Given a structure* $\mathcal{A}$ *and* $\overline{e},\overline{f} \in A^k$,
(1) $\mathcal{A} \models [\overset{k}{\mathsf{TP}}{}^{m,1}_{\overline{x};\overline{y}}\varphi(\overline{x},\overline{y})][\overline{e},\overline{f}]$ *iff* $\mathcal{A} \models \mathrm{LFP}_{\overline{x},\overline{y},X}(\overline{x}=\overline{y}\vee\exists\overline{w}X\overline{x}\overline{w}_0\wedge\cdots\wedge X\overline{x}\overline{w}_{m-1}\wedge \varphi(\overline{w},\overline{y}))[\overline{e},\overline{f}]$, *where* $\overline{w}=\overline{w}_0,\ldots,\overline{w}_{m-1}$.
(2) $\mathcal{A} \models [\mathrm{LFP}_{\overline{y},Y}(\varphi_0(\overline{y}) \vee \exists\overline{x}(Y\overline{x}_0 \wedge \ldots \wedge Y\overline{x}_{m-1} \wedge \varphi_1(\overline{x}_0\ldots\overline{x}_{m-1},\overline{y}))][\overline{e}]$, *iff* $\mathcal{A} \models [\neg \overset{k}{\mathsf{P}}{}^{m,1}_{\overline{x};\overline{y}}\neg\varphi_1(\overline{x},\overline{y}) \wedge \overline{x} \neq \overline{u} \wedge \neg\varphi_0(\overline{y})][\overline{e}]$, *where* $\overline{x}=\overline{x}_0,\ldots,\overline{x}_{m-1}$.

Proof. Routine. □

The LFP-formulae in above proposition indeed fall into Bounded Fixed-Point Logic, FO(BFP)[3], which allows the LFP operator only if there is bounded $m \geq 1$ such that the tuple in a new stage is already witnessed by a set of at most m many tuples of the preceding stage, i.e. all fixed-point formulae are of the form: $\mathrm{LFP}_{\overline{y},Y}(\varphi_0(\overline{y})\vee\exists\overline{x}_0\ldots\exists\overline{x}_{m-1}(Y\overline{x}_0\wedge\ldots\wedge Y\overline{x}_{m-1}\wedge\varphi_1(\overline{x}_0\ldots\overline{x}_{m-1},\overline{y}))$.
Thus by Proposition 1 and the equivalence between $\overset{k}{\mathsf{P}}^{m,1}$ and $\overset{k}{\mathsf{TP}}^{m,1}$,

Theorem 2. ([8]) $\quad \mathrm{FO}(\overset{\omega}{\mathsf{P}}^{\omega,1}) = \mathrm{FO(BFP)}$.

We know that FO(TC) < FO(BFP) < FO(LFP)[3], and 3-colorability can separate $\mathrm{FO}(\overset{\omega}{\mathsf{P}}^{\omega,\omega})$ and $\mathcal{L}^{\omega}_{\infty\omega}$, henceforth

Corollary 1.

$$\mathrm{FO}(\overset{\omega}{\mathsf{P}}^{1,1}) < \mathrm{FO}(\overset{\omega}{\mathsf{P}}^{\omega,1}) < \mathrm{FO}(\overset{\omega}{\mathsf{P}}^{\omega,\omega}),$$
$$\mathrm{FO}(\overset{\omega}{\mathsf{P}}^{\omega,1}) < \mathrm{FO(LFP)}.$$

4 Characterizing NP Machine

In this section, we will characterize **NP** Turing machines by $\mathrm{FO}(\overset{\omega}{\mathsf{P}}^{\omega,\omega})$ on finite ordered structures. The main technique and convention used here follow [3]. First a simple warming-up example is given, then we sketch the proof idea of the main theorem enlightened by this example, while full-length proof is omitted due to the lack of space.

Example 3. Let τ be the vocabulary for binary tree, i.e. $\tau = \{\epsilon, S_1, S_2\}$, where ϵ is the constant symbol for the root node, and S_1,S_2 are respectively the left and right successor relation symbols. The sentence Θ defined left below asserts that there exists a path from the root such that if a node in the path satisfies the formula $\varphi(x)$ then its right successor must lie in this path too, which may be depicted by the figure right below, where the solid nodes are those satisfying φ in the path.

$$\begin{array}{lll} \Theta := & \mathsf{P}^{2,2}_{x,x';y,y'}\ \neg(\epsilon = y) & (1) \\ & \quad \wedge\ \neg(S_1yx \vee S_2yx) & (2) \\ & \quad \wedge\ \neg(S_1xy \wedge S_2xy') & (3) \\ & \quad \wedge\ \neg\exists z(S_1zx \wedge S_2zx') & (4) \\ & \quad \wedge\ \neg(\varphi(x) \wedge S_2xy) & (5) \end{array}$$

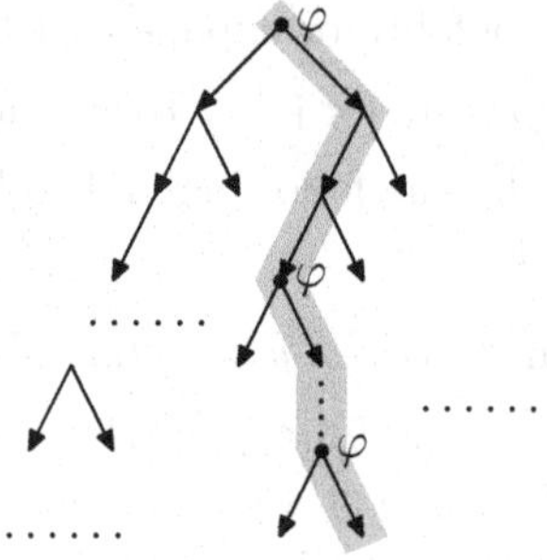

Now for any finite tree structure $\mathcal{A} = \langle A, \epsilon^{\mathcal{A}}, S_1^{\mathcal{A}}, S_2^{\mathcal{A}}\rangle \models \Theta$, there exists a partition $U|V$ of A such that the inner formula is satisfied homogeneously on this $U|V$. We claim that U is the required path. First by (1), $\epsilon^{\mathcal{A}} \in U$, since for any $y \in V$, $\epsilon^{\mathcal{A}} \neq y$. To verify that U is closed under predecessor, note if there is some node x in U such that its predecessor node is not in U, i.e. some $y \in V$ with $S_1^{\mathcal{A}}yx$ or $S_2^{\mathcal{A}}yx$, then these x and y will refute the second conjunct. The subformula (3) implies that for any $x \in U$, at least one of its successors will also lie in U, assume contrarily one x's left successor y and right successor y' are both in V, then (3) can not be satisfied homogeneously. On the other hand, for any node $z \in U$, at most one of its successors is inside U, which is ensured by (4). Henceforth, (1)-(4) guarantee that the first partition subset U is indeed a path of $\mathcal{A}$. Finally by (5), no element $y \in V$ is the right successor of an $x \in U$ satisfying φ. The reverse direction is trivial.

Next we fix a vocabulary $\tau = \tau_0 \dot{\cup} \tau_1$, where $\tau_0 = \{<, S, \min, \max\}$ and $\tau_1 = \{R_1, \ldots, R_k\}$, while all τ-structures under consideration have the interpretation of $<$, S, min and max as the ordering, successor relation, minimum and maximum elements, i.e. ordered structures. For an **NP** Turing machine M which is time-bounded by n^d to accept τ-structures using $k+1$ input tapes for coding the structure, and some other m work tapes for intermediate computation, let br be the maximum number of choices that M can face each time, and note $br = 1$ if and only if M is deterministic. It is a standard technique to code any computation run of M by a $(2d+2)$-ary relation on the input structure: the first d-ary part is the "time stamp", the rest will code the actual configuration of M at a particular time, that is to say, if $|t|$ is the n-adic representation of time t, then the $(d+2)$-ary relation $R|t|$ fully describes M's configuration at time t, including the state, the inscriptions of each cell on each input or work tape, and also the position of each reading head on those tapes.

The Fagin's theorem relies on the observation that it is possible to define a first order formula $\varphi(X, Y)$ saying that Y is a valid configuration after M makes a move from the original configuration X, and in the meantime we can introduce two simple formulae $\psi_{\text{init}}(X)$ and $\psi_{\text{end}}(X)$ expressing X is the initial configuration and a final configuration with an accepting state respectively. Thus a Σ_1^1 sentence

$$\begin{aligned}\Theta = \exists R[\psi_{\text{init}}(R|t_{\text{init}}|) \wedge \psi_{\text{end}}(R|t_{\text{end}}|) \\ \wedge\forall|t_0|\ |t_1|(\text{“}t_1 = t_0 + 1\text{''} \rightarrow \varphi(R|t_0|, R|t_1|))]\end{aligned}$$

where $|t_{\text{init}}|$ and $|t_{\text{end}}|$ are FO definable constant vectors representing the initial and the final times of the computation run, can be constructed. Clearly for any finite structure $\mathcal{A} \models \Theta$ iff R can be interpreted as a accepting run of M on $\mathcal{A}$, i.e. M recognizes $\mathcal{A}$.

Shifted to partition logic, we aim to devise a sentence $\overset{2d+2}{\mathsf{P}}\ \chi$, such that once it is satisfied in a structure $\mathcal{A}$, the first partition subset will be interpreted as an accepting run. Surely the above φ, ψ_{init} and ψ_{end} can not be directly applied, because we are deprived of the explicit use of second order variables. Though the overall idea is similar to the previous tree example, much more deliberation is needed. Our description of one step computation is divided into two phases: firstly M chooses an instruction according to the current configuration, then M changes its configuration following the instruction. For the first phase, we add a new element into each $R|t|$ to indicate which instruction will be actually carried out concerning all nondeterministic choices that M can make over $R|t|$. Note the set of all possible instructions is fully determined by the state of M at time t together with the symbols read by those heads on input and work tapes at that moment, which are reflected by a finite number of elements in $R|t|$. Therefore we are in a similar situation like Example 3 which must regulate any nonterminal point of the path has one and only one successor also lying in the path, while the choice between left and right successor can be nondeterministic. Once the instruction is chosen for time t, the configuration of time $t+1$ is totally determined. It is crucial that each element in $R|t+1|$ only depends on a bounded number of elements in $R|t|$: the new state and the new inscriptions on those positions originally the heads were pointing at are determined by the chosen instruction; the new head positions are determined by the instruction and the heads' original positions; the inscriptions of rest positions remain unchanged, thus determined by their original inscriptions. So $R|t+1|$ can be characterized in the same fashion as the tree example requires those nodes satisfying φ must have right successor in the path. Thus a careful and tedious elaboration will yield,

Theorem 3. *if a class of finite ordered τ-structures K is accepted by M, then K is axiomatizable in* $\mathrm{FO}(\overset{\omega}{\mathsf{P}}{}^{\omega,\omega})$ *by a sentence* $\Theta = \overset{2d+2}{\mathsf{P}}{}^{3+k+2m,br}\chi$, *where χ is a quantifier-free formula.*

Conversely, we can effectively construct a **NP** machine for any given $\varphi \in \mathrm{FO}(\text{pos } \overset{\omega}{\mathsf{P}}{}^{\omega,\omega})$ to check whether $\mathcal{A} \models \varphi$ for each $\mathcal{A}$, so

Theorem 4. *if K is a* $\mathrm{FO}(\text{pos } \overset{\omega}{\mathsf{P}}{}^{\omega,\omega})$ *definable τ-structure class, then $K \in$* **NP**.

Combining Theorems 2, 3 and 4, we obtain

Corollary 2. *On finite ordered structures,*

$$\begin{aligned} \mathrm{FO}(\mathrm{pos}\ \overset{\omega}{\mathsf{P}}{}^{\omega,\omega}) &= \mathbf{NP}, \\ \mathrm{FO}(\overset{\omega}{\mathsf{P}}{}^{\omega,1}) &= \mathbf{P}. \end{aligned}$$

5 0-1 Law for Partition Logic

In this section, we are only interested in the labeled 0-1 law under the uniform probability measure, i.e. each structure of cardinality n has $\{0, \ldots, n-1\}$ as its underlying universe, and with same probability. To prove the 0-1 law for partition logic, we would rather focus on TP instead of the original quantifier P, making substantial use of its preservation over extensions, i.e. Lemma 1.2. There have been several results concerning logics that deal with properties closed under extensions, or equivalently closed under substructures[5,2,12]. In particular [2] proved that the 0-1 law is retained in FO augmented with those generalized quantifiers defined over classes of structures that are closed under extensions. Later we will see TP can be regarded as such generalized quantifier. First we revise the definition of generalized quantifiers. Fix a vocabulary $\sigma = \{R_1, \ldots, R_s\}$, where each R_i is r_i-ary relation symbol. Now upon a class of σ-structures K which is closed under isomorphisms, if for $1 \leq i \leq s$, $\psi_i(\overline{x}_i, \overline{y})$ is a formula with $\mathsf{FV}(\psi_i) \subseteq \{\overline{x}_i, \overline{y}\}$, then $Q_K \overline{x}_1, \ldots, \overline{x}_s[\psi_1, \ldots, \psi_s]$ is also a formula which bounds all $\overline{x}_i$. For any τ-structure $\mathcal{A}$ and $\overline{a} \in A^{|\overline{y}|}$,

$$\mathcal{A} \models Q_K \overline{x}_1, \ldots, \overline{x}_s[\psi_1, \ldots, \psi_s][\overline{a}], \text{ iff } (A, \psi_1^{\mathcal{A}}(\_\_, \overline{a}), \ldots, \psi_s^{\mathcal{A}}(\_\_, \overline{a})) \in K,$$

where each $\psi_i^{\mathcal{A}}(\_\_, \overline{a})$ stands for $\{\overline{b} \in A^{r_i} \mid \mathcal{A} \models \psi_i[\overline{b}, \overline{a}]\}$. For a set of generalized quantifiers $\mathcal{Q}$, $\mathrm{FO}(\mathcal{Q})$ and $\mathcal{L}^t_{\infty\omega}(\mathcal{Q})$ denote the extension of FO and t-variable infinitary logic with $\mathcal{Q}$ respectively, while $\mathcal{L}^{\omega}_{\infty\omega}(\mathcal{Q}) = \bigcup_{t<\omega} \mathcal{L}^t_{\infty\omega}(\mathcal{Q})$.

Next we detail some notions concerning the class of structures that are used to define generalized quantifiers of which we will prove the 0-1 law.

Definition 3. *Let K be a class of structures as defined above, K is said to be* closed under extensions *iff whenever $\mathcal{A} \in K$ and $\mathcal{A} \subseteq \mathcal{A}'$, we have $\mathcal{A}' \in K$. While a* finitely witnessed *K means for any infinite $\mathcal{A}$, if $\mathcal{A} \in K$, then there exists a finite $\mathcal{A}'$ such that $\mathcal{A}' \subset \mathcal{A}$ and $\mathcal{A}' \in K$, and we can say $\mathcal{A}'$ finitely witnesses $\mathcal{A} \in K$. K is* finitely based *if it is both closed under extensions and finitely witnessed. A Q_K is called closed under extensions, finitely witnessed or finitely based if K is respectively so.*

Example 4.
(1) To define first order quantifier $\exists$, assume $K = \{(A, U) \mid U \subseteq A \text{ and } U \neq \emptyset\}$, then $\models \exists x \varphi(x) \leftrightarrow Q_K x[\varphi(x)]$. Counting quantifier $\exists^{\geq l}$ can be regarded as Q_K where $K = \{(A, U) \mid |U| \geq l\}$.

(2) For each $\overset{k}{\mathsf{TP}}{}^{m,n}$, let $\sigma = \{R, P\}$ where R and P are respectively $k(m+n)$-ary

and $2k$-ary relation symbols, set $K = \{\mathcal{A} \mid \mathcal{A} \models \exists \overline{uv}\{[\overset{k}{\mathsf{TP}}{}^{m,n}_{\overline{x};\overline{y}} R\overline{xy}](\overline{uv}) \wedge P\overline{uv}\}\}$. We have

$$\models [\overset{k}{\mathsf{TP}}{}^{m,n}_{\overline{x};\overline{y}} \psi(\overline{x},\overline{y})](\overline{uv}) \leftrightarrow Q_K\overline{xy}\overline{x}'\overline{y}'[\psi, \overline{x}' = \overline{u} \wedge \overline{y}' = \overline{v}],$$

$$\models Q_K\overline{xy}\overline{x}'\overline{y}'[\psi_1(\overline{x},\overline{y}), \psi_2(\overline{x}',\overline{y}')] \leftrightarrow \exists \overline{uv}\{[\overset{k}{\mathsf{TP}}{}^{m,n}_{\overline{x};\overline{y}} \psi_1(\overline{x},\overline{y})](\overline{uv}) \wedge \psi_2(\overline{u},\overline{v})\}.$$

In the above examples, all Q_Ks are finitely based. Specifically, Lemma 1.2 guarantees TP is closed under extensions, while in Lemma 1.3 $\mathcal{A} \restriction D_{\overline{e},\overline{f}}$ finitely witnesses $\mathcal{A} \models [\overset{k}{\mathsf{TP}}{}^{m,n}_{\overline{x};\overline{y}} \mathrm{R}\overline{x}\,\overline{y}][\overline{e},\overline{f}]$. As a consequence, one result in [2] implies that $\mathrm{FO}(\overset{\omega}{\mathsf{P}}{}^{\omega,\omega})$ has the 0-1 law. Furthermore in the following we will strengthen it to

Theorem 5. *$\mathcal{L}^{\omega}_{\infty\omega}(\overset{\omega}{\mathsf{P}}{}^{\omega,\omega})$ has the labeled 0-1 law.*

The proof method we adopt here is rather traditional, i.e. based on a *transfer property* (Theorem 6) , compared with [2] of which we see no easy extension to $\mathcal{L}^{\omega}_{\infty\omega}$.

Let ϵ_i be the conjunction of finitely many r-extension axioms with $r \leq i$, and $\mathrm{T_{rand}}$ is the set of all extension axioms, i.e. $\mathrm{T_{rand}} = \bigwedge_{i>0} \epsilon_i$. Furthermore $\mathcal{A}_{\mathrm{rand}}$ is the unique countable random structure up to isomorphism, i.e. $\mathcal{A}_{\mathrm{rand}} \models \mathrm{T_{rand}}$. We will rely on the following lemma heavily later.

Lemma 2. *Given two structures $\mathcal{A}, \mathcal{B}$, particularly $\mathcal{A} \models \epsilon_i$ and $h : \overline{a} \longmapsto \overline{b}$ is a partial isomorphism between $\mathcal{A}$ and $\mathcal{B}$ with finite domain $|\{\overline{a}\}| \leq i$, then for any finite subset $S \subseteq B$ with $|S| \leq i - |\{\overline{a}\}|$, there exists a finite subset $S' \subseteq A$, such that h can be extended to some larger partial isomorphism $h' : S', \overline{a} \longmapsto S, \overline{b}$.*

Proof. Easy. □

For any structure $\mathcal{A}$ and $\overline{a} \in A^*$, define a first order formula

$$\varphi^0_{\mathcal{A},\overline{a}} = \bigwedge\{\psi(\overline{x}) \mid \psi \text{ atomic or negated atomic, and } \mathcal{A} \models \psi[\overline{a}]\},$$

and it follows that $\models \varphi^0_{\mathcal{A},\overline{a}} \leftrightarrow \varphi^0_{\mathcal{B},\overline{b}}$ iff $\overline{a} \longmapsto \overline{b}$ is a partial isomorphism between $\mathcal{A}$ and $\mathcal{B}$. Later on we will write $\varphi^0_{\mathcal{A},\overline{a}} = \varphi^0_{\mathcal{B},\overline{b}}$ instead of $\models \varphi^0_{\mathcal{A},\overline{a}} \leftrightarrow \varphi^0_{\mathcal{B},\overline{b}}$, if no ambiguity arises. Obviously an equivalence relation over A^* can be induced: $\overline{a} \equiv^0_{\mathcal{A}} \overline{b}$ iff $\varphi^0_{\mathcal{A},\overline{a}} = \varphi^0_{\mathcal{A},\overline{b}}$, so each $\varphi^0_{\mathcal{A},\overline{a}}$ defines an equivalence class. Next lemma will show in $\mathcal{A}_{\mathrm{rand}}$, this equivalence relation holds for arbitrary higher level formulae.

Lemma 3. *For any $\psi \in \mathcal{L}[\tau]$ and $\overline{a}, \overline{b} \in A^{|\mathsf{FV}(\psi)|}_{\mathrm{rand}}$ with $\varphi^0_{\mathcal{A}_{\mathrm{rand}},\overline{a}} = \varphi^0_{\mathcal{A}_{\mathrm{rand}},\overline{b}}$, $\mathcal{A}_{\mathrm{rand}} \models \psi[\overline{a}]$ iff $\mathcal{A}_{\mathrm{rand}} \models \psi[\overline{b}]$. The above $\mathcal{L}$ could be any logical system whose satisfaction relation is closed under isomorphisms and permits substitution.*

Proof. By $\varphi^0_{\mathcal{A}_{\text{rand}},\overline{a}} = \varphi^0_{\mathcal{A}_{\text{rand}},\overline{b}}$, $h : \overline{a} \longmapsto \overline{b}$ is a partial automorphism over $\mathcal{A}_{\text{rand}}$. Then the fact of $\mathcal{A}_{\text{rand}}$ satisfying T_{rand} and being countable ensures that h can be enlarged to an automorphism h' on $\mathcal{A}_{\text{rand}}$ via a back and forth process. Thereby

$$\begin{aligned} & \mathcal{A}_{\text{rand}} \models \psi[\overline{a}] \\ \text{iff } & h'(\mathcal{A}_{\text{rand}}) \models \psi[h'(\overline{a})] \\ \text{iff } & \mathcal{A}_{\text{rand}} \models \psi[h(\overline{a})] \\ \text{iff } & \mathcal{A}_{\text{rand}} \models \psi[\overline{b}]. \end{aligned}$$

□

So it makes sense to introduce the following canonization function δ on $\mathcal{A}_{\text{rand}}$. Let θ be a choice function over the equivalence classes of $\equiv^0_{\mathcal{A}_{\text{rand}}}$, i.e. $\theta([\overline{a}]_{\equiv^0_{\mathcal{A}_{\text{rand}}}}) \in [\overline{a}]_{\equiv^0_{\mathcal{A}_{\text{rand}}}}$, define $\delta : A^*_{\text{rand}} \longrightarrow A^*_{\text{rand}}$ with $\delta(\overline{a}) = \theta([\overline{a}]_{\equiv^0_{\mathcal{A}_{\text{rand}}}})$. Surely $\varphi^0_{\mathcal{A}_{\text{rand}},\overline{a}} = \varphi^0_{\mathcal{A}_{\text{rand}},\delta(\overline{a})}$, so by Lemma 3, $\mathcal{A}_{\text{rand}} \models \psi[\overline{a}]$ iff $\mathcal{A}_{\text{rand}} \models \psi[\delta(\overline{a})]$ for arbitrary ψ. Note for any fixed i, $|\{\delta(\overline{a}) \mid \overline{a} \in A^i_{\text{rand}}\}| = |\{[\overline{a}]_{\equiv^0_{\mathcal{A}_{\text{rand}}}} \mid \overline{a} \in A^i_{\text{rand}}\}| < \omega$ due to the finiteness of τ.

Clearly Lemma 3 exhibits the extreme symmetry of $\mathcal{A}_{\text{rand}}$, furthermore it implies the following technical result which gives a bound on the finite witness of any Q_K over $\mathcal{A}_{\text{rand}}$.

Lemma 4. *Given a finitely witnessed* Q_K *and* $t \in \mathbb{N}$, *we can find a fixed* n *such that: for any* $\varphi = Q_K\overline{x}_1, \ldots, \overline{x}_s[\psi_1, \ldots, \psi_s] \in \mathcal{L}[\tau]$ *with* $|\mathsf{FV}(\varphi)| \leq t$ *and* $\overline{a} \in A^{|\mathsf{FV}(\varphi)|}_{\text{rand}}$, *if* $\mathcal{A}_{\text{rand}} \models \varphi[\overline{a}]$, *then there exists a finite set* $D \subset A_{\text{rand}}$ *with* $|D| \leq n$ *such that*

$$(D, \psi_1^{\mathcal{A}_{\text{rand}}}(\_\_, \overline{a}) \upharpoonright D, \ldots, \psi_s^{\mathcal{A}_{\text{rand}}}(\_\_, \overline{a}) \upharpoonright D) \in K.$$

Note $\mathcal{L}$ *is the same as in Lemma 2.*

Proof. First observe that for any $\psi \in \mathcal{L}[\tau]$,

$$\mathcal{A}_{\text{rand}} \models \psi \leftrightarrow \bigvee_{\mathcal{A}_{\text{rand}} \models \psi[\delta(\overline{a})], \overline{a} \in A^{|\mathsf{FV}(\psi)|}_{\text{rand}}} \varphi^0_{\mathcal{A}_{\text{rand}},\delta(\overline{a})}$$

by Lemma 3, and the finiteness of $\delta(\overline{a})$ ensures the above conjunction is finite, i.e. in FO. So we can define an equivalent translation $[\![\_\_]\!] : \mathcal{L}[\tau] \to \text{FO}[\tau]$ over $\mathcal{A}_{\text{rand}}$,

$$[\![\psi]\!] = \bigvee_{\mathcal{A}_{\text{rand}} \models \psi[\delta(\overline{a})], \overline{a} \in A^{|\mathsf{FV}(\psi)|}_{\text{rand}}} \varphi^0_{\mathcal{A}_{\text{rand}},\delta(\overline{a})}.$$

It is important that the number of possible $\bigvee_{\mathcal{A}_{\text{rand}} \models \psi[\delta(\overline{a})], \overline{a} \in A^{|\mathsf{FV}(\psi)|}_{\text{rand}}} \varphi^0_{\mathcal{A}_{\text{rand}},\delta(\overline{a})}$ is finite, when ψ ranges over all $\mathcal{L}[\tau]$-formulae with a bounded number of free variables.

Given $\varphi = Q_K\overline{x}_1, \ldots, \overline{x}_s[\psi_1, \ldots, \psi_s] \in \mathrm{FO}(\{Q_K\})[\tau]$ and $\overline{a} \in A_{\mathrm{rand}}^{|\mathsf{FV}(\varphi)|}$, if $\mathcal{A}_{\mathrm{rand}} \models \varphi[\overline{a}]$, since Q_K is finitely witnessed, there exists a finite $D \subset A_{\mathrm{rand}}$ to fulfill

$$(D, \psi_1^{\mathcal{A}_{\mathrm{rand}}}(\_\_, \overline{a}) \restriction D, \ldots, \psi_s^{\mathcal{A}_{\mathrm{rand}}}(\_\_, \overline{a}) \restriction D) \in K.$$

Such D may not be unique, so fix one specific $D^{\overline{a}}_{\psi_1,\ldots,\psi_s}$.

Then we set

$$\begin{aligned} n = \max\{|D^{\delta(\overline{a})}_{[\![\psi_1]\!],\ldots,[\![\psi_s]\!]}| \mid\ & \mathcal{A}_{\mathrm{rand}} \models Q_K\overline{x}_1, \ldots, \overline{x}_s[[\![\psi_1]\!], \ldots, [\![\psi_s]\!]][\delta(\overline{a})], \\ & \text{where } \varphi = Q_K\overline{x}_1, \ldots, \overline{x}_s[\psi_1, \ldots, \psi_s] \in \mathcal{L}[\tau] \\ & \text{with } |\mathsf{FV}(\varphi)| \le t, \text{ hence } |\mathsf{FV}(\psi_i)| \le t + r_i, \\ & \text{and } \overline{a} \in A_{\mathrm{rand}}^{|\mathsf{FV}(\varphi)|} \ \}. \end{aligned}$$

By the discussion in the beginning and the finiteness of $\delta(\overline{a})$, the right-hand set is also finite, so n is well defined.

Now for any $\varphi = Q_K\overline{x}_1, \ldots, \overline{x}_s[\psi_1, \ldots, \psi_s] \in \mathcal{L}[\tau]$ with $|\mathsf{FV}(\varphi)| \le t$ and $\overline{a} \in A_{\mathrm{rand}}^{|\mathsf{FV}(\varphi)|}$, assume $\mathcal{A}_{\mathrm{rand}} \models \varphi[\overline{a}]$, by $[\![\_\_]\!]$ is an equivalent translation and Lemma 3, we have $\mathcal{A}_{\mathrm{rand}} \models Q_K\overline{x}_1, \ldots, \overline{x}_s[[\![\psi_1]\!], \ldots, [\![\psi_s]\!]][\delta(\overline{a})]$. Hence $D^{\delta(\overline{a})}_{[\![\psi_1]\!],\ldots,[\![\psi_s]\!]}$ exists, i.e.

$$\begin{aligned} (D^{\delta(\overline{a})}_{[\![\psi_1]\!],\ldots,[\![\psi_s]\!]}, [\![\psi_1]\!]^{\mathcal{A}_{\mathrm{rand}}}(\_\_, \delta(\overline{a})) \restriction D^{\delta(\overline{a})}_{[\![\psi_1]\!],\ldots,[\![\psi_s]\!]}, \\ \ldots, [\![\psi_s]\!]^{\mathcal{A}_{\mathrm{rand}}}(\_\_, \delta(\overline{a})) \restriction D^{\delta(\overline{a})}_{[\![\psi_1]\!],\ldots,[\![\psi_s]\!]}) \in K. \end{aligned}$$

Moreover $|D^{\delta(\overline{a})}_{[\![\psi_1]\!],\ldots,[\![\psi_s]\!]}| \le n$. Like the proof of Lemma 3, $\delta(\overline{a}) \longmapsto \overline{a}$ can be extended to an automorphism $h : A_{\mathrm{rand}} \to A_{\mathrm{rand}}$ with $h(\delta(\overline{a})) = \overline{a}$. Then we deduce

$$\begin{aligned} & (D^{\delta(\overline{a})}_{[\![\psi_1]\!],\ldots,[\![\psi_s]\!]}, \psi_1^{\mathcal{A}_{\mathrm{rand}}}(\_\_, \delta(\overline{a})) \restriction D^{\delta(\overline{a})}_{[\![\psi_1]\!],\ldots,[\![\psi_s]\!]}, \\ & \qquad \ldots, \psi_s^{\mathcal{A}_{\mathrm{rand}}}(\_\_, \delta(\overline{a})) \restriction D^{\delta(\overline{a})}_{[\![\psi_1]\!],\ldots,[\![\psi_s]\!]}) \in K, \\ & \qquad \text{by } [\![\_\_]\!] \text{ is an equivalent translation,} \\ \Rightarrow\ & (h(D^{\delta(\overline{a})}_{[\![\psi_1]\!],\ldots,[\![\psi_s]\!]}), \psi_1^{\mathcal{A}_{\mathrm{rand}}}(\_\_, \overline{a}) \restriction h(D^{\delta(\overline{a})}_{[\![\psi_1]\!],\ldots,[\![\psi_s]\!]}), \\ & \qquad \ldots, \psi_s^{\mathcal{A}_{\mathrm{rand}}}(\_\_, \overline{a}) \restriction h(D^{\delta(\overline{a})}_{[\![\psi_1]\!],\ldots,[\![\psi_s]\!]})) \in K, \\ & \qquad \text{as } L \text{ is closed under isomorphisms.} \end{aligned}$$

So $\mathcal{A}_{\mathrm{rand}} \restriction h(D^{\delta(\overline{a})}_{[\![\psi_1]\!],\ldots,[\![\psi_s]\!]})$ witnesses $\mathcal{A}_{\mathrm{rand}} \models Q_K\overline{x}_1, \ldots, \overline{x}_s[\psi_1, \ldots, \psi_s][\overline{a}]$ with required $|h(D^{\delta(\overline{a})}_{[\![\psi_1]\!],\ldots,[\![\psi_s]\!]})| \le n$. □

Now we say a set of generalized quantifiers $\mathcal{Q}$ is closed under extensions, finitely witnessed or finitely based if each $Q_K \in \mathcal{Q}$ is respectively so.

Theorem 6. *Assume a finite set of generalized quantifiers $\mathcal{Q}$ is finitely based, and $t \in \mathbb{N}$. There exists an $i \in \mathbb{N}$ such that for any $\varphi \in \mathcal{L}^t_{\infty\omega}(\mathcal{Q})[\tau]$ and $\overline{a} \in A_{\mathrm{rand}}^{|\mathsf{FV}(\varphi)|}$, if $\mathcal{A}_{\mathrm{rand}} \models \varphi[\overline{a}]$, then for any finite $\mathcal{A} \models \epsilon_i$ and $\overline{a}' \in A^{|\mathsf{FV}(\varphi)|}$ with $\varphi^0_{\mathcal{A},\overline{a}'} = \varphi^0_{\mathcal{A}_{\mathrm{rand}},\overline{a}}$, we have $\mathcal{A} \models \varphi[\overline{a}']$.*

Proof. Let n be the maximum such as in Lemma 4, when Q_K ranges over the finite set $\mathcal{Q}$. Take $i = n + t$. First for any φ, we can use De Morgan Law to push all its negation symbols either the atomic level or right before a Q_K. Then we proceed by induction on the structure of such transformed φ. Note as mentioned before $\exists$ can be treated as a finitely based Q_K, so the cases of first order quantifiers are absorbed in the discussion of general finitely based Q_K.
(i) φ is atomic or negated atomic, trivial.
(ii) The proof for $\varphi = \bigwedge_{j\in J} \varphi_j$ or $\bigvee_{j\in J} \varphi_j$ is an easy induction argument.
(iii) $\varphi = Q_K\overline{x}_1, \ldots, \overline{x}_s[\psi_1, \ldots, \psi_s]$ where $Q_K \in \mathcal{Q}$ and clearly $|\mathsf{FV}(\varphi)| \leq t$. By definition $\mathcal{A}_{\text{rand}} \models \varphi[\overline{a}]$ is equivalent to

$$(A_{\text{rand}}, \psi_1^{\mathcal{A}_{\text{rand}}}(\_\_, \overline{a}), \ldots, \psi_s^{\mathcal{A}_{\text{rand}}}(\_\_, \overline{a})) \in K.$$

As Q_K is finitely witnessed, for some finite $D \subset A_{\text{rand}}$, $(D, \psi_1^{\mathcal{A}_{\text{rand}}}(\_\_, \overline{a}) \restriction D, \ldots, \psi_s^{\mathcal{A}_{\text{rand}}}(\_\_, \overline{a}) \restriction D) \in K$. Moreover by Lemma 4, D can be chosen in a way such that $|D| \leq n$.

Now for any finite $\mathcal{A} \models \epsilon_i$ and $\overline{a}' \in A^{|\mathsf{FV}(\varphi)|}$ with $\varphi^0_{\mathcal{A},\overline{a}'} = \varphi^0_{\mathcal{A}_{\text{rand}},\overline{a}}$, since $|\{\overline{a}\}| \leq |\mathsf{FV}(\varphi)| \leq t$, so $i = n + t \geq |D| + |\{\overline{a}\}|$, thus we can enlarge the partial isomorphism $\overline{a}' \longmapsto \overline{a}$ to $h : S, \overline{a}' \longmapsto D, \overline{a}$ between $\mathcal{A}$ and $\mathcal{A}_{\text{rand}}$ by Lemma 2 for some $S \subseteq A$. We claim

$$\begin{aligned} &h((S, \psi_1^{\mathcal{A}}(\_\_, \overline{a}') \restriction S, \ldots, \psi_s^{\mathcal{A}}(\_\_, \overline{a}') \restriction S)) \\ &= (D, \psi_1^{\mathcal{A}_{\text{rand}}}(\_\_, \overline{a}) \restriction D, \ldots, \psi_s^{\mathcal{A}_{\text{rand}}}(\_\_, \overline{a}) \restriction D). \end{aligned} \tag{1}$$

Then it follows that $(A, \psi_1^{\mathcal{A}}(\_\_, \overline{a}'), \ldots, \psi_s^{\mathcal{A}}(\_\_, \overline{a}')) \in K$, since K is closed under isomorphisms and extensions, that is to say $\mathcal{A} \models \varphi[\overline{a}']$.

Indeed (1) is equivalent to for any $1 \leq j \leq s$ and $\overline{e} \in S^{r_j}$,

$$\mathcal{A} \models \psi_j[\overline{e}\overline{a}'] \text{ iff } \mathcal{A}_{\text{rand}} \models \psi_j[h(\overline{e})\overline{a}] \quad \text{i.e.} \quad \mathcal{A}_{\text{rand}} \models \psi_j[h(\overline{e}\overline{a}')].$$

If $\mathcal{A}_{\text{rand}} \models \psi_j[h(\overline{e}\overline{a}')]$, observe $\varphi^0_{\mathcal{A},\overline{e}\overline{a}'} = \varphi^0_{\mathcal{A}_{\text{rand}},h(\overline{e}\overline{a}')}$, so by induction hypothesis, $\mathcal{A} \models \psi_j[\overline{e}\overline{a}']$. The case for $\mathcal{A}_{\text{rand}} \models \neg\psi_j[h(\overline{e}\overline{a}')]$ is the same.
(iv) $\varphi = \neg Q_K\overline{x}_1, \ldots, \overline{x}_s[\psi_1, \ldots, \psi_s]$, $\mathcal{A}_{\text{rand}} \models \varphi[\overline{a}]$ i.e.

$$(A_{\text{rand}}, \psi_1^{\mathcal{A}_{\text{rand}}}(\_\_, \overline{a}), \ldots, \psi_s^{\mathcal{A}_{\text{rand}}}(\_\_, \overline{a})) \notin K.$$

Given any finite $\mathcal{A}$ with $\mathcal{A} \models \epsilon_i$ and $\overline{a}' \in A^{|\mathsf{FV}(\varphi)|}$ with $\varphi^0_{\mathcal{A},\overline{a}'} = \varphi^0_{\mathcal{A}_{\text{rand}},\overline{a}}$, i.e. $\overline{a}' \longmapsto \overline{a}$ is a partial isomorphism between $\mathcal{A}$ and $\mathcal{A}_{\text{rand}}$, moreover it can be extended to an isomorphic embedding $h : A \to A_{\text{rand}}$ with $h(\overline{a}') = \overline{a}$, by Lemma 2 for $\mathcal{A}_{\text{rand}} \models \epsilon_{|A|}$. Similar to (iii), we shall prove

$$\begin{aligned} &h((A, \psi_1^{\mathcal{A}}(\_\_, \overline{a}'), \ldots, \psi_s^{\mathcal{A}}(\_\_, \overline{a}'))) \\ &\subset (A_{\text{rand}}, \psi_1^{\mathcal{A}_{\text{rand}}}(\_\_, \overline{a}), \ldots, \psi_s^{\mathcal{A}_{\text{rand}}}(\_\_, \overline{a})). \end{aligned} \tag{2}$$

Then assume contrarily $\mathcal{A} \models \neg\varphi[\overline{a}']$, i.e. $(A, \psi_1^{\mathcal{A}}(\_\_, \overline{a}'), \ldots, \psi_s^{\mathcal{A}}(\_\_, \overline{a}')) \in K$. Because K is closed under isomorphisms and extensions, we would have

$$(A_{\text{rand}}, \psi_1^{\mathcal{A}_{\text{rand}}}(\_\_, \overline{a}), \ldots, \psi_s^{\mathcal{A}_{\text{rand}}}(\_\_, \overline{a})) \in K$$

by (2), a contradiction. Therefore $\mathcal{A} \models \varphi[\overline{a}']$.

To establish (2), it amounts to show that for any $1 \leq j \leq s$ and $\overline{e} \in A^{r_j}$,

$$\mathcal{A} \models \psi_j[\overline{ea}'] \text{ iff } \mathcal{A}_{\text{rand}} \models \psi_j[h(\overline{e})\overline{a}] \quad \text{i.e.} \quad \mathcal{A}_{\text{rand}} \models \psi_j[h(\overline{ea})],$$

of which the proof is identical to the last part of (iii). □

Corollary 3. *For a finite set of generalized quantifiers $\mathcal{Q}$, $\mathcal{L}^{\omega}_{\infty\omega}(\mathcal{Q})$ satisfies the labeled 0-1 law, if $\mathcal{Q}$ is closed under extensions.*

Proof.
To make Theorem 6 applicable, $\mathcal{Q}$ must be also finitely witnessed. Set

$$\mathcal{Q}^{\uparrow}_{\text{fin}} = \{Q_{K^{\uparrow}_{\text{fin}}} \mid Q_K \in \mathcal{Q}, K^{\uparrow}_{\text{fin}} = \{\mathcal{A} \mid \text{for some finite } \mathcal{A}' \in K, \mathcal{A}' \subseteq \mathcal{A}\}\}.$$

It is easy to verify that $\mathcal{Q}^{\uparrow}_{\text{fin}}$ is finitely based for any $\mathcal{Q}$, and more importantly $\mathcal{Q}^{\uparrow}_{\text{fin}}$ behaves exactly the same as $\mathcal{Q}$ on all finite structures provided $\mathcal{Q}$ is closed under extensions. So we can safely assume $\mathcal{Q}$ is finitely based. Then the result follows from the fact that each ϵ_i has the asymptotic probability 1 and Theorem 6. □

When $\mathcal{Q}$ is an infinite set of generalized quantifiers closed under extensions, for any logic $\mathcal{L} < \mathcal{L}^{\omega}_{\infty\omega}$ with finitary syntax like FO, LFP and PFP, that is, each sentence in $\mathcal{L}(\mathcal{Q})$ only involves finitely many Q_Ks, we can argue in the same way as if $\mathcal{Q}$ were finite like the above corollary, so the 0-1 law still holds in $\mathcal{L}(\mathcal{Q})$. But for $\mathcal{L}^{\omega}_{\infty\omega}$ itself, consider the sentence

$$\varphi = \bigvee_{l \text{ is even.}} \exists^{\geq l} x(x = x) \wedge \neg\exists^{\geq l+1} x(x = x),$$

surely it defines the Parity property which has no asymptotic probability, while each $\exists^{\geq l}$ is finitely based, so Theorem 6 and Corollary 3 can by no means be extended to infinite $\mathcal{Q}$. Nevertheless finite many of generalized quantifiers usually do not suffice. One typical situation of infinite many quantifiers is the *vectorization*, which extends a given quantifier to finite Cartesian product of the universe of the original structure. For instance, $\overset{k}{\mathsf{TP}}$ can be viewed as the k-vectorization of monadic TP. Another extension of much interest is the *relativization* which closes the Lindström logic under definable unary set. Combine them together, we have for any Q_K, its relativized k-vectorization $\overset{k}{Q}{}^{\text{rel}}_K$ is the $Q_{K'}$ for $K' = \{(\mathcal{A}, U) \mid U \subseteq A^k \text{ and } (U, R^{\mathcal{A}}_1 \upharpoonright U, \ldots, R^{\mathcal{A}}_s \upharpoonright U) \in K\}$. It is easy to prove being closed under extensions, finitely witnessed and finitely based are all preserved under relativization and vectorization. Observe that each $\overset{k}{Q}{}^{\text{rel}}_K$ must consume $k + \sum_{1 \leq i \leq s} kr_i$ distinct variables, so there are only finite many types of $\overset{k}{Q}{}^{\text{rel}}_K$ that are valid to appear in a $\mathcal{L}^{t}_{\infty\omega}$ formula for any given t, thus we have

Corollary 4. *For a finite set $\mathcal{Q}$ of Q_K closed under extensions, $\mathcal{L}^{\omega}_{\infty\omega}(\overset{\omega}{\mathcal{Q}}{}^{\text{rel}})$ has the labeled 0-1 law, where $\overset{\omega}{\mathcal{Q}}{}^{\text{rel}} = \{\overset{k}{Q}{}^{\text{rel}}_K \mid Q_K \in \mathcal{Q}, k \in \mathbb{N}\}$.*

A similar argument can be applied to $\mathcal{L}^{\omega}_{\infty\omega}(\overset{\omega}{\mathsf{P}}^{\omega,\omega})$, so Theorem 5 holds. Note one of its immediate consequences is that Hamiltonicity can not be defined in partition logic, for FO[Ham], the minimal regular logic capturing Hamiltonicity does not have a 0-1 law[2].

Acknowledgments

We are grateful to H.-D. Ebbinghaus for suggestions on an earlier draft of the paper. An anonymous referee's comments also help to improve the presentation.

References

1. A. Dawar. Generalized quantifiers and logical reducibilities. *Journal of Logic and Computation*, 5(1995), 213-226.
2. A. Dawar and E. Grädel. Generalized quantifiers and 0-1 laws. in Proceedings of 10th IEEE Symposium on Logic in Computer Science, 54-64, 1995.
3. H.-D. Ebbinghaus and J. Flum, *Finite Model Theory*. Springer-Verlag, 1995.
4. R. Fagin. Generalized first-order spectra and polynomial-time recognizable sets. In R.M. Karp, editor, *Complexity of Computation, SIAM-AMS Proceedings, Vol. 7*, 43-73, 1974.
5. G. Fayolle, S. Grumbach, and C. Tollu. Asymptotic probabilities of languages with generalized quantifiers, in Proceedings of 8th IEEE Symposium on Logic in Computer Science, 199-207, 1993.
6. E. Grädel. Capturing complexity classes by fragments of second-order logic. *Theoretical Computer Science*, 101(1992), 35–57.
7. E. Grädel, *et al.* Problems in finite model theory. Available at: http://www-mgi.informatik.rwth-aachen.de/FMT/, June 2000.
8. H. Imhof. Logiken mit Partitionsquantoren. *Diplomarbeit*, Freiburg University, 1992.
9. N. Immerman. Languages that capture complexity classes. *SIAM Journal on Computing*, 16(1987), 760-778.
10. N. Immerman. Nondeterministic space is closed under complementation. *SIAM Journal on Computing*, 17(1988), 935-938.
11. N. Immerman. *Descriptive Complexity*. Graduate Texts in Computer Science, Springer-Verlag, 1999.
12. T. Lacoste. 0-1 laws by preservation. *Theoretical Computer Science*, 184(1-2):237-245, 30, 1997.
13. J. Malitz. Downward transfer of satisfiability for sentences of $L^{1,1}$. *Journal Symbolic Logic*, 48(1984), 1146–1150.
14. E. Shen. Partition logics and transitive closure logics (Abstract). *Bulletin of Science* (Chinese), 38(1993), 1271–1272.
15. E. Shen. Partition logics and colored pebble games. *Prog. Maths.* (Chinese), 24(6)(1996), 540–546.
16. E. Shen and Y. Chen. The downward transfer of elementary satisfiability of partition logics. *Mathematical Logic Quarterly*, 46, 477-487, No. 4, 2000.
17. E. Shen and Q. Tian. Monadic partition logics and finite automata. *Theoretical Computer Science*, 166(1996), 63–81.
18. I. A. Stewart. Using the Hamiltonian path operator to capture NP. *Journal of Computer and System Sciences*, 45, 127-151(1992).

An Existential Locality Theorem

Martin Grohe[1] and Stefan Wöhrle[2]

[1] Laboratory for Foundations of Computer Science, University of Edinburgh, Edinburgh EH9 3JZ, Scotland, UK
grohe@dcs.ed.ac.uk

[2] Lehrstuhl für Informatik VII, RWTH Aachen, 52056 Aachen, Germany
woehrle@informatik.rwth-aachen.de

Abstract. We prove an existential version of Gaifman's locality theorem and show how it can be applied algorithmically to evaluate existential first-order sentences in finite structures.

1 Introduction

Gaifman's locality theorem [12] states that every first-order sentence is equivalent to a Boolean combination of sentences saying: There exist elements $a_1, \ldots, a_k$ that are far apart from one another, and each a_i satisfies some local condition described by a first-order formula whose quantifiers only range over a fixed-size neighborhood of an element of a structure. We prove that every *existential* first-order sentence is equivalent to a *positive* Boolean combination of sentences saying: There exist elements $a_1, \ldots, a_k$ that are far apart from one another, and each a_i satisfies some local condition described by an *existential* first-order formula.

The locality of first-order logic can be explored to prove that certain properties of finite structures are not expressible in first-order logic, and it seems that this was Gaifman's main motivation. More recently, Libkin and others considered this technique of proving inexpressibility results using locality in a complexity theoretic context (see, e.g., [5,15,14,16]).

A completely different application of Gaifman's theorem has been proposed in [11]: It can be used to evaluate first-order sentences in certain finite structures quite efficiently. In general, it takes time $n^{\Theta(l)}$ to decide whether a structure of size n satisfies a first-order sentence of size l, and under complexity theoretic assumptions, it can be proved that no real improvement is possible: The problem of deciding whether a given structure satisfies a given first-order sentence is PSPACE-complete [18,20], and if parameterized by the size of the input sentence, it is complete for the parameterized complexity class AW[*] [7]. The latter result implies that it is unlikely that the problem is fixed-parameter tractable (cf. [6]), i.e., that it can be solved in time $f(l) \cdot n^c$, for a function f and a constant c.

Gaifman's theorem reduces the question of whether a first-order sentence holds in a structure to the question of whether the structure contains elements that are far apart from one another and satisfy some local condition expressed by a first-order formula. In certain structures, it is much easier to decide whether

L. Fribourg (Ed.): CSL 2001, LNCS 2142, pp. 99–114, 2001.

an element satisfies a local first-order formula than to decide whether the whole structure satisfies a first-order sentence. An example are graphs of bounded degree: Local neighborhoods of vertices in such graphs have a size bounded by a constant only depending on the radius of the neighborhoods, so the time needed to check whether a vertex satisfies a local condition does not depend on the size of the graph. Another, less obvious example are planar graphs. To evaluate local conditions in planar graphs, we can exploit the fact that in planar graphs neighborhoods of fixed radius have bounded tree-width [17]. In general, such a locality based approach to evaluating first-order sentences in finite structures works for classes of structures that have a property called bounded local tree-width; the class of planar graphs and all classes of structures of bounded degree are examples of classes having this property. It has been proved in [11] that for each class C of structures of bounded local tree-width there is an algorithm that, given a structure $\mathcal{A} \in \mathrm{C}$ and a first-order sentence φ, decides whether $\mathcal{A}$ satisfies φ in time near linear in the size of the structure $\mathcal{A}$ (the precise statement is Theorem 7).

While a linear dependence on the size of the input structure is optimal, the dependence of these algorithms on the size of the input sentence leaves a lot to be desired: There is not even an elementary upper bound for the runtime in terms of the size of the sentence. Although the dependence of the algorithm on the structure size matters much more than the dependence on the size of the sentence, because usually we are evaluating small sentences in large structures,[1] it would be desirable to have a dependence on the size of the sentence that is not worse than exponential. Of course, since we are dealing with a PSPACE complete problem, we cannot expect the runtime of an algorithm to be polynomial in both the size of the input structure and the size of the input sentence.

We have observed that one of the main factors contributing to the enormous runtime of the locality based algorithms in terms of the formulas size is the number of quantifier alternations in the formula. This has motivated the present paper. We can use a variant of our existential locality theorem to improve the algorithms described above to algorithms whose runtime "only" depends doubly exponentially on the size of the input sentence.

In this paper we concentrate on the proof of our existential locality theorem, which is surprisingly complicated. This proof is presented in Section 3. The algorithmic application is outlined in Section 4.

2 Preliminaries

A *vocabulary* is a finite set of relation symbols. Associated with every relation symbol R is a positive integer called the *arity* of R. In the following, τ always denotes a vocabulary.

[1] The generic example is the problem of evaluating SQL database queries against finite relational databases, which can be modeled by the problem of evaluating first-order sentences in finite structures.

A *τ-structure* $\mathcal{A}$ consists of a non-empty set A, called the *universe* of $\mathcal{A}$, and a relation $R^{\mathcal{A}} \subseteq A^r$ for each r-ary relation symbol $R \in \tau$. For instance, we consider *graphs* as $\{E\}$-structures $\mathcal{G} = (G, E^{\mathcal{G}})$, where the binary relation $E^{\mathcal{G}}$ is symmetric and anti-reflexive (i.e. graphs are undirected and loop-free). If $\mathcal{A}$ is a τ-structure and $B \subseteq A$, then $\langle B \rangle^{\mathcal{A}}$ denotes the substructure induced by $\mathcal{A}$ on B, that is, the τ-structure $\mathcal{B}$ with universe B and $R^{\mathcal{B}} := R^{\mathcal{A}} \cap B^r$ for every r-ary $R \in \tau$.

The formulas of *first-order logic* are build up from *atomic formulas* using the usual Boolean connectives and existential and universal quantification over the elements of the universe of a structure. Remember that an *atomic formula*, or *atom*, is a formula of the form $x = y$ or $R(x_1, \ldots, x_r)$, where R is an r-ary relation symbol. The set of all variables of a formula φ is denoted by $\text{var}(\varphi)$. A *free variable* in a first-order formula is a variable x not in the scope of a quantifier $\exists x$ or $\forall x$. The set of all free variables of a formula φ is denoted by $\text{free}(\varphi)$. A *sentence* is a formula without free variables. The notation $\varphi(x_1, \ldots, x_k)$ indicates that all free variables of the formula φ are among $x_1, \ldots, x_k$; it does not necessarily mean that the variables $x_1, \ldots, x_k$ all appear in φ. For a formula $\varphi(x_1, \ldots, x_k)$, a structure $\mathcal{A}$, and $a_1, \ldots, a_k \in A$ we write $\mathcal{A} \models \varphi(a_1, \ldots, a_k)$ to say that $\mathcal{A}$ satisfies φ if the variables $x_1, \ldots, x_k$ are interpreted by the vertices $a_1, \ldots, a_k$, respectively.

The *weight* of a first-order formula φ is the number of quantifiers $\exists x$ and $\forall x$ occurring in φ.

A first-order formula is *existential* if it contains no universal quantifiers and if every existential quantifier occurs in the scope of an even number of negation symbols. A *literal* is an atom or a negated atom. A *conjunctive query with negation* is a formula of the form $\exists \bar{x} \bigwedge_{i=1}^{m} \lambda_i$, where each λ_i is a literal. Every existential formula φ of weight w and length l is equivalent to a disjunction of at most 2^l conjunctive queries with negation, each of which is of weight at most w and length at most l.

We often denote tuples $a_1 \ldots a_k$ of elements of a set A by $\bar{a}$, and we write $\bar{a} \in A$ instead of $\bar{a} \in A^k$. Similarly, we denote tuples of variables by $\bar{x}$.

Our underlying model of computation is the standard RAM-model with addition and subtraction as arithmetic operations (cf. [1,19]). In our complexity analysis we use the uniform cost measure. Structures are represented on a RAM in a straightforward way by listing all elements of the universe and then all tuples in the relations. For details we refer the reader to [10]. We define the *size* of a τ-structure $\mathcal{A}$ to be $||\mathcal{A}|| := |A| + \sum_{R \in \tau \text{ } r\text{-ary}} r \cdot |R^{\mathcal{A}}|$; this is the length of a reasonable representation of $\mathcal{A}$ (if we suppress details that are inessential for us). We fix some reasonable encoding for first-order formulas and denote by $||\varphi||$ the size of the encoding of a formula φ.

2.1 Gaifman's Locality Theorem

The *Gaifman graph* of a τ-structure $\mathcal{A}$ is the graph $\mathcal{G}_{\mathcal{A}}$ with vertex set A and an edge between two vertices $a, b \in A$ if there exists an $R \in \tau$ and a tuple

$a_1 \dots a_k \in R^{\mathcal{A}}$ such that $a, b \in \{a_1, \dots, a_k\}$. The *distance* $d^{\mathcal{A}}(a,b)$ between two elements $a, b \in A$ of a structure $\mathcal{A}$ is the length of the shortest path in $\mathcal{G}_{\mathcal{A}}$ connecting a and b. For $r \geq 1$ and $a \in A$, we define the *r-neighborhood* of a in $\mathcal{A}$ to be $N_r^{\mathcal{A}}(a) := \{b \in A \mid d^{\mathcal{A}}(a,b) \leq r\}$. For a subset $B \subseteq A$ we let $N_r^{\mathcal{A}}(B) := \bigcup_{b \in B} N_r^{\mathcal{A}}(b)$.

For every $r \geq 0$ there is an existential first-order formula $\delta_r(x,y)$ such that for all τ-structures $\mathcal{A}$ and $a, b \in A$ we have $\mathcal{A} \models \delta_r(a,b)$ if, and only if, $d^{\mathcal{A}}(a,b) \leq r$. In the following, we write $d(x,y) \leq r$ instead of $\delta_r(x,y)$ and $d(x,y) > r$ instead of $\neg\delta_r(x,y)$.

If $\varphi(x)$ is a first-order formula, then $\varphi^{N_r(x)}(x)$ is the formula obtained from $\varphi(x)$ by relativizing all quantifiers to $N_r(x)$, that is, by replacing every subformula of the form $\exists y \psi(x,y,\bar{z})$ by $\exists y(d(x,y) \leq r \wedge \psi(x,y,\bar{z}))$ and every subformula of the form $\forall y \psi(x,y,\bar{z})$ by $\forall y(d(x,y) \leq r \rightarrow \psi(x,y,\bar{z}))$. We usually write $\exists y \in N_r(x)\ \psi$ instead of $\exists y(d(x,y) \leq r \wedge \psi)$ and $\forall y \in N_r(x)\ \psi$ instead of $\forall y(d(x,y) \leq r \rightarrow \psi)$.

A formula $\psi(x)$ of the form $\varphi^{N_r(x)}(x)$, for some $\varphi(x)$, is called *r-local.* The basic property of r-local formulas $\psi(x)$ is that it only depends on the r-neighborhood of x whether they hold at x or not, that is, for all structures $\mathcal{A}$ and $a \in A$ we have $\mathcal{A} \models \psi(a)$ if, and only if, $\langle N_r^{\mathcal{A}}(a) \rangle \models \psi(a)$. Observe that if $\psi(x)$ is r-local and $s > r$, then $\psi(x)$ is equivalent to the s-local formula $\psi^{N_s(x)}(x)$. We often use this observation implicitly when considering r-local formulas as s-local for some $s > r$.

Sentences can never be local in the sense just defined. As a substitute, we say that a *local sentence* is a sentence of the form

$$\exists x_1 \dots \exists x_k \Big(\bigwedge_{1 \leq i < j \leq k} d(x_i, x_j) > 2r \wedge \bigwedge_{1 \leq i \leq k} \psi(x_i) \Big),$$

where $r, k \geq 1$ and $\psi(x)$ is r-local.

Theorem 1 (Gaifman [12]). *Every first-order sentence is equivalent to a Boolean combination of local sentences.*

3 The Existential Locality Theorems

If $\psi(x)$ is an existential first-order formula, then for every $r \geq 1$ the r-local formula $\psi^{N_r(x)}(x)$ obtained from ψ is also existential. We define a local sentence

$$\exists x_1 \dots \exists x_k \Big(\bigwedge_{1 \leq i < j \leq k} d(x_i, x_j) > 2r \wedge \bigwedge_{1 \leq i \leq k} \psi(x_i) \Big)$$

to be *existential* if the formula ψ is existential and r-local. Let us remark that, in general, an existential local sentence is *not* equivalent to an existential first-order sentence, because the formula $d(x_i, x_j) > s$ is not existential for any $s \geq 2$.

Theorem 2. *Every existential first-order sentence is equivalent to a positive Boolean combination of existential local sentences.*

Unfortunately, neither Gaifman's original proof of his locality theorem (based on quantifier elimination) nor Ebbinghaus and Flum's [8] model theoretic proof can be adapted to prove this existential version of Gaifman's theorem. Compared to these proofs, our proof is very combinatorial, which is not surprising, because there is not much "logic" left in existential sentences.

We illustrate the basic idea by a simple example:

Example 3. Let

$$\varphi := \exists x \exists y \big(\neg E(x,y) \wedge \textsc{Red}(x) \wedge \textsc{Blue}(y)\big)$$

(here E is a binary relation symbol and $\textsc{Red}$, $\textsc{Blue}$ are unary relation symbols). Although the syntactical form of φ is close to that of an existential local sentence, it is not obvious how to find a positive Boolean combination of existential local sentences equivalent to φ. Here is one:

$$\begin{aligned}&\Big(\exists x \exists y \big(d(x,y) > 2 \wedge (\textsc{Red}(x) \vee \textsc{Blue}(y)) \wedge (\textsc{Red}(y) \vee \textsc{Blue}(y))\big) \\ &\quad \wedge \exists x\, \textsc{Red}(x) \wedge \exists x\, \textsc{Blue}(x)\Big) \\ &\vee \exists x\, \exists x' \in N_2(x) \exists y \in N_2(x) \big(\neg E(x',y) \wedge \textsc{Red}(x') \wedge \textsc{Blue}(y)\big).\end{aligned}$$

To understand the following proof it is worthwhile trying to extend the idea of this example to the sentence

$$\exists x \exists y \exists z \big(\neg E(x,y) \wedge \neg E(x,z) \wedge \neg E(y,z) \wedge \textsc{Red}(x) \wedge \textsc{Blue}(y) \wedge \textsc{Green}(z)\big)$$

(although it is very complicated to actually write down an equivalent positive Boolean combination of existential local sentences). Indeed, it is the main difficulty of the proof to handle sentences saying "there is an independent set of points $x_1, \ldots, x_k$ of colors $c_1, \ldots, c_k$, respectively." Playing with such sentences leads to the crucial observation that the basic combinatorial problem can be handled by the marriage theorem (as it is done in Step 4 of the proof of Lemma 4).

The proof requires some preparation. We define the *rank* of a local sentence

$$\exists x_1 \ldots \exists x_k \Big(\bigwedge_{1 \le i < j \le k} d(x_i, x_j) > 2r \wedge \bigwedge_{1 \le i \le k} \psi(x_i)\Big),$$

to be the pair $(k + w, r)$, where w is the weight of ψ. We partially order the ranks by saying that $(q, r) \le (q', r')$ if $q \le q'$ and $r \le r'$.

Lemma 4. *Let $k \ge 2$, $r \ge 1$, $w \ge 0$, and let $\mathcal{A}$, $\mathcal{B}$ be structures such that every existential local sentence of rank at most $(k \cdot (w+1), 2^{k^2} r)$ that holds in $\mathcal{A}$ also holds in $\mathcal{B}$. Let*

$$\varphi := \exists x_1 \ldots \exists x_k \Big(\bigwedge_{1 \le i < j \le k} d(x_i, x_j) > 2^{k^2} r \wedge \bigwedge_{i=1}^{k} \psi_i(x_i)\Big),$$

where for $1 \leq i \leq k$, the formula $\psi_i(x_i)$ is r-local, existential and of weight at most w. Suppose that $\mathcal{A} \models \varphi$.

Then

$$\mathcal{B} \models \exists x_1 \ldots \exists x_k \Big(\bigwedge_{1 \leq i < j \leq k} d(x_i, x_j) > 2r \wedge \bigwedge_{i=1}^{k} \psi_i(x_i) \Big).$$

Proof: We prove the lemma in four steps.

Step 1. We show that if for some $l, 1 \leq l \leq k$, say $l = k$, there are $b_1, \ldots, b_k \in B$ such that $d(b_i, b_j) > 4r$ for $1 \leq i < j \leq k$, and $\mathcal{B} \models \psi_l(b_i)$ for $1 \leq i \leq k$, then it suffices to prove that

$$\mathcal{B} \models \exists x_1 \ldots \exists x_{k-1} \Big(\bigwedge_{1 \leq i < j \leq k-1} d(x_i, x_j) > 2r \wedge \bigwedge_{i=1}^{k-1} \psi_i(x_i) \Big).$$

To see this, suppose that we have such $b_1, \ldots, b_k$ and we find $c_1, \ldots, c_{k-1}$ such that $d(c_i, c_j) > 2r$ for all $1 \leq i < j \leq k-1$, and $\mathcal{B} \models \psi_i(c_i)$ for $1 \leq i \leq k-1$. Then there will be at least one $i, 1 \leq i \leq k$ such that b_i has distance greater than $2r$ from c_j for all $j, 1 \leq j \leq k-1$. Thus $c_1, \ldots, c_{k-1}, b_i$ witness that

$$\mathcal{B} \models \exists x_1 \ldots \exists x_k \Big(\bigwedge_{1 \leq i < j \leq k} d(x_i, x_j) > 2r \wedge \bigwedge_{i=1}^{k} \psi_i(x_i) \Big).$$

So without loss of generality, in the following we assume that for $1 \leq i \leq k$, there are at most $(k-1)$ elements of B of pairwise distance greater than $4r$ satisfying ψ_i.

Step 2. We let $K := \{1, \ldots, k\}$, and for every set $I \subseteq K$ we let $\psi_I(x) := \bigvee_{i \in I} \psi_i(x)$. Note that ψ_I is a formula of weight at most $k \cdot w$. Let $C := \{c \in B \mid \mathcal{B} \models \psi_K(c)\}$. By the assumption we made at the end of Step 1, there exist at most $k(k-1)$ elements of C of pairwise distance greater than $4r$.

Claim: There are $p, l, 1 \leq p \leq k(k-1)+1, 1 \leq l \leq k(k-1)$, and elements $c_1, \ldots, c_l \in C$ such that $d^{\mathcal{B}}(c_i, c_j) > 2^{p+1}r$ for $1 \leq i < j \leq l$, and for all $c \in C$ there exists an $i \leq l$ such that $d^{\mathcal{B}}(c, c_i) \leq 2^p r$.

Proof: We construct $c_1, \ldots, c_l$ inductively: As the inductive basis, let c_1 be an arbitrary element of C. If $c_1, \ldots, c_i$ are constructed, we choose $c_{i+1} \in C$ such that for $1 \leq j \leq i$ we have $d^{\mathcal{B}}(c_{i+1}, c_j) > 2^{k(k-1)+1-(i-1)}r$. If no such c_{i+1} exists, we let $l := i$, $p := k(k-1)+1-(l-1)$ and stop.

Our construction guarantees that for $1 \leq i < j \leq l$ we have

$$d^{\mathcal{B}}(c_i, c_j) > 2^{k(k-1)+1-(j-2)}r. \tag{1}$$

For $j \leq k(k-1)+1$, this implies $d^{\mathcal{B}}(c_i, c_j) > 4r$. Since there are at most $k(k-1)$ elements of C of pairwise distance greater than $4r$, this guarantees that $l \leq k(k-1)$. (1) also guarantees that for $1 \leq i < j \leq l$ we have $d^{\mathcal{B}}(c_i, c_j) > 2^{k(k-1)+1-(l-2)}r = 2^{p+1}r$.

Since we stopped at $l = i$, for all $c \in C$ there exists an $i \leq l$ such that $d^{\mathcal{B}}(c, c_i) \leq 2^{k(k-1)+1-(l-1)} = 2^p r$. This proves the claim.

Step 3. Let p, l, $c_1, \dots, c_l$ be as stated in the claim in Step 2. For $I \subseteq K$, let

$$\varphi_I := \exists x_1 \dots \exists x_k \Big(\bigwedge_{\substack{i,j \in I \\ i<j}} d(x_i, x_j) > 2^{p+1} r \wedge \bigwedge_{i \in I} \psi_I(x_i) \Big).$$

Since $\mathcal{A} \models \varphi$, we have $\mathcal{A} \models \varphi_I$. Thus, since φ_I is an existential local sentence of rank at most $(k \cdot (w+1), 2^{k^2} r)$, we also have $\mathcal{B} \models \varphi_I$.

Step 4. Let $L := \{1, \dots, l\}$. We define a relation $R \subseteq K \times L$ as follows: For $i \in K, j \in L$ we let iRj if there is a $b \in B$ such that $\mathcal{B} \models \psi_i(b)$ and $d^{\mathcal{B}}(b, c_j) \leq 2^p r$.

Claim: For every $I \subseteq K$ the set $R(I) := \{j \in L \mid \exists i \in I : iRj\}$ contains at least as many elements as I.

Proof: Recall that $\mathcal{B} \models \varphi_I$. For $i \in I$, let $b_i \in B$, such that for all $i, j \in I$ with $i < j$ we have $d^{\mathcal{B}}(b_i, b_j) > 2^{p+1} r$ and for all $i \in I$ we have $\mathcal{B} \models \psi_I(b_i)$. Then $b_i \in C$, and thus there exist a $j \in L$ such that $d^{\mathcal{B}}(b_i, c_j) \leq 2^p r$. Since $d^{\mathcal{B}}(b_i, b_j) > 2^{p+1} r$, for every $j \in L$ there can be at most one $i \in I$ such that $d^{\mathcal{B}}(b_i, c_j) \leq 2^p r$. This proves the claim.

By the marriage theorem, there exists a one-to-one mapping f of K into L such that for all $i \in K$ we have $iRf(i)$. In other words, there exist $b_1, \dots, b_k$ such that for $1 \leq i \leq k$ we have $\mathcal{B} \models \psi_i(b_i)$ and $d^{\mathcal{B}}(b_i, c_{f(i)}) \leq 2^p r$. Since $d^{\mathcal{B}}(c_{f(i)}, c_{f(j)}) > 2^{p+1} r$, the latter implies $d^{\mathcal{B}}(b_i, b_j) > 2r$. Thus

$$\mathcal{B} \models \exists x_1 \dots \exists x_k \Big(\bigwedge_{1 \leq i < j \leq k} d(x_i, x_j) > 2r \wedge \bigwedge_{i=1}^{k} \psi_i(x_i) \Big).$$

□

Lemma 5. *There is a function $f(k)$, such that the following holds for all $k \geq 1$: Let $\mathcal{A}$, $\mathcal{B}$ be structures such that every existential local sentence of rank $(k(k+1), f(k))$ that holds in $\mathcal{A}$ also holds in $\mathcal{B}$. Then every existential sentence of weight at most k that holds in $\mathcal{A}$ also holds in $\mathcal{B}$.*

Proof: Since every existential sentence is equivalent to a disjunction of conjunctive queries with negation of the same weight, it suffices to prove that every conjunctive query with negation of weight k that holds in $\mathcal{A}$ also holds in $\mathcal{B}$. Let

$$\varphi := \exists x_1 \dots \exists x_k \psi(x_1, \dots, x_k)$$

with

$$\psi(x_1, \dots, x_k) := \Big(\bigwedge_{i=1}^{p} \alpha_i \wedge \bigwedge_{i=1}^{q} \beta_i \Big),$$

where all the α_i are atoms and the β_i are negated atoms. Suppose that $\mathcal{A} \models \varphi$. We shall prove that $\mathcal{B} \models \varphi$.

We define the *positive graph of* φ to be the graph $\mathcal{G}$ with universe $G := \text{var}(\varphi) = \{x_1, \dots, x_k\}$ and

$$E^{\mathcal{G}} := \{xy \mid \exists i, 1 \leq i \leq p : \ x, y \in \text{var}(\alpha_i)\}.$$

Let $\mathcal{H}_1, \dots, \mathcal{H}_r$ be the connected components of $\mathcal{G}$. Without loss of generality, we may assume that for $1 \leq i \leq r$ we have $x_i \in H_i$. Then we know that $H_i \subseteq N_k^{\mathcal{G}}(x_i)$. If $r = 1$, then this means that $\text{var}(\varphi) \subseteq N_k^{\mathcal{G}}(x_1)$, and φ is equivalent to the k-local sentence

$$\exists x_1 \, \exists x_2 \in N_k(x_1) \dots \exists x_k \in N_k(x_1)\psi$$

of rank (k, k). If we choose f such that $f(k) \geq k$, then $\mathcal{A} \models \varphi$ implies $\mathcal{B} \models \varphi$. In the following, we assume that $r \geq 2$.

Let $c_0 := 0$ and $c_{i+1} := 2^{k^2}(c_i + k + 1)$ for $i \geq 0$. We let $R := \{\{i,j\} \mid 1 \leq i < j \leq r\}$, $h := |R| = \binom{r}{2}$ and

$$f(k) = 2^{k^2}(c_h + k + 1). \tag{2}$$

For $\bar{a} = a_1 \dots a_r \in A^r$, the *distance pattern* of $\bar{a}$ is the mapping $\Delta_{\bar{a}} : R \to \{0, \dots, h\}$ defined by

$$\Delta_{\bar{a}}(\{i,j\}) := \begin{cases} 0 & \text{if } d^{\mathcal{A}}(a_i, a_j) = 0 \\ t & \text{if } c_t < d^{\mathcal{A}}(a_i, a_j) \leq c_{t+1} \text{ for some } t \text{ such that } 0 \leq t < h \\ h & \text{if } d^{\mathcal{A}}(a_i, a_j) > c_h \end{cases}$$

By the pigeonhole principle, for every distance pattern Δ there is an integer $\text{gap}(\Delta)$ such that $0 \leq \text{gap}(\Delta) \leq h$ and $\Delta(\{i,j\}) \neq \text{gap}(\Delta)$ for all $\{i,j\} \in R$.

Let $\bar{a} = a_1 \dots a_k \in A^k$ such that $\mathcal{A} \models \psi(\bar{a})$. Let $\Delta := \Delta_{a_1 \dots a_r}$, and $g := \text{gap}(\Delta)$. Then for all $\{i,j\} \in R$ we either have $d(a_i, a_j) \leq c_g$ or $d(a_i, a_j) > 2^{k^2}(c_g + k + 1)$. This implies that the relation on $\{a_1, \dots, a_r\}$ defined by $d^{\mathcal{A}}(a_i, a_j) \leq c_g$ is an equivalence relation. Without loss of generality, we may assume that $a_1, \dots, a_s$ form a system of representatives of the equivalence classes.

We let $l := c_g + k$. For $1 \leq i \leq s$, we let $I_i := \{j \mid 1 \leq j \leq k, d^{\mathcal{A}}(a_i, a_j) \leq l\}$. Then $(I_i)_{1 \leq i \leq s}$ is a partition of $\{1, \dots, k\}$. To see this, first recall that for $1 \leq j \leq r$ there is an $i, 1 \leq i \leq s$ such that $d^{\mathcal{A}}(a_i, a_j) \leq c_g$. For t with $r + 1 \leq t \leq k$ there exist a $j, 1 \leq j \leq r$ such that $x_t \in H_j$, the connected component of x_j in the positive graph of φ. Since $\mathcal{A} \models \psi(\bar{a})$, this implies that $d^{\mathcal{A}}(a_j, a_t) \leq k$. Thus there exists an $i, 1 \leq i \leq s$ such that $d^{\mathcal{A}}(a_i, a_t) \leq c_g + k$.

For $1 \leq i \leq s$, we let

$$\psi_i(x_i) := \exists \bar{x}^i \in N_l(x_i) \bigwedge_{\text{var}(\alpha_i) \subseteq I_i} \alpha_i \wedge \bigwedge_{\text{var}(\beta_i) \subseteq I_i} \beta_i,$$

where $\bar{x}^i$ consists of all variables x_j with $j \in I_i \setminus \{i\}$. Then for $1 \leq i \leq s$ we have $\mathcal{A} \models \psi_i(a_i)$, because $\mathcal{A} \models \psi(\bar{a})$. Thus

$$\mathcal{A} \models \exists x_1 \ldots \exists x_s \Big(\bigwedge_{1 \leq i < j \leq s} d(x_i, x_j) > 2^{k^2}(l+1) \wedge \bigwedge_{1 \leq i \leq s} \psi_i(x_i) \Big).$$

Since $f(k) = 2^{k^2}(c_h + k + 1) \geq 2^{k^2}(l+1)$, by Lemma 4, this implies

$$\mathcal{B} \models \exists x_1 \ldots \exists x_s \Big(\bigwedge_{1 \leq i < j \leq k} d(x_i, x_j) > 2(l+1) \wedge \bigwedge_{1 \leq i \leq s} \psi_i(x_i) \Big).$$

Thus there exist $b_1, \ldots, b_s \in B$ such that for $1 \leq i < j \leq s$ we have $d^{\mathcal{B}}(b_i, b_j) > 2(l+1)$ and for $1 \leq i \leq s$ we have $\mathcal{B} \models \psi_i(b_i)$. Since $I_1, \ldots, I_s$ is a partition of $\{1, \ldots, k\}$, there are $b_{s+1}, \ldots, b_k \in B$ such that:

(i) $d^{\mathcal{B}}(b_i, b_j) \leq l$ for all $j \in I_i$.
(ii) $\mathcal{B} \models \alpha_j(\bar{b})$ for all $j, 1 \leq j \leq p$ such that $\text{var}(\alpha_j) \subseteq I_j$.
(iii) $\mathcal{B} \models \beta_j(\bar{b})$ for all $j, 1 \leq j \leq q$ such that $\text{var}(\beta_j) \subseteq I_j$.

We claim that $\mathcal{B} \models \psi(\bar{b})$. Since for each connected component H_j of the positive graph of φ there is an $i, 1 \leq i \leq s$ such that $t \in I_i$ whenever $x_t \in H_j$, (ii) implies that $\mathcal{B} \models \alpha_j(\bar{b})$ for $1 \leq j \leq p$. It remains to prove that $\mathcal{B} \models \beta_j(\bar{b})$ for $1 \leq j \leq q$. If $\text{var}(\beta_j) \subseteq I_i$ for some i, then $\mathcal{B} \models \beta_j(\bar{b})$ by (iii). Otherwise, β_j has variables x_u, x_v such that there exist $i \neq i'$ with $x_u \in I_i, x_v \in I_{i'}$. Then by (i), $d^{\mathcal{B}}(b_i, b_u) \leq l$ and $d^{\mathcal{B}}(b_{i'}, b_v) \leq l$. Since $d^{\mathcal{B}}(b_i, b_{i'}) > 2l + 1$, this implies $d^{\mathcal{B}}(b_u, b_v) > 1$. Since β_j is a negated atom, this implies $\mathcal{B} \models \beta_j(\bar{b})$.

Thus $\mathcal{B} \models \varphi$. □

Proof (of Theorem 2): Let φ be an existential sentence of weight k and $\mathcal{K} := \{\mathcal{A} | \mathcal{A} \models \varphi\}$ the class of all finite structures satisfying φ. Let Ψ be the set of all existential local sentences of rank at most $(k(k+1), f(k))$, where f is the function from Lemma 5. Let

$$\varphi' := \bigvee_{\mathcal{A} \in \mathcal{K}} \bigwedge_{\substack{\psi \in \Psi \\ \mathcal{A} \models \psi}} \psi.$$

We claim that φ is equivalent to φ'. The forward implication is trivial, and the backward implication follows from Lemma 5. Since up to logical equivalence, the set Ψ is finite and therefore φ' contains at most $2^{|\Psi|}$ non-equivalent disjuncts, this proves the theorem. □

Our proof of the existential version of Gaifman's theorem does not give us good bounds on the size and rank of the local formulas to which we translate a given existential formula. Therefore, for the algorithmic applications, it is preferable to work with the following weaker version of Theorem 2, which gives us better bounds.

An *asymmetric local sentence* is a sentence φ of the form

$$\exists x_1 \dots \exists x_k \Big(\bigwedge_{1 \le i < j \le k} d(x_i, x_j) > 2r \wedge \bigwedge_{1 \le i \le k} \psi_i(x_i) \Big),$$

where $r, k \ge 1$ and $\psi_1(x), \dots, \psi_k(x)$ are r-local. φ is an *existential asymmetric local sentence*, if in addition $\psi_1(x), \dots, \psi_k(x)$ are existential.

An *r-local conjunctive query with negation*, for some $r \ge 1$, is a formula $\psi(x)$ of the form $\exists y_1 \in N_r(x) \dots \exists y_n \in N_r(x) \bigwedge_{i=1}^{m} \lambda_i$, where each λ_i is a literal.

Theorem 6. *Every existential first-order sentence φ is equivalent to a disjunction φ' of existential asymmetric local sentences.*

More precisely, if k is the weight of φ and l its size, then φ' is a disjunction of $2^{O(l+k^4)}$ asymmetric local sentences of the form

$$\exists x_1 \dots \exists x_k \Big(\bigwedge_{1 \le i < j \le k} d(x_i, x_j) > 2r \wedge \bigwedge_{1 \le i \le k} \psi_i(x_i) \Big),$$

where $\psi_1, \dots, \psi_k$ are r-local conjunctive queries with negation. The rank of each of these local sentences is at most $(k, 2^{k^2+1})$, and their size is in $O(l)$.

Furthermore, there is a polynomial p and an algorithm translating φ to φ' in time $O(2^{p(l)})$.

Proof: We first assume that φ is a conjunctive query with negation, say,

$$\varphi := \exists x_1 \dots \exists x_k \psi(x_1, \dots, x_k)$$

with

$$\psi(x_1, \dots, x_k) := \Big(\bigwedge_{i=1}^{p} \alpha_i \wedge \bigwedge_{i=1}^{q} \beta_i \Big),$$

where all the α_i are atoms and the β_i are negated atoms. Without loss of generality, we may assume that $k \ge 2$, because for $k = 1$ there is nothing to prove. We define the *positive graph of* φ to be the graph $\mathcal{G}$ with $G := \mathrm{var}(\varphi) = \{x_1, \dots, x_k\}$ and

$$E^{\mathcal{G}} := \{xy \mid \exists i, 1 \le i \le p : \; x, y \in \mathrm{var}(\alpha_i)\}.$$

Let $\mathcal{H}_1, \dots, \mathcal{H}_r$ be the connected components of $\mathcal{G}$. Without loss of generality, we may assume that $r \ge 2$, and that for $1 \le i \le r$ we have $x_i \in H_i$. Then we know that $H_i \subseteq N_k^{\mathcal{G}}(x_i)$.

Let $c_0 := 0$ and $c_{i+1} := 2(c_i + k + 1)$ for $i \ge 0$. Let $R := \big\{\{i, j\} \bigm| 1 \le i < j \le r\big\}$ and $h := |R| = \binom{r}{2}$. It is not difficult to prove that $c_h + k + 1 \le 2^{k^2+1}$.

Let $\mathcal{A}$ be a structure and $\bar{a} = a_1 \dots a_r \in A^r$. The *distance pattern* of $\bar{a}$ is the mapping

$$\Delta_{\bar{a}} : R \to \{0, \dots, h\}$$

defined by

$$\Delta_{\bar{a}}(\{i,j\}) := \begin{cases} 0 & \text{if } d^{\mathcal{A}}(a_i,a_j) = 0 \\ t & \text{if } c_t < d^{\mathcal{A}}(a_i,a_j) \leq c_{t+1} \text{ for some } t \text{ such that } 0 \leq t < h \\ h & \text{if } d^{\mathcal{A}}(a_i,a_j) > c_h \end{cases}$$

By the pigeonhole principle, for every distance pattern Δ there is a number $\text{gap}(\Delta), 0 \leq \text{gap}(\Delta) \leq h$ such that $\Delta(\{i,j\}) \neq \text{gap}(\Delta)$ for all $\{i,j\} \in R$.

Let $\bar{a} \in A^r$, $\Delta := \Delta_{\bar{a}}$, and $g := \text{gap}(\Delta)$. Then for all $\{i,j\} \in R$ we either have $d(a_i,a_j) \leq c_g$ or $d(a_i,a_j) > 2(c_g + k + 1)$. This implies that the relation on $\{a_1,\dots,a_r\}$ defined by $d^{\mathcal{A}}(a_i,a_j) \leq c_g$ is an equivalence relation. Without loss of generality, we may assume that $a_1,\dots,a_s$ form a system of representatives of the equivalence classes.

Now suppose that we extend $a_1 \dots a_r$ to a k-tuple $\bar{a} = a_1 \dots a_k \in A^k$ such that $\mathcal{A} \models \psi(\bar{a})$. We let $l := c_g + k$. For $1 \leq i \leq s$, we let $I_i := \{j \mid d^{\mathcal{A}}(a_i,a_j) \leq l\}$. Then $(I_i)_{1 \leq i \leq s}$ is a partition of $\{1,\dots,k\}$. For $1 \leq i \leq s$, we let

$$\psi_i(x_i) := \exists \bar{x}^i \in N_l(x_i) \bigwedge_{\text{var}(\alpha_i) \subseteq I_i} \alpha_i \wedge \bigwedge_{\text{var}(\beta_i) \subseteq I_i} \beta_i,$$

where $\bar{x}^i$ consists of all variables x_j with $j \in I_i \setminus \{i\}$, and

$$\psi_\Delta(x_1,\dots,x_s) := \bigwedge_{1 \leq i < j \leq s} d(x_i,x_j) > 2(l+1) \wedge \bigwedge_{1 \leq i \leq s} \psi_i(x_i).$$

Then $\mathcal{A} \models \psi_\Delta(a_1,\dots,a_s)$. Furthermore, for every tuple $a'_1 \dots a'_s \in A^s$ with $\mathcal{A} \models \psi_\Delta(a'_1,\dots,a'_s)$ there exists an extension $\bar{a}' := a'_1 \dots a'_k$ such that $\mathcal{A} \models \psi(\bar{a}')$. To see this, observe that every positive literal α_j occurs in an I_i and thus in ψ_Δ, and every negative literal β_j either occurs in an I_i or has variables with indices in two distinct $I_i, I_{i'}$ and is thus automatically satisfied, because the variables are forced to be far apart.

The formula $\psi_\Delta(x_1,\dots,x_k)$ only depends on the distance pattern Δ and not on the tuple $\bar{a}$ realizing it. So for every distance pattern Δ we obtain a formula $\psi_\Delta(\bar{x}^\Delta)$, whose free variables $\bar{x}^\Delta$ are among $x_1,\dots,x_r$, with the following properties:

- $\exists \bar{x}^\Delta \psi_\Delta$ is an existential asymmetric local sentence of rank at most $(k, 2^{k^2+1})$.
- For every tuple $\bar{a} \in A^k$ with $\mathcal{A} \models \psi(\bar{a})$ and $\Delta_{\bar{a}} = \Delta$ we have $\mathcal{A} \models \psi_\Delta(\bar{a}^\Delta)$, where $\bar{a}^\Delta$ consists of the same entries of $\bar{a}$ as $\bar{x}^\Delta$ of $\bar{x}$.
- Every tuple $\bar{a}^\Delta$ with $\mathcal{A} \models \psi_\Delta(\bar{a}^\Delta)$ can be extended to a tuple $\bar{a} = a_1 \dots a_k$ such that $\mathcal{A} \models \psi(\bar{a})$.

The last two items imply that φ is equivalent to the formula

$$\varphi' := \bigvee_{\Delta \text{ distance pattern}} \exists \bar{x}^\Delta \psi_\Delta.$$

It is not hard to see that the number of distance pattern is in $2^{O(k^4)}$, thus φ is a disjunction of $2^{O(k^4)}$ existential asymmetric local sentences of rank at most $(k, 2^{k^2+1})$ and size in $O(l)$ (where l denotes the length of φ).

If φ is an arbitrary existential sentence, we first transform it to a disjunction of at most 2^l conjunctive queries with negation of the same weight as φ.

Finally, we observe that the translation from φ to the disjunction of asymmetric local formulas is effective within the desired time bound: Given φ, we first translate it to a disjunction of conjunctive queries with negation. This is possible in time $2^{O(l)}$. Then we treat each of the conjunctive queries with negation separately. We compute the positive graph and all possible patterns. For each of pattern Δ, we compute the gap and then the formula φ_Δ. Since $k \leq l$, this is clearly possible in time $2^{p(l)}$ for a suitable polynomial p. □

4 An Algorithmic Application

The appropriate structural notion for the algorithmic applications of locality is *bounded local tree-width.* We assume that the reader is familiar with the definition of *tree-width* of graphs (see e.g. [4]). The tree-width of a structure $\mathcal{A}$, denoted by $\text{tw}(\mathcal{A})$, is the tree-width of its Gaifman graph. The *local tree-width* of a structure $\mathcal{A}$ is the function $\text{ltw}_\mathcal{A} : \mathbb{N} \to \mathbb{N}$ defined by

$$\text{ltw}_\mathcal{A}(r) := \max\Big\{\text{tw}(\langle N_r^\mathcal{A}(a)\rangle) \;\Big|\; a \in A\Big\}.$$

A class C of structures has *bounded local tree-width* if there is a function $\lambda : \mathbb{N} \to \mathbb{N}$ such that $\text{ltw}_\mathcal{A}(r) \leq \lambda(r)$ for all $\mathcal{A} \in \text{C}, r \in \mathbb{N}$. Many well-known classes of structures have bounded local tree-width, among them the class of planar graphs and all classes of structures of bounded degree.

Theorem 7 (Frick and Grohe [11]). *Let* C *be a class of structures of bounded local tree-width. Then there is a function f and, for every $\epsilon > 0$, an algorithm deciding in time $O(f(||\varphi||)|A|^{1+\epsilon})$ whether a given structure $\mathcal{A} \in \text{C}$ satisfies a given first-order sentence φ.*

If the class C is *locally tree-decomposable*, which is a slightly stronger requirement than having bounded local tree-width, then there is a function f and an algorithm deciding whether a given structure $\mathcal{A} \in \text{C}$ satisfies a given first-order sentence φ in time $O(f(||\varphi||)|A|)$.

These algorithms proceed as follows: Given a structure $\mathcal{A}$ and a sentence φ, they first translate φ to a Boolean combination of local sentences. Then they evaluate each local sentence and combine the results. To evaluate a local sentence, say,

$$\exists x_1 \ldots \exists x_k \Big(\bigwedge_{1 \leq i < j \leq k} d(x_i, x_j) > 2r \wedge \bigwedge_{1 \leq i \leq k} \psi(x_i) \Big),$$

they first compute the set $\psi(\mathcal{A})$ of all $a \in A$ such that $\mathcal{A} \models \psi(a)$. Since ψ is local and the class C has bounded local tree-width or even is locally tree-decomposable, this is possible quite efficiently. (In the special case of structures of bounded degree, this is easy to see, because ψ only has to be evaluated in substructures of $\mathcal{A}$ of bounded size.) Finally, the algorithms test whether there are $a_1, \dots, a_k \in \psi(\mathcal{A})$ of pairwise distance greater than $2r$. This is possible in linear time by the following lemma:

Lemma 8 (Frick and Grohe [11]). *Let* C *be a class of structures of bounded local tree-width. Then there is a function* g *and an algorithm that, given a structure* $\mathcal{A}$*, a subset* $P \subseteq A$*, and integers* k, r*, decides in time* $O(g(k,r)|A|)$ *whether there are* $a_1, \dots, a_k \in P$ *of pairwise distance greater than* $2r$*.*

The drawback of these algorithms is that we cannot even give an elementary upper bound for the function f in Theorem 7. The main reason for the enormous runtime of the algorithms in terms of the formula size is that to evaluate the local formulas, they translate them to tree-automata, and in the worst case the size of these automata grows exponentially with each quantifier alternation. Therefore, it is a natural idea to bound the number of quantifier alternations in order to obtain smaller automata. But this would require that the translation of first-order sentences into local sentences preserves the quantifier structure. Unfortunately, the known proofs of Gaifman's theorem do not preserve the quantifier structure of the input formula.

These considerations motivated the present paper. Indeed, Theorem 2 shows that existential first-order sentences can be translated into Boolean combinations of existential local formulas. The price we pay for this is that these Boolean combinations of existential local formulas can get enormously large. Therefore, we use Theorem 6, because this theorem at least gives us an exponential upper bound on the size of the resulting formula. To evaluate an asymmetric local sentence, say

$$\exists x_1 \dots \exists x_k \Big(\bigwedge_{1 \le i < j \le k} d(x_i, x_j) > 2r \wedge \bigwedge_{1 \le i \le k} \psi_i(x_i) \Big),$$

where the ψ_i are conjunctive queries with negation, we first compute the sets $\psi_1(\mathcal{A}), \dots, \psi_k(\mathcal{A})$. This can be done as in the algorithms described above, but is actually faster since the ψ_i are conjunctive queries with negation. We use Lemma 9. Then we have to decide whether there are $a_1 \in \psi_1(\mathcal{A}), \dots, a_k \in \psi_k(\mathcal{A})$ of pairwise distance greater than $2r$. Lemma 10 is an analogue of Lemma 8 for this more general situation.

Lemma 9. *There is a polynomial* p *and an algorithm that solves the following problem in time* $O(2^{p(||\varphi|| + \mathrm{tw}(\mathcal{A}))} \cdot |A|)$*.*

Input: Structure $\mathcal{A}$, conjunctive query with negation φ. *Problem:* Decide if $\mathcal{A} \models \varphi$.

Details of the proof of Lemma 9 and the following Lemma 10 can be found in the full version of this paper [13] and in the second author's Diploma thesis [21].

Lemma 10. *There is a polynomial p and an algorithm that solves the following problem in time $O(2^{p(\mathrm{ltw}_{\mathcal{A}}((k+1)r)+r+k)} \cdot |A|)$:*

Input: Structure $\mathcal{A}$, sets $P_1, \dots, P_k \subseteq A$, integer $r \geq 1$.
Problem: Decide if there are $a_1 \in P_1, \dots, a_k \in P_k$ of pairwise distance greater than k.

If we combine these two lemmas together with Theorem 6 and plug them in the algorithms described in [11], we obtain the following theorem.

Theorem 11. *Let C be a class of structures whose local tree-width is bounded by a function $\lambda : \mathbb{N} \to \mathbb{N}$ (i.e., for all $\mathcal{A} \in \mathrm{C}$ and $r \geq 0$ we have $\mathrm{ltw}_{\mathcal{A}}(r) \leq \lambda(r)$). Then there are polynomials p, q such that for every $\epsilon > 0$ there is an algorithm that, given a structure $\mathcal{A}$ and an existential first-order sentence φ, decides if $\mathcal{A} \models \varphi$ in time*

$$O\Big(2^{2^{p(\lambda(q(||\varphi||+(1/\epsilon)))+||\varphi||+(1/\epsilon))}} \cdot |A|^{1+\epsilon}\Big),$$

i.e., in time doubly exponential in $||\varphi||, (1/\epsilon), \lambda(q(||\varphi|| + (1/\epsilon)))$ and near linear in $|A|$.

For many interesting classes of structures of bounded local tree-width, such as planar graphs, the local tree-width is bounded by a linear function λ.

5 Conclusions

Our main result is an existential version of Gaifman's locality theorem. It would be interesting to see if there are similar structure preserving locality theorems for other classes of first-order formulas, such as formulas monotone in some relation symbol or Σ_2-formulas. The combinatorial techniques we use in our proof seem to be specific to existential formulas; we do not see how to apply them to other classes of formulas. With the algorithmic applications in mind, it would be nice to get better bounds on the size and rank of the Boolean combinations of local sentences the locality theorems give us, both in the existential and in the general case.

In the second part of the paper, we show how a variant of our locality theorem can be applied to evaluate existential first-order sentences in structures of bounded local tree-width by improving an algorithm of [11] for the special case of existential sentences. We are able to prove a doubly exponential upper bound for the dependence of the runtime of the algorithm on the size of the input sentence. Though not really convincing, it is much better than what we have for arbitrary first-order sentences — recall that no elementary bound is known there — and it shows that quantifier alternation really is an important factor contributing to the large complexity. It might be possible to further improve the algorithm to obtain a (singly) exponential dependence on the size of the input sentence. But even then we would probably not get a practical algorithm, because the hidden constant would still be too large.

The best chance to get practical algorithms might be to concentrate on particular classes of graphs, such as graphs of bounded degree or planar graphs, and use their specific properties. For example, the local tree-width of planar graphs is bounded by the function $r \mapsto 3 \cdot r$, and it is quite easy to compute tree-decompositions of neighborhoods in planar graphs [2,9]. This already eliminates certain very expensive parts of our algorithms. The algorithms can also be improved by using weaker forms of locality. We have taken a step in this direction by admitting asymmetric local sentences. Further improvement might be possible by admitting "weak" asymmetric sentences stating that there are elements of pairwise distance greater than s satisfying some r-local condition, where s is no longer required to be $2r$. For the algorithms, it does not really matter if the local neighborhoods are disjoint, and relaxing this condition may give us smaller formulas.

References

1. A.V. Aho, J.E. Hopcroft, and J.D. Ullman. *The Design and Analysis of Computer Algorithms.* Addison-Wesley, 1974.
2. H.L. Bodlaender. NC-algorithms for graphs with small treewidth. In J. van Leeuwen, editor, *Proceedings of the 14th International Workshop on Graph theoretic Concepts in Computer Science WG'88*, volume 344 of *Lecture Notes in Computer Science*, pages 1–10. Springer-Verlag, 1988.
3. H.L. Bodlaender. A linear-time algorithm for finding tree-decompositions of small treewidth. *SIAM Journal on Computing*, 25:1305–1317, 1996.
4. R. Diestel. *Graph Theory.* Springer-Verlag, second edition, 2000.
5. G. Dong, L. Libkin, and L. Wong. Local properties of query languages. In *Proceedings of the 5th International Conference on Database Theory*, volume 1186 of *Lecture Notes in Computer Science*, pages 140–154. Springer-Verlag, 1997.
6. R.G. Downey and M.R. Fellows. *Parameterized Complexity.* Springer-Verlag, 1999.
7. R.G. Downey, M.R. Fellows, and U. Taylor. The parameterized complexity of relational database queries and an improved characterization of $W[1]$. In Bridges, Calude, Gibbons, Reeves, and Witten, editors, *Combinatorics, Complexity, and Logic – Proceedings of DMTCS '96*, pages 194–213. Springer-Verlag, 1996.
8. H.-D. Ebbinghaus and J. Flum. *Finite Model Theory.* Springer-Verlag, second edition, 1995.
9. D. Eppstein. Diameter and treewidth in minor-closed graph families. *Algorithmica*, 27:275–291, 2000.
10. J. Flum, M. Frick, and M. Grohe. Query evaluation via tree-decompositions. In Jan van den Bussche and Victor Vianu, editors, *Proceedings of the 8th International Conference on Database Theory*, volume 1973 of *Lecture Notes in Computer Science*, pages 22–38. Springer Verlag, 2001.
11. M. Frick and M. Grohe. Deciding first-order properties of locally tree-decomposable structures. Currently available at `http://www.math.uic.edu/~grohe`. A preliminary version of the paper appeared in *Proceedings of the 26th International Colloquium on Automata, Languages and Programming*, LNCS 1644, Springer-Verlag, 1999.

12. H. Gaifman. On local and non-local properties. In *Proceedings of the Herbrand Symposium, Logic Colloquium '81*. North Holland, 1982.
13. M. Grohe and S. Wöhrle. An existential locality theorem. Technical Report AIB-2001-07, RWTH Aachen, 2001.
14. L. Hella, L. Libkin, J. Nurmonen, and L. Wong. Logics with aggregate operators. In *Proceedings of the 14th IEEE Symposium on Logic in Computer Science*, 1999.
15. L. Hella, L. Libkin, and Y. Nurmonen. Notions of locality and their logical characterizations over finite models. *Journal of Symbolic Logic*, 64:1751–1773, 1999.
16. L. Libkin. Logics with counting and local properties. *ACM Transaction on Computational Logic*, 1:33–59, 2000.
17. N. Robertson and P.D. Seymour. Graph minors III. Planar tree-width. *Journal of Combinatorial Theory, Series B*, 36:49–64, 1984.
18. L.J. Stockmeyer. *The Complexity of Decision Problems in Automata Theory.* PhD thesis, Department of Electrical Engineering, MIT, 1974.
19. P. van Emde Boas. Machine models and simulations. In J. van Leeuwen, editor, *Handbook of Theoretical Computer Science*, volume 1, pages 1–66. Elsevier Science Publishers, 1990.
20. M.Y. Vardi. The complexity of relational query languages. In *Proceedings of the 14th ACM Symposium on Theory of Computing*, pages 137–146, 1982.
21. S. Wöhrle. *Lokalität in der Logik und ihre algorithmischen Anwendungen.* Diploma thesis, Faculty of Mathematics, Albert–Ludwigs–University Freiburg, 2000.

Actual Arithmetic and Feasibility

Jean-Yves Marion

Loria, B.P. 239, 54506 Vandœuvre-lès-Nancy Cedex, France
Jean-Yves.Marion@loria.fr
http://www.loria.fr/~marionjy

Abstract. This paper presents a methodology for reasoning about the computational complexity of functional programs. We introduce a first order arithmetic $\mathbf{AT}^0$ which is a syntactic restriction of Peano arithmetic. We establish that the set of functions which are provably total in $\mathbf{AT}^0$, is exactly the set of polynomial time functions.The cut-elimination process is polynomial time computable.
Compared to others feasible arithmetics, $\mathbf{AT}^0$ is conceptually simpler. The main feature of $\mathbf{AT}^0$ concerns the treatment of the quantification. The range of quantifiers is restricted to the set of *actual terms* which is the set of constructor terms with variables. The inductive formulas are restricted to conjunctions of atomic formulas.

1 Introduction

1.1 Motivation

We investigate feasible logics, that is systems in which the class of provably total functions is exactly the set of polynomial time functions. There are three main motivations:

1. Proof development environments, which are based on the proofs-as-programs principle, such as Alf, Coq and Nuprl, synthetise correct programs. The efficiency of a program is not guaranteed, though it is a crucial property of a running implementation. Benzinger [4] has developed a prototype to determine the runtime of Nuprl-extracted programs. Here, we propose instead a proof theoretical method to analyse the runtime complexity of extracted programs.
2. Computational Complexity Theory (CCT) delineates classes of functions, which are computable within bounded resources. CCT characterisations are extensional, that is all functions of a given class are captured, but most of the efficients algorithms are missing. Runtime analysis of programs necessitates to reason on programs, or on proofs in the "proofs-as-programs" context. For this, we need to develop logics to study algorithmic contents of proofs.
3. It seems worthwhile to develop feasible logics in order to reason about "polynomial-time mathematics", analogous to constructive mathematics.

L. Fribourg (Ed.): CSL 2001, LNCS 2142, pp. 115–129, 2001.

1.2 This Arithmetic

We propose a sub-system of first order Peano arithmetic **AT** in which, intuitively, we quantify over terms that actually exist. The range of quantifiers is limited to terms, that we call *actual terms.* The set of actual terms is the free word algebra with variables. This restriction implies *the actual elimination quantifier principle* which stipulates that from $\forall x.A$, we derive $A[x \leftarrow p]$ *where* p is an actual term,*i.e.* $p \in \mathcal{T}(\mathcal{W}, \mathcal{X})$. The system **AT** is confluent and strongly normalizable.

1.3 First Order Logic with Actual Quantification

We present the natural deduction calculus **AT**. We assume some familiarity with natural deduction, see Prawitz [24] or Girard, Lafont and Taylor [9]. Terms are divided into different categories which are listed in Figure 1. *Actual terms* are built up from constructors of $\mathcal{W}$, and variables of $\mathcal{X}$, and forms the set $\mathcal{T}(\mathcal{W}, \mathcal{X})$. The logical rules of **AT** are written in Figure 2.

The difference with the $\{\to, \wedge, \forall\}$-fragment of the minimal logic is *the actual elimination quantifier principle* which is obtained by the $\forall E^{\mathbf{S}}$-rule.

1.4 Arithmetic over Words with Actual Quantification

We extend **AT** to an arithmetic **AT**(**W**) in order to reason about the free word algebra $\mathcal{T}(\mathcal{W})$. The set of words is denoted by a unary predicate **W** together with the rules displayed in Figure 3. Throughout, we shall make no distinction between the set of words $\{0, 1\}^*$, and the set $\mathcal{T}(\mathcal{W})$ of constructor terms.

Following Martin-Löf [22] and Leivant [18,20], the introduction rules indicate the construction of words, and the elimination rules specify the computational behaviour associated with them. Both elimination rules, that is the induction rule and the selection rule, are necessary, because of the actual elimination quantifier principle. Indeed, the induction rule schema Ind(**W**) corresponds to the usual induction. However, the range of the universal quantifier is restricted to actual terms. So, the last quantifier of the induction filters the instantiation through the $\forall E^{\mathbf{S}}$-rule. Roughly speaking, an induction is guarded by a universal quantifier, like a proof-net box in linear logic.

(Constructors)	$\mathcal{W} \ni \mathbf{c}$	$::= \epsilon \mid \mathbf{s}_0 \mid \mathbf{s}_1$	
(Function symbols)	$\mathcal{F} \ni \mathbf{f}$	$::= \mathbf{f} \mid \mathbf{g} \mid \mathbf{h} \mid \ldots$	with fixed arities
(Variables)	$\mathcal{X} \ni x$	$::= x \mid y \mid z \mid \ldots$	
(Words)	$\mathcal{T}(\mathcal{W}) \ni \mathbf{w}$	$::= \epsilon \mid \mathbf{s}_0(\mathbf{w}) \mid \mathbf{s}_1(\mathbf{w})$	
(Terms)	$\mathcal{T}(\mathcal{W}, \mathcal{F}, \mathcal{X}) \ni t$	$::= \epsilon \mid \mathbf{s}_0(t) \mid \mathbf{s}_1(t) \mid \mathbf{f}(t_1, \ldots, t_n) \mid x$	
(Actual terms)	$\mathcal{T}(\mathcal{W}, \mathcal{X}) \ni p$	$::= \epsilon \mid \mathbf{s}_0(p) \mid \mathbf{s}_1(p) \mid x$	

Fig. 1. Categories of terms

Premiss : A (Predicate A)

Introduction rules / *Elimination rules*

$$\frac{\begin{matrix}\{A\}\\ \vdots\\ B\end{matrix}}{A \to B} \to I \qquad \frac{A \to B \quad A}{B} \to E$$

$$\frac{A_1 \quad A_2}{A_1 \wedge A_2} \wedge I \qquad \frac{A_1 \wedge A_2}{A_j} \wedge E$$

$$\frac{A}{\forall x.A} \forall I \qquad \frac{\forall x.A}{A[x \leftarrow p]} \forall E^{\mathbf{S}}, \text{ where } p \in \mathcal{T}(\mathcal{W}, \mathcal{X})$$

Restrictions on the rules

- In $\forall I$-rule, x is not free in any premiss.
- In $\forall E^{\mathbf{S}}$-rule, p is an actual term, *i.e.* $p \in \mathcal{T}(\mathcal{W}, \mathcal{X})$.

Fig. 2. Logical rules of **AT**

On the other hand, the selection rule expresses that a word t is either the empty word ϵ, or $\mathbf{s_i}(y)$ for some term y. We shall employ the selection rule to perform definitions by cases over words. Unlike the induction rule, the term t in the conclusion of the selection rule can be any term. It is worth noticing that the application of the selection rule is restricted. There is no application of $\forall E^{\mathbf{S}}$-rule in the derivations $\pi_{\mathbf{s_0}}$ and $\pi_{\mathbf{s_1}}$. Thus, we prohibit nested applications of induction rule, inside the selection rule. Otherwise it would be possible to unguard an induction.

2 Reasoning over Programs

2.1 First Order Functional Programs

An equational program $\mathtt{f}$ is a set of (oriented) equations $\mathcal{E}$. Each equation is of the form $\mathtt{f}(p_1, \cdots, p_n) \to t$ where each p_i is an actual term, and corresponds to a pattern. The term t is in $\mathcal{T}(\mathcal{W}, \mathcal{F}, \mathcal{X})$ and each variable of t also appears in $\mathtt{f}(p_1, \cdots, p_n)$.

The semantics is based on term rewriting. One might consult [7] about general references on rewrite systems. A set of equations $\mathcal{E}$ induces a rewriting rule $u \to v$ if the term v is obtained from u by applying an equation of $\mathcal{E}$. We write $t \xrightarrow{!} s$ to mean that s is a normal form of t. A program is confluent if the rewriting rule $\to$ is confluent, *i.e.* has the Church-Rosser property.

Introduction rules

$$\frac{}{\mathbf{W}(\epsilon)}\,\epsilon I \qquad \frac{\mathbf{W}(t)}{\mathbf{W}(\mathbf{s}_0(t))}\,\mathbf{s}_0 I \qquad \frac{\mathbf{W}(t)}{\mathbf{W}(\mathbf{s}_1(t))}\,\mathbf{s}_1 I$$

Elimination rules

$$\text{Selection}\quad \frac{\begin{array}{c}\vdots\,\pi_{\mathbf{s}_0}\\ \mathbf{W}(y)\rightarrow A[\mathbf{s}_0(y)]\end{array}\quad \begin{array}{c}\vdots\,\pi_{\mathbf{s}_1}\\ \mathbf{W}(y)\rightarrow A[\mathbf{s}_1(y)]\end{array}\quad A[\epsilon]\quad \mathbf{W}(t)}{A[t]}\,\text{Sel}(\mathbf{W})$$

$$\text{Induction}\quad \frac{\forall y.A[y],\mathbf{W}(y)\rightarrow A[\mathbf{s}_0(y)]\quad \forall y.A[y],\mathbf{W}(y)\rightarrow A[\mathbf{s}_1(y)]\quad A[\epsilon]}{\forall x.\mathbf{W}(x)\rightarrow A[x]}\,\text{Ind}(\mathbf{W})$$

Restrictions on the rules :

- In Sel($\mathbf{W}$)-rule, derivations of $\pi_{\mathbf{s}_0}$ and $\pi_{\mathbf{s}_1}$ do not use the rule $\forall E^{\mathbf{S}}$. The variable y must not occur in any assumption on which $A[t]$ depends.

Fig. 3. Rules for word reasonning in $\mathbf{AT}(\mathbf{W})$

Definition 1. *A confluent equational program f computes a function $[\![f]\!]$ over $\mathcal{T}(\mathcal{W})$ which is defined as follows.*
For each $\mathbf{w}_i, \mathbf{v} \in \mathcal{T}(\mathcal{W})$, $[\![f]\!](\mathbf{w}_1, \cdots, \mathbf{w}_n) = \mathbf{v}$ iff $f(\mathbf{w}_1, \cdots, \mathbf{w}_n) \xrightarrow{!} \mathbf{v}$, otherwise $[\![f]\!](\mathbf{w}_1, \cdots, \mathbf{w}_n)$ is undefined.

2.2 Feasible Provably Total Functions

Let $\mathtt{f}$ be an equational program. We define $\mathbf{AT}(\mathtt{f})$ as the calculus $\mathbf{AT}(\mathbf{W})$ extended with the replacement rule below,

$$\frac{A[u\theta]}{A[v\theta]}\,\text{R}$$

where $(v \rightarrow u) \in \mathcal{E}$ and θ is a substitution $\mathcal{X} \rightarrow \mathcal{T}(\mathcal{W}, \mathcal{X})$.

Throughout, we abbreviate $\tau_1, \cdots, \tau_n \rightarrow \tau$ by $\tau_1 \rightarrow (\ldots(\tau_n \rightarrow \tau)\ldots)$ We write $A[t]$ to express that the term t occurs in the formula A. We write $\mathbf{W}(t)$ to express that the term t is the argument of the unary predicate $\mathbf{W}$.

Definition 2. *A function ϕ of arity n is provably total in $\mathbf{AT}(\mathbf{W})$ iff there are an equational program f such that $\phi = [\![f]\!]$ and a derivation in $\mathbf{AT}(f)$ of*

$$\text{Tot}(f) \equiv \forall x_1 \cdots x_n.\mathbf{W}(x_1), \cdots, \mathbf{W}(x_n) \rightarrow \mathbf{W}(f(x_1, \cdots, x_n))$$

Definition 3. *A formula $A[x]$ is an induction formula if $\forall x.\mathbf{W}(x) \to A[x]$ is the conclusion of an induction. Define $\mathbf{AT}^0(\mathbf{W})$ as the restriction of $\mathbf{AT}(\mathbf{W})$ in which induction formulas are just conjunctions of predicates (i.e. atomic formulas).*

Theorem 1 (Main result). *A function ϕ is polynomial time computable if and only if the function ϕ is provably total in $\mathbf{AT}^0$.*

Remark 1. By removing the restriction on the elimination rule of the universal quantifier in $\mathbf{AT}(\mathbf{W})$, we obtain a system which is equivalent to the system $\mathbf{IT}(\mathbf{W})$ developed by Leivant in [20]. The set of provably total functions in $\mathbf{IT}(\mathbf{W})$ is exactly the set of provably total functions in Peano arithmetic. Similarly, the provably total functions of $\mathbf{AT}^0(\mathbf{W})$, without the restriction on universal quantifiers, are the primitive recursive functions.

It is not difficult to modify $\mathbf{AT}$ to reason about any sorted algebra. In particular, a consequence of Theorem 1 is that the set of provably total functions is exactly the set of functions computable in linear space.

Example 1. We begin with the word concatenation whose equations are

$$\begin{aligned} &\mathtt{cat}(\epsilon, w) \to w \\ &\mathtt{cat}(\mathtt{s_i}(x), w) \to \mathtt{s_i}(\mathtt{cat}(x, w)) \qquad \mathbf{i} = \mathbf{0}, \mathbf{1} \end{aligned}$$

The derivation $\pi_{\mathtt{cat}}$, below, shows that the concatenation is a provably total function of $\mathbf{AT}(\mathtt{cat})$.

$$\dfrac{\dfrac{\dfrac{\{\mathbf{W}(z)\} \quad \dfrac{\dfrac{\{\mathbf{W}(\mathtt{cat}(z,w))\}}{\mathbf{W}(\mathtt{s_i}(\mathtt{cat}(z,w)))}\,\mathtt{s_i}I}{\mathbf{W}(\mathtt{cat}(\mathtt{s_i}(z),w))}\,\mathrm{R}}{\mathbf{W}(z), \mathbf{W}(\mathtt{cat}(z,w)) \to \mathbf{W}(\mathtt{cat}(\mathtt{s_i}(z),w))}\to I}{\forall z.\mathbf{W}(z), \mathbf{W}(\mathtt{cat}(z,w)) \to \mathbf{W}(\mathtt{cat}(\mathtt{s_i}(z),w))}\,\forall I \qquad \dfrac{\mathbf{W}(w)}{\mathbf{W}(\mathtt{cat}(\epsilon,w))}\,\mathrm{R}}{\forall x.\mathbf{W}(x) \to \mathbf{W}(\mathtt{cat}(x,w))}\,\mathrm{Ind}(\mathbf{W})$$

Notice that the term w is any term, and so w can be substituted by a non-actual term. Let us investigate the word multiplication whose equations are

$$\begin{aligned} &\mathtt{mul}(\epsilon, x) \to \epsilon \\ &\mathtt{mul}(\mathtt{s_i}(y), x) \to \mathtt{cat}(x, \mathtt{mul}(y, x)) \qquad \mathbf{i} = \mathbf{0}, \mathbf{1} \end{aligned}$$

The word multiplication is a provably total function as the derivation below shows it.

$$\dfrac{\dfrac{\dfrac{\dfrac{\dfrac{\dfrac{\begin{matrix}\{\mathbf{W}(\mathtt{mul}(z,x))\}\\ \vdots\ \pi_{\mathtt{cat}}[w \leftarrow \mathtt{mul}(z,x)]\\ \forall x.\mathbf{W}(x)\to\mathbf{W}(\mathtt{cat}(x,\mathtt{mul}(z,x)))\end{matrix}}{\mathbf{W}(x)\to\mathbf{W}(\mathtt{cat}(x,\mathtt{mul}(z,x)))}\forall E^{\mathbf{S}} \quad \{\mathbf{W}(x)\}}{\mathbf{W}(\mathtt{cat}(x,\mathtt{mul}(z,x)))}\to E}{\mathbf{W}(\mathtt{mul}(\mathtt{s_i}(z),x))}\mathrm{R}}{\mathbf{W}(z),\mathbf{W}(\mathtt{mul}(z,x))\to\mathbf{W}(\mathtt{mul}(\mathtt{s_i}(z),x))}\to I}{\forall z.\mathbf{W}(z),\mathbf{W}(\mathtt{mul}(z,x))\to\mathbf{W}(\mathtt{mul}(\mathtt{s_i}(z),x))}\forall I \quad \dfrac{\dfrac{}{\mathbf{W}(\epsilon)}\epsilon I}{\mathbf{W}(\mathtt{mul}(\epsilon,x))}\mathrm{R}}{\dfrac{\forall y.\mathbf{W}(y)\to\mathbf{W}(\mathtt{mul}(y,x))}{\forall x.\forall y.\mathbf{W}(x),\mathbf{W}(y)\to\mathbf{W}(\mathtt{mul}(y,x))}(\forall E^{\mathbf{S}};\to I;\forall I;\forall I)}$$

Now, consider the equations defining the exponential :

$$\begin{aligned} \mathtt{exp}(\epsilon) &\to \mathtt{s_0}(\epsilon) \\ \mathtt{exp}(\mathtt{s_i}(y)) &\to \mathtt{cat}(\mathtt{exp}(y),\mathtt{exp}(y)) && \mathbf{i}=\mathbf{0},\mathbf{1} \end{aligned}$$

In order to establish that the program exp defines a provably total function, we have to make an induction. At the induction step, under the assumptions $\mathbf{W}(\mathtt{exp}(y))$ and $\mathbf{W}(y)$, we have to prove $\mathbf{W}(\mathtt{cat}(\mathtt{exp}(y),\mathtt{exp}(y)))$. However, $\mathtt{exp}(y)$ is not an actual term, and so we can not "plug in" the derivation $\pi_{\mathtt{cat}}$ to conclude.

2.3 Comments

These examples illustrate that actual terms play a role similar to terms of higher tier (safe) used in ramified recursions, as defined by Bellantoni and Cook [2], and Leivant in [19]. Intuitively, we do not assume that two terms are equal just because they have the same value. We are not concerned by term denotations, but rather by the resource necessary to evaluate a term, or in other words, by term intention. From this point of view, a non-actual term is unsafe. So, we have no justification to quantify over non-actual terms. On the other hand, there are no computation rules associated to actual terms, so they are safe with respect to polynomial-time computation. In a way, this idea is similar to "read-only" programs of Jones [12].

The concept arising from the work of Simmons [27], Bellantoni and Cook [2] and Leivant [19], is the ramification of the domain of computation and the ramification of recursion schemata. One usually compares this solution with Russell's type theory. One unattractive feature is that objects are duplicated at different tiers. This drawback is eliminated here. It is amazing to see that this solution seems related to Zermelo or Quine answers to Russell's type theory.

Lastly, the actual elimination quantifier principle reminds one of logic with existence predicate, in which quantifiers are supposed to range only over existing terms. The motivation is to take into account undefined terms. Such logics have their roots in works of Weyl [28] and Heyting [10], and have since extensively studied and are related to *free logic.*

2.4 Others Feasible Logics

Theories of feasible mathematics originate with Buss [5] on bounded arithmetic. Subsequently, Leivant [17] established that the functions provably total in second order arithmetic with the comprehension axiom restricted to positive existential formulas, are exactly the polynomial time functions. Leivant [18] also translated his characterisation [19] of feasible functions by mean of ramified recursion. For this, he has introduced a sequence of predicate $\mathbf{N}_0, \mathbf{N}_1, \ldots$ corresponding to copies of $\mathbf{N}$ with increasing computational potential. Çağman, Ostrin and Wainer [6] defined a two sorted Peano arithmetic PA(;) in the spirit of Bellantoni and Cook [2]. They characterize the functions computable in linear space, and the elementary functions. Predicates have two kinds of arguments : safe and normal. Quantifiers are allowed only over safe terms and range over hereditary basic terms.

In a recent article [21], Leivant suggests a new direction by giving some structural conditions on proof hipothesis and on inductive formulas.

There are also theories of feasible mathematics which are affiliated to linear logic. Girard, Scedrov and Scott in [11] have introduced bounded linear logic, in which resources are explicitly counted. Then, Girard [8] constructed light linear logic which is a second order logic with a new modality which controls safely the resources. See also the works of Asperti [1] and Roversi [25]. Lastly, Bellantoni and Hofmann [3] and Schwichtenberg [26], have proposed feasible arithmetics based on linear logic with extra counting modalities.

2.5 Strong Normalisation and Confluence

Detours and conversions are listed in Figures 4 and 5. Suppose that π is a derivation of $\mathbf{AT}(\mathtt{f})$, and that $\pi \rhd \pi'$. Then, π' is a derivation of $\mathbf{AT}(\mathtt{f})$.

Theorem 2. *The proof reduction relation* $\rhd$ *is confluent and terminating.*

Proof. A derivation π of $\mathbf{AT}(\mathtt{f})$ is also a derivation of Peano arithmetic. From the observation above, we conclude $\mathbf{AT}(\mathtt{f})$ is confluent and terminating because Peano arithmetic enjoys both properties.

3 Extraction of Polynomial Time Programs

We follow the program construction methodology behind AF_2 of Leivant [15], see also the works of Krivine and Parigot [13,14,23]. An equational program $\mathtt{f}$ is seen as a specification of the function $[\![\mathtt{f}]\!]$ which is compiled in two steps. First, we establish that $\mathrm{Tot}(\mathtt{f})$ holds in $\mathbf{AT}^0(\mathtt{f})$. Second, we extract a lambda-term which computes $[\![\mathtt{f}]\!]$.

Detours *Conversions*

β

$$\dfrac{\dfrac{\begin{matrix}\{A\}\\ \vdots\ \pi_0\\ B\end{matrix}}{A\to B}\to I \qquad \begin{matrix}\vdots\ \pi_1\\ A\end{matrix}}{B}\to E \qquad \rhd \qquad \begin{matrix}\vdots\ \pi_1\\ A\\ \vdots\ \pi_0\\ B\end{matrix}$$

$\wedge_i$

$$\dfrac{\dfrac{\begin{matrix}\vdots\ \pi_1\\ A_1\end{matrix}\qquad\begin{matrix}\vdots\ \pi_2\\ A_2\end{matrix}}{A_1\wedge A_2}\wedge I}{A_i}\wedge E \qquad \rhd \qquad \begin{matrix}\vdots\ \pi_i\\ A_i\end{matrix}$$

$\forall$

$$\dfrac{\dfrac{\begin{matrix}\vdots\ \pi\\ A\end{matrix}}{\forall x.A}\forall I}{A[x\leftarrow p]}\forall E^{\mathbf{S}},\ p\in\mathcal{T}(\mathcal{W},\mathcal{X}) \qquad \rhd \qquad \begin{matrix}\vdots\ \pi[x\leftarrow p]\\ A[x\leftarrow p]\end{matrix}$$

Sel

$$\dfrac{\begin{matrix}\vdots\ \pi_{\mathbf{s_0}}\\ \mathbf{W}(x)\to A[\mathbf{s_0}(x)]\end{matrix}\quad\begin{matrix}\vdots\ \pi_{\mathbf{s_1}}\\ \mathbf{W}(x)\to A[\mathbf{s_1}(x)]\end{matrix}\quad\begin{matrix}\vdots\ \pi_{\epsilon}\\ A[\epsilon]\end{matrix}\quad \mathbf{W}(\epsilon)}{A[\epsilon]}\mathrm{Sel}(\mathbf{W}) \qquad \rhd \qquad \begin{matrix}\vdots\ \pi_{\epsilon}\\ A[\epsilon]\end{matrix}$$

$$\dfrac{\begin{matrix}\vdots\ \pi_{\mathbf{s_0}}\\ \mathbf{W}(x)\to A[\mathbf{s_0}(x)]\end{matrix}\quad\begin{matrix}\vdots\ \pi_{\mathbf{s_1}}\\ \mathbf{W}(x)\to A[\mathbf{s_1}(x)]\end{matrix}\quad\begin{matrix}\vdots\ \pi_{\epsilon}\\ A[\epsilon]\end{matrix}\quad \dfrac{\begin{matrix}\vdots\ \pi_t\\ \mathbf{W}(t)\end{matrix}}{\mathbf{W}(\mathbf{s_i}(t))}\mathbf{s_i}I}{A[\mathbf{s_i}(t)]}\mathrm{Sel}(\mathbf{W})$$

$$\rhd \qquad \dfrac{\begin{matrix}\vdots\ \pi_{\mathbf{s_i}}[x\leftarrow t]\\ \mathbf{W}(t)\to A[\mathbf{s_i}(t)]\end{matrix}\qquad\begin{matrix}\vdots\ \pi_t\\ \mathbf{W}(t)\end{matrix}}{A[\mathbf{s_i}(t)]}\to E$$

Fig. 4. Detours and conversions of **AT(W)**

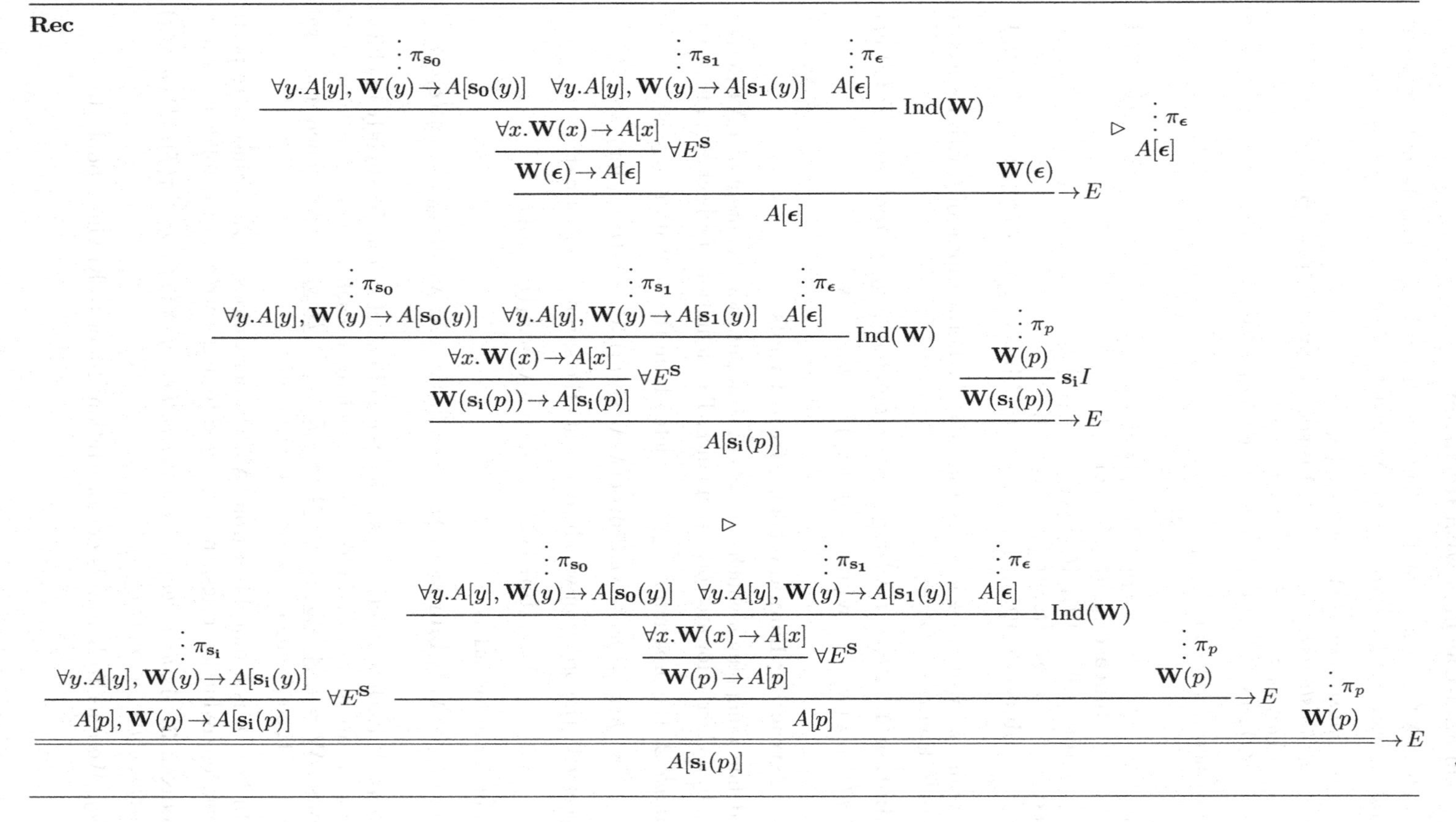

Fig. 5. Detours and conversions for the induction rule of **AT**(**W**)

3.1 Lambda-Calculus with Recurrence

We define $\boldsymbol{\lambda}^{\mathcal{W}}$ as the lambda-calculus with pairing, selection and recurrence operators over words. Terms of $\boldsymbol{\lambda}^{\mathcal{W}}$ are generated by

$$\begin{array}{llll} \textit{(Constructors)} & \mathcal{W} \ni \mathbf{c} & ::= & \boldsymbol{\epsilon} \mid \mathbf{s_0} \mid \mathbf{s_1} \\ \textit{(Constants)} & \mathbf{r} & ::= & \mathbf{binrec} \mid \mathbf{bincase} \mid \mathbf{pr}_1 \mid \mathbf{pr}_2 \mid \langle \_, \_ \rangle \\ \textit{(Variables)} & \mathcal{X} \ni x & ::= & x \mid y \mid z \mid \dots \\ \textit{(Terms)} & t & ::= & \mathbf{c} \mid \mathbf{r} \mid \lambda x.t \mid t(s) \end{array}$$

The reduction rules are

$$\begin{array}{llll} \beta & (\lambda x.t)s & \to & t[x \leftarrow s] \\ \wedge_i & \mathbf{pr}_i(\langle t_1, t_2 \rangle) & \to & t_i \qquad\qquad \mathbf{i} = \mathbf{0}, \mathbf{1} \\ \mathbf{Sel} & \mathbf{bincase}(t_{\mathbf{s_0}})(t_{\mathbf{s_1}})(t_\epsilon)(\epsilon) & \to & t_\epsilon \\ & \mathbf{bincase}(t_{\mathbf{s_0}})(t_{\mathbf{s_1}})(t_\epsilon)(\mathbf{s_i}(u)) & \to & t_{\mathbf{s_i}}(u) \\ \mathbf{Rec} & \mathbf{binrec}(t_{\mathbf{s_0}})(t_{\mathbf{s_1}})(t_\epsilon)(\epsilon) & \to & t_\epsilon \\ & \mathbf{binrec}(t_{\mathbf{s_0}})(t_{\mathbf{s_1}})(t_\epsilon)(\mathbf{s_i}(u)) & \to & t_{\mathbf{s_i}}(\mathbf{binrec}(t_{\mathbf{s_i}})(t_\epsilon)(u))(u) \qquad \mathbf{i} = \mathbf{0}, \mathbf{1} \end{array}$$

Again, we write $t \xrightarrow{!} s$ to say that s is the normal form of t with respect to the reduction rules above.

Definition 4. *A term t $\boldsymbol{\lambda}^{\mathcal{W}}$-represents a function ϕ iff for each $\mathbf{w}_i \in \mathcal{T}(\mathcal{W})$ and $\mathbf{v} \in \mathcal{T}(\mathcal{W})$, $\phi(\mathbf{w}_1, \cdots, \mathbf{w}_n) = \mathbf{v}$ iff $t(\mathbf{w}_1) \cdots (\mathbf{w}_n) \xrightarrow{!} \mathbf{v}$.*

3.2 Extracting Lambda-Terms from Proofs

We define a mapping κ which extracts the computational content of a **AT** derivation, by a Curry-Howard correspondance. The definition of κ is given in Figure 6.

Example 2. the program $\kappa(\pi_{\mathtt{cat}})$ extracted from the derivation $\pi_{\mathtt{cat}}$ is

$$\kappa(\pi_{\mathtt{cat}})[w] \equiv \lambda x.\mathbf{binrec}(\lambda z \lambda v.\mathbf{s_0}(v))(\lambda z \lambda v.\mathbf{s_1}(v))(w)(x)$$

Henceforth, the program which represents the specification of `cat`, is

$$\lambda w \lambda x.\mathbf{binrec}(\lambda z \lambda v.\mathbf{s_0}(v))(\lambda z \lambda v.\mathbf{s_1}(v))(w)(x)$$

The program for `mul` is

$$\kappa(\pi_{\mathtt{mul}}) \equiv \lambda x \lambda y.\mathbf{binrec}(\lambda z.\lambda w.\kappa(\pi_{\mathtt{cat}})[w](x))(\lambda z.\lambda w.\kappa(\pi_{\mathtt{cat}})[w](x))(\epsilon)(y)$$

Remark 2. Each term $\kappa(\pi)$ of $\boldsymbol{\lambda}^{\mathcal{W}}$ is a term of Gödel system T, slightly modified to reason about words. Indeed, we associate to each formula A a type $\underline{A}$ as follows : $\underline{\mathbf{W}} = \iota$, $\underline{A \to B} = \underline{A} \to \underline{B}$, $\underline{A \wedge B} = \underline{A} \wedge \underline{B}$, $\underline{\forall x.A} = \underline{A}$. Then, the term $\kappa(\pi)$ is of type $\underline{A}$ where A is the conclusion of π.

In fact, κ is a morphism from $\mathbf{AT}^0(\mathtt{f})$ derivations to $\boldsymbol{\lambda}^{\mathcal{W}}$. Since κ respects conversion/reduction rules, the extracted term is correct in the following sense.

Theorem 3. *Assume that π is a derivation of $\mathbf{AT}(f)$ of* Tot(f). *Then $\kappa(\pi)$ $\boldsymbol{\lambda}^{\mathcal{W}}$-represents the function $[\![f]\!]$.*

Proof. We refer to Leivant's demonstration [16] or to Krivine's book [13].

$$\text{Derivation}\,\pi \quad \xrightarrow{\kappa} \quad \text{Terms}\,\kappa(\pi)$$

$$A \xrightarrow{\kappa} x \qquad\qquad \dfrac{\overset{\vdots\,\pi}{A[u\theta]}}{A[v\theta]}\,\mathrm{R} \xrightarrow{\kappa} \kappa(\pi)$$

$$\dfrac{\overset{\{A\}}{\overset{\vdots\,\pi}{B}}}{A\to B}\to I \xrightarrow{\kappa} \lambda x.\kappa(\pi) \qquad\qquad \dfrac{\overset{\vdots\,\pi_1}{A\to B}\quad \overset{\vdots\,\pi_2}{A}}{B}\to E \xrightarrow{\kappa} \kappa(\pi_1)(\kappa(\pi_2))$$

$$\dfrac{\overset{\vdots\,\pi_1}{A_1}\quad \overset{\vdots\,\pi_2}{A_2}}{A_1\wedge A_2}\wedge I \xrightarrow{\kappa} \langle\kappa(\pi_1),\kappa(\pi_2)\rangle \qquad\qquad \dfrac{\overset{\vdots\,\pi}{A_1\wedge A_2}}{A_i}\wedge E \xrightarrow{\kappa} \mathbf{pr}_i(\kappa(\pi))$$

$$\dfrac{\overset{\vdots\,\pi}{A}}{\forall x.A}\forall I \xrightarrow{\kappa} \kappa(\pi) \qquad\qquad \dfrac{\overset{\vdots\,\pi}{\forall x.A}}{A[t/x]}\forall E^{\mathbf{S}} \xrightarrow{\kappa} \kappa(\pi)$$

$$\dfrac{}{\mathbf{W}(\epsilon)}\epsilon I \xrightarrow{\kappa} \epsilon \qquad\qquad \dfrac{\overset{\vdots\,\pi}{\mathbf{W}(t)}}{\mathbf{W}(\mathbf{s_i}(t))}\mathbf{s_i}I \xrightarrow{\kappa} \mathbf{s_i}(\kappa(\pi))$$

$$\dfrac{\overset{\vdots\,\pi_{\mathbf{s_0}}}{\mathbf{W}(y)\to A[\mathbf{s_0}(y)]}\quad \overset{\vdots\,\pi_{\mathbf{s_1}}}{\mathbf{W}(y)\to A[\mathbf{s_1}(y)]}\quad \overset{\vdots\,\pi_{\epsilon}}{A[\epsilon]}\quad \overset{\vdots\,\pi_t}{\mathbf{W}(t)}}{A[t]}\,\mathrm{Sel}(\mathbf{W})$$

$$\xrightarrow{\kappa} \quad \mathbf{bincase}(\kappa(\pi_{\mathbf{s_0}}))(\kappa(\pi_{\mathbf{s_1}}))(\kappa(\pi_{\epsilon}))(\kappa(\pi_t))$$

$$\dfrac{\overset{\vdots\,\pi_{\mathbf{s_0}}}{\forall y.A[y],\mathbf{W}(y)\to A[\mathbf{s_0}(y)]}\quad \overset{\vdots\,\pi_{\mathbf{s_1}}}{\forall y.A[y],\mathbf{W}(y)\to A[\mathbf{s_1}(y)]}\quad \overset{\vdots\,\pi_{\epsilon}}{A[\epsilon]}}{\forall x.\mathbf{W}(x)\to A[x]}\,\mathrm{Ind}(\mathbf{W})$$

$$\xrightarrow{\kappa} \quad \mathbf{binrec}(\kappa(\pi_{\mathbf{s_0}}))(\kappa(\pi_{\mathbf{s_1}}))(\kappa(\pi_{\epsilon}))$$

Fig. 6. Program extraction by Curry-Howard correspondance κ

3.3 Analysis of the Complexity of Extracted Programs

The size $|\mathbf{w}|$ of a term is the number of constructors which makes up $\mathbf{w}$.

Theorem 4. *Let ϕ be a provably total function of $\mathbf{AT}^0(\mathbf{W})$. The function ϕ is computable in polynomial time with respect to input sizes.*

Proof. There is an equational program $\mathtt{f}$ which computes ϕ, that is $\phi = [\![\mathtt{f}]\!]$ and such that $\text{Tot}(\mathtt{f})$ is derivable in $\mathbf{AT}^0(\mathtt{f})$. In order to avoid needlessly difficulties, we shall assume that the derivation of $\text{Tot}(\mathtt{f})$ does not use the rule $\wedge E$.

Before proceeding with the demonstration of the theorem, it is helpful to say that an occurrence of a predicate $\mathbf{W}(p)$ in a derivation π, is critical if

(a) Either the occurrence of $\mathbf{W}(p)$ is in the negative part of the conclusion of an induction rule.
(b) Or the occurrence $\mathbf{W}(p)$ is the minor of a detour of an induction.

A typical example is written below. Here, $\mathbf{W}(x)$ satisfies (a) and the occurrence of $\mathbf{W}(t)$, which is the conclusion of π_t, satisfies (b).

$$\dfrac{\dfrac{\dfrac{}{\forall x.\mathbf{W}(x)\to A[x]}\,\text{Ind}(\mathbf{W})}{\mathbf{W}(t)\to A[t]}\,\forall E^{\mathbf{S}} \qquad \begin{matrix}\vdots\ \pi_t \\ \mathbf{W}(t)\end{matrix}}{A[t]}\to E$$

Next, say that a predicate $\mathbf{W}(t)$ is critical in a derivation π if one of its occurrence in π is critical. It is important to notice that if $\mathbf{W}(t)$ is critical then, the term t is necessary actual. Now, the demonstration of the theorem is a consequence of the following claim.

Claim. Let π be a normal derivation, without the rule $\wedge E$, of

$$\forall \boldsymbol{x}.\mathbf{W}(s_{m+1}), \dots, \mathbf{W}(s_{m+n}) \to \mathbf{W}(t)$$

under the premises $\mathbf{W}(s_1), \dots, \mathbf{W}(s_m)$ and where t is not an actual term. There is a polynomial p_π such that for every $\mathbf{w}_1, \cdots, \mathbf{w}_{m+n} \in \mathcal{T}(\mathcal{W})$, the computation of

$$\kappa(\pi)(\mathbf{w}_{m+1})\dots(\mathbf{w}_{m+n})[s_1 \leftarrow \mathbf{w}_1, \dots, s_m \leftarrow \mathbf{w}_m]$$

is performed in time bounded by $p_\pi(M)$ and constant in N where

$$M = \max(|\mathbf{w}_i| : \mathbf{W}(s_i) \text{ is critical in } \pi)$$
$$N = \max(|\mathbf{w}_i| : \mathbf{W}(s_i) \text{ is not critical in } \pi)$$

The demonstration goes by induction on the size of the normal derivation π. We shall just consider case where the las rule is an induction rule. The other cases are similar or immediate.

Following Figure 5 the term $\kappa(\pi)$ is $\lambda x.\mathbf{binrec}(\kappa(\pi_{\mathbf{s_0}}))(\kappa(\pi_{\mathbf{s_1}}))(\kappa(\pi_\epsilon))(x)$. The computation of $\kappa(\pi)\mathbf{w}_{m+1}[s_1 \leftarrow \mathbf{w}_1, \dots, s_m \leftarrow \mathbf{w}_m]$ consists in a recursion of length $|\mathbf{w}_{m+1}|$. Terms $\kappa(\pi_\epsilon)$ and $\kappa(\pi_{\mathbf{s_i}})_{i=0,1}$ satisfy the induction hypothesis. By claim assumption, $t[y]$ is not an actual terms. So, $\mathbf{W}(t[y])$ is not critical. It follows that $|\mathbf{w}_{m+1}|$-long recursion is evaluated in time bounded by $|\mathbf{w}_{m+1}| \cdot \max(P_{\pi_{\mathbf{s_0}}}(M), P_{\pi_{\mathbf{s_1}}}(M)) + P_{\pi_\epsilon}(M) + O(1)$ and constant in N.

4 Polynomial-Time Functions Are Provably Total

Theorem 5. *Each polynomial time computable function ϕ is provably total in* $\mathbf{AT}^0$.

Proof. Suppose that the function ϕ is computed by a single tape Turing machine M within $k \cdot n^k$ steps, for some constant k where n is the input size. The tape alphabet is $\{0,1\}$ and the set of states is $Q \subset \{0,1\}^+$. A description of M is a triplet (q,u,v) where q is the current state, u is the left hand side of the tape, and v the right hand side of the tape. The head scans the first letter of v. Let $\delta : Q \times \{0,1\} \mapsto Q \times \{0,1\} \times \{l,r\}$ be the transition function of M. The initial description is (q_0, ϵ, v). The result of the computation is written on the right hand side of the tape.

We construct an equational program $\mathtt{f}$, which simulates the computation of M and which is provably total in $\mathbf{AT}^0(\mathbf{W})$. We begin by defining three programs $\mathtt{D}_0, \mathtt{D}_1$ and $\mathtt{D}_2$, which determine the next description. From the transition function δ, those programs indicate, respectively, the next state, the word on the left, and the word on the right, as follows.

$$\begin{array}{rll}
\mathtt{D}_0(q,u,\mathtt{s_i}(v)) \to q' & \text{if } \delta(q,i) = (q',k,*) & * \in \{l,r\} \\
\mathtt{D}_1(q,\mathtt{s_j}(u),\mathtt{s_i}(v)) \to u & \text{if } \delta(q,i) = (q',k,l) & \\
\mathtt{D}_1(q,u,\mathtt{s_i}(v)) \to \mathtt{s_k}(u) & \text{if } \delta(q,i) = (q',k,r) & \\
\mathtt{D}_2(q,\mathtt{s_j}(u),\mathtt{s_i}(v)) \to \mathtt{s_j}(\mathtt{s_k}(v)) & \text{if } \delta(q,i) = (q',k,l) & \\
\mathtt{D}_2(q,u,\mathtt{s_i}(v)) \to v & \text{if } \delta(q,i) = (q',k,r) &
\end{array}$$

In all others cases, the result of $\mathtt{D}_i$ is ϵ. Thus, (c_0,c_1,c_2) yields, in one step, the description $(\mathtt{D}_0(c_0,c_1,c_2), \mathtt{D}_1(c_0,c_1,c_2), \mathtt{D}_2(c_0,c_1,c_2))$. Each $\mathtt{D}_i$ is defined by cases. The current state and the neighbour letter of the head are determined by nested applications of Sel($\mathbf{W}$)-rules. So, we can build a derivation $\pi_{\mathtt{D}_i}$ of $\wedge_{i=0,1,2}\mathbf{W}(\mathtt{D}_i(q,u,v))$, under the assumptions $\mathbf{W}(q), \mathbf{W}(u), \mathbf{W}(v)$ and without using the induction rule. Consequently, we shall be able to replace the free variables q, u et v by any term.
Next, we define the sequence $(\mathtt{L}^i_m(t,x,c_0,c_1,c_2))_{i=0,1,2}$, by mutual recursion:

$$\begin{array}{rll}
\mathtt{L}^i_0(t,x,c_0,c_1,c_2) \to \mathtt{D}_i(c_0,c_1,c_2) & & \\
\mathtt{L}^i_{m+1}(\epsilon,x,c_0,c_1,c_2) \to c_i & & \\
\mathtt{L}^i_{m+1}(\mathtt{s_j}(t),x,c_0,c_1,c_2) \to \mathtt{L}^i_m(x,x,\Delta^0,\Delta^1,\Delta^2) & & \mathtt{s_j} = \mathtt{s_0},\mathtt{s_1}
\end{array}$$

where $\Delta^i = \mathtt{L}^i_{m+1}(t,x,c_0,c_1,c_2)$, for $i = 0,1,2$.

The description of M after $|t| \cdot |x|^k$ steps, starting form the description (c_0,c_1,c_2), is $(\mathtt{L}^0_{k+1}(t,x,c_0,c_1,c_2), \mathtt{L}^1_{k+1}(t,x,c_0,c_1,c_2), \mathtt{L}^2_{k+1}(t,x,c_0,c_1,c_2))$. The function ϕ, computed by the Turing machine M, is represented by $\mathtt{f}$:

$$\mathtt{f}(v) \to \mathtt{L}^2_{k+1}(\mathtt{s_0}^k(\epsilon), v, q_0, \epsilon, v)$$

It remains to establish that $\mathtt{f}$ is provably total in $\mathbf{AT}^0(\mathbf{W})$. For this, we construct for every m, a derivation π_m of $\forall z.\mathbf{W}(z) \to \wedge_{i=0,1,2}\mathbf{W}(\mathtt{L}^i_m(z,x,c_0,c_1,c_2))$

under the premises $\mathbf{W}(x), \mathbf{W}(c_0), \mathbf{W}(c_1), \mathbf{W}(c_2)$ where c_0, c_1 and c_2 are variables which can be substituted for any term, unlike x.

References

1. A. Asperti. Light affine logic. In *Thirteenth Annual IEEE Symposium on Logic in Computer Science(LICS'98)*, pages 300–308, 1998.
2. S. Bellantoni and S. Cook. A new recursion-theoretic characterization of the poly-time functions. *Computational Complexity*, 2:97–110, 1992.
3. S. Bellantoni and M. Hofmann. A new "feasible" arithmetic. *Journal of symbolic logic*, 2000. to appear.
4. R. Benzinger. *Automated complexity analysis of NUPRL extracts.* PhD thesis, Cornell University, 1999.
5. S. Buss. *Bounded arithmetic.* Bibliopolis, 1986.
6. N. Çağman, G. Ostrin, and S. Wainer. Proof theoretic complexity of low subrecursive classes. In F. L. Bauer and R. Steinbrueggen, editors, *Foundations of Secure Computation*, Foundations of Secure Computation, pages 249–286. IOS Press Amsterdam, 2000.
7. N. Dershowitz and J-P Jouannaud. *Handbook of Theoretical Computer Science vol.B*, chapter Rewrite systems, pages 243–320. Elsevier Science Publishers B. V. (North-Holland), 1990.
8. J.-Y. Girard. Light linear logic. *Information and Computation*, 143(2):175–204, 1998. présenté à LCC'94, LNCS 960.
9. J.-Y. Girard, Y. Lafont, and P. Taylor. *Proofs and Types*, volume 7 of *Cambridge tracts in theoretical computer science.* Cambridge university press, 1989.
10. A. Heyting. Die formalen regeln der intuitionischen logik. *Sitzungsberichte der preussischen akademie von wissenschaften*, pages 57–71, 1930.
11. P. Scott J.-Y. Girard, A. Scedrov. A modular approach to polynomial-time computability. *Theoretical Computer Science*, 97(1):1–66, 1992.
12. N. Jones. LOGSPACE and PTIME characterized by programming languages. *Theoretical Computer Science*, 228:151–174, 1999.
13. J.-L. Krivine. *Lambda-calcul.* Masson, 1990. English translation by R. Cori: Lambda-Calculus, Types and Models (Ellis Horwood Series) 1993.
14. J.L. Krivine and M. Parigot. Programming with proofs. In *SCT'87*, pages 145–159, 1987.
15. D. Leivant. Reasoning about functional programs and complexity classes associated with type disciplines. In *Twenty Fourth Symposium on Foundations of Computer Science*, pages 460–469, 1983.
16. D. Leivant. Contracting proofs to programs. In P. Odifreddi, editor, *Logic and Computer Science*, pages 279–327. Academic Press, London, 1990.
17. D. Leivant. A foundational delineation of computational feasiblity. In *Proceedings of the Sixth IEEE Symposium on Logic in Computer Science (LICS'91)*, 1991.
18. D. Leivant. Intrinsic theories and computational complexity. In *Logical and Computational Complexity*, volume 960 of *Lecture Notes in Computer Science*, pages 177–194. Springer, 1994.
19. D. Leivant. Predicative recurrence and computational complexity I: Word recurrence and poly-time. In Peter Clote and Jeffery Remmel, editors, *Feasible Mathematics II*, pages 320–343. Birkhäuser, 1994.

20. D. Leivant. Intrinsic reasoning about functional programs I: first order theories. *Annals of Pure and Applied Logic*, 2000. to appear.
21. D. Leivant. Termination proofs and complexity certification. In *Proceedings of the third international workshop on Implicit Computational Complexity*, 2001.
22. P. Martin-Löf. *Intuitionistic Type Theory.* Bibliopolis, 1984.
23. M. Parigot. Recursive programming with proofs. *Theoretical Computer Science*, 94:335–356, 1992.
24. D. Prawitz. *Natural Deduction.* Almqvist and Wiskel, Uppsala, 1965.
25. L. Roversi. A P-Time completeness proof for light logics. In *Computer Science Logic, 13th International Workshop, CSL '99*, volume 1683 of *Lecture Notes in Computer Science*, pages 469–483, 1999.
26. H. Schwichtenberg. Feasible programs from proofs. http://www.mathematik.uni-muenchen.de/~schwicht/.
27. H. Simmons. The realm of primitive recursion. *Archive for Mathematical Logic*, 27:177–188, 1988.
28. H. Weyl. über die neue grundlagenkrise der mathematik. Mathematische zeitschrift, 1921.

The Natural Order-Generic Collapse for ω-Representable Databases over the Rational and the Real Ordered Group

Nicole Schweikardt

Institut für Informatik / FB 17
Johannes Gutenberg-Universität, D-55099 Mainz
nisch@informatik.uni-mainz.de
http://www.informatik.uni-mainz.de/~nisch/homepage.html

Abstract. We consider *order-generic* queries, i.e., queries which commute with every order-preserving automorphism of a structure's universe. It is well-known that first-order logic has the natural order-generic collapse over the rational and the real ordered group for the class of dense order constraint databases (also known as *finitely representable databases*). I.e., on this class of databases over $\langle \mathbb{Q}, < \rangle$ or $\langle \mathbb{R}, < \rangle$, *addition* does not add to the expressive power of first-order logic for defining order-generic queries. In the present paper we develop a *natural generalization* of the notion of finitely representable databases, where an *arbitrary* (i.e. possibly infinite) number of regions is allowed. We call these databases ω*-representable*, and we prove the natural order-generic collapse over the rational and the real ordered group for this larger class of databases.

Keywords: Logic in Computer Science, Database Theory, Constructive Mathematics

1 Introduction and Main Results

In relational database theory a database is modelled as a relational structure over a fixed, possibly infinite universe $\mathbb{U}$. A k-ary *query* is a mapping $\mathcal{Q}$ which assigns to each database $\mathcal{A}$ a k-ary relation $\mathcal{Q}(\mathcal{A}) \subseteq \mathbb{U}^k$. In many applications the elements in $\mathbb{U}$ only serve as identifiers which are exchangeable. If this is the case, one demands that queries commute with every permutation of $\mathbb{U}$. Such queries are called *generic*. If $\mathbb{U}$ is *linearly ordered*, a query may refer to the ordering. In this setting it is more appropriate to consider queries which commute with every order-preserving (i.e. strictly increasing) mapping of $\mathbb{U}$. Such queries are called *order-generic*.

A basic way of expressing order-generic queries is by first-order formulas that make use of the linear ordering and of the database relations. Database theorists distinguish between two different semantics: *active* semantics, where quantifiers only range over database elements, and the (possibly) stronger *natural* semantics, where quantifiers range over all of $\mathbb{U}$. In the present paper we always consider *natural* semantics.

L. Fribourg (Ed.): CSL 2001, LNCS 2142, pp. 130–144, 2001.

It is a reasonable question whether the use of additional, e.g. arithmetical, predicates on $\mathbb{U}$ allows first-order logic to express more order-generic queries than with linear ordering alone. In some situations this question can be answered "yes" (e.g. if $\mathbb{U}$ is the set of natural numbers with $+$ and $\times$ as additional predicates, cf. [3]). In other situations the question must be answered "no" (e.g. if $\mathbb{U}$ is the set of natural numbers with $+$ alone, cf. [8]) — such results are then called *collapse results*, because first-order logic with the additional predicates collapses to first-order logic with linear ordering alone. A recent overview of this area of research is given in [3].

In classical database theory, attention usually is restricted to *finite* databases. In this setting Benedikt et al. [2] have obtained a strong collapse result: *First-order logic has the natural order-generic collapse for finite databases over o-minimal structures.* This means that if the universe $\mathbb{U}$ together with the additional predicates, has a certain property called *o-minimality*, then for every order-generic first-order formula φ which uses the additional predicates, there is a formula with linear ordering alone which is equivalent to φ on all *finite* databases.

Belegradek et al. [1] have extended this result: Instead of o-minimality they consider *quasi o-minimality*, and instead of finite databases they consider *finitely representable* databases (also known as *dense order constraint databases*). Many structures interesting to database theory, including $\langle \mathbb{N}, <, + \rangle$, $\langle \mathbb{Q}, <, + \rangle$, $\langle \mathbb{R}, <, + \rangle$, and $\langle \mathbb{R}, <, +, \times, e^x \rangle$, are indeed *o-minimal* or at least *quasi o-minimal.* A database is called *finitely representable* if each of its relations can be explicitly defined by a first-order formula which makes use of the linear ordering and of finitely many constants in $\mathbb{U}$. For $\mathbb{U} \in \{\mathbb{Q}, \mathbb{R}\}$, finitely representable databases are exactly those databases where every relation is defined by a Boolean combination of order-constraints over $\mathbb{U}$. I.e., those database relations essentially consist of a *finite* number of multidimensional rectangles in $\mathbb{U}$.

A reasonable question is whether such collapse results hold for even larger classes of databases. In [8] it was shown that over $\langle \mathbb{N}, <, + \rangle$ the natural order-generic collapse does indeed hold for *arbitrary* databases. However, this result cannot be carried over to *dense* linear orders: Belegradek et al. have shown (cf. [1, Theorem 3.2]) that e.g. over $\langle \mathbb{Q}, <, + \rangle$ the natural order-generic collapse does *not* hold for *arbitrary* databases. This result draws a borderline between finite and finitely representable databases on the one side and arbitrary databases on the other. In the present paper we extend that borderline. We develop a *natural generalization* of the notion of finitely representable databases. We call these databases *ω-representable*, and we obtain the following

Main Theorem 1. *First-order logic has the natural order-generic collapse for ω-representable databases over $\langle \mathbb{Q}, <, + \rangle$ and $\langle \mathbb{R}, <, + \rangle$.* □

We call a database *ω-representable* if each of its relations can be explicitly defined by a formula in *infinitary logic* which makes use of the linear ordering and of a countable, unbounded sequence of constants $s_1 < s_2 < \cdots$ in $\mathbb{U}$. For $\mathbb{U} \in \{\mathbb{Q}, \mathbb{R}\}$, ω-representable databases turn out to be exactly those databases where every relation is defined by an *infinitary* Boolean combination of order-constraints over $\langle \mathbb{U}, (s_n)_{n \geqslant 1} \rangle$. I.e., those database relations essentially consist of a finite or a *countable* number of multidimensional rectangles in $\mathbb{U}$.

In particular, the theorem above shows that there is a natural class that contains "essentially infinite" databases, to which the collapse results of Benedikt et al. and Belegradek et al. can be generalized, for the special case of $\langle \mathbb{Q}, <, + \rangle$ or $\langle \mathbb{R}, <, + \rangle$ as underlying structures.

The two main tools for proving Main Theorem 1 are

(1.) a result of [8] that implies, for $\mathbb{U} \in \{\mathbb{Q}, \mathbb{R}\}$, the natural order-generic collapse over $\langle \mathbb{U}, <, + \rangle$ for the class of *ω-databases* (these are the databases whose active domain is either finite or consists of an unbounded sequence $s_1 < s_2 < \cdots$ of elements in $\mathbb{U}$), and

(2.) the following Main Theorem 2, which allows us to lift collapse results for *ω-databases* to collapse results for *ω-representable databases.*

Main Theorem 2. Let $\langle \mathbb{U}, <, \cdots \rangle$ be an extension of $\langle \mathbb{U}, < \rangle$ with arbitrary additional predicates. *If first-order logic has the natural order-generic collapse over $\langle \mathbb{U}, <, \cdots \rangle$ for the class of ω-databases, then it also has the natural order-generic collapse over $\langle \mathbb{U}, <, \cdots \rangle$ for the class of ω-representable databases.* □

Structure of the Paper. In section 2 we provide the notation used throughout the paper. In section 3 we give an outline of the proof and we point out analogies and differences compared with related papers which use a similar proof method. In section 4 we explain the collapse result of [8] which gives us the collapse for ω-databases. In section 5 we examine infinitary logic and give a characterization of ω-representable relations. In section 6 we explain how an ω-representable database can be represented by a ω-database. In section 7 we show that there are first-order interpretations that map an ω-representable database to an ω-database, and vice versa. In section 8 we prove the two main theorems. In section 9 we conclude the paper by pointing out further questions and a potential application.

2 Preliminaries

We use $\mathbb{Q}$ for the set of rationals, $\mathbb{R}$ for the set of reals, and ω for the set of non-negative integers. For $r, s \in \mathbb{R}$ we write $int[r,s]$ to denote the closed interval $\{x \in \mathbb{R} : r \leqslant x \leqslant s\}$. Analogously, we write $int[r,s)$ for the halfopen interval $int[r,s] \setminus \{s\}$, and $int(r,s)$ for the open interval $int[r,s] \setminus \{r,s\}$.

Depending on the particular context, we use $\vec{x}$ as abbreviation for a sequence $x_1,..,x_m$ or a tuple $(x_1,..,x_m)$. Accordingly, if q is a mapping defined on all elements in $\vec{x}$, we write $q(\vec{x})$ to denote the sequence $q(x_1),..,q(x_m)$ or the tuple $(q(x_1),..,q(x_m))$. If R is an m-ary relation on the domain of q, we write $q(R)$ to denote the relation $\{q(\vec{x}) : \vec{x} \in R\}$. Instead of $\vec{x} \in R$ we often write $R(\vec{x})$. For two disjoint sets A and B we write $A \uplus B$ to denote the disjoint union of A and B.

First-Order Logic *FO*(τ). A *signature* τ consists of finitely many relation and constant symbols. Each relation symbol $R \in \tau$ has a fixed arity $ar(R) \in \omega$. Whenever we refer to some "$c \in \tau$", we implicitly assume that c is a *constant* symbol in τ. Analogously, "$R \in \tau$" always means that R is a *relation* symbol in τ. We use $x_1, x_2, ..$ as variable symbols. *Atomic* τ-formulas are $y_1 = y_2$ and $R(y_1, .., y_m)$, where $R \in \tau$ is of arity, say, m and $y_1, .., y_m$ are constant symbols in τ or variable symbols. *FO*(τ)*-formulas* are built up as usual from the atomic τ-formulas and the logical connectives $\wedge$, $\vee$, $\neg$, the variable symbols $x_1, x_2, ..$, the existential quantifier $\exists$, and the universal quantifier $\forall$. We write $\text{qd}(\varphi)$ to denote the *quantifier depth* of a formula φ, i.e., the maximum number of nested quantifiers that occurs in φ. We sometimes write $\varphi(x_1, .., x_k)$ to indicate that $x_1, .., x_k$ are the free variables of φ, i.e., those variables that are not bound by a quantifier. We say that φ is a *sentence* if it has no free variables. If we insert additional constant or relation symbols, e.g. $<$ and $+$, into a signature τ, then we simply write $FO(\tau, <, +)$ instead of $FO(\tau \cup \{<, +\})$.

Structures. Let τ be a signature. A τ*-structure* $\mathcal{A} = \langle \mathbb{U}, \tau^{\mathcal{A}} \rangle$ consists of an arbitrary set $\mathbb{U}$ which is called the *universe* of $\mathcal{A}$, and a set $\tau^{\mathcal{A}}$ that contains

- an interpretation $R^{\mathcal{A}} \subseteq \mathbb{U}^{ar(R)}$, for each $R \in \tau$, and
- an interpretation $c^{\mathcal{A}} \in \mathbb{U}$, for each $c \in \tau$.

The *active domain* of $\mathcal{A}$ is the set of all constants of $\mathcal{A}$, together with the set of all elements in $\mathbb{U}$ that belong to one of $\mathcal{A}$'s relations.

Sometimes we explicitly want to specify the universe $\mathbb{U}$ of a τ-structure $\mathcal{A}$. In these cases we say that $\mathcal{A}$ is a $\langle \mathbb{U}, \tau \rangle$-structure. In the present paper, we only consider structures with universe $\mathbb{U} \in \{\mathbb{R}, \mathbb{Q}\}$.

For a $FO(\tau)$-sentence φ we say that $\mathcal{A}$ models φ and write $\mathcal{A} \models \varphi$ to indicate that φ is satisfied when interpreting each symbol in τ by its interpretation in $\tau^{\mathcal{A}}$. We write $\mathcal{A} \not\models \varphi$ to indicate that $\mathcal{A}$ does *not* model φ. For a $FO(\tau)$-formula $\varphi(x_1, .., x_k)$ and for elements $a_1, .., a_k$ in the universe of $\mathcal{A}$ we write $\mathcal{A} \models \varphi(a_1, .., a_k)$ to indicate that the $(\tau \cup \{x_1, .., x_k\})$-structure $\langle \mathcal{A}, a_1, .., a_k \rangle$ models the $FO(\tau \cup \{x_1, .., x_k\})$-sentence φ.

Since it is more convenient for our proof, we will talk about *structures* instead of *databases*. A structure can be viewed as a database whose database schema may contain not only relation symbols but also constant symbols. This allows us to restrict ourselves to boolean queries (which are formulated by *sentences*) instead of considering the general case of k-ary queries for arbitrary k (which are formulated by formulas with k free variables).

Order-Generic Collapse. Let $\mathbb{U} \in \{\mathbb{R}, \mathbb{Q}\}$. A mapping $\alpha : \mathbb{U} \to \mathbb{U}$ is called an *order-automorphism* of $\mathbb{U}$ if it is bijective and strictly increasing. For a $\langle \mathbb{U}, \tau \rangle$-structure $\mathcal{A}$ we write $\alpha(\mathcal{A})$ to denote the $\langle \alpha(\mathbb{U}), \tau \rangle$-structure with $R^{\alpha(\mathcal{A})} = \alpha(R^{\mathcal{A}})$ for all $R \in \tau$ and $c^{\alpha(\mathcal{A})} = \alpha(c^{\mathcal{A}})$ for all $c \in \tau$.

Let $\langle \mathbb{U}, <, \cdots \rangle$ be an extension of $\langle \mathbb{U}, < \rangle$ with arbitrary additional predicates. A $FO(\tau, <, \cdots)$-sentence φ is called *order-generic* on $\mathcal{A}$ iff for every order-automorphism α of $\mathbb{U}$ it is true that "$\langle \mathcal{A}, <, \cdots \rangle \models \varphi$ iff $\langle \alpha(\mathcal{A}), <, \cdots \rangle \models \varphi$".

Let $\mathcal{C}$ be a class of structures. We say "*first-order logic has the natural order-generic collapse over* $\langle \mathbb{U}, <, \cdots \rangle$ *on structures in* $\mathcal{C}$" to express that the following is valid for every signature τ: Let φ be a $FO(\tau, <, \cdots)$-sentence, and let $\mathcal{K}$ be the class of all $\langle \mathbb{U}, \tau \rangle$-structures in $\mathcal{C}$ on which φ is order-generic. There exists a $FO(\tau, <)$-sentence ψ which is equivalent to φ on $\mathcal{K}$, i.e., "$\langle \mathcal{A}, <, \cdots \rangle \models \varphi$ iff $\langle \mathcal{A}, < \rangle \models \psi$" is true for all $\mathcal{A} \in \mathcal{K}$.

Infinitary Logic $L_{\infty\omega}(<, S)$. Infinitary logic is defined in the same way as first-order logic, except that *arbitrary* (i.e. possibly infinite) disjunctions and conjunctions are allowed. Only in the context of infinitary logic we allow a signature to contain *infinitely* many symbols. What we need in the present paper is the following: Let S be a possibly infinite set of constant symbols. The logic $L_{\infty\omega}(<, S)$ is given by the following clauses: It contains all atomic formulas $x{=}y$ and $x{<}y$, where x and y are variable symbols or elements in S. If it contains φ, then it contains also $\neg\varphi$. If it contains φ and if x is a variable symbol, then it contains also $\exists x \varphi$ and $\forall x \varphi$. If Φ is a (possibly infinite) set of $L_{\infty\omega}(<, S)$-formulas, then $\bigvee \Phi$ and $\bigwedge \Phi$ are formulas in $L_{\infty\omega}(<, S)$.

The semantics is a direct extension of the semantics of first-order logic, where $\bigvee \Phi$ is true if there is some $\varphi \in \Phi$ which is true; and $\bigwedge \Phi$ is true if every $\varphi \in \Phi$ is true.

In the present paper we use infinitary logic only for the universe $\mathbb{U} = \mathbb{R}$ or $\mathbb{U} = \mathbb{Q}$, where the constant symbols are interpreted by numbers in $\mathbb{U}$. Consequently, we identify the set S of constant symbols with a set $S \subseteq \mathbb{U}$.

Sets of Type at Most ω, ω-Structures, and ω-Representable Structures. Let $\mathbb{U} \in \{\mathbb{R}, \mathbb{Q}\}$. We say that $S \subseteq \mathbb{U}$ is *of type* ω if $\langle \mathbb{U}, <, S \rangle$ is isomorphic to $\langle \mathbb{U}, <, \omega \rangle$. One can easily see that S is of type ω if and only if $S = \{s_1 < s_2 < \cdots\}$, where the sequence $(s_n)_{n \geqslant 1}$ is strictly increasing and unbounded. Accordingly, we say that S is *of type at most* ω if S is finite or of type ω.

We say that a $\langle \mathbb{U}, \tau \rangle$-structure $\mathcal{A}$ is an *ω-structure* if the active domain of $\mathcal{A}$ is of type at most ω.

A relation $R \subseteq \mathbb{U}^m$ is called *ω-representable* if there is a set $S \subseteq \mathbb{U}$ of type at most ω such that R is definable in $L_{\infty\omega}(<, S)$, i.e. there is a $L_{\infty\omega}(<, S)$-formula $\varphi_R(x_1, .., x_m)$ with $R = \{\vec{a} \in \mathbb{U}^m : \mathbb{U} \models \varphi_R(\vec{a})\}$. Accordingly, a $\langle \mathbb{U}, \tau \rangle$-structure $\mathcal{A}$ is called ω-representable if each of $\mathcal{A}$'s relations is ω-representable.

For better readability, we formulate the rest of the paper only for the case $\mathbb{U} = \mathbb{R}$. However, all statements remain correct if one replaces $\mathbb{R}$ by $\mathbb{Q}$.

3 Outline of the Proof – The Lifting Method

It is by now quite a common method in database theory to *lift* results from one class of databases to another. This *lifting method* can be described as follows:

Known: *A result for a class of "easy" databases.*

Wanted: *The analogous result for a class of "complicated" databases.*

Method:

(1.) Show that all the relevant information about a "complicated" database can be represented by an "easy" database.

(2.) Show that the translation from the "complicated" to the "easy" database (and vice versa) can be performed in an appropriate way (e.g. via an efficient algorithm or via FO-formulas).

(3.) Use this to translate the known result for the "easy" databases into the desired result for the "complicated" databases.

In the literature the "easy" database which represents a "complicated" database is usually called *the invariant* of the "complicated" database. Table 1 gives a listing of recent papers in which the lifting method has been used.

Table 1. Some papers using the lifting method.

	"compl." dbs	**"easy" dbs**	**result ("easy" dbs)**	**result ("compl." dbs)**
[9]	planar spatial dbs	finite dbs	evaluation of fixpoint+counting queries	evaluation of top. *FO*($\mathbb{R}, <$)-queries
[7]	region dbs	finite dbs	order-generic collapse over $\langle \mathbb{R}, <, +, \times \rangle$ (cf. [2])	collapse from top. *FO*($\mathbb{R}, <, +, \times$)-queries to top. *FO*($\mathbb{R}, <$)-queries
[5]	finitely rep. dbs	finite dbs	logical characterization of complexity classes	complexity of query evaluation
[1]	finitely rep. dbs	finite dbs	order-generic collapse over quasi o-minimal structures	order-generic collapse over quasi o-minimal structures
[here]	ω-rep. dbs	ω-dbs	order-generic collapse over $\langle \mathbb{R}, <, + \rangle$	order-generic collapse over $\langle \mathbb{R}, <, + \rangle$

In particular, Belegradek, Stolboushkin, and Taitslin [1] and Grädel and Kreutzer [5] show that all the relevant information about a finitely representable database (i.e. a database defined by a finite Boolean combination of order-constraints) can be represented by a finite database, and that the translation from finitely representable to finite (and vice versa, in [1]) can be done by a first-order interpretation.

Grädel and Kreutzer use this translation to carry over logical characterizations of complexity classes to results on the data complexity of query evaluation. They lift, e.g., the well-known logical characterization "PTIME = FO+LFP on ordered finite structures" to the result stating that the polynomial time computable queries against finitely representable databases are exactly the FO+LFP-definable queries.

Belegradek, Stolboushkin, and Taitslin use their FO-translations from finitely representable databases to finite databases (and vice versa) to lift collapse results for finite databases to collapse results for finitely representable databases.

In the present paper the same is done for ω-representable databases and ω-databases (instead of finitely representable databases and finite databases, respectively). I.e.:

(1.) We show how all the relevant information about an ω-representable database can be represented by an ω-database (cf. sections 5 and 6).
The representation here is considerably different from the representations of [1] and [5]. It is, as the author feels, more natural for the context considered in the present paper.

(2.) We show that the translation from the ω-representable to the ω-database (and vice versa) can be done by a first-order interpretation (cf. section 7).

(3.) We use this translation to carry over a collapse result for ω-databases from [8] to a collapse result for ω-representable databases (cf. section 8).

4 The Collapse Result for ω-Structures

In [8] a structure $\mathcal{A}$ is called *nicely representable* if it satisfies the following conditions:

(1) There is an infinite sequence $(I_n)_{n\in\omega}$ of intervals $I_n = int[l_n, r_n]$, such that $l_n \leqslant r_n < l_{n+1}$, and the sequence $(r_n)_{n\in\omega}$ is unbounded,

(2) $\biguplus_{n\in\omega} I_n$ is the active domain of $\mathcal{A}$,

(3) every relation $R^{\mathcal{A}}$ of $\mathcal{A}$ is constant on the multi-dimensional rectangles $I_{n_1} \times \cdots \times I_{n_{ar(R)}}$ (for all $n_1, .., n_{ar(R)} \in \omega$). I.e., either all elements in $I_{n_1} \times \cdots \times I_{n_{ar(R)}}$ belong to $R^{\mathcal{A}}$, or no element in $I_{n_1} \times \cdots \times I_{n_{ar(R)}}$ belongs to $R^{\mathcal{A}}$.

Theorem 1 ([8], Theorem 4). *First-order logic has the natural order-generic collapse over $\langle \mathbb{R}_{\geqslant 0}, <, + \rangle$ for nicely representable structures.* □

Let us mention that the class of ω-representable structures (considered in the present paper) properly contains both, the class of finitely representable and the class of nicely representable structures, whereas the class of nicely representable structures does *not* contain the class of finitely representable structures.

The proof of Theorem 1 presented in [8] even shows the slightly stronger result which states that first-order logic has the natural order-generic collapse over $\langle \mathbb{R}, <, + \rangle$ for structures that satisfy the conditions (1), (2'), and (3), where the condition (2') says that there is a set $N \subseteq \omega$ such that $\biguplus_{n\in N} I_n$ is the active domain of $\mathcal{A}$. In particular ω-structures, i.e. structures whose active domain is of type at most ω, *do* satisfy the conditions (1), (2'), and (3). This gives us the following

Corollary 1. *First-order logic has the natural order-generic collapse over $\langle \mathbb{R}, <, + \rangle$ for ω-structures.* □

5 Infinitary Logic and ω-Representable Relations

It is well-known that $FO(<, S)$ allows quantifier elimination over $\mathbb{R}$, for every set of constants $S \subseteq \mathbb{R}$. In this section we show that also $L_{\infty\omega}(<, S)$ allows quantifier elimination over $\mathbb{R}$, provided that S is *of type at most* ω. Recall from section 2 that $S \subseteq \mathbb{R}$ is *of type* ω if and only if $S = \{s_1 < s_2 < \cdots\}$, where the sequence $(s_n)_{n \geqslant 1}$ is strictly increasing and unbounded. Accordingly, S is *of type at most* ω if S is finite or of type ω.

However, our aim is not only to show that $L_{\infty\omega}(<, S)$ allows quantifier elimination, but to give an *explicit characterization* of the quantifier free formulas. This characterization will give us full understanding of what ω-representable relations look like.

Before giving the formalization of the quantifier elemination let us fix some notation. For the rest of this paper let $S \subseteq \mathbb{R}$ always be of type at most ω. We write $S(i)$ to denote the i-th smallest element in S. For infinite S we define $S(0) := -\infty$ and $N(S) := \omega$, and we obtain $\mathbb{R} = \biguplus_{i \in N(S)} int[S(i), S(i{+}1))$.

For finite S we define $S(0) := -\infty$, $N(S) := \{0, .., |S|\}$, and $S(|S|{+}1) := +\infty$; and, as before, we obtain $\mathbb{R} = \biguplus_{i \in N(S)} int[S(i), S(i{+}1))$.

For $m \geqslant 1$ and $\vec{\imath} = (i_1, .., i_m) \in N(S)^m$ we define $S(\vec{\imath}) := (S(i_1), .., S(i_m))$, and

$$Cube_{S;\vec{\imath}} := int[S(i_1), S(i_1{+}1)) \times \cdots \times int[S(i_m), S(i_m{+}1)) .$$

We say that $S(\vec{\imath})$ are the *coordinates* of the cube $Cube_{S;\vec{\imath}}$. Obviously,

$$\mathbb{R}^m = \biguplus_{\vec{\imath} \in N(S)^m} Cube_{S;\vec{\imath}} .$$

Let $\vec{a} = (a_1, .., a_m) \in \mathbb{R}^m$. The type $type_{\vec{a};S;\vec{\imath}}$ of $\vec{a}$ with respect to $Cube_{S;\vec{\imath}}$ is the conjunction of all atoms in $\{y_i{=}x_i,\ y_i{<}x_i,\ x_i{=}x_j,\ x_i{<}x_j\ :\ i,j \in \{1.., m\},\ i \neq j\}$ which are satisfied if one interprets the variables $x_1, .., x_m, y_1, .., y_m$ by the numbers $a_1, .., a_m, S(i_1), .., S(i_m)$.

We define $\boldsymbol{types}_m$ to be the set of all complete conjunctions of atoms in $\{y_i{=}x_i,\ y_i{<}x_i,\ x_i{=}x_j,\ x_i{<}x_j\ :\ i,j \in \{1.., m\},\ i \neq j\}$, i.e., the set of all conjuctions t where, for all $i, j \in \{1, .., m\}$ with $i \neq j$, either $y_i{=}x_i$ or $y_i{<}x_i$ occurs in t, and either $x_i{=}x_j$ or $x_i{<}x_j$ or $x_j{<}x_i$ occurs in t. Of course, $\boldsymbol{types}_m$ is finite, and $type_{\vec{a};S;\vec{\imath}} \in \boldsymbol{types}_m$. Analogously, we define $\boldsymbol{Types}_m$ to be the set of all subsets of $\boldsymbol{types}_m$, i.e., $\boldsymbol{Types}_m = \{T : T \subseteq \boldsymbol{types}_m\}$. Of course, $\boldsymbol{Types}_m$ is finite.

For a relation $R \subseteq \mathbb{R}^m$ we define $Type_{R;S;\vec{\imath}} := \{type_{\vec{a};S;\vec{\imath}} : \vec{a} \in R \cap Cube_{S;\vec{\imath}}\}$ to be the set of all types occurring in the restriction of R to $Cube_{S;\vec{\imath}}$. We say that $Type_{R;S;\vec{\imath}}$ is the *type of* $Cube_{S;\vec{\imath}}$ *in* R. Of course, $Type_{R;S;\vec{\imath}} \in \boldsymbol{Types}_m$.

In the formalization of the quantifier elimination we further use the following notation: If φ is a $L_{\infty\omega}(<, S)$-formula with free variables $\vec{x} := x_1, .., x_k$ and $\vec{y} := y_1, .., y_m$, we write $\varphi(\vec{y}/S(\vec{\imath}))$ to denote the formula one obtains by replacing the variables $y_1, .., y_m$ by the real numbers $S(i_1), .., S(i_m)$.

Proposition 1 (Quantifier Elimination). *Let $S \subseteq \mathbb{R}$ be of type at most ω and let $m \geqslant 1$. Every formula $\varphi(x_1,..,x_m)$ in $L_{\infty\omega}(<,S)$ is equivalent over $\mathbb{R}$ to the formula*

$$\tilde{\varphi}(\vec{x}) \quad := \quad \bigvee_{\vec{\imath} \in N(S)^m} \ \bigvee_{t \in Type_{R;S;\vec{\imath}}} \Big(t(\vec{y}/S(\vec{\imath})) \wedge \bigwedge_{j=1}^{m} S(i_j) \leqslant x_j < S(i_j{+}1) \Big)$$

where $R \subseteq \mathbb{R}^m$ is the relation defined by $\varphi(\vec{x})$.
I.e., $R \;=\; \{\vec{a} \in \mathbb{R}^m : \mathbb{R} \models \varphi(\vec{a})\} \;=\; \{\vec{a} \in \mathbb{R}^m : \mathbb{R} \models \tilde{\varphi}(\vec{a})\}.$ □

The proof is similar to the quantifier elimination for *FO*$(<,S)$ over $\mathbb{R}$. Due to space limitations it is omitted here.

Recall from section 2 that a relation $R \subseteq \mathbb{R}^m$ is called ω-representable iff there is a set $S \subseteq \mathbb{R}$ of type at most ω such that R is definable in $L_{\infty\omega}(<,S)$. From Proposition 1 we know what R looks like: It is defined by an infinitary boolean combination of order-constraints over S, and it essentially consists of a finite or a countable number of multidimensional rectangles. (Note, however, that also certain triangles are allowed, e.g. via the constraint $S(i) \leqslant x_1 < x_2 < S(i{+}1)$). An ω-representable *binary* relation is illustrated in Figure 1.

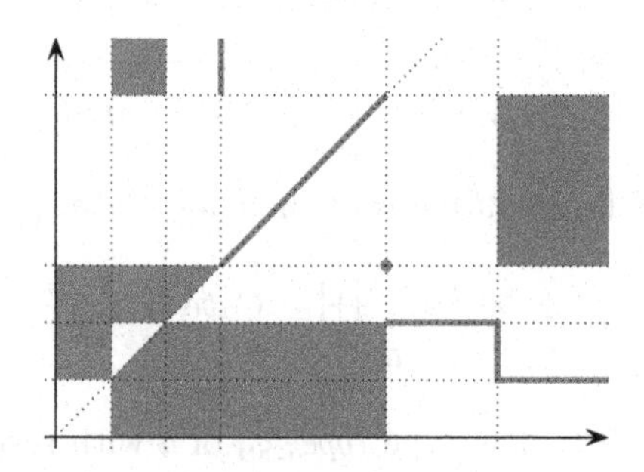

Fig. 1. An ω-rep. binary relation R. The grey regions are those that belong to R.

6 ω-Representations of Relations and Structures

Definition 1. *Let $R \subseteq \mathbb{R}^m$. A set $S \subseteq \mathbb{R}$ is called* sufficient for defining R *if S is of type at most ω and R is definable in $L_{\infty\omega}(<,S)$.* □

Remark 1. We say that a relation $R \subseteq \mathbb{R}^m$ is *constant* on a set $M \subseteq \mathbb{R}^m$ if either all elements of M belong to R or no element of M belongs to R.

From Proposition 1 we obtain that a set $S \subseteq \mathbb{R}$ of type at most ω is sufficient for defining R if and only if R is constant on the sets

$$Cubes_{S;\vec{\imath};t} \ := \ \{\vec{b} \in Cubes_{S;\vec{\imath}} \, : \, type_{\vec{b};S;\vec{\imath}} = t\},$$

for all $\vec{\imath} \in N(S)^m$ and all $t \in \boldsymbol{types}_m$. □

Let $R \subseteq \mathbb{R}^m$ be ω-representable and let $S \subseteq \mathbb{R}$ be sufficient for defining R. From Remark 1 we know, for all $\vec{\imath} \in N(S)^m$ and all $t \in \textbf{\textit{types}}_m$, that either $R \cap Cubes_{S;\vec{\imath};t} = \emptyset$ or $R \supseteq Cubes_{S;\vec{\imath};t}$. This means that if we know, for each $\vec{\imath} \in N(S)^m$ and each $t \in \textbf{\textit{types}}_m$, whether or not R contains an element of $Cubes_{S;\vec{\imath};t}$, then we can reconstruct the entire relation R.

For $i_j \neq 0$ we represent the interval $int[S(i_j), S(i_j+1)) \subseteq \mathbb{R}$ by the number $S(i_j)$. Consequently, for $\vec{\imath} \in (N(S) \setminus \{0\})^m$, we can represent $Cubes_{S;\vec{\imath};t} \subseteq \mathbb{R}^m$ by the tuple $S(\vec{\imath}) \in S^m$. The information whether or not R contains an element of $Cubes_{S;\vec{\imath};t}$ can be represented by the relation $R_{S;t} := \{S(\vec{\imath}) : \vec{\imath} \in (N(S) \setminus \{0\})^m$ and $R \cap Cubes_{S;\vec{\imath};t} \neq \emptyset\}$.

In general, we would like to represent every $Cubes_{S;\vec{\imath};t}$, for *every* $\vec{\imath} \in N(S)^m$, by a tuple in S^m. Unfortunately, the case where $i_j = 0$ must be treated separately, because $S(0) = -\infty \notin S$. There are various possibilities for solving this technical problem. Here we propose the following solution: Use $S(1)$ to represent the interval $int[S(0), S(1))$. With every tuple $\vec{\imath} \in N(S)^m$ we associate a *characteristic tuple* $char(\vec{\imath}) := (c_1, .., c_m) \in \{0,1\}^m$ and a tuple $\vec{\imath'} \in (N(S) \setminus \{0\})^m$ via $c_j := 0$ and $i'_j := 1$ if $i_j = 0$, and $c_j := 1$ and $i'_j := i_j$ if $i_j \neq 0$. Now $Cubes_{S;\vec{\imath};t}$ can be represented by the tuple $S(\vec{\imath'}) \in S^m$. The information whether or not R contains an element of $Cubes_{S;\vec{\imath};t}$ can be represented by the relations $R_{S;t;\vec{u}} := \{S(\vec{\imath'}) : \vec{\imath} \in N(S)^m,\ char(\vec{\imath}) = \vec{u}$, and $R \cap Cubes_{S;\vec{\imath};t} \neq \emptyset\}$ (for all $\vec{u} \in \{0,1\}^m$). This leads to

Definition 2 (ω-Representation of a Relation).
Let $R \subseteq \mathbb{R}^m$ be ω-representable, and let $S \subseteq \mathbb{R}$ be sufficient for defining R.

(a) We represent the m-ary relation R over $\mathbb{R}$ by a finite number of m-ary relations over S as follows: The ω-representation of R with respect to S is the collection

$$rep_S(R) \ := \ \langle\, R_{S;t;\vec{u}} \,\rangle_{t \in \textbf{\textit{types}}_m,\, \vec{u} \in \{0,1\}^m},$$

where $R_{S;t;\vec{u}} := \{S(\vec{\imath'}) : \vec{\imath} \in N(S)^m,\ char(\vec{\imath}) = \vec{u}$, and $R \cap Cubes_{S;\vec{\imath};t} \neq \emptyset\}$. Here, for $\vec{\imath} \in N(S)^m$ we define $\vec{\imath'}$ and $char(\vec{\imath})$ via $i'_j := 1$ and $\big(char(\vec{\imath})\big)_j := 0$ if $i_j = 0$, and $i'_j := i_j$ and $\big(char(\vec{\imath})\big)_j := 1$ if $i_j \neq 0$.

(b) For $\vec{x} \in Cubes_{S;\vec{\imath};t}$ we say that $\vec{u} := char(\vec{\imath})$ is the characteristic tuple *of $\vec{x}$ w.r.t. S, $\vec{y} := S(\vec{\imath'})$ is* the representative *of $\vec{x}$ w.r.t. S, and t is* the type *of $\vec{x}$ w.r.t. S. From Remark 1 we obtain that $\vec{x} \in R$ iff $\vec{y} \in R_{S;t;\vec{u}}$.* □

We will now tranfer the notion of "ω-representation" from relations to τ-structures.

Recall from section 2 that a $\langle \mathbb{R}, \tau \rangle$-structure $\mathcal{A}$ is called ω-representable iff each of $\mathcal{A}$'s relations is ω-representable.

Definition 3. *Let $\mathcal{A}$ be a $\langle \mathbb{R}, \tau \rangle$-structure. A set $S \subseteq \mathbb{R}$ is called* sufficient for defining $\mathcal{A}$ *if*

- *S is of type at most ω,*
- *$c^{\mathcal{A}} \in S$, for every constant symbol $c \in \tau$, and*

– *S is sufficient for defining $R^{\mathcal{A}}$, for every relation symbol $R \in \tau$.* □

Let $\mathcal{A}$ be a $\langle \mathbb{R}, \tau \rangle$-structure, and let S be a set sufficient for defining $\mathcal{A}$. According to Definition 2, each of $\mathcal{A}$'s relations $R^{\mathcal{A}}$ of arity, say, m can be represented by a finite collection $reps_S(R^{\mathcal{A}}) = \langle R^{\mathcal{A}}_{S;t;\vec{u}} \rangle_{t \in \textbf{types}_m,\, \vec{u} \in \{0,1\}^m}$ of relations over S. I.e. $\mathcal{A}$ can be represented by a structure $reps_S(\mathcal{A})$ with active domain S as follows:

Definition 4 (ω-Representation of a Structure). *Let τ be a signature.*

(a) The type extension *τ' of τ is the signature which consists of*
  - *the same constant symbols as τ,*
  - *a unary relation symbol S, and*
  - *a relation symbol $R_{t;\vec{u}}$ of arity, say, m, for every relation symbol $R \in \tau$ of arity m, every $t \in \textbf{types}_m$, and every $\vec{u} \in \{0,1\}^m$.*

(b) Let $\mathcal{A}$ be an ω-representable $\langle \mathbb{R}, \tau \rangle$-structure and let S be a set sufficient for defining $\mathcal{A}$. We represent $\mathcal{A}$ by the $\langle \mathbb{R}, \tau' \rangle$-structure $reps_S(\mathcal{A})$ which satisfies
  - $c^{reps_S(\mathcal{A})} = c^{\mathcal{A}}$ *(for each $c \in \tau'$),*
  - $S^{reps_S(\mathcal{A})} = S$ *(for the unary relation symbol $S \in \tau'$), and*
  - $R^{reps_S(\mathcal{A})}_{t;\vec{u}} = R^{\mathcal{A}}_{S;t;\vec{u}}$ *(for each $R \in \tau$, each $t \in \textbf{types}_{ar(R)}$, and each $\vec{u} \in \{0,1\}^{ar(R)}$).* □

7 FO-Interpretations

The concept of first-order interpretations (or, reductions) is well-known in mathematical logic (cf., e.g. [4]). In the present paper we consider the following easy version:

Definition 5 (FO-Interpretation of σ in τ). *Let σ and τ be signatures. A* FO-interpretation of σ in τ *is a collection*

$$\Phi = \Big\langle \big(\varphi_c(x)\big)_{c \in \sigma},\ \big(\varphi_R(x_1, \ldots, x_{ar(R)})\big)_{R \in \sigma} \Big\rangle$$

of FO(τ)-formulas. For every $\langle \mathbb{U}, \tau \rangle$-structure $\mathcal{A}$, the $\langle \mathbb{U}, \sigma \rangle$-structure $\Phi(\mathcal{A})$ is given via

- $\{c^{\Phi(\mathcal{A})}\} = \{a \in \mathbb{U} : \mathcal{A} \models \varphi_c(a)\}$, *for each constant symbol $c \in \sigma$,*
- $R^{\Phi(\mathcal{A})} = \{\vec{a} \in \mathbb{U}^{ar(R)} : \mathcal{A} \models \varphi_R(\vec{a})\}$, *for each relation symbol $R \in \sigma$.* □

Making use of a FO-interpretation of σ in τ, one can translate FO(σ)-formulas into FO(τ)-formulas (cf., [4, Exercise 11.2.4]):

Lemma 1. *Let σ and τ be signatures, let Φ be a FO-interpretation of σ in τ, and let d be the maximum quantifier depth of the formulas in Φ.*

For every FO(σ)-sentence χ there is a FO(τ)-sentence χ' with $qd(\chi') \leqslant qd(\chi)+d$, such that "$\mathcal{A} \models \chi'$ iff $\Phi(\mathcal{A}) \models \chi$" is true for every $\langle \mathbb{U}, \tau \rangle$-structure $\mathcal{A}$. □

Proof. χ' is obtained from χ by replacing every atomic formula $R(\vec{x})$ (resp. $x{=}c$) by the formula $\varphi_R(\vec{x})$ (resp. $\varphi_c(x)$). ■

The following lemma shows that $\mathcal{A}$ is first-order definable in $reps_S(\mathcal{A})$, i.e.: all relevant information about $\mathcal{A}$ can be reconstructed from $reps_S(\mathcal{A})$ (if $\mathcal{A}$ is ω-representable and if S is sufficient for defining $\mathcal{A}$).

Lemma 2. *There is a FO-interpretation Φ of τ in $\tau' \cup \{<\}$ such that $\Phi(\langle reps_S(\mathcal{A}), <\rangle) = \mathcal{A}$, for every ω-representable $\langle \mathbb{R}, \tau\rangle$-structure $\mathcal{A}$ and every set S which is sufficient for defining $\mathcal{A}$.* □

Proof (sketch). For every constant symbol $c \in \tau$ we define $\varphi_c(x) := x{=}c$.

For every relation symbol $R \in \tau$ of arity, say, m we construct a formula $\varphi_R(\vec{x})$ which expresses that $\vec{x} \in R$. From Definition 2(b) we know that $\vec{x} \in R$ iff $\vec{y} \in R_{S;t;\vec{u}}$, where $\vec{y}$, t, and $\vec{u}$ are the representative, the type, and the characteristic tuple, respectively, of $\vec{x}$ w.r.t. S.

It is straightforward to construct, for fixed $t \in \boldsymbol{types}_m$ and $\vec{u} \in \{0,1\}^m$, a $FO(\tau', <)$-formula $\psi_{t,\vec{u}}(\vec{x})$ which expresses that

- $\vec{x}$ has type t w.r.t. S,
- $\vec{u}$ is the characteristic tuple of $\vec{x}$ w.r.t. S, and
- for the representative $\vec{y}$ of $\vec{x}$ w.r.t. S it holds that $R_{t;\vec{u}}(\vec{y})$.

The disjunction of the formulas $\psi_{t;\vec{u}}(\vec{x})$, for all $t \in \boldsymbol{types}_m$ and all $\vec{u} \in \{0,1\}^m$, gives us the desired formula $\varphi_R(\vec{x})$ which expresses that $\vec{x} \in R$. ■

We now want to show the converse of Lemma 2, i.e., we want to show that the ω-representation of $\mathcal{A}$ is first-order definable in $\mathcal{A}$. Up to now the ω-representation $reps_S(\mathcal{A})$ was parameterized by a set S which is sufficient for defining $\mathcal{A}$. For the current step we need the existence of a *canonical*, first-order definable set S. For this canonization we can use the following result of Grädel and Kreutzer [5, Lemma 8]:

Lemma 3 (Canonical set sufficient for defining R). *Let $R \subseteq \mathbb{R}^m$ be ω-representable and let S_R be the set of all elements $s \in \mathbb{R}$ which satisfy the following condition $(*)$:*

> *There are $a_1, .., a_m, \varepsilon \in \mathbb{R}$, $\varepsilon > 0$, such that one of the following holds:*
> - *For all $s' \in int(s{-}\varepsilon, s)$ and for no $s' \in int(s, s{+}\varepsilon)$ we have $R(\vec{a}[s/s'])$. Here $\vec{a}[s/s']$ means that all components $a_j{=}s$ are replaced by s'.*
> - *For no $s' \in int(s{-}\varepsilon, s)$ and for all $s' \in int(s, s{+}\varepsilon)$ we have $R(\vec{a}[s/s'])$.*
> - *$R(\vec{a}[s/s'])$ holds for all $s' \in int(s{-}\varepsilon, s{+}\varepsilon) \setminus \{s\}$, but not for $s' = s$.*
> - *$R(\vec{a}[s/s'])$ holds for $s' = s$, but not for any $s' \in int(s{-}\varepsilon, s{+}\varepsilon) \setminus \{s\}$.*

The following holds true:

(1.) S_R is included in every set $S \subseteq \mathbb{R}$ which is sufficient for defining R.
(2.) S_R is sufficient for defining R.

The set S_R is called the canonical set sufficient for defining R. *It is straightforward to formulate a $FO(R, <)$-formula $\zeta_R(x)$ which expresses condition $(*)$, such that $S_R = \{s \in \mathbb{R} : \langle \mathbb{R}, R, <\rangle \models \zeta_R(s)\}$ for every ω-representable m-ary relation R.* □

Definition 6 (Canonical Representation of a Structure). *Let τ be a signature and let $\mathcal{A}$ be an ω-representable $\langle \mathbb{R}, \tau \rangle$-structure. The set*

$$S_{\mathcal{A}} := \{c^{\mathcal{A}} : c \in \tau\} \cup \bigcup_{R \in \tau} S_{R^{\mathcal{A}}}$$

is called the canonical set sufficient for defining $\mathcal{A}$. *Similarly, the representation* $canrep(\mathcal{A}) := reps_{S_{\mathcal{A}}}(\mathcal{A})$ *is called* the canonical representation of $\mathcal{A}$. □

Remark 2. It is straightforward to see that $\alpha\big(canrep(\mathcal{A})\big) = canrep\big(\alpha(\mathcal{A})\big)$ is true for every ω-representable $\langle \mathbb{R}, \tau \rangle$-structure $\mathcal{A}$ and every order-automorphism α of $\mathbb{R}$. □

We are now ready to prove the converse of Lemma 2.

Lemma 4. *There is a FO-interpretation Φ' of τ' in $\tau \cup \{<\}$ such that $\Phi'(\langle \mathcal{A}, <\rangle) = canrep(\mathcal{A})$, for every ω-representable $\langle \mathbb{R}, \tau \rangle$-structure $\mathcal{A}$.* □

Proof (sketch). For every constant symbol $c \in \tau'$ we define $\varphi_c(x) := x{=}c$.

For every relation symbol $R \in \tau$ let $\zeta_R(x)$ be the formula from Lemma 3 describing the canonical set sufficient for defining $R^{\mathcal{A}}$. Obviously, the formula $\varphi_S(x) := \bigvee_{c \in \tau} x{=}c \;\vee\; \bigvee_{R \in \tau} \zeta_R(x)$ describes the canonical set sufficient for defining $\mathcal{A}$.

For every relation symbol $R_{t;\vec{u}} \in \tau'$ of arity, say, m we construct a formula $\varphi_{R_{t;\vec{u}}}(\vec{y})$ which expresses that $\vec{y} \in R_{t;\vec{u}}$. We make use of Definition 2(b). I.e., $\varphi_{R_{t;\vec{u}}}$ states that $y_1, .., y_m$ satisfy φ_S and that there is some $\vec{x}$ such that

- $\vec{y}$ is the representative of $\vec{x}$ w.r.t. $S_{\mathcal{A}}$,
- $R(\vec{x})$,
- $\vec{x}$ has type t w.r.t. $S_{\mathcal{A}}$, and
- $\vec{u}$ is the characteristic tuple of $\vec{x}$ w.r.t. $S_{\mathcal{A}}$.

It is straightforward to formalize this in first-order logic. ■

8 The Main Theorems and Their Proofs

We first show the

Main Theorem 2. Let $\langle \mathbb{R}, <, \cdots \rangle$ be an extension of $\langle \mathbb{R}, < \rangle$ with arbitrary additional predicates. *If first-order logic has the natural order-generic collapse over $\langle \mathbb{R}, <, \cdots \rangle$ for the class of ω-structures, then it has the natural order-generic collapse over $\langle \mathbb{R}, <, \cdots \rangle$ for the class of ω-representable structures.* □

Proof. Let τ be a signature, let φ be a $FO(\tau, <, \cdots)$-sentence, and let $\mathcal{K}$ be the class of all ω-representable $\langle \mathbb{R}, \tau \rangle$-structures on which φ is order-generic. We need to find a $FO(\tau, <)$-sentence ψ such that "$\langle \mathcal{A}, <, \cdots \rangle \models \varphi$ iff $\langle \mathcal{A}, < \rangle \models \psi$" is valid for all $\mathcal{A} \in \mathcal{K}$.

Let τ' be the type extension of τ. We first make use of Lemma 2: Let Φ be the FO-interpretation of τ in $\tau' \cup \{<\}$ which is obtained in Lemma 2. In particular, we have $\Phi(\langle canrep(\mathcal{A}), < \rangle) = \mathcal{A}$, for all $\mathcal{A} \in \mathcal{K}$. From Lemma 1 we obtain a $FO(\tau', <, \cdots)$-sentence φ' such that "$\langle canrep(\mathcal{A}), <, \cdots \rangle \models \varphi'$ iff $\langle \Phi(\langle canrep(\mathcal{A}), < \rangle), <, \cdots \rangle \models \varphi$ iff $\langle \mathcal{A}, <, \cdots \rangle \models \varphi$" is true for all $\mathcal{A} \in \mathcal{K}$.

From our assumption we know that first-order logic has the natural order-generic collapse over $\langle \mathbb{R}, <, \cdots \rangle$ for the class of ω-structures. Of course $canrep(\mathcal{A})$ is an ω-structure. Furthermore, with Remark 2 we obtain that φ' is order-generic on $canrep(\mathcal{A})$ for all $\mathcal{A} \in \mathcal{K}$.

Hence there must be a $FO(\tau', <)$-sentence ψ' such that "$\langle canrep(\mathcal{A}), <, \cdots \rangle \models \varphi'$ iff $\langle canrep(\mathcal{A}), < \rangle \models \psi'$" is true for all $\mathcal{A} \in \mathcal{K}$.

We now make use of Lemma 4: Let Φ' be the FO-interpretation of τ' in $\tau \cup \{<\}$ which is obtained in Lemma 4. In particular, we have $\Phi'(\langle \mathcal{A}, < \rangle) = canrep(\mathcal{A})$, for all $\mathcal{A} \in \mathcal{K}$. According to Lemma 1, we can transform ψ' into a $FO(\tau, <)$-sentence ψ such that "$\langle \mathcal{A}, < \rangle \models \psi$ iff $\langle \Phi'(\langle \mathcal{A}, < \rangle), < \rangle \models \psi'$ iff $\langle canrep(\mathcal{A}), < \rangle \models \psi'$" is true for all $\mathcal{A} \in \mathcal{K}$. Obviously, ψ is the desired sentence, and hence our proof is complete. ■

Main Theorem 2 and Corollary 1 directly give us the following

Main Theorem 1. *First-order logic has the natural order-generic collapse over* $\langle \mathbb{R}, <, + \rangle$ *for the class of* ω*-representable structures.* □

9 Conclusion

We have developed the notion of ω-representable databases, which is a natural generalization of the notion of finitely representable (i.e. dense order constraint) databases. We have shown that any collapse result for ω-databases can be lifted to the analogous collapse result for ω-representable databases. In particular, this implies that first-order logic has the natural order-generic collapse over $\langle \mathbb{R}, <, + \rangle$ and $\langle \mathbb{Q}, <, + \rangle$ for ω-representable databases.

Recursive Databases. In theoretical computer science one is often interested in things that can be represented in the *finite*. This is not a priori true for ω-representable databases. sHowever, there is a line of research considering *recursive structures* (cf. [6]). In this setting a database is called *recursive* if there is, for each of its relations, an algorithm which effectively decides whether or not an input tuple belongs to that relation. The results of the present paper are, in particular, true for the class of ω*-representable recursive databases*, which still is a rather natural extension of the class of finitely representable (i.e. dense order constraint) databases.

Open Questions. It is an obvious question if the collapse results discussed in the present paper also hold for $\mathbb{Z}$-databases (i.e. databases whose active domain is of type at most $\mathbb{Z}$) and for $\mathbb{Z}$-representable databases. It should be straightforward to transform the proof of Main Theorem 2 in such a way that it is valid for these databases. However, we do not know if the corresponding analogue to Corollary 1 is valid.

Another question is whether such a collapse result *for ω-representable databases* is valid also over structures other than $\langle \mathbb{R}, <, + \rangle$ and $\langle \mathbb{Q}, <, + \rangle$. E.g.: Is it valid over $\langle \mathbb{R}, <, +, \times \rangle$, or even over all *(quasi) o-minimal* structures? (This would then fully generalize the results of Belegradek et al. [1].)

We also want to mention a potential application concerning *topological queries*: Kuijpers and Van den Bussche [7] used the theorem of Benedikt et al. [2] to obtain a collapse result for *topological* first-order definable queries. One step of their proof was to encode spatial databases (of a certain kind) by *finite* databases, to which the result of [2] can be applied. Here the question arises whether there is an interesting class of spatial databases that can be encoded by ω-representable (but not by finite) databases in such a way that our main theorem helps to obtain some collapse result for topological queries.

Acknowledgements

I want to thank Luc Segoufin for pointing out to me the connection to topological queries. Furthermore, I thank Clemens Lautemann for helpful discussions on the topics of this paper.

References

1. O.V. Belegradek, A.P. Stolboushkin, and M.A. Taitslin. Extended order-generic queries. *Annals of Pure and Applied Logic*, 97:85–125, 1999.
2. M. Benedikt, G. Dong, L. Libkin, and L. Wong. Relational expressive power of constraint query languages. *Journal of the ACM*, 45:1–34, 1998.
3. M. Benedikt and L. Libkin. Expressive power: The finite case. In G. Kuper, L. Libkin, and J. Paredaens, editors, *Constraint Databases*, pages 55–87. Springer, 2000.
4. H.D. Ebbinghaus and J. Flum. *Finite Model Theory.* Springer, 1999.
5. E. Grädel and S. Kreutzer. Descriptive complexity theory for constraint databases. In *Proc. CSL 1999*, volume 1683 of *Lecture Notes in Computer Science*, pages 67–81. Springer, 1999.
6. D. Harel. Towards a theory of recursive structures. In *Proc. MFCS 1998*, volume 1450 of *Lecture Notes in Computer Science*, pages 36–53. Springer, 1998.
7. B. Kuijpers and J. Van den Bussche. On capturing first-order topological properties of planar spatial databases. In *Proc. ICDT 1999*, volume 1540 of *Lecture Notes in Computer Science*, pages 187–198. Springer, 1999.
8. C. Lautemann and N. Schweikardt. An Ehrenfeucht-Faïssé approach to collapse results for first-order queries over embedded databases. In *Proc. STACS 2001*, volume 2010 of *Lecture Notes in Computer Science*, pages 455–466. Springer, 2001.
9. L. Segoufin and V. Vianu. Querying spatial databases via topological invariants. *JCSS*, 61(2):270–301, 2000.

An Algebraic Foundation for Higraphs

John Power* and Konstantinos Tourlas**

Division of Informatics, The University of Edinburgh
King's Buildings, Edinburgh EH9 3JZ
United Kingdom
{ajp,kxt}@dcs.ed.ac.uk
Tel. +44 131 650 5159, Fax. +44 131 667 7209

Abstract. Higraphs, which are structures extending graphs by permitting a hierarchy of nodes, underlie a number of diagrammatic formalisms popular in computing. We provide an algebraic account of higraphs (and of a mild extension), with our main focus being on the mathematical structures underlying common operations, such as those required for understanding the semantics of higraphs and Statecharts, and for implementing sound software tools which support them.

1 Introduction

Recent years have witnessed a rapid, ongoing popularisation of diagrammatic notations in the specification, modelling and programming of computing systems. Most notable among them are Statecharts [3], a notation for modelling reactive systems, and the Unified Modelling Language (UML) [10], a family of diagrammatic notations for object-based modelling. As the popularity of diagrammatic languages in computing and software engineering increases, so does the need of supporting best practice in terms of a sound theory accounting for the multitude of syntactic, semantic and pragmatic issues involved.

A major difficulty in achieving this goal becomes evident when one begins to appreciate the intricate structural complexities of the diagrams typically found in practice, which consist of a multitude of largely heterogeneous and interacting features, the combinations and interactions among them often being ad hoc and poorly understood. Our approach to dealing with this problem is to investigate how diagrams may be decomposed into elementary, underlying structures and features, the properties and interpretations of which we study in mathematical and cognitive terms, and to formulate principles and techniques for sensibly combining them in the design of improved diagrammatic notations. Thus our approach is to first uncover underlying, fundamental structure, which serves for diagrams a role akin in spirit to the role played by various λ-calculi in the study of conventional programming languages.

* This work has been done with the support of grant GR/M56333 and a British Council grant, and the COE budget of STA Japan.

** Support of grant GR/N12480/01 and the COE budget of STA Japan is gratefully acknowledged.

L. Fribourg (Ed.): CSL 2001, LNCS 2142, pp. 145–159, 2001.

This paper presents a step in this direction by developing a category-theoretic framework for *higraphs* [4,5]. The latter are an extension of graphs which underlie a number of sophisticated diagrammatic formalisms including, most prominently, Statecharts, the state diagrams of UML, and the domain-specific language Argos [8] for programming reactive systems. The feature of higraphs we consider as definitive is that of *depth*, the containment of nodes inside other nodes. This feature is systematically exploited in applications to produce concise, economical representations of complex state-transition systems, such as those underlying realistic reactive systems.

The first operation we study is an essential device in understanding the concurrency features of Statecharts. Another operation we develop accounts in a natural way for the connection between higraphs and the state-transition structures they represent.

Even diagrams which are designed with economy and compactness in mind may still grow impractically large, or simply become too detailed to be effective. One therefore still needs effective mechanisms, and tools to support them, for re-organising, abstracting and filtering the information present in diagrams [9].

The leading example studied here is a filtering operation on higraphs, introduced briefly and motivated by Harel in [4] under the name of *zooming out.* Further, we generalise Harel's operation to include other cases of practical interest and show its precise correspondence to a pushout construction in a suitable category of higraphs.

A promising approach to understanding the meaning of certain modelling diagrams (but not presently higraphs) in terms of sketches [1] has been the product of recent work by Diskin *et al.* [2]. Our emphasis here is instead on accounting for the algebraic structure of higraphs (and, in future, of other kinds of diagrams in computing) as a basis for making precise the semantic import of common syntactic operations on them. The rationale is that, in practice, diagrams are first-class objects which are constantly manipulated and altered in the course of design and reasoning about computing systems.

Many of the mathematical structures developed here seem to generalise smoothly to graph-based constructions other than higraphs. In [11] we explore such relevant, deeper category-theoretic structures, which include internal categories, symmetric monoidal adjunctions, and the so-called "other" symmetric monoidal closed structure on **Cat** (the category of all small categories and functors). As such, [11] is only accessible to an audience of category theory experts, aiming at detailed mathematical investigation and relying only on few motivating examples and terse explanation to support the abstract development. Our objective here is instead to present and study lucidly the concrete case of higraphs; and to do so in a way accessible to a wide audience of computer scientists, who have immediate scientific and practical interest in higraphs and their applications in UML and Statecharts, but only minimal knowledge of categories and functors.

Section 2 introduces higraphs and their applications in computing, followed by the development in Section 3 of a category of higraphs. The latter is endowed with a symmetric monoidal closed structure in Section 4 which underpins the concurrency feature of Statecharts. An exposition of Harel's zoom-out operation and its generalisation is the subject of Section 5. A completion operation underlying the state-transition interpretation of higraphs is developed and studied in Section 6. Finally, Section 7 accommodates into our framework a mild extension of higraphs.

2 Higraphs

Higraphs, originally developed by Harel [4] as a foundation for Statecharts [3], are diagrammatic ("visual") objects which extend graphs by permitting spatial containment among the nodes. Figure 1 illustrates the pictorial representation

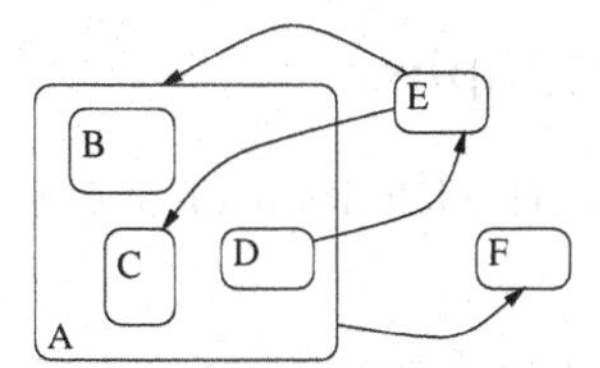

Fig. 1. A simple higraph.

of a simple higraph consisting of six nodes and four edges, with the nodes labelled B, C and D being spatially contained within the node labelled A. It is therefore common, and we shall hereafter adhere to convention, to call the nodes of a higraph *blobs*, as an indication of their pictorial representation by convex contours on the plane. A blob is called *atomic* if no other blobs are contained in it. The feature of spatial containment is often referred to as *depth*, leading to an expression of the relationship of higraphs to graphs in terms of Harel's "equation": higraphs = graphs + depth[1].

The main application of higraphs has been in the specification and visualisation of complex state-transition systems, manifested mainly in Statecharts and, more recently, in the state diagrams of UML. In such applications, depth is used both as a conceptual device, in decomposing the overall system into meaningful subsystems, and as an economical and effective representation of *interrupts*. In terms of our example higraph in Figure 1, the edge emanating from blob A may

[1] Higraph is a term coined-up by Harel[4] as short for *hierarchical graph*, but often used quite liberally to include several variants. The view taken here is that depth is the most distinguishing, definitive feature of higraphs, common to all variants. Harel's original definition includes an extra feature which he called *orthogonality* and which is not treated here. It is our conviction, supported by preliminary results outside the scope of the present paper, that orthogonality can, at least mathematically, be regarded as an extension to the basic, "depth-only" higraphs considered here.

be regarded as a *higher-level* transition interrupting the operation of the subsystem comprising states (i.e. atomic blobs) B, C and D. When applied at multiple levels, depth therefore facilitates the concise representation of large systems by drastically reducing the number of edges required to specify the transition relation among states. Thus, for instance, the higraph on the left of " : " in

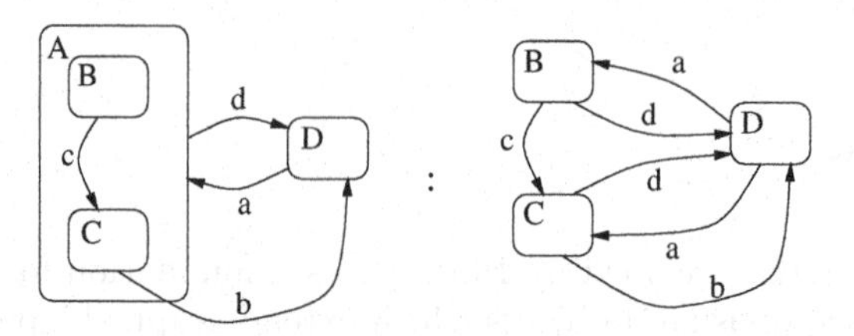

concisely represents the transition system on the right[2].

3 Categories of Higraphs

We begin with a set-theoretic definition of our notion of higraphs, based on Harel's [4]:

Definition 1. *A* higraph *is a 5-tuple* $(B, \leq_B, E, \leq_E, s, t)$*, where* B *and* E *are respectively the sets of* blobs *and* edges*,* $\leq_B$ *is a partial order on* B*,* $\leq_E$ *is a partial order on* E*, and* $s, t : E \to B$ *are monotone functions giving, for each edge* $e \in E$*, its* source *blob* $s(e)$ *and target blob* $t(e)$*.* □

In practice, a higraph typically arises as a graph (B, E, s, t) together with a partial order $\leq_B$ on B. In that case, the poset structure on E may be taken to be the discrete one. However, other choices of orders on E may be useful, e.g. for encoding the conflict resolution schemes [6] adopted in Statecharts.

Thus, each higraph χ is, essentially, a pair of "parallel" monotone functions $\chi_s = s : \chi_E \to \chi_B$ and $\chi_t = t : \chi_E \to \chi_B$, with common domain the poset $\chi_E = (E, \leq_E)$ and codomain $\chi_B = (B, \leq_B)$. By taking into account the two implicit identity functions on B and E, which are trivially monotone, every such pair is exactly a functor χ from the category $\cdot \rightrightarrows \cdot$, consisting of two objects and two non-identity arrows as shown, to the category **Poset** having all (small) posets as objects and monotone functions as arrows. We have therefore arrived at a categorical formulation in which a higraph is regarded as a functor from $\cdot \rightrightarrows \cdot$ to **Poset**.

Notation 1 *Hereafter we shall denote such functors* χ*,* χ'*,* χ'' *... , and implicitly decompose them as* $\chi = (s, t : E \to B)$*,* $\chi' = (s', t' : E' \to B')$ *and so on, unless specifically indicated otherwise.*

[2] In Statecharts, however, either B or C would normally be designated as the *default* state within A, thus resulting in a less general interpretation. We have chosen not to add such a device to higraphs in the present paper, for reasons of generality and simplicity of exposition.

3.1 Morphisms of Higraphs

Definition 2. *Given any two such functors* χ *and* χ', *a* natural transformation τ *from* χ *to* χ', *denoted* $\tau : \chi \to \chi'$, *consists of two monotone functions* $\tau_B : B \to B'$ *and* $\tau_E : E \to E'$ *such that* $\tau_B \circ s = s' \circ \tau_E$ *and* $\tau_B \circ t = t' \circ \tau_E$. □

In pictorial terms, a natural transformation $\tau : \chi \to \chi'$ provides an image of χ into χ' while preserving the two pertinent visual relations in higraphs (containment of blobs and attachment of edges). As such, natural transformations among functors from $\cdot \rightrightarrows \cdot$ to **Poset** provide an intuitive account of *morphisms* between higraphs.

Notation 2 *Let, as usual,* $[\cdot \rightrightarrows \cdot, \mathbf{Poset}]$ *denote the category having all functors from* $\cdot \rightrightarrows \cdot$ *to* **Poset** *as objects, and all natural transformations among them as arrows. Hereafter we shall abbreviate this category as* $\mathcal{H}$ *and regard it as our base category of higraphs and their morphisms.*

4 A Binary Operation on Higraphs

Diagrammatic representations of complex reactive systems directly in terms of simple (or "flat") state-transition diagrams become impractical owing to the large number of states involved. Statecharts deal with this problem by allowing the modelling of reactive systems directly in terms of their identifiable concurrent subsystems: consider for instance the (very simple) Statechart

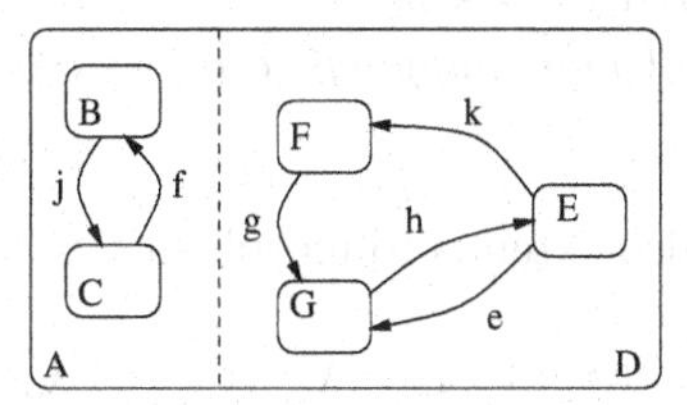

representing two subsystems A and D operating concurrently. Assuming an interleaving model of concurrency, as is the case with Statecharts, the meaning of this picture is captured precisely by the operation

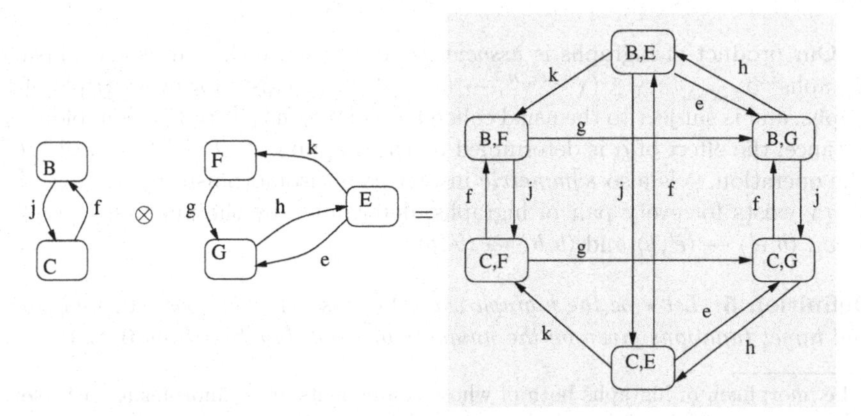

where the resulting transition system is exactly the intended behaviour of the complete system. Thus the diagrammatic device employed enables the representation of a system having $n \times m$ states using only $n + m$ blobs.

This operation, which in [4] is referred to as "a sort of product of automata", is shown here to extend smoothly to higraphs. This is an essential step in pinpointing the precise mathematical structures underpinning the semantics of Statecharts. For, more generally, the specifications of the subsystems (such as A and D in our example Statechart) typically bear higraph structure.

4.1 Monoidal Product of Higraphs

Notation 3 *Given posets A and B, write $A \times B$ for their (cartesian) product partially ordered pointwise, and $A+B$ for their disjoint union. $\mathbf{0}$ will denote the empty poset, whereas $\mathbf{1}$ will stand for the poset over the singleton $\{\star\}$. For any poset A, $A \times \mathbf{0}$ will henceforth be regarded as identical to $\mathbf{0}$, whereas $A + \mathbf{0}$ will be identified with A.*

Definition 3. *Given higraphs $\chi = s, t : E \to B$ and $\chi = s', t' : E' \to B'$, their* monoidal product *$\chi \otimes \chi'$ is defined as having*

- *poset of blobs the cartesian product $B \times B'$*
- *poset of edges $(E \times B') + (B \times E')$*
- *source function given by the mappings $\langle e, b' \rangle \mapsto \langle s(e), b' \rangle$, and target function given similarly.* □

The definition of $\otimes$ also extends to morphisms:

Definition 4. *Given arrows $h : \chi_1 \to \chi_2$ and $h' : \chi_1' \to \chi_2'$ in $\mathcal{H}$, where $\chi_i = s_i, t_i : E_i \to B_i$ and $\chi_i' = s_i', t_i' : E_i' \to B_i'$, let $h \otimes h' : \chi_1 \otimes \chi_2 \to \chi_1' \otimes \chi_2'$ be the natural transformation with components determined by the mapping $\langle b_1, b_1' \rangle \mapsto \langle h(b_1), h'(b_1') \rangle$ on blobs and $\langle e_1, b_1' \rangle \mapsto \langle h(e_1), h'(b_1') \rangle$, $\langle b, e_1' \rangle \mapsto \langle h(b), h'(e_1') \rangle$ on edges.* □

Our product of higraphs is *associative*, in the sense that an isomorphism of higraphs[3] $\alpha_{\chi,\chi',\chi''} : \chi \otimes (\chi' \otimes \chi'') \to (\chi \otimes \chi') \otimes \chi''$ exists for every triple of higraphs, and is subject to the usual coherence conditions [7] (p. 161). On blobs, for instance, the effect of α is determined by the mapping $\langle b, \langle b', b'' \rangle \rangle \mapsto \langle \langle b, b' \rangle, b'' \rangle$. The operation $\otimes$ is also *symmetric* insofar as an isomorphism $\gamma_{\chi,\chi'} : \chi \otimes \chi' \to \chi' \otimes \chi$ exists for every pair of higraphs, determined by the mappings $\langle e, b' \rangle \mapsto \langle b', e \rangle$, $\langle b, e' \rangle \mapsto \langle e', b \rangle$ and $\langle b, b' \rangle \mapsto \langle b', b \rangle$.

Definition 5. *Let $\star$ be the higraph with blob poset $\mathbf{1}$, edge poset $\mathbf{0}$, and source and target functions given by the unique monotone function from $\mathbf{0}$ to $\mathbf{1}$.* □

[3] I.e. morphism of higraphs both of whose components are isomorphisms in **Poset**.

The higraph $\star$ acts as a *unit* for $\otimes$ in the sense that isomorphisms $\rho_\chi : \chi \otimes \star \to \chi$ and $\lambda_\chi : \star \otimes \chi \to \chi$ exist for every χ. Routine calculation confirms that α, λ, ρ and γ satisfy the coherence conditions [7] (p. 158, 159) necessary to show the following

Theorem 1. *$(\mathcal{H}, \otimes, \star, \alpha, \lambda, \rho, \gamma)$ is a symmetric monoidal category.*

Given this symmetric monoidal structure on $\mathcal{H}$, every pair χ, χ' of higraphs determines very naturally an intuitive higraph $[\chi, \chi']$ in which the blobs correspond to all morphisms in $\mathcal{H}$ from χ to χ' and the edges capture all transformations[4] between them:

Definition 6. *Define the higraph $[\chi', \chi'']$ as having*

- *blobs B all morphisms $h : \chi' \to \chi''$ in $\mathcal{H}$ ordered pointwise as pairs of arrows in* **Poset***;*
- *poset E of edges consisting of all $\langle h, \tau, h' \rangle \in B \times [B', E''] \times B$, where $[B', E'']$ is the poset of all monotone functions from B' to E'', such that $s''(\tau(b')) = h(b')$ and $t''(\tau(b')) = h'(b')$;*
- *source and target functions given by the evident first and third projections from E into B.* □

Moreover, an arrow $\mathrm{eval}_{\chi,\chi'} : [\chi, \chi'] \otimes \chi \to \chi'$ always exists subject to the following universal property: for every other arrow $f : \chi'' \otimes \chi \to \chi'$ in $\mathcal{H}$ a unique arrow $f' : \chi'' \to \chi$ exists in $\mathcal{H}$ such that $\mathrm{eval}_{\chi,\chi'} \circ (f' \otimes \mathrm{id}_\chi) = f$. In this case one says that the symmetric monoidal structure on $\mathcal{H}$ is *closed* [7] (p. 180). Indeed:

Theorem 2. *The symmetric monoidal category $(\mathcal{H}, \otimes, \alpha, \lambda, \rho, \gamma)$ is closed.*

Proof. *One straightforwardly calculates that $\mathrm{eval}_{\chi,\chi'}$ with components given by the mappings $\langle h, b \rangle \mapsto h(b)$ and $\langle \langle h, \tau, h' \rangle, b \rangle \mapsto \tau(b)$ (where h, h' are arrows from χ to χ' in $\mathcal{H}$ and τ is a transformation from h to h') has the universal property required.*

5 Zooming Out

We begin our analysis with the simplest, and most frequently occurring in practice, instance of a zooming operation on higraphs: the selection of a single blob and the subsequent removal from view of all blobs contained in it. An example is illustrated in the transition from the left to the right half of Figure 2.

To capture the notion of selecting a blob in a higraph we introduce the following:

[4] Given any two morphisms $h, k : \chi \to \chi'$ in $\mathcal{H}$, a *transformation* from h to k is a monotone function $\sigma : B \to E'$ such that $s''(\tau(b)) = h(b)$ and $t''(\tau(b)) = k(b)$ for all $b \in B$.

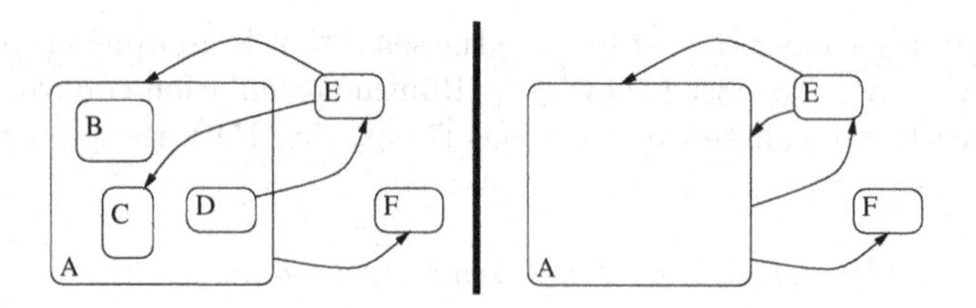

Fig. 2. Zooming out of a blob in a higraph.

Definition 7. *A* pointed higraph ψ *consists of an ordinary higraph* χ *together with a distinguished blob in* χ *called the* point *of* ψ. *This is tantamount to defining a pointed higraph as an arrow* $\psi : \star \to \chi$ *in* $\mathcal{H}$ *from* $\star$, *the higraph consisting of a single blob and no edges, to* χ. □

Definition 8. *A morphism from a pointed higraph* $\psi : \star \to \chi$ *to a pointed higraph* $\psi' : \star \to \chi'$ *is a morphism* $m : \chi \to \chi'$ *such that* $\psi' = m \circ \psi$. □

That is, m sends the point of ψ to the point of ψ'. Thus, the category $\mathcal{H}_\star$ of pointed higraphs is the comma category [7] (p. 46) $\star \downarrow \mathcal{H}$

Notation 4 *Let* $\mathcal{H}_{\star,min}$ *be the full subcategory of* $\mathcal{H}_\star$ *consisting of all objects (pointed higraphs) in which the point is minimal wrt. the partial order on blobs; in other words, the point is an atomic blob. Let* I *be the full functor including* $\mathcal{H}_{\star,min}$ *into* $\mathcal{H}_\star$.

The operation of zooming out may thus be approached at first as a function Z from the objects of $\mathcal{H}_\star$ to the objects of $\mathcal{H}_{\star,min}$ since, in essence, it reduces the point (selected blob) of ψ to a minimal point in $Z(\psi)$.

Definition 9. *Let* $\psi : \star \to \chi$ *be a pointed higraph with* $\chi = (s, t : E \to B)$ *and point, say,* $p \in B$. *Formally,* $Z(\psi)$ *is determined by the following data:*

- *blobs:* $B' = B \setminus \{b \mid b < p\}$ *(ordered by the restriction to* B' *of the partial order on* B*);*
- *edges:* E, *with the source and target functions being* $q \circ s$ *and* $q \circ t$ *respectively, where* $q : B \to B'$ *is the (obviously monotone) function mapping each* $b \not< p$ *in* B *to* $b \in B'$ *and each* $b < p$ *to* $p \in B'$*;*
- *point:* p □

Thus, any edges emanating from or targeting sub-blobs of the point p in ψ have their source or target fixed accordingly to the new, minimal point in $Z(\psi)$[5].

One now observes that, while the essence of Z is to turn the point of ψ to a minimal point, it does so "least disruptively" wrt. the structure of ψ. This latter

[5] Thus, edges contained entirely within p in ψ become endo-edges on p in $Z(\psi)$ and may be subsequently removed, if required, by means of a straightforward operation on pointed higraphs.

observation suggests that the effect of Z is precisely captured by a universal property corresponding to an *adjunction* of functors.

The universal property in question is formulated in terms of a family $\eta_\psi : \psi \to I(Z(\psi))$ of arrows in $\mathcal{H}_\star$, one for each object ψ. It states that for each pointed higraph of the form $I(\psi')$, that is one with minimal point, and arrow $f : \psi \to I(\psi')$ there is a *unique* arrow $f' : Z(\psi) \to \psi'$ in $\mathcal{H}_{\star,min}$ making

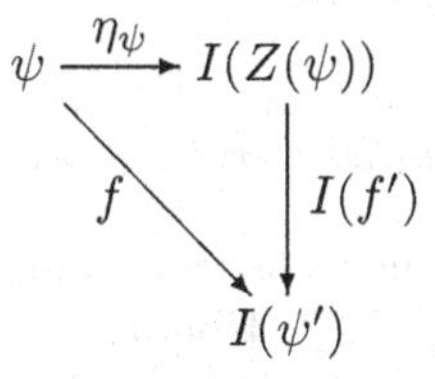

commute. If this universal property holds, Z may be extended to a functor which is called a *left adjoint* to I, with the family η being referred to as the *unit* of the adjunction [7]. Indeed, this is so in our case:

Theorem 3. *The function Z extends to a functor from $\mathcal{H}_\star$ to $\mathcal{H}_{\star,min}$ which is left adjoint to the inclusion functor I.*

Proof. *Let $\psi = (s, t : E \to B)$. It is routine to verify that $\eta_\psi = \langle \mathrm{id}_E, q\rangle$ (with q as given in Definition 9) is an arrow in $\mathcal{H}_\star$ from ψ to $I(Z(\psi)) = (q \circ s, q \circ t : E \to B')$. Given any other arrow $f = \langle f_E, f_B\rangle : \psi \to I(\psi)$ in $\mathcal{H}_\star$, the component f_B induces a unique monotone function $f'_B : B \to B'$ (the one mapping each $b \in B'$ to $f_B(b)$) such that $f'_B \circ q = f_B$. Then $f' = \langle f_E, f'_B\rangle : \psi \to I(\psi)$ is the unique arrow in $\mathcal{H}_{\star,min}$ such that $I(f') \circ \eta_\psi = f$ as required. It now follows from Theorem IV.2(ii) of [7] that the function Z extends to a functor which is left adjoint to the inclusion I.*

5.1 Generalising Zoom-Outs

We now seek to generalise the zoom-out operation on higraphs so that detail is selectively suppressed throughout a chosen part of a given higraph, rather than only within a single blob. In particular, this allows the precise selection of which specific sub-blobs are to disappear from view. In the example of Figure 3, the

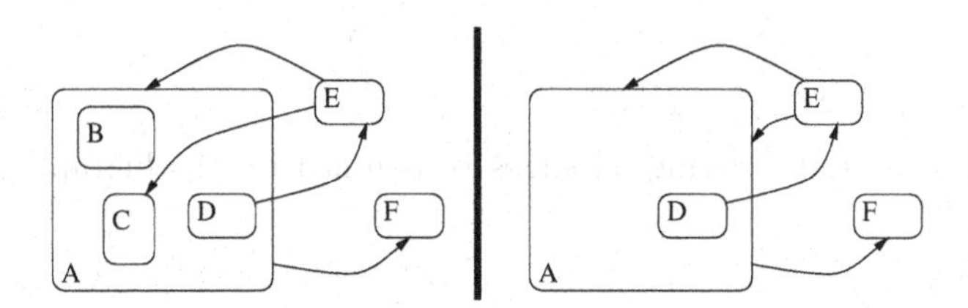

Fig. 3. Example of generalised zoom-out on higraphs.

sub-blobs B and C of A in higraph on the left are selectively discarded from view resulting in the higraph pictured on the right.

Connected Components in Partial Orders. Consider the functor $\delta : \mathbf{Set} \to \mathbf{Poset}$ taking each set to the discrete partial order over the same elements, and the functor $\pi : \mathbf{Poset} \to \mathbf{Set}$ giving for each poset A its set of *connected components*. Explicitly, $\pi(A)$ is the quotient set $A/\!\sim$, where $\sim$ is the equivalence relation generated by $a \leq_A a'$ or $a' \leq_A a$.

Proposition 1. *π is left adjoint to δ.*

Proof. Routine calculation reveals $\eta_A : A \to \delta(\pi(A))$, defined as mapping each $a \in A$ to $[a]_\sim$, to have the required universal property.

The functors π and δ induce corresponding functors $\Pi : [\cdot \rightrightarrows \cdot, \mathbf{Poset}] \to [\cdot \rightrightarrows \cdot, \mathbf{Set}]$ and $D : [\cdot \rightrightarrows \cdot, \mathbf{Set}] \to [\cdot \rightrightarrows \cdot, \mathbf{Poset}]$.

Definition 10. *Given any object g of $[\cdot \rightrightarrows \cdot, \mathbf{Set}]$ (that is, a graph), $D(g)$ is the higraph produced by the discrete partial orderings on both the edges and vertices of g. On the other hand, $\Pi(\chi)$, where $\chi = (s, t : E \to B)$, is the graph having:*

- *sets of edges $\pi(E)$, the connected components of E;*
- *set of nodes $\pi(B)$, the connected components of B;*
- *source function the unique arrow $s' : \pi(E) \to \pi(B)$ in $\mathbf{Set}$, asserted by the universal property of the adjunction in Proposition 1, such that $\delta(s') \circ \eta_E = \eta_B \circ s$ in $[\cdot \rightrightarrows \cdot, \mathbf{Poset}]$ (where η is the unit of the adjunction in Proposition 1);*
- *target function given similarly.* □

Proposition 2. *Π is left adjoint to D.*

Proof. A routine verification shows that each arrow $\eta_\chi : \chi \to D(\Pi(\chi))$ in $\mathcal{H} = [\cdot \rightrightarrows \cdot, \mathbf{Poset}]$ with components $\langle \eta_E, \eta_B \rangle$ (given by the unit of the adjunction in Proposition 1) has the requisite universal property.

As an example consider the mapping,

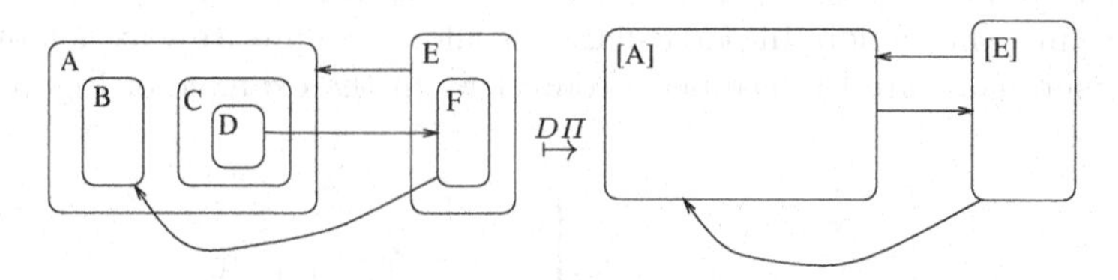

in which the partial ordering on edges is assumed for the higraph pictured on the left of $\overset{D\Pi}{\mapsto}$.

Zooming-Out as a Pushout Construction. Recalling that we are generalising zooming-out to part of a given higraph χ, one first selects the required part (sub-higraph) by means of a monomorphism[6], $m : \sigma \to \chi$. In the example of Figure 3, σ would be the higraph

[6] I.e. morphism of higraphs both of whose components are injective functions.

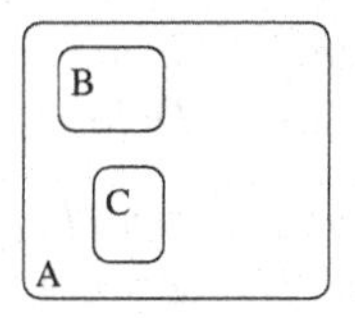

and m would be its inclusion into the left-hand side of Figure 3.

Rewriting the selected part involves the reduction of (the image in χ of) σ to its connected components wrt. the partial orders involved. Thus the rewrite step is captured precisely as the morphism $\eta_\sigma : \sigma \to D\Pi(\sigma)$ mapping the blobs and edges in σ to the connected components in which they belong. Specifically, η_σ is the component at σ of the unit of the adjunction in Proposition 2.

Assume now $Z(m)$ to be a candidate higraph for the result of our zooming-out operation given m. In $Z(m)$, the occurrence of σ in χ specified by m must be rewritten as $D\Pi(\sigma)$. Thus, a morphism $f : D\Pi(\sigma) \to Z(m)$ must exist witnessing this and, moreover, a morphism $g : \chi \to Z(m)$ must also exist making the square

$$\begin{array}{ccc} \sigma & \xrightarrow{\eta_\sigma} & D\Pi(\sigma) \\ {\scriptstyle m}\downarrow & & \downarrow{\scriptstyle f} \\ \chi & \xrightarrow[g]{} & Z(m) \end{array} \tag{1}$$

in $\mathcal{H}$ commute.

One now observes that $Z(m)$ is the required higraph if it satisfies a particular universal property: given the morphisms m and η_σ, any other pair of morphisms $f' : D\Pi(\sigma) \to z'$ and $g' : \chi \to z'$ into another candidate z' such that $f' \circ \eta_\sigma = g' \circ m$ induces a *unique* morphism $u : Z(m) \to z'$ such that both $f' = u \circ f$ and $g' = u \circ g$.

A square such as (1) above with this property is called a *pushout* square. The universal property in our case expresses precisely that $Z(m)$ contains no edges or blobs which are not in χ or $D\Pi(\sigma)$ and that it is obtained in the "least disruptive" way wrt. the structure of χ.

The following definition provides an explicit description of the required higraph $Z(m)$:

Definition 11. *Let $m : \sigma \to \chi$ be a monomorphism in $\mathcal{H}$, where $\chi = (s, t : E \to B)$ and $\sigma = (s', t' : E' \to B')$. Define the higraph $Z(m)$ as having:*

- *blobs: $B/{\sim_m}$, where $\sim_m$ is the least equivalence relation on (the set underlying) B containing all pairs $\langle m(b'_1), m(b'_2)\rangle$ such that $\eta_\sigma(b'_1) = \eta_\sigma(b'_2)$, partially ordered by $[b_1] \leq [b_2]$ iff $b_1 \leq_B b_2$;*
- *edges defined similarly to the case for blobs above;*
- *source and target functions sending each $[e]$ to $[s(e)]$ and $[t(e)]$ respectively.*

□

Theorem 4. *For every monomorphism $m : \sigma \to \chi$ there exist arrows f and g such that square (1), with $Z(m)$ as in Definition 11, is a pushout square in $\mathcal{H}$.*

Proof. *Outline: g is defined to map every element (i.e. blob or edge) x in χ to its equivalence class $[x]$ in $Z(m)$. The arrow f maps every element $[s]$, which by definition of $D\Pi(\sigma)$ is an equivalence class of elements in σ, to $[m(s)]$ in $Z(m)$. The universal property follows from the properties of the quotients involved in the definition of $Z(m)$.*

The Special Case of Single-Blob Zoom-Out. We conclude this section by recasting single-blob zoom-out as an instance of the generalised zoom-out operation.

Proposition 3. *Let $\psi : \star \to \chi$ be a pointed higraph with point p, and $\chi_{\leq p}$ be the higraph with no edges and blobs all $b \leq p$ in χ. Denote by m the inclusion of $\chi_{\leq p}$ into χ and by $z(\psi)$ be the ordinary higraph underlying $Z(\psi)$ as in Definition 9. Then*

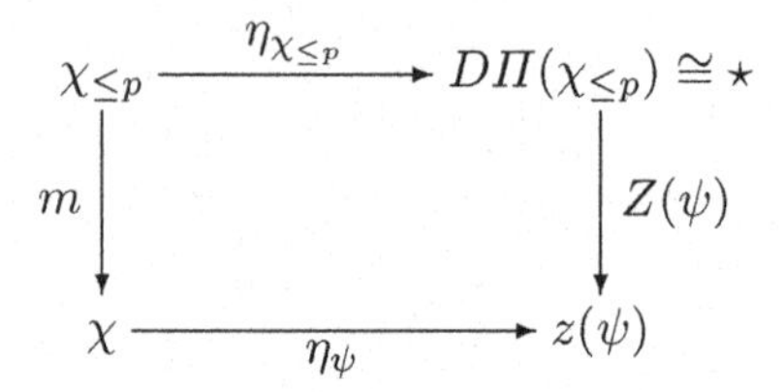

$$\begin{array}{ccc} \chi_{\leq p} & \xrightarrow{\eta_{\chi_{\leq p}}} & D\Pi(\chi_{\leq p}) \cong \star \\ {\scriptstyle m}\downarrow & & \downarrow{\scriptstyle Z(\psi)} \\ \chi & \xrightarrow[\eta_\psi]{} & z(\psi) \end{array}$$

where $\eta_{\chi_{\leq p}}$ is the instance at $\chi_{\leq p}$ of the unit of the adjunction in Theorem 3, is a pushout square in $\mathcal{H}$.

Proof. *For conciseness, we have identified $Z(\psi) \circ i$, where i is the isomorphism $D\Pi(\chi_{\leq p}) \cong \star$, with $Z(\psi)$. Taking this convention into account the square is easily seen to commute by the definition of η_ψ.*

Assume now arrows $f : D\Pi(\chi_{\leq p}) \to \chi'$ and $g = \langle g_E, g_B \rangle : \chi \to \chi'$ in $\mathcal{H}$ to be such that $g \circ \psi = f \circ \eta_{\chi_{\leq p}}$. Form the arrow $u : z(\psi) \to \chi'$ mapping each edge $[e] = \{e\}$ in $z(\psi)$ to $g_E(e)$ and each blob $[b]$ in $z(\psi)$ to $g_B(b)$. This is easily seen to be the unique such arrow satisfying $u \circ \eta_\psi = g$ and $u \circ Z(\psi) = f$.

6 Completion of a Higraph

Another operation, useful in understanding the semantics of higraphs, is to explicate all edges which are understood as being implicitly present in a higraph. Recall that the intuition underlying the interpretation of higraphs is that any edge between blobs b and b' *implies* the presence of "lower-level", implicit edges from all blobs contained in b to all blobs contained in b'. The effect of our "completion" operation is illustrated in Figure 4.

We now proceed to formalise this construction and formulate the universal property from which it arises.

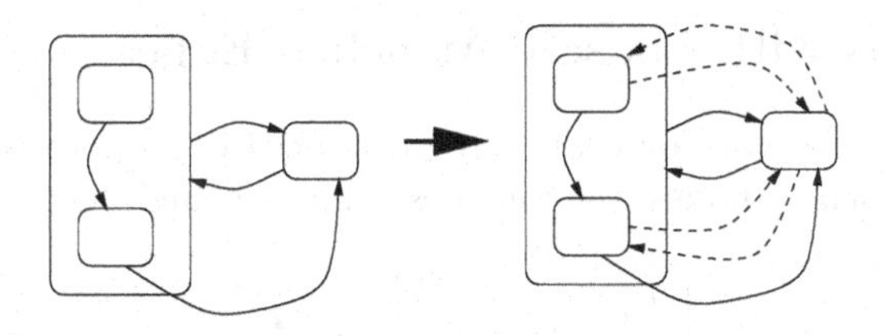

Fig. 4. Completion of a simple higraph, where the added edges are shown dashed.

Definition 12. *Let* $\chi = s, t : E \to B$ *be a higraph. The higraph* $T(\chi)$*, called the* completion *of* χ*, is determined by the following data:*

- *edges: the subset of* $E \times (B \times B)$ *consisting of those pairs* $\langle e, \langle b, b' \rangle \rangle$ *such that* $b \leq_B s(e)$ *and* $b' \leq_B t(e)$*, partially ordered by* $\langle e_1, \langle b_1, b_1' \rangle \rangle \leq \langle e_2, \langle b_2, b_2' \rangle \rangle$ *iff* $e_1 \leq_E e_2$*,* $b_1 \leq_B b_2$ *and* $b_1' \leq_B b_2'$*;*
- *blobs:* B*;*
- *source and target functions given respectively by the (monotone) mappings* $\langle e, \langle b, b' \rangle \rangle \mapsto b$ *and* $\langle e, \langle b, b' \rangle \rangle \mapsto b'$*.* □

Notice, in particular, how each "added" or "new" edge in $T(\chi)$ becomes "less than" the one from which it is derived. In order to formalise this observation, we require a non-strict notion of morphism capable of mapping $T(\chi)$ into χ:

Definition 13. *An* oplax *natural transformation* τ *from* χ *to* χ' *(functors from* $\cdot \rightrightarrows \cdot$ *to* **Poset***) consists of the same data as a (strict) natural transformation except that the naturality condition requires that*

$$\tau_B \circ s \sqsubseteq s' \circ \tau_E \quad \text{and} \quad \tau_B \circ t \sqsubseteq t' \circ \tau_E \ .$$

Here $\sqsubseteq$ *is the usual partial order on the set* $[A, B]$ *of all monotone functions from poset* A *to poset* B *whereby* $f \sqsubseteq g$ *iff* $f(a) \leq_B g(a)$ *for all* $a \in A$*.* □

Notation 5 *Hereafter we abbreviate* Oplax$[\cdot \rightrightarrows \cdot, \mathbf{Poset}]$*, the category having all functors from* $\cdot \rightrightarrows \cdot$ *to* **Poset** *as objects with oplax natural transformations as arrows, as* $\mathcal{H}_{\sqsubseteq}$*.*

Clearly, as every (strict) natural transformation is also an oplax one, an inclusion functor $J : \mathcal{H} \to \mathcal{H}_{\sqsubseteq}$ exists.

Theorem 5. *The function* T *extends to a functor* $T : \mathcal{H}_{\sqsubseteq} \to \mathcal{H}$ *which is right adjoint to the inclusion* $J : \mathcal{H} \to \mathcal{H}_{\sqsubseteq}$*.*

Proof. (Sketch) For every χ *with edges* E *and blobs* B*, let* $\varepsilon_\chi : J(T(\chi)) \to \chi$*, where* J *is the inclusion of* $\mathcal{H}$ *into* $\mathcal{H}_{\sqsubseteq}$*, be the oplax morphism sending each edge* $\langle e, \langle b, b' \rangle \rangle$ *to* e *and acting like the identity on the common set of blobs* B*. The resulting family* ε *of morphisms satisfies a universal property dual to that of the unit in an adjunction: every arrow* $f : J(\chi') \to \chi$ *in* $\mathcal{H}_{\sqsubseteq}$ *induces a unique arrow* $f' : \chi' \to T(\chi')$ *in* $\mathcal{H}$ *such that* $f = \varepsilon_\chi \circ J(f')$*. One now appeals to Theorem IV.2.(iv) of [7].*

7 Higraphs with Loosely Attached Edges

A mild extension to higraphs is briefly introduced in [4] permitting edges to be "loosely" attached to nodes, the four possibilities being illustrated in

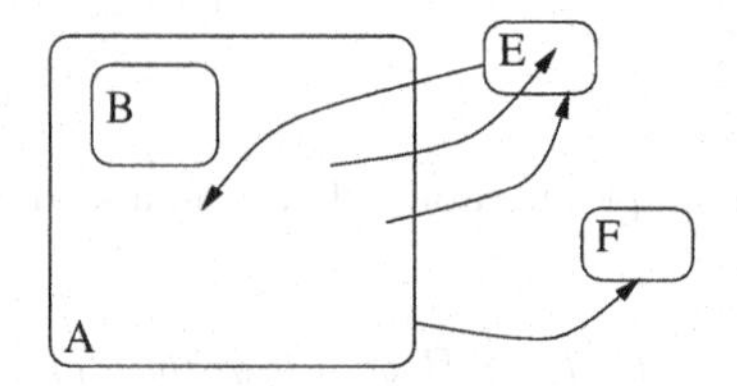

The rationale is to indicate transitions or relations between some as yet unspecified parts of the represented system.

We cast such an extended higraph with blobs B as an ordinary one having the same edges but containing two distinct copies $\langle 0, b\rangle$ and $\langle 1, b\rangle$ of each $b \in B$, tagged with 0's and 1's. In the pictorial representation of such extended higraphs the convention is that blobs tagged with 0 are not shown at all and that, for instance, an edge with target of the form $\langle 0, b\rangle$ has its endpoint lying *inside* the contour picturing b.

Moreover one stipulates that $\langle 0, b\rangle < \langle 1, b\rangle$ for all b, to capture the intuition underlying the pictorial representation. Any edge $\langle i, b\rangle \to \langle j, b'\rangle$ will be called *non-firm* if $i = 0$ or $j = 0$.

Definition 14. *The category* $\mathcal{LH}$ *of* higraphs with loosely attached edges *has*

- *objects: all pairs of parallel arrows* $\phi = s, t : E \to (\cdot \to \cdot \times B)$ *in* **Poset**, *where* $\cdot \to \cdot$ *is the poset* $0 \leq 1$;
- *arrows: from* $\phi = (s, t : E \to \cdot \to \cdot \times B)$ *to* $\phi' = (s', t' : E' \to \cdot \to \cdot \times B')$ *all pairs* $m = \langle m_0 : B \to B', m_1 : E \to E'\rangle$ *such that*

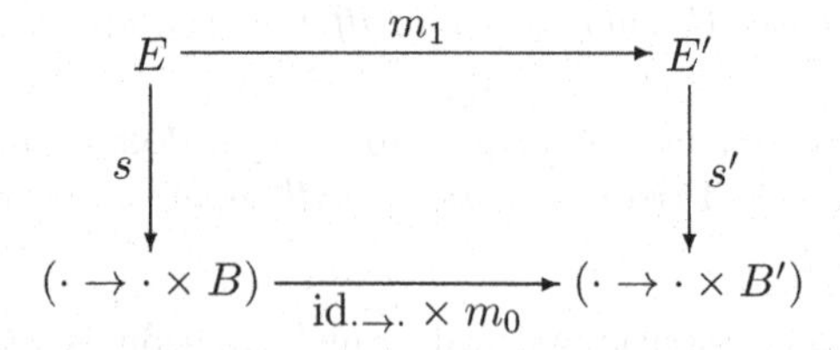

$$\begin{array}{ccc} E & \xrightarrow{m_1} & E' \\ {\scriptstyle s}\downarrow & & \downarrow{\scriptstyle s'} \\ (\cdot \to \cdot \times B) & \xrightarrow[\mathrm{id}_{\cdot \to \cdot} \times m_0]{} & (\cdot \to \cdot \times B') \end{array}$$

commutes, and similarly for the corresponding square with sides t *and* t';
- *evident identities and composition defined componentwise.* □

Let $U : \mathcal{LH} \to \mathcal{H}$ be the functor sending each ϕ in $\mathcal{LH}$ with blobs $\cdot \to \cdot \times B$ and edges E to the higraph obtained by "forgetting" the non-firm edges of ϕ. That is, $U(\phi)$ has blobs B and an edge $e : b \to b'$ for each edge $e : \langle 1, b\rangle \to \langle 1, b'\rangle$ in ϕ. Consider now the function J_1 sending each $\chi = (s, t : E \to B)$ in $\mathcal{H}$ to $J(\chi)$ with blobs $\cdot \to \cdot \times B$ and an edge $e : \langle 1, b\rangle \to \langle 1, b'\rangle$ for each edge $e : b \to b'$ in χ.

Proposition 4. *The function* J_1 *extends to a functor which is left adjoint to* U.

Proof. It is routine to verify that the arrows $\eta_\chi : \chi \to U(J_1(\chi))$ with components given by the identities have the required universal property.

8 Outlook and Future Work

This work is part of a project, drawing on cognitive, computational and mathematical views of diagrams, to research principles which will improve the design of diagrammatic, domain-specific programming languages.

Typically the diagrams used in practice contain a multitude of subtly interacting features. This situation neccesitates an analytic approach to identify suitable primitive structures and ways of combining them. Higraphs, featuring only an underlying graph structure and depth (hierarchy), appear as being an excellent point of departure in the study of diagrammatic features and their interaction.

Specifically with respect to higraphs and their applications, we aim to consider additional features (such as labelling and Harel's "orthogonality") towards obtaining a structured account of simple Statecharts.

References

1. M. Barr and C. Wells. *Category Theory for Computing Science.* Prentice-Hall, 1990.
2. Zinovy Diskin, Michael Johnson, Boris Kadish, and Frank Piessens. Universal arrow foundations for visual modelling. In *Proceedings of Diagrams 2000*, number 1889 in Lecture Notes in Artificial Intelligence, pages 345–360, 2000.
3. David Harel. Statecharts: A visual approach to complex systems. *Science of Computer Programming*, 8(3):231–275, 1987.
4. David Harel. On visual formalisms. *Communications of the ACM*, 31(5), 1988.
5. David Harel. On visual formalisms. In J. Glasgow, N.H. Narayanan, and B. Chandrasekaran, editors, *Diagrammatic Reasoning: Cognitive and Computational Perspectives*, pages 235–272. AAAI Press/The MIT Press, 1995.
6. David Harel and Amnon Naamad. The STATEMATE semantics of Statecharts. *ACM Transactions on Software Engineering Methodology*, 5(4), October 1996.
7. Saunders MacLane. *Categories for the Working Mathematician*, volume 5 of *Graduate Texts in Mathematics.* Springer-Verlag, 1971.
8. F. Maraninchi. The Argos language: Graphical representation of automata and description of reactive systems. In *Proceedings of the IEEE Workshop on Visual Languages*, 1991.
9. Bonnie M. Nardi. *A Small Matter of Programming: Perspectives on End-User Computing.* MIT Press, 1993.
10. Rob Pooley and Perdita Stevens. *Using UML.* Addison Wesley, 1999.
11. John Power and Konstantinos Tourlas. An algebraic foundation for graph-based diagrams in computing. In *Proceedings of the 17th Conference on the Mathematical Foundations of Programming Semantics (MFPS)*, 2001. To appear.

Semantic Characterisations of Second-Order Computability over the Real Numbers*

M.V. Korovina[1] and O.V. Kudinov[2]

[1] Institute of Informatics Systems, Lavrent'ev pr., 6,
Novosibirsk, Russia
rita@inet.ssc.nsu.ru
[2] Institute of Mathematics, Koptug pr., 4,
Novosibirsk, Russia
kud@math.nsc.ru

Abstract. We propose semantic characterisations of second-order computability over the reals based on Σ-definability theory. Notions of computability for operators and real-valued functionals defined on the class of continuous functions are introduced via domain theory. We consider the reals with and without equality and prove theorems which connect computable operators and real-valued functionals with validity of finite Σ-formulas.

1 Introduction

To investigate semantic properties of computable operators and real-valued functionals we use the concept of generalised computability firstly proposed in [20]. This concept is a result of development of two well-known non-equivalent approaches to computability over the real numbers.

The first one is related to abstract machines and scheme of computations (e.g. [5,26,14]). The result of this computation is defined by a finite algorithm. Semantic characterisations of computable functions have been given in [25,6,10] from the point of view of definability. In this approach equality is usually used as a basic relation so a computable function can be discontinuous. It diverges from the situation in concrete computability over the reals, in particularly, in computable analysis.

The second approach (e.g. [17,28,29,30,16,7,8,11,35,36,37]) is closely related to computable analysis. In this approach computation is an infinite process which produces approximations closer and closer to the result. We work in the framework of the second approach. The main result of our paper is an application of definability theory, which was originally used in the first approach, to characterisation of an infinite computational process via validity of finite Σ-formulas. This paper structured as follows.

* This research was supported in part by the RFBR (grants N 99-01-00485, N 00-01-00810) and by the Siberian Division of RAS (a grant for young researchers, 2000)

L. Fribourg (Ed.): CSL 2001, LNCS 2142, pp. 160–172, 2001.

In Section 2, we give basic definitions and tools. We study properties of operators and functionals considering their generalised computability relative either to the ordered reals with equality, or to the strictly ordered reals without equality.

In Section 3 we introduce notions of second-order computability via domain theory. In this work, to construct computational models for operators and functionals we will use continuous domains. Continuous domains, e.g. [29,30,16,7,8,11,35,36,37], are generalisation of algebraic domains, e.g. [2,27,32,33]. The continuous domain (more precisely, the interval domain) for the reals was first proposed by Dana Scott [29] and later was applied to mathematics, physics and real number computation in [7,8,36,37,27] and other publications. In this section we propose continuous domains, named as function domains to construct a computational model of operators and real-valued functionals defined on the set of continuous real-valued functions.

In Section 4 we give semantic characterisations of computable operators and real-valued functionals.

In Section 5 we provide some concluding remarks.

2 Generalised Computability

2.1 Terminology

Throughout the article we consider two models of the real numbers, $< \mathbb{R}, \sigma_1 > \rightleftharpoons < \mathbb{R}, 0, 1, +, \cdot, <, -x, \frac{x}{2} >$ is the model of the reals without equality, and $< \mathbb{R}, \sigma_2 > \rightleftharpoons < \mathbb{R}, 0, 1, +, \cdot, \leq >$ is the model of the reals with equality. Below if statements concern the languages σ_1 and σ_2 we will write σ for a language. Denote $\mathbf{D}_2 = \{z \cdot 2^{-n} | z \in \mathbb{Z},\ n \in \mathbb{N}\}$. Let us use $\bar{r}$ to denote $r_1, \ldots, r_m$.

2.2 Basic Definitions

To recall the notion of generalised computability, let us construct the set of hereditarily finite sets $\mathrm{HF}(M)$ over a model $\mathbf{M}$. This structure is rather well studied in the theory of admissible sets [3] and permits us to define the natural numbers and to code and store information via formulas. Let $\mathbf{M}$ be a model of a language σ whose carrier set is M. We construct the set of hereditarily finite sets, $\mathrm{HF}(M) = \bigcup_{n \in \omega} \mathrm{S}_n(M)$, where $\mathrm{S}_0(M) \rightleftharpoons M$, $\mathrm{S}_{\mathrm{n}+1}(M) \rightleftharpoons \mathcal{P}_\omega(\mathrm{S}_\mathrm{n}(M)) \cup \mathrm{S}_\mathrm{n}(M)$, where $n \in \omega$ and for every set B, $\mathcal{P}_\omega(B)$ is the set of all finite subsets of B.

We define $\mathbf{HF}(\mathbf{M}) \rightleftharpoons \langle \mathrm{HF}(M), M, \sigma, \emptyset_{\mathbf{HF}(\mathbf{M})}, \in_{\mathbf{HF}(\mathbf{M})} \rangle$, where the unary predicate $\emptyset$ singles out the empty set and the binary predicate symbol $\in_{\mathbf{HF}(\mathbf{M})}$ has the set-theoretic interpretation.

Below we consider $M \rightleftharpoons \mathbb{R}$, the language without equality $\sigma_1^* = \sigma_1 \cup \{\in,\ \emptyset\}$, and the language with equality $\sigma_2^* = \sigma_2 \cup \{\in,\ \emptyset\}$.

Below if statements concern the languages σ_1^* and σ_2^* we will write σ^* for a language.

To introduce the notions of terms and atomic formulas we use variables of two sorts. Variables of the first sort range over $\mathbb{R}$ and variables of the second sort range over $\mathrm{HF}(\mathbb{R})$.

The terms in the language σ_1^* are defined inductively as follows: 1. the constant symbols 0 and 1 are terms; 2. the variables of the first sort are terms; 3. if t_1, t_2 are terms then $t_1 + t_2$, $t_1 \cdot t_2$, $-t_1$, $\frac{t_2}{2}$ are terms. The notions of a term in the language σ_2^* can be given in a similar way.

The following formulas in the language σ_1^* are atomic: $t_1 < t_2$, $t \in s$ and $s_1 \in s_2$ where t_1, t_2, t are terms and s_1, s_2 are variables of the second sort. The following formulas in the language σ_2^* are atomic: $t_1 < t_2$, $t \in s$, $s_1 \in s_2$ and $t_1 = t_2$ where t_1, t_2, t are terms and s_1, s_2 are variables of the second sort.

The set of Δ_0-*formulas* in the language σ^* is the closure of the set of atomic formulas in the language σ^* under $\wedge, \vee, \neg, (\exists x \in s)$ and $(\forall x \in s)$, where $(\exists x \in s)\varphi$ denotes $\exists x(x \in s \wedge \varphi)$ and $(\forall x \in s)$ φ denotes $\forall x(x \in s \to \varphi)$ and x, s are variables of second type.

The set of Σ-*formulas* in the language σ^* is the closure of the set of Δ_0 formulas in the language σ^* under $\wedge, \vee, (\exists x \in s), (\forall x \in s)$, and $\exists$. The natural numbers $0, 1, \ldots$ are identified with $\emptyset$, $\{\emptyset, \{\emptyset\}\}, \ldots$ so that, in particular, $n+1 = n \cup \{n\}$ and the set ω is a subset of $\mathrm{HF}(\mathbb{R})$.

Definition 1. *A relation $B \subseteq \mathrm{HF}(\mathbb{R}^n)$ is Σ-definable in σ^*, if there exists a Σ-formula $\Phi(\bar{x})$ in the language σ^* such that $\bar{x} \in B \leftrightarrow \mathbf{HF}(\mathbb{R}) \models \Phi(\bar{x})$. A function is Σ-definable if its graph is Σ-definable.*

Note that the set $\mathbb{R}$ is Δ_0–definable in the language σ^*. This fact makes $\mathbf{HF}(\mathbb{R})$ a suitable domain for studying relations in $\mathbb{R}^n$ and functions from $\mathbb{R}^n$ to $\mathbb{R}$ where $n \in \omega$. To introduce the definition of generalised computability, we use the class of Σ-definable sets as a basic class. So, we recall some useful properties of Σ-definable subsets of $\mathbb{R}^n$.

Proposition 1. *1. If a set A is Σ-definable in the language σ_2^* then A is Σ-definable in the language σ_1^*.*
2. The sets $\mathbf{HF}(\emptyset)$, ω and the predicate of equality on $\mathbf{HF}(\emptyset)$ are Σ-definable in the language σ^*.
3. The set $\{\langle n, r\rangle \mid n$ is a Gödel number of a Σ-formula Φ, $r \in \mathbb{R}$, and $\mathbf{HF}(\mathbb{R}) \models \Phi(r)\}$ is Σ-definable in the language σ^*.
4. A set $B \subseteq \mathbb{R}^n$ is Σ-definable in the language σ^* if and only if there exists an effective sequence of quantifier free formulas in the language σ, $\{\Phi_s(x)\}_{s\in\omega}$, such that $x \in B \leftrightarrow \mathbb{R} \models \bigvee_{s\in\omega} \Phi_s(x)$.

Proof. The parts *1.-3.* can be easy proved by technique developed in [3,10,20]. The parts *4.* immediately follow from the part *3.* □

Below we will write $(\exists n \in \mathbb{N})\, \Phi(n, x)$ instead of $\bigvee_n \Phi(\underline{n}, x)$ and $(\exists n \in \mathrm{D}_2)\, \Phi(n, x)$ instead of $\bigvee_{n,m} \left(\Phi(\frac{n}{2^m}, x) \vee \Phi(\frac{-n}{2^m}, x)\right)$, where $\underline{0} = 0, \ldots,$ $\underline{n+1} = \underline{n} + 1$.

Without loss of generality we consider the set of continuous functions defined on compact intervals with endpoints which are computable numbers in the sense of computable analysis (e.g. [28]).

To introduce generalised computability of operators and functionals we extend the languages σ_1^* and σ_2^* by two 3-ary predicates U_1 and U_2.

The following technical defintinion turns out to be rather clear in the framework of Deninition 3 and Definition 4.

Definition 2. *Let $\varphi_1(U_1, U_2, x_1, x_2, c)$, $\varphi_2(U_1, U_2, x_1, x_2, c)$ be formulas in the language σ^*. We suppose that U_1, U_2 occur positively in φ_1, φ_2 and the predicates U_1, U_2 define open sets on $\mathbb{R}^3$. The formulas φ_1, φ_2 are said to satisfy joint continuity property if the following formulas are valid in* $\mathbf{HF}(\mathbb{R})$.

1. $\forall x_1 \forall x_2 \forall x_3 \forall x_4 \forall z\, ((x_1 \leq x_3) \wedge (x_4 \leq x_2) \wedge \varphi_i(U_1, U_2, x_1, x_2, z)) \to$ $\varphi_i(U_1, U_2, x_3, x_4, z)$, *for* $i = 1, 2$
2. $\forall x_1 \forall x_2 \forall c \forall z\, ((z < c) \wedge \varphi_1(U_1, U_2, x_1, x_2, c)) \to \varphi_1(U_1, U_2, x_1, x_2, z)$,
3. $\forall x_1 \forall x_2 \forall c \forall z\, ((z > c) \wedge \varphi_2(U_1, U_2, x_1, x_2, c)) \to \varphi_2(U_1, U_2, x_1, x_2, z)$,
4. $\forall x_1 \forall x_2 \forall x_3 \forall z\, (\varphi_i(U_1, U_2, x_1, x_2, z) \wedge \varphi_i(U_1, U_2, x_2, x_3, z)) \to$ $\varphi_i(U_1, U_2, x_1, x_3, z)$, *for* $i = 1, 2$,
5. $(\forall y_1 \forall y_2 \exists z \forall z_1 \forall z_2 (U_1(y_1, y_2, z_1) \wedge U_2(y_1, y_2, z_1) \to (z_1 < z < z_2))) \to$ $(\forall x_1 \forall x_2 \exists c \forall c_1 \forall c_2 (\varphi_1(U_1, U_2, x_1, x_2, c_1) \wedge \varphi_2(U_1, U_2, x_1, x_2, c_2) \to$ $(c_1 < c < c_2)))$.

Definition 3. *A partial operator $F : C[a, b] \to C[c, d]$ is said to be shared by two Σ-formulas φ_1 and φ_2 in the language σ^* if the following assertions hold. For every $u \in C([a, b])$ and $h \in C([c, d])$, $F(u) = h$ holds if and only if*

$$h|_{[x_1,x_2]} > z \leftrightarrow \mathbf{HF}(\mathbb{R}) \models \varphi_1(U_1, U_2, x_1, x_2, z) \text{ and}$$
$$h|_{[x_1,x_2]} < z \leftrightarrow \mathbf{HF}(\mathbb{R}) \models \varphi_2(U_1, U_2, x_1, x_2, z),$$

where $U_1(x_1, x_2, c) \rightleftharpoons u|_{[x_1,x_2]} > c$, $U_2(x_1, x_2, c) \rightleftharpoons u|_{[x_1,x_2]} < c$ and the predicates U_1 and U_2 occur positively in φ_1, φ_2.

Definition 4. *A partial operator $F : C[a, b] \to C[c, d]$ is said to be generalised computable in the language σ^*, if F is shared by two Σ-formulas in the language σ^* which satisfy the joint continuity property.*

Definition 5. *A partial functional $F : C[a, b] \times [c, d] \to \mathbb{R}$ is said to be generalised computable in the language σ^*, if there exists an operator $F^* : C[a, b] \to C[c, d]$ generalised computable in the language σ^* such that $F(f, x) = F^*(f)(x)$.*

Definition 6. *A partial functional $F : C[a, b] \times \mathbb{R} \to \mathbb{R}$ is said to be generalised computable in the language σ^*, if there exists an effective sequence $\{F_n^*\}_{n \in \omega}$ of operators generalised computable in the language σ^* of the types $F_n^* : C[a, b] \to C[-n, n]$ such that $F(f, x) = y \leftrightarrow \forall n\, (-n \leq x \leq n \to F_n^*(f)(x) = y)$.*

2.3 Generalised Computability in the Various Languages

Now we prove the main theorem which connects generalised computabilities in the various languages.

Theorem 1. *A continuous total operator $F : C[a,b] \to C[c,d]$ is generalised computable in the language with equality if and only if it is generalised computable in the language without equality.*

Proof. We outline the basic elements of the proof only. Without loss of generality we consider a continuous operator $F : C[0,1] \to C[0,1]$. Assume that F is generalised computable in the language with equality. By definition, there exist Σ-formulas φ_1, φ_2 in the language with equality which satisfy the joint continuity property and share F. Let us construct new Σ-formulas φ_1''', φ_2''' in the language without equality which satisfy the joint continuity property and share F. By properties of Σ-definable sets we have the following equivalence:

$$\varphi_1(U_1, U_2, x, y, z) \Leftrightarrow \bigvee_{i \in \omega} \Theta_i(U_1, U_2, x, y, z).$$

In the language with equality, for all $i \in \omega$ the formula Θ_i can be written in the form:

$$\Theta_i \rightleftharpoons \exists \bar{r}_1 \in U_1 \exists \bar{r}_2 \in U_2 \psi_i(\bar{r}_1, \bar{r}_2, x, y, z),$$

where ψ_i is a quantifier free formula. By the definition of U_1, U_2, the tuples $\bar{r}_1, \bar{r}_2$ can be represented in the following way

$$\bar{r}_1 = \langle \alpha_1, \beta_1, \gamma_1 \rangle, \dots, \langle \alpha_s, \beta_1, \gamma_s \rangle, \; \bar{r}_2 = \langle u_1, v_1, w_1 \rangle, \dots, \langle u_s, v_1, w_s \rangle.$$

For fixed x, y, $z \in \mathbf{D_2}$, using properties of open sets definable in $\mathbb{R}$, we can effectively construct the set of $6s$-tuples

$$A^{x,y,z} = \{\langle \bar{r}_1, \bar{r}_2 \rangle\} = \{\langle \alpha_1, \beta_1, \gamma_1, u_1, v_1, w_1, \dots, \alpha_s, \beta_s, \gamma_s, u_s, v_s, w_s \rangle\}$$

with the following properties:

1. $\alpha_1, \beta_1, \gamma_1, u_1, v_1, w_1, \dots \alpha_s, \beta_s, \gamma_s, u_s, v_s, w_s \in \tilde{\mathbb{Q}} \cap \mathbb{R}$,
2. for each $\langle \bar{r}_1, \bar{r}_2 \rangle \in A^{x,y,z}$ there exists $i \in \omega$ such that $\mathbf{HF}(\mathbb{R}) \models \psi_i(\bar{r}_1, \bar{r}_2, x, y, z)$,
3. for all $I \subseteq \{1, \dots, s\}$ and $J \subseteq \{1, \dots, s\}$ such that $\bigcap_{i \in I} [\alpha_i, \beta_i] \cap \bigcap_{j \in J} [u_j, v_j] \neq \emptyset$ we have $\max_{i \in I} \gamma_i < \min_{j \in J} w_j$.

By construction, for each $x, y, z \in \mathbf{D_2}$ the set $A^{x,y,z}$ is computably enumerable. Let us fix some numbering and denote $6s$-typle numbered by j as $\left\langle \bar{r}_1^j, \bar{r}_2^j \right\rangle$. Let us construct $\psi'_{i,j}$ in the following way:

$$\psi'_{i,j}(\bar{r}_1^j, \bar{r}_2^j, x, y, z) = \begin{cases} 0 > 1 & \text{if } \neg\psi_i(\bar{r}_1^j, \bar{r}_2^j, x, y, z), \\ \left(\bar{r}_1^j \in U_1\right) \wedge \left(\bar{r}_2^j \in U_2\right) & \text{if } \psi_i(\bar{r}_1^j, \bar{r}_2^j, x, y, z). \end{cases}$$

Using Tarski's quantifier elimination theorem for real closed fields, Σ-definability of $\mathbf{D_2}$ and the following equivalences,

$$\langle \alpha,\beta,\gamma\rangle \in U_1 \Leftrightarrow \exists a,b,c \in \mathbf{D_2}(((a<\alpha<\beta<b)\wedge(c>\gamma)\wedge U_1(a,b,c))\vee ((b>\beta)\wedge(c>\gamma)\wedge U_1(0,b,c))\vee((a<\alpha)\wedge(c>\gamma)\wedge U_1(a,1,c))\vee((c>\gamma)\wedge U_1(0,1,c))),$$
$$\langle \alpha,\beta,\gamma\rangle \in U_2 \Leftrightarrow \exists a,b,c \in \mathbf{D_2}(((a<\alpha<\beta<b)\wedge(c<\gamma)\wedge U_2(a,b,c))\vee ((b>\beta)\wedge(c<\gamma)\wedge U_2(0,b,c))\vee((a<\alpha)\wedge(c<\gamma)\wedge U_2(a,1,c))\vee((c<\gamma)\wedge U_2(0,1,c))),$$

we can construct a Σ-formula $\psi''_{i,j}$ in the language without equality which is equivalent to $\psi'_{i,j}$ with respect to the predicates U_1 and U_2 such that $U_1(x_1,x_2,c) \rightleftharpoons f|_{[x_1,x_2]} > c, U_2(x_1,x_2,c) \rightleftharpoons f|_{[x_1,x_2]} < c$ for a continuous function f. Put

$$\varphi'_1(U_1,U_2,x,y,z) \rightleftharpoons \bigvee_{i,j\in\omega} \psi''_{i,j}(U_1,U_2,x,y,z).$$

Let us define formula, which is equivalent to $\varphi_1(U_1,U_2,x,y,z)$ for arbitrary $x,y,z \in \mathbb{R}$, the predicates U_1 and U_2 such that $U_1(x_1,x_2,c) \rightleftharpoons f|_{[x_1,x_2]} > c, U_2(x_1,x_2,c) \rightleftharpoons f|_{[x_1,x_2]} < c$ for a continuous function f:

$$\varphi''_1(U_1,U_2,x,y,z) \Leftrightarrow \exists x',y',z' \in \mathbf{D_2}(((x'<x<y<y')\wedge(z'>z)\wedge \varphi'_1(U_1,U_2,x',y',z'))\vee((y'>y)\wedge(z'>z)\wedge\varphi'_1(U_1,U_2,0,y',z'))\vee ((x'<x)\wedge(z'>z)\wedge\varphi'_1(U_1,U_2,x',1,z'))\vee((z'>z)\wedge\varphi'_1(U_1,U_2,0,1,z'))).$$

Note that $\mathbf{D_2}$ and $\mathbb{R}$ are Σ-definable in the language without equality. So the formula $\varphi''_1(U_1,U_2,x,y,z)$ is equivalent to a Σ-formula $\varphi'''_1(U_1,U_2,x,y,z)$ in the language without equality for arbitrary $x,y,z\in\mathbb{R}$, the predicates U_1 and U_2 such that $U_1(x_1,x_2,c) \rightleftharpoons f|_{[x_1,x_2]} > c, U_2(x_1,x_2,c) \rightleftharpoons f|_{[x_1,x_2]} < c$ for a continuous function f. Similarly we can construct a Σ-formula in the language without equality φ'''_2 which is equivalent to φ_2. The formulas are the required ones. It follows from continuity of the operator F. By definition, the operator F is generalised computable in the language without equality. □

Proposition 2. *Let $F : C[a,b]\times\mathbb{R}\to\mathbb{R}$ be a continuous total functional. Then the functional F is generalised computable in the language with equality if and only if it is generalised computable in the language without equality.*

3 Second Order Computability over the Reals

In this section, we will use continuous domains to construct computational models for operators and real-valued functionals. Continuous domains, e.g. [29,30,16,7,8,11,35,36,37], are generalisation of algebraic domains, e.g. [2,27,32,33]. The continuous domain (more precisely, the interval domain) for the reals was first proposed by Dana Scott [29] and later was applied to mathematics, physics and real number computation in [7,8,36,37,27] and others. In this section we propose continuous domains, named as function domains to construct a computational model of operators and real-valued functionals defined on the set of continuous real-valued functions.

3.1 Interval Domain for the Reals

By *the interval domain for the reals* we mean the set of compact intervals of $\mathbb{R}$, partially ordered with reversed subset inclusion and endowed with the least element.
We recall the definition of the *interval domain* $\mathcal{I}$ proposed in [8]:
$\mathcal{I} = \{[a,b] \subseteq \mathbb{R} \mid a,b \in \mathbb{R},\ a \leq b\} \cup \{\bot\}$.
The order is reversed subset inclusion, i.e. $\bot \sqsubseteq I$ for all $I \in \mathcal{I}$ and $[a,b] \sqsubseteq [c,d]$ iff $a \leq c$ and $d \leq b$ in the usual ordering of the reals. One can consider the least element $\bot$ as the set $\mathbb{R}$. Directed suprema are filtered intersections of intervals. The way-below relation is given by $I \ll J$ iff $J \subseteq \mathrm{int}(I)$, where $\mathrm{int}(I)$ denotes the interior of I. For the relation $\ll$ we have the following properties: $\bot \ll J$ for all $J \in \mathcal{I}$ and $[a,b] \ll [c,d]$ if and only if $a < c$ and $b > d$. Note that $\mathcal{I}$ is an effective ω-continuous domain. A countable basis $\mathcal{I}_0$ is given by the collection of all intervals with rational endpoints together with the least element $\bot$. Similarly, we can define the interval domain $\mathcal{I}_{[a,b]}$ for an interval $[a,b]$.

The maximal elements are the intervals $[a,a]$ denoted as $\{a\}$. We denote the set of maximal elements as $\max(\mathcal{I})$. It is easy to see that the maximal elements with the subspace topology of Scott topology on $\mathcal{I}$ is homeomorphic to the real line with the standard topology. That is why we can identify a real number r with $\{r\}$.

3.2 Function Domains

In this subsection we introduce effective function domains which are effective ω-continuous domains. Based on the notion of computability of mapping between two domains we propose computability of operators and functionals defined on $C[a,b]$. The main feature of this approach is related to the fact that continuous operators and functionals defined on continuous real-valued functions can be extended to continuous operators and functionals defined on the corresponding function domain. Moreover, we propose a semantic characterisation of computable operators and functionals via validity of finite Σ-formulas.

Let $\mathcal{I}$ be equipped with Scott topology (for the definition we refer to [29,30,16]). We consider the set of continuous functions $f : [a,b] \to \mathcal{I}$ defined on a compact interval $[a,b]$ with computable endpoints. Let $\mathbb{R}^-$ denote $\mathbb{R} \cup \{-\infty\}$ and $\mathbb{R}^+$ denote $\mathbb{R} \cup \{+\infty\}$.

Definition 7. *A function $f : [a,b] \to \mathbb{R}^-$ is said to be lower semicontinuous if the set $Y_f^- = \{x | f(x) \neq -\infty\}$ is open w.r.t the standard topology and*

$$\left(\forall x_0 \in Y_f^-\right) (\forall y < f(x_0)) \exists \delta \left(|x_0 - x| < \delta \to y < f(x)\right).$$

A function $f : [a, b] \to \mathbb{R}^+$ *is said to be upper semicontinuous if the set* $Y_f^+ = \{x | f(x) \neq +\infty\}$ *is open w.r.t the standard topology and*

$$\left(\forall x_0 \in Y_f^+\right)(\forall y > f(x_0))\, \exists \delta\, (|x_0 - x| < \delta \to y > f(x)).$$

For the classical theory of semicontinuous functions the reader should refer a standard textbook (e.g. [4]). The reader can also find some properties of computability on continuous and semicontinuous real functions in [38]. It is easy to see that a continuous function $f : [a, b] \to \mathcal{I}$ is closely related to the pair of functions $\langle f^1 : [a, b] \to \mathbb{R}^-, f^2 : [a, b] \to \mathbb{R}^+ \rangle$, where $f^1(x) = \inf\ f(x)$ is lower semicontinuous and $f^2(x) = \sup\ f(x)$ is upper semicontinuous (see [11]). The function f^1 is called a *lower bound* of f and f^2 is called a *upper bound* of f. Below we denote $Y_f = \{x | f(x) \neq \bot\}$ for $f : [a, b] \to \mathcal{I}$.

To introduce our notions of computable operators and real-valued functionals, we introduce functional domains which are effective ω-continuous domains.

Definition 8. *Let* a, b *be computable real numbers. A function domain* $\mathcal{I}_f([a, b])$ *is the collection of all continuous functions* $f : [a, b] \to \mathcal{I}$ *with the least element* $\bot_{[a,b]}$ *partially ordered by the following relation:* $f \sqsubseteq g$ *iff* $(\forall x \in [a, b])\, (f(x) \sqsubseteq g(x))$ *and* $\bot_{[a,b]} \sqsubseteq I$ *for all* $I \in \mathcal{I}_f([a, b])$.

The way-below relation $\ll$ is induced by $\sqsubseteq$ in the standard way.

Proposition 3. *For each compact interval* $[a, b]$ *with computable endpoints the function domain* $\mathcal{I}_f([a, b])$ *is an effective* ω*-continuous domain.*

Proof. The existence of $\bigvee^\uparrow A$ for each directed subset $A \subseteq \mathcal{I}_f([a, b])$ follows from the properties of semicontinuous functions. Indeed, $\bigvee^\uparrow A = \langle \sup_{f \in A} f^1, \inf_{f \in A} f^2 \rangle$, where f^1 is the lower bound and f^2 is the upper bound of f.

Let us prove that $f = \vee^\uparrow (\Downarrow f)$ for $f \in \mathcal{I}_f([a, b])$, where $\Downarrow f$ denotes the set $\{g \in \mathcal{I}_f([a, b]) | g \ll f\}$. Let U be open and $\mathrm{cl}U = \bar{U} \subset Y_f$. The set $\Downarrow f$ contains all functions of the type $g_U^n = \langle \mathbf{a}_U^n, \mathbf{c}_U^n \rangle$, where

$$\mathbf{a}_U^n(x) = \begin{cases} -\infty & \text{if } x \notin U, \\ \inf_{z \in \bar{U}} f^1(z) - \frac{1}{n} & \text{if } x \in U, \end{cases}$$

$$\mathbf{c}_U^n(x) = \begin{cases} +\infty & \text{if } x \notin U, \\ \sup_{z \in \bar{U}} f^2(z) + \frac{1}{n} & \text{if } x \in U, \end{cases}$$

By the properties of semicontinuous functions, $\vee^\uparrow \{g_U^n | \bar{U} \subset Y_f,\ n \in \omega\} = f$, so $\vee^\uparrow \Downarrow f = f$.

It is obvious that the function domain $\mathcal{I}_f([a, b])$ is ω-continuous. An example for a countable basis is the set $\mathcal{I}_{f,0}([a, b]) = \{\mathbf{b}_n\}_{n \in \omega} \cup \{\bot_{[a,b]}\}$, where the lower bound $\mathbf{b}_n^1$ and the upper bound $\mathbf{b}_n^2$ of $\mathbf{b}_n$ satisfy the following conditions: there exist $a = a_0 \ldots \leq a_i \leq \ldots \leq a_n = b$ such that

1. for all $x \in (a_i, a_{i+1})$ $\mathbf{b}_n^1(x) = -\infty$ and $\mathbf{b}_n^2(x) = +\infty$ or $\mathbf{b}_n^1(x) = \alpha_i x + \beta_i$ and $\mathbf{b}_n^2(x) = \gamma_i x + \zeta_i$;

2. if for $x \in (a_i, a_{i+1}) \cup (a_{i+1}, a_{i+2})$ $\mathbf{b}_n^1$ and $\mathbf{b}_n^2$ are finite then $\mathbf{b}_n^1(a_{i+1}) = \alpha_i a_{i+1} + \beta_i = \alpha_{i+1} a_{i+1} + \beta_{i+1}$ and $\mathbf{b}_n^2(a_{i+1}) = \gamma_i a_{i+1} + \zeta_i = \gamma_{i+1} a_{i+1} + \zeta_{i+1}$;
3. if $\mathbf{b}_n^1$ and $\mathbf{b}_n^2$ are infinite on (a_i, a_{i+1}) then $\mathbf{b}_n^1(a_i) = \mathbf{b}_n^1(a_{i+1}) = -\infty$ and $\mathbf{b}_n^2(a_i) = \mathbf{b}_n^2(a_{i+1}) = +\infty$, where a_i, α_i, β_i, γ_i, $\zeta_i \in \mathrm{D}_2$.

Using the standard numbering of the set of piecewise linear functions with coefficients from D_2, it is easy to prove that $\mathcal{I}_{f,0}([a,b])$ is countable and effective. □

3.3 Second Order Computability

Now we introduce notions of computable operators and computable functionals defined on total continuous real-valued functions. Below we use the standard notion of continuity of a total operator $F : \mathcal{I}_f([a,b]) \to \mathcal{I}_f([c,d])$ w.r.t. the Scott topologies on $\mathcal{I}_f([a,b])$ and $\mathcal{I}_f([c,d])$.

Definition 9. *Let $\mathcal{I}_f([a,b])$, $\mathcal{I}_f([c,d])$ be some function domains and $\mathcal{I}_{f,0}([a,b]) = \{\mathbf{b}_i\}_{i\in\omega}$, $\mathcal{I}_{f,0}([c,d]) = \{\mathbf{c}_i\}_{i\in\omega}$ be their effective bases constructed as in Proposition 3. A continuous total operator $F : \mathcal{I}_f([a,b]) \to \mathcal{I}_f([c,d])$ is computable, if the relation $\mathbf{c}_m \ll F(\mathbf{b}_n)$ is computably enumerable in n and m, where $\mathbf{b}_n \in \mathcal{I}_{f,0}([a,b])$ and $\mathbf{c}_m \in \mathcal{I}_{f,0}([c,d])$.*

Definition 10. *A partial operator $F : C[a,b] \to C[c,d]$ is computable, if there exists a computable operator $F^* : \mathcal{I}_f([a,b]) \to \mathcal{I}_f([c,d])$ such that*

$$F(f) = g \leftrightarrow F^*(\hat{f}) = \hat{g}, \;\; where\;\; \hat{f}(x) = \{f(x)\},\; \hat{g}(x) = \{g(x)\}.$$

Proposition 4. *For a computable partial operator $F : C[a,b] \to C[c,d]$, $\mathrm{dom}(F)$ is the countable intersection of open sets.*

Definition 11. *A partial functional $F : C[a,b] \times [c,d] \to \mathbb{R}$ is computable, if there exists a computable operator $F^* : C[a,b] \to C[c,d])$ such that*

$$F(f,x) = y \leftrightarrow F^*(f)(x) = y.$$

To introduce computability of a functional $F : C[a,b] \times \mathbb{R} \to \mathbb{R}$, we use an effective sequence $\{\mathcal{I}_f([-n,n])\}_{n\in\omega}$ of domains $\mathcal{I}_f([-n,n])$ with coordinated bases in the following sense. We consider a sequence of bases $\{\mathcal{I}_{f,0}([-n,n])\}_{n\in\omega} = \{\{\mathbf{b}_i^n\}_{i\in\omega}\}_{n\in\omega}$ with the homomorphisms $\mathrm{res}_{m,n} : \mathcal{I}_f([-m,m]) \to \mathcal{I}_f([-n,n])$ of restrictions for $m > n$ defined by the natural rules $\mathrm{res}_{m,n}(\mathbf{b}_i^m) = \mathbf{b}_i^m|_{[-n,n]} = \mathbf{b}_i^n$ and $\mathrm{res}_{m,n}(\bot_{[-m,m]}) = \bot_{[-n,n]}$.

Definition 12. *A sequence $\{F_k\}_{k\in\omega}$ of computable operators $F_k : \mathcal{I}_f[a,b] \to \mathcal{I}_f[-k,k]$ is uniformly computable, if $\mathbf{b}_m^k \ll F_k(\mathbf{b}_n^k)$ is computably enumerable in k, n and m.*

Definition 13. *A sequence $\{F_k\}_{k\in\omega}$ of computable operators $F_k : C[a,b] \to C[-k,k]$ is uniformly computable, if there exists a uniformly computable sequence $\{F_k^*\}_{k\in\omega}$ of computable operators $F_k : \mathcal{I}_f[a,b] \to \mathcal{I}_f[-k,k]$ such that for $k \in \omega$*

$$F_k(f) = g \leftrightarrow F_k^*(\hat{f}) = \hat{g}, \text{ where } \hat{f}(x) = \{f(x)\},\ \hat{g}(x) = \{g(x)\},\ x \in [a,b].$$

Definition 14. *A functional $F : C[a,b] \times \mathbb{R} \to \mathbb{R}$ is computable, if there exists a uniformly computable sequence $\{F_k^*\}_{k\in\omega}$ of computable operators $F_k^* : C[a,b] \to C[-k,k]$ such that*

$$F(f,x) = y \leftrightarrow \forall k\, (x \in [-k,k] \to F_k^*(f)(x) = y)\,.$$

Note that for $m > n$ the condition $\mathrm{res}_{m,n}(F_m^*(f)) = F_n^*(f)$ holds by construction.

4 Generalised Computability and Second Order Computability

Now we prove the theorem which connects computable operators and real-valued functionals with validity of finite Σ-formulas.

Theorem 2. *An operator $F : C[a,b] \to C[c,d]$ is computable if and only if it is generalised computable in the language without equality.*

Proof. Let $F : C[a,b] \to C[c,d]$ be computable. Let us consider its corresponding operator $F^* : \mathcal{I}_f[a,b] \to \mathcal{I}_f[c,d]$. We construct two Σ-formulas $\varphi_1,\ \varphi_2$ satisfying the conditions of Definition 2. Let $\mathcal{I}_{f,0}([a,b]) = \{\mathbf{b}_i\}_{i\in\omega}$ and $\mathcal{I}_{f,0}([c,d]) = \{\mathbf{c}_i\}_{i\in\omega}$ be effective bases constructed as in Proposition 3 for $\mathcal{I}_f([a,b])$ and $\mathcal{I}_f([c,d])$. Suppose $u \in \mathcal{I}_f([a,b])$. It is easy to see that the relation $\mathbf{b}_n \ll u$ is definable by Σ-formulas in the language without equality with positive occurrences of U_1 and U_2, where $U_1(r_1,r_2,c) \rightleftharpoons u^1|_{[r_1,r_2]} > c$, $U_2(r_1,r_2,c) \rightleftharpoons u^2|_{[r_1,r_2]} < c$. Therefore the set $\{(n,m) | u \gg \mathbf{b}_n \wedge F^*(\mathbf{b}_n) \gg \mathbf{c}_m\}$ is definable by some Σ-formula $\Phi(n,m,U_1,U_2)$.
Then $F^*(u) \gg \mathbf{c}_m \leftrightarrow \mathbf{HF}(\mathbb{R}) \models \exists n \Phi(n,m,U_1,U_2)$.

Put

$$\varphi_1(U_1,U_2,x_1,x_2,z) \rightleftharpoons \exists m \exists n \left(b_m^1|_{[x_1,x_2]} > z\right) \wedge \Phi(n,m,U_1,U_2),$$

$$\varphi_2(U_1,U_2,x_1,x_2,z) \rightleftharpoons \exists m \exists n \left(b_m^2|_{[x_1,x_2]} < z\right) \wedge \Phi(n,m,U_1,U_2).$$

Clearly, $\varphi_1,\ \varphi_2$ are required formulas.

Let $F : C[a,b] \to C[c,d]$ be generalised computable. It is not hard to see that the formulas $\varphi_1,\ \varphi_2$ define operator $F^* : \mathcal{I}_f[a,b] \to \mathcal{I}_f[c,d]$. Monotonicity of F^* follows from positive occurrences of U_1 and U_2 in the formulas φ_1 and φ_2.

Because $\mathcal{I}_f[a,b]$ and $\mathcal{I}_f[c,d]$ are ω-continuous domains, it is enough to prove that F^* preserves suprema of countable directed sets.

Let $A = \{< u_n^1, u_n^2 >\}_{n\in\omega}$ and $\bigvee^{\uparrow} A =< u^1, u^2 >$. Put $U_{1n}(x_1,x_2,c) \rightleftharpoons u_n^1|_{[x_1,x_2]} > c$ and $U_{2n}(x_1,x_2,c) \rightleftharpoons u_n^2|_{[x_1,x_2]} < c$ for $n \in \omega$ and $U_1(x_1,x_2,c) \rightleftharpoons u^1|_{[x_1,x_2]} > c$, $U_2(x_1,x_2,c) \rightleftharpoons u^2|_{[x_1,x_2]} < c$.

Note that if A is a directed set of lower semicontinuous functions and $\lim_{a\in A} a(x) = g(x)$, then for a compact V and every $c \in \mathbb{R}$ the following assertion holds: if $g(x) > c$ for all $x \in V$, then there exists $a \in A$ such that $a(x) > c$ for all $x \in V$.

So, if $u^1|_{[x_1,x_2]} > c$ then there exists n such that $u^1_n|_{[x_1,x_2]} > c$, and if $u^2|_{[x_1,x_2]} < c$ then there exists n such that $u^2_n|_{[x_1,x_2]} < c$.

So

$$U_1(x_1,x_2,c) = \bigvee_{n\in\omega} U_{1n}(x_1,x_2,c) \text{ and } U_2(x_1,x_2,c) = \bigvee_{n\in\omega} U_{2n}(x_1,x_2,c).$$

By the properties of Σ-formulas and positive occurrences of U_1 and U_2 in φ_1 and φ_2,

$$\varphi_1(U_1,U_2,x_1,x_2,c) \leftrightarrow \bigvee_{n\in\omega} \varphi_{1n}(U_1,U_2,x_1,x_2,c),$$
$$\varphi_2(U_1,U_2,x_1,x_2,c) \leftrightarrow \bigvee_{n\in\omega} \varphi_{2n}(U_1,U_2,x_1,x_2,c).$$

Hence it is clear that $F^*(\bigvee^{\uparrow} A) = \bigvee^{\uparrow} F^*(A)$.

Now we show that the set $\{(n,m)|F^*(\mathbf{b}_n) \gg \mathbf{c}_m\}$ is Σ-definable and, as a consequence, is computable enumerable in n and m. Let $F^*(<\mathbf{b}^1_n, \mathbf{b}^2_n>) = <h^1, h^2>$. Since $\mathbf{b}^1_n$, $\mathbf{b}^2_n$, $\mathbf{c}^1_m$ and $\mathbf{c}^2_m$ are piecewise linear, it is obvious that the sets $\mathbf{b}^1_n|_{[x_1,x_2]} > c$, $\mathbf{b}^2_n|_{[x_1,x_2]} < c$ and $\mathbf{c}^1_m|_{[x_1,x_2]} > c$, $\mathbf{c}^2_m|_{[x_1,x_2]} < c$ are Σ-definable. As is evident from the definition of F^*, the sets $h^1|_{[x_1,x_2]} > c$, $h^2|_{[x_1,x_2]} < c$ are Σ-definable too. By properties of semicontinuous functions, there exist upper semicontinuous step functions s^1 and s^2 such that $\mathbf{c}^1_m(x) < s^1(x) < h^1(x)$ and $\mathbf{c}^2_m(x) > s^2(x) > h^2(x)$ for $x \in [c,d]$.

As one can see, the following Σ-formula

$$\exists x_0 \ldots \exists x_n \exists y_1 \ldots \exists y_n \exists z_1 \ldots \exists z_n \bigwedge_{i\leq n} \left((\mathbf{c}^1_m|_{[x_i,x_{i+1}]} < y_i) \wedge \right.$$
$$\left. (h^1|_{[x_i,x_{i+1}]} > y_i) \wedge (\mathbf{c}^2_m|_{[x_i,x_{i+1}]} > z_i) \wedge (h^2|_{[x_i,x_{i+1}]} < z_i) \right)$$

defines the set $\{(n,m)|F^*(\mathbf{b}_n) \gg \mathbf{c}_m\}$. As a consequence this set is computable enumerable in n and m. □

Note that using the previous theorem one can elegantly prove computability of such functions as $\sup_{x\in[x_1,x_2]} f(x)$, $\inf_{x\in[x_1,x_2]} f(x)$ and Riemann integral on $[x_1,x_2]$. A couple of corresponding examples and counterexamples can be found in [27,28,39].

Corollary 1. *A functional $F : C[a,b] \times [c,d] \to \mathbb{R}$ is computable if and only if it is generalised computable in the language without equality.*

Corollary 2. *A functional $F : C[a,b] \times \mathbb{R} \to \mathbb{R}$ is computable if and only if it is generalised computable in the language without equality.*

Corollary 3. *Let $F : C[0,1] \times \mathbb{R} \to \mathbb{R}$ be an continuous functional. Then the functional F is computable in the sense of computable analysis if and only if F is generalised computable with equality.*

5 Conclusions

In this work we have analysed computability of operators and functionals defined on the class of continuous functions. Using the effective ω-continuous domains presented here we introduced a notion of second-order computability in the framework of domain theory. We took into consideration semantic characterisations of computable objects. It was shown that definability theory can be useful for analysing computability of higher order objects over the reals.

References

1. S. Abramsky, A. Jung, Domain theory, Handbook of Logic in Computer Science, v. 3, Clarendon Press, 1994.
2. J. Blanck, Domain representability of metric space, Annals of Pure and Applied Logic, 83, 1997, pages 225–247.
3. J. Barwise, Admissible sets and structures, Berlin, Springer–Verlag, 1975.
4. A. Brown, C. Pearcy, Introduction to Analysis, Springer-Verlag, Berlin, 1989.
5. L. Blum and M. Shub and S. Smale, On a theory of computation and complexity over the reals: NP-completeness, recursive functions and universal machines, Bull. Amer. Math. Soc., (N.S.) , v. 21, no. 1, 1989, pages 1–46.
6. C. E. Gordon, Comparisons between some generalizations of recursion theory, Composition Mathematica, v. 22, N 3, 1970, pages 333–346.
7. A. Edalat, Domain Theory and integration, Theoretical Computer Science, 151, 1995, pages 163–193.
8. A. Edalat, P. Sünderhauf, A domain-theoretic approach to computability on the real line, Theoretical Computer Science, 210, 1998, pages 73–98.
9. Yu. L. Ershov, Computable functionals of finite types, Algebra and Logic, 11(4), 1996 pages 367-437.
10. Yu. L. Ershov, Definability and computability, Plenum, New York, 1996.
11. M. H. Escardó, PCF extended with real numbers: a domain-theoretic approach to hair-order exact real number computation, PhD thesis, Imperial College, University of London, London, 1997.
12. T. Erker, M. H. Escardó,K. Keimel, The way-below relation of function spaces over semantic domain, Topology and its Applications, 89(1-2), pages 61-74, 1998.
13. M. H. Escardó, Function-space compactifications of function spaces, To appear in Topology and its Applications, 2000.
14. H. Friedman, Algorithmic procedures, generalized Turing algorithms, and elementary recursion theory, Logic colloquium 1969, C.M.E., Hollang, Amsterdam, 1971, pages 361–390.
15. H. Friedman and K. Ko, Computational complexity of real functions, Theoret. Comput. Sci. , v. 20, 1982, pages 323–352.
16. G. Gierz, K.H. Hofmann, K. Keimel, J.D. Lawson, M.W. Mislove, D.S. Scott, A Compendium Of Continuous Lattices, Springer Verlag, Berlin, 1980.
17. A. Grzegorczyk, On the definitions of computable real continuous functions, Fund. Math., N 44, 1957, pages 61–71.
18. A. Jung, Cartesian Closed Categories of Domains, CWI Tract. Centrum voor Wiskunde en Informatica, Amsterdam v. 66, 1989.
19. M. Korovina, Generalized computability of real functions, Siberian Advance of Mathematics, v. 2, N 4, 1992, pages 1–18.

20. M. Korovina, O. Kudinov, A New Approach to Computability over the Reals, SibAM, v. 8, N 3, 1998, pages 59–73.
21. M. Korovina, O. Kudinov, Characteristic Properties of Majorant-Computability over the Reals, Proc. of CSL'98, LNCS, 1584, 1999, pages 188–204.
22. M. Korovina, O. Kudinov, Computability via Approximations, Bulletin of Symbolic Logic, v. 5, N 1, 1999
23. M. Korovina, O. Kudinov, A Logical approach to Specifications of Hybrid Systems, Proc. of PSI'99, LNCS 1755, 2000, pages 10–16.
24. M. Korovina, O. Kudinov, Formalisation of Computability of Operators and Real-Valued Functionals via Domain Theory, Proceedings of CCA-2000, LNCS, to appear.
25. R. Montague, Recursion theory as a branch of model theory, Proc. of the third international congr. on Logic, Methodology and the Philos. of Sc., 1967, Amsterdam, 1968, pages 63–86.
26. Y. N. Moschovakis, Abstract first order computability, Trans. Amer. Math. Soc., v. 138, 1969, pages 427–464.
27. Pietro Di Gianantonio, Real number computation and domain theory, Information and Computation, N 127, 1996, pages 11-25.
28. M. B. Pour-El, J. I. Richards, Computability in Analysis and Physics, Springer-Verlag, 1988.
29. D. Scott, Outline of a mathematical theory of computation, In 4th Annual Princeton Conference on Information Sciences and Systems, 1970, pages 169–176.
30. D. Scott, Continuous lattices, Lecture Notes in Mathematics, 274, Toposes, Algebraic geometry and Logic, Springer-Verlag, 1972, pages 97–136.
31. E.Schechter, Handbook of Analysis and Its Foundations, Academic Pressbook, 1996.
32. V. Stoltenberg-Hansen and J. V. Tucker, Complete local rings as domains, Journal of Symbolic Logic, 53, 1988, pages 603-624.
33. V. Stoltenberg-Hansen and J. V. Tucker, Effective algebras, Handbook of Logic in computer Science, v. 4, Clarendon Press, 1995, pages 375–526.
34. H. Tong, Some characterizations of normal and perfectly normal space, Duke Math. J. N 19, 1952, pages 289-292.
35. K. Weihrauch, Computability, volume 9 of EATCS Monographs on Theoretical Computer Science, Springer, Berlin, 1987.
36. K. Weihrauch, A simple introduction to computable analysis, Informatik Berichte 171, FernUniversitat, Hagen, 1995, 2-nd edition.
37. C. Kreitz, K. Weihrauch, Complexity Theory on Real Numbers and Functions, LNCS, 145, 1983, pages 165-175.
38. K. Weihrauch, X. Zheng Computability on Continuous, Lower Semi-Continuous and Upper Semi-Continuous real Functions, LNCS, 1276, 1997, pages 166-186.
39. K. Weihrauch, Computable Analysis. An Introduction, Springer-Verlag, 2000.

An Abstract Look at Realizability

Edmund Robinson[1,*] and Giuseppe Rosolini[2]

[1] Queen Mary, University of London
[2] Università di Genova

Abstract. This paper is about the combinatorial properties necessary for the construction of realizability models with certain type-theoretic properties. We take as our basic construction a form of tagging in which elements of sets are equipped with tags, and functions must operate constructively on tags. To complete the construction we allow a form of closure under quotients by equivalence relations. In this paper we analyse first the condition for a natural monoidal structure to be product structure, and then investigate necessary conditions for the realizability model to be locally cartesian closed and to have a subobject classifier.

Introduction

Realizability is a technique for constructing models in which all operations of a given type are computable, according to a given notion of computation. It extends the naive approach of enumerating elements and requiring that operations be computable with respect to the enumerations, in particular by allowing the construction of higher-order types. It produces extensional models which validate various forms of constructive reasoning, e.g. [10,17,19], and forms the basis for PER models of polymorphic lambda calculi e.g. [11]. All this work uses traditional intensional models of untyped computation, such as the Kleene algebra of partial recursive functions. However there is recent interest in extending this, for example to process models [1] or to the typed setting [13,12].

These approaches tend to take quite a concrete approach, giving structures and building combinators into the definition. For example Longley's notion of typed pca assumes function spaces and application, and then uses them to construct a locally cartesian closed category (the category-theorists analogue of a type theory with dependent products). The purpose of the present paper is to attempt to reverse this. One of our results is that, modulo a condition to do with the way pairs are represented in the realizability model, if the realizability model is locally cartesian closed, then the model of computation has a weak form of function space, though not quite Longley's. This to some extent validates the use of combinatorial structures which have function spaces built in, and is typical of the form of our results. Broadly, they say that for the realizability model to support extensional forms of type structure, i.e. with both β

* The authors wish to acknowledge the support of the EPSRC, EU Working Group 26142 APPSEM, and MURST

L. Fribourg (Ed.): CSL 2001, LNCS 2142, pp. 173–187, 2001.

and η laws, the underlying model of computation has to interpret corresponding combinators, but in a weak sense. This holds both for products and for function spaces. There is an exception to this pattern in the result which discusses what happens when the realizability model has a subobject classifier: in this case the model of computation must have a universal object, again in a weak sense, and thus that from the point of view of the model a typed form of realizability gives no extra generality over an untyped form.

We have chosen to use categorical technology and to couch our results in categorical terms. Thus, for us, a model of computation will be a category (for example the category with a single object, to be thought of as N, where the morphisms are partial recursive functions), and the existence of combinators will be given by structure on that category. There are two reasons for this choice. The first is that our account of the construction of a realizability model is essentially categorical. It is of course possible to give the construction in more set-theoretic language, and indeed this appears quite natural for the first part of the construction. However, set-theoretic constructions can be overly concrete. Our category-theoretic framework applies immediately to pointed cpo's, where there are at least two possible ways of assigning a set (include bottom or not). Moreover, if one uses a set-theoretic presentation, the second part of the construction (freely adjoining quotients of equivalence relations) is poorly motivated. It would not be clear why that particular definition should be chosen over a number of possible variants. Our second reason is that the categorical formulation gives a fairly clear idea of what the minimal supporting structure might be. Set-theoretic formulations have not.

In these senses the paper contrasts with recent work particularly by Longley [13] and Lietz and Streicher [12], in which the basis is taken as a typed generalisation of a partial combinatory algebra. We, like they, will be interested in when the construction yields a topos, and hence gives a full interpretation of higher-order logic. This is also a theme of Birkedal's work, see [2,3], and his joint work in [4].

We present realizability toposes as the product of two constructions. First one takes a category (which corresponds to the typed partial combinatory algebra), and then one glues **Set** to it in a variant of the comma construction. This step is the categorical equivalent of forming a category in which objects are sets whose elements are tagged by possible realizers, e.g. natural numbers. The result should be a category with finite products, and we study the conditions under which it is so, or rather we study the conditions under which a natural monoidal structure gives finite products. In this event, it has long been known [6,16] that in the examples derived from standard realizability the associated realizability topos is the exact completion, i.e. it is obtained essentially by freely adjoining quotients of equivalence relations. In the general case which we study, we do not necessarily get a cartesian closed category, still less a topos. We produce necessary conditions for *local* cartesian closure (dependent products, not just function spaces) and the existence of a subobject classifier (an object of truth values).

Our study of finite limits depends on the initial category being monoidal. This is the level of product structure exhibited by multiplicatives in linear logic. We show that finite products demand in addition combinators corresponding to diagonal and projections. These results can be read as saying that Birkedal was correct to use categories of partial maps as a basis for his theory, nothing significantly more general would have worked. In the case of function spaces, however, we get something slightly weaker than Birkedal's condition. Birkedal's condition is an analogue of the standard partial function space, in that every partial function is representable. Our results suggest that this is too strong for the current purpose, and all that is required is that some extension of any partial function be representable. This is a new notion of partial function space, which, to our knowledge, has not previously been encountered. Finally, we present our version of a result given independently by Birkedal and Lietz and Streicher, that if the realizability category has a subobject classifier, then the original category has a form of universal object. Our result is slightly more general than theirs, since it is independent of questions of cartesian closure inherent in their frameworks, and we give an explicit account of how it relates to untyped realizability.

The motivation for this work came from two directions. The first was to provide a general categorical account of traditional work on realizability. Our results show limitations on the use of typed forms of standard realizability in terms of the models they produce. There remains, however, modified realizability. It is possible to read the sets of "possible realizers" in modified realizability as a form of type, and hence to think of modified realizability as a form of typed realizability. Alas, our results show that this can not be if by typed realizability we mean either the construction given here, or, more particularly Longley's setting.

Our second motivation was to provide a case study giving the limitations of what could be achieved using these structures, but admitting the possibility of starting out with a very different model of computation, as in Abramsky's work on process realizability [1]. Here we believe that our results and techniques could be useful in narrowing down the design space.

A longer version of this paper is available from the authors. It contains a more substantial introduction as well as those proofs which have been excised for reasons of space.

We would like to acknowledge useful discussions with Lars Birkedal, Peter Lietz, Thomas Streicher and particularly Federico de Marchi.

1 The $\mathcal{F}$-Construction

There is a simple categorical generalisation of the construction of the category of partitioned assemblies, given by a variant of the standard comma construction.

We write **Ptl** for the category of sets and partial functions. The standard cartesian product of sets is no longer a categorical product, but it does provide a monoidal structure, which we shall use later.

Suppose $U : \mathbf{C} \longrightarrow \mathbf{Ptl}$ is a functor. Let $\mathcal{F}(\mathbf{C}, U)$ be the category whose objects are triples $(C, S, \sigma : S \longrightarrow U(C))$, where σ is total, and a map

$f : (C, S, \sigma) \longrightarrow (C', S', \sigma')$ is a (total) function $f : S \longrightarrow S'$ such that there exists a map $\phi : C \longrightarrow C'$ in $\mathbf{C}$ for which

$$\begin{array}{ccc} S & \xrightarrow{f} & S' \\ \sigma \downarrow & & \downarrow \sigma' \\ U(C) & \xrightarrow{U(\phi)} & U(C') \end{array}$$

commutes.

Notation. We shall always write $\mathcal{F}(\mathbf{C})$ leaving U understood. Instead of $(C, S, \sigma : S \longrightarrow U(C))$, we shall write a typical object of $\mathcal{F}(\mathbf{C})$ as $\sigma : S \longrightarrow U(C)$, using the fact that we can recover C from the notation $U(C)$. Finally, we shall write morphisms as pairs (f, ϕ). This is redundant in that the equality between morphisms is based only on the first component, but we shall need to use the second in some of our constructions.

We can think of this as a category of tagged sets. $\sigma : S \longrightarrow U(C)$ represents the tagging of the elements of S by realizers taken from $U(C)$. The functions are functions at the level of sets which can be traced by a function on tags.

Example 1. As part of the construction of a standard realizability topos, $\mathbf{C}$ can be taken to be the monoid of representable partial endo-functions on the partial combinatory algebra in use. In this case $\mathcal{F}(\mathbf{C})$ is the category of partitioned assemblies (the projective objects) in the associated realizability topos. In particular, if we take $\mathbf{C}$ to be the monoid of partial recursive functions on N, then we will get the projective objects of the classical effective topos.

The category $\mathcal{F}(\mathbf{C})$ always has equalizers, and we shall see that weak conditions on $\mathbf{C}$ ensure products in $\mathcal{F}(\mathbf{C})$. Similarly, weak conditions on $\mathbf{C}$ ensure that the exact completion $\mathcal{F}(\mathbf{C})_{\mathrm{ex}}$ is locally cartesian closed.

Like a comma category, $\mathcal{F}(\mathbf{C})$ comes equipped with a number of functors.

Let $\mathbf{C}_{\mathrm{t}}$ be the inverse image along U of the subcategory of total functions, then there is a full functor $\mathsf{Y} : \mathbf{C}_{\mathrm{t}} \longrightarrow \mathcal{F}(\mathbf{C})$ defined by

$$[C \xrightarrow{f} C'] \longmapsto [(C, U(C), \mathrm{id}) \xrightarrow{(U(f), f)} (C', U(C'), \mathrm{id})]$$

or

$$[C \xrightarrow{f} C'] \longmapsto [(C \xrightarrow{\mathrm{id}} U(C)) \xrightarrow{(U(f), f)} (C' \xrightarrow{\mathrm{id}} U(C'))]$$

This becomes full and faithful when U is faithful.

Because of the existence condition in the definition of morphisms, there is no forgetful functor $\mathcal{F}(\mathbf{C}) \to \mathbf{C}$, however there is one $\mathcal{F}(\mathbf{C}) \to \mathbf{Set}$. More significantly, let C be an arbitrary object of $\mathbf{C}$, and $x : 1 \longrightarrow U(C)$ an arbitrary element of $U(C)$, then there is a full embedding $\nabla_{C,x} : \mathbf{Set} \longrightarrow \mathcal{F}(\mathbf{C})$ defined by

$$\nabla : [S \xrightarrow{f} S'] \longmapsto [(S \xrightarrow{x \circ !} U(C)) \xrightarrow{(f, U(\mathrm{id}))} (S' \xrightarrow{x \circ !} U(C))]$$

This definition is quite robust. If there are morphisms $\phi : C \longrightarrow C'$ and $\psi : C' \longrightarrow C$ such that $\phi x = x'$ and $\psi x' = x$, then $\nabla_{C,x}$ is naturally isomorphic to $\nabla_{C',x'}$.

Because of the existence condition in the definition of maps of $\mathcal{F}(\mathbf{C})$, it is clear that $\mathcal{F}(\mathbf{C})$ is equivalent to $\mathcal{F}(U[\mathbf{C}])$ where $U[\mathbf{C}]$ is the quotient of $\mathbf{C}$ with two maps identified when they have the same value under U (in other words, the category sitting in the middle of the (full and identity on objects)/faithful factorisation of U).

Notation. In order to make things less cluttered, from now on we shall write $\overline{(\ )}$ for the functor U, so $U(C) = \overline{C}$ and $U(\delta) = \overline{\delta}$.

2 Exact Completions

Our construction proceeds in two stages. We begin by constructing a base category using the $\mathcal{F}$-construction, and then we construct a better-behaved category from that using an exact completion. In other words our final category is a free exact category on a category obtained by means of the $\mathcal{F}$-construction.

Our results, then, rely on the fundamental property of an exact completion (cf. [7]): Given an exact category $\mathbf{A}$, let $\mathbf{P}$ be the full subcategory on the regular projectives of $\mathbf{A}$. Then $\mathbf{A}$ is an exact completion of a category with finite limits if and only if $\mathbf{P}$ is closed under finite limits and each object in $\mathbf{A}$ is covered by a regular projective (i.e. for every A in $\mathbf{A}$ there is a regular epi $P \longrightarrow\!\!\!\!\triangleright\, A$ from a regular projective). When this is the case, $\mathbf{A}$ is the exact completion of $\mathbf{P}$.

The crucial point here is that the base category of projectives, $\mathbf{P}$, which in our case is going to be $\mathcal{F}(\mathbf{C})$ must be left exact, and in the next section we explore conditions under which this is so.

3 Finite Limits

First, we observe that $\mathcal{F}(\mathbf{C})$ always has equalisers. This reduces the question to when $\mathcal{F}(\mathbf{C})$ has products. It is fairly easy to see when $\mathcal{F}(\mathbf{C})$ has a terminal object, though the condition seems both delicate and a little unnatural. However, characterising products seems more difficult.

Fortunately, in the cases we know about $\mathbf{C}$ can be taken to be a monoidal category, and $\overline{(\ )}$ a monoidal functor (cf. [9]). This means that $\mathcal{F}(\mathbf{C})$ has a candidate for a monoidal structure. The unit is given by $\psi : 1 \longrightarrow \overline{I}$, and the tensor by $(f : X \longrightarrow \overline{C}) \otimes (g : Y \longrightarrow \overline{D}) = \theta(f \times g) : X \times Y \longrightarrow \overline{C \otimes D}$, where ψ and $\theta : \overline{C} \times \overline{D} \longrightarrow \overline{C \otimes D}$ are the maps given by the monoidal structure of $\overline{(\ )}$. These definitions give valid objects of $\mathcal{F}(\mathbf{C})$ if and only if ψ and θ are total. In this case, the resulting structure is indeed monoidal. The verification is straightforward category theory, except that at some points we have to use the totality of various morphisms.

This allows us to ask a simpler question: when is this monoidal structure actually a product? This simplification is not without cost. We noted above that $\mathcal{F}(\mathbf{C})$ is equivalent to $\mathcal{F}(\overline{\mathbf{C}})$ where $\overline{\mathbf{C}}$ is the quotient of $\mathbf{C}$ with two maps identified when they have the same value under $\overline{(\;)}$. This suggests that without loss of generality we can take $\overline{(\;)}$ to be faithful. This is not, unfortunately, the case. The problem is that the monoidal structure on $\mathbf{C}$ does not necessarily transfer to one on $\overline{\mathbf{C}}$. The reason is that unless θ is iso, the monoidal tensor does not necessarily respect equivalence of maps. But, unless $\overline{(\;)}$ is faithful we cannot completely reflect properties of $\mathcal{F}(\mathbf{C})$ back into properties of $\mathbf{C}$. This explains why in general we prove properties up to the functor $\overline{(\;)}$, leaving the cleaner and perhaps more interesting case where $\overline{(\;)}$ is faithful to corollaries. This is first evident in the characterisation of when the monoidal unit on $\mathcal{F}(\mathbf{C})$ is terminal.

Lemma 1. *$\psi : 1 \longrightarrow \overline{I}$ is terminal in $\mathcal{F}(\mathbf{C})$ if and only if for each object C of $\mathbf{C}$ there is a map $t_C : C \longrightarrow I$ such that $\overline{t_C} = \psi \circ !$.*

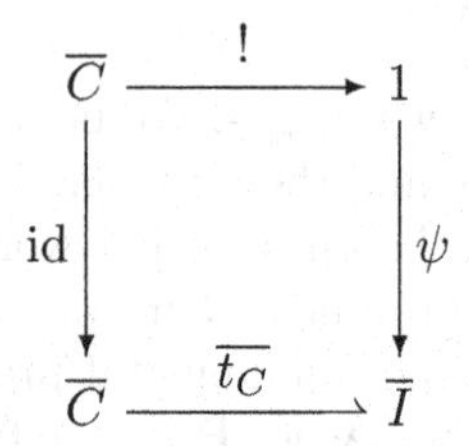

Proof. If $\psi : 1 \longrightarrow \overline{I}$ is terminal, then we obtain t_C by considering the terminal map from $\mathrm{id} : \overline{C} \longrightarrow \overline{C}$ to ψ (the diagram is as above). Conversely, given such a family of maps, ψ is weakly terminal because for any $f : X \longrightarrow \overline{C}$, we have the following diagram (note that the upper triangle commutes because f is total).

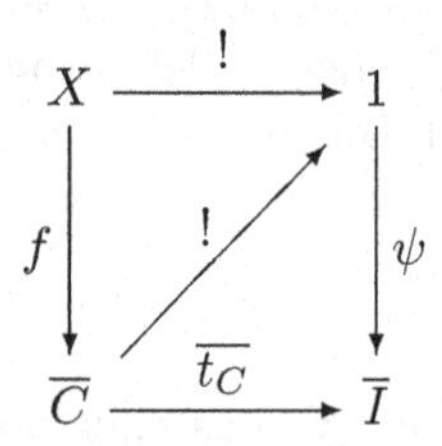

However, maps into ψ are unique, when they exist, because maps into 1 are. This establishes that ψ is terminal. □

Now, if the unit of a monoidal category is terminal, then there are candidates for left and right projections from the tensor:

$$\pi_{0,XY} = \rho_X(\mathrm{id}_X \otimes t_Y) : X \otimes Y \longrightarrow X \qquad \pi_{1,XY} = \lambda_Y(t_X \otimes \mathrm{id}_Y) : X \otimes Y \longrightarrow Y$$

This allows us to ask the question of when the monoidal tensor is a product, in the precise sense that these projections together form a product cone.

Lemma 2. *In the case that $\psi : 1 \longrightarrow \overline{I}$ is terminal in $\mathcal{F}(\mathbf{C})$, then the candidates for projections above form product cones if and only if for each object c of $\mathbf{C}$ there is a map $d_C : C \longrightarrow C \otimes C$ such that $\overline{d_C} = \theta \circ \Delta_{\overline{C}}$, where $\Delta_{\overline{C}} : \overline{C} \longrightarrow \overline{C} \times \overline{C}$ is the ordinary cartesian diagonal.*

Proof. (Sketch) First, suppose that the tensor is cartesian product. Then the tensor of $\mathrm{id} : \overline{C} \longrightarrow \overline{C}$ with itself is $\theta : \overline{C} \times \overline{C} \longrightarrow \overline{C \otimes C}$. This must have a diagonal

$$\begin{array}{ccc} \overline{C} & \xrightarrow{D} & \overline{C} \times \overline{C} \\ {\scriptstyle \mathrm{id}}\downarrow & & \downarrow{\scriptstyle \theta} \\ \overline{C} & \xrightarrow{\overline{d_C}} & \overline{C \otimes C} \end{array}$$

Composing with the projections we see that D must be the diagonal $\Delta_{\overline{C}}$, and the square then yields $\overline{d_C} = \theta \circ \Delta_{\overline{C}}$, as required.

For the converse, suppose we have two maps

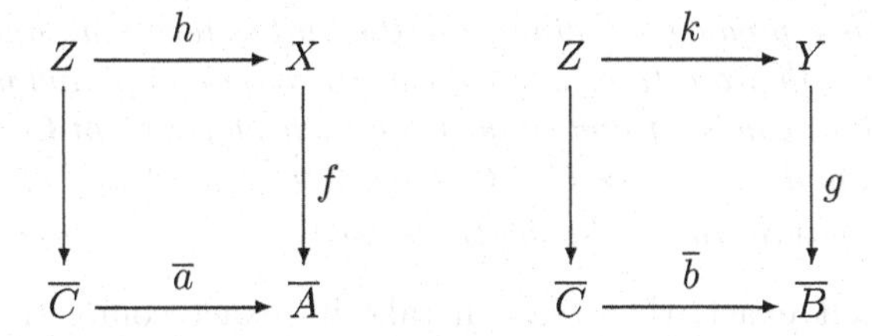

then we can form the pairing

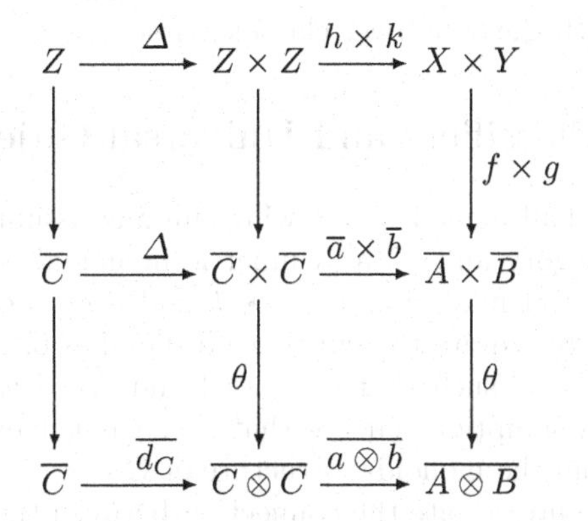

given by composing the obvious "diagonal" on $Z \longrightarrow \overline{C}$ (the left-hand half of the diagram) with the tensor product. □

Note that Δ and θ are natural considered as transformations between functors $\mathbf{C} \longrightarrow \mathbf{Ptl}$, hence $\overline{d_C}$ is natural in C. Thus, if $\overline{(\ )}$ is faithful, then d_C itself is natural in C. However, although ψ is natural in C, $\overline{!}$ is only natural in the subcategory of total maps. Whence $\overline{t_C}$ (and hence t_C, if $\overline{(\ )}$ is faithful) is natural only in the category of total maps in $\mathbf{C}$.

Moreover, it is not necessarily the case that $\overline{I}$ is isomorphic to 1, or that $\overline{X \otimes Y}$ is isomorphic to $\overline{X} \times \overline{Y}$. However:

Lemma 3. *There is a map* $e : X \otimes Y \longrightarrow X \otimes Y$ *such that* $\overline{e}$ *is an idempotent split by*

$$\overline{X \otimes Y} \xrightarrow{(\overline{\pi_0}, \overline{\pi_1})} \overline{X} \times \overline{Y} \xrightarrow{\theta} \overline{X \otimes Y}$$

where $\pi_0 = \rho(\mathrm{id} \otimes t)$ *and* $\pi_1 = \lambda(t \otimes \mathrm{id})$.

Similarly t_I is an endomorphism on I, such that $\overline{t_I}$ is split by

$$\overline{I} \xrightarrow{\quad ! \quad} 1 \xrightarrow{\quad \psi \quad} \overline{I}$$

In summary:

Lemma 4. *If* $\mathbf{C}$ *is a symmetric monoidal category and* $\overline{(\ )}$ *a faithful symmetric monoidal functor* $\mathbf{C} \longrightarrow \mathbf{Ptl}$, *for which the structural maps* $\psi : 1 \longrightarrow \overline{I}$ *and* $\theta_{C,D} : \overline{C} \times \overline{D} \longrightarrow \overline{C \otimes D}$ *are total, then* $\mathcal{F}(\mathbf{C})$ *carries a symmetric monoidal structure. This is a product structure, i.e. the unit is terminal, and the monoidal product together with projections defined from terminal maps and monoidal structure forms product cones, if and only if for each object* C *of* $\mathbf{C}$ *there are maps* $t_C : C \longrightarrow I$ *and* $d_C : C \longrightarrow C \otimes C$ *such that* $\overline{t_C} = \psi!$ *and* $\overline{d_C} = \theta_{C,C}\Delta_{\overline{C}}$. *In addition,* $\overline{d_C}$ *is natural in* C *(though* $\overline{t_C}$ *is not).*

If $\mathcal{F}(\mathbf{C})$ is left exact, then we can take its exact completion. This is our candidate for a topos. In the next two sections we see what we can say about $\mathbf{C}$ when this category is (locally) cartesian closed or has a subobject classifier. It is simpler to deal with the subobject classifier first.

4 Subobject Classifiers and Universal Objects

From this point we shall make the following running assumptions: $\mathbf{C}$ is a symmetric monoidal category and $\overline{(\ )}$ a symmetric monoidal functor $\mathbf{C} \longrightarrow \mathbf{Ptl}$, for which the structural maps $\psi : 1 \longrightarrow \overline{I}$ and $\theta_{C,D} : \overline{C} \times \overline{D} \longrightarrow \overline{C \otimes D}$ are total. Moreover we require the existence of families of maps $t_C : C \longrightarrow I$ and $d_C : C \longrightarrow C \otimes C$ such that $\overline{t_C} = \psi \circ !$ and $\overline{d_C} = \theta_{C,C} \circ \Delta_{\overline{C}}$, as in the last section. These assumptions ensure that $\mathcal{F}(\mathbf{C})$ is left exact, with cartesian structure derived from the monoidal structure of $\mathbf{C}$.

In this section we investigate the connection between the existence of a subobject classifier in $\mathcal{F}(\mathbf{C})_{\mathrm{ex}}$ and universal objects in $\mathbf{C}$. As before, our main result takes its cleanest form when U is faithful, but can be deduced immediately from a more technical statement which holds in general.

Definition 1. *The category* $\mathbf{C}$ *has a* universal object W *if each object* C *of* $\mathbf{C}$ *is a retract of* W.

Proposition 1. *If the category* $\mathcal{F}(\mathbf{C})_{\mathrm{ex}}$ *has a subobject classifier, then there is an object* W *of* $\mathbf{C}$, *such that for each object* C *of* $\mathbf{C}$ *there are morphisms* $\gamma : C \longrightarrow W$ *and* $\delta : W \longrightarrow C$ *such that* $\overline{\delta\gamma}$ *is the identity on* $\overline{C}$.

Intuitively, modulo U, W is universal in $\mathbf{C}$. If U is faithful, then this immediately implies that $\delta\gamma = \mathrm{id}_C$.

Corollary 1. *If the category $\mathcal{F}(\mathbf{C})_{\mathrm{ex}}$ has a subobject classifier, and the functor U is faithful, then the category $\mathbf{C}$ has a universal object.*

This result is closely connected to one in [12,2] obtained for the subcategory of an exact completion as above which is the regular completion of $\mathbf{P}$, see [5].

Our proof builds on previous analysis of subobject classifiers in exact completions. The following is a slight variant of Menni [14].

Definition 2. *A map $u : W \longrightarrow V$ is a (weakly) weak proof classifier if every map in the category appears as weakly equivalent to a (weak) pullback of u: i.e. for every map $a : X \longrightarrow A$ there is a diagram*

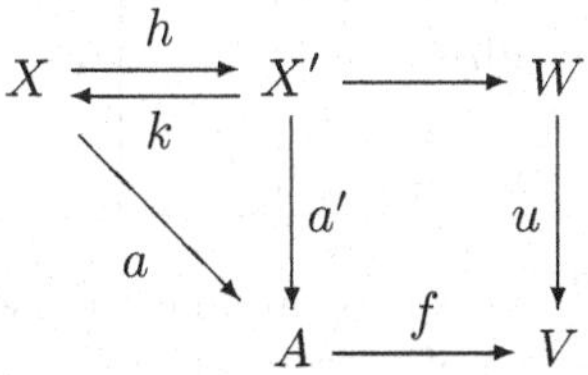

$$\begin{array}{ccccc} X & \underset{k}{\overset{h}{\rightleftarrows}} & X' & \longrightarrow & W \\ & \searrow^{a} & \downarrow{a'} & & \downarrow{u} \\ & & A & \xrightarrow{f} & V \end{array} \tag{1}$$

where the square is a (weak) pullback, and the triangles commute.

In an exact category $\mathbf{A}$ where every object is covered by a regular projective, a weakly weak proof classifier is what can be traced directly in the full subcategory $\mathbf{P}$ of projectives when $\mathbf{A}$ has a subobject classifier. If in addition $\mathbf{P}$ is a left exact subcategory (as when $\mathbf{A}$ is its exact completion), then any weakly weak proof classifier is actually a weak proof classifier in the sense of Menni (the weak pullback in the definition can always be taken to be a pullback).

We will also need a further technical result, establishing a factorisation property which generalises a standard lemma for subobject classifiers, and is best seen in the abstract:

Lemma 5. *Suppose that $u : W \longrightarrow V$ is a (weakly) weak proof classifier, and that $a : X \longrightarrow A$ is an arbitrary map. The (weakly) weak proof classifier produces a diagram*

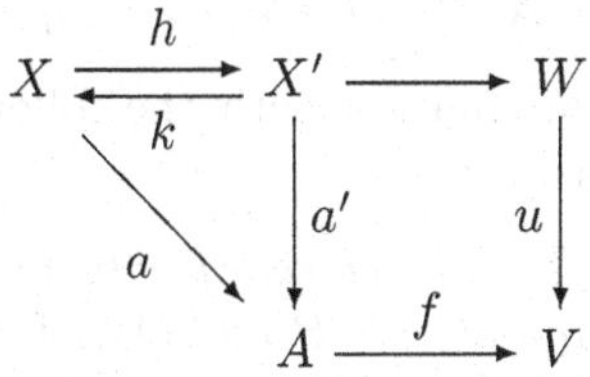

$$\begin{array}{ccccc} X & \underset{k}{\overset{h}{\rightleftarrows}} & X' & \longrightarrow & W \\ & \searrow^{a} & \downarrow{a'} & & \downarrow{u} \\ & & A & \xrightarrow{f} & V \end{array}$$

Suppose, now that $b : Y \longrightarrow A$ makes f true, in the sense that $f \circ b$ factors through u, then b factors through a.

Proof (Proposition 1, sketch). Taking a weak proof classifier

$$\begin{array}{ccc} Q & \xrightarrow{f} & P \\ {\scriptstyle w}\downarrow & & \downarrow{\scriptstyle v} \\ \overline{W} & \xrightarrow{\overline{\phi}} & \overline{V} \end{array} \qquad (2)$$

we prove that $\overline{W}$ is "universal" (quotation marks indicate that this holds modulo $\overline{(\ )}$).

To prove that C is a "retract", classify

$$\begin{array}{ccc} \overline{C} & \xrightarrow{\mathrm{id}_{\overline{C}}} & \overline{C} \\ {\scriptstyle \overline{\mathrm{id}_C}}\downarrow & & \downarrow{\scriptstyle \overline{t_C}} \\ \overline{C} & \xrightarrow{\overline{t_C}} & \overline{I} \end{array} \qquad (3)$$

giving a map $\overline{\gamma} : \overline{C} \longrightarrow \overline{W}$. This is total, and we use it as an object of $\mathcal{F}(\mathbf{C})$ in order to establish the existence of a "retraction". □

5 Function Spaces

In this section we deal with conditions for the cartesian closure of $\mathcal{F}(\mathbf{C})_{\mathrm{ex}}$. We continue with the running assumption made at the start of section 4: that $\mathbf{C}$ is a symmetric monoidal category, and $U : \mathbf{C} \longrightarrow \mathbf{Ptl}$ a symmetric monoidal functor satisfying certain conditions so that $\mathcal{F}(\mathbf{C})$ is a left exact category with product structure constructed from the monoidal structure on $\mathbf{C}$. We shall abuse the structure and refer to a map f in $\mathbf{C}$ as *total* just when its image under U, $\overline{f}$, is total.

As in section 4, our work builds heavily on previous work on properties of exact completions. One of the major lessons of [8] is that in this context it is easier to deal with local cartesian closure, than simple cartesian closure. So it is an important fact that exact completion is a local construction.

Lemma 6. *Let* $\mathbf{P}$ *be a left exact category with exact completion* $\mathbf{A}$*. Then for any object* P *of* $\mathbf{P}$*, the slice* $\mathbf{A}/P$ *is the exact completion of* $\mathbf{P}/P$*.*

We shall use this in combination with the following facts about cartesian closure of exact completions.

Lemma 7. *Let* $\mathbf{P}$ *be a left exact category with exact completion* $\mathbf{A}$*. If* $\mathbf{A}$ *is cartesian closed, then for any objects* P *and* Q *of* $\mathbf{P}$*, there is a weak evaluation* $\epsilon : F \times P \longrightarrow Q$ *from* P *to* Q *in* $\mathbf{P}$*, i.e. any map* $\phi : X \times P \longrightarrow Q$ *can be expressed as* $\epsilon \circ (f \times \mathrm{id}_P)$ *for some* $f : X \longrightarrow F$ *(here all of the last part of the statement takes place in* $\mathbf{P}$*).*

We shall need to deal with a monoidal structure that approximates a product. First some notation.

Notation. The structure on $\mathbf{C}$ gives operations which we can loosely think of as pairing and projections, and which we shall write:

$$\langle f, g \rangle = (f \otimes g) \circ d_Z : Z \longrightarrow X \otimes Y$$

$$\pi_0 = \rho_X \circ (\mathrm{id}_X \otimes t_Y) : X \otimes Y \longrightarrow X \qquad \pi_1 = \lambda_Y \circ (t_X \otimes \mathrm{id}_Y) : X \otimes Y \longrightarrow Y$$

We shall use (a, b), p_0 and p_1 for the usual pairing and projections from a categorical product.

Returning to our main interest, if

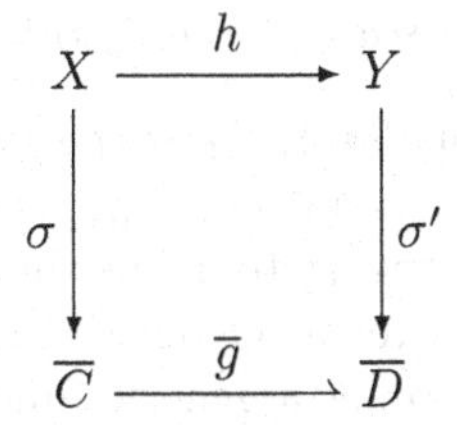

a morphism in $\mathcal{F}(\mathbf{C})$, then we can replace g by any morphism f such that $\overline{f}$ extends $\overline{g}$. It follows that if we have constructed some morphism f in $\mathbf{C}$, then we will not be able to prove that $\overline{f}$ is a particular *partial* function h, only that it is an extension of h. Moreover, we have seen that our monoidal structure is not a product, but the product is related by retraction. This motivates the following definition:

Definition 3. *Suppose* $\mathbf{C}$ *is a monoidal category equipped with a functor* $\overline{(\ )}$ *into* **Ptl**, *together with families of maps* $t_C : C \longrightarrow I$ *and* $d_C : C \longrightarrow C \otimes C$, *as in our standard structure. Then we say that a morphism* $f : A \longrightarrow B$ extends *a morphism* $g : A \longrightarrow B$ $(g \subseteq f : A \longrightarrow B)$ *if* $\overline{g} \subseteq \overline{f}$. *We now say that a map* $\epsilon : F \otimes C \longrightarrow C'$ *is a weak partial evaluation from* C *to* C' *if for every map* $\phi : X \otimes C \longrightarrow C'$ *there is a total* $f : X \longrightarrow F$, *such that* $\epsilon \circ (f \otimes \mathrm{id}_C) \circ e$ *extends* $\phi \circ e$, *where* $e : X \otimes C \longrightarrow X \otimes C$ *is the "η-retraction" for pairing* $e = \langle \pi_0, \pi_1 \rangle$.

This differs from a standard definition of partial function space in that it does not demand that arbitrary partial functions be represented, only that some extension of them be, and also in that the equation unexpectedly passes through the "η-retraction" for pairing. This can be viewed as saying that the equation does not have to hold on the whole of $X \otimes C$, but only on those elements which are actually ordered pairs.

Moreover, the definition we have given depends upon U to give notions of totality and extension for morphisms in $\mathbf{C}$. However, instead of deriving these notions directly from U, we could instead use the "diagonal" and "terminal" maps in $\mathbf{C}$ to give internal definitions. This is a standard trick in p-categories, and fortunately agrees with our other definition. It follows that if we regard the "diagonal" and "terminal" maps as part of our structure, then we can reasonably suppress mention of this dependence on U.

Proposition 2. *Suppose* $\mathbf{C}$ *and* $U : \mathbf{C} \longrightarrow \mathbf{Ptl}$ *satisfy our running assumptions, then if the exact completion of* $\mathcal{F}(\mathbf{C})$ *is locally cartesian closed, then for any pair of objects* C *and* C' *of* $\mathbf{C}$*, there is an object* F *of* $\mathbf{C}$ *and a map* $\epsilon : F \otimes C \longrightarrow C'$*, such that for any map* $\phi : X \otimes C \longrightarrow C'$*, there is a map* $f : X \longrightarrow F$ *such that* $\overline{f}$ *is total and* $\overline{\epsilon \circ (f \otimes \mathrm{id}_C) \circ e}$ *extends* $\overline{\phi \circ e} : X \otimes C \longrightarrow C'$*, where* $e : X \otimes C \longrightarrow X \otimes C$ *is the* η*-retraction for pairing* $e = \langle \pi_0, \pi_1 \rangle$*, as before.*

Corollary 2. *If in the above the functor* U *is faithful, then the exact completion of* $\mathcal{F}(\mathbf{C})$ *is locally cartesian closed if and only if* $\mathbf{C}$ *has weak partial evaluations.*

The corollary follows immediately from the proposition, which, however, is technically the most demanding result in the paper. The proof depends on the use of cartesian closure in a slice category to define the weak partial evaluation. More specifically we work in a slice over a set derived from the possible subfunctions of the identity on C in order to get a generic function space. Details are in the full version of this paper.

6 Consequences for Realizability

In this section we draw out the consequences of our previous results in the case that most interests us. We shall suppose that $\mathbf{C}$ is a category of sets and functions and that $\overline{(\ )}$ is the underlying set functor. Thus $\overline{(\ )}$ is faithful. What we have in mind is that $\mathbf{C}$ is the category obtained from some form of typed partial applicative structure, as in Longley [13], but part of the game is to see how much of that structure we can reconstruct from properties of the resulting realizability category.

In section 3, we examined the case when a monoidal structure induced a product on $\mathcal{F}(\mathbf{C})$. In lemma 4 we showed that in this case we had a diagonal $d_C : C \longrightarrow C \otimes C$ and collection of maps into the unit $t_C : C \longrightarrow I$, satisfying certain properties. We have seen that these induce projections $\pi_0 : X \otimes Y \longrightarrow X$ and $\pi_1 : X \otimes Y \longrightarrow Y$, and a form of pairing $\langle a, b\rangle : z \longrightarrow X \otimes Y$. This pairing satisfies the beta laws:

$$\pi_0 \circ \langle a, b\rangle = a \quad \text{and} \quad \pi_1 \circ \langle a, b\rangle = b$$

but not necessarily the eta law

$$\langle \pi_0, \pi_1\rangle = \mathrm{id} : X \otimes Y \longrightarrow X \otimes Y$$

The result is that we have something which is almost, but not quite, a category of partial maps on a category with finite products. It is interesting to compare with the formalisms given in [15], and to check when the equations listed there are satisfied. It turns out that the transformations have the correct naturality properties, but equations whose domains are tensors $X \otimes Y$ are valid only when composed with the retraction on $X \otimes Y$. We can therefore obtain a category of partial maps by splitting suitable idempotents. Since idempotents split in **Ptl**, $\overline{(\ )}$ extends to the resulting category (though it is not obviously still faithful). Now

Lemma 8. *If in* **C**, C *is a retract of* D

$$C \underset{i}{\overset{r}{\rightleftarrows}} D$$

then an object $f : X \longrightarrow \overline{C}$ *of* $\mathcal{F}(\mathbf{C})$ *is isomorphic to* $\overline{i} \circ f : X \longrightarrow \overline{D}$.

Corollary 3. *If* $\overline{(\ )} : \mathbf{C} \longrightarrow \mathbf{Ptl}$ *and* **D** *is a category obtained from* C *by splitting idempotents, then* $\overline{(\ )}$ *extends to* **D**, *and* $\mathcal{F}(\mathbf{D})$ *is equivalent to* $\mathcal{F}(\mathbf{C})$.

So this process does not affect the resulting category.

This means that if $\mathcal{F}(\mathbf{C})$ is a left exact category (or more exactly if it is lex and that structure is obtained from monoidal structure on **C**), then **C** must already have interpretations of the combinators for pairing and unpairing satisfying similar properties to the pairing and unpairing in **Ptl**. At this level, then we parallel very closely the structure used by Birkedal [3], with only the minor details of certain equations holding only up to η.

Suppose now, that $\mathcal{F}(\mathbf{C})_{\mathrm{ex}}$ is locally cartesian closed. Then by corollary 2, **C** has weak partial evaluations. This means that for any pair C, D of objects of **C**, there is an object which we can call $[C \rightharpoonup D]$ together with an evaluation map $\epsilon : [C \rightharpoonup D] \otimes C \longrightarrow D$. This generates an "application" in **Ptl**: $\overline{\epsilon} \circ \theta : \overline{[C \rightharpoonup D]} \times \overline{C} \longrightarrow \overline{D}$. This is more general than the structure used by Birkedal. We use it to construct a partial combinatory type structure in the sense of Longley [13].

The type world T is the set $\mathbf{C}_0$ of objects of **C**, the binary product operation $C \times D$ is tensor product $C \otimes D$, and the arrow type is given by the weak partial evaluations $[C \rightharpoonup D]$. The associated family of sets is $(A_C | C \in \mathbf{C}_0) = (\overline{C} | C \in \mathbf{C}_0)$, and the application functions $\overline{[C \rightharpoonup D]} \times \overline{C} \longrightarrow \overline{D}$ are as above.

Longley's structure also requires s and k combinators, along with combinators for pairing and first and second projections. These are obtained by currying corresponding maps in **C**. For example, the combinator $k \in \overline{[C \rightharpoonup D \rightharpoonup C]}$ is obtained from $\pi_0 : C \otimes D \longrightarrow C$. We first curry to get a map $k_1 : C \longrightarrow [D \rightharpoonup C]$, and then again to get $k_2 : I \longrightarrow [C \rightharpoonup D \rightharpoonup C]$, apply $\overline{(\ )}$ to get a (total) function $\overline{I} \longrightarrow \overline{[C \rightharpoonup D \rightharpoonup C]}$, and finally compose this with $\psi : 1 \longrightarrow \overline{I}$ to get k. The construction of s is similar, this time starting with

$$\begin{aligned}(C \rightharpoonup D \rightharpoonup E) \otimes (C \rightharpoonup D) \otimes C &\xrightarrow{\mathrm{id} \otimes \mathrm{id} \otimes d} (C \rightharpoonup D \rightharpoonup E) \otimes (C \rightharpoonup D) \otimes C \otimes C \\ &\xrightarrow{\ \sim\ } (C \rightharpoonup D \rightharpoonup E) \otimes C \otimes (C \rightharpoonup D) \otimes C\end{aligned}$$

$$\xrightarrow{\epsilon \otimes \epsilon} \quad (D \rightharpoonup E) \otimes D$$
$$\xrightarrow{\epsilon} \quad E$$

Pairing and unpairing combinators are also obtained in this way, and satisfy the requisite equations. This can be seen from the following lemma.

Lemma 9. *Suppose* $f : C \longrightarrow D$ *is curried to give* $F : I \longrightarrow [C \rightharpoonup D]$, *then for all* $c \in \overline{C}$, *if* $\overline{f}(c)$ *is defined, then so is* $(\overline{\epsilon} \circ \theta \circ (\overline{F} \times \mathrm{id}_{\overline{C}})(\psi, c)$, *and they are equal.*

Proof. $(\overline{\epsilon} \circ \theta \circ (\overline{F} \times \mathrm{id}_{\overline{C}})(\psi, c) = (\overline{\epsilon \circ (F \otimes \mathrm{id}_C)} \circ \theta)(\psi, c)$, and the result follows from corollary 2. □

We thus have a partial combinatory type structure. This in turn generates a graph $\mathbf{C}'$ equipped with a graph morphism into **Ptl**: the vertices are the same as the objects of **C**, and the edges are the partial functions induced from the arrow types by application. It is irritating that $\mathbf{C}'$ is not necessarily a category, but it may fail to be closed under composition (we only know that the composite of two partial functions in the graph can be extended to a third). But we only need this lax structure, not a full category, to define $\mathcal{F}(\mathbf{C}')$. Since any partial function obtained from **C** is extended by a partial function obtained from $\mathbf{C}'$, there is an embedding $\mathcal{F}(\mathbf{C}) \longrightarrow \mathcal{F}(\mathbf{C}')$. Unfortunately, this is not necessarily an equivalence. The problem is that there is no reason why a partial function obtained from $\mathbf{C}'$ should extend to one obtained from **C**. One way of viewing the problem is that in a typed pca, as a consequence of the K combinator, every element of a type is named by a constant function. This is not the case for us. The K combinator corresponds to a projection. Suppose, however, that **C** is concrete in the sense that every element of $\overline{C}$ is obtained from a morphism $I \longrightarrow C$. In this case every partial function in $\mathbf{C}'$ is already in **C**, and the realizability structure obtained categorically is identical to that obtained from the partial combinatory type structure.

We now turn our attention to the case when $\mathcal{F}(\mathbf{C})_{\mathrm{ex}}$ has a subobject classifier. In that case we have that **C** has a universal object V, and applying lemma 8, we get that $\mathcal{F}(\mathbf{C})$ is equivalent to $\mathcal{F}(\mathbf{M})$ where **M** is the monoid of endomorphisms on V.

Corollary 4. *If* $\mathcal{F}(\mathbf{C})$ *is a topos, then it is equivalent to the topos constructed using the monoid of endomorphisms of the universal object in* **C**.

Putting these observations together, we can see that if $\mathcal{F}(\mathbf{C})_{\mathrm{ex}}$ is a topos, then, much as in Scott [18], $\overline{V}$ is a partial combinatory algebra, and if **C** is concrete, then the topos obtained is the conventional realizability topos from this algebra.

References

1. S. Abramsky. Process realizability. Unpublished notes available at `http://web.comlab.ox.ac.uk/oucl/work/samson.abramsky/pr209.ps.gz`.

2. L. Birkedal. Developing theories of types and computability via realizability. *Electronic Notes in Theoretical Computer Science*, 34, 2000. Available at `http://www.elsevier.nl/locate/entcs/volume34.html`. The pdf version has active hyperreferences and is therefore the preferred version for reading online.
3. L. Birkedal. A general notion of realizability. In *Proceedings of the 15th Annual IEEE Symposium on Logic in Computer Science*, Santa Barbara, California, June 2000. IEEE Computer Society.
4. L. Birkedal, A. Carboni, G. Rosolini, and D.S. Scott. Type theory via exact categories. In *Proceedings of the 13th Annual IEEE Symposium on Logic in Computer Science*, pages 188–198, Indianapolis, Indiana, June 1998. IEEE Computer Society Press.
5. A. Carboni. Some free constructions in realizability and proof theory. *Journal of Pure and Applied Algebra*, 103:117–148, 1995.
6. A. Carboni, P.J. Freyd, and A. Scedrov. A categorical approach to realizability and polymorphic types. In M. Main, A. Melton, M. Mislove, and D.Schmidt, editors, *Mathematical Foundations of Programming Language Semantics*, volume 298 of *Lectures Notes in Computer Science*, pages 23–42, New Orleans, 1988. Springer-Verlag.
7. A. Carboni and R. Celia Magno. The free exact category on a left exact one. *Journal of Australian Mathematical Society*, 33(A):295–301, 1982.
8. A. Carboni and G. Rosolini. Locally cartesian closed exact completions. *J.Pure Appl. Alg.*, 154:103–116, 2000.
9. S. Eilenberg and G.M. Kelly. Closed categories. In S. Eilenberg, D.K. Harrison, S. Mac Lane, and H. Röhrl, editors, *Categorical Algebra (LaJolla, 1965)*. Springer-Verlag, 1966.
10. J.M.E. Hyland. The effective topos. In A.S. Troelstra and D. Van Dalen, editors, *The L.E.J. Brouwer Centenary Symposium*, pages 165–216. North Holland Publishing Company, 1982.
11. J.M.E. Hyland, E.P. Robinson, and G. Rosolini. The discrete objects in the effective topos. *Proceedings of the London Mathematical Society*, 60:1–36, 1990.
12. Peter Lietz and Thomas Streicher. Impredicativity entails untypedness. Submitted for publication, 2000.
13. J.R. Longley. Unifying typed and untyped realizability. Electronic note, available at `http://www.dcs.ed.ac.uk/home/jrl/unifying.txt`, 1999.
14. M. Menni. A characterization of the left exact categories whose exact completions are toposes. Submitted to *Journ.Pure Appl.Alg.*, 1999.
15. E.P. Robinson and G. Rosolini. Categories of partial maps. *Inform. and Comput.*, 79:95–130, 1988.
16. E.P. Robinson and G. Rosolini. Colimit completions and the effective topos. *Journal of Symbolic Logic*, 55:678–699, 1990.
17. G. Rosolini. *Continuity and Effectiveness in Topoi.* PhD thesis, University of Oxford, 1986.
18. D.S. Scott. Relating theories of the λ-calculus. In R. Hindley and J. Seldin, editors, *To H.B. Curry: Essays in Combinatory Logic, Lambda Calculus and Formalisms*, pages 403–450. Academic Press, 1980.
19. J. van Oosten. History and developments. In L. Birkedal, J. van Oosten, G. Rosolini, and D.S. Scott, editors, *Tutorial Workshop on Realizability Semantics, FLoC'99, Trento, Italy, 1999*, volume 23 of *Electronic Notes in Theoretical Computer Science*. Elsevier, 1999.

The Anatomy of Innocence

Vincent Danos[1] and Russell Harmer[2]

[1] Université Paris 7
Équipe Preuves, Programmes, Systèmes
2 place Jussieu 75251 Paris Cedex 05
France
[2] University of Sussex
School of Cognitive and Computing Sciences
Brighton BN1 9QH
UK

Abstract. We reveal a symmetric structure in the HO/N games model of innocent strategies, introducing rigid strategies, a concept dual to bracketed strategies. We prove a direct definability theorem of general innocent strategies with respect to a simply typed language of extended Böhm trees, which gives an operational meaning to rigidity in call-by-name. A corresponding factorization of innocent strategies into rigid ones with some form of conditional as an oracle is constructed.

1 Introduction

Game models have swept over the semantic grounds because of their accuracy in describing computations. They're so precise indeed that reading them as a sort of infinite and glorified syntax seems reasonable to a certain extent. A lot is known about the variety of sequential behaviors they can express, which is a good thing. But our mathematical understanding of them is comparatively embarassingly poor. And this is especially true of the HO/N kind of models [HO94,Nic94], which at the same time are certainly the most successful [HM99,DH00,McC96,AM97,AHM98,Lai97,MH99], *i.e.* the ones we'd most like to understand.

The whole family of related Cartesian closed categories (Ccc) known as HO-models, are organized around a common stable kernel Ccc of *innocent* strategies. We bring in this paper what we think is a significant clarification of this basic structure, by giving a clean decomposition of this kernel based upon the distinction between Questions and Answers. These were introduced in games to sort out which innocent strategies are necessarily (defined by programs) using global control. The ones that don' t are called *bracketed.* When suitably abstracted this property has a dual which we call being *rigid* (co-bracketed seemed somewhat too heavy).

This splits the innocent Ccc in two sub-Cccs with a dualizing endofunctor running in between. Subject to a technical condition about Answers, we prove that

L. Fribourg (Ed.): CSL 2001, LNCS 2142, pp. 188–202, 2001.

case a suitable form of conditional is a universal oracle with respect to rigidity — a result mirroring Laird's proof that a form of 'taster' known as catch is an oracle for bracketedness [Lai97]. This brings to the fore an interesting duality between these oracles, namely that they form a retract. Moving to syntax we show that Herbelin's language of classical Böhm trees from [Her97] defines any compact innocent strategy (in arenas interpreting simple types over Integers). This second result gives an operational interpretation of rigidity in a call-by-name world.

Future Work. Concerning the factorization result we know an abstract treatment that relies on catch and case forming a retract; there are other such (non-innocent) retracts in the games universe and these are very likely to also have something to say about the wider model of single-threaded strategies. Concerning definability we give an original internal category-theoretic treatment which should be somewhat strengthened using more abstract tools and extended to cope with full polarized linear logic. That could also bring some deeper understanding on the mathematical side of affairs.

On the logical side any kind of proof-theoretical analysis of rigid strategies would be welcome, *e.g.* a tautology they would depend on in the same way that global control is related to Peirce's law. Finally, on the programming side, it's a big question mark. Does rigid programming have any interest to simplify flow analysis, or to program in time- or security-constrained environments, how does it merge with imperative programming, or with call-by-value, is it even a complete computability model. We don't know, but we'd be surprised if a structurally cogent concept had no computational interest.

2 Game Semantics

We proceed first to the standard presentation of HO-style games.

2.1 Arenas & Plays

The basic "playing area" of a game is described as an ***arena***. Formally, this is a triple $\langle M_A, \lambda_A, \vdash_A \rangle$ where

- M_A is a countable set of ***moves***.
- $\lambda_A : M_A \to \{\mathsf{O}, \mathsf{P}\} \times \{\mathsf{Q}, \mathsf{A}\}$ is an application determining for each $m \in M_A$ whether it's an O***pponent*** or a P***layer*** move and a Q***uestion*** or an A***nswer***. We denote by λ_A^{OP} and λ_A^{QA} composition with 1st and 2nd projection respectively.
- $\vdash_A$ is a binary ***enabling*** relation on M_A formalizing the *temporal* nature of arenas whereby the possibility of one move being played may be contingent on some other "enabling" move having already been played.

An ***initial*** move of A is a move that has no enabler; we write I_A for the set of such moves. To begin with, we ask two conditions on arenas:

(a1) if $m \vdash_A n$ then $\lambda_A^{\mathsf{OP}}(m) \neq \lambda_A^{\mathsf{OP}}(n)$;
(a2) if $m \in I_A$ then $\lambda_A^{\mathsf{OP}}(m) = \mathsf{O}$;

i.e. the protagonists enable each other's moves, not their own, and initial moves are Opponent moves. We record two further conditions on the enabling relation that will be 'switched on' later in the paper:

(a3) if $m \in I_A$ then $\lambda_A^{\mathsf{QA}}(m) = \mathsf{Q}$;
(a4) if $m \vdash_A n$ then $\lambda_A^{\mathsf{QA}}(m) = \mathsf{Q}$;

i.e. initial moves can only be Questions, and Answers can't be pointed at. Moves enabling no moves will be called ***terminal***; thus **(a4)** can be rephrased as 'Answers are terminal'.

We'll frequently use shorthand notation such as PQ for a move m such that $\lambda_A(m) = \mathsf{PQ}$.

Examples of Arenas. The unique arena with no moves at all is denoted by **1**. Another useful arena is $\bot$ where $M_\bot = \{\mathsf{q}\}$ and $\lambda_\bot$ sends this to OQ. Any countable set X generates the ***discrete*** arena $\mathbf{X}$ with $M_{\mathbf{X}} = X$ and $\lambda_{\mathbf{X}}(m) = \mathsf{OA}$ for all $m \in X$. In particular, we denote by **C**, **B** and **N** respectively the discrete arenas generated by $\{\mathsf{a}\}$, $\{\mathtt{tt}, \mathtt{ff}\}$ and $\{0, 1, 2, \ldots\}$. Of course, none satisfy **(a3)**.

Plays & Legal Plays. A **legal play** in arena A consists of an OP-alternating string s equipped with a "pointer" from each non-initial move in s to some earlier *enabling* move in s. The set of all legal plays of A is written $\mathcal{L}_A$. The set of **plays** of A, written $\mathcal{P}_A$, is defined to be the set of all *suffixes* of legal plays of A beginning with an O-move. In particular, $\mathcal{L}_A \subseteq \mathcal{P}_A$, because of **(a2)**. Note that a play of A may have "missing" pointers and may also begin with a non-initial move.

Various Views. The **view** of a (non-empty) play $s \in \mathcal{P}_A$ is defined inductively as follows.

(v1) $\mathsf{V}(sm) = m$, if m is an O-move with no pointer;
(v2) $\mathsf{V}(sntm) = \mathsf{V}(s) \cdot nm$, if m is an O-move pointing to n;
(v3) $\mathsf{V}(sm) = \mathsf{V}(s) \cdot m$, if m is a P-move;

i.e. we follow back pointers from O-moves, skipping all intervening moves, and "step over" P-moves—until we reach a pointerless O-move.

Note that the view of a legal play might not itself be legal because in the last clause, when m is a P-move, one might lose m's pointer. If s is legal in A, then $\mathsf{V}(s)$ first move must be an initial O-move; if A has only terminal Answers and a PA-move occurs in $\mathsf{V}(s)$, then it is its last move.

We define now an even more stringent notion of partial information. The ***R-view*** or the rigid view of a (non-empty) play $s \in \mathcal{P}_A$ is defined inductively as follows.

(r1) $\mathsf{R}(sm) = m$, if m is an O-move with no pointer or an OA-move;
(r2) $\mathsf{R}(sntm) = \mathsf{R}(s) \cdot nm$, if m is an OQ-move pointing to n;
(r3) $\mathsf{R}(sm) = \mathsf{R}(s) \cdot m$, if m is a P-move;

i.e. again we follow back pointers from Opponent Questions until we either we "run out of pointers" or we reach an Opponent Answer. Obviously the rigid view of a play will be a suffix of its (normal) view.

A dual of the rigid view, called the ***B-view*** or the bracket view, is obtained by switching the roles of Questions and Answers, *i.e.* we stop on pointerless O-moves or when we reach an Opponent Question (sometimes called the 'pending question').

Product & Arrow. If A and B are arenas, we define their product $A \times B$ by:

- $M_{A\times B} = M_A + M_B$, the disjoint union;
- $\lambda_{A\times B} = [\lambda_A, \lambda_B]$, the copairing;
- $m \vdash_{A\times B} n$ iff $m \vdash_A n$ or $m \vdash_B n$.

This places A and B "side by side" with no chance of any interaction between them. The "empty arena" **1**, defined above, is the unit for this constructor. This construction easily generalizes to countable products and we write A^ω for the product of countably many copies of A.

Our other constructor is the arrow, defined by:

- $M_{A\Rightarrow B} = M_A + M_B$;
- $\lambda_{A\Rightarrow B} = [\langle \overline{\lambda}_A^{\mathsf{OP}}, \lambda_A^{\mathsf{QA}} \rangle, \lambda_B]$, where $\overline{\lambda}_A^{\mathsf{OP}}(m) = \mathsf{O}$ iff $\lambda_A^{\mathsf{OP}}(m) = \mathsf{P}$;
- $m \vdash_{A\Rightarrow B} n$ iff $m \vdash_A n$ or $m \vdash_B n$ or $m \in I_B \wedge n \in I_A$;

i.e. the roles of Opponent and Player are reversed in A and the (formerly) initial moves of A are now enabled by the (still) initial moves of B.

Conditions (**a1**) to (**a3**) are preserved by product and arrow, while (**a4**) is preserved by the arrow iff (**a3**) holds of B, of course.

More Arenas. If $\mathbf{X}$ is the hyperflat arena generated by X, we define the ***flat*** arena over X to be $\mathbf{X} \Rightarrow \bot$. The effect of this is to make the Answers of X non-initial P-moves, all enabled by the unique initial Question.

We will sometimes write $\neg A$ as a shorthand for $A \Rightarrow \bot$, and $\mathbb{C}$, $\mathbb{B}$ and $\mathbb{N}$ for $\neg\mathbf{C}$, $\neg\mathbf{B}$ and $\neg\mathbf{N}$. An arena of the form $\neg A$ is known as a ***pointed*** arena. By construction, any flat arena is pointed. Moreover, if B is pointed, *i.e.* of the form $B' \Rightarrow \bot$ for some B', then $A \Rightarrow B$ is pointed too, regardless of A, as can be easily seen since it's isomorphic to $\neg(A \times B')$.

2.2 Innocent Strategies

A ***strategy*** σ for arena A is a non-empty set of even-length legal plays of A satisfying

(s1) if $sab \in \sigma$ then $s \in \sigma$;
(s2) if $sab \in \sigma$ and $sac \in \sigma$ then $sab = sac$;

i.e. σ is deterministic and closed under even-length prefixes. For the purposes of this paper, we'll spend most, but not all, of the time in the more restricted class of ***innocent*** strategies, defined to be those strategies σ further satisfying

(s3) if $sab \in \sigma$ then b points to a move in $\mathsf{V}(sa)$;
(s4) if $sab \in \sigma$, $t \in \sigma$, $ta \in \mathcal{L}_A$ and $\mathsf{V}(sa) = \mathsf{V}(ta) = v$ then $tab \in \sigma$ where both bs point to the same move in v;

Condition (**s4**) implies (**s3**), by just taking $s = t$; (**s3**) is known as P-***visibility*** and is exactly saying that σ's views are legal plays, while (**s4**) says that these views are enough to completely describe σ's behaviour.

No two O-moves in a view are pointing to the same P-move and the view of a play is a *linearization* of O's behavior so far. Constraining strategies to play innocently, enforces a particular way they access and handle information which is strongly related to *linear head reduction* as shown in [DHR96,Her97,Lev98].

The innocent strategies on arena A form an ω-algebraic CPO when ordered by subset inclusion; the ***compact*** strategies are precisely those with a finite number of distinct views.

Interactions. Let u be a finite string of moves from arenas A, B and C equipped with pointers. We define $u \upharpoonright B, C$ to be the subsequence of u where we delete all moves from and pointers to A; $u \upharpoonright A, B$ is defined similarly. Next, we define $u \upharpoonright A, C$ by removing all moves from and pointers to B and additionally, in the case where $a \in M_A$ points to $b \in M_B$ which, in turn, points to $c \in M_C$, we 'compose' these pointers, *i.e.* make a point directly to c in $u \upharpoonright A, C$.

In the sequel, we will refer to these as respectively the two inner projections and the outer projection.

A ***legal interaction*** of A, B and C is such a u satisfying $u \upharpoonright A, B \in \mathcal{L}_{A \Rightarrow B}$, $u \upharpoonright B, C \in \mathcal{L}_{B \Rightarrow C}$ and $u \upharpoonright A, C \in \mathcal{L}_{A \Rightarrow C}$ (*i.e.* all three projections are legal). The set of all legal interactions (of A, B and C) is written $\mathsf{int}(A, B, C)$. (An example is given in Fig. 1.)

Composition of Strategies. With this definition in place, we define the composite of strategies $\sigma : A \Rightarrow B$ and $\tau : B \Rightarrow C$ by:

$$\sigma ; \tau = \{u \upharpoonright A, C \mid u \in \mathsf{int}(A, B, C) \wedge u \upharpoonright A, B \in \sigma \wedge u \upharpoonright B, C \in \tau\}.$$

This is easily seen to define a set of even-length legal plays. A bit more work is needed to show it does define a valid strategy for $A \Rightarrow C$. In fact, one can define a symmetric closed category with arenas as objects, strategies on $A \Rightarrow B$ as arrows from A to B, and the tensor and exponential structure defined from the arena constructors $\times$ and $\Rightarrow$ introduced above. Specializing to innocent strategies makes the product a cartesian product indeed and yields a Cartesian closed category.

2.3 Rigid and Bracketed Strategies

We now make use of our truncated views to define two more conditions on strategies. The second is already well-known from the original work of Hyland and Ong. The purpose of the paper is to introduce and analyze the first.

(s5) if $sab \in \sigma$ and $\lambda_A^{\mathsf{QA}}(b) = \mathsf{Q}$, b points in $\mathsf{R}(sa)$;
(s6) if $sab \in \sigma$ and $\lambda_A^{\mathsf{QA}}(b) = \mathsf{A}$, b points in $\mathsf{B}(sa)$.

A strategy σ for arena A is said to satisfy ***R-visibility***, or to be ***rigid*** when satisfying (**s5**) and to satisfy ***B-visibility***, or to be ***bracketed*** when satisfying (**s6**). Note that condition (**s5**) only bites on Questions and dually (**s6**) only on Answers. Their conjunction implies (**s3**).

Any rigid strategy on a 'function' type can inspect at most one of its arguments, and only once, and then return immediately. Rigid strategies tend not to question too long once they get an answer, while bracketed ones tend not to answer too fast.

Rigidity is preserved by composition only subject to a certain condition.

Proposition 1 Let B be an arena *without initial answers* and $\sigma : A \Rightarrow B$, $\tau : B \Rightarrow C$ be rigid strategies, then $\sigma;\tau$ is rigid as well.

The proof follows the general method developed for bracketedness and P-visibility by McCusker [McC98]. In Fig. 1 a shortest counterexample is given when initial Answers are allowed in the middle arena B. Both inner projections $\mathsf{a}_2\mathsf{q}_3$ and $\mathsf{q}_1\mathsf{q}_2\mathsf{a}_1$ are legal rigid plays and the outer one $\mathsf{q}_1\mathsf{q}_2\mathsf{a}_1\mathsf{q}_3$ is legal too but *not* rigid because $\mathsf{R}(\mathsf{q}_1\mathsf{q}_2\mathsf{a}_1\mathsf{q}_3) = \mathsf{a}_1\mathsf{q}_3$ and q_3 points past a_1 to q_1.

Proposition 2 Let A be an arena *without initial answers* and $\sigma : A \Rightarrow B$, $\tau : B \Rightarrow C$ be bracketed strategies, then $\sigma;\tau$ is bracketed as well.

A slight variation also gives a counterexample for bracketedness when the left-most arena A has an initial Answer.

If we now restrict to arenas where initial moves can only be questions, *i.e.* if $m \in I_A$ then $\lambda_A^{\mathsf{QA}}(m) = \mathsf{Q}$ which was condition (**a3**), then both our constraints pass the 'composition test' and generate sub-Cccs.

Before looking for a syntactic materialization of rigidity, we look for strategies seriously violating the constraint (to be used in the factorization section).

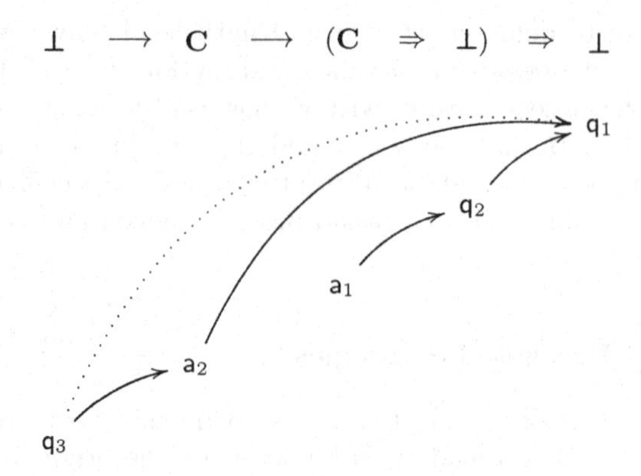

Fig. 1. Losing rigidity with a middle initial Answer.

2.4 Oracles

Since the strategies we want to describe now are all innocent, it is enough to describe their views, and by prefix closure it is even enough to define their maximal views. Thus case is uniquely defined on arena $\mathbb{N} \Rightarrow \mathbb{N}^\omega \Rightarrow \mathbb{N}$ as including views represented in Fig. 2, where i, j range over the integers and q_i denotes the initial question in the ith factor of $\mathbb{N}^\omega$; and for all $n > 0$, case_n on $\mathbb{N} \Rightarrow \mathbb{N}^n \Rightarrow \mathbb{N}$ is defined as case except it responds to i with q_i only when $i < n$. All these strategies are innocent, bracketed but none are rigid (because of Questions q_i).

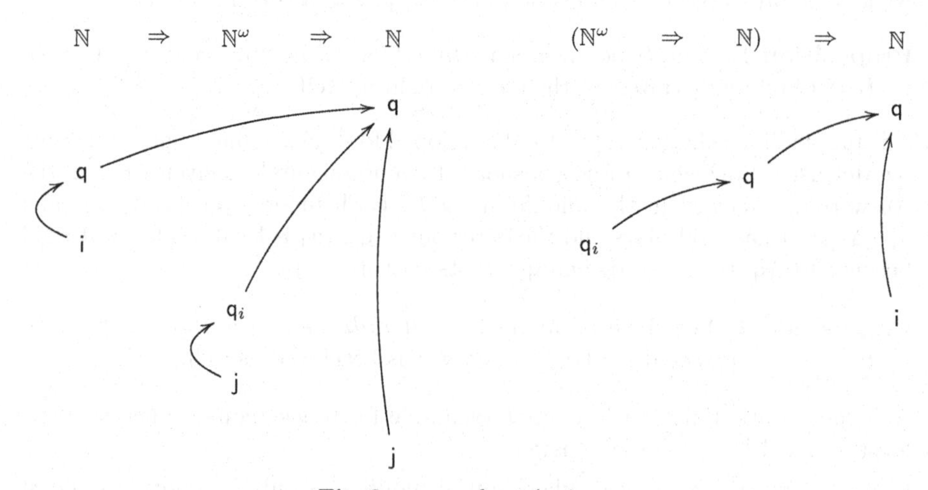

Fig. 2. case and catch views.

On the dual side of the matter, we have catch defined on the 'converse' arena $(\mathbb{N}^\omega \Rightarrow \mathbb{N}) \Rightarrow \mathbb{N}$, and described in Fig. 2, and similarly the miniaturized versions

catch_n on arenas $(\mathbb{N}^n \Rightarrow \mathbb{N}) \Rightarrow \mathbb{N}$, which are all innocent, rigid but not bracketed (because of the premature Answers i).

Proposition 3 (retract) The pair $\mathsf{catch} : (\mathbb{N}^\omega \Rightarrow \mathbb{N}) \Rightarrow \mathbb{N}$ and $\mathsf{case} : \mathbb{N} \Rightarrow \mathbb{N}^\omega \Rightarrow \mathbb{N}$ defines a retract, *i.e.* $\mathsf{case} \,;\, \mathsf{catch} = \mathsf{id}_\mathbb{N}$.

The proof is an easy exercise in composition.

The functor '?' of which the object part is $?A = A^{\mathsf{Q}/\mathsf{A}} \Rightarrow \bot$ (*i.e.* it flips Questions and Answers and 'negates' the arena, so that $?\bot = \neg\mathbf{C}$, $?(\bot \times \bot) = \neg\mathbf{B}$ and $?(\bot^\omega) = \neg\mathbf{N}$) maps the rigid sub-Ccc to the bracketed one and vice-versa. In particular condition (**a3**) is always preserved (but (**a4**) is not).

Perhaps we have here the beginnings of a control and co-control category pair as axiomatised in [Sel99]. A deeper analysis of this functor should clarify—in a suitably abstract sense—the duality between Questions and Answers, maybe leading to a model of polarized linear logic [Lau99].

3 Definability

3.1 Classical Böhm Trees

It's time by now that we find a syntax expressive enough to define our compact strategies and home in on a precise picture of what it means in *this* language to be rigid. The tersest solution at hand seems to be CBT, Herbelin's classical Böhm trees (documented in [Her97]).

We assume an infinite supply of λ- and μ-variables respectively ranged over by x_js and α_ks. Terms come in two flavours, the ***executables*** written E, and the ***functionals*** written F, and we also single out C, a sub-species of E, for reasons to be clear at the end of the section (suspense !).

$$C ::= \Omega \mid [\alpha]\mathtt{n}$$
$$E ::= C \mid \mathtt{case}\ (x)\boldsymbol{F}\ \mathtt{of}\ 0 \mapsto E, \cdots, \mathtt{n} \mapsto E$$
$$F ::= \lambda \boldsymbol{x}.\mu\alpha.\ E$$

Types are usual simple types built over some base types using implication only. Here we'll be content with `Nat` as the only base type. Typing judgments also come in two flavors: $\Gamma\,;\,\Delta \vdash E$ (for executables) and $\Gamma\,;\,\Delta \vdash F : A$ (for functionals), where A is any type, Γ is any finite set of types annotated by λ-variables and Δ is any finite set of *base* types annotated by μ-variables. These judgments are derived using the rules in Fig. 3. Maybe the simplest instructive example is $\mathtt{catch}_2 = \mathtt{case}\ (f\,\mu\delta.[\alpha]0\,\mu\delta.[\alpha]1)\ \mathtt{of}\ 0 \mapsto [\alpha]0, 1 \mapsto [\alpha]1$ which can be typed in the context $f : \mathtt{Nat} \Rightarrow \mathtt{Nat} \Rightarrow \mathtt{Nat}; \alpha : \mathtt{Nat}$.

The evaluation of a functional term applied to a series of functional arguments is governed by the following principle: a $\mu\alpha$ catches any integer thrown by the

$$\frac{}{\Gamma\,;\Delta,\alpha:\texttt{Nat}\vdash\Omega}\qquad\frac{}{\Gamma\,;\Delta,\alpha:\texttt{Nat}\vdash[\alpha]\texttt{n}}$$

$$\frac{\Gamma,\boldsymbol{x}:\boldsymbol{A}\,;\Delta,\alpha:\texttt{Nat}\vdash E}{\Gamma\,;\Delta\vdash\lambda\boldsymbol{x}.\mu\alpha.E:\boldsymbol{A}\Rightarrow\texttt{Nat}}$$

$$\frac{(\Gamma\,;\Delta\vdash F_i:A_i)_{i=1\cdots p}\qquad(\Gamma\,;\Delta\vdash E_j)_{j=0\cdots n}}{\Gamma,x:\boldsymbol{A}\Rightarrow\texttt{Nat}\,;\Delta\vdash\texttt{case}\,(x)\boldsymbol{F}\,\texttt{of}\,0\mapsto E_0,\cdots,\texttt{n}\mapsto E_n}$$

Fig. 3. CBT typing rules.

$[\alpha]$s it statically binds, this integer is then matched against the continuation $0 \mapsto E_0, \cdots, \texttt{n} \mapsto E_n$ offered by the case that $\mu\alpha$ was substituted in (if any), and the matching executable is run. For instance evaluating $\texttt{catch}_2$ in a context where $f = \lambda x\lambda y.\mu\beta.\texttt{case}\,x\ \texttt{of}\,0 \mapsto \Omega$ will end directly throwing 0 to the free continuation α, whereas if $f = \lambda x\lambda y.\mu\beta.[\beta]0$, the result will be the same but going through the continuation part.

3.2 Some Useful Strategies

We need two auxiliary strategies to perform our decomposition. For all natural numbers n, we have a strategy $\mathsf{a2q}_n$ in $\mathbf{N} \Rightarrow \bot^n$ which responds to the initial Question in the ith copy, $i < n$, of $\bot^n$ with the Answer i in $\mathbf{N}$. In other words, it performs a trivial (and partial) mapping from data to space.

For any arena A, we also have a 'linearizing' strategy lin_A in $(\neg A_1 \Rightarrow \bot) \Rightarrow \neg A_2 \Rightarrow A_3$ (indices are there to distinguish occurrences of the *same* arena A) which separates out the first "thread of activity" (or "copy") of A_1 into A_3, removing the two opening Questions (in $\bot$ and $\neg A_1$) in the process. The remaining threads of A_1 are transposed into A_2. Thus, a typical play looks as in Fig. 4, with the second thread pictured as dotted pointers. Note that lin_A is not innocent since at each opening Question by Opponent in $\neg A_1$, it has the same view $\mathsf{m}_1\mathsf{qq}$, and by definition responds differently the first time, thus violating (**s4**); in general, it is not bracketed either, because the second m_2 can be an Answer, and it points past the pending q played by Opponent in its view $\mathsf{m}_1\mathsf{qqm}_1\mathsf{m}_2$. It does satisfy all the rest yet, namely P-visibility and rigidity.

3.3 Decomposition

In a moment we're going to give a decomposition argument for definability of compact innocent strategies. That is, we explain how any such strategy can be obtained as the interpretation of some term. As said, the language we're going to use is CBT. The argument itself follows the general form of the Hyland-Ong decomposition, adapted to the unbracketed case, and also *internalizing* the vital 'separation of head occurrence' step.

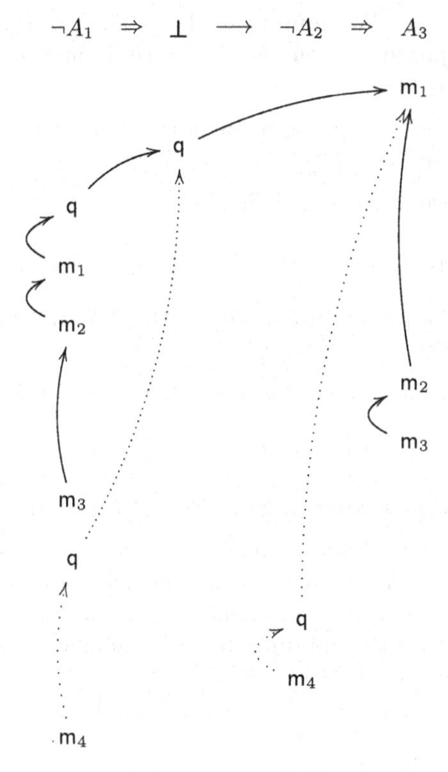

Fig. 4. A play of lin_A.

Our only base type Nat is interpreted by the flat arena $\mathbb{N}$, and thus every simple type over the Integers has a natural arena associated; conversely we say an arena B is a ***simple arena*** if of one of the two following forms

$$A_1 \times \cdots \times A_n \times b_1 \times \cdots \times b_m \Rightarrow \bot,$$
$$A_1 \times \cdots \times A_n \times b_1 \times \cdots \times b_m \Rightarrow A,$$

where $A_1, \ldots, A_n, A$ are arenas interpreting types, and $b_1, \ldots, b_m$ are discrete arenas coming from base types (integers here).

Proposition 4 (definability) Any compact innocent strategy σ on some simple arena is CBT definable.

If σ's arena is not of the first form, we uncurry (and munge around) σ, arriving at a strategy for

$$A_1 \times \cdots \times A_p \times b_1 \times \cdots \times b_m \;\Rightarrow\; B,$$

with B a flat arena. Since B is of the form $b_{m+1} \Rightarrow \bot$ for some discrete arena b_{m+1}, we can uncurry one more time, yielding σ' for

$$A_1 \times \cdots \times A_p \times b_1 \times \cdots \times b_{m+1} \;\Rightarrow\; \bot,$$

an arena of the first form. We abbreviate the product of the A_is by Γ and the 'μ-context', *i.e.* the product of the b_js by Δ (it is just a product of discrete natural number arenas $\mathbf{N}$).

And now, let's look at σ''s response to the initial Question in $\perp$ (we know there is at most one response from (**s2**)) .

— If it has no response, then σ' is defined by

$$x_1 : T_1, \ldots, x : T_p \,; \alpha_1 : U_1, \ldots, \alpha_{m+1} : U_{m+1} \vdash \Omega$$

where the T_is are the types corresponding to the A_is and the U_js are the types corresponding to the $\neg b_j$s.

— If it responds with some Answer a in b_j, then σ' is defined by

$$x_1 : T_1, \ldots, x : T_p; \alpha_1 : U_1, \ldots, \alpha_{m+1} : U_{m+1} \vdash [\alpha_j]\mathsf{a}$$

where a is the language constant corresponding to the move a.

— If it responds with a Question q_i in $A_i = A_{i,1} \Rightarrow \cdots A_{i,k_i} \Rightarrow B_i$, then writing $A_i^- = A_{i,1} \times \cdots \times A_{i,k_i} \times b_i$, we know A_i is isomorphic to $A_i^- \Rightarrow \perp$, and we may decompose the strategy into its "arguments" to A_i and its "continuations" from A_i by a composition with the appropriate $\mathsf{lin}_{A_i^-}$, namely:

$$\lambda_{A_i}(\sigma') \,; \mathsf{lin}_{A_i^-} : (\Gamma / A_i) \times \Delta \to A_i \Rightarrow A_i^-$$

and thus, uncurrying A_i back:

$$\lambda_{A_i}^{-1}(\lambda_{A_i}(\sigma') \,; \mathsf{lin}_{A_i^-}) : \Gamma \times \Delta \to A_i^-$$

Note that the resulting strategy is still rigid if σ' was in the first place (by proposition 1, since the middle arena here $\neg A_i$, being pointed, has no initial Answers). It can also be shown that it is innocent.

- each "argument" strategy $\sigma'_{i,j} : \Gamma \times \Delta \Rightarrow A_{i,j}$ has a strictly smaller set of views than σ' and can therefore be defined by some functionals F_j by inductive hypothesis;
- since σ' is compact, the "continuation" strategy $\sigma_c : \Gamma \times \Delta \Rightarrow b_i$ with $b_i = \mathbf{N}$ has no response to initial moves $m > n$ for some n. Hence we extract $n+1$ continuations by forming the composite $\sigma_c \,; \mathsf{a2q}_{n+1} : \Gamma \times \Delta \Rightarrow \perp^{n+1}$ and, again by the inductive hypothesis, these $n+1$ continuations are definable by some executables E_i for $0 \leq i \leq n$.

The strategy σ' can now be seen as the semantics of the executable E:

$$t\,\mathtt{case}\,(x_i)F_1 \cdots F_{k_i}\ \mathtt{of}\ 0 \mapsto E_0, \cdots, \mathtt{n} \mapsto E_n.$$

and we finally get a functional defining σ by λ-abstracting the appropriate x_is and μ-abstracting α_{m+1}, *i.e.* $\lambda x_{n+1} \cdots \lambda x_p.\mu\alpha_{m+1}.\, E$.

3.4 The Rigid Case

The constraint of being rigid has dramatic effect on the shape of the functional extracted from the above argument. If we examine the views of σ' that contribute to the continuation strategy σ_c, we find that they all begin as in:

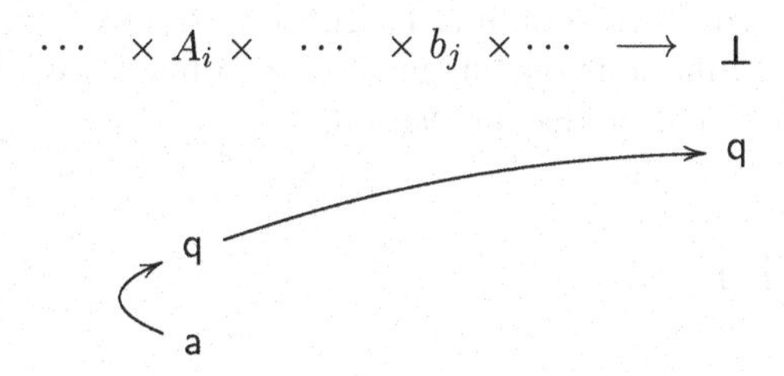

But at this point, the rigid view is just the last move, namely the Answer a. Since, in a simple type, all Answers are terminal, this means that we cannot play a Question: if we did, it would have to point at the Answer a which is impossible. So, if we play anything at all, we're forced to play an Answer—which is only possible in one of the b_js.

In more operational terms, the continuations, *i.e.* the E_is, have to be of the form $[\alpha]$a or Ω. In other words, we can't "nest" **case** statements. We can easily give a precise grammar of "rigid" Böhm trees, say RBT, matching exactly this syntactic constraint:

$$
\begin{aligned}
C &::= \Omega \mid [\alpha]\mathbf{n} \\
E &::= C \mid \texttt{case}\ (x)\boldsymbol{F}\ \mathbf{of}\ 0 \mapsto C, \cdots, \mathbf{n} \mapsto C \\
F &::= \lambda \boldsymbol{x}.\mu\alpha.\ E
\end{aligned}
$$

Note that the use of **case** in this language is restricted to the definition of unary functions: no local control flow is possible. We could equally define a version of the language with no **case** whatsoever—but we would still need the facility to define unary functions (and **case** seems as good a syntax for that as any).

Proposition 5 (rigid definability) Any compact innocent rigid strategy σ on some simple arena is RBT definable.

Indeed **catch$_2$** is in CBT and gives one possible definition of catch$_2$. On the other hand applying the definability machinery to a strategy interpreting a left-and for instance (upcasting Booleans to Integers) yields:

$$
\begin{aligned}
&\lambda x \lambda y.\mu\alpha. \\
&\texttt{case}\ (x)\ \texttt{of}\ \{\ 0 \mapsto [\alpha]0, \\
&\qquad\qquad 1 \mapsto \texttt{case}\ (y)\ \texttt{of}\ \{\ 0 \mapsto [\alpha]0, \\
&\qquad\qquad\qquad\qquad\qquad\qquad 1 \mapsto [\alpha]1\}\}
\end{aligned}
$$

which as expected is not in RBT (because upon receiving 1 for x the term launches a new computation instead of returning immediately).

Bracketedness also has some considerable effect: compact innocent bracketed strategies always throw to the μ just above. Therefore one can dispose in the language of $\mu\alpha$s and $[\alpha]$s altogether, and this brings back the Hyland-Ong language of 'finite canonical forms' [HO94].

It is possible to *combine* both constraints: being rigid and being bracketed. It is easy to see that the corresponding language is simply typed λ-calculus with unary finite partial functions on Integers. This forms the core of the 'canonical forms' echoing our semantic decomposition.

4 Factorization

We can rework the decomposition, but from within the model, in a more intrinsic way. We're only giving an informal argument that one can 'extrude' case out of an innocent strategy and obtain something rigid, much in the same way as Laird proved in [Lai97] that one can extrude catch out of an innocent strategy and obtain something bracketed. To do this, and for reasons explained in the proof sketch below, we have to restrict to arenas where Answers are *terminal*, *i.e.* condition (**a4**).

Typical arenas where this is not the case are sums or lifted arenas as used in modeling sums and call-by-value. So this is a strong assumption.

Let σ be a strategy on arena A violating rigidity. This means in some views of σ, there are Player Questions, PQ, pointing past Opponent Answers, OA, as in:

$$v = \cdots \ \mathsf{OQ} \ \cdots \ \mathsf{PQ} \ \mathsf{OA} \ \cdots \ \mathsf{PQ} \ \cdots$$

We'd like to 'mend' this by hiding OA from the view of any further P-move. If OAs are not terminal, this is hopeless, since removing them from the view would cause visibility violations in case P would target them. So we suppose they are. Having said this, we need a suitable simple side arena, say S so as to play in the extended arena $S \Rightarrow A$, and a helper strategy on S. Now we have room for the following manoeuvre: each time σ is about to move in A, we first play an opening move P_1 in S and wait for Opponent to respond with O_1; only then we play as σ would; now, if Opponent responds with a question we simply rearm the trap and prepare the next σ-move in A; if Opponent responds with an answer OA, which is the case that matters, we play $\mathsf{P}_2^{\mathsf{OA}}$ pointing back to O_1 in S, encoding the answer, and hope for Opponent to play $\mathsf{O}_2^{\mathsf{OA}}$ and hide OA in response. As in:

$$v' = \cdots \ \mathsf{P}_1 \ \mathsf{O}_1 \ \mathsf{PQ} \ \mathsf{OA} \ \mathsf{P}_2^{\mathsf{OA}} \ \mathsf{O}_2^{\mathsf{OA}}$$

Thus the obstacle OA is excised from the view and yet enough information is retained, through the encoding $\mathsf{O}_2^{\mathsf{OA}}$, so that the modified strategy is still innocent.

Once this pattern of four side moves P_1, O_1, P_2 and O_2 is fixed, one has *no choice* left regarding their Q/A status. First $\mathsf{P}_2^{\mathsf{OA}}$ has to be an Answer, otherwise it reintroduces a rigidity violation; just for the same reason O^{OA} must be a question; and the moves they point to must themselves be Questions, since Answers are assumed terminal. Finally there must be Questions enough in S to encode OA moves. Actually case turns out to be a convenient embodiment of this pattern and we may choose $S = \mathbb{N} \Rightarrow \mathbb{N}^\omega \Rightarrow \mathbb{N}$.

Proposition 6 (rigid factorization) Let σ be an innocent strategy on an arena A with only terminal answers, then there exists a rigid innocent strategy $\mathrm{RIG}(\sigma) : (\mathbb{N} \Rightarrow \mathbb{N}^\omega \Rightarrow \mathbb{N}) \Rightarrow A$ such that $\sigma = \mathsf{case}\,;\mathrm{RIG}(\sigma)$.

One interesting question here would be to understand if 'rigidification' has any functorial property. Another issue is whether this can be combined with 'bracketifiction' and yes it can be done when rigidification is applied first.

References

[AHM98] Samson Abramsky, Kohei Honda, and Guy McCusker. A fully abstract game semantics for general references. In *Proceedings, Thirteenth Annual IEEE Symposium on Logic in Computer Science*, pages 334–344, 1998.

[AM97] Samson Abramsky and Guy McCusker. Linearity, sharing and state: a fully abstract game semantics for Idealized Algol with active expressions. In P. W. O'Hearn and R. D. Tennent, editors, *Algol-like languages*. Birkhaüser, 1997.

[DH00] Vincent Danos and Russell Harmer. Probabilistic games semantics. In *Proceedings of the 15th Symposium on Logic in Computer Science*, Santa Barbara, 2000. IEEE.

[DHR96] Vincent Danos, Hugo Herbelin, and Laurent Regnier. Games semantics and abstract machines. In *Proceedings of the 11th Symposium on Logic in Computer Science*, pages 394–405, New Brunswick, 1996. IEEE.

[Her97] Hugo Herbelin. Games and weak head reduction for classical PCF. In ?, editor, *Proceedings of TLCA'97*, number 1210 in Lecture Notes in Computer Science, pages ?–?, ?, 1997. Springer Verlag.

[HM99] R. S. Harmer and G. A. McCusker. A fully abstract game semantics for finite nondeterminism. In *Proceedings, Fourteenth Annual IEEE Symposium on Logic in Computer Science*. IEEE Computer Society Press, 1999.

[HO94] J. M. E. Hyland and C.-H. L. Ong. On full abstraction for PCF: I, II and III. To appear in Information and Computation, 200?, 1994.

[Lai97] J. Laird. Full abstraction for functional languages with control. In *Proceedings, Twelfth Annual IEEE Symposium on Logic in Computer Science*. IEEE Computer Society Press, 1997.

[Lau99] Olivier Laurent. Polarized proof nets and $\lambda\mu$-calculus. Submitted to Theoretical Computer Science, 1999.

[Lev98] P. B. Levy. Game semantics as continuation passing (extended abstract). unpublished, 1998.

[McC96] Guy McCusker. Games and full abstraction for FPC. In *Proceedings, Eleventh Annual IEEE Symposium on Logic in Computer Science*, pages 174–183. IEEE Computer Society Press, 1996.

[McC98] Guy McCusker. *Games and Full Abstraction for a Functional Metalanguage with Recursive Types.* Distinguished Dissertations in Computer Science. Springer-Verlag, 1998.

[MH99] P. Malacaria and C. Hankin. Non-deterministic games and program analysis: An application to security. In *Proceedings, Fourteenth Annual IEEE Symposium on Logic in Computer Science.* IEEE Computer Society Press, 1999.

[Nic94] Hanno Nickau. Hereditarily sequential functionals. In *Proceedings, Logical Foundations of Computer Science*, volume 813 of *Lecture Notes in Computer Science.* Springer-Verlag, 1994.

[Sel99] Peter Selinger. Control categories and duality: on the categorical semantics of the $\lambda\mu$-calculus. To appear, 1999.

An Improved Extensionality Criterion for Higher-Order Logic Programs

Marc Bezem

Institute for Informatics, University of Bergen, Norway
bezem@ii.uib.no

Abstract. Extensionality means, very roughly, that the semantics of a logic program can be explained in terms of the set-theoretic extensions of the relations involved. This allows one to reason about the program by ordinary extensional logic. First-order logic programming is extensional. Due to syntactic equality tests in the unification procedure, higher-order logic programming is generally not extensional. Extensionality is a highly undecidable property. We give a decidable extensionality criterion for simply typed logic programs, improving both on Wadge's definitional programs from [9] and on our good programs from [2].

1 The Problem of Extensionality in Higher-Order Logic Programming

Consider the following Prolog program defining the usual ordering of the natural numbers as the transitive closure of the successor relation.

```
succ(X,s(X)).                   % assume also constant 0
less(X,Y):- succ(X,Y).
less(X,Z):- succ(X,Y),less(Y,Z).
```

It is better to acknowledge that transitive closure is a generic operation:

```
tc(R,X,Y):- R(X,Y).             % R: D->D->A
tc(R,X,Z):- R(X,Y),tc(R,Y,Z).   % tc = transitive closure
```

and to define `less` as `tc(succ)`. We are then in the realm of higher-order logic programming, by the type of the variable `R`.[1] The semantics of `tc` is still perfectly extensional: `tc` relates extensions of binary predicates to their transitive closures. So far so good, but now consider:

```
p(a).                           % p: D->A
q(a).                           % q: D->A
p_named(p).                     % p_named: (D->A)->A
```

[1] In the examples we stretch the Prolog syntax a bit, in the underlying theory everything is fully curried. Moreover we shall identify subsets with their characteristic function.

L. Fribourg (Ed.): CSL 2001, LNCS 2142, pp. 203–216, 2001.

Here p and q have the same extension {a}, and an extensional interpretation of p_named should contain that extension. As a consequence, p_named(q) would also be true, for which the program provides no justification. In general, comparing infinite extensions is computationally hopeless (>RE), but even in the finite case the extensional approach is not semantically tenable either. In order to see this, add the following clause to the above program:

```
p(b):- p_named(q).
```

Denoting interpretations by [ ], we have for the extended program:

```
NON-extensional model: [p]=[q]={a}, [p_named]={p}
NOT a model: [p]=[q]={a}, [p_named]={{a}}
minimal model 1: [p]={a,b}, [q]={a}, [p_named]={{a,b}}
minimal model 2: [p]={a}, [q]={a,b}, [p_named]={{a}}
```

Note that in both minimal models the constant b makes the extensions of p and q distinct, but lacks a proper explanation. Even worse, the two minimal models disagree on the extension of p_named.

The relevance of extensionality for software engineering has been explained in [9] and [2]. For example, the way in which Pascal handles procedures and functions as parameters is extensional. In short, extensional programs are easier to understand and to maintain, since plugging in a new component with a different name but with the same semantics does not change the overall semantics of the program. This paper gives a rigorous definition of the extensionality of a simply-typed logic program and provides a decidable criterion which captures a large class of extensional programs.

2 Terms and Types

In order to get a clear picture of the notion of extensionality itself, we keep the terms and the types as simple as possible. For this reason we do not yet consider lambda abstraction, higher-order unification, polymorphism, subtyping and so on, although we are well aware of their importance. We take terms to be built up from constants and variables by application. We adopt a fully curried version of the syntax, so that we can do with the type constructor $\to$ only, associating to the right, such as in $D{\to}D{\to}A$.

For compliance with the usual syntax we practise a liberal way of currying. For example, the following denotations are all identified:

$$pxa \equiv (px)a \equiv p(x,a) \equiv p(x)a \equiv p(x)(a)$$

This means that brackets and argument lists can be used freely, but serve only as alternative denotations of curried terms.

The (untyped) terms of logic programming are given by the following abstract syntax:

$$\mathcal{T} ::= \mathcal{C} \mid \mathcal{V} \mid (\mathcal{T}\mathcal{T})$$

Here $\mathcal{C}$ and $\mathcal{V}$ are sets of (typed) constants and variables, respectively, and $\mathcal{TT}$ are terms obtained by application. In principle $\mathcal{C}$ and $\mathcal{V}$ contain all correct Prolog constants and variables, respectively, but in most examples we will tacitly assume that $\mathcal{C}$ contains only those constants that are explicitly mentioned in the example. We use x, y, z to denote variables, a, b, c, p, q as constants, and r, s, t to denote arbitrary terms. Application is taken to be left associative, so $p(xa)$ is notably different from the denotations above. The *head symbol* of a term is its leftmost symbol, either a variable or a constant.

A *clause* is an expression of the form $t_0 \leftarrow t_1, \ldots, t_n$, where all t_i are terms $(0 \leq i \leq n)$. A program is a finite set of clauses. We use C for clauses and P for programs. In the examples we use Prolog notation for clauses and programs.

In order to single out the well-typed higher-order logic programs we use types. The fragment consisting of first-order terms can be typed with types given by the following abstract syntax:

$$\mathcal{D} ::= D \mid (D \to \mathcal{D})$$

Here D is the base type of individuals and $(D \to D)$, $(D \to (D \to D))$, ... are the types of unary functions, binary functions, and so on. We let $\to$ associate to the right and drop outermost brackets.

The types for the higher-order objects are given by the following abstract syntax, allowing predicates to depend on predicates and on individuals:

$$\mathcal{A} ::= A \mid (\mathcal{A} \to \mathcal{A}) \mid (D \to \mathcal{A})$$

Here A is the base type of atoms. The types $D \to A$, $D \to D \to A$, ... are the types of unary predicates, of binary predicates, and so on. The types $A \to A$, $(D \to A) \to A$, ... are the types of predicates on atoms, on unary predicates, and so on. More complicated types can easily be constructed, for example, $((A \to A) \to A) \to (D \to D \to A) \to A$. Note that all types in $\mathcal{A}$ end in A and not in D, so that individuals do not depend on predicates.

The set of types for higher-order logic programming is the union of $\mathcal{D}$ and $\mathcal{A}$. The former will only play a minor role in this paper, and we will focus attention on the latter. We use τ, τ' to denote arbitrary types. By induction one proves that every type $\tau \in \mathcal{A}$ is of the form

$$\tau \equiv \tau_1 \to \cdots \to \tau_k \to A$$

for some $k \geq 0$, with τ_i either D or in $\mathcal{A}$, for every $1 \leq i \leq k$. The types τ_i are called the *argument types* of τ.

Due to the absence of bound variables within terms, the typing system for terms is very simple and consists of one single typing rule:

$$\frac{t : \tau \to \tau' \quad s : \tau}{ts : \tau'} \ (\to\text{-elim})$$

We will now formally define when a higher-order logic program is typable.

Definition 1. *A* declaration *is an expression either of the form $x : \tau$ with x a variable, or of the form $c : \tau$ with c a constant, stating that x (respectively c) has type τ. The variable (constant) on the left hand side is called the* declarandum *of the declaration.*

A context *is a finite list of declarations with different declaranda. Contexts are denoted by Γ.*

The typing relation *$\Gamma \vdash t : \tau$ is defined inductively as the smallest relation which holds whenever $t : \tau$ is a declaration in Γ and which is closed under the typing rule ($\rightarrow$-elim) above.*

A term t is typable *by Γ if there exists τ such that $\Gamma \vdash t : \tau$.*

A clause $t_0 \leftarrow t_1, \ldots, t_n$ is typable *by Γ if $\Gamma \vdash t_i : A$ for all $0 \leq i \leq n$. In that case we write $\Gamma \vdash t_0 \leftarrow t_1, \ldots, t_n : A$.*

A program P consisting of clauses $C_1, \ldots, C_n$ is typable *by Γ if $\Gamma \vdash C_i : A$ for all $1 \leq i \leq n$. In that case we write $\Gamma \vdash P : A$.*

We call a term (clause, program) typable *if it is typable by Γ for suitable Γ. If we speak of a clause (term) in relation to a typable program, then we implicitly assume the clause (term) to be typable by the same context.* □

Intuitively, $\Gamma \vdash P : A$ means that the declarations in Γ ensure that each atom in P is of the base type A. Note that one and the same variable may occur in different clauses of P, but always with the same type as declared in Γ. In cases in which different types are required, the program clauses should be standardized apart.

An alternative characterization of the typing relation is the following: $\Gamma \vdash t : \tau$ holds if and only if there exists a ($\rightarrow$-elim) derivation tree with root $t : \tau$ and leaves in Γ.

Due to the absence of abstraction, the typing system is in fact a subsystem of that for simply typed combinatory logic. We rely on the well-established techniques on principal type schemes for type checking and type synthesis. For the purpose of this paper it is not necessary to enter this important subject, instead we refer to the original source [4] and to the ML literature.

To give the reader at least some idea, the principal type scheme of the term xy is $x : \tau\rightarrow\tau'$, $y : \tau$, with τ, τ' arbitrary types. The principal type scheme of the atom xy is $x : \tau\rightarrow A$, $y : \tau$, and of the clause $xy \leftarrow y$ it is the context $x : A\rightarrow A$, $y : A$.

3 Operational Semantics

The operational semantics is in fact an extension of the usual SLD-resolution procedure for first-order logic programming [1]. We treat only some key points needed for a proper understanding of the sequel, namely unification, well-typed substitution and the immediate consequence operator. The latter will play a role in inductive proofs.

Consider an arbitrary first-order unification algorithm, e.g. by Martelli and Montanari, see [1]. We will sketch how to extend it to the terms here. Unification of two terms can only succeed if they have the same type. In unifying them, we first write the terms in the form $t_0 t_1 \ldots t_k$ and $s_0 s_1 \ldots s_l$, where t_0 and s_0 are the respective head symbols. Second we make the hidden application functions explicit for $k, l > 1$. That is, we write the terms in the form $\alpha(\alpha(\ldots\alpha(t_0, t_1), \ldots), t_k)$ and $\alpha(\alpha(\ldots\alpha(s_0, s_1), \ldots), s_l)$, where α is a new binary function symbol. Next we apply the first-order unification algorithm to the terms written with explicit application functions. In recursive calls of the unification procedure hidden applications must again first be made explicit. In the end we return to implicit applications by discarding the α symbols properly from the resulting substitution.

Type persistence under well-typed substitution (and hence under resolution) is ensured by the following lemma.

Lemma 1. *If $\Gamma, x : \tau \vdash t : \tau'$ and $\Gamma, \Gamma' \vdash s : \tau$, then $\Gamma, \Gamma' \vdash t[x/s] : \tau'$.*

Proof: By induction on the derivation of $\Gamma, x : \tau \vdash t : \tau'$. □

We recall the familiar notions of Herbrand Base, Herbrand Universe, immediate consequence operator and its least fixed point. These notions are now slightly more general as the terms involved stem from the higher-order syntax. By convention all terms are assumed to be typable in the context of the program.

Definition 2. *Let P be a typable higher-order logic program. We define the* Herbrand Base B_P *(resp. the* Herbrand Universe U_P*) to be the set of all closed terms of type A (resp. D). For every $S \subseteq B_P$ we define $T_P(S) \subseteq B_P$ by $t \in T_P(S)$ iff there exists a closed instance of a program clause in P with head t and all body atoms in S. The operator T_P is called the* immediate consequence operator *of P. As usual, $T_P{\uparrow}0 = \emptyset$, $T_P{\uparrow}(n{+}1) = T_P(T_P{\uparrow}n)$ and $M_P = T_P{\uparrow}\omega = \bigcup_{n \geq 0} T_P{\uparrow}n$ is the least fixed point of T_P.* □

4 Declarative Semantics

In this section we propose a notion of model for higher-order logic programs. The idea is to separate the applicative behaviour of higher-order objects from the logical behaviour. Although this semantical framework may seem overly general at first sight, there are strong reasons in favour of this generality:

- The framework covers all relevant approaches to higher-order logic.
- The extensional collapse, to be introduced in the next section, can be carried out within the model class.
- The larger the model class is, the more applications there are.

Definition 3. *A* type structure *consists of sets D_τ for every type τ and application mappings*

$$ap_{\tau,\tau'} : D_{\tau \to \tau'} \times D_\tau \to D_{\tau'}$$

for all types τ, τ'. We denote application by juxtaposition and associate to the left. A type structure is functionally extensional *if $D_{\tau \to \tau'} \subseteq D_\tau \to D_{\tau'}$ and $ap_{\tau,\tau'}$*

is set-theoretic function application, for all τ, τ'. *It is* logically extensional *if it is functionally extensional and moreover* $\emptyset \neq D_A \subseteq \{\mathbf{T}, \mathbf{F}\}$. *(The cases in which* D_A *consists of just one truth value are borderline cases in which everything is true or everything is false.)*

A type structure is extended to an interpretation for higher-order logic programs in the following way. First we add an interpretation function I *which assigns an element of* D_τ *to every constant of type* τ. *What follows now is a standard development of the interpretation of terms, but with application according to the given type structure. An* assignment *is a function mapping variables to domain elements of the corresponding types. Given an assignment* α, *the interpretation function* I *can be extended to an* interpretation $[\![t]\!]_\alpha$ *for all terms* t *in the following way:*

- $[\![c]\!]_\alpha = I(c)$ *for every constant* c,
- $[\![x]\!]_\alpha = \alpha(x)$ *for every variable* x, *and*
- $[\![tt']\!]_\alpha = [\![t]\!]_\alpha [\![t']\!]_\alpha$.

Thus interpretation is homomorphic with respect to syntactic and semantic application.

Next we add a valuation function V *assigning a truth value to every element of* D_A *(in the case of logical extensionality we take* $V(\mathbf{T}) = \mathbf{T}$ *and/or* $V(\mathbf{F}) = \mathbf{F}$*). A term* t *of type* A *(that is, an atom) is* true *(*false*) under an assignment* α *if* $V([\![t]\!]_\alpha) = \mathbf{T}$ $(\mathbf{F})$. *The valuation* V *is extended to formulas according to the usual meaning of the logical connectives and quantifiers.*

A type structure with interpretation/valuation functions I *and* V *is called a* model *of higher-order logic program* P *if it makes true every clause of* P *under any assignment* α. *A model is called* extensional *if it is logically extensional.* □

The *general* models of Henkin [3] are logically extensional models as above. The *standard* model [3] satisfies moreover $D_{\tau\to\tau'} = D_\tau \to D_{\tau'}$. The model class of Nadathur and Miller [6], who allow D_o to be a set of *labeled* truth values, is covered by the functionally extensional type structures. We continue by showing that the closed term model of a well-typed program P is functionally extensional up to isomorphy, and initial in the model class.

Definition 4. *Let* P *be a typable higher-order logic program. By convention we assume that the language of* P *is such that we do have closed terms of base types* D *and* A. *In addition we assume that if some term* t *of type* τ *occurs in* P, *then there exist closed terms of type* τ *and of all argument types of* τ. *(If necessary, assume some default constants in the signature.) We use* $|\tau|_P$ *to denote the set of closed terms of type* τ. *Let* $\mathcal{M}_P$ *be the type structure defined by sets* $|\tau|_P$ *and syntactic application mappings, with the interpretation* $I_P(c) = c$ *for any constant* c. *The valuation function* V_P *on* $|A|_P$ *is defined by* $V_P(t) = \mathbf{T}$ *if* $t \in M_P$, *and* $V_P(t) = \mathbf{F}$ *otherwise.* □

The following lemma is obvious, initiality is proved for $t \in T_P\!\uparrow\! n$ by induction on n.

Lemma 2. *Let P be a typable higher-order logic program. Then we have:*

1. *Every closed term $t : \tau \to \tau'$ defines through application a unique mapping $|\tau|_P \to |\tau'|_P$.*
2. *The type structure $\mathcal{M}_P$ is functionally extensional up to isomorphy.*
3. *$\mathcal{M}_P$ is a model of P which is initial in the following sense: for every closed term t of type A, if t is true in $\mathcal{M}_P$, then t is true in every model of P.*

The operational semantics from the previous section can be related to the declarative semantics by generalizing the usual soundness and completeness results for SLD-resolution [1]. Space limitations prevent us from giving more details here.

5 The Extensional Collapse

In this section we define a notion of extensionality for higher-order logic programs and show that for extensional programs the semantics can be considerably simplified. This so-called extensional collapse originates from the model theory of finite type arithmetic and is described and attributed to Zucker in [8].

Definition 5. *Let P be a typable higher-order logic program. We define relations $\approx_P^\tau$ on $|\tau|_P$, expressing extensional equality of type τ.*

$\mathcal{D}$: We put $\approx_P^\tau$ to be $\equiv$, syntactic equality on $|\tau|_P$, for every $\tau \in \mathcal{D}$.
$\mathcal{A}$: By induction on $\tau \in \mathcal{A}$. For the base type A we put $t \approx_P^A s$ if and only if $V_P(t) = V_P(s)$, that is, either $t, s \in M_P$, or $t, s \notin M_P$.
For the induction steps $\tau' \to \tau$ with $\tau' \equiv D$ or $\tau' \in \mathcal{A}$ we define $t \approx_P^{\tau' \to \tau} s$ if and only if $tt' \approx_P^\tau ss'$ for all t', s' such that $t' \approx_P^{\tau'} s'$.

We will often omit type superscripts and the subscript P. A closed term t is called extensional *if $t \approx t$. We call P* extensional *if all closed terms over the signature of P are extensional. In that case the relations $\approx$ will be called* extensional equalities. □

We give some examples and counterexamples of extensional programs. Proofs are postponed till after Corollary 1.

Example 1. The following clauses form an extensional program:

```
R(a,b).                    % (a,b) in every binary relation D->D->A
call(X):- X.
or(X,Y):- X.
or(X,Y):- Y.
tc_T(R,X,Y):- R(X,Y).              % tc_T = transitive closure with
tc_T(R,X,Z):- R(X,Y),tc_T(R,Y,Z).                          % R: T->T->A
...
sort(R,L1,L2):- ...          % sorting parameterized by R: D->D->A
```

Counterexamples, i.e., examples of non-extensionality, are:

```
eq(Y,Y).                    % Y: A
apply(F,Z,F(Z)).            % F: (D->D->A)->D->D->A
```

For example, we have `p` $\approx^A$ `or(p,p)` , but not `eq(p,p)` $\approx^A$ `eq(p,or(p,p))`, and hence not `eq` $\approx^{A \to A \to A}$ `eq`. The non-extensionality of the second clause arises when, for example, one considers a transitive relation `r`, so that `r` has the same extension as `tc(r)`, and `apply(tc,r,tc(r))` holds, whereas `apply(tc,r,r)` does not hold. □

Lemma 3. *Let P be a typable higher-order logic program. Then $\approx^\tau_P$ is a partial equivalence relation for every type τ. More precisely, for every type τ, the relation $\approx^\tau_P$ is symmetric and transitive (and hence reflexive where defined: $t \approx^\tau_P s \Rightarrow t \approx^\tau_P t$ for all closed terms $t, s : \tau$). If the relations $\approx$ are everywhere reflexive, then the usual form of extensionality holds: if $tr \approx sr$ for all r, then $t \approx s$.*

Proof: Symmetry and transitivity are proved simultaneously by induction on τ. If reflexivity also holds, then $tr \approx sr$ and $r \approx r'$ imply $tr \approx sr'$. □

Definition 6. *Let P be a typable, extensional, higher-order logic program, with $\mathcal{M}_P$ as in Definition 4 and $\approx^\tau_P$ as in Definition 5. The extensional equalities $\approx^\tau_P$ are by definition congruences with respect to application. Let $[t]_P$ denote the equivalence class of t with respect to $\approx^\tau_P$, for any closed term $t : \tau$. We define $\mathcal{M}_P/\approx$ as the type structure defined by sets $|\tau|_P/\approx^\tau_P$ and application mappings satisfying $[t]_P[t']_P = [tt']_P$ for all closed terms of appropriate types. The interpretation and valuation function are also collapsed into $I/\approx(c) = [c]_P$ and $V/\approx([t]_P) = V_P(t)$. Note that $V/\approx$ is well-defined. The quotient structure $\mathcal{M}_P/\approx$ is called the* extensional collapse *of $\mathcal{M}_P$.*

Theorem 1. *Let P be a typable,* extensional, *higher-order logic program. Then we have the following:*

1. *Every closed term $t : \tau \to \tau'$ defines through application a mapping from $|\tau|_P/\approx^\tau_P$ to $|\tau'|_P/\approx^{\tau'}_P$, and two such terms are extensionally equal if and only if they define the same mapping.*
2. *The quotient structure $\mathcal{M}_P/\approx$ is logically extensional up to isomorphy.*
3. *The quotient structure $\mathcal{M}_P/\approx$ is a model of P that is elementarily equivalent to $\mathcal{M}_P$ with respect to the clausal language of P.*

Proof: Let conditions be as above, in particular P is extensional. Obviously, $t : \tau \to \tau'$ can be viewed to map any $[t']_P$ to $[tt']_P$. Moreover, extensionally equal terms define the same mapping. Conversely, if t and s define the same mapping, then $[tt']_P = [ss']_P$ whenever $[t']_P = [s']_P$, so $t \approx s$. This proves 1 and implies that the quotient structure is functionally extensional up to isomorphy. It is logically extensional since $|A|_P/\approx^A_P$ consists of at most two classes, namely M_P and $B_P - M_P$, the true and the false closed atoms. This proves 2. The elementary equivalence of the quotient structure with the original structure follows easily, since both structures make true the same closed atoms. This proves 3. □

6 The Extensionality Criterion

In this section we develop a decidable, syntactic criterion which is sufficient for extensionality. The criterion is not necessary, and cannot be so given the undecidability of extensionality in general.

Conceptually the criterion is very simple and satisfying. Note that the unification process may operate on arguments in an intensional way, for example, by comparing arguments syntactically. The idea is that a program is extensional provided higher-order arguments (hoa) are only passed (p), applied (a) and/or thrown away (ta). For this it suffices that all higher-order arguments in the head of any program clause are distinct variables, so that the unification for the higher-order arguments reduces to matching.

It may come as a surprise that the elaboration of this very simple idea requires all the technicalities that are to follow, as well as those that are omitted for reasons of space. We do not see how the proofs below can be simplified, but some readers may be challenged.

Definition 7. *Let P be a typable higher-order logic program. We call P a* hoapata *program if the higher order arguments in the head of any program are all distinct variables.*

Observe that, in the definition of a hoapata program, arguments of type D do not play a role. As a consequence, arguments of type D may be used without any restriction in hoapata programs, but we will ignore them in the considerations below. Thus the head of a clause of a hoapata program has one of the following two forms: $a\,\vec{x}$ or $x\,\vec{x}$, where all variables x_i in $\vec{x}$ are distinct (for typing reasons we also have $x \not\equiv x_i$). For atoms in the body of a clause of a hoapata program we distinguish the following four typical forms: $b\,\vec{c}(x,\vec{x},\vec{y})$, $x\,\vec{d}(x,\vec{x},\vec{y})$, $x_i\,\vec{e}(x,\vec{x},\vec{y})$, or $y_j\,\vec{f}(x,\vec{x},\vec{y})$. Here $\vec{y}$ are the local variables, i.e. variables that occur in the body but not in the head of the program clause. Furthermore, a, b are constants and $\vec{c}(x,\vec{x},\vec{y}), \vec{d}(x,\vec{x},\vec{y}), \vec{e}(x,\vec{x},\vec{y}), \vec{f}(x,\vec{x},\vec{y})$ are sequences of terms containing zero or more occurrences of the variables $x, \vec{x}, \vec{y}$. By convention everything is assumed to be well-typed.

The four forms above represent the cases in which the head symbol of the body atom is a constant, the head variable x, one of the arguments x_i and one of the local variables y_j, respectively. In the first and the fourth case all higher-order variables from the head of the program clause are passed and/or thrown away. In the second and the third case they are also applied.

Examples of hoapata programs are: all definitional programs from [9], all good programs from [2], all programs claimed to be extensional in Example 1. Also the following fixed point clauses for different types T, which can capture all recursions, form a hoapata program.

```
fp_T(X,Y1,...Yk) :- X(fp_T(X),Y1,...Yk).    % fp_T = fixed point
% X: T->T, T = T1->...->Tk->A, Yi: Ti for i=1,...,k
```

The following lemma states an essential property of hoapata programs, namely that all higher-order arguments that do play a role in a derivation are eventually applied. Here the higher-order arguments are closed instantiations of the variables $\vec{z}$, and they are applied in closed instantiations of atoms of the form $z_i\,\vec{e}$. Playing a role in the derivation is expressed in terms of the layered structure of M_P induced by T_P.

Lemma 4. *Let P be a hoapata program and $\vec{z}$ be a sequence of distinct higher-order variables. Let $t : A$ be an atom headed by a constant and with free variables among $\vec{z}$, and σ a closed substitution with domain $\vec{z}$ such that $t^\sigma \in T_P{\uparrow}(n{+}1)$. Then there is a conjunction G of atoms, all starting with a head variable from $\vec{z}$ and with free variables among $\vec{z}$, such that M_P is closed under $t \leftarrow G$ and $G^\sigma \subseteq T_P{\uparrow}n$. The former means that for any closed instance $t' \leftarrow G'$ we have $t' \in M_P$ whenever $G' \subseteq M_P$ and the latter means that all atoms in G^σ are in a lower stratum than t^σ. In particular the conjunction G is empty if $n = 0$.*

Proof: Let P be a hoapata program. We proceed by induction on n, ignoring arguments of type D. For the base case $n = 0$, let $t \equiv a\,\vec{t}$ with σ as above. Now $a\,\vec{t^\sigma} \in T_P{\uparrow}1$ implies that there is a clause $a\,\vec{x}$ or $x\,\vec{x}$ in P, with all variables in $\vec{x}$ distinct, such that $a\,\vec{t^\sigma}$ is an instantiation. But then also $a\,\vec{t}$ and all its instantiations are in M_P and we can take G empty. (Note that the program clause could also take the form $y\,\vec{y}$ with the type of y smaller than the type of a; in that case y is instantiated by, say, $a\,t_1 t_2$, but again we can take G empty.)

For the induction step, assume the lemma holds for n and let again $t \equiv a\,\vec{t}$ with σ as above such that $a\,\vec{t}^\sigma \in T_P{\uparrow}(n{+}2)$. Recall that we have the following two typical forms of hoapata program clauses:

$$a\,\vec{x} \leftarrow b\,\vec{c}(\vec{x},\vec{y}),\ldots,x_i\,\vec{e}(\vec{x},\vec{y}),\ldots,y_j\,\vec{f}(\vec{x},\vec{y})$$
$$x\,\vec{x} \leftarrow b\,\vec{c}(x,\vec{x},\vec{y}),\ldots,x\,\vec{d}(x,\vec{x},\vec{y}),\ldots,x_i\,\vec{e}(x,\vec{x},\vec{y}),\ldots,y_j\,\vec{f}(x,\vec{x},\vec{y})$$

We will start with first form which is somewhat simpler. Assume there is a program clause of the first form with closed instance

$$a\,\vec{t^\sigma} \leftarrow b\,\vec{c}(\vec{t^\sigma},\vec{s}),\ldots,t_i^\sigma\,\vec{e}(\vec{t^\sigma},\vec{s}),\ldots,s_j\,\vec{f}(\vec{t^\sigma},\vec{s})$$

all whose body atoms are in $T_P{\uparrow}(n{+}1)$. Since all higher-order variables in the head $a\,\vec{x}$ are distinct, we also have the following instance of the same program clause:

$$a\,\vec{t} \leftarrow b\,\vec{c}(\vec{t},\vec{s}),\ldots,t_i\,\vec{e}(\vec{t},\vec{s}),\ldots,s_j\,\vec{f}(\vec{t},\vec{s})$$

Observe that $\vec{s}$ is closed. We apply the induction hypothesis to all the body atoms $g_1,\ldots,g_m$ of the clause above that do not (yet) have the desired form $z_i\,\vec{e}(\vec{t},\vec{s})$. For each such body atom g_k there exists a conjunction G_k such that M_P is closed under $g_k \leftarrow G_k$ and $G_k^\sigma \subseteq T_P{\uparrow}n$ and all atoms in G_k have the desired form $(1 \le k \le m)$. Now take G to be the conjunction of all G_k together with all body atoms that do have already the desired form $z_i\,\vec{e}(\vec{t},\vec{s})$. One verifies easily that G has also the other desired properties: M_P is closed under $a\,\vec{t} \leftarrow G$ as M_P is closed under the program clause involved in the induction step, and $G^\sigma \subseteq T_P{\uparrow}(n{+}1)$.

The proof for the second form can now be presented as a refinement of the argument for the first form. Assume there is a program clause of the second form with closed instance and instance respectively:

$$
\begin{aligned}
a\,\vec{t^\sigma} &\leftarrow b\,\vec{c}(a,\vec{t^\sigma},\vec{s}),\ldots,a\,\vec{d}(a,\vec{t^\sigma},\vec{s}),\ldots,t_i^\sigma\,\vec{e}(a,\vec{t^\sigma},\vec{s}),\ldots,s_j\,\vec{f}(a,\vec{t^\sigma},\vec{s})\\
a\,\vec{t} &\leftarrow b\,\vec{c}(a,\vec{t},\vec{s}),\ldots,a\,\vec{d}(a,\vec{t},\vec{s}),\ldots,t_i\,\vec{e}(a,\vec{t},\vec{s}),\ldots,s_j\,\vec{f}(a,\vec{t},\vec{s})
\end{aligned}
$$

All body atoms of the closed instance are in $T_P{\uparrow}(n{+}1)$. Here we have assumed that the head variable x has the same type as a. It is not difficult to see that the same inductive argument as in the treatment of the first form can be applied. It remains to consider a slightly more difficult subcase of the second form, in which the head variable of the program clause has a type smaller than that of a. Let $\vec{t} \equiv \vec{u}\vec{v}$ and assume that the head variable matches $a\,\vec{u}$. Thus we assume there is a program clause of the first form with closed instance and instance as above, with $a\,\vec{u}$ instead of a and $\vec{v}$ instead of $\vec{t}$. In this case t_i may belong to either $\vec{u}$ or $\vec{v}$. Although this looks more complicated, again the same inductive argument applies.

It takes some reflection to see that we have justifiably ignored arguments of type D in the above proof. □

Observe that the head constant plays an essential role in the above lemma. If, for example, the atom t is of the form $z_i\,\vec{s}$, then instantiations of z_i can be decomposed by applying program clauses, and nothing can be concluded.

Theorem 2. *If P is a hoapata program, then $t \approx t$ for every closed $t : \tau$.*

Proof: Let P be a hoapata program. We proceed by double induction, first on τ and then on $T_P{\uparrow}n$ using the previous lemma. For the base types A and D the lemma holds by definition. For types $\tau \in \mathcal{D}$ the lemma is trivial. For the induction step, let $\tau \in \mathcal{A}$ and assume the lemma has been proved for all types smaller than τ.

Let $\vec{z}$ be a sequence of distinct higher-order variables of types smaller than τ. Again we ignore arguments of type D. Let $P(n)$ be the following property: for all terms $t : A$ headed by a constant and with free variables among $\vec{z}$, for all closed substitutions σ, σ' such that $\vec{z}^\sigma \approx \vec{z}^{\sigma'}$ holds componentwise, if $t^\sigma \in T_P{\uparrow}n$, then $t^{\sigma'} \in M_P$. By induction we prove $P(n)$ for all n. It follows that every closed $t : \tau$ is extensional, by considering $t\,\vec{z} : A$.

The base case $n = 0$ holds vacuously, since $T_P{\uparrow}0 = \emptyset$. For the induction step, assume $P(n)$ as secondary induction hypothesis, the primary induction hypothesis being that all closed terms of type smaller than τ are extensional. We have to prove $P(n+1)$, so assume t and σ, σ' as above, with $t^\sigma \in T_P{\uparrow}(n{+}1)$. By Lemma 4 there is a (possibly empty) conjunction G of atoms, all starting with a head variable from $\vec{z}$, such that M_P is closed under $t \leftarrow G$ and $G^\sigma \subseteq T_P{\uparrow}n$. We want to infer $t^{\sigma'} \in M_P$, which by the above property of G reduces to $G^{\sigma'} \subseteq M_P$. If G is empty, for example if $n = 0$, then we are done. Otherwise, we would like to apply the secondary induction hypothesis to all atoms in G, which are all of the form $z_i\,\vec{s}$ with $z_i^\sigma\,\vec{s}^\sigma \in T_P{\uparrow}n$. Unfortunately $P(n)$ does not allow us to conclude directly that $z_i^{\sigma'}\,\vec{s}^{\sigma'} \in M_P$, since z_i is a variable (cf. the remark just after Lemma 4). We have $z_i^\sigma \approx z_i^{\sigma'}$ by assumption. However, in this stage of the proof we do not know whether or not $\vec{s}^\sigma \approx \vec{s}^{\sigma'}$ holds. There is exactly one way out. Note that z_i^σ starts with a constant. By the secondary induction hypothesis $P(n)$ we get that $z_i^\sigma\,\vec{s}^{\sigma'} \in M_P$. From the primary induction hypothesis we get that all $\vec{s}^{\sigma'}$ are extensional, since the types are smaller than τ. Combining those two facts with $z_i^\sigma \approx z_i^{\sigma'}$ yields $z_i^\sigma\,\vec{s}^{\sigma'} \approx z_i^{\sigma'}\,\vec{s}^{\sigma'} \in M_P$. □

Corollary 1. *Every hoapata program is extensional.*

The above corollary justifies all extensionality claims in Example 1. The non-extensional programs there are indeed not hoapata. However, the criterion does not (and cannot) capture all extensional programs, as shown by the following example.

Example 2. The following program is obtained from the example program in Section 1 by leaving out `q(a)`. Then there is only one unary predicate and the program is extensional from poverty, although the second clause is not hoapata.

```
p(a).                            % p: D->A
p_named(p).                      % p_named: (D->A)->A
```

7 Related and Future Work

We took over the non-extensional example program in Sec. 1 from Wadge [9], but with head `p(b)` instead of `q(b)` in the fourth clause `p(b):- p_named(q)`. Thus we avoid the introduction of a new constant `c` in the explanation and we get the dramatic disagreement of the two minimal models on `[p_named]`. The result of [9] may be rendered as follows: every *definitional* higher-order logic program has a minimal *standard* model. Here 'definitional' is a syntactic criterion which is strictly stronger than 'hoapata', since it disallows local variables of types other than D. Although this excludes, for example, the transitive closure of a binary predicate of unary predicates from being definitional, the main difference in view between Wadge and us is on the semantical level. In our opinion standard models are less suitable as models of extensional higher-order logic programs, as standard models are uncountable if the domain of individuals is infinite. For example, a slight variation of the program in Example 2, with `p(s(0))` instead of `p(a)`, is still extensional, but the interpretation of by `p_named` in the minimal standard model is a non-computational object, which must single out `{s(0)}` in an uncountable power set. In our declarative semantics the (collapsed) closed term model consists only of `p` at type $D{\to}A$ and the interpretation of `p_named` is computationally tractable.

The main technical improvement of this paper over [2] is the argument in Section 6 consisting of the double induction proving Theorem 2 and the syntactical analysis behind Lemma 4. Hoapata programs strictly extend the good programs from [2], since they impose no restrictions on neither local variables, nor on the use of the head variable of the head of a clause.

There seems to be no connection with Miller's [5] condition for the decidability of higher-order unification. First of all, unification is first-order here, since we have no bound variables within terms. Second, the hoapata criterion, unlike the condition in [5], does not impose restrictions on the first-order arguments of a higher-order object.

We expect our results to carry over to other symbolic forms of simultaneous inductive definitions. Normann proposes to study the relation between hoapata programs and PCF [7].

For the future we plan to explore the following connection. The rewrite rules $K_\tau XY \to X$ and $S_\tau XYZ \to XZ(YZ)$ in typed Combinatory Logic are hoapata. The same holds for the rules $R_\tau XY0 \to X$ and $R_\tau XY(Z{+}1) \to YZ(R_\tau XYZ)$ for the primitive recursor in Gödel's T. The latter rules are hoapata since the patterns 0 and $Z{+}1$ are of base type, i.c. the natural numbers. Also the rules for some forms of transfinite recursion are hoapata. It is known that the closed term models of those calculi can be collapsed extensionally, but the proofs rely on confluence and termination and are ad hoc. We expect that, along the lines of the hoapata criterion, a general extensionality result in the context of higher-order rewriting can be obtained, unifying the results on the calculi above with the result of this paper.

In order to get a clear picture of the notion of extensionality we have kept the syntax of the higher-order logic programs as simple as possible. In particular we have left out important features such as lambda abstraction, higher-order unification, polymorphism, subtyping and so on. For the future we also plan to extend the notion of extensionality to a richer language and to develop an extensionality criterion for that language.

Conclusion

We have developed a notion of extensionality and a decidable extensionality criterion for simply typed logic programs. Thus we have captured the phenomenon of extensionality for a natural class of higher-order logic programs which supports predicate abstraction. We have shown that under the criterion the closed term model of a simply typed logic program can be collapsed into a computationally tractable, extensional set-theoretic model, which is initial in the model class.

References

1. K. R. Apt. Logic programming. In J. van Leeuwen (ed.) *Handbook of Theoretical Computer Science*, Vol. B, pp. 493–574. Elsevier, Amsterdam, 1990.
2. M.A. Bezem. Extensionality of Simply Typed Logic Programs. In D. de Schreye (ed.) *Proceedings ICLP'99*, pp. 395–410. MIT Press, Cambridge, Massachusetts, 1999.

3. L. Henkin. Completeness in the theory of types. *Journal of Symbolic Logic*, 15:81–91, 1950.
4. J.R. Hindley. The principal type-scheme of an object in combinatory logic. *Transactions of the AMS*, 146:29–60, 1969.
5. D.A. Miller. A logic programming language with lambda-abstraction, function variables, and simple unification. *Journal of logic and computation*, 2(4):497–536, 1991.
6. G. Nadathur and D.A. Miller. Higher-order logic programming. In D. Gabbay e.a. (eds.) *Handbook of logic in artificial intelligence*, Vol. 5, pp. 499–590. Clarendon Press, Oxford, 1998.
7. G. Plotkin. LCF considered as a programming language. *Theoretical Computer Science*, 5:223–256, 1977.
8. A.S. Troelstra. *Metamathematical Investigation of Intuitionistic Arithmetic and Analysis.* Number 344 in Lecture Notes in Mathematics. Springer-Verlag, Berlin, 1973.
9. W.W. Wadge. Higher-order Horn logic programming. In V. Saraswat and K. Ueda (eds.) *Proceedings of the 1991 International Symposium on Logic Programming*, pp. 289–303. MIT Press, Cambridge, Massachusetts, 1991.

A Logic for Abstract State Machines

Robert F. Stärk and Stanislas Nanchen

Computer Science Department, ETH Zürich, CH-8092 Zürich, Switzerland
{staerk,nanchen}@inf.ethz.ch

Abstract. We introduce a logic for sequential, non distributed Abstract State Machines. Unlike other logics for ASMs which are based on dynamic logic, our logic is based on atomic propositions for the function updates of transition rules. We do not assume that the transition rules of ASMs are in normal form, for example, that they concern distinct cases. Instead we allow structuring concepts of ASM rules including sequential composition and possibly recursive submachine calls. We show that several axioms that have been proposed for reasoning about ASMs are derivable in our system and that the logic is complete for hierarchical (non-recursive) ASMs.

Keywords: Abstract State Machines, dynamic logic, modal logic, logical foundations of specification languages.

1 Introduction

Gurevich's Abstract State Machines (ASMs) [5] are widely used for the specification of software, hardware, algorithms, and the semantics of programming languages [1]. Most logics that have been proposed for ASMs are based on variants of dynamic logic. There are, however, fundamental differences between the imperative programs of dynamic logic and ASMs. In dynamic logic, states are represented with variables. In ASMs, states are represented with dynamic functions. The fundamental program constructs in dynamic logic are non-deterministic iteration (star operator) and sequential composition. The basic transition rules of ASMs consist of parallel function updates. Since parallel function updates may conflict, a logic for ASMs must have a clear notion of consistency of transition rules. Therefore, rather than to encode ASMs into imperative programs of dynamic logic or to extend the axioms and rules of dynamic logic, we propose new axioms and rules which are directly based on an update predicate for transition rules. Since ASMs are special instances of transition systems, our logic contains a modal operator, too [14].

What comes closest to our system is known as dynamic logic with array assignments [6,7]. The substitution principle which is used in its axiomatization is derivable in our system (Lemma 9). The dynamic logic with array assignments, however, is not concerned with parallel execution of assignments and therefore does not need a notion of consistency.

Groenboom and Renardel de Lavalette introduce in [4] the *Formal Language for Evolving Algebras* (FLEA), a system for formal reasoning about abstract

L. Fribourg (Ed.): CSL 2001, LNCS 2142, pp. 217–231, 2001.

state machines. Their system is in the tradition of dynamic logic and contains for every rule R a modal operator $[R]\varphi$ with the intended meaning that φ holds always after the execution or R. The logic of their formal language contains besides *true* and *false* a third truth-value which stands for *undefined.* Although they consider parallel composition of transition rules, they have no formal notion of consistency in their system. We adopt their modal operator $[R]\varphi$ such that the basic axioms of their system are derivable in our logic (Lemma 5). We use, however, the two-valued logic of the classical predicate logic.

Our system is designed to deal with Börger and Schmid's *named parameterized ASM rules* [2] which include also recursive definitions of transition rules. Recursive rule definitions in combination with sequential compositions of transition rules are maybe too powerful and not in the spirit of the basic ASM concept [5]. Nevertheless we include them in our logic and solve the technical problems that arise with an explicit definedness predicate.

Schönegge extends in [11] the dynamic logic of the KIV system (Karlsruhe Interactive Verifier) to turn it into a tool for reasoning about abstract state machines. The transition rules of ASMs are considered as imperative programs of dynamic logic. While-programs of dynamic logic are used as an interpreter for abstract state machines; loop-programs are used to apply an ASM a finite number of times. Schönegge's rules for simultaneous function updates and parallel composition of transition rules are derivable in our system (Lemma 7). His sequent calculus is mainly designed as a practical extension of the KIV system and not as a foundation for reasoning about transition rules and ASMs.

Schellhorn and Ahrendt simulate in [10] abstract state machines in the KIV system by formalizing dynamic functions as association lists, serializing parallel updates and transforming transition rules into flat imperative program of the underlying dynamic logic. Schellhorn is able to fully mechanize a correctness proof of the Prolog to WAM compilation in his dissertation in [9]. He argues that the inconsistency of an ASM (clash in simultaneous updates) can only be detected when the ASM is in normal form and that the transformation of the ASM into normal form by a pre-processor is more efficient than a formalization of consistency in terms of logical axioms as we do it in our system. We do think that a suitable theorem prover can automatically process and simplify our consistency conditions (Lemma 3 and Table 5) without problems.

Poetzsch-Heffter introduces in [8] a basic logic for a class of ASMs consisting of simultaneous updates of 0-ary functions (dynamic constants) and if-then-else rules only. His basic axiom states that the truth of the *weakest backwards transformer* of a formula implies the truth of the formula in the next state. He then derives partial correctness logics for a class of simple imperative programming languages by specifying their semantics with ASMs of his restricted class. His basic axiom is derivable in our system (Lemma 10).

Gargantini and Riccobene show in [3] how the PVS theorem prover can provide tool support for ASMs. They show how ASMs can be encoded in PVS (and hence in the underlying formal system which is Church's simple theory of types). Functions are encoded as PVS functions and an interpreter for ASMs is implemented in PVS. The parallel rule application is serialized. The abstraction level of the abstract states is preserved by assuming properties of certain static functions rather than by implementing (explicitly defining) them in PVS. They do not provide a technique for proving consistency of ASMs as we do in our logic. We think that for verification of large ASMs a theorem prover like PVS should directly provide support for transition rules such that the overhead introduced by the encoding is avoided (cf. [12]).

The plan of this paper is as follows. In Sect. 2 we give a short overview on ASMs with sequential composition and (possibly recursive) rule definitions. After some considerations on formalizing the consistency of transition rules in Sect. 3, we introduce in Sect. 4 the basic axioms and rules of our logic and show that several useful principles are derivable. In Sect. 5 we prove that the logic is complete for hierarchical (non-recursive) ASMs.

2 ASM Rules and Update Sets

The notion of an *abstract state* is the classical notion of a mathematical structure $\mathfrak{A}$ for a vocabulary Σ consisting of a a non-empty set $|\mathfrak{A}|$ and of functions $f^{\mathfrak{A}}$ from $|\mathfrak{A}|^n$ to $|\mathfrak{A}|$ for each n-ary function name f of Σ. The terms s, t and the first-order formulas φ, ψ of the vocabulary Σ are interpreted as usual in the structure $\mathfrak{A}$ with respect to a variable assignment ζ. The value of a term t in the structure $\mathfrak{A}$ under ζ is denoted by $[\![t]\!]^{\mathfrak{A}}_{\zeta}$; the truth value of a formula φ in $\mathfrak{A}$ under ζ is denoted by $[\![\varphi]\!]^{\mathfrak{A}}_{\zeta}$ (see Table 1). The variable assignment which is obtained from ζ by assigning the element a to the variable x is denoted by $\zeta\frac{a}{x}$.

Abstract State Machines (ASMs) are systems of finitely many *transition rules* which update some of the functions of the vocabulary Σ in a given state at some arguments. The functions of the vocabulary Σ are divided into *static* functions which cannot be updated by an ASM and *dynamic* ones which typically do change as a consequence of updates by the ASM. The transition rules R, S of an ASM are syntactic expressions generated as follows (the function arguments can be read as vectors):

1. *Skip Rule:* **skip**
   Meaning: Do nothing.
2. *Update Rule:* $f(t) := s$
   Syntactic condition: f is a dynamic function name of Σ
   Meaning: In the next state, the value of f at the argument t is updated to s.
3. *Block Rule:* $R\ S$
   Meaning: R and S are executed in parallel.
4. *Conditional Rule:* **if** φ **then** R **else** S
   Meaning: If φ is true, then execute R, otherwise execute S.
5. *Let Rule:* **let** $x = t$ **in** R
   Meaning: Assign the value of t to x and execute R.
6. *Forall Rule:* **forall** x **with** φ **do** R
   Meaning: Execute R in parallel for each x satisfying φ.

7. *Sequence Rule:* $R\,;S$
   Meaning: R and S are executed sequentially, first R and then S.
8. *Call Rule:* $\rho(t)$
   Meaning: Call ρ with parameter t.

A *rule definition* for a rule name ρ is an expression $\rho(x) = R$, where R is a transition rule in which there are no free occurrences of variables except of x. The calling convention is *lazy.* This means that in a call $\rho(t)$ the variable x is replaced in the body R of the rule by the parameter t. The parameter t is not evaluated in the state where the rule is called but only later when it is used in the body (maybe in different states due to sequential compositions). Call-by-value evaluation of rule calls can be simulated as follows:

$$\rho(y) = \textbf{let } x = y \textbf{ in } R$$

Then upon calling $\rho(t)$ the parameter t is evaluated in the same state.

Definition 1 (ASM). An *abstract state machine* M consists of a vocabulary Σ, an initial state $\mathfrak{A}$ for Σ, a rule definition for each rule name, and a distinguished rule name of arity zero called the *main rule name* of the machine.

The semantics of transition rules is given by sets of updates. Since due to the parallelism (in the Block and the Forall rules), a transition rule may prescribe to update the same function at the same arguments several times, such updates are required to be consistent. The concept of consistent update sets is made more precise by the following definitions.

Definition 2 (Update). An *update* for $\mathfrak{A}$ is a triple $\langle f, a, b\rangle$, where f is a dynamic function name, and a and b are elements of $|\mathfrak{A}|$.

The meaning of the update is that the interpretation of the function f in $\mathfrak{A}$ has to be changed at the argument a to the value b. The pair of the first two components of an update is called a *location.* An update specifies how the function table of a dynamic function has to be updated at the corresponding location. An *update set* is a set of updates.

Definition 3 (Consistent update set). An update set U is called *consistent,* if it satisfies the following property: If $\langle f, a, b\rangle \in U$ and $\langle f, a, c\rangle \in U$, then $b = c$.

This means that a consistent update set contains for each function and each argument at most one value. If an update set U is consistent, it can be fired in a given state. The result is a new state in which the interpretations of dynamic function names are changed according to U. The interpretations of static function names are the same as in the old state.

Definition 4 (Firing of updates). The result of firing a consistent update set U in a state $\mathfrak{A}$ is a new state $U(\mathfrak{A})$ with the same universe as $\mathfrak{A}$ satisfying the following two conditions for the interpretations of function names f of Σ:

Table 1. The semantics of formulas.

$$[\![s = t]\!]^{\mathfrak{A}}_{\zeta} := \begin{cases} \textit{true}, & \text{if } [\![s]\!]^{\mathfrak{A}}_{\zeta} = [\![t]\!]^{\mathfrak{A}}_{\zeta}; \\ \textit{false}, & \text{otherwise.} \end{cases}$$

$$[\![\neg\varphi]\!]^{\mathfrak{A}}_{\zeta} := \begin{cases} \textit{true}, & \text{if } [\![\varphi]\!]^{\mathfrak{A}}_{\zeta} = \textit{false}; \\ \textit{false}, & \text{otherwise.} \end{cases}$$

$$[\![\varphi \wedge \psi]\!]^{\mathfrak{A}}_{\zeta} := \begin{cases} \textit{true}, & \text{if } [\![\varphi]\!]^{\mathfrak{A}}_{\zeta} = \textit{true} \text{ and } [\![\psi]\!]^{\mathfrak{A}}_{\zeta} = \textit{true}; \\ \textit{false}, & \text{otherwise.} \end{cases}$$

$$[\![\varphi \vee \psi]\!]^{\mathfrak{A}}_{\zeta} := \begin{cases} \textit{true}, & \text{if } [\![\varphi]\!]^{\mathfrak{A}}_{\zeta} = \textit{true} \text{ or } [\![\psi]\!]^{\mathfrak{A}}_{\zeta} = \textit{true}; \\ \textit{false}, & \text{otherwise.} \end{cases}$$

$$[\![\varphi \to \psi]\!]^{\mathfrak{A}}_{\zeta} := \begin{cases} \textit{true}, & \text{if } [\![\varphi]\!]^{\mathfrak{A}}_{\zeta} = \textit{false} \text{ or } [\![\psi]\!]^{\mathfrak{A}}_{\zeta} = \textit{true}; \\ \textit{false}, & \text{otherwise.} \end{cases}$$

$$[\![\forall x\, \varphi]\!]^{\mathfrak{A}}_{\zeta} := \begin{cases} \textit{true}, & \text{if } [\![\varphi]\!]^{\mathfrak{A}}_{\zeta\frac{a}{x}} = \textit{true} \text{ for all } a \in |\mathfrak{A}|; \\ \textit{false}, & \text{otherwise.} \end{cases}$$

$$[\![\exists x\, \varphi]\!]^{\mathfrak{A}}_{\zeta} := \begin{cases} \textit{true}, & \text{if there exists an } a \in |\mathfrak{A}| \text{ with } [\![\varphi]\!]^{\mathfrak{A}}_{\zeta\frac{a}{x}} = \textit{true}; \\ \textit{false}, & \text{otherwise.} \end{cases}$$

1. If $\langle f, a, b\rangle \in U$, then $f^{U(\mathfrak{A})}(a) = b$.
2. If there is no b with $\langle f, a, b\rangle \in U$ or if f is static, then $f^{U(\mathfrak{A})}(a) = f^{\mathfrak{A}}(a)$.

Since U is consistent, the state $U(\mathfrak{A})$ is determined in a unique way. Notice that only those locations can have a new value in state $U(\mathfrak{A})$ with respect to state $\mathfrak{A}$ for which there is an update in U.

The composition '$U\,;V$' of two update sets U and V is defined such that the following equation is true for any state $\mathfrak{A}$:

$$(U\,;V)(\mathfrak{A}) = V(U(\mathfrak{A}))$$

The equation says that applying the update set '$U\,;V$' to state $\mathfrak{A}$ should be the same as first applying U and then V. Hence, '$U\,;V$' is the set of updates obtained from U by adding the updates of V and overwriting updates in U which are redefined in V. If U and V are consistent, then $U\,;V$ is consistent, too.

Definition 5 (Composition of update sets). The composition of two update sets U and V is defined by $U\,;V := \{\langle f, a, b\rangle \in U \mid \neg\exists c\,\langle f, a, c\rangle \in V\} \cup V$.

In a given state, a transition rule of an ASM produces for each variable assignment an update set. Since the rule can contain recursive calls to other rules, it is also possible that the rule does not terminate and has no semantics at all. Therefore, the semantics of ASM rules is given by an inductive definition of a predicate $\mathcal{S}(R, \mathfrak{A}, \zeta, U)$ with the meaning 'rule R yields in state $\mathfrak{A}$ under the variable assignment ζ the update set U.' Instead of $\mathcal{S}(R, \mathfrak{A}, \zeta, U)$ we write $[\![R]\!]^{\mathfrak{A}}_{\zeta} \rhd U$

Table 2. Inductive definition of the semantics of ASM rules.

$\dfrac{}{[\![\mathbf{skip}]\!]^{\mathfrak{A}}_{\zeta} \rhd \emptyset}$		(skip)
$\dfrac{}{[\![f(s) := t]\!]^{\mathfrak{A}}_{\zeta} \rhd \{\langle f, a, b\rangle\}}$	if $a = [\![s]\!]^{\mathfrak{A}}_{\zeta}$ and $b = [\![t]\!]^{\mathfrak{A}}_{\zeta}$	(upd)
$\dfrac{[\![R]\!]^{\mathfrak{A}}_{\zeta} \rhd U \qquad [\![S]\!]^{\mathfrak{A}}_{\zeta} \rhd V}{[\![R\ S]\!]^{\mathfrak{A}}_{\zeta} \rhd U \cup V}$		(par)
$\dfrac{[\![R]\!]^{\mathfrak{A}}_{\zeta} \rhd U}{[\![\mathbf{if}\ \varphi\ \mathbf{then}\ R\ \mathbf{else}\ S]\!]^{\mathfrak{A}}_{\zeta} \rhd U}$	if $[\![\varphi]\!]^{\mathfrak{A}}_{\zeta} = \mathit{true}$	(if$_1$)
$\dfrac{[\![S]\!]^{\mathfrak{A}}_{\zeta} \rhd U}{[\![\mathbf{if}\ \varphi\ \mathbf{then}\ R\ \mathbf{else}\ S]\!]^{\mathfrak{A}}_{\zeta} \rhd U}$	if $[\![\varphi]\!]^{\mathfrak{A}}_{\zeta} = \mathit{false}$	(if$_2$)
$\dfrac{[\![R]\!]^{\mathfrak{A}}_{\zeta\frac{a}{x}} \rhd U}{[\![\mathbf{let}\ x = t\ \mathbf{in}\ R]\!]^{\mathfrak{A}}_{\zeta} \rhd U}$	if $a = [\![t]\!]^{\mathfrak{A}}_{\zeta}$	(let)
$\dfrac{[\![R]\!]^{\mathfrak{A}}_{\zeta\frac{a}{x}} \rhd U_a \qquad \text{for each } a \in I}{[\![\mathbf{forall}\ x\ \mathbf{with}\ \varphi\ \mathbf{do}\ R]\!]^{\mathfrak{A}}_{\zeta} \rhd \bigcup_{i \in I} U_i}$	if $I = \{a \in \lvert\mathfrak{A}\rvert : [\![\varphi]\!]^{\mathfrak{A}}_{\zeta\frac{a}{x}} = \mathit{true}\}$	(forall)
$\dfrac{[\![R]\!]^{\mathfrak{A}}_{\zeta} \rhd U \qquad [\![S]\!]^{U(\mathfrak{A})}_{\zeta} \rhd V}{[\![R\,;S]\!]^{\mathfrak{A}}_{\zeta} \rhd U\,;V}$	if U is consistent	(seq$_1$)
$\dfrac{[\![R]\!]^{\mathfrak{A}}_{\zeta} \rhd U}{[\![R\,;S]\!]^{\mathfrak{A}}_{\zeta} \rhd U}$	if U is inconsistent	(seq$_2$)
$\dfrac{[\![R\frac{t}{x}]\!]^{\mathfrak{A}}_{\zeta} \rhd U}{[\![\rho(t)]\!]^{\mathfrak{A}}_{\zeta} \rhd U}$	if $\rho(x) = R$ is a rule definition	(def)

and present the inductive definition as a (possibly infinitary) calculus in Table 2. We say that a rule R is defined in state $\mathfrak{A}$ under the variable assignment ζ, if there exists an update set U such that $[\![R]\!]^{\mathfrak{A}}_{\zeta} \rhd U$ is derivable in the calculus in Table 2. Note that for each state $\mathfrak{A}$ and variable assignment ζ there exists at most one update set U such hat $[\![R]\!]^{\mathfrak{A}}_{\zeta} \rhd U$ is derivable in the calculus. Hence transition rules are deterministic.

The notion of ASM run is the classical notion of computation of transition systems. A computation step in a given state consists in executing *simultaneously* all updates of the main transition rule of the ASM, if these updates are consistent. The run stops if the main transition rule is not defined or yields an inconsistent update set. If the update set is empty, then the ASM produces an infinite run (stuttering, never changing anymore the state). We do not allow that so-called *monitored* functions change during a computation.

Definition 6. (Run of an ASM) Let M be an ASM with vocabulary Σ, initial state $\mathfrak{A}$ and main rule name ρ. Let ζ be a variable assignment. A *run* of M is a finite or infinite sequence $\mathfrak{B}_0, \mathfrak{B}_1, \ldots$ of states for Σ such that the following conditions are satisfied:

1. $\mathfrak{B}_0 = \mathfrak{A}$.
2. If $[\![\rho]\!]_\zeta^{\mathfrak{B}_n}$ is not defined or inconsistent, then $\mathfrak{B}_n$ is the last state.
3. Otherwise, $\mathfrak{B}_{n+1} = U(\mathfrak{B}_n)$, where $[\![\rho]\!]_\zeta^{\mathfrak{B}_n} \triangleright U$.

Runs are deterministic and independent of the variable assignment ζ, since we forbid global variables in rule definitions.

Remark 1. In the presence of sequential composition the following two rules are not equivalent:

$$\textbf{let } x = t \textbf{ in } R \quad \not\equiv \quad R\tfrac{t}{x}.$$

For a counter example, consider the following transition rule:

$$\textbf{let } x = f(0) \textbf{ in } (f(0) := 1\,;\, f(1) := x)$$

If we substitute the term $f(0)$ for x, then we obtain:

$$f(0) := 1\,;\, f(1) := f(0)$$

In general, the two rules are not the same, because $f(0)$ is evaluated in different states. The following substitution property, however, is true for *static* terms t: If t is static and $[\![t]\!]_\zeta^{\mathfrak{A}} = a$, then

$$[\![R]\!]_{\zeta\frac{a}{x}}^{\mathfrak{A}} \triangleright U \iff [\![R\tfrac{t}{x}]\!]_\zeta^{\mathfrak{A}} \triangleright U.$$

If the term t contains dynamic functions, then the equivalence is not necessarily true, because t could be evaluated in different states on the right-hand side (due to sequential compositions).

3 Formalizing the Consistency of ASMs

Following Groenboom and Renardel de Lavalette [4] we extend the language of first-order predicate logic by a modal operator $[R]$ for each rule R. The intended meaning of a formula $[R]\varphi$ is that the formula φ is true after firing R. More precisely, the formula $[R]\varphi$ is true iff one of the following conditions is satisfied:

1. R is not defined or the update set of R is inconsistent, or
2. R is defined, the update set of R is consistent and φ is true in the next state after firing the update set of R.

Equivalently we can say that the formula $[R]\varphi$ is true in state $\mathfrak{A}$ under the variable assignment ζ iff for each set U such that $[\![R]\!]_\zeta^{\mathfrak{A}} \triangleright U$ is derivable and U is consistent, the formula φ is true in the state $U(\mathfrak{A})$ under ζ (see Table 3).

In order to express the definedness and the consistency of transition rules we extend the set of formulas by atomic formulas $\mathrm{def}(R)$ and $\mathrm{upd}(R, f, x, y)$. The semantics of these formulas is defined in Table 3 and the basic properties are

Table 3. The semantics of modal formulas and basic predicates.

$$[\![[R]\varphi]\!]^{\mathfrak{A}}_{\zeta} := \begin{cases} \mathit{true}, & \text{if } [\![\varphi]\!]^{U(\mathfrak{A})}_{\zeta} = \mathit{true} \text{ for each consistent } U \text{ with } [\![R]\!]^{\mathfrak{A}}_{\zeta} \rhd U; \\ \mathit{false}, & \text{otherwise.} \end{cases}$$

$$[\![\mathrm{def}(R)]\!]^{\mathfrak{A}}_{\zeta} := \begin{cases} \mathit{true}, & \text{if there exists an update set } U \text{ with } [\![R]\!]^{\mathfrak{A}}_{\zeta} \rhd U; \\ \mathit{false}, & \text{otherwise.} \end{cases}$$

$$[\![\mathrm{upd}(R, f, s, t)]\!]^{\mathfrak{A}}_{\zeta} := \begin{cases} \mathit{true}, & \text{if ex. } U \text{ with } [\![R]\!]^{\mathfrak{A}}_{\zeta} \rhd U \text{ and } \langle f, [\![s]\!]^{\mathfrak{A}}_{\zeta}, [\![t]\!]^{\mathfrak{A}}_{\zeta} \rangle \in U; \\ \mathit{false}, & \text{otherwise.} \end{cases}$$

Table 4. Axioms for definedness.

D1. $\mathrm{def}(\mathbf{skip})$

D2. $\mathrm{def}(f(s) := t)$

D3. $\mathrm{def}(R\ S) \leftrightarrow \mathrm{def}(R) \wedge \mathrm{def}(S)$

D4. $\mathrm{def}(\mathbf{if}\ \varphi\ \mathbf{then}\ R\ \mathbf{else}\ S) \leftrightarrow (\varphi \wedge \mathrm{def}(R)) \vee (\neg\varphi \wedge \mathrm{def}(S))$

D5. $\mathrm{def}(\mathbf{let}\ x = t\ \mathbf{in}\ R) \leftrightarrow \exists x\,(x = t \wedge \mathrm{def}(R))$ if $x \notin \mathrm{FV}(t)$

D6. $\mathrm{def}(\mathbf{forall}\ x\ \mathbf{with}\ \varphi\ \mathbf{do}\ R) \leftrightarrow \forall x\,(\varphi \to \mathrm{def}(R))$

D7. $\mathrm{def}(R\,;S) \leftrightarrow \mathrm{def}(R) \wedge [R]\mathrm{def}(S)$

D8. $\mathrm{def}(\rho(t)) \leftrightarrow \mathrm{def}(R\frac{t}{x})$ if $\rho(x) = R$ is a rule definition of M

Table 5. Axioms for updates.

U1. $\neg\mathrm{upd}(\mathbf{skip}, f, x, y)$

U2. $\mathrm{upd}(f(s) := t, f, x, y) \leftrightarrow s = x \wedge t = y,\quad \neg\mathrm{upd}(f(s) := t, g, x, y)$ if $f \neq g$

U3. $\mathrm{upd}(R\ S, f, x, y) \leftrightarrow \mathrm{def}(R\ S) \wedge (\mathrm{upd}(R, f, x, y) \vee \mathrm{upd}(S, f, x, y))$

U4. $\mathrm{upd}(\mathbf{if}\ \varphi\ \mathbf{then}\ R\ \mathbf{else}\ S, f, x, y) \leftrightarrow (\varphi \wedge \mathrm{upd}(R, f, x, y)) \vee (\neg\varphi \wedge \mathrm{upd}(S, f, x, y))$

U5. $\mathrm{upd}(\mathbf{let}\ z = t\ \mathbf{in}\ R, f, x, y) \leftrightarrow \exists z\,(z = t \wedge \mathrm{upd}(R, f, x, y))$ if $z \notin \mathrm{FV}(t)$

U6. $\mathrm{upd}(\mathbf{forall}\ z\ \mathbf{with}\ \varphi\ \mathbf{do}\ R, f, x, y) \leftrightarrow$
$\mathrm{def}(\mathbf{forall}\ z\ \mathbf{with}\ \varphi\ \mathbf{do}\ R) \wedge \exists z\,(\varphi \wedge \mathrm{upd}(R, f, x, y))$

U7. $\mathrm{upd}(R\,; S, f, x, y) \leftrightarrow$
$(\mathrm{upd}(R, f, x, y) \wedge [R](\mathrm{def}(S) \wedge \mathrm{inv}(S, f, x))) \vee$
$(\mathrm{Con}(R) \wedge [R]\mathrm{upd}(S, f, x, y))$

U8. $\mathrm{upd}(\rho(t), f, x, y) \leftrightarrow \mathrm{upd}(R\frac{t}{z}, f, x, y)$ if $\rho(z) = R$ is a rule definition of M

listed in Tables 4 and 5. The formula $\mathrm{def}(R)$ expresses that the rule R is defined. The formula $\mathrm{upd}(R, f, x, y)$ expresses that rule R is defined and yields an update set which contains an update for f at x to y. The formula $\text{Con}(R)$ used in U7 to characterize an update of a sequential composition is defined as follows:

$$\text{Con}(R) := \text{def}(R) \wedge \bigwedge_{f \text{ dyn.}} \forall x, y, z\, (\text{upd}(R, f, x, y) \wedge \text{upd}(R, f, x, z) \rightarrow y = z)$$

It is true in a state iff the rule R is defined and yields a consistent update set:

$$[\![\text{Con}(R)]\!]_\zeta^{\mathfrak{A}} = \mathit{true} \iff \text{there exists a consistent } U \text{ with } [\![R]\!]_\zeta^{\mathfrak{A}} \triangleright U.$$

The formula $\text{inv}(R, f, x)$ in U7 expresses that the rule R does not update the function f at the argument x. It is a simple abbreviation defined as follows:

$$\text{inv}(R, f, x) := \forall y\, \neg\text{upd}(R, f, x, y)$$

Note, that it would be wrong to define the predicate $\text{upd}(R, f, x, y)$ by saying that $f(x)$ is different from y in the present state but equal to y in the next state after firing rule R:

$$\text{upd}(R, f, x, y) := f(x) \neq y \wedge [R] f(x) = y \qquad \text{(wrong definition)}$$

Using this definition, the predicate $\text{upd}(f(0) := 1, f, 0, 1)$ would be false in a state where $f(0)$ is equal to 1, although the rule $f(0) := 1$ does update the function f at the argument 0 to 1.

4 Basic Axioms and Rules of the Logic

The formulas of the logic for abstract state machines are generated by the following grammar:

$$\begin{aligned} \varphi, \psi \ ::= \ & s = t \mid \neg\varphi \mid \varphi \wedge \psi \mid \varphi \vee \psi \mid \varphi \rightarrow \psi \mid \forall x\, \varphi \mid \exists x\, \varphi \mid \\ & \text{def}(R) \mid \text{upd}(R, f, s, t) \mid [R]\varphi \end{aligned}$$

A formula is called *pure* (or *first-order*), if it contains neither the predicate 'def' nor 'upd' nor the modal operator $[R]$. A formula is called *static*, if it does not contain dynamic function names. The formulas used in If-Then-Else and Forall rules must be pure formulas.

The semantics of formulas is given by the definitions in Table 1 and Table 3. The equivalence $\varphi \leftrightarrow \psi$ is defined by $(\varphi \rightarrow \psi) \wedge (\psi \rightarrow \varphi)$. A formula φ is called *valid*, if $[\![\varphi]\!]_\zeta^{\mathfrak{A}} = \mathit{true}$ for all states $\mathfrak{A}$ and variable assignments ζ. The substitution of a term t for a variable x in a formula φ is denoted by $\varphi\frac{t}{x}$ and is defined as usual. Variables bound by a quantifier, a **let** or a **forall** have to be renamed when necessary. The substitution is also performed inside of transition rules that occur in formulas. The following substitution property holds.

Lemma 1 (Substitution). *If t is static and $a = [\![t]\!]_\zeta^{\mathfrak{A}}$, then $[\![\varphi]\!]_{\zeta\frac{a}{x}}^{\mathfrak{A}} = [\![\varphi\frac{t}{x}]\!]_\zeta^{\mathfrak{A}}$.*

We define two transition rules R and S to be *equivalent*, if they are equiconsistent and produce the same next state when they are fired.

Definition 7 (Equivalence). The formula $R \simeq S$ is defined as follows:

$$\begin{aligned} R \simeq S \;:\Longleftrightarrow\; & (\mathrm{Con}(R) \lor \mathrm{Con}(S)) \to \mathrm{Con}(R) \land \mathrm{Con}(S) \land \\ & \bigwedge\!\!\!\bigwedge_{f\ \mathrm{dyn.}} \forall x, y\, (\mathrm{upd}(R, f, x, y) \to \mathrm{upd}(S, f, x, y) \lor f(x) = y) \land \\ & \bigwedge\!\!\!\bigwedge_{f\ \mathrm{dyn.}} \forall x, y\, (\mathrm{upd}(S, f, x, y) \to \mathrm{upd}(R, f, x, y) \lor f(x) = y) \end{aligned}$$

The formula $R \simeq S$ has the intended meaning:

Lemma 2. *The formula $R \simeq S$ is true in $\mathfrak{A}$ under ζ iff the following two conditions are true:*

1. $[\![\mathrm{Con}(R)]\!]_\zeta^{\mathfrak{A}} = true$ *iff* $[\![\mathrm{Con}(S)]\!]_\zeta^{\mathfrak{A}} = true$.
2. *If* $[\![R]\!]_\zeta^{\mathfrak{A}} \rhd U$, $[\![S]\!]_\zeta^{\mathfrak{A}} \rhd V$ *and* U *and* V *are consistent, then* $U(\mathfrak{A}) = V(\mathfrak{A})$.

We already know that the axioms D1–D8 and U1–U8 are valid for a given abstract state machine M. Together with the following principles they will be the basic axioms and rules of our logic $\mathcal{L}(M)$.

I. Classical logic with equality: We use the axioms and rules of the classical predicate calculus with equality. The quantifier axioms, however, are restricted.

II. Restricted quantifier axioms:

1. $\forall x\, \varphi \to \varphi\frac{t}{x}$ if t is static or φ is pure
2. $\varphi\frac{t}{x} \to \exists x\, \varphi$ if t is static or φ is pure

III. Modal axioms and rules:

3. $[R](\varphi \to \psi) \land [R]\varphi \to [R]\psi$
4. $\dfrac{\varphi}{[R]\varphi}$
5. $\neg\mathrm{Con}(R) \to [R]\varphi$
6. $\neg[R]\varphi \to [R]\neg\varphi$

IV. The Barcan axiom:

7. $\forall x[R]\varphi \to [R]\forall x \varphi$, if $x \notin \mathrm{FV}(R)$.

V. Axioms for pure static formulas:

8. $\varphi \to [R]\varphi$ if φ is pure and static
9. $\mathrm{Con}(R) \land [R]\varphi \to \varphi$ if φ is pure and static

VI. Axioms for def **and** upd**:**

10. D1–D8 in Table 4
11. U1–U8 in Table 5

VII. Update axioms for transition rules:

12. $\mathrm{upd}(R, f, x, y) \to \mathrm{def}(R)$
13. $\mathrm{upd}(R, f, x, y) \to [R]f(x) = y$
14. $\mathrm{inv}(R, f, x) \land f(x) = y \to [R]f(x) = y$

VIII. Extensionality axiom for transition rules:

15. $R \simeq S \rightarrow ([R]\varphi \leftrightarrow [S]\varphi)$

IX. Axioms from dynamic logic:

16. $[\mathbf{skip}]\varphi \leftrightarrow \varphi$
17. $[R\,;S]\varphi \leftrightarrow [R][S]\varphi$

The principles I–IX are valid, therefore the logic is sound.

Theorem 1 (Soundness). *If a formula is derivable with the axioms and rules I–IX, then it is valid.*

The formula $\forall x\,\varphi \rightarrow \varphi\frac{t}{x}$ is not valid for non-static terms t. Consider the following tautology:

$$\forall x\,(x = 0 \rightarrow [f(0) := 1]x = 0).$$

If we substitute the term $f(0)$ for x, then we obtain the formula

$$f(0) = 0 \rightarrow [f(0) := 1]f(0) = 0.$$

This formula is not valid. Hence, the quantifier axioms must be restricted.

Lemma 3. *The following consistency properties are derivable:*

18. $\mathrm{Con}(\mathbf{skip})$
19. $\mathrm{Con}(f(s) := t)$
20. $\mathrm{Con}(R\ S) \leftrightarrow \mathrm{Con}(R) \wedge \mathrm{Con}(S) \wedge \mathrm{joinable}(R, S)$
21. $\mathrm{Con}(\mathbf{if}\ \varphi\ \mathbf{then}\ R\ \mathbf{else}\ S) \leftrightarrow (\varphi \wedge \mathrm{Con}(R)) \vee (\neg\varphi \wedge \mathrm{Con}(S))$
22. $\mathrm{Con}(\mathbf{let}\ x = t\ \mathbf{in}\ R) \leftrightarrow \exists x\,(x = t \wedge \mathrm{Con}(R))$ *if* $x \notin \mathrm{FV}(t)$
23. $\mathrm{Con}(\mathbf{forall}\,x\,\mathbf{with}\,\varphi\,\mathbf{do}\,R) \leftrightarrow \forall x\,(\varphi \rightarrow \mathrm{Con}(R) \wedge \forall y(\varphi\frac{y}{x} \rightarrow \mathrm{joinable}(R, R\frac{y}{x}))$
24. $\mathrm{Con}(R\,;S) \leftrightarrow \mathrm{Con}(R) \wedge [R]\mathrm{Con}(S)$
25. $\mathrm{Con}(\rho(t)) \leftrightarrow \mathrm{Con}(R\frac{t}{x})$ *if* $\rho(x) = R$ *is a rule definition of* M

The predicate joinable(R, S) which is used in 20 to reduce the consistency of a parallel composition $R\ S$ into consistency properties of R and S is defined as follows (where x, y, z are not free in R):

$$\mathrm{joinable}(R, S) := \bigwedge_{f\ \mathrm{dyn.}} \forall x, y, z\,(\mathrm{upd}(R, f, x, y) \wedge \mathrm{upd}(S, f, x, z) \rightarrow y = z),$$

It expresses that the update sets of R and S do not conflict. This means, whenever R and S both update a function f at the same argument x, then the new values of f at x are the same.

Lemma 4. *The following principles are derivable:*

26. $\mathrm{Con}(R) \wedge [R]f(x) = y \rightarrow \mathrm{upd}(R, f, x, y) \vee (\mathrm{inv}(R, f, x) \wedge f(x) = y)$
27. $\mathrm{Con}(R) \wedge [R]\varphi \rightarrow \neg[R]\neg\varphi$
28. $[R]\exists x\,\varphi \leftrightarrow \exists x\,[R]\varphi,$ *if* $x \notin \mathrm{FV}(R)$.

Groenboom and Renardel de Lavalette introduce in [4] different axioms for transition rules. Their axioms FM1, FM2, AX1, AX2 are derivable in our system using the update axioms 13 and 14.

Lemma 5. *The following principles of [4] are derivable:*

29. $s = x \to (y = t \leftrightarrow [f(s) := t]f(x) = y)$
30. $s \neq x \to (y = f(x) \leftrightarrow [f(s) := t]f(x) = y)$
31. $[R]f(x) = y \wedge [S]f(x) = y \to [R\ S]f(x) = y$
32. $f(x) \neq y \wedge ([R]f(x) = y \vee [S]f(x) = y) \to [R\ S]f(x) = y.$

The following inverse implication of 31 and 32 is not mentioned in [4] (maybe because of the lack of a consistency notion), but is derivable in our system:

$$\begin{aligned}&\mathrm{Con}(R\ S) \wedge [R\ S]f(x) = y \to \\ &\quad ([R]f(x) = y \wedge [S]f(x) = y) \vee (f(x) \neq y \wedge ([R]f(x) = y \vee [S]f(x) = y))\end{aligned}$$

Several principles known from dynamic logic are derivable using the extensionality axiom 15.

Lemma 6. *The following principles are derivable:*

33. $[\textbf{if}\ \varphi\ \textbf{then}\ R\ \textbf{else}\ S]\psi \leftrightarrow (\varphi \wedge [R]\psi) \vee (\neg\varphi \wedge [S]\psi)$
34. $[\textbf{let}\ x = t\ \textbf{in}\ R]\varphi \leftrightarrow \exists x\,(x = t \wedge [R]\varphi),$ *if* $x \notin \mathrm{FV}(t) \cup \mathrm{FV}(\varphi)$.
35. $[\rho(t)]\varphi \leftrightarrow [R\frac{t}{x}]\varphi,$ *if* $\rho(x) = R$ *is a rule definition of* M.

Schönegge uses in his sequent calculus for the extended dynamic logic in [11] new rules that express the commutativity, the associativity and similar properties of the parallel combination of transition rules. In our system, these properties are derivable.

Lemma 7. *The following principles of [11] are derivable:*

36. $(R\ \textbf{skip}) \simeq R$
37. $(R\ S) \simeq (S\ R)$
38. $((R\ S)\ T) \simeq (R\ (S\ T))$
39. $(R\ R) \simeq R$
40. $(\textbf{if}\ \varphi\ \textbf{then}\ R\ \textbf{else}\ S)\ T \simeq \textbf{if}\ \varphi\ \textbf{then}\ (R\ T)\ \textbf{else}\ (S\ T)$
41. $T\ (\textbf{if}\ \varphi\ \textbf{then}\ R\ \textbf{else}\ S) \simeq \textbf{if}\ \varphi\ \textbf{then}\ (T\ R)\ \textbf{else}\ (T\ S)$

If we can derive $R \simeq S$, then we immediately obtain the principle $[R]\varphi \leftrightarrow [S]\varphi$ using the extensionality axiom 15. It is not clear to us, whether for example the commutativity of the parallel composition, $[R\ S]\varphi \leftrightarrow [S\ R]\varphi$, could be derived in the formal system of [4].

Lemma 8. *The following properties of the sequential composition are derivable:*

42. $(R\,;\textbf{skip}) \simeq R$
43. $(\textbf{skip}\,; R) \simeq R$
44. $((R\,; S)\,; T) \simeq (R\,; (S\,; T))$
45. $(\textbf{if}\ \varphi\ \textbf{then}\ R\ \textbf{else}\ S)\,; T \simeq \textbf{if}\ \varphi\ \textbf{then}\ (R\,; T)\ \textbf{else}\ (S\,; T)$

The dynamic logic with array assignments (see [6]) uses a substitution principle which is derivable in our system. Let φ be a pure (first-order) formula. Then by $\varphi\frac{t}{f(s)}$ we denote the formula which is obtained in the following way. First, φ is transformed into an equivalent formula

$$\varphi \leftrightarrow \exists \boldsymbol{x}\, \exists \boldsymbol{y} \left(\bigwedge_{i=1}^{n} f(x_i) = y_i \wedge \psi \right),$$

where $\boldsymbol{x} = x_1, \ldots, x_n$, $\boldsymbol{y} = y_1, \ldots, y_n$ and ψ does not contain f. Then we define:

$$\varphi\frac{t}{f(s)} := \exists \boldsymbol{x}\, \exists \boldsymbol{y} \left(\bigwedge_{i=1}^{n} ((x_i = s \wedge y_i = t) \vee (x_i \neq s \wedge f(x_i) = y_i)) \wedge \psi \right)$$

Lemma 9. *For any first-order formula φ, the following substitution principle is derivable: $\varphi\frac{t}{f(s)} \leftrightarrow [f(s) := t]\varphi$.*

The If-Then rule can be defined in terms of If-Then-Else in the standard way:

if φ **then** $R :=$ **if** φ **then** R **else skip**

An ASM is called *simple*, if it is defined by a single rule R, which has the following form:

if φ_1 **then** $f(s_1) := t_1$
if φ_2 **then** $f(s_2) := t_2$
⋮
if φ_n **then** $f(s_n) := t_n$

Simple ASMs have the obvious properties formulated in the following lemma. Property 49 is a variant of the basic axiom of Poetzsch-Heffter [8]. It can easily be extended to disjoint If-Then rules with simultaneous function updates.

Lemma 10. *Let R be the rule of a simple AMS. Then,*

46. $\mathrm{Con}(R) \leftrightarrow \bigwedge_{i<j} (\varphi_i \wedge \varphi_j \wedge s_i = s_j \to t_i = t_j)$

47. $\mathrm{upd}(R, f, x, y) \leftrightarrow \bigvee_{i=1}^{n} (\varphi_i \wedge x = s_i \wedge y = t_i)$

48. $\mathrm{inv}(R, f, x) \leftrightarrow \bigwedge_{i=1}^{n} (\varphi_i \to x \neq s_i)$

49. $\bigvee_{i=1}^{n} \varphi_i \wedge \bigwedge_{i<j} \neg(\varphi_i \wedge \varphi_j) \wedge \bigwedge_{i=1}^{n} (\varphi_i \to \psi\frac{t_i}{f(s_i)}) \to [R]\psi,$ *if ψ is first-order.*

Iteration can be reduced to recursion. We can define the While rule recursively, as follows:

while φ **do** $R =$ **if** φ **then** $(R\,;$ **while** φ **do** $R)$

The expression **while** φ **do** R has to be read as a rule call $\rho(\boldsymbol{x})$, where $\boldsymbol{x}$ are the free variables of φ and R. So the above equation stands for the following rule definition:

$\rho(\boldsymbol{x}) =$ **if** φ **then** $(R\,; \rho(\boldsymbol{x}))$

Lemma 11. *The following properties of the While rule are derivable:*
50. $\mathrm{Con}(\textbf{while } \varphi \textbf{ do } R) \leftrightarrow (\varphi \rightarrow \mathrm{Con}(R) \wedge [R]\mathrm{Con}(\textbf{while } \varphi \textbf{ do } R))$
51. $[\textbf{while } \varphi \textbf{ do } R]\psi \leftrightarrow (\varphi \wedge [R][\textbf{while } \varphi \textbf{ do } R]\psi) \vee (\neg\varphi \wedge \psi)$

Several properties of ASMs can be expressed using the basic logic (where M is the distinguished rule name of the ASM and φ_{init} is a formula characterizing initial states):

- The formula ψ ensures consistency: $(\varphi_{\mathrm{init}} \rightarrow \psi) \wedge (\psi \rightarrow \mathrm{Con}(M) \wedge [M]\psi)$.
- The formula ψ is an invariant: $(\varphi_{\mathrm{init}} \rightarrow \psi) \wedge (\psi \rightarrow [M]\psi)$.

The statement in [13] for the correctness of the compiler from Java to the JVM can be formulated as follows:

$$(\varphi_{\mathrm{init}} \rightarrow \varphi_{\mathrm{eqv}}) \wedge (\varphi_{\mathrm{eqv}} \rightarrow [J](\varphi_{\mathrm{eqv}} \vee [V]\varphi_{\mathrm{eqv}} \vee [V][V]\varphi_{\mathrm{eqv}} \vee [V][V][V]\varphi_{\mathrm{eqv}}))$$

Here, J is an ASM that specifies the semantics of a Java source level program according to the *Java Language Specification*; V is an ASM that specifies the *Java Virtual Machine.* The two ASMs have disjoint dynamic function names and use the same static functions. The formula φ_{eqv} expresses that two dynamic states of the two ASMs are equivalent for a given Java program and its compiled bytecode program. The above formula says, that if two states are equivalent, then for each step of J the ASM V has to make zero, one, two, or three steps to reach an equivalent state again. The proof in [13] which comprises 83 cases could be carried out in the basic system with appropriate structural induction principles for lists and abstract syntax trees (which are encoded using static functions).

5 Completeness for Hierarchical ASMs

An ASM is called *hierarchical*, if the call graph of the rule definitions does not contain cycles, in other words, if the ASM does not contain recursive rule definitions. An ASM is hierarchical iff it is possible to assign levels to the rule names such that in a rule definition $\rho(x) = R$ the levels of rule names in R are less than the level of ρ. Transition rules of hierarchical ASMs are always defined. If R is a transition rule which uses rules from a hierarchical machine M, then $\mathrm{def}(R)$ is derivable in $\mathcal{L}(M)$.

Theorem 2 (Completeness). *If M is a hierarchical ASM and φ is valid, then φ is derivable in $\mathcal{L}(M)$.*

The proof of the completeness theorem follows the traditional Henkin-style completeness proof and uses the fact that for a maximal consistent set Φ of formulas, if $\mathrm{Con}(R) \in \Phi$, then the set $\{\psi \mid [R]\psi \in \Phi\}$ is also maximal consistent. The extensionality axiom 15, Axiom 16 for **skip** and Axiom 17 for the sequential compositions are not used in the completeness proof. Since the axioms are valid, by the completeness theorem they must be derivable for hierarchical ASMs.

Acknowledgment

We are grateful to Egon Börger for helpful comments on an earlier version of the article. Because of his remarks we decided not to use Con(R), upd(R, f, x, y) and inv(R, f, x) as basic notions of the logic, but def(R) and upd(R, f, x, y) instead.

References

1. E. Börger and J. Huggins. Abstract State Machines 1988–1998: Commented ASM Bibliography. *Bulletin Europ. Assoc. Theoret. Comp. Science*, 64:105–127, 1998. Updated bibliography available at `http://www.eecs.umich.edu/gasm`.
2. E. Börger and J. Schmid. Composition and submachine concepts. In P. Clote and H. Schwichtenberg, editors, *Computer Science Logic (CSL 2000)*, pages 41–60. Springer-Verlag, Lecture Notes in Computer Science 1862, 2000.
3. A. Gargantini and E. Riccobene. Encoding abstract state machines in PVS. In Y. Gurevich, P. Kutter, M. Odersky, and L. Thiele, editors, *Abstract State Machines: Theory and Applications*, pages 303–322. Springer-Verlag, Lecture Notes in Computer Science 1912, 2000.
4. R. Groenboom and G. R. Renardel de Lavalette. A formalization of evolving algebras. In *Proceedings of Accolade 95*. Dutch Research School in Logic, 1995.
5. Y. Gurevich. Evolving algebras 1993: Lipari guide. In E. Börger, editor, *Specification and Validation Methods*, pages 9–36. Oxford University Press, 1993.
6. D. Harel. Dynamic logic. In D. M. Gabbay and F. Guenthner, editors, *Handbook of Philosophical Logic*, pages 497–604. Reidel, Dordrecht, 1983.
7. D. Harel, D. Kozen, and J. Tiuryn. *Dynamic Logic*. The MIT Press, 2000.
8. A. Poetzsch-Heffter. Deriving partial correctness logics from evolving algebras. In B. Pehrson and I. Simon, editors, *IFIP 13th World Computer Congress*, volume I: Technology/Foundations, pages 434–439, Elsevier, Amsterdam, 1994.
9. G. Schellhorn. *Verification of Abstract State Machines*. PhD thesis, Universität Ulm, 1999.
10. G. Schellhorn and W. Ahrendt. Reasoning about abstract state machines: the WAM case study. *J. of Universal Computer Science*, 3(4):377–413, 1997.
11. A. Schönegge. Extending dynamic logic for reasoning about evolving algebras. Technical Report IRATR–1995–49, Universität Karlsruhe, Institut für Logik, Komplexität und Deduktionssysteme, 1995.
12. N. Shankar. Symbolic analysis of transition systems. In Y. Gurevich, P. Kutter, M. Odersky, and L. Thiele, editors, *Abstract State Machines: Theory and Applications*, pages 287–302. Springer-Verlag, Lecture Notes in Computer Science 1912, 2000.
13. R. F. Stärk, J. Schmid, and E. Börger. *Java and the Java Virtual Machine—Definition, Verification, Validation*. Springer-Verlag, 2001.
14. J. van Benthem and J. A. Bergstra. Logic of transition systems. *J. of Logic, Language, and Information*, 3(4):247–283, 1995.

Constrained Hyper Tableaux

Jan van Eijck

CWI and ILLC, Amsterdam, Uil-OTS, Utrecht
jve@cwi.nl

Abstract. Hyper tableau reasoning is a version of clausal form tableau reasoning where all negative literals in a clause are resolved away in a single inference step. Constrained hyper tableaux are a generalization of hyper tableaux, where branch closing substitutions, from the point of view of model generation, give rise to constraints on satisfying assignments for the branch. These variable constraints eliminate the need for the awkward 'purifying substitutions' of hyper tableaux. The paper presents a non-destructive and proof confluent calculus for constrained hyper tableaux, together with a soundness and completeness proof, with completeness based on a new way to generate models from open tableaux. It is pointed out that the variable constraint approach applies to free variable tableau reasoning in general.

1 Introduction

Hyper tableau reasoning was introduced in [2]; like (positive) hyper resolution [9] it resolves away all negative literals of a clause in a single inference step, but it combines this with the notion of a tableau style search for counterexamples. Hyper tableau reasoning, in the improved version proposed in [1], allows local universally quantified variables. The key element in hyper tableau reasoning, the use of purifying substitutions to get rid of variable distribution over different head literals (or, in the improved version, the generation of proper clause instantiations by means of a *Link* rule) is replaced in constrained hyper tableau reasoning by the generation of constraints on the interpretation of the variables that get distributed. Constrained hyper tableaux solve the problem of model generation from open tableaux with free variables in a general way.

2 Basic Definitions

Language. Let Σ be a first order signature. A $\mathcal{L}_\Sigma$ literal is an $\mathcal{L}_\Sigma$ atom or its negation, and an $\mathcal{L}_\Sigma$ clause is a multiset of $\mathcal{L}_\Sigma$ literals, written as $\neg A_1 \vee \cdots \vee \neg A_m \vee B_1 \vee \cdots \vee B_n$ $(m, n \geq 0)$. If $m, n > 0$ the clause is *mixed*; if $m = 0, n > 0$ the clause is *positive*; if $m > 0, n = 0$ the clause is *negative*, and if $m = n = 0$ the clause is *empty*. A mixed clause $\neg A_1 \vee \cdots \vee \neg A_m \vee B_1 \vee \cdots \vee B_n$ may be written as $A_1 \wedge \cdots \wedge A_m \Rightarrow B_1 \vee \cdots \vee B_n$, and a negative clause as $\neg(A_1 \wedge \cdots \wedge A_m)$. The empty clause is written as $\bot$. We write $\top$ for the formula that is always true.

L. Fribourg (Ed.): CSL 2001, LNCS 2142, pp. 232–246, 2001.

Substitutions. A *substitution* σ is a function $V \to T_\Sigma$ that makes only a finite number of changes, i.e., σ has the property that $\mathrm{dom}\,(\sigma) = \{v \in V \mid \sigma(v) \neq v\}$ is finite. We use ϵ for the substitution with domain $\emptyset$ (the identity substitution). We represent a substitution σ in the standard way, as a list $\{v_1 \mapsto \sigma(v_1), \ldots, v_n \mapsto \sigma(v_n)\}$, where $\{v_1, \ldots, v_n\}$ is $\mathrm{dom}\,(\sigma)$. Write substitution application in post-fix notation, and write $\sigma\theta$ for 'θ after σ'.

If σ, θ are substitutions, then $\sigma \preceq \theta$ if σ is less general than θ, i.e., if there is a ρ with $\sigma = \theta\rho$. The relation $\preceq$ is a pre-order (transitive and reflexive), and its poset reflection is a partial order. For this, put $\sigma \sim \theta$ if $\sigma \preceq \theta$ and $\theta \preceq \sigma$, and consider substitutions modulo renaming, i.e., put $|\sigma| = \{\theta \mid \sigma \sim \theta\}$, and put $|\sigma| \sqsubseteq |\theta|$ if $\sigma \preceq \theta$. A renaming is a substitution that is a bijection on the set of variables. For convenience we continue to write σ for $|\sigma|$.

Extend the set of substitutions (modulo renaming) with the improper substitution $\perp\!\!\!\perp$, the substitution with the property that $\perp\!\!\!\perp \sqsubseteq \sigma$ for every substitution σ. Now for every pair of substitutions σ and θ, $\sigma \sqcap \theta$, the greatest common instance of σ and θ, and $\sigma \sqcup \theta$, the least common generalization of σ and θ, exist. If $\sigma \sqcap \theta = \perp\!\!\!\perp$ we say that σ and θ do not unify. We get that ϵ, the substitution that is more general than any, is the top of the lattice given by $\sqsubseteq$, and $\perp\!\!\!\perp$ its bottom. The grounding substitutions are the least general proper substitutions; In the lattice of substitutions, they are just above $\perp\!\!\!\perp$. Note that this hinges on the fact that substitutions have finite domains. If $\sigma \sqsubseteq \rho$, and $\sigma \neq \perp\!\!\!\perp$, we call σ an instance of ρ. A clause ϕ is a proper instance of a clause ψ if for some substitution σ that is not a renaming it is the case that $\phi = \psi\sigma$.

A variable map is a function in $V \to T_\Sigma$ (i.e., we drop the finite domain restriction of substitutions). Variable maps modulo renaming form a *complete lattice* under the 'less general than' ordering. A grounding is a variable map that maps every variable to a closed term.

Substitutions as Formulas; Variable Constraints. Associate with a substitution

$$\sigma = \{v_1 \mapsto \sigma(v_1), \ldots, v_n \mapsto \sigma(v_n)\}$$

the formula $v_1 \approx \sigma(v_1) \wedge \cdots \wedge v_n \approx \sigma(v_n)$. We can then say what it means that assignment α satisfies substitution σ in model $\mathcal{M}$ in the usual way. Notation $\mathcal{M} \models_\alpha \sigma$. A *variable constraint* is the negation of a substitution as formula, i.e., a variable constraint is a multiset of inequalities $v \not\approx t$, with $t \in T_\Sigma$, written as $v_1 \not\approx t_1 \vee \cdots \vee v_n \not\approx t_n$. From a substitution σ we derive a variable constraint $\overline{\sigma}$ by complementation, as follows:

$$\overline{\sigma} = \bigvee\{v \not\approx \sigma(v) \mid v \in \mathrm{dom}\,(\sigma)\}.$$

E.g., the complement $\overline{\sigma}$ of $\sigma = \{x \mapsto a, y \mapsto b\}$ is $x \not\approx a \vee y \not\approx b$. Note that $\overline{\epsilon} = \perp$.

Tableaux, Branches. A hyper tableau over Σ is a finitely branching tree with nodes labeled by positive $\mathcal{L}_\Sigma$ literals, or by variable constraints. A branch in a tableau $\boldsymbol{T}$ is a maximal path in $\boldsymbol{T}$. We occasionally identify a branch $\boldsymbol{B}$ with the set of its atomic facts and constraints. The variables of a tableau branch are the variables that occur in a literal or a constraint along the branch. A variable v distributes over branches $\boldsymbol{B}, \boldsymbol{B}'$ if v occurs in constraints or literals on both sides of a split point, as follows:

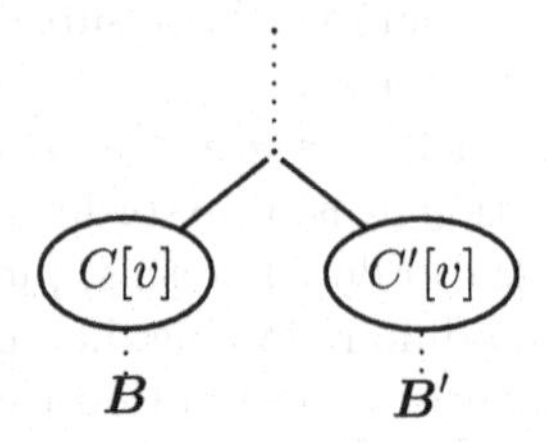

The rigid variables of a branch $\boldsymbol{B}$ are the variables of $\boldsymbol{B}$ that are distributed over $\boldsymbol{B}$ and some other branch. The tableau construction rules will ensure that every rigid variable in a tableau has a unique split point (highest point where it gets distributed).

A hyper tableau for a set Φ of $\mathcal{L}_\Sigma$ formulas in clause form is a finite or infinite tree grown according to the following instructions.

Initialize. Put $\top$ at the root node of the tableau.

Expand. Branches of a hyper tableau for clause set Φ are expanded by the only inference rule of Constrained Hyper Tableau (CHT) reasoning, the rule *Expand*, in the following manner.

$$\frac{C_1, \dots, C_m, \quad \neg A_1 \vee \dots \vee \neg A_m \vee B_1 \vee \dots \vee B_n}{B_1\sigma \quad | \quad \cdots \quad | \quad B_n\sigma \quad | \quad \overline{\theta}},$$

where

- $\neg A_1 \vee \dots \vee \neg A_m \vee B_1 \vee \dots \vee B_n$ is fresh copy of a clause in Φ (fresh with respect to the tableau),
- the C_i are positive literals from the current branch,
- σ is a most general substitution such that $A_i\sigma = C_i\sigma \quad (1 \leq i \leq m)$, and, moreover, σ *does not rename any rigid branch variables*,
- θ is the restriction of σ to the *rigid* variables of the branch.

An application of *Expand* to a branch expands the branch with an instance of a literal from the list $B_1, \dots, B_n$, or with a variable constraint.

Remark. It is convenient to use mgu's σ in *Expand* that do not rename any rigid branch variables. Suppose Px is a positive literal on a branch, with x rigid. Then a match with the rule $Py \Rightarrow Qy$ can rename either x or y. If x is renamed, the application of *Expand* branches, and two leafs are created, one with constraint $x \not\approx y$, the other with literal Qy. If x is not renamed, only a single leaf Qx is created, for in this case the constraint leaf extension carries constraint $\overline{\epsilon}$, and can be suppressed.

Here is an example application of *Expand.* Here and below, uppercase characters are used for predicates, $x, y, z, u, \ldots$ for variables, $a, b, c, \ldots$ for individual constants (skolem constants), $f, g \ldots$ for skolem functions. In the example, it is assumed that x is rigid and y is not:

$$\frac{Pxy, \quad Qb, \quad \neg Paz \vee \neg Qz \vee Raz}{Rab \quad | \quad x \not\approx a}.$$

Note that in the case of a positive clause, no branch literals are involved, and the substitution that is produced is ϵ, with corresponding constraint $\overline{\epsilon}$, i.e., $\perp$. In this case the rule boils down to:

$$\frac{B_1 \vee \cdots \vee B_n}{B_1 \mid \quad \cdots \quad \mid B_n}.$$

If there are no positive clauses $B_1 \vee \cdots \vee B_n$ in the clause set Φ, the set Φ cannot be refuted since in this case we can always build a model for Φ from just negative facts.

In case *Expand* is applied with a negative clause, the rule boils down to the following:

$$\frac{C_1, \ldots, C_m, \neg(A_1 \wedge \cdots \wedge A_m)}{\overline{\theta}},$$

where the C_i are as before, there is a most general σ such that $A_i\sigma = C_i\sigma$ $(1 \leq i \leq m)$ and no rigid variables get renamed, and θ is the restriction of σ to the rigid variables of the branch.

History Conditions on Expand. To avoid superfluous applications of *Expand*, a *history list* is kept of all clause instances that were applied to a branch. For this we need a preliminary definition. We say that a literal B reaches a k-fold in clause set Φ if either there is a clause in Φ in which the predicate of B has at least k negative occurrences, or there is a clause $\ldots B \ldots \Rightarrow \ldots C \ldots$ in Φ, and C reaches a k-fold in Φ. E.g., if $Qa \Rightarrow Pa, Px \wedge Py \Rightarrow Rxy$ in Φ, then Qx reaches a 2-fold in Φ. If Qx is also in Φ, we should generate two copies Qx', Qx'', which in turn will yield two copies of Pa, so that Raa can be derived.

If a clause is applied with substitution σ, the conditions on the application, in a tableau for clause set Φ, are:

1. $\neg A_1\sigma \vee \cdots \vee \neg A_m\sigma \vee B_1\sigma \vee \cdots \vee B_n\sigma$ is not a proper instance of any of the instances of $\neg A_1 \vee \cdots \vee \neg A_m \vee B_1 \vee \cdots \vee B_n$ that were applied to the branch before;
2. if $\neg A_1\sigma \vee \cdots \vee \neg A_m\sigma \vee B_1\sigma \vee \cdots \vee B_n\sigma$ is the k-th variant of any of the instances of $\neg A_1 \vee \cdots \vee \neg A_m \vee B_1 \vee \cdots \vee B_n$ that were applied to the branch before, then at least one of the B_i must reach a k-fold in Φ.

If these two conditions are fulfilled, we say that the instance of the clause is *fresh to the branch.* All clause instances used on a branch are kept in a branch history list. The *history conditions* on *Expand* are fair, for application of proper instances of previously applied clause instances to a branch is spurious, and generation of alphabetic variants only makes sense if they (eventually) lead to the generation of alphabetic variants that can be matched simultaneously against a single clause in Φ.

Constraint Merge for Closure. To check a tableau consisting of n branches for closure, apply the following *constraint merge for closure.* It is assumed that the $\overline{\sigma_i}$ are constraints on the different branches.

$$\frac{\overline{\sigma_1}, \quad \cdots \quad , \overline{\sigma_n}}{\text{closure by: } \sigma_1 \sqcap \cdots \sqcap \sigma_n} \sigma_1 \sqcap \cdots \sqcap \sigma_n \neq \perp\!\!\!\perp.$$

The idea of the constraint merge for closure is that if σ, θ each close a branch and can be unified, then $\sigma \sqcap \theta$ closes both branches, and so on, until the whole tableau is closed.

Open and Closed Tableaux. A hyper tableau is *open* if one of the following two conditions holds, otherwise it is *closed*:

- some branch in the tableau carries no constraint,
- all branches in the tableau carry constraints, but there is no way to pick constraints from individual branches and merge their corresponding substitutions into a single substitution (in the sense of: pick a finite initial stage $\boldsymbol{T}$, and pick $\overline{\sigma_i}$ on each B_i of $\boldsymbol{T}$ such that $\sigma_1 \sqcap \cdots \sqcap \sigma_n \neq \perp\!\!\!\perp$).

Fair Tableaux. A hyper tableau $\boldsymbol{T}$ for clause set Φ is *fair* if on every open branch B of $\boldsymbol{T}$, *Expand* is applied to each clause in Φ as many times as is compatible with the history conditions on the branch.

Tableau Bundles; Herbrand Universes for Open Tableaux. A pair of different branches in a tableau is *connected* if some variable distributes over the two branches. Since connectedness is symmetric, the reflexive transitive closure of this relation (connected*) is an equivalence. A tableau *bundle* is an equivalence class of connected* branches.

We will consider term models built from Herbrand universes of ground terms. The Herbrand universe of a bundle $\mathcal{B}$ in a tableau is the set of terms built from the skolem constants and functions that occur in $\mathcal{B}$, or, if no skolem constants are present, the set of terms built from the constant c and the skolem functions that occur in $\mathcal{B}$. If $\mathcal{B}$ contains no skolem functions and $\mathcal{B}$ is finite, the Herbrand universe of $\mathcal{B}$ is finite; if $\mathcal{B}$ contains skolem functions it is infinite. The models over such a Herbrand universe are completely specified by a set of ground positive literals. We use $H_{\mathcal{B}}$ for the Herbrand universe of $\mathcal{B}$, and we call a variable map σ with $\text{dom}(\sigma) = \text{vars}(\mathcal{B})$ and $\text{rng}(\sigma) \subseteq H_{\mathcal{B}}$ a *grounding* for $\mathcal{B}$ in $H_{\mathcal{B}}$, and a ground instance of a clause under a grounding for $\mathcal{B}$ in $H_{\mathcal{B}}$ an $H_{\mathcal{B}}$ instance. Note that a grounding need not be a substitution, as the set $\text{vars}(\mathcal{B})$ may be infinite.

3 Refutation Proof Examples

Let us agree on some conventions for tableau representation. To represent an application of extension in the tableau, we just have to write the rule instance $B_1\sigma \wedge \cdots \wedge B_n\sigma \Rightarrow A_1\sigma \vee \cdots \vee A_m\sigma$, and the branch extensions with the list of daughters $A_1\sigma$, ..., $A_m\sigma$, $\overline{\theta}$, as follows:

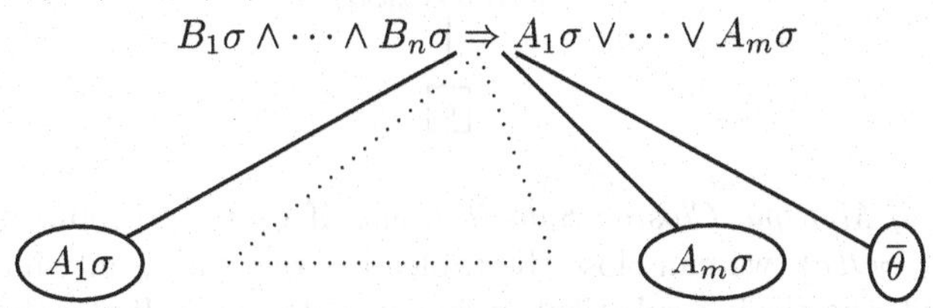

In case the constraint $\overline{\theta}$ that is generated is $\perp$, we suppress that leaf, unless it is the single leaf that closes the branch. If a constraint gives rise to a substitution that closes the whole tableau, then the substitution will be put in a box, like this (note that $\boxed{\theta}$ should be read as $\overline{\theta}$):

$$\neg(B_1\sigma \wedge \cdots \wedge B_n\sigma)$$

$$\boxed{\theta}$$

Reasoning about Relations. To prove that every transitive and irreflexive relation is asymmetric, we refute the clause form of its negation:

$$\{Rxy \wedge Ryz \Rightarrow Rxz, \neg Ruu, Rab, Rba\},$$

where the Rab, Rba provide the witnesses of non-asymmetry.

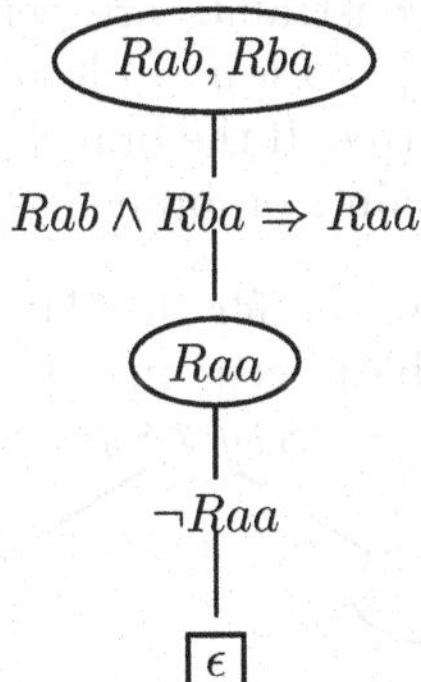

To apply the negative clause $\neg Ruu$, we use the substitution $\{u \mapsto a\}$. The restriction of that substitution to the rigid tableau variables is ϵ, so ϵ is the closing substitution of the tableau.

Closure by Renaming. To refute the clause set $\{Rxy, \neg Rab \vee \neg Rba\}$, two applications of *Expand* to the clause Rxy are needed. The second application uses fresh variables. Since none of the variables is distributed in the tableau, the closing substitution is ϵ.

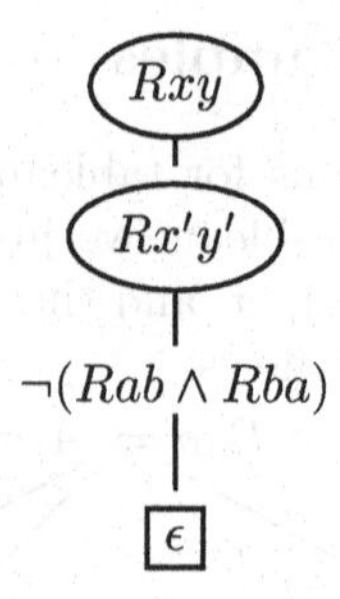

Generation of Multiple Closing Substitutions. If we try to refute the clause set $\{Oxy, \neg Oab, \neg Obc\}$, we can close the tableau in two ways, but since no variable is distributed, the closing substitution is ϵ in both cases. If the clauses are used to expand a tableau branch in which x and y are distributed, the following two constraints are generated on the branch.

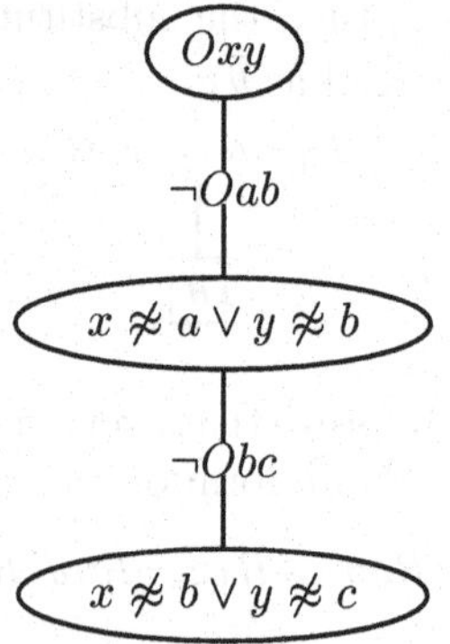

The order in which the constraints are generated does not matter. Both substitutions $\{x \mapsto a, y \mapsto b\}, \{x \mapsto b, y \mapsto c\}$ are candidates for use in the merge check for closure of the whole tree. If the branch is part of an open tableau, then both constraints act as constraints on branch satisfaction.

Closure by Merge. A hyper tableau for the clause set $\{Sxy \vee Syx, \neg Sab, \neg Sba\}$ has x, y rigid, so these variables occur in the constraints that are generated.

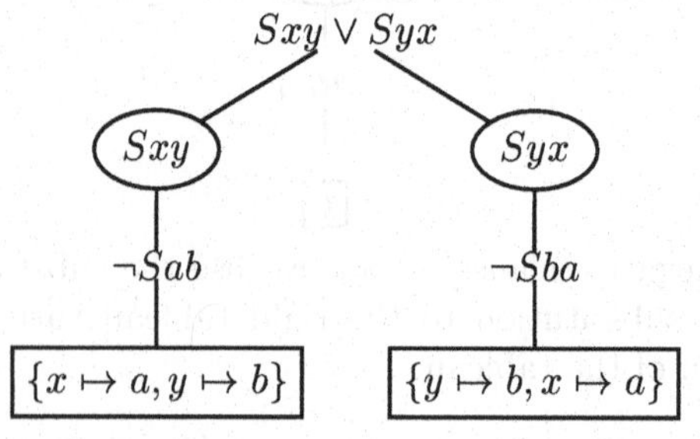

The substitutions unify (are, in fact, identical), so the tableau closes.

An AI Puzzle. If a is green, a is on top of b, b is on top of c, and c is not green, then there is a green object on top of an object that is not green. For a

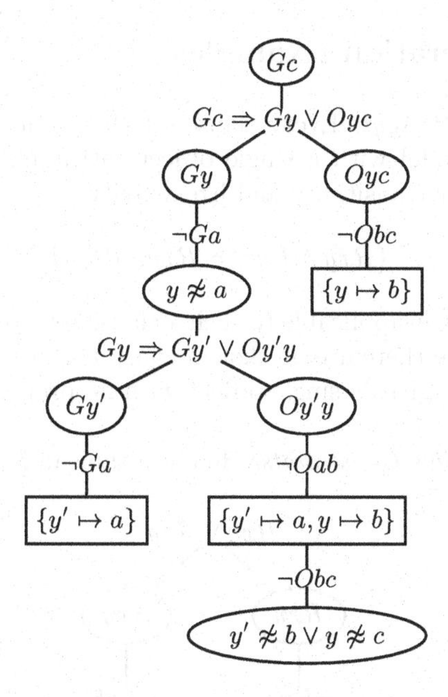

Fig. 1. Tableau for AI Puzzle

hyper tableau refutation proof, refute the clause form of the negation of this: $\{Ga, \neg Gc, Oab, Obc, Gx \wedge Oxy \Rightarrow Gy\}$. To make for a more interesting example, we swap positive and negative literals, as follows.

$$\{\neg Ga, Gc, \neg Oab, \neg Obc, Gx \Rightarrow Gy \vee Oyx\}.$$

This is not in the Horn fragment of FOL, so beyond Prolog (except through Horn renaming; for the present example, a Horn renaming is a swap of O and $\neg O$, and of G and $\neg G$). In the tableau for this example, in Fig. 1, note that when the rule $Gx \Rightarrow Gy \vee Oyx$ is used for the second time, its variables are first renamed. The variable y gets distributed at the first tableau split, the variable y' at the second split. The tableau of Figure 1 closes, for the substitution $\{y' \mapsto a, y \mapsto b\}$ closes every branch. This closing substitution is found by an attempt to merge closing substitutions of the individual branches. The branch constraints for which this works are boxed. Other possibilities fail. In particular, the substitution $\{y' \mapsto b, y \mapsto c\}$ closes the middle branch all right, but it clashes with both substitutions that close the left hand branch, either on the y or on the y' value, and with the substitution that closes the rightmost branch on the y value.

4 Model Generation Examples

No Positive Clause Present. Every clause set that contains no positive clauses is satisfiable in a model with a single object satisfying no atomic predicates. Example: a model for transitivity and irreflexivity.

$$\{Rxy \wedge Ryz \Rightarrow Rxy, \neg Ruu\}.$$

No hyper tableau rule is applicable to such a clause set, so we get no further than the top node $\top$. Since there are no skolem constants, we generate the Herbrand universe from c. This gives a single object with no properties.

Disjunctively Satisfiable Constraints. Here is a tableau for the clause set $\{Rxy \vee Sxy, \neg Rza, \neg Sub\}$:

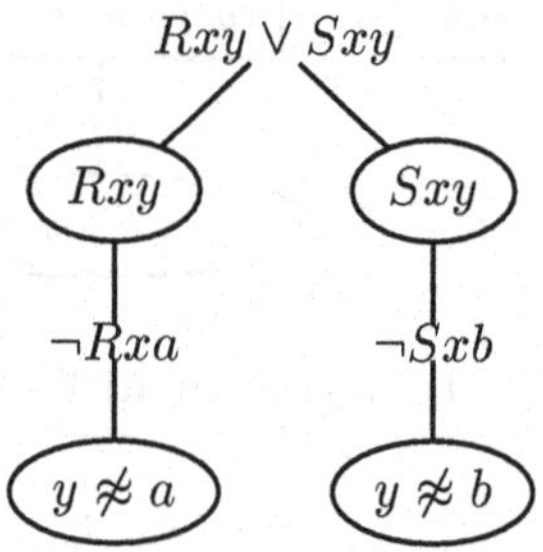

This tableau does not close, for the two substitutions disagree on the value for y. Or, put differently, the two constraints can be satisfied disjunctively. There are no further rule applications, so we have an open tableau. A model for the clause set is not generated by a single branch in this case, as the two branches share a constrained variable. The domain of a model generated from this tableau is the set of closed terms of the tableau, i.e., the set $\{a, b\}$. The set of groundings in this domain consists of $\theta_1 = \{x \mapsto a, y \mapsto a\}$, $\theta_2 = \{x \mapsto a, y \mapsto b\}$, $\theta_3 = \{x \mapsto b, y \mapsto a\}$, $\theta_4 = \{x \mapsto b, y \mapsto b\}$. θ_1 satisfies only the right branch, so it generates the fact Saa. θ_2 satisfies only the left branch, so it generates the fact Rab. θ_3 satisfies only the right branch, so it generates the fact Sba. Finally, θ_4 satisfies only the left branch, so it generates the fact Rbb. The model is given by the set of facts $\{Saa, Rab, Sba, Rbb\}$.

Infinitary Tableau Development. There are relations that are transitive and serial. The attempt to refute this combination of properties should lead to an open hyper tableau. In fact, the model that is generated for the clause set $\{Rxy \wedge Ryz \Rightarrow Rxz, Ruf(u)\}$ is infinite. The step from $Ruf(u)$ to $Rwf(w)$, in Fig. 2, is an application of *Expand* that generates an alphabetic variant. This agrees with the history condition, since there is a clause in the clause set with two negative R occurrences. The tableau will not close, and tableau development will not be stopped by the check on instantiations, for new instances of the rule $Rxy \wedge Ryz \Rightarrow Rxz$ will keep turning up. The corresponding model is isomorphic to $\mathbb{N}, <$. Although finite models for the clause set exist (a single reflexive point

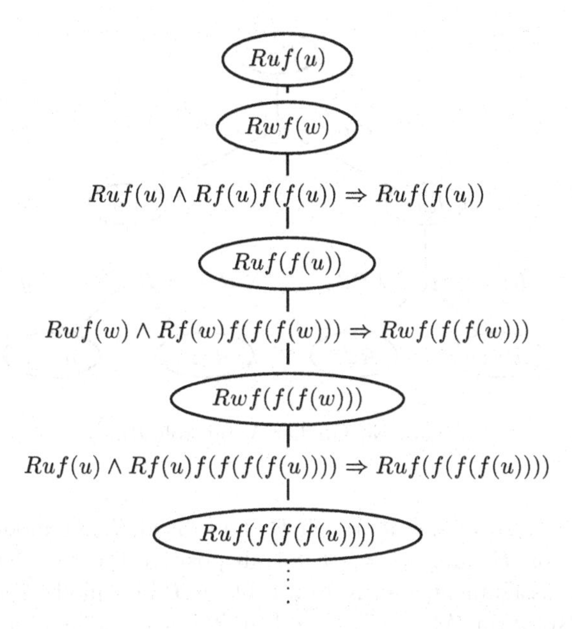

Fig. 2. Infinitary Tableau Development

also constitutes a model for this example) the calculus needs to be modified to generate them. For finite model generation, we need a slightly more sophisticated treatment of literals that introduce new skolem terms to a branch. This is beyond the scope of the present paper.

Open Tableau; No Further Rules Applicable. In the tableau for clause set $\{Rxy \Rightarrow Rxz \vee Rzy, Ruu\}$, given in Fig. 3, no further branch extensions are generated, as on all branches the next instance of $Rxy \Rightarrow Rxz \vee Rzy$ is a variant of an instance that has already been used on the branch. Generation of variants of the R predicate on a branch is spurious, because no clause in the clause set has more that a single negative occurrence of the R predicate. Note that variables are renamed in the second and the third application of the rule $Rxy \Rightarrow Rxz \vee Rzy$. Since there are no skolem constants, we generate the Herbrand model from a fresh c. This gives a model consisting of a single reflexive point, from any of the branches.

5 Soundness, Model Generation, Completeness

An assignment α in a model $\mathcal{M}$ meets a constraint $\overline{\sigma}$ if $\mathcal{M} \models_\alpha \overline{\sigma}$. Let $[\![\cdot]\!]^{\mathcal{M}}_\alpha$ give the term interpretation in the model with respect to α. Then we have:

Theorem 1. *$\mathcal{M} \models_\alpha \overline{\sigma}$ iff there is a $v \in \mathrm{dom}\,(\sigma)$ with $\alpha(v) \neq [\![v\sigma]\!]^{\mathcal{M}}_\alpha$.*

The idea of the constraints is to *forbid* certain variable interpretations!

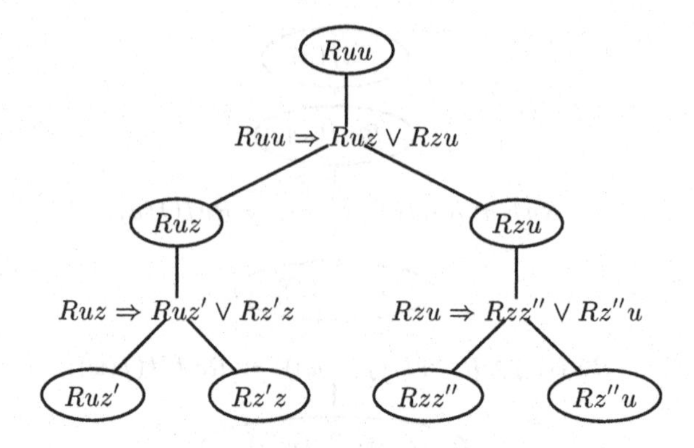

Fig. 3. No Further Applicable Rules

An assignment α satisfies a branch $\boldsymbol{B}$ of a tableau $\boldsymbol{T}$ in a model $\mathcal{M}$ if α meets all constraints on $\boldsymbol{B}$, and $\mathcal{M} \models_\alpha L$ for all positive literals L on $\boldsymbol{B}$. Notation: $\mathcal{M} \models_\alpha \boldsymbol{B}$. An assignment α satisfies a tableau $\boldsymbol{T}$ in a model $\mathcal{M}$ if α satisfies a branch of $\boldsymbol{T}$. Notation: $\mathcal{M} \models_\alpha \boldsymbol{T}$. A tableau $\boldsymbol{T}$ is satisfiable if for some model $\mathcal{M}$ it is the case that all assignments α for $\mathcal{M}$ satisfy $\boldsymbol{T}$ in $\mathcal{M}$. Notation: $\mathcal{M} \models \boldsymbol{T}$.

Theorem 2 (Satisfiability). *If Φ is a satisfiable set of clauses, then any tableau for Φ is satisfiable.*

Proof. Let $\boldsymbol{T}$ be a tableau for Φ. Since a tableau for Φ is a any tree grown from the seed $\top$ with the rule *Expand*, either there is a finite tableau sequence $\boldsymbol{T}_1, \dots, \boldsymbol{T}_n = \boldsymbol{T}$, or there is an infinite sequence $\boldsymbol{T}_1, \dots$, with $\boldsymbol{T} = \bigcup_{i=1}^{\infty} \boldsymbol{T}_i$. In any case, $\boldsymbol{T}_1$ consists of a single node $\top$, and $\boldsymbol{T}_{i+1}$ is constructed from $\boldsymbol{T}_i$ by an application of *Expand.* To prove by induction on n that a finite $\boldsymbol{T}$ is satisfiable, we have to check that satisfiability is preserved by each of these steps. Take some $\mathcal{M}$ with $\mathcal{M} \models \Phi$. Assume that $\mathcal{M} \models \boldsymbol{T}_i$, and $\boldsymbol{T}_{i+1}$ is the result of applying *Expand* to $\boldsymbol{T}_i$. Assume the branch to which *Expand* is applied is $\boldsymbol{B}$, the clause is $B_1 \wedge \cdots \wedge B_k \Rightarrow A_1 \vee \cdots \vee A_m$, the branch literals used in the rule are $C_1, \dots, C_k$, the matching substitution is σ, and the restriction of σ to the rigid branch variables is θ.

Consider an assignment α that satisfies $\boldsymbol{T}_i$ in $\mathcal{M}$. In case α satisfies a branch different from $\boldsymbol{B}$ then the application of *Expand* will not affect this, and α will satisfy $\boldsymbol{T}_{i+1}$ in $\mathcal{M}$. Suppose, therefore, that α satisfies only $\boldsymbol{B}$. We have to show that α satisfies at least one of the branch extensions, with $A_1\sigma$, with ..., with $A_m\sigma$, or with $\overline{\theta}$. From $\mathcal{M} \models \Phi$ we get that

$$\mathcal{M} \models B_1 \wedge \cdots \wedge B_k \Rightarrow A_1 \vee \cdots \vee A_m,$$

and therefore, since $B_i\sigma = C_i\sigma$,

$$\mathcal{M} \models C_1\sigma \wedge \cdots \wedge C_k\sigma \Rightarrow A_1\sigma \vee \cdots \vee A_m\sigma,$$

so in particular

$$\mathcal{M} \models_\alpha C_1\sigma \wedge \cdots \wedge C_k\sigma \Rightarrow A_1\sigma \vee \cdots \vee A_m\sigma.$$

In case $\mathcal{M} \models_\alpha C_1\sigma \wedge \cdots \wedge C_k\sigma$, it follows from the above that $\mathcal{M} \models_\alpha A_1\sigma \vee \cdots \vee A_m\sigma$, and we are done. In case $\mathcal{M} \not\models_\alpha C_1\sigma \wedge \cdots \wedge C_k\sigma$ we have to show that $\mathcal{M} \models_\alpha \overline{\theta}$. In this case, there is an i with $\mathcal{M} \models_\alpha C_i$ and $\mathcal{M} \not\models_\alpha C_i\sigma$. Let assignment α' be given by $\alpha'(v) = [\![v\sigma]\!]^{\mathcal{M}}_\alpha$. Then $\mathcal{M} \models_\alpha C_i$ and $\mathcal{M} \not\models_{\alpha'} C_i$. Thus $\mathcal{M} \not\models_{\alpha'} \boldsymbol{B}$, and by the satisfiability of $\boldsymbol{T}_i$, there has to be a $\boldsymbol{B}'$ with $\mathcal{M} \models_{\alpha'} \boldsymbol{B}'$. Since $\boldsymbol{B}$ is the only branch with $\mathcal{M} \models_\alpha \boldsymbol{B}$, $\mathcal{M} \not\models_\alpha \boldsymbol{B}'$. So there has to be a variable v that is both on $\boldsymbol{B}$ and $\boldsymbol{B}'$ with the property that $\alpha(v) \neq \alpha'(v)$. But this means that $v \in \text{dom}(\sigma)$ and v is rigid in $\boldsymbol{T}_i$. It follows that $v \in \text{dom}(\theta)$, and that α does meet $\overline{\theta}$, i.e., $\mathcal{M} \models_\alpha \overline{\theta}$.

Satisfiability in $\mathcal{M}$ for an infinite $\boldsymbol{T} = \bigcup_{i=1}^\infty \boldsymbol{T}_i$ follows from the fact that satisfiability in $\mathcal{M}$ is a universal property (it has the form 'for all literals and all constraints on the branch ... '), and is therefore by standard model-theoretic reasoning preserved under limit constructions. □

Theorem 3 (Merge). *If a hyper tableau $\boldsymbol{T}$ closes by constraint merge, then $\boldsymbol{T}$ is not satisfiable.*

Proof. If $\boldsymbol{T}$ closes by constraint merge then there is a way to pick a finite initial stage $\boldsymbol{T}'$ of $\boldsymbol{T}$, and pick constraints $\overline{\sigma_1}, \ldots, \overline{\sigma_n}$, one on each tableau branch of $\boldsymbol{T}'$, such that $\sigma_1 \sqcap \cdots \sqcap \sigma_n \neq \perp\!\!\!\perp$. Thus, there is a ground substitution θ with $\theta \sqsubseteq \sigma_1 \sqcap \cdots \sqcap \sigma_n$. Note that we can associate with each ground substitution an assignment in a model, as follows. If α is an assignment for $\mathcal{M}$, and θ is a ground substitution, then the assignment $\theta\alpha$ is given by $\theta\alpha(v) = [\![v\theta]\!]^{\mathcal{M}}_\alpha$. Thus, for any model $\mathcal{M}$ and any assignment α for $\mathcal{M}$ it will be the case that $\mathcal{M} \models_{\theta\alpha} \sigma_1 \sqcap \cdots \sqcap \sigma_n$. So for any $\mathcal{M}$ there is an assignment α' with $\mathcal{M} \models_{\alpha'} \sigma_1$ and ... and $\mathcal{M} \models_{\alpha'} \sigma_n$, i.e., with $\mathcal{M} \not\models_{\alpha'} \overline{\sigma_1}$ and ... and $\mathcal{M} \not\models_{\alpha'} \overline{\sigma_n}$. In other words, for any $\mathcal{M}$ there has to be an assignment that does not meet any of the constraints $\overline{\sigma_1}, \ldots, \overline{\sigma_n}$. □

Theorem 4 (Soundness). *If there is a hyper tableau refutation for a clause set Φ, then Φ is unsatisfiable.*

Proof. Immediate from the Satisfiability Theorem and the Merge Theorem. □

A variable map θ meets a constraint $\overline{\sigma}$ if $\theta \sqcap \sigma = \perp\!\!\!\perp$; θ is *compatible* with a branch $\boldsymbol{B}$ if θ meets all constraints $\overline{\sigma}$ on $\boldsymbol{B}$; θ is *compatible* with a tableau $\boldsymbol{T}$ if θ is compatible with at least one branch $\boldsymbol{B}$ of $\boldsymbol{T}$.

Theorem 5 (Compatibility). *If a tableau $\boldsymbol{T}$ is open, then every ground variable map θ for vars($\boldsymbol{T}$) is compatible with $\boldsymbol{T}$.*

Proof. Assume $\boldsymbol{T}$ consists of (open) branches $\{\boldsymbol{B}_i\}_{i\geq 0}$. We have to show that every grounding is compatible with at least one $\boldsymbol{B}_i$. Suppose θ is a grounding for vars($\boldsymbol{T}$) that is not compatible with any $\boldsymbol{B} \in \boldsymbol{T}$. Note that θ need not be a substitution, as the set vars($\boldsymbol{T}$) may be infinite. Then for each of the $\boldsymbol{B}_i$ there is a constraint $\overline{\sigma_i}$ on $\boldsymbol{B}_i$ such that $\sigma_i \sqcap \theta \neq \perp\!\!\!\perp$. Since θ is grounding, $\sigma_i \sqcap \theta = \theta$, i.e., $\theta \sqsubseteq \sigma_i$. Since variable maps modulo renaming form a complete lattice under $\sqsubseteq$, it follows that $\theta \sqsubseteq \sqcap_{(i \geq 0)} \sigma_i$. Now, since any tableau is finitely branching, and since any constraint is at finite distance from the root of the tableau, by König's lemma there has to be a *finite* set of constraints $\overline{\sigma_1}, \ldots, \overline{\sigma_n}$ with $\forall i \geq 0\ \exists j \leq n$ such that $\overline{\sigma_j}$ occurs on $\boldsymbol{B}_i$. But then $\sigma_1 \sqcap \cdots \sqcap \sigma_n \neq \perp\!\!\!\perp$, and contradiction with the assumption that $\boldsymbol{T}$ is open. □

Theorem 6 (Model Generation). *Every fair open tableau for Φ has a model $\mathcal{M}$ with $\mathcal{M} \models \Phi$.*

Proof. Since in a Herbrand universe groundings play the role of assignments, all we have to do to satisfy a tableau $\boldsymbol{T}$ in a Herbrand model is look at all the ground instances of the tableau. To generate a model from an open hyper tableau, proceed as follows. Pick an open bundle $\mathcal{B}$, and consider groundings for $\mathcal{B}$ in $H_{\mathcal{B}}$.

- If there is an unconstrained $\boldsymbol{B} \in \mathcal{B}$, the set of all $H_{\mathcal{B}}$ instances of the positive literals along $\boldsymbol{B}$ constitutes a model for the tableau. By the fairness of the tableau construction process, this model also satisfies Φ.
- If all branches in $\mathcal{B}$ are constrained, then generate $H_{\mathcal{B}}$ instances from groundings for $\mathcal{B}$ in $H_{\mathcal{B}}$, as follows. For every grounding θ for $\mathcal{B}$ in $H_{\mathcal{B}}$, we can pick, according to the Compatibility Theorem, a branch $\boldsymbol{B}$ in $\mathcal{B}$ that is compatible with θ. Collect the ground instances of the positive literals of $\boldsymbol{B}$. The union, for all groundings θ, of the sets of ground positive literals collected from branches compatible with θ, constitutes a model for the tableau. Again, by the fairness of the tableau construction process, the model also satisfies Φ. □

Theorem 7 (Completeness). *If a clause set Φ is unsatisfiable, then there exists a hyper tableau refutation for Φ.*

Proof. Immediate from the Model Generation Theorem. □

6 Fair Computation

A tableau calculus is *non-destructive* if all tableaux that can be constructed with the help of its rules from a given tableau $\boldsymbol{T}$ contain $\boldsymbol{T}$ as an initial sub-tree [6]. The usual versions of free variable tableaux are all destructive. Clearly, the CHT calculus is non-destructive. A tableau calculus is *proof confluent* if every tableau for an unsatisfiable clause set Φ can be expanded to a closed tableau [6]. Again, it is clear that the CHT calculus is proof-confluent.

Because of its non-destructiveness and proof-confluence, fair computation with constrained hyper tableaux is easy. We give a mere sketch. First apply

Expand to every all positive clause, on every branch. Next, use the list of positive literals on a branch to select the candidates from the clause set Φ for a match. For a given P on the branch and candidate clause ϕ, determine whether P needs to be copied for a match. If so, apply *Expand* again to generate the appropriate number of alphabetic variants. Next apply *Expand* to the mixed and the negative clauses. As the applications of *Expand* are non-destructive, no backtracking is ever needed in the merge check for closure.

As one of the referees pointed out, there is scope for further redundancy tests. E.g., for the clause set $\{Pa, Pb, Px \Rightarrow Q\}$ we would get Q twice on the branch. This can be avoided by adding a check like 'never expand a branch with a clause instance if it is already true in all models of the branch'. In the same spirit, if a branch is expanded with a literal A, but a proper instance $A\sigma$ is already present, then $A\sigma$ may be deleted from the branch, on condition of course that none of the variables in A are rigid.

We are experimenting with an implementation of CHT reasoning in Haskell [7], with merge checks for closure performed on tableau branches represented as lazy lists. Since the method is essentially breadth-first, space consumption is an issue, and it remains to be seen what the practical merits of the approach are.

7 Related Work

The standard reference for free variable reasoning in first order tableaux is [3]. With the introduction of free variables in tableaux, easy model generation from open tableaux got lost. Working with variable constraints in the manner explained above restores this delightful property of tableau reasoning.

The research for this paper was sparked off by a suggestion from [4] to do tableau proof search by merging closing substitutions for tableau branches into a closing substitution for the whole tableau. This suggestion is worked out in [5]. The difference between that approach and the present one is that we use disunification constraints rather than unification constraints. In our approach, the negations of the substitutions that close a branch are viewed as constraints on branch satisfiability, and a tableau remains open along as there is *no* way to unify a list of constraints selected from each branch. Since this is done in a setting where open branches only contain positive literals, it is ensured that a constraint can never clash with a branch literal.

The idea to enrich tableau branches with history lists for keeping track of the clause instances used in the construction of the branch is from [1]. As is mentioned there, this bookkeeping stratagem makes hyper tableaux a decision engine for satisfiability of the Bernays Schönfinkel class (relational $\exists^*\forall^*$ sentences without equality). The clause form of such sentences may have skolem constants, but since there are no skolem functions, any clause has only a finite number of instances. Thus, the history conditions ensure that tableau developments for Bernays Schönfinkel sentences are always finite.

One of the referees drew my attention to [8], an earlier proposal for handling rigid variables in a hyper tableau setting (with no variable constraints involved, however), for which completeness unfortunately remained open. The present paper settles this issue.

As far as I know, the idea to use constraints on the interpretation of rigid tableau variables for model generation from open free variable tableaux is new. This idea, by the way, applies to free variable tableau reasoning in general. Instead of using a closure rule that applies a most general unifying substitution σ for A and $\neg A'$ to a whole tableau, generate the constraint $\overline{\theta}$, where θ is the restriction of σ to the rigid variables of the current branch, and add a constraint merge for closure check: see [10] for details.

Acknowledgments

Many thanks to the members of the *Dynamo team* (Balder ten Cate, Juan Heguiabehere and Breanndán Ó Nualláin), to Bernhard Beckert, Martin Giese, Maarten Marx and Yde Venema, and to three anonymous referees.

References

1. BAUMGARTNER, P. Hyper tableaux — the next generation. In *Proceedings International Conference on Automated Reasoning with Analytic Tableaux and Related Methods* (1998), H. d. Swart, Ed., no. 1397 in Lecture Notes in Computer Science, Springer, pp. 60–76.
2. BAUMGARTNER, P., FRÖHLICH, P., AND NIEMELÄ, I. Hyper tableaux. In *Proceedings JELIA 96* (1996), Lecture Notes in Artificial Intelligence, Springer.
3. FITTING, M. *First-order Logic and Automated Theorem Proving; Second Edition.* Springer Verlag, Berlin, 1996.
4. GIESE, M. Proof search without backtracking using instance streams, position paper. In *Proc. Int. Workshop on First-Order Theorem Proving, St. Andrews, Scotland* (2000). Available online at `http://i12www.ira.uka.de/~key/doc/2000/giese00.ps.gz`.
5. GIESE, M. Incremental closure of free variable tableaux. In *IJCAR 2001 Proceedings* (2001).
6. HÄHNLE, R. Tableaux and related methods. In *Handbook of Automated Reasoning*, A. Robinson and A. Voronkov, Eds. Elsevier Science Publishers, to appear, 2001.
7. JONES, S. P., HUGHES, J., ET AL. Report on the programming language Haskell 98. Available from the Haskell homepage: `http://www.haskell.org`, 1999.
8. KÜHN, M. Rigid hypertableaux. In *KI-97: Advances in Artificial Intelligence* (1997), G. Brewka, C. Habel, and B. Nebel, Eds., pp. 87–98.
9. ROBINSON, J. Automated deduction with hyper-resolution. *International Journal of Computer Mathematics 1* (1965), 227–234.
10. VAN EIJCK, J. Model generation from constrained free variable tableaux. In *IJCAR 2001 - Short Papers* (Siena, 2001), R. Goré, A. Leitsch, and T. Nipkov, Eds., pp. 160–169.

Modal Logic and the Two-Variable Fragment

Carsten Lutz[1], Ulrike Sattler[1], and Frank Wolter[2]

[1] LuFG Theoretical Computer Science, RWTH Aachen, Ahornstr. 55, 52074 Aachen, Germay
[2] Institut für Informatik, Universität Leipzig, Augustus-Platz 10-11, 04109 Leipzig, Germany

Abstract. We introduce a modal language L which is obtained from standard modal logic by adding the difference operator and modal operators interpreted by boolean combinations and the converse of accessibility relations. It is proved that L has the same expressive power as the two-variable fragment FO^2 of first-order logic but speaks less succinctly about relational structures: if the number of relations is bounded, then L-satisfiability is ExpTime-complete but FO^2 satisfiability is NExpTime-complete. We indicate that the relation between L and FO^2 provides a general framework for comparing modal and temporal languages with first-order languages.

1 Introduction

Ever since it became common knowledge that many modal logics can be regarded as fragments of first-order logics, the exploration of the relation between those two families of languages has been a major research topic: Kamp's result [18] that modal logic with binary operators *Since* and *Until* has the same expressive power as monadic first-order logic over structures like $\langle \mathbb{N}, < \rangle$ and $\langle \mathbb{R}, < \rangle$ was the starting point. Van Benthem [27,28] provided a systematic model theoretic analysis of the relation between families of modal logics and predicate logics and Gabbay [10,9] extended Kamp's result to a systematic investigation of the possibilities of designing expressively complete modal logics. As part of his investigation Gabbay made the basic observation that often modal languages are contained in *finite variable fragments* of first-order logics. The basic modal language with unary operators only, for example, lies embedded even in the two-variable fragment FO^2 of first-order logic. In the early 1990s, this observation was regarded as an explanation for the decidability of many modal logics: the decidability of FO^2 (cf. [22,24,14]) explains the decidability of standard modal logics simply because they are contained in it.[1] The situation is different as soon as our concern is computational complexity: while most standard modal logics

[1] More recently it has been argued that some "modal phenomena" are better explained by their *tree-model-property* [29] (i.e., they are determined by tree-like structures) and/or by embedding them in bounded (or guarded) fragments of first-order logic [1,13]. The logics we consider here do not have those properties.

L. Fribourg (Ed.): CSL 2001, LNCS 2142, pp. 247–261, 2001.

are decidable in ExpTime or even PSpace and NP (see e.g. [19,2,26]), the two-variable fragment is NExpTime-complete [14]. A question naturally arising is why modal logics, in general, are of a lower complexity than the two-variable fragment. It is worth noting that this phenomenon is not due to the fact that modal logics have a fixed number of modal operators (alias accessibility relations interpreting them), whereas the two-variable fragments allows for arbitrarily many binary relations: Even without relation symbols of arity > 1, the two-variable fragment is NExpTime-hard [8,3]. There are two possible explanations for this phenomenon:

1. Explanation: any "standard modal logic" contained in FO^2 has strictly less expressive power than FO^2 itself, or

2. Explanation: although the expressive power of the two-variable logic coincides with the expressive power of a standard modal logic, the way it speaks about relational structures is strictly more succinct than the way modal languages do.

The main contribution of this paper is to show that (2.) is the case: to this end, we define a natural modal logic L, prove that it has the *same expressive power as* FO^2, and show that, as soon as we allow for a bounded number of relation symbols only, L-satisfiability is only ExpTime-complete. The logic L extends basic multi-modal logic $\mathbf{K}_m$ by means of taking the closure under forming (1) Boolean combinations of accessibility relations (2) the converse of accessibility relations, and (3) the identity relation. All those ingredients have been investigated and applied intensively: see [11,17,20] for (1), [5,12,31] for (2), and [6] for (3). Hence L can certainly be regarded as a standard member of the modal family.

The usefulness of our result, i.e., the expressive completeness of L for FO^2, is demonstrated by showing that it provides a *general framework* for comparing the expressive power and complexity of modal and first-order logic. For example, as soon as our concern are weak temporal logics (with the operators 'always in the future' and 'always in the past' only) interpreted over strict linear orderings, the Boolean operations and the identity relation become definable, which means that weak temporal logics have the same expressive power as FO^2 over strict linear orderings (and without further binary relation symbols). For the strict linear ordering $\langle \mathbb{N}, < \rangle$, this was first proved in [7]. In this case the complexity-gap is even wider: Over $\langle \mathbb{N}, < \rangle$, weak temporal logic is in NP [25] while FO^2 is NExpTime-complete [7,15]. In the present paper we show that this holds for $\langle \mathbb{Q}, < \rangle$ and $\langle \mathbb{R}, < \rangle$ as well.

2 Expressivity

We start with definitions of the languages under consideration. FO^2 comprises exactly those first-order formulas without constants and function symbols but with equality whose only variables are x and y and whose relation symbols have arity ≤ 2. The unary predicates are denoted by $P_1, \ldots$ while the binary ones are $R_1, \ldots$. For $m \leq \omega$ we denote by FO^2_m the fragment of FO^2 consisting of formulas containing only the first m binary relations. FO^2 is interpreted in the standard manner in structures of the form $\langle W, \mathcal{P}_1, \ldots, \mathcal{R}_1, \ldots \rangle$ in which the $\mathcal{P}_i$ interpret the P_i and the $\mathcal{R}_i$ interpret the R_i.

The modal language $\mathcal{ML}^{\neg,\cap,\cup,-,id}$ is Boolean modal logic [11,20] enriched with a converse constructor and the identity relation.

Definition 1. *A* complex modal parameter *is an expression built up from atomic modal parameters $R_1, \ldots$, the identity parameter id, and the operators $\neg, \cap, \cup$, and $\cdot^-$. For $m \leq \omega$ we denote by $\mathcal{ML}_m^{\neg,\cap,\cup,-,id}$ the modal language defined inductively as follows:*

- *all propositional variables $p_1, p_2, \ldots$ belong to $\mathcal{ML}_m^{\neg,\cap,\cup,-,id}$;*
- *if $\varphi, \psi \in \mathcal{ML}_m^{\neg,\cap,\cup,-,id}$ and S is a complex modal parameter built from the first m atomic modal parameters $R_1, \ldots, R_m$ and id, then $\neg\psi$, $\varphi \wedge \psi$, and $\langle S \rangle \varphi$ belong to $\mathcal{ML}_m^{\neg,\cap,\cup,-,id}$.*

We abbreviate $\top = p_1 \vee \neg p_1$ and $\bot = \neg\top$. The box operator $[S]\varphi$ and other Boolean connectives are defined as abbreviations in the standard manner.

A Kripke-model is a structure $\mathcal{M} = \langle W, \pi, \mathcal{R}_1, \ldots, \rangle$ in which π associates with every variable p a subset $\pi(p)$ of W. Let S be a (possibly complex) modal parameter. Then the extension $\mathcal{E}(S)$ is inductively defined as follows:

if $S = R_i$ (i.e., S is atomic)	then $\mathcal{E}(S) = \mathcal{R}_i$
if $S = id$	then $\mathcal{E}(S) = \{(w, w) \mid w \in W\}$
if $S = \neg S'$	then $\mathcal{E}(S) = (W \times W) \setminus \mathcal{E}(S')$
if $S = S_1 \cap S_2$	then $\mathcal{E}(S) = \mathcal{E}(S_1) \cap \mathcal{E}(S_2)$
if $S = S_1^-$	then $\mathcal{E}(S) = \{(w, w') \mid (w', w) \in \mathcal{E}(S_1)\}$

The semantics of formulas is defined inductively in the standard way, e.g. for the diamond operator we have

$$\mathcal{M}, w \models \langle S \rangle \varphi \text{ iff } \exists w' \in W \text{ with } (w, w') \in \mathcal{E}(S) \text{ and } \mathcal{M}, w' \models \varphi$$

Given a Kripke-model $\mathcal{M} = \langle W, \pi, \mathcal{R}_1, \ldots \rangle$, define a corresponding first-order model $\mathcal{M}_\sigma = \langle W, \mathcal{P}_1, \ldots, \mathcal{R}_1, \ldots \rangle$ by setting $\mathcal{P}_i = \pi(p_i)$.

We start our investigation of the relationship between FO_m^2 and $\mathcal{ML}_m^{\neg,\cap,\cup,-,id}$ by showing that these logics are equally expressive. If we write $\varphi(x)$, $\varphi(y)$ for formulas, we assume that at most the displayed variable occurs free in φ.

Theorem 1 (Expressive completeness for 2-variable-logic). *For every $\varphi \in \mathcal{ML}_m^{\neg,\cap,\cup,-,id}$ there exists a formula $\varphi^\sharp(x) \in FO_m^2$ whose length is linear in the length of φ such that the following holds for all Kripke-models $\mathcal{M}$ and all $a \in W$:*

$$\mathcal{M}, a \models \varphi \Leftrightarrow \mathcal{M}_\sigma \models \varphi^\sharp(a).$$

Conversely, given $\varphi(x) \in FO_m^2$ there exists a formula $\varphi^{\sigma_x} \in \mathcal{ML}_m^{\neg,\cap,\cup,-,id}$ whose length is exponential in the length of φ such that the following holds for all Kripke-models $\mathcal{M}$ and all $a \in W$:

$$\mathcal{M}, a \models \varphi^{\sigma_x} \Leftrightarrow \mathcal{M}_\sigma \models \varphi(a).$$

Proof. The proof of the first claim is standard [27,28], so we concentrate on the second one, whose proof is rather similar to the proof provided in [7] for temporal logics.

An FO^2-formula $\rho(x,y)$ is called a *binary atom* if it is an atom of the form $R_i(x,y)$, $R_i(y,x)$, or $x = y$. A *binary type* t for a formula ψ is a set of FO^2-formulas containing (i) either χ or $\neg\chi$ for each binary atom χ occurring in ψ, (ii) either $x = y$ or $x \neq y$, and (iii) no other formulas than these. The set of binary types for ψ is denoted by $\mathcal{R}_\psi$. A formula ξ is called a *unary atom* if it is of the form $R_i(x,x)$, $R_i(y,y)$, $A_i(x)$, or $A_i(y)$.

Let $\varphi(x) \in FO^2_m$. We assume $\varphi(x)$ is built using $\exists$, $\wedge$, and $\neg$ only. We inductively define two mappings $\cdot^{\sigma_x}$ and $\cdot^{\sigma_y}$ where the former one takes each FO^2_m-formula $\varphi(x)$ to the corresponding $\mathcal{ML}_m^{\neg,\cap,\cup,-,id}$-formula φ^{σ_x} and the latter does the same for FO^2_m-formulas $\varphi(y)$. We only give the details of $\cdot^{\sigma_x}$ since $\cdot^{\sigma_y}$ is defined analogously by switching the roles of x and y.

Case 1. If $\varphi(x) = P_i(x)$, then put $(\varphi(x))^{\sigma_x} = p_i$.

Case 2. If $\varphi(x) = R_i(x,x)$, then put $(\varphi(x))^{\sigma_x} = \langle id \cap R_i \rangle \top$.

Case 3. If $\varphi(x) = \chi_1 \wedge \chi_2$, then put, recursively, $(\varphi(x))^{\sigma_x} = \chi_1^{\sigma_x} \wedge \chi_2^{\sigma_x}$.

Case 4. If $\varphi(x) = \neg\chi$, then put, recursively, $(\varphi(x))^{\sigma_x} = \neg(\chi)^{\sigma_x}$.

Case 5. If $\varphi(x) = \exists y \chi(x,y)$, then $\chi(x,y)$ can clearly be written as

$$\chi(x,y) = \gamma[\rho_1, \dots, \rho_r, \gamma_1(x), \dots, \gamma_l(x), \xi_1(y), \dots, \xi_s(y)],$$

i.e., as a Boolean combination γ of ρ_i, $\gamma_i(x)$, and $\xi_i(y)$; the ρ_i are binary atoms; the $\gamma_i(x)$ are unary atoms or of the form $\exists y \gamma_i'$; and the $\xi_i(y)$ are unary atoms or of the form $\exists x \xi_i'$. We may assume that x occurs free in $\varphi(x)$. Our first step is to move all formulas without a free variable y out of the scope of $\exists$: obviously, $\varphi(x)$ is equivalent to

$$\bigvee_{\langle w_1,\dots,w_\ell\rangle \in \{\top,\bot\}^\ell} \Big(\bigwedge_{1 \le i \le \ell} (\gamma_i \leftrightarrow w_i) \wedge \exists y \gamma(\rho_1, \dots, \rho_r, w_1, \dots, w_l, \xi_1, \dots, \xi_s) \Big). \quad (1)$$

For every binary type $t \in \mathcal{R}_\varphi$ and binary atom α from φ, we have $t \models \alpha$ or $t \models \neg\alpha$—hence we can "guess" a binary type t and then replace all binary atoms by either true or false. For $t \in \mathcal{R}_\varphi$, let $\rho_i^t = \top$ if $t \models \rho_i$, and $\rho_i^t = \bot$, otherwise. Then $\varphi(x)$ is equivalent to

$$\begin{aligned} &\textstyle\bigvee_{\langle w_1,\dots,w_\ell\rangle \in \{\top,\bot\}^\ell} (\bigwedge_{1 \le i \le \ell} (\gamma_i \leftrightarrow w_i) \wedge \\ &\qquad \textstyle\bigvee_{t \in \mathcal{R}_\varphi} \exists y ((\bigwedge_{\alpha \in t} \alpha) \wedge \gamma(\rho_1^t, \dots, \rho_r^t, w_1, \dots, w_l, \xi_1, \dots, \xi_s))). \end{aligned} \quad (2)$$

Define, for every negated and unnegated binary atom α, a complex modal parameter α^{σ_x} as follows:

$$\begin{array}{ll} (x = y)^{\sigma_x} = id & (\neg(x = y))^{\sigma_x} = \neg id \\ (R_i(x,y))^{\sigma_x} = R_i & (\neg R_i(x,y))^{\sigma_x} = \neg R_i \\ (R_i(y,x))^{\sigma_x} = R_i^- & (\neg R_i(y,x))^{\sigma_x} = \neg R_i^- . \end{array}$$

Put, for every binary type $t \in \mathcal{R}_\varphi$, $t^{\sigma_x} = \bigcap_{\alpha \in t} \alpha^{\sigma_x}$. Now compute, recursively, $\gamma_i^{\sigma_x}$ and $\xi_i^{\sigma_y}$, and define $\varphi(x)^\sigma$ as

$$\bigvee_{\langle w_1,\ldots,w_\ell\rangle \in \{\top,\bot\}^\ell} \Big(\bigwedge_{1\le i\le \ell} (\gamma_i^{\sigma_x} \leftrightarrow w_i) \wedge \bigvee_{t\in\mathcal{R}_\varphi} \langle t^{\sigma_x}\rangle \gamma(\rho_1^t,\ldots,\rho_r^t, w_1,\ldots,w_l,\xi_1^{\sigma_y},\ldots,\xi_s^{\sigma_y})\Big).$$

❑

Note that φ^{σ_x} can be computed in polynomial time in the length of φ^{σ_x}. We should like to stress that the existence of formalisms with some 'modal flavour' and the same expressive power as FO^2 is known [4,10]. However, these formalisms have a number of purely technical constructs which did not find applications in modal or description logic. In [4], for example, Borgida constructs a counterpart L' of FO^2 in which accessibility relations $\mathcal{R}$ can be defined as products of extensions of formulas: for any two formulas φ_1, φ_2 one can form $\mathcal{R} = \{w \in W : w \models \varphi_1\} \times \{w \in W : w \models \varphi_2\}$. The expressive completeness result for L' becomes rather straightforward. In fact, the translation provided by Borgida is polynomial so that L' is speaking about relational structures as succinctly as FO^2 does.

3 Complexity

We show that, for $0 < m < \omega$, $\mathcal{ML}_m^{\neg,\cap,\cup,-,id}$-satisfiability is ExpTime-complete and hence in a lower complexity class than FO_m^2-satisfiability which is known to be NExpTime-complete [14].[2] Together with the expressivity result obtained in the previous section, this shows that FO_m^2 speaks about relational structures more succinctly than $\mathcal{ML}_m^{\neg,\cap,\cup,-,id}$ does (if ExpTime $\neq$ NExpTime, to be precise). The ExpTime lower bound for $\mathcal{ML}_m^{\neg,\cap,\cup,-,id}$-satisfiability is an immediate consequence of the fact that $\mathcal{ML}_m^{\neg}$ is ExpTime-hard even if $m = 1$ [20]. Hence, we concentrate on the upper bound. It is established by first (polynomially) reducing $\mathcal{ML}_m^{\neg,\cap,\cup,-,id}$-satisfiability to a certain variant of $\mathcal{ML}_k^{\neg id}$-satisfiability—$\mathcal{ML}_k^{\neg id}$ is multi-modal **K** enriched with the difference modality [6]—and then showing that this variant of $\mathcal{ML}_k^{\neg id}$-satisfiability can be decided in ExpTime.

3.1 Reducing $\mathcal{ML}_m^{\neg,\cap,\cup,-,id}$ to $\mathcal{ML}_{s,t,n}^{\neg id}$

In this section, we generally assume that $0 < m < \omega$. The following languages are used in the reduction:

Definition 2 (Languages). *(1) By $\mathcal{ML}_{s,t,n}^{\neg id}$ we denote the modal language $\mathcal{ML}_k^{\neg id}$ with $k = 2s + t + n$ modal parameters*

$$\mathfrak{P} = \{K_1,\ldots,K_s, I_1,\ldots,I_s, X_1,\ldots,X_t, Y_1,\ldots,Y_n\}$$

[2] Throughout this paper, we assume that ExpTime is defined as $\cup_{k\ge 0}\mathrm{DTIME}(2^{n^k})$ and NExpTime as $\cup_{k\ge 0}\mathrm{NTIME}(2^{n^k})$.

and the difference modality $\langle d \rangle$, *where* d *is an abbreviation for* $\neg id$.

(2) $\mathcal{ML}_m^{(\neg),\cap,-,id}$ *is* $\mathcal{ML}_m^{\neg,\cap,\cup,-,id}$ *with negation of modal parameters restricted to atomic modal parameters and without union of modal parameters.*

(3) By $\mathcal{ML}_{s,t,n}^{-}$ *we denote the modal language* $\mathcal{ML}_k^{-}$ *with converse and* $k = s+t+n$ *modal parameters* $K_1,\dots,K_s,X_1,\dots,X_t,Y_1,\dots,Y_n$.

Definition 3 (Semantics). *A structure*

$$\mathcal{M} = \langle W, \mathcal{K}_1, \dots \mathcal{K}_s, \mathcal{X}_1, \dots, \mathcal{X}_t, \mathcal{Y}_1, \dots, \mathcal{Y}_n \rangle$$

is called a c-frame *iff*

1. *the relations* $\mathcal{K}_i$ *are irreflexive and antisymmetric,*
2. *the relations* $\mathcal{X}_i$ *are irreflexive and symmetric,*
3. *the relations* $\mathcal{Y}_i$ *are subsets of* $\{(w,w) \mid w \in W\}$,
4. *for all* $w, w' \in W$ *with* $w \neq w'$, *there exists a unique*

$$\mathcal{S} \in \{\mathcal{K}_1, \dots, \mathcal{K}_s, \mathcal{K}_1^{-1}, \dots, \mathcal{K}_s^{-1}, \mathcal{X}_1, \dots, \mathcal{X}_t\}$$

   such that $(w, w') \in \mathcal{S}$, *and*

5. *for each* $w \in W$, *there exists a unique* i *with* $1 \leq i \leq n$ *such that* $(w,w) \in \mathcal{Y}_i$,

where $\mathcal{R}_i^{-1}$ *is used to denote the converse of a binary relation* $\mathcal{R}_i$. *An* $\mathcal{ML}_{s,t,n}^{-}$*-formula is called* c-satisfiable *iff it has a model which is based on a c-frame. Such a model is called a c-model.*

A structure $\mathcal{M} = \langle W, \mathcal{K}_1, \dots \mathcal{K}_s, \mathcal{I}_1, \dots, \mathcal{I}_s, \mathcal{X}_1, \dots, \mathcal{X}_t, \mathcal{Y}_1, \dots, \mathcal{Y}_n \rangle$ *is called an* s-frame *iff there exists a c-frame*

$$\mathcal{M}' = \langle W, \mathcal{K}_1', \dots \mathcal{K}_s', \mathcal{X}_1', \dots, \mathcal{X}_t', \mathcal{Y}_1', \dots, \mathcal{Y}_n' \rangle$$

such that $\mathcal{K}_i \subseteq \mathcal{K}_i'$, $\mathcal{I}_i \subseteq \mathcal{K}_i'^{-1}$, $\mathcal{X}_i \subseteq \mathcal{X}_i'$, *and* $\mathcal{Y}_i \subseteq \mathcal{Y}_i'$. *An* $\mathcal{ML}_{s,t,n}^{\neg id}$*-formula is called* s-satisfiable *iff it is satisfiable in a model based on an s-frame. Such a model is called an s-model.*

A *literal* is a modal parameter that matches one of the following descriptions:

- an atomic parameter or the negation thereof,
- the inverse of an atomic parameter or the negation thereof,
- the identity parameter or the negation of the identity parameter.

The reduction is comprised of a series of polynomial reduction steps. Let φ be a $\mathcal{ML}_m^{\neg,\cap,\cup,-,id}$-formula.

Step 1. Exhaustively apply the following rewrite rules to modal parameters in φ:

$$(\neg S)^- \rightsquigarrow \neg(S^-) \qquad (S_1 \cup S_2)^- \rightsquigarrow S_1^- \cup S_2^- \qquad id^- \rightsquigarrow id$$
$$S^{--} \rightsquigarrow S \qquad (S_1 \cap S_2)^- \rightsquigarrow S_1^- \cap S_2^- \qquad \neg id^- \rightsquigarrow \neg id$$

In the resulting formula φ_1, all modal parameters are Boolean combinations of literals.

Step 2. Convert all modal parameters in φ_1 to disjunctive normal form over literals using a truth table (as, e.g., described in [23], page 20). If the "empty disjunction" is obtained when converting a modal parameter S, then replace every occurrence of $\langle S\rangle\psi$ with $\bot$. Call the result of the conversion φ_2. The conversion can be done in linear time since the number m of atomic modal and we use a truth table for the conversion (instead of applying equivalences). It is easy to see that φ_2 is satisfiable iff φ_1 is satisfiable. Since the conversion to DNF was done using a truth table, each disjunct occurring in a modal parameter in φ_2 is a *relational type*, i.e., of the form $S_0 \cap S_1 \cap \cdots \cap S_m \cap S_1' \cap \cdots \cap S_m'$,where

1. $S_0 = id$ or $S_0 = \neg id$,
2. $S_i = R_i$ or $S_i = \neg R_i$ for $1 \leq i \leq m$, and
3. $S_i' = R_i^-$ or $S_i' = \neg(R_i^-)$ for $1 \leq i \leq m$.

Let $\Gamma_=$ be the set of all relational types with $S_0 = id$, $\Gamma_{\neq}$ be the set of all relational types with $S_0 = \neg id$, and $\Gamma = \Gamma_= \cup \Gamma_{\neq}$.

Step 3. We reduce satisfiability of $\mathcal{ML}_m^{\neg,\cap,\cup,-,id}$-formulas of the form of φ_2 (i.e, the modal parameters are disjunctions of relational types) to the satisfiability of $\mathcal{ML}_m^{(\neg),\cap,-,id}$-formulas in which all modal parameters are relational types. As the first step, recursively apply the following substitution to φ_2 from the inside to the outside (i.e., no union on modal parameters occurs in φ)

$$\langle S_1 \cup \cdots \cup S_k\rangle\, \varphi \leadsto \langle S_1\rangle\, p_\varphi \vee \cdots \vee \langle S_k\rangle\, p_\varphi$$

where p_φ is a new propositional variable. Call the result of these substitutions φ_2'. Secondly, define

$$\varphi_3 \quad := \quad \varphi_2' \wedge \bigwedge_{p_\varphi \text{ occurs in } \varphi_2'} \bigwedge_{S \in \Gamma} [S](p_\varphi \leftrightarrow \varphi).$$

φ_3 is an $\mathcal{ML}_m^{(\neg),\cap,-,id}$-formula as required.[3] Furthermore, φ_2 is satisfiable iff φ_3 is satisfiable, and the reduction is linear.

Step 4. It is not hard to see that the set $\Gamma_{\neq}$ (from Step 3) can be partitioned into three sets $\Gamma_{\neq}^s$, $\Gamma_{\neq}^1$, and $\Gamma_{\neq}^2$ such that there exists a bijection F from $\Gamma_{\neq}^1$ onto $\Gamma_{\neq}^2$ and, for every Kripke structure $\mathcal{M}$ with set of worlds W, and $w, w' \in W$, the following holds:

1. for all $S \in \Gamma_{\neq}^s$: $\mathcal{M}, (w, w') \models S$ iff $\mathcal{M}, (w', w) \models S$ and
2. for all $S \in \Gamma_{\neq}^1$: $\mathcal{M}, (w, w') \models S$ iff $\mathcal{M}, (w', w) \models F(S)$.

Given this, it is easy to reduce satisfiability of $\mathcal{ML}_m^{(\neg),\cap,-,id}$-formulas of the form of φ_3 to c-satisfiability of $\mathcal{ML}_{s,t,n}^{-}$-formulas, where $s = |\Gamma_{\neq}^1|$, $t = |\Gamma_{\neq}^s|$, and $n = |\Gamma_=|$. Let r be some bijection between $\Gamma_{\neq}^1$ and the set $\{K_1, \ldots, K_s\}$, r' some bijection between $\Gamma_{\neq}^s$ and the set $\{X_1, \ldots, X_t\}$, and r'' some bijection between $\Gamma_=$ and the set $\{Y_1, \ldots, Y_n\}$. The formula φ_4 is obtained from φ_3 by replacing

[3] We use $\bigwedge_{S\in\Gamma}[S](p_\varphi \leftrightarrow \varphi)$ instead of the more natural $[R](p_\varphi \leftrightarrow \varphi) \wedge [\neg R](p_\varphi \leftrightarrow \varphi)$ (for some atomic R) to ensure that all modal parameters in φ_3 are still relational types after the application of Step 3.

(1.) each element S of $\Gamma^1_{\neq}$ that appears in φ_3 with $r(S)$, (2.) each element S of $\Gamma^2_{\neq}$ with $r(S)^-$, (3.) each element S of $\Gamma^s_{\neq}$ with $r'(S)$, and (4.) each element S of $\Gamma_=$ with $r''(S)$. It can be proved that φ_3 is satisfiable iff φ_4 is c-satisfiable. Furthermore, the reduction is obviously linear.

Step 5. We reduce c-satisfiability of $\mathcal{ML}^-_{s,t,n}$-formulas to s-satisfiability of $\mathcal{ML}^{\neg id}_{s,t,n}$-formulas. W.l.o.g., we assume that φ_3 does not contain modal parameters of the form X_i^- and Y_i^-: since these parameters are interpreted by symmetric relations, X_i^- (resp. Y_i^-) can be replaced by X_i (resp. Y_i). For $\chi \in \mathcal{ML}^-_{s,t,n}$ (without X_i^- and Y_i^-), denote by $\chi^* \in \mathcal{ML}^{\neg id}_{s,t,n}$ the formula obtained from χ by replacing all occurrences of K_i^- with I_i.

For each $S \in \mathfrak{P}$ (see Definition 2), we use $S^\smile$ to denote (i) I_i if $S = K_i$, (ii) K_i if $S = I_i$, (iii) X_i if $S = X_i$, and (iv) Y_i if $S = Y_i$. For convenience, we define two more sets

$$\mathfrak{P}_1 = \{K_1, \ldots, K_s, I_1, \ldots, I_s, X_1, \ldots, X_t\} \text{ and } \mathfrak{P}_2 = \{Y_1, \ldots, Y_n\}.$$

Define φ_5 as the conjunction of φ_4^* with all formulas $\vartheta \wedge [d]\vartheta$, where ϑ can be obtained from the following formulas by replacing ψ and all ψ_S with subformulas of φ_4^*.

$$\chi_1 := \Big(\bigwedge_{S \in \mathfrak{P}_2} [S]\psi_S\Big) \rightarrow \Big(\bigvee_{S \in \mathfrak{P}_2} \psi_S\Big)$$

$$\chi_2 := \bigwedge_{\mathcal{P} \subseteq \mathfrak{P}_1} \Big[\Big(\bigwedge_{S \in \mathcal{P}} [S]\neg\psi_S \wedge \langle d\rangle\Big(\bigwedge_{S \in \mathcal{P}} \psi_S \wedge \bigwedge_{S \in \mathfrak{P}_1 \setminus \mathcal{P}} [S^\smile]\neg\psi_S\Big)\Big) \rightarrow \bigvee_{S \in \mathfrak{P}_1 \setminus \mathcal{P}} \neg\psi_S\Big)\Big]$$

$$\chi_3 := \bigwedge_{S \in \mathfrak{P}} \psi \rightarrow [S]\langle S^\smile\rangle\psi$$

Obviously, φ_5 is an $\mathcal{ML}^{\neg id}_{s,t,n}$-formula. The formula χ_1 deals with Item 5 from the definition of c-frames, χ_2 with Item 4, and χ_3 with symmetry from Item 2 and with the semantics of the converse operator. Note that the length of φ_5 is polynomial in the length $|\varphi_4|$ of φ_4 since the set of modal parameters is fixed.

Lemma 1. *φ_4 is c-satisfiable iff φ_5 is s-satisfiable.*

Proof: The "only if" direction is straightforward: Let

$$\mathcal{M} = \langle W, \pi, \mathcal{K}_1, \ldots, \mathcal{K}_s, \mathcal{X}_1, \ldots, \mathcal{X}_t, \mathcal{Y}_1, \ldots, \mathcal{Y}_n\rangle$$

be a c-model for φ_4. It is readily checked that

$$\mathcal{M}' = \langle W, \pi, \mathcal{K}_1, \ldots, \mathcal{K}_s, \mathcal{K}_1^{-1}, \ldots, \mathcal{K}_s^{-1}, \mathcal{X}_1, \ldots, \mathcal{X}_t, \mathcal{Y}_1, \ldots, \mathcal{Y}_n\rangle$$

is an s-frame and that the ϑ formulas from above are true in $\mathcal{M}'$. Hence, by the semantics of converse, $\mathcal{M}'$ is obviously a model for φ_5.

It remains to prove the "if" direction. Let

$$\mathcal{M} = \langle W, \pi, \mathcal{K}_1, \ldots, \mathcal{K}_s, \mathcal{I}_1, \ldots, \mathcal{I}_s, \mathcal{X}_1, \ldots, \mathcal{X}_t, \mathcal{Y}_1, \ldots, \mathcal{Y}_n\rangle$$

be an s-model for φ_5. In particular, this implies that all formulas derived from χ_1 to χ_3 are true in $\mathcal{M}$. Before we construct the c-model for φ_4, we prove two claims:

Claim 1. For each $w, w' \in W$ with $w \neq w'$, there exists an $S \in \mathfrak{P}_1$ such that, for all subformulas ψ of φ_4^*, we have that $\mathcal{M}, w \not\models \langle S \rangle \psi$ implies $\mathcal{M}, w' \not\models \psi$ and $\mathcal{M}, w' \not\models \langle S^\smile \rangle \psi$ implies $\mathcal{M}, w \not\models \psi$.

Proof: Assume that the claim does not hold. Fix $w, w' \in W$ with $w \neq w'$ that do not have the property from the claim. This means that, for each $S \in \mathfrak{P}_1$,

(i) there is a subformula ψ_S^1 of φ_4^* such that $\mathcal{M}, w \models [S]\neg\psi_S^1$ and $\mathcal{M}, w' \models \psi_S^1$ or

(ii) there is a subformula ψ_S^2 of φ_4^* such that $\mathcal{M}, w' \models [S^\smile]\neg\psi_S^2$ and $\mathcal{M}, w \models \psi_S^2$.

Let $\mathcal{P}$ be the subset of $\mathfrak{P}_1$ such that $S \in \mathcal{P}$ iff S satisfies (i) and let $\psi_S = \psi_S^1$ if $S \in \mathcal{P}$ and $\psi_S = \psi_S^2$ otherwise. Let ϑ be the instantiation of χ_2 with $\mathcal{P}$ and the ψ_S.[4] Since all formulas derived from χ_2 are true in $\mathcal{M}$, we have $\mathcal{M}, w \models \vartheta$. It is straightforward to verify that this is a contradiction to the properties of the ψ_S as stated under (i) and (ii). ❑

Claim 2. For each $w \in W$, there exists an $S \in \mathfrak{P}_2$ such that, for all subformulas ψ of φ_4^*, we have that $\mathcal{M}, w \models [S]\psi$ implies $\mathcal{M}, w \models \psi$.

Proof: Similar to the previous claim, only simpler using χ_1 in place of χ_2. ❑

Construct a Kripke model $\mathcal{M}' = \langle W, \pi, \mathcal{K}_1', \dots, \mathcal{K}_s', \mathcal{X}_1', \dots, \mathcal{X}_t', \mathcal{Y}_1', \dots, \mathcal{Y}_n' \rangle$ as follows: Initially, set $\mathcal{K}_i' := \mathcal{K}_i \cup \mathcal{I}_i^{-1}$, $\mathcal{X}_i' := \mathcal{X}_i \cup \mathcal{X}_i^{-1}$, and $\mathcal{Y}_i' := \mathcal{Y}_i$. Then, augment the relations as follows:

1. For each $w, w' \in W$ with $w \neq w'$, if

$$(w, w') \notin \bigcup_{1 \leq i \leq s} \mathcal{K}_i' \cup \bigcup_{1 \leq i \leq s} (\mathcal{K}_i')^{-1} \cup \bigcup_{1 \leq i \leq t} \mathcal{X}_i'$$

then choose an $S \in \mathfrak{P}_1$ as in Claim 1 and set
   - $\mathcal{K}_i' := \mathcal{K}_i' \cup \{(w, w')\}$ if $S = K_i$,
   - $\mathcal{K}_i' := \mathcal{K}_i' \cup \{(w', w)\}$ if $S = I_i$, and
   - $\mathcal{X}_i' := \mathcal{X}_i' \cup \{(w, w'), (w', w)\}$ if $S = X_i$.
2. For each $w \in W$, if $(w, w) \notin \bigcup_{1 \leq i \leq n} \mathcal{Y}_i'$ then choose a $Y_i \in \mathfrak{P}_2$ as in Claim 2 and set $\mathcal{Y}_i' := \mathcal{Y}_i' \cup \{(w, w)\}$.

It is not hard to check that $\mathcal{M}'$ is a c-model, i.e., that the properties from Definition 2 are satisfied. It hence remains to prove that

$$\mathcal{M}, w \models \psi^* \text{ iff } \mathcal{M}', w \models \psi$$

for all subformulas ψ of φ_4. The proof is by a straightforward induction and can be found in the full version of this paper [21]. Since $\mathcal{M}$ is a model for φ_4^*, we have that $\mathcal{M}'$ is a model for φ_4. ❑

[4] For the cases $\mathcal{P} = \emptyset$ and $\mathcal{P} = \mathfrak{P}_1$, we assume that the "empty" conjunction is equivalent to $\top$ and the "empty" disjunction equivalent to $\bot$.

3.2 An ExpTime upper Bound for $\mathcal{ML}^{\neg id}_{s,t,n}$

We show that s-satisfiability of $\mathcal{ML}^{\neg id}_{s,t,n}$-formulas can be decided in deterministic exponential time. Consider an $\mathcal{ML}^{\neg id}_{s,t,n}$-formula φ with modal parameters from $\{d\} \cup \mathfrak{P}_1 \cup \mathfrak{P}_2$ as defined above. Denote by $\mathsf{cl}(\varphi)$ the closure under single negation of the set of all subformulas of φ. In what follows we identify $\neg\neg\psi$ with ψ. A φ-type t is a subset of $\mathsf{cl}(\varphi)$ with

- $\neg\chi \in \mathsf{t}$ iff $\chi \notin \mathsf{t}$, for all $\neg\chi \in \mathsf{cl}(\varphi)$;
- $\chi_1 \wedge \chi_2 \in \mathsf{t}$ iff $\chi_1, \chi_2 \in \mathsf{t}$, for all $\chi_1 \wedge \chi_2 \in \mathsf{cl}(\varphi)$.

Given a world w in a model, the set of formulas in $\mathsf{cl}(\varphi)$ which are realized in w is a (φ-)type. We use the following notation:

- for $R \in \mathfrak{P}$ we write $\mathsf{t}_1 \rightarrow_R \mathsf{t}_2$ iff $\{\neg\chi \mid \neg \langle R \rangle \chi \in \mathsf{t}_1\} \subseteq \mathsf{t}_2$;
- a φ-type t is called a χ-*singleton type* if $\{\chi, \neg \langle d \rangle \chi\} \subseteq \mathsf{t}$.

Intuitively, singleton types are types which cannot be realized by two different worlds in a model. A *candidate* for φ is a maximal set (w.r.t. $\subseteq$) $\mathcal{T}$ of φ-types with the following properties:

(C1) for all $\mathsf{t} \in \mathcal{T}$: if $\langle Y_i \rangle \chi_1 \in \mathsf{t}$ and $\langle Y_j \rangle \chi_2 \in \mathsf{t}$, then $i = j$;
(C2) for all $\mathsf{t} \in \mathcal{T}$: if for some $i \leq n$, $\langle Y_i \rangle \chi \in \mathsf{t}$, then $\mathsf{t} \rightarrow_{Y_i} \mathsf{t}$ and $\{\chi \mid \langle Y_i \rangle \chi \in \mathsf{t}\} \subseteq \mathsf{t}$;
(C3) if $\mathcal{T}$ contains a χ-singleton type t, then $\neg\chi \in \mathsf{t}'$, for all $\mathsf{t}' \in \mathcal{T} - \{\mathsf{t}\}$,
(C4) for every $\langle d \rangle \chi \in \mathsf{cl}(\varphi)$ and $\mathsf{t}, \mathsf{t}' \in \mathcal{T}$: $\neg\chi, \neg \langle d \rangle \chi \in \mathsf{t}$ iff $\neg\chi, \neg \langle d \rangle \chi \in \mathsf{t}'$.

Intuitively, (C1) says that it suffices to add at most a single reflexive edge Y_i to each world of type t which is necessary since we are heading for s-models. By (C2), for each $\langle Y_i \rangle$-formula in t we find a witness in t itself. (C3) states that, for every $\langle d \rangle \chi \in \mathsf{cl}(\varphi)$, $\mathcal{T}$ does not contain more than one χ-singleton type (C3). (C4) should be obvious by the semantics of $\langle d \rangle$. We have an exponential upper bound of $n_\varphi = 2^{(|\mathsf{cl}(\varphi)|+1)^2}$ for the number of candidates (see [21]).

A *relational candidate* is a triple $\langle \mathcal{T}, \mathcal{F}, \mathcal{I} \rangle$ consisting of

- a candidate $\mathcal{T}$ for φ;
- a function $\mathcal{F} : \{1, \ldots, k\} \rightarrow \mathcal{T}_N$ with $k \leq |\mathsf{cl}(\varphi)|^2$ (in what follows we often use $\mathcal{F} = \{(1, \mathcal{F}(1)), \ldots, (k, \mathcal{F}(k))\}$);
- and a function $\mathcal{I}$ mapping each modal parameter $R \in \mathfrak{P}$ to a relation $R^{\mathcal{I}} \subseteq (\mathcal{T}_S \cup \mathcal{F}) \times (\mathcal{T}_S \cup \mathcal{F})$ such that

  (R1) $\langle \mathcal{T}_S \cup \mathcal{F}, (R^{\mathcal{I}} : R \in \mathfrak{P}) \rangle$ is an s-frame;
  (R2) for all $R \in \mathfrak{P}$, $m, m' \leq k$ and types t, t': if $\mathsf{t} R^{\mathcal{I}} \mathsf{t}'$, $\mathsf{t} R^{\mathcal{I}} (m, \mathsf{t}')$, $(m, \mathsf{t}) R^{\mathcal{I}} \mathsf{t}'$, or $(m, \mathsf{t}) R^{\mathcal{I}} (m', \mathsf{t}')$, then $\mathsf{t} \rightarrow_R \mathsf{t}'$;
  (R3) for all $R \in \mathfrak{P}$, $\langle R \rangle \chi \in \mathsf{cl}(\varphi)$, and $\mathsf{t} \in \mathcal{T}_S$ with $\langle R \rangle \chi \in \mathsf{t}$ we find $\mathsf{t}' \in \mathcal{T}_S$ with $\mathsf{t} R^{\mathcal{I}} \mathsf{t}'$ and $\chi \in \mathsf{t}'$ or we find $(m, \mathsf{t}') \in \mathcal{F}$ with $\mathsf{t} R^{\mathcal{I}} (m, \mathsf{t}')$ and $\chi \in \mathsf{t}'$;
  (R4) for all $R \in \mathfrak{P}_2$ and $(m, \mathsf{t}) \in \mathcal{F}$, if $\langle R \rangle \chi \in \mathsf{t}$, then $(m, \mathsf{t}) R^{\mathcal{I}} (m, \mathsf{t})$.

Intuitively, $\mathcal{T}_S$ is the set of worlds realizing singleton types, $\mathcal{F}$ is the set of worlds providing witnesses for diamond formulas in singleton types, and $\mathcal{I}$ fixes the extension of the modal parameters on $\mathcal{T}_S \cup \mathcal{F}$. Note that $\mathcal{F}$ need not contain more than $|\mathsf{cl}(\varphi)|^2$ worlds since each candidate contains at most $|\mathsf{cl}(\varphi)|$ singleton types (one for each $\langle d \rangle \chi \in \mathsf{cl}(\varphi)$, see above) and each type may contain at most $|\mathsf{cl}(\varphi)|$ diamond formulas. (R2) ensures that the relations fixed satisfy all box formulas. (R3) guarantees that diamond-formulas in $\mathsf{t} \in \mathcal{T}_S$ with parameters $R \in \mathfrak{P}$ have witnesses in $\mathcal{T}_S \cup \mathcal{F}$. And (R4) says that relations from $\mathfrak{P}_2$ are interpreted by $\mathcal{I}$ as enforced by the diamond formulas. We need not consider types from $\mathcal{T}_S$ in (R4) since the corresponding claim already follows from (R1) and (R3). The number of relational candidates is bounded by $n_\varphi \cdot 2^{|\mathsf{cl}(\varphi)|^3} \cdot |\mathsf{cl}(\varphi)|^{6 \cdot |\mathfrak{P}|+2}$ [21].

Our algorithm enumerates all (exponentially many) relational candidates and performs, for each such candidate, an elimination procedure that checks whether the candidate under consideration induces a model or not. Concerning the enumeration of relational candidates, note that it can be checked in polynomial time whether some $\mathcal{I}$ defines an s-frame as required by (R1) above: It is tedious but straightforward to write down explicit conditions that determine s-frames. We now describe the elimination procedure. Inituitively, we remove those non-singleton types whose diamond formulas are not witnessed: for a given relational candidate $\langle \mathcal{T}, \mathcal{F}, \mathcal{I} \rangle$ we can form a sequence $\mathcal{T} = \mathcal{T}_0 \supseteq \mathcal{T}_1 \supseteq \cdots$ inductively as follows: put $\mathcal{T}_0 = \mathcal{T}$. Suppose $\mathcal{T}_i$ is defined. Then delete non-singleton types $\mathsf{t} \in \mathcal{T}_i$ which are not in the range of $\mathcal{F}$ whenever

(E1) there are no pairwise disjoint relations $R^{\mathcal{I}} \subseteq \{\mathsf{t}\} \times \mathcal{T}_S$ for all $R \in \mathfrak{P}_1$, such that (i) $\mathsf{t} \rightarrow_R \mathsf{t}'$ whenever $\mathsf{t} R^{\mathcal{I}} \mathsf{t}'$, and (ii) for all $\langle R \rangle \chi \in \mathsf{t}$, $R \in \mathfrak{P}_1$, there exists t' with $\mathsf{t} R^{\mathcal{I}} \mathsf{t}'$ and $\chi \in \mathsf{t}'$ or there exists $\mathsf{t}' \in \mathcal{T}_i - \mathcal{T}_S$ with $\mathsf{t} \rightarrow_R \mathsf{t}'$ and $\chi \in \mathsf{t}'$, or

(E2) there is $\langle d \rangle \chi \in \mathsf{t}$ but no $\mathsf{t}' \in \mathcal{T}_i$ with $\chi \in \mathsf{t}'$

and denote the result by $\mathcal{T}_{i+1}$. Clearly, $\mathcal{T}_i = \mathcal{T}_{i+1}$ after at most $2^{|\mathsf{cl}(\varphi)|}$ rounds. We denote the result of the elimination procedure started on $\mathcal{T}$ with $\widehat{\mathcal{T}}$. Obviously, for each non-singleton type t in $\widehat{\mathcal{T}}$ which is not in the range of $\mathcal{F}$, each diamond formula in t is witnessed by some type in $\widehat{\mathcal{T}}$ such that at most one "edge" from t to any $\mathsf{t}' \in \mathcal{T}_S$ is required (this is crucial for building s-models). Together with (R3), (C2), and (R4), this implies that the only diamond formulas not witnessed in $\widehat{\mathcal{T}}$ are either $\langle R \rangle$-formulas in types from the range of $\mathcal{F}$ with $R \in \{d\} \cup \mathfrak{P}_1$, or are $\langle d \rangle$-formulas in types from $\mathcal{T}_S$. Since we are building s-models, we must be careful choosing singleton types as witnesses for these formulas:

Lemma 2. *φ is s-satisfiable iff there exists a relational candidate $\langle \mathcal{T}, \mathcal{F}, \mathcal{I} \rangle$ such that $\left\langle \widehat{\mathcal{T}}, \mathcal{F}, \mathcal{I} \right\rangle$ has the following properties:*

- *there exists $\mathsf{t} \in \widehat{\mathcal{T}}$ with $\varphi \in \mathsf{t}$,*
- *for every $(m, \mathsf{t}) \in \mathcal{F}$ and all $\langle R \rangle \chi \in \mathsf{t}$ with $R \in \mathfrak{P}_1$: (i) there exists $\mathsf{t}' \in \widehat{\mathcal{T}} - \mathcal{T}_S$ with $\mathsf{t} \rightarrow_R \mathsf{t}'$ and $\chi \in \mathsf{t}'$ or (ii) there exists $\mathsf{t}' \in \mathcal{T}_S$ with $(m, \mathsf{t}) R^{\mathcal{I}} \mathsf{t}'$ and $\chi \in \mathsf{t}'$.*
- *for every $(m, \mathsf{t}) \in \mathcal{F}$ and $\langle d \rangle \chi \in \mathsf{t}$ we find $\mathsf{t}' \in \widehat{\mathcal{T}}$ with $\chi \in \mathsf{t}'$.*
- *for every $\mathsf{t} \in \mathcal{T}_S$ and $\langle d \rangle \chi \in \mathsf{t}$ we find a $\mathsf{t}' \in \widehat{\mathcal{T}}$ with $\mathsf{t} \neq \mathsf{t}'$ such that $\chi \in \mathsf{t}'$.*

Proof. Suppose φ is s-satisfiable. Take a witness $\langle W, \pi, \mathcal{R}_1, \ldots, \mathcal{R}_k \rangle$. Let, for $w \in W$,

$$\mathsf{t}(w) = \{\chi \in \mathsf{cl}(\varphi) \mid w \models \chi\},$$

and $\mathcal{T} = \{\mathsf{t}(w) \mid w \in W\}$. Due to the semantics of the modal operator $\langle d \rangle$, for each singleton type $\mathsf{t} \in \mathcal{T}$, we find precisely one w_t with $\mathsf{t}(w_\mathsf{t}) = \mathsf{t}$. Select, for each singleton type $\mathsf{t} \in \mathcal{T}$ and each $\langle R \rangle \chi \in \mathsf{t}$, $R \in \mathfrak{P}_1$, a world $v_{\mathsf{t}, \langle R \rangle \chi} \in W$ such that $w_\mathsf{t} \mathcal{R} v_{\mathsf{t}, \langle R \rangle \chi}$ and $v_{\mathsf{t}, \langle R \rangle \chi} \models \chi$. Let $v_1, \ldots, v_r$ be an enumeration of those $v_{\mathsf{t}, \langle R \rangle \chi}$ for which $\mathsf{t}(v_{\mathsf{t}, \langle R \rangle \chi})$ is not a singleton-type and put

$$\mathcal{F} = \{(i, \mathsf{t}(v_i)) \mid 1 \le i \le r\}.$$

Note that $r \le |\mathsf{cl}(\varphi)|^2$. Let, for $x, y \in \mathcal{T}_S \cup \mathcal{F}$ and $R \in \mathfrak{P}$:

$$xR^{\mathcal{I}}y \Leftrightarrow \begin{cases} w_\mathsf{t} \mathcal{R} v_m & : \quad x = \mathsf{t}, y = (m, \mathsf{t}'), \\ v_m \mathcal{R} w_\mathsf{t} & : \quad x = (m, \mathsf{t}'), y = \mathsf{t}, \\ w_\mathsf{t} \mathcal{R} w_{\mathsf{t}'} & : \quad x = \mathsf{t}, y = \mathsf{t}', \\ v_n \mathcal{R} v_m & : \quad x = (n, \mathsf{t}), y = (m, \mathsf{t}'). \end{cases}$$

Now take a candidate $\mathcal{S} \supseteq \mathcal{T}$ ($\mathcal{T}$ itself may violate the maximality condition) containing precisely the singleton types from $\mathcal{T}$, and $\mathcal{I}$ and $\mathcal{F}$ as defined above. It is easy to see that the elimination procedure applied to $\langle \mathcal{S}, \mathcal{F}, \mathcal{I} \rangle$ terminates with a structure satisfying the four properties in Lemma 2.

Conversely, suppose the elimination procedure terminates with $\langle \mathcal{T}, \mathcal{F}, \mathcal{I} \rangle$ that satisfies all four conditions in Lemma 2. We define an s-model satisfying φ as follows: W consists of $\mathcal{T}_S \cup \mathcal{F}$ and the set W_S of finite sequences

$$(\mathsf{t}_{i_0}, R_{i_0}, \mathsf{t}_{i_1}, R_{i_1}, \ldots, \mathsf{t}_{i_k}),$$

where $\mathsf{t}_{i_j} \in \mathcal{T}_N$, $R_{i_j} \in \mathfrak{P}_1$, and $k \ge 0$.

Note that adding the *paths* from W_S instead of elements of $\mathcal{T}_N$ to $\mathcal{T}_S \cup \mathcal{F}$ allows us to make sure that the same type reached via different paths yields different worlds. Like in standard unravelling, a path represents its last element, and will therefore be interpreted according to its last type. So, define a valuation π into W as follows:

$$x \in \pi(p) \Leftrightarrow \begin{cases} p \in x & : \quad x \in \mathcal{T}_S, \\ p \in \mathsf{t} & : \quad x = (m, \mathsf{t}) \in \mathcal{F}, \\ p \in \mathsf{t}_{i_k} & : \quad x = (\mathsf{t}_{i_0}, R_{i_0}, \mathsf{t}_{i_1}, R_{i_1}, \ldots, \mathsf{t}_{i_k}) \in W_S \end{cases}$$

It remains to define the relational structure of our s-model. Intuitively, we start with the relational structure provided by $\mathcal{I}$ and then, for $R \in \mathfrak{P}_1$ and each non-singleton type $\mathsf{t} \in \mathcal{T}_N$ which is not in the range $ran(\mathcal{F})$ of $\mathcal{F}$, take $R^{\mathcal{I}} \subseteq \{\mathsf{t}\} \times \mathcal{T}_S$ supplied by (E1). For every non-singleton type $\mathsf{t} \in ran(\mathcal{F})$, take an m_t with $\mathcal{F}(m_\mathsf{t}) = \mathsf{t}$. Define, for $x, y \in W$ and $R \in \mathfrak{P}_1$:

$$x\mathcal{R}y \Leftrightarrow \begin{cases} xR^{\mathcal{I}}y & : \quad x, y \in \mathcal{T}_S \cup \mathcal{F}, \\ \mathsf{t} \to_R \mathsf{t}_{i_k} & : \quad x = (m, \mathsf{t}), y = (\mathsf{t}_{i_0}, \ldots, R_{i_{k-1}}, \mathsf{t}_{i_k}), R = R_{i_{k-1}}, \\ \mathsf{t}_{i_n} \to_R \mathsf{t}_{i_{n+1}}, & : \quad x = (\cdots, \mathsf{t}_{i_n}), y = (\cdots, \mathsf{t}_{i_n}, R_{i_n}, \mathsf{t}_{i_{n+1}}), R = R_{i_n}, \\ \mathsf{t}_{i_k} R^{\mathcal{I}} \mathsf{t} & : \quad x = (\mathsf{t}_{i_0}, \ldots, R_{i_{k-1}}, \mathsf{t}_{i_k}), y = \mathsf{t} \in \mathcal{T}_S, \mathsf{t}_{i_k} \in ran(\mathcal{F}), \\ (m_{\mathsf{t}_{i_k}}, \mathsf{t}_{i_k}) R^{\mathcal{I}} \mathsf{t} & : \quad x = (\mathsf{t}_{i_0}, \ldots, R_{i_{k-1}}, \mathsf{t}_{i_k}), y = \mathsf{t} \in \mathcal{T}_S, \mathsf{t}_{i_k} \notin ran(\mathcal{F}). \end{cases}$$

Define, for $x, y \in W$ and $R \in \mathfrak{P}_2$:

$$x\mathcal{R}y \Leftrightarrow \begin{cases} xR^{\mathcal{I}}y & : \quad x, y \in \mathcal{T}_S \cup \mathcal{F}, \\ \exists\psi.\, \langle R\rangle\,\psi \in \mathsf{t}_{i_k} & : \quad x = y = (\mathsf{t}_{i_0}, \ldots, R_{i_{k-1}}, \mathsf{t}_{i_k}) \in W_S. \end{cases}$$

It is left to the reader to check that $\langle W, \pi, (\mathcal{R} : R \in \mathfrak{P})\rangle$ is an s-model satisfying φ. ❑

Obviously, the conditions listed in Lemma 2 can be checked in exponential time and we have obtained an ExpTime upper bound for $\mathcal{ML}^{\neg id}_{s,t,n}$-satisfiability. The reduction of $\mathcal{ML}^{\neg,\cap,\cup,-,id}_m$ to $\mathcal{ML}^{\neg id}_{s,t,n}$ given in Section 3.1 immediately yields an ExpTime upper bound for the satisfiability of $\mathcal{ML}^{\neg,\cap,\cup,-,id}_m$-formulas.

Theorem 2. *For $0 < m < \omega$, satisfiability of $\mathcal{ML}^{\neg,\cap,\cup,-,id}_m$-formulas is* ExpTime*-complete.*

Note that, for $m = \omega$, satisfiability of $\mathcal{ML}^{\neg,\cap,\cup,-,id}_m$-formulas is NExpTime-complete: In [20], it is proved that satisfiability in $\mathcal{ML}^{\neg,\cap,\cup}_\omega$ is NExpTime-hard and the upper bound follows from Theorem 1 and the NExpTime upper bound for FO^2_ω. So, in the modal language, the complexity depends on whether we have a bounded number of accessibility relations or not, while FO^2 does not "feel" this difference.

4 The Temporal Case

We briefly indicate that the expressive completeness result presented in this paper provides a general framework for comparing the expressivity of modal languages with first-order languages.

Fix a class $\mathcal{K}$ of frames of the form $\mathfrak{F} = \langle W, \mathcal{R}_1, \ldots, \mathcal{R}_m\rangle$. Denote by $\mathcal{E}_{\mathfrak{F}}$ the mapping which determines the extension of any complex modal parameter in $\mathfrak{F}$. A set $\mathcal{S}$ of complex modal parameters over $\{R_1, \ldots, R_m, id\}$ is called *exhaustive for $\mathcal{K}$* if for every complex modal parameter S, such that there exists $\mathfrak{F} \in \mathcal{K}$ with $\mathcal{E}_{\mathfrak{F}}(S) \neq \emptyset$, we find $S_1, \ldots, S_k \in \mathcal{S}$ such that $\mathcal{E}_{\mathfrak{F}}(S) = \mathcal{E}_{\mathfrak{F}}(S_1 \cup \cdots \cup S_k)$ for all $\mathfrak{F} \in \mathcal{K}$. Denote by $\mathcal{ML}(\mathcal{S})$ the modal language with operators $\langle S\rangle$, $S \in \mathcal{S}$.

Theorem 3. *Let $\mathcal{K}$ be a class of frames and $\mathcal{S}$ a set of complex modal parameters which is exhaustive for $\mathcal{K}$. Then $\mathcal{ML}(\mathcal{S})$ is expressively complete for the two-variable fragment over $\mathcal{K}$; i.e., for every $\varphi \in FO^2$ we find a $\varphi^\sigma \in \mathcal{ML}(\mathcal{S})$ such that for all $\mathcal{M} = \langle W, \pi, \mathcal{R}_1, \ldots, \mathcal{R}_m\rangle$ with $\langle W, \mathcal{R}_1, \ldots, \mathcal{R}_m\rangle \in \mathcal{K}$ and all $a \in W$:*

$$\mathcal{M}, a \models \varphi^\sigma \Leftrightarrow \mathcal{M}_\sigma \models \varphi(a).$$

Moreover, given φ the formula φ^σ is exponential in the size of φ and can be computed in polynomial time in the size of φ^σ.

The proof of this theorem is similar to the proof of Theorem 1. We provide two examples from temporal logic:

(i) Let $\mathcal{K}$ be a class of strict linear orderings $\langle W, \mathcal{R}\rangle$. Then $\mathcal{S} = \{R, R^-, id\}$ is exhaustive for $\mathcal{K}$. Hence, $\mathcal{ML}(\mathcal{S})$ is expressively complete for the two-variable fragment over $\mathcal{K}$. It is not hard to see that any $\mathcal{ML}(\mathcal{S})$-formula ψ can be translated into an equivalent $\mathcal{ML}(\{R, R^-\})$-formula ψ' whose length is linear in the length of ψ. In other words, the language of temporal logic with operators 'always in the future' and 'always in the past' [5,12] is expressively complete for the two-variable fragment over any class of strict linear orderings.

(ii) Consider again a class $\mathcal{K}$ of strict linear orderings $\langle W, \mathcal{R}\rangle$. Let, for every $\mathfrak{F} = \langle W, \mathcal{R}\rangle \in \mathcal{K}$, $\mathfrak{Int}(\mathfrak{F}) = \langle \mathcal{I}(\mathfrak{F}), \mathcal{R}_1, \ldots, \mathcal{R}_{13}\rangle$, where $\mathcal{I}(\mathfrak{F})$ is the set of intervals in $\mathfrak{F}$ and $\mathcal{R}_1, \ldots, \mathcal{R}_{13}$ is the list of Allen's relations over $\mathcal{I}(\mathfrak{F})$. $\mathcal{S} = \{R_1, \ldots, R_{13}\}$ is exhaustive for $\{\mathfrak{Int}(\mathfrak{F}) \mid \mathfrak{F} \in \mathcal{K}\}$ and so $\mathcal{ML}(\mathcal{S})$ is expressively complete for $\{\mathfrak{Int}(\mathfrak{F}) \mid \mathfrak{F} \in \mathcal{K}\}$. This interval-based temporal logic was introduced in [16].

Using (i), we obtain the following complexity result for the two-variable fragment interpreted in strict linear orderings:

Theorem 4. *Suppose $\mathcal{K}$ is a class of strict linear orderings such that satisfiability of temporal propositional formulas with operators 'always in the future' and 'always in the past' in $\mathcal{K}$ is in* NP *and $\mathcal{K}$ contains an infinite ordering. Then satisfiability of FO^2 with one binary relation interpreted by the strict linear ordering is* NExpTime*-complete.*

Proof. NExpTime-hardness follows from the condition that $\mathcal{K}$ contains an infinite structure and that FO^2 without binary relation symbols is NExpTime-hard already. Conversely, the following algorithm is in NExpTime: given φ, compute φ^σ (in exponential time) and check whether φ^σ is satisfiable in $\mathcal{K}$. ❑

This Theorem applies to e.g. (i) the class of all strict linear orderings, (ii) $\{\langle \mathbb{N}, <\rangle\}$, (iii) $\{\langle \mathbb{Q}, <\rangle\}$, and (iv) $\{\langle \mathbb{R}, <\rangle\}$, see [25,30].

References

1. H. Andréka, I. Németi, and J. van Benthem. Modal languages and bounded fragments of predicate logic. *Journal of Philosophical Logic*, 27:217–274, 1998.
2. P. Blackburn, M. de Rijke, and Y. Venema. Modal logic. In print.
3. E. Börger, E. Grädel, and Yu. Gurevich. *The Classical Decision Problem*. Perspectives in Mathematical Logic. Springer, 1997.
4. A. Borgida. On the relative expressivness of description logics and predicate logics. *Artificial Intelligence*, 82(1 - 2):353–367, 1996.
5. J.P. Burgess. Basic tense logic. In D.M. Gabbay and F. Guenthner, editors, *Handbook of Philosophical Logic*, volume 2, pages 89–133. Reidel, Dordrecht, 1984.
6. Maarten de Rijke. The modal logic of inequality. *The Journal of Symbolic Logic*, 57(2):566–584, June 1992.
7. K. Etessami, M. Vardi, and T. Wilke. First-order logic with two variables and unary temporal logic. In *Proceedings of 12th. IEEE Symp. Logic in Computer Science*, pages 228–235, 1997.
8. M. Fürer. The computational complexity of the unconstrained limited domino problem (with implications for logical decision problems). In *Logic and Machines: Decision problems and complexity*, pages 312–319. Springer, 1984.

9. D. Gabbay, I. Hodkinson, and M. Reynolds. *Temporal Logic: Mathematical Foundations and Computational Aspects, Volume 1.* Oxford University Press, 1994.
10. D.M. Gabbay. Expressive functional completeness in tense logic. In U. Mönnich, editor, *Aspects of Philosophical Logic*, pages 91–117. Reidel, Dordrecht, 1981.
11. G. Gargov and S. Passy. A note on boolean modal logic. In D. Skordev, editor, *Mathematical Logic and Applications*, pages 253–263, New York, 1987. Plenum Press.
12. R.I. Goldblatt. *Logics of Time and Computation.* Number 7 in CSLI Lecture Notes, Stanford. CSLI, 1987.
13. E. Grädel. Why are modal logics so robustly decidable? *Bulletin of the European Association for Theoretical Computer Science*, 68:90–103, 1999.
14. E. Grädel, P. Kolaitis, and M. Vardi. On the Decision Problem for Two-Variable First-Order Logic. *Bulletin of Symbolic Logic*, 3:53–69, 1997.
15. E. Grädel and M. Otto. On Logics with Two Variables. *Theoretical Computer Science*, 224:73–113, 1999.
16. J. Halpern and Y. Shoham. A propositional modal logic of time intervals. *Journal of the ACM*, 38:935–962, 1991.
17. I. L. Humberstone. Inaccessible worlds. *Notre Dame Journal of Formal Logic*, 24(3):346–352, 1983.
18. H. Kamp. *Tense Logic and the Theory of Linear Order.* Ph. D. Thesis, University of California, Los Angeles, 1968.
19. R.E. Ladner. The computational complexity of provability in systems of modal logic. *SIAM Journal on Computing*, 6:467–480, 1977.
20. C. Lutz and U. Sattler. The complexity of reasoning with boolean modal logics. In Frank Wolter, Heinrich Wansing, Maarten de Rijke, and Michael Zakharyaschev, editors, *Advances in Modal Logics Volume 3.* CSLI Publications, Stanford, 2001.
21. C. Lutz, U. Sattler, and F. Wolter. Modal logic and the two-variable fragment. LTCS-Report 01-04, LuFG Theoretical Computer Science, RWTH Aachen, Germany, 2001. See http://www-lti.informatik.rwth-aachen.de/Forschung/Reports.html.
22. M. Mortimer. On languages with two variables. *Zeitschrift für Mathematische Logik und Grundlagen der Mathematik*, 21:135–140, 1975.
23. Anil Nerode and Richard A. Shore. *Logic for Applications.* Springer Verlag, New York, 1997.
24. D. Scott. A decision method for validity of sentences in two variables. *Journal of Symbolic Logic*, 27(377), 1962.
25. A. Sistla and E. Clarke. The complexity of propositional linear temporal logics. *Journal of the Association for Computing Machinery*, 32:733–749, 1985.
26. E. Spaan. *Complexity of Modal Logics.* PhD thesis, Department of Mathematics and Computer Science, University of Amsterdam, 1993.
27. J. van Benthem. *Modal Logic and Classical Logic.* Bibliopolis, Napoli, 1983.
28. J. van Benthem. Correspondence theory. In D.M. Gabbay and F. Guenthner, editors, *Handbook of Philosophical Logic*, volume 2, pages 167–247. Reidel, Dordrecht, 1984.
29. M. Vardi. Why is modal logic so robustly decidable? In *Descriptive Complexity and Finite Models*, pages 149–184. AMS, 1997.
30. F. Wolter. Tense logics without tense operators. *Mathematical Logic Quarterly*, 42:145–171, 1996.
31. F. Wolter. Completeness and decidability of tense logics closely related to logics containing $K4$. *Journal of Symbolic Logic*, 62:131–158, 1997.

A Logic for Approximate First-Order Reasoning

Frédéric Koriche

LIRMM, UMR 5506, Université Montpellier II CNRS
161, rue Ada. 34392 Montpellier Cedex 5, France
koriche@lirmm.fr

Abstract. In classical approaches to knowledge representation, reasoners are assumed to derive all the logical consequences of their knowledge base. As a result, reasoning in the first-order case is only semi-decidable. Even in the restricted case of finite universes of discourse, reasoning remains inherently intractable, as the reasoner has to deal with two independent sources of complexity: *unbounded chaining* and *unbounded quantification*. The purpose of this study is to handle these difficulties in a logic-oriented framework based on the paradigm of *approximate reasoning*. The logic is semantically founded on the notion of resource, an accuracy measure, which controls at the same time the two barriers of complexity. Moreover, a stepwise technique is included for improving approximations. Finally, both sound approximations and complete ones are covered. Based on the logic, we develop an approximation algorithm with a simple modification of classical instance-based theorem provers. The procedure yields approximate proofs whose precision increases as the reasoner has more resources at her disposal. The algorithm is interruptible, improvable, dual, and can be exploited for anytime computation. Moreover, the algorithm is flexible enough to be used with a wide range of propositional satisfiability methods.

Keywords: approximate reasoning, first-order logic, multi-modal logics, resource-bounded algorithms.

1 Introduction

A widely accepted framework for studying intelligent agents is the knowledge representation approach. Knowledge is described is some logical formalism and stored into a knowledge base. This component is coupled with an inference algorithm, the reasoner, which determines whether a given query is entailed from the knowledge base. One of the main challenges of knowledge representation lies in the computational tradeoff between expressiveness of representation languages and their complexity [21]. On the one hand, knowledge needs to be represented in a very expressive language, such as first-order logic and, on the other, reasoning has to be very efficient, especially for knowledge bases of the size required for human level common-sense. Unfortunately, it is well-known that first-order reasoning is only *semi-decidable*. In other words, if the base and the query are represented in first-order logic, there is no guaranteed way to determine in finite time whether the query is entailed, or not, from the knowledge base.

L. Fribourg (Ed.): CSL 2001, LNCS 2142, pp. 262–276, 2001.

Since real agents are constrained by finite resources, it seems appropriate to examine first-order reasoning in the setting of *finite universes of discourses.* This assumption has received increasing attention in the communities of database systems [23, 26], planning [11, 12] and theorem proving [13, 25, 28, 29]. This can be expressed by a domain closure axiom or, less restrictively, through a constraint expressing a finite upper limit on the cardinality of the domain of any interpretation. In this setting, every first-order formula can be rewritten to a finite "propositional" formula which, however, is in general exponentially larger. Each atom with m free variables gives rise to (n^m) ground instances for an universe of size n. As a result the complexity of first-order reasoning is (at most) exponentially higher than the complexity of propositional reasoning. Intuitively, this means that a first-order reasoner is confronted with two sources of complexity: *unbounded chaining* and *unbounded quantification.* The first one is related to propositional entailment, which is known to be intractable, while the second one is related to the exponential number of ground instances generated by a first-order formula. So, even in finite universes of discourse, first-order reasoning remains very much demanding from a computational point of view.

Approximate reasoning is an approach advocated in many areas of artificial intelligence to deal with the computational intractability of problems. The motivation behind this paradigm stems from the fact that practical agents have limited time to solve problems and limited memory to remember information. As suggested by Lakemeyer in [16], an approximate reasoning system provides something of middle ground between what is explicit or evident and can be retrieved using few resources and what is implicit and should be inferred given enough time and memory. In case the answer of the reasoner is not satisfactory, one can still decide to continue reasoning. The simplest way is to switch to a conventional reasoning technique. A better way is to use the resource-bounded reasoner to produce better and better answers in a cumulative fashion.

Many such inference algorithms have been proposed, and these algorithms are generally reasonably easy to understand procedurally. For example, one may take an existing theorem prover and bound its execution time and memory in some way. However, understanding inference procedurally is no substitute for understanding what sort of "semantics" and "axiomatics" underlie the inference. This kind of deeper understanding is the domain of *logic.* A logic for a resource-bounded reasoner gives a clear picture to the notion of resource and tells us what the inference algorithm is, and is not, able to deduce from its knowledge base.

There have been a number of attempts at devising logics for approximate reasoning either proof-theoretically or model-theoretically. On the proof-theoretic side, for example, Dalal defines tractable forms of reasoning by eliminating certain inference rules from propositional logic [5]. The starting point of its framework relies on unit resolution which is tractable but weaker than propositional deduction. Based on this inference rule, the author obtains better and better approximations by imposing incremental bounds on the size of clauses used as lemmas. This technique has been further studied, for example, in [6, 10].

On the model-theoretic side, one of the first framework proposed in the literature is that of Levesque [20]. Based on a multi-valued logic, especially a fragment of Belnap's relevance logic [1], the author introduces a notion of inference which is weaker than propositional deduction and that captures a tractable form of reasoning. This study has been further extended by Schaerf and Cadoli in [24]. Their framework considers a subset S of the propositional variables, which are deserved a classical interpretation, while the rest of propositions is given a multi-valued interpretation. By increasing the parameter S, the reasoner can regain full logical deduction in an incremental fashion. This treatment of approximate reasoning has been applied to a wide range of reasoning problems [2, 14, 15, 22].

To the best of our knowledge, most of the studies in approximate reasoning have concentrated to propositional logic. On the proof-theoretic camp, Crawford and Etherington in [10] have recently attempted to extend Dalal's approach to first-order logic but, as they pointed it out, unit resolution alone is undecidable. On the model-theoretic camp, Schaerf and Cadoli have extended their semantics to the description logics $\mathcal{ALE}$ and $\mathcal{ALC}$ but these languages remain restricted fragments of first-order logic. A more general framework has been investigated by Lakemeyer in [16, 17]. Based on Levesque's multi-valued logic, the author defines a inference relation which is weaker than classical first-order implication and that captures a decidable form of reasoning. However, Lakemeyer's framework is only "one-shot": if an approximate solution is wrong, the sole thing to do is to switch to a general-purpose reasoning technique. From this perspective, it seems appropriate to pursue investigations in the direction to "improvable" reasoners that would produce better and better solutions in an incremental fashion.

In this paper, we introduce a model-theoretic framework for approximate first-order reasoning. The framework is based on a multi-modal logic which contains a well-founded semantics and a correct and complete axiomatization. To some extent, our logic combines ideas from Schaerf and Cadoli's approximation technique and Lakemeyer's system for limited reasoning. In essence, our framework integrates the following features.

- The logic is founded on the notion of *resource*, an accuracy measure which semantically captures bounded approximations of first-order inference. The measure reflects both the quality and the cost of the approximations.
- The framework enables *improvable reasoning*: the quality of approximations is an increasing function of the resources that have been spent.
- The framework covers *dual reasoning*: both sound but incomplete and complete but unsound solutions are returned at any computation step.

The rest of the paper is organized as follows. In section 2, we define the syntax, the semantics, and a sound and complete axiomatization for the logic. In section 3, we investigate the semantical properties of approximate reasoning. In section 4, we show how to transform a traditional instance-based theorem prover into a resource-bounded algorithm which is interruptible, improvable, dual, and that can be exploited for anytime computation [30]. The approximation schema is flexible enough to be used with a wide range of efficient satisfiability methods. Finally, in section 5, we suggest some topics for future research.

2 The Logic

In this section we present a logic, named **AFOR**, for approximate first-order reasoning. We begin to define the syntax, next we examine the semantics in detail, and then we present a sound and complete axiomatization for the logic.

2.1 Syntax

The basic building block of our framework is Levesque's first-order logic with standard names presented in [19] and further examined by Lakemeyer in [16, 17]. A *first-order signature* consists of denumerable sets P, F, N of symbols called *predicates*, *functions* and *standard names* respectively. Each predicate and function symbol has a fixed *arity* which is defined by the number of its arguments. Function symbols with arity zero are called *constants*. The set of standard names corresponds to the universe of discourse over which quantifiers range.

A *term* is either a variable, a standard name, or a function symbol whose arguments are themselves terms. A *ground term* is a term not containing any variable. A *primitive term* is either a constant or a function symbol whose arguments are standard names. An *atom* is a predicate whose arguments are terms and a *literal* is an atom or its negation. A *primitive atom* (resp. *primitive literal*) is an atom (resp. literal) with standard names as arguments. The sets of primitive atoms and primitive literals generated from the signature are denoted A and L, respectively. A *standard formula* is either an atom or can be obtained by the usual rules for the connectives $\neg$ and $\wedge$, and the quantifier $\forall$. Other connectives such as $\vee$, $\supset$ and $\equiv$, and the quantifier $\exists$ are defined in the usual way.

We now turn to the concept of resource. The key point behind this notion is to control the two aforementioned sources of complexity of first-order reasoning, namely unbounded chaining and unbounded quantification. To this end, a *resource parameter* is defined as a pair $S = (P_S, N_S)$ where P_S is a finite subset of P and N_S is a nonempty finite subset of N. A parameter can be seen as the collection of primitive atoms and standard names which are relevant for chaining and quantification for a given problem instance. The sets of primitive atoms and primitive literals generated from S are denoted A_S and L_S, respectively. The "empty" parameter, which doesn't contain any predicate, is denoted S_0.

A *formula* is either a standard formula or can be obtained by the following rules: if α is a formula, then $\neg\,\alpha$ is a formula, if α and β are formulas then $\alpha \wedge \beta$ is a formula, and if α is a standard formula and S a resource parameter then $\Box_S\, \alpha$ is a formula. Notice that the syntax does not allow quantifying-in or nested modalities. The modality $\Diamond_S$ is used as an abbreviation of $\neg\Box_S\,\neg$. A formula such as $\Box_S\, \alpha$ is read "the reasoner necessarily infers α given the resources S"; dually $\Diamond_S\, \alpha$ is read "the reasoner possibly infers α given the resources S".

A *sentence* is a closed formula, a *declaration* is a closed standard formula, and a *ground declaration* is a quantifier free declaration. In the following, sequences of terms are written in vector notion. For example, $(t_1, \cdots, t_k)$ is abbreviated as $\boldsymbol{t}$. If a formula α contains the free variables $x_1, \cdots, x_k$, then the notation $\alpha[\boldsymbol{x}/\boldsymbol{t}]$ will denote the result of replacing each occurrence of x_i by t_i in α.

2.2 Semantics

The semantics of **AFOR** combines ideas from Belnap's four-valued logic [1] with possible world interpretations that allow varying domains of quantification [7].

We first assign semantics to ground terms. To this point, the logic makes the assumption that the universe of discourse is isomorphic to the set of standard names, that is, a ground term is identified with a unique name. A *denotation function* is defined as a mapping d from primitive terms to standard names. A denotation function determines a unique map, also denoted d, on the set of all ground terms, according to the following conditions: $d(n) = n$, where n is a standard name, and $d(f(\boldsymbol{t})) = d(f(\boldsymbol{n}))$, where $n_i = d(t_i)$.

We now examine the semantics for all non-logical symbols. Our approach rests on an extension of the standard notion of possible world which we call *valuations.* While worlds use fixed domains of quantification and assign a truth value to every primitive atoms, valuations, in contrast, allow varying domains of quantification and assign truth values to all the literals. In formal terms, a *valuation* is a structure $v = (N_v, L_v, d_v)$, where N_v is a subset of N, L_v is a subset of L, and d_v is a denotation function. A *world* is a valuation w, where N_w is the set of all standard names N and L_w is a subset of L such that for every primitive atom a, $a \in L_w$ if and only if $\neg a \notin L_w$. We say that a valuation v is more *specific* than v', and write $v \subseteq v'$, if $N_v = N_{v'}$, $L_v \subseteq L_{v'}$ and $d_v = d_{v'}$. We remark that the specificity relation is a partial order on the space of all valuations. Moreover it induces a lattice structure on each subspace of valuations defined over the same domain of quantification and the same denotation function.

The concept of resource parameter S is semantically captured by an accessibility relation on valuations $\mathcal{R}_S$. Given two valuations v and v' and any resource parameter S, $v' \in \mathcal{R}_S(v)$, if $N_{v'} = N_S$, $L_{v'} \cap L_S = L_v \cap L_S$ and $d_{v'} = d_v$. In other words, $\mathcal{R}_S(v)$ is the set of valuations that share the same domain of quantification N_S, that assign the same truth value as v to each literal in L_S, and that use the same denotation functions. We remark that, for any valuation v, $\mathcal{R}_S(v)$ is a complete lattice under the specificity ordering $\subseteq$. The smallest and the largest valuations of $\mathcal{R}_S(v)$ are denoted $\cap\mathcal{R}_S(v)$ and $\cup\mathcal{R}_S(v)$, respectively.

We now turn to the semantics of sentences. Since valuations assign independent truth values to literals and their complements, the semantic rules for sentences must define truth support for both sentences and their negation.

$$v \models p(\boldsymbol{t}) \quad \text{iff} \quad p(\boldsymbol{n}) \in L_v \text{ and } \boldsymbol{n} = d_v(\boldsymbol{t}), \tag{1}$$

$$v \models \neg p(\boldsymbol{t}) \quad \text{iff} \quad \neg p(\boldsymbol{n}) \in L_v \text{ and } \boldsymbol{n} = d_v(\boldsymbol{t}), \tag{2}$$

$$v \models \neg\neg\alpha \quad \text{iff} \quad v \models \alpha, \tag{3}$$

$$v \models \alpha \wedge \beta \quad \text{iff} \quad v \models \alpha \text{ and } v \models \beta, \tag{4}$$

$$v \models \neg(\alpha \wedge \beta) \quad \text{iff} \quad v \models \neg\alpha \text{ or } v \models \neg\beta, \tag{5}$$

$$v \models (\forall x)\, \alpha \quad \text{iff} \quad \text{for all } n \in N_v,\ v \models \alpha[x/n], \tag{6}$$

$$v \models \neg(\forall x)\, \alpha \quad \text{iff} \quad \text{for some } n \in N_v,\ v \models \neg\alpha[x/n], \tag{7}$$

$$v \models \Box_S\, \alpha \quad \text{iff} \quad \text{for all } v' \in \mathcal{R}_S(v),\ v' \models \alpha, \tag{8}$$

$$v \models \neg\Box_S\, \alpha \quad \text{iff} \quad v \not\models \Box_S\, \alpha. \tag{9}$$

A sentence α is called *satisfiable* iff $w \models \alpha$ for some world w. We say that a sentence α is *valid*, and write $\models \alpha$, iff $w \models \alpha$ holds for all worlds w. Finally, given two sentences α and β, we say that β is a *logical consequence* of α iff $\models \alpha \supset \beta$ holds. The following lemmas capture important structural properties of the semantics. They will be frequently used in the remaining sections.

Lemma 1. *For any declaration α and any valuations v, v' such that $v \subseteq v'$,*

$$\text{if } v \models \alpha \text{ then } v' \models \alpha.$$

Lemma 2. *For any declaration α and any world w,*

$$w \models \Box_S \alpha \text{ iff } \cap\mathcal{R}_S(w) \models \alpha, \quad (1)$$

$$w \models \Diamond_S \alpha \text{ iff } \cup\mathcal{R}_S(w) \models \alpha. \quad (2)$$

Proof (1). Suppose that $w \models \Box_S \alpha$. By semantic rule 8, we obtain $v \models \alpha$ for all $v \in \mathcal{R}_S(w)$. It follows that $\cap\mathcal{R}_S(w) \models \alpha$. Dually, suppose that $w \not\models \Box_S \alpha$. By semantic rule 9, we obtain $v \not\models \alpha$ for some $v \in \mathcal{R}_S$. Since $\cap\mathcal{R}_S(w) \subseteq v$, by contraposition of lemma 1, it follows that $\cap\mathcal{R}_S(w) \not\models \alpha$.

Proof (2). Suppose that $w \models \Diamond_S \alpha$. By semantic rule 9, we obtain $v \not\models \neg\alpha$ for some $v \in \mathcal{R}_S(w)$. If $v \models \alpha$ then by lemma 1 we have $\cup\mathcal{R}_S(w) \models \alpha$. Otherwise, by contraposition of lemma 1 we obtain $\cap\mathcal{R}_S(w) \not\models \alpha \vee \neg\alpha$. By induction on the structure of α, it follows that $\cup\mathcal{R}_S(w) \models \alpha \wedge \neg\alpha$. So $\cup\mathcal{R}_S(w) \models \alpha$. Now suppose that $w \not\models \Diamond_S \alpha$. By semantic rule 8, we have $v \models \neg\alpha$ for all $v \in \mathcal{R}_S(w)$. If $\cup\mathcal{R}_S(w) \models \alpha$ then by lemma 1 we obtain $\cup\mathcal{R}_S(w) \models \alpha \wedge \neg\alpha$. By induction on the structure of α, it follows that $\cap\mathcal{R}_S(w) \not\models \alpha \vee \neg\alpha$, but this contradicts the former hypothesis. Therefore, we must obtain $\cup\mathcal{R}_S(w) \not\models \alpha$.

2.3 Axiomatization

We now focus on obtaining a sound and complete axiomatization for our logic. An *axiom system* consists of a collection of *axioms* and *inferences rules.* A *proof* in an axiom system is a finite sequence of sentences, each of which is either an instance of an axiom or follows by an application of an inference rule. Finally, we say that a sentence α is a *theorem* of the axiom system and write $\vdash \alpha$ if there exists a proof of α in the system. The axiom system of **AFOR** is the following.

Axioms:

All tautologies of first-order logic (A1)

$\Box_S \neg\neg\alpha \equiv \Box_S \alpha$ (A2)

$\Box_S (\alpha \wedge \beta) \equiv \Box_S \alpha \wedge \Box_S \beta$ (A3)

$\Box_S \neg(\alpha \wedge \beta) \equiv \Box_S \neg\alpha \vee \Box_S \neg\beta$ (A4)

$\Box_S (\forall x)\, \alpha \equiv \Box_S \bigwedge_{n \in N_s} \alpha[x/n]$ (A5)

$\Box_S \neg(\forall x)\, \alpha \equiv \Box_S \bigvee_{n \in N_s} \neg\alpha[x/n]$ (A6)

$\Box_S (a \vee \neg a)$, where $a \in A_S$ (A7)

$\Diamond_S (a \wedge \neg a)$, where $a \notin A_S$ (A8)

$\Box_S \alpha \supset \alpha$ where α is a ground declaration (A9)

Inference rules:

$$\text{From } \vdash \alpha \text{ and } \vdash \alpha \supset \beta \text{ infer } \vdash \beta \quad \text{(R1)}$$

$$\text{From } \vdash \alpha[x/n] \text{ infer } \vdash (\forall x)\,\alpha \quad \text{(R2)}$$

The axiom system may be divided into two categories. The first one is concerned by axiom (A1) and rules (R1) and (R2) which come from standard first-order logic. Hence, the first-order fragment of **AFOR** is correctly handled. The specificity of our logic lies on the second category. Axioms (A2)-(A4) capture the properties of double negation, conjunction and disjunction, respectively. Axioms (A5)-(A8) are the key point of resource-bounded reasoning. The first two axioms introduce a limitation on the *quantification* capabilities of the reasoner. Specifically, they state that any universal (existential) quantifier can be rewritten to a finite number of conjunctions (disjunctions) which is bounded by the size of N_S. From an orthogonal point of view, the last two axioms impose a limitation on the *chaining* capabilities of the reasoner. Axiom (A7) says that the system necessarily infers the tautology $a \vee \neg a$, whenever a is in A_S. Dually axiom (A8) says that the system can infer the antilogy $a \wedge \neg a$, if a is not in A_S. Finally, axiom (A9) claims that reasoning under the scope of $\Box_S$ is sound, provided that the declaration α is quantifier-free.

It is interesting to analyse the axiomatization from the standpoint of the so-called *logical omniscience problem* [9]. A reasoner is called logically omniscient if its inference capabilities are closed under logical consequence. By inference rule (R1) and axioms (A2)-(A4), we remark that $\Box_S\,\alpha \supset (\Box_S\,(\alpha \supset \beta) \supset \Box_S\,\beta)$ is a theorem of the axiom system. So, the inference capabilities of the reasoner are closed under *material* implication. However, we also remark that the sentence $\Box_S\,\alpha \supset ((\alpha \supset \beta) \supset \Box_S\,\beta)$ is *not* a theorem of the axiom system. Hence, the inference capabilities of the reasoner are not closed under *logical* implication. The following result gives soundness and completeness for the axiom system.

Theorem 1 (Soundness and Completeness). *For any sentence* α,

$$\vdash \alpha \text{ iff } \models \alpha.$$

Proof (sketch). The soundness of the axiom system is easily demonstrated from the semantic rules and lemma 2. The proof of completeness is based on the technique of saturated sets presented in [7]. A set of sentences is called *saturated* if it is ω-complete and consistent. A set of sentences E is ω-complete if $E \cup p[x/n]$ is consistent whenever $E \cup \{\neg(\forall x)\ p\}$ is consistent. Completeness follows if we can show that any saturated set is satisfiable. We begin by extending a given ω-complete set E to a maximally consistent set using the Lindenbaum procedure. Then we build a world w_E as follows. d_E is an injective morphism from the ground terms in E to N, and $a \in L_E$ iff $a \in E$, for every primitive atom a. The central lemma in the proof shows that for any sentence α, we have $\alpha \in E$ iff $w_E \models \alpha$. The only difficulty is the case where α is of the form $\Box_S\,\beta$. We begin to rewrite β to an equivalent ground declaration into conjunctive normal form, by using (A2)-(A6). The "if" part of the proof is based on axioms (A8) and (A9). Dually, the "only if" part of the proof is built from axioms (A7) and (A9).

3 Approximate Reasoning

After an excursion into the logic **AFOR**, we now apply our results to the formalization of approximate first-order reasoning. In the knowledge representation paradigm, the main task for a reasoner is to decide whether a query is entailed, or not, from the knowledge base. In general, this task is divided into two steps. First, convert the knowledge base and the negation of the query into clausal form and next, determine whether the resulting declaration is satisfiable or not. In this study, we concentrate on the second step of the reasoning process.

From this perspective, we specify a resource-bounded reasoner as an "abstract type" that takes in input a clausal declaration α and an increasing sequence of resources $(S_0, \cdots, S_n)$, and that approximates the problem of deciding whether α is satisfiable, or not, by means of two dual families of modal operators $(\Box_{S_0}, \cdots, \Box_{S_n})$ and $(\Diamond_{S_0}, \cdots, \Diamond_{S_n})$. If we prove that $\Box_{S_i} \alpha$ is satisfiable for any index i, then we have proved that α is satisfiable. Dually, if we prove that $\Diamond_{S_i} \alpha$ is unsatisfiable for any i, then we have proved that α is unsatisfiable. This stepwise process has the important advantage that the iteration may be stopped when a confirming answer is already obtained for a small index i.

Before examining into detail the properties of approximate reasoning, we introduce some useful definitions. An *Herbrand signature* is a first-order signature such that the set of function symbols F contains at least one constant symbol, and the set of standard names N is the set of all ground terms built from F. In other words, N is the *Herbrand universe* of the signature. Such a context greatly simplifies the technical aspects of the semantics. Specifically, in the language defined over a Herbrand signature, there exists exactly one denotation function d from ground terms to standard names, namely the identity function. Thus, any valuation v is uniquely determined by its components N_v and L_v.

A *clause* is a disjunction of literals. A *clausal declaration* is a declaration in prenex normal form containing only universal quantifiers, whose matrix is a disjunction of clauses. When clear from the context, such sentences will be respectively modeled as sets of literals and sets of clauses. With each clausal declaration α, we can uniquely associate a Herbrand signature whose function and predicate symbols are those occurring in α, with the additional condition that if α does not contain any constant, then we introduce a new constant in its signature. To avoid some complicated notations, we assume from now that the underlying representation language of a clausal declaration α is the language built from the Herbrand signature of α.

With these notions in hand, we can now examine the semantical properties of approximations. First, we show that resource-bounded reasoning is *improvable* and *dual.* The quality of approximations improves as we increase the resources. Moreover, both sound approximations and complete ones are improvable.

Theorem 2 (Monotonicity). *For any clausal declaration α and any resource parameters R and S such that $R \subseteq S$,*

$$\text{if } \Box_R \alpha \text{ is satisfiable, then } \Box_S \alpha \text{ is satisfiable,} \tag{1}$$

$$\text{if } \Diamond_R \alpha \text{ is unsatisfiable, then } \Diamond_S \alpha \text{ is unsatisfiable.} \tag{2}$$

Proof (1). Suppose that $\Box_R\,\alpha$ is satisfiable. Then $w \models \Box_R\,\alpha$ for some world w. Let v denotes $\cap\mathcal{R}_R(w)$. By construction, $N_v = N_R$ and $L_v = L_w \cap L_R$. Moreover, by lemma 2, it follows that $v \models \alpha$. Now, let us define a total mapping τ from N_S to N_R such that $\tau(n) = n$ for every $n \in N_R$. Let v' be a new valuation where $N_{v'} = N_S$ and $L_{v'}$ is the set of all literals $l(\boldsymbol{n})$ such that $l(\tau(\boldsymbol{n})) \in L_v$. Suppose that $v' \not\models \alpha$. Then there exists at least one ground clause $\gamma(\boldsymbol{n})$ of α generated from N_S and such that $\gamma(\boldsymbol{n}) \cap L_{v'} = \varnothing$. It follows that $\gamma(\tau(\boldsymbol{n})) \cap L_v = \varnothing$. Since $\gamma(\tau(\boldsymbol{n}))$ is a ground clause of α generated from N_R, it follows that $v \not\models \alpha$, but this contradicts the former hypothesis. Hence, $v' \models \alpha$. Let w' be a new world such that $L_{w'} = L_{v'} \cup (L_w / L_S)$ and let v'' denotes $\cap\mathcal{R}_S(w)$. By construction, $N_{v''} = N_S$ and $L_{v''} = L_{v'} \cap L_R$. Thus $v' \subseteq v''$. By lemma 1, it follows that $v'' \models \alpha$. Hence, $w' \models \Box_S\,\alpha$. Therefore, $\Box_S\,\alpha$ is satisfiable.

Proof (2). Suppose that $\Diamond_S\,\alpha$ is satisfiable. Then $w \models \Diamond_S\,\alpha$ for some world w. Let v and v' denote $\cup\mathcal{R}_S(w)$ and $\cup\mathcal{R}_R(w)$, respectively. By application of lemma 2, we have $v \models \alpha$. Now, let us define a new valuation v'' such that $N_{v''} = N_v$ and $L_{v''} = L_v \cup (L - L_R)$. By construction, $v \subseteq v''$. Thus, by lemma 1 it follows that $v'' \models \alpha$. Moreover, it is clear that $L_{v''} = L_{v'}$. Suppose that $v' \not\models \alpha$. Then, there exists at least one ground clause γ of α generated from N_R and such that $\gamma \cap L_{v'} = \varnothing$. Since $N_R \subseteq N_S$, it follows that γ is a ground clause of α generated from N_S. Moreover, since $L_{v''} = L_{v'}$, it follows that $\gamma \cap L_{v''} = \varnothing$. Hence $v'' \not\models \alpha$, but this contradicts the former hypothesis. So, $v' \not\models \alpha$ and by lemma 2, it follows that $w \models \Diamond_R\,\alpha$. Therefore, $\Diamond_R\,\alpha$ is satisfiable.

Second, we demonstrate that there exists a systematic adequacy relationship between approximate reasoning and classical first-order reasoning.

Theorem 3 (Adequacy). *For any clausal declaration α and parameter S,*

$$\text{if } \Box_S\,\alpha \text{ is satisfiable, then } \alpha \text{ is satisfiable,} \tag{1}$$

$$\text{if } \Diamond_S\,\alpha \text{ is unsatisfiable, then } \alpha \text{ is unsatisfiable.} \tag{2}$$

Proof (1). Let R be a new parameter such that $N_R = N_R$ and P_R is the set of all predicates occurring in the formula α. Suppose that $\Box_S\,\alpha$ is satisfiable. By theorem 2 it follows that $\Box_R\,\alpha$ is satisfiable. Then $w \models \Box_R\,\alpha$ for some world w. Let τ be a function from N to N_R such that $\tau(n) = n$ for every $n \in N_R$. Let w' be a world where $L_{w'} = \{l(\boldsymbol{n}) : l(\tau(\boldsymbol{n})) \in L_w\}$. Suppose that $w' \not\models \alpha$. Then there exists a ground clause $\gamma(\boldsymbol{n})$ of α such that $w' \not\models \gamma(\boldsymbol{n})$. It follows that $w \not\models \gamma(\tau(\boldsymbol{n}))$. However, by contraposition of axiom (A9) we obtain $w \not\models \Box_R\,\gamma(\tau(\boldsymbol{n}))$. Since $\gamma(\tau(\boldsymbol{n}))$ is a ground clause generated from N_R, it follows that $w \not\models \Box_R\,\alpha$, hence contradiction. So, $w' \models \alpha$ and therefore α is satisfiable.

Proof (2). We use the parameter R defined in part (1). Suppose that $\Diamond_S\,\alpha$ is unsatisfiable. By theorem 2 it follows that $\Diamond_R\,\alpha$ is unsatisfiable. Let E be the set of all ground clauses of α generated by N_R. Clearly, $\Diamond_R\,E$ is unsatisfiable. Moreover, from axiom (A9) and rule (R1) we infer that $E \supset \Diamond_R\,E$ is a theorem of our logic. By contraposition of this theorem, it follows that E is unsatisfiable. However, E is a finite subset of all ground instances of clauses of α. By application of Herbrand's theorem [3] (only if part), it follows that α is unsatisfiable.

Third and finally, we guarantee the convergence of approximate unsatisfiability: if a declaration is unsatisfiable, then using enough resources, we are guaranteed to find the correct solution. Notice that we cannot hope obtaining an analogue result for the dual part: since first-order unsatisfiability is recursively but *not* co-recursively denumerable, a infinite amount of resources may be necessary for determining satisfiability of a first-order clausal declaration.

Corollary 1 (Convergence). *For any clausal declaration α, if α is unsatisfiable then there exists a resource parameter S such that $\Diamond_S\, \alpha$ is unsatisfiable.*

Proof. Suppose that α is unsatisfiable. Then, by application of Herbrand's theorem [3] (if part), there must exist a finite unsatisfiable set E of ground clauses of α. Let R_S and N_S be the set of all predicates and ground terms that occur in E. Clearly enough, $\Diamond_S\, E$ is unsatisfiable. Since E is a subset of the set of all ground clauses of α generated by N_S, it follows that $\Diamond_S\, \alpha$ is unsatisfiable.

Example 1. Suppose we are given the following declaration:

$$\alpha = \{\{p(x,y), r(x)\}, \{\neg q(f(b)), r(x)\}, \{\neg r(a), q(f(x))\}, \{\neg p(a,b), q(x)\}\}.$$

The Herbrand signature of α is defined by the sets $P = \{p,q,r\}$, $F = \{a,b,f\}$ and $N = \{a, b, f(a), f(b), \ldots\}$. We want to show that α is satisfiable. Hence, we need to find a parameter S such that $\Box_S\, \alpha$ is satisfiable. In fact, this happens with $P_S = \{q, r\}$ and $N_S = \{a, f(a)\}$ which are restricted subparts of the signature.

Example 2. Suppose we are given the following declaration:

$$\alpha = \{\{p(x)\}, \{\neg p(a), q(x)\}, \{r(g(x), y), q(f(a))\}, \{\neg p(b), \neg q(x)\}, \{\neg r(x, f(y))\}\}.$$

The Herbrand signature of the formula is defined by $P = \{p, q, r\}$, $F = \{a, b, f, g\}$ and $N = \{a, b, f(a), g(a), \ldots\}$. We want to show that α is unsatisfiable. So we need to find a parameter S such that $\Diamond_S\, \alpha$ is unsatisfiable. In fact this holds with $P_S = \{p, q\}$ and $N_S = \{a, b\}$ which are restricted subparts of the signature.

4 Approximate Computation

In this section, we investigate the computational aspects of approximate reasoning. We present an original algorithm, named AFOS, for approximate first-order satisfiability. We begin to specify the algorithm, next we prove its soundness and completeness and then we analyse its computational complexity.

Before exploring the algorithm into detail, we introduce some additional definitions. A *substitution* is a mapping θ from variables to terms. Given a resource parameter S, a S-*substitution* is a substitution θ such that the range of θ is a subset of N_S. Given two parameters R and S such that $R \subseteq S$, a (R,S)-*substitution* is a S-substitution θ such that the range of θ contains a nonempty subset of $N_S - N_R$. A clause δ is called a S-*instance* (resp. (R,S)-*instance*) of a clause γ if $\delta = \gamma\theta$ for some S-substitution (resp. (R,S)-*substitution*) θ. In the following, α_S denotes the set of all S-instances of α generated by S.

The algorithm AFOS, presented in figure 1, can be thought as an iterative instance-based theorem prover. The three major parts of the algorithm are: first the choice of new resources, second the resource-bounded instance generation, and third the satisfiability test by a standard propositional prover. The algorithm basically carries out these three steps until a proof for satisfiability or unsatisfiability is found or a time-space limit (i.e. interruption) is reached. It is important to remark that the procedure both returns the solution and the resources that have been spent for computing the solution. For example, suppose that AFOS(α) returns ($true, S$) for some clausal declaration α. The intuitive reading of this result is "α can be shown satisfiable using the resources S". Such an information not only provides knowledge about the solution but also meta-knowledge about the resources needed to compute the solution.

Input : a clausal declaration α;
Output: a resource parameter S and the truth-value *true* if $\Box_S\, \alpha$ is satisfiable, *false* if $\Diamond_S\, \alpha$ is unsatisfiable and *unknown* otherwise;

$S \longleftarrow S_0$;
$\alpha_S^\Box \longleftarrow \varnothing$;
$\alpha_S^\Diamond \longleftarrow \varnothing$;
while *not interruption*() **do**
  Choice of resource parameter;
  $R \longleftarrow S$;
  $S \longleftarrow choose()$;
  Instantiation;
  if $P_R = P_S$ **then**
    foreach (R, S)*-instance* γ *of* α **do**
      $\alpha_S^\Box \longleftarrow \alpha_S^\Box \cup \{\gamma \cap L_S\}$;
      if $\gamma \subseteq L_S$ **then** $\alpha_S^\Diamond \longleftarrow \alpha_S^\Diamond \cup \{\gamma\}$;
  else
    $\alpha_S^\Box \longleftarrow \varnothing$;
    $\alpha_S^\Diamond \longleftarrow \varnothing$;
    foreach S*-instance* γ *of* α **do**
      $\alpha_S^\Box \longleftarrow \alpha_S^\Box \cup \{\gamma \cap L_S\}$;
      if $\gamma \subseteq L_S$ **then** $\alpha_S^\Diamond \longleftarrow \alpha_S^\Diamond \cup \{\gamma\}$;
  Satisfiability;
  if $\alpha_S^\Box$ *is satisfiable* **then** return ($true, S$);
  if $\alpha_S^\Diamond$ *is unsatisfiable* **then** return ($false, S$);
return ($unknown, S$);

Fig. 1: Approximate First-Order Satisfiability (AFOS)

Interestingly, our algorithm incorporates several major features. First, the algorithm is *interruptible*: it can be stopped at any time and provide some answer. Second, the procedure is *dual*: it can compute at the same time the satisfiability and the unsatisfiability of a generic formula. Third, the instance generation step can be shown *progressive*. Specifically, if R is a subset of S, then any clause in the declaration $\alpha_R^\Box$ is a subset of some clause in the declaration $\alpha_S^\Box$; furthermore, the declaration $\alpha_R^\Diamond$ is a subset of the declaration $\alpha_S^\Diamond$. Fourth and finally, our algorithm can be shown *incremental* and *anytime* [30]. In particular, the reasoner can decide to fix the choice of predicates P_S for a certain number of iterations. During these iterations, the set of ground terms N_S is progressively increased and the declarations $\alpha_S^\Box$ and $\alpha_S^\Diamond$ are progressively expanded in an incremental fashion. If a solution is not found then the reasoner can choose a new set P_S, reinitialize the set N_S, and apply again the same strategy.

The approximation schema is general enough to be combined with a wide range of satisfiability testers. The underlying interest is to use methods which are appropriate for the problem at hand and that have been shown powerful enough for solving large size instances of the problem. Complete methods such as depth first search enumeration [27] can be used to compute at the same time the satisfiability of $\alpha_S^\Box$ and the unsatisfiability of $\alpha_S^\Diamond$. On the other hand, incomplete methods such as local search algorithms [8, 12] can be exploited if we concentrate on the satisfiability of $\alpha_S^\Box$.

The two following results clarify the interest of approximate computation. The first theorem gives soundness and completeness for the algorithm. The second theorem states its computational complexity. To this point, the last result claims that the two barriers of complexity in first order reasoning, that is "quantification" and "chaining", are bounded by the resource parameter S.

Theorem 4 (Soundness and Completeness for AFOS). *Given a clausal declaration α a resource parameter S and no interruption of the algorithm,*

$$\text{AFOS}(\alpha) \textit{ returns } (\textit{true}, S) \textit{ iff } \Box_S\, \alpha \textit{ is satisfiable,} \tag{1}$$

$$\text{AFOS}(\alpha) \textit{ returns } (\textit{false}, S) \textit{ iff } \Diamond_S\, \alpha \textit{ is unsatisfiable.} \tag{2}$$

Proof. We only examine part (1) as a dual strategy applies to part (2). we know that $\Box_S\, \alpha$ is satisfiable iff $w \models \Box_S\, \alpha$ for some world w. By lemma 2, $w \models \Box_S\, \alpha$ iff $\cap\mathcal{R}_S(w) \models \alpha$. Let v denotes $\cap\mathcal{R}_S(w)$. By semantic rules (4) and (6), $v \models \alpha$ iff $v \models \gamma$ for every S-instance γ of α. By semantic rules (1) and (5), $v \models \gamma$ iff $\gamma \cap L_v \neq \varnothing$. So, $v \models \alpha$ iff $v \models \alpha_S^\Box$ and hence, $\Box_S\, \alpha$ is satisfiable iff $\Box_S\, \alpha_S^\Box$ is satisfiable. Now suppose that AFOS(α) returns (*true*, S). Then $\alpha_S^\Box$ is satisfiable. Since the relations and terms that occur in $\alpha_S^\Box$ are subsets of R_S and N_S it follows that $\Box_S\, \alpha_S^\Box$ is satisfiable. Therefore, $\Box_S\, \alpha$ is satisfiable. Dually, assume that $\Box_S\, \alpha$ is satisfiable. So, $\Box_S\, \alpha_S^\Box$ is satisfiable. By axiom (A9), it follows that $\alpha_S^\Box$ is satisfiable. Provided that no interruption occurred, AFOS(α) returns (*true*, S).

Theorem 5 (Complexity). *For any clausal declaration α and any resource parameter S, deciding whether $\Box_S\, \alpha$ is satisfiable and $\Diamond_S\, \alpha$ is satisfiable can be computed in $O(|\alpha_S| \cdot 2^{|P_S|})$ time.*

Proof. Let us examine the sentence $\Box_S\, \alpha$. By application of theorem 4, $\Box_S\, \alpha$ is satisfiable iff AFOS(α) returns (*true*, S). The instantiation part of the algorithm must, at most, generate all S-instances of α which is $O(|\alpha_S|)$. So, the time complexity of this part is $O(|\alpha_S|)$. In the worst case, all the clauses of α are stored in $\alpha_S^\Box$, so the size of $\alpha_S^\Box$ is $O(|\alpha_S|)$. Moreover, the cardinality of all distinct ground atoms in $\alpha_S^\Box$ is $O(|P_S)|$. So, the worst-case time complexity of the satisfiability part is $O(|\alpha_S| \cdot 2^{|P_S|})$. A dual argument applies to the sentence $\Diamond_S\, \alpha$.

Example 3. We consider again example 1. Suppose that AFOS chooses the resource parameter S defined by $P_S = \{q, r\}$ and $N_S = \{a, f(a)\}$. We obtain $\alpha_S^\Box = \{\{r(a)\}, \{r(f(a))\}, \{\neg r(a), q(f(a))\}, \{q(a)\}, \{q(f(a)\}\}$. Clearly, $\alpha_S^\Box$ is satisfiable. Therefore, it follows that $\Box_S\, \alpha$ is satisfiable.

Example 4. Now consider again example 2 and suppose that AFOS selects the resource parameter S with $P_S = \{p, q\}$ and $N_S = \{a, b\}$. We obtain: $\alpha_S^\Diamond = \{\{p(a)\}, \{p(b)\}, \{\neg p(a), q(a)\}, \{\neg p(a), q(b)\}, \{\neg p(b), \neg q(a)\}, \{\neg p(b), \neg q(b)\}\}$. It is clear that $\alpha_S^\Diamond$ is unsatisfiable. Hence, it follows that $\Diamond_S\, \alpha$ is unsatisfiable.

Example 5. We assume that the following knowledge base E is part of a very large ontology of group theory and other algebraic structures. E uses a small signature which we assume to be part of a larger vocabulary.

$$E = \begin{cases} \{p(x, y, f(x, y))\}, \\ \{p(e, x, x)\}, \\ \{p(x, e, x)\}, \\ \{p(i(x), x, e)\}, \\ \{p(x, i(x), e)\}, \\ \{\neg p(x, y, v), \neg p(y, z, v), \neg p(x, w, u), p(v, z, u)\}, \\ \{\neg p(x, y, v), \neg p(y, z, v), \neg p(v, z, u), p(x, w, u)\}, \\ \{\neg s(x), \neg s(y), \neg p(x, i(y), z), s(z)\} \end{cases}$$

Suppose we are given the following query: for any given element in a subgroup, show that its inverse is still in the subgroup. This can be denoted by $(\forall x)(s(x) \supset s(i(x))$. Let $\alpha = E \cup \{\{s(a)\}, \{\neg s(i(a))\}\}$. We want to show that α is unsatisfiable. So the algorithm needs to find a resource parameter S such that $\alpha_S^\Diamond$ is unsatisfiable. In fact, this holds with $R_S = \{s, p\}$ and $N_S = \{a, i(a), e\}$. We remark that the number of clauses and atoms in $\alpha_S^\Diamond$ is 1475 and 30, respectively. By combining iterative instantiation with an improved version of the Davis-Putnam procedure [27], unsatisfiability should be inferred in short time.

5 Conclusion

This main motivation behind this work has been to obtain a model approximate reasoning that defines a computationally more attractive reasoner than classical first order logic. We have stressed on a multi-modal logic which contains a well-founded semantics and a correct and complete axiomatization. Based on this logic, we have shown that our framework integrates several major features: bounded resources, improvability and dual reasoning. Finally, we designed a resource-bounded algorithm with a simple modification of classical instance-based theorem provers, and we have discussed the quality and complexity guarantees of this approximate deduction mechanism.

There are various avenues of research that come out of this work. On the logic side, a first investigation is to extend the language with equality. In particular, the question whether first-order logic with equality can be embedded in our framework should be settled. A secondary, but important, investigation is to examine decidable sub-fragments of first-order logic such as, for instance, Schönfinkel-Bernays expressions. To this point, it would be interesting to obtain a convergence result for the satisfiability problem of these sub-fragments. On the algorithmic side, a number of open problems remain to be explored. In particular, an important issue is the development of "intelligent" strategies for the incremental choice of resources. This choice may be heuristic; search strategies advocated in the literature of instance-based theorem proving should play a major role in the global efficiency of the framework [4, 13, 18, 29]. For example, a well-known principle is to iteratively choose predicates and terms according to their increasing arity. More sophisticated heuristics can be developed by combining unification techniques used for instantiation and strategies employed in propositional testers. Alternatively, the choice of the resources may be guided by *control knowledge*; it is a commonplace that knowledge about the structure of a base is important for efficient inferences [24]. To this very point, we recall that AFOS communicates at the same time knowledge about the solution but also meta-knowledge about the computation. In the query-answering process this meta-knowledge should be automatically learned to perform an appropriate choice of the resources which guarantees a high degree of confidence. These criteria are under study and we hope that an intelligent control strategy will be possible for approximate first-order reasoning.

References

1. N. D. Belnap. A useful four-valued logic. In *Modern Uses of Multiple-Valued Logic*, pages 8–37. Reidel, Dordrecht, 1977.
2. M. Cadoli and M. Schaerf. Approximate inference in default reasoning and circumscription. *Fundamenta Informaticae*, 2:123–143, 1995.
3. C. Chang and R. C. Lee. *Symbolic Logic and Mechanical Theorem Proving.* Academic Press Inc., 1973.
4. H. Chu and D. A. Plaisted. CLIN-S: A semantically guided first-order theorem prover. *Journal of Automated Reasoning*, 18(2):183–188, 1997.
5. M. Dalal. Anytime clausal reasoning. *Annals of Mathematics and Artificial Intelligence*, 22(3-4):297–318, 1998.
6. A. del Val. Tractable databases: How to make propositional unit resolution complete through compilation. In *Proceedings of the 4th International Conference on Principles of Knowledge Representation and Reasoning*, pages 551–561, 1994.

7. J. W. Garson. Quantification in modal logic. In D. Gabbay and F. Guenthner, editors, *Handbook of Philosophical Logic, Volume II: Extensions of Classical Logic*, pages 249–307. D. Reidel Publishing Co., 1984.
8. J. Gu, P. W. Purdom, J. Franco, and B. W. Wah. Algorithms for satisfiability problem: A survey. In *Satisfiability Problem: Theory and Applications*, volume 35, pages 19–152. American Mathematical Society, 1997.
9. J. Hintikka. *Knowledge and Belief.* Cornell University Press, 1962.
10. D. W. Etherington J. M. Crawford. A non-deterministic semantics for tractable inference. In *Proceedings of the 15th National Conference on Artificial Intelligence*, pages 286–291, 1998.
11. H. Kautz, D. McAllester, and B. Selman. Encoding plans in propositional logic. In *Principles of Knowledge Representation and Reasoning*, pages 374–384, 1996.
12. H. Kautz and B. Selman. Pushing the envelope: Planning, propositional logic and stochastic search. In *Proceedings of the 13th National Conference on Artificial Intelligence*, pages 1194–1201, 1996.
13. S. Kim and H. Zhang. ModGen: Theorem proving by model generation. In *Proc. of the 12th National Conference on Artificial Intelligence*, pages 162–167, 1994.
14. F. Koriche. Approximate reasoning about combined knowledge. In *Intelligent Agents Volume IV*, volume 1365 of *LNAI*, pages 259–273. Springer Verlag, 1998.
15. F. Koriche. A logic for anytime deduction and anytime compilation. In *Logics in Artificial Intelligence*, volume 1489 of *LNAI*, pages 324–342. Springer Verlag, 1998.
16. G. Lakemeyer. Limited reasoning in first-order knowledge bases. *Artificial Intelligence*, 71:213–255, 1994.
17. G. Lakemeyer. Limited reasoning in first-order knowledge bases with full introspection. *Artificial Intelligence*, 84:209–255, 1996.
18. S. J. Lee and D. A. Plaisted. Problem solving by searching for models with a theorem prover. *Artificial Intelligence*, 69(1-2):205–233, 1994.
19. H. J. Levesque. Foundations of a Functional Approach to Knowledge Representation. *Artificial Intelligence*, 23:125–212, 1984.
20. H. J. Levesque. A logic of implicit and explicit belief. In *Proceedings of the 6th National Conference on Artificial Intelligence*, pages 198–202, 1984.
21. H. J. Levesque and R. J. Brachman. Expressiveness and tractability in knowledge representation and reasoning. *Computational Intelligence*, 3(2):78–93, 1987.
22. F. Massacci. Anytime approximate modal reasoning. In *Proceedings of 15th National Conference on Artificial Intelligence*, pages 274–279, 1998.
23. R. Reiter. What should a database know? In *Proceedings of the 7th ACM Symposium on Principles of Database Systems*, pages 302–304, 1988.
24. M. Schaerf and M. Cadoli. Tractable reasoning via approximation. *Artificial Intelligence*, 74:249–310, 1995.
25. J. Slaney. SCOTT: A model-guided theorem prover. In *Proceedings of the 13th International Joint Conference on Artificial Intelligence*, pages 109–115, 1993.
26. M. Y. Vardi. On the complexity of bounded-variable queries. In *Proceedings of the 14th ACM Symposium on Principles of Database Systems*, pages 266–276, 1995.
27. H. Zhang and M. E. Stickel. Implementing the davis-putnam method. *Journal of Automated Reasoning*, 24(1/2):277–296, 2000.
28. J. Zhang and H. Zhang. SEM: a system for enumerating models. In *Proc. of the 14th International Joint Conference on Artificial Intelligence*, pages 298–303, 1995.
29. Y. Zhu. *Efficient First-Order Semantic Deduction Techniques.* PhD thesis, University of North Carolina at Chapel Hill (USA), 1998.
30. S. Zilberstein. Using anytime algorithms in intelligent systems. *AI Magazine*, 17(3):73–83, 1996.

Inflationary Fixed Points in Modal Logic

A. Dawar[1,*], E. Grädel[2], and S. Kreutzer[2]

[1] University of Cambridge Computer Laboratory, Cambridge CB2 3QG, UK
anuj.dawar@cl.cam.ac.uk
[2] Mathematische Grundlagen der Informatik, RWTH Aachen, D-52065 Aachen
{graedel,kreutzer}@informatik.rwth-aachen.de
http://www-mgi.informatik.rwth-aachen.de

Abstract. We consider an extension of modal logic with an operator for constructing inflationary fixed points, just as the modal μ-calculus extends basic modal logic with an operator for least fixed points. Least and inflationary fixed point operators have been studied and compared in other contexts, particularly in finite model theory, where it is known that the logics IFP and LFP that result from adding such fixed point operators to first order logic have equal expressive power. As we show, the situation in modal logic is quite different, as the modal iteration calculus (MIC) we introduce has much greater expressive power than the μ-calculus. Greater expressive power comes at a cost: the calculus is algorithmically much less manageable.

1 Introduction

The modal μ-calculus L_μ is an extension of multi-modal logic with an operator for forming least fixed points. This logic has been extensively studied, having acquired importance for a number of reasons. In terms of expressive power, it subsumes a variety of modal and temporal logics used in verification, in particular LTL, CTL, CTL*, PDL and also many logics used in other areas of computer science, for instance description logics. On the other hand, L_μ has a rich theory, and is well-behaved in model-theoretic and algorithmic terms.

The logic L_μ is only one instance of a logic with an explicit operator for forming least fixed points. Indeed, in recent years, a number of fixed point extensions of first order logic have been studied in the context of finite model theory. It may be argued that fixed point logics play a central role in finite model theory, more important than first order logic itself. The best known of these fixed point logics is LFP, which extends first order logic with an operator for forming the least fixed points of positive formulae, defining monotone operators. In this sense, it relates to first order logic in much the same way as L_μ relates to propositional modal logic. However, a number of other fixed point operators have been extensively studied in finite model theory, including inflationary, partial, nondeterministic and alternating fixed points. All of these have in common that they allow the construction of fixed points of operators that are not necessarily monotone.

* Research supported by EPSRC grants GR/L69596 and GR/N23028.

L. Fribourg (Ed.): CSL 2001, LNCS 2142, pp. 277–291, 2001.

Furthermore, a variety of fragments of the fixed point logics formed have been studied, such as existential and stratified fragments, bounded fixed point logics, transitive closure logic and varieties of Datalog. Thus, there is a rich theory of the structure and expressive power of fixed point logics on finite relational structures and, to a lesser extent, on infinite structures.

In the present paper, we take a first step in the study of extensions of propositional modal logic by operators that allow us to form fixed points of non-monotone formulae. We focus on the simplest of these, that is the inflationary fixed point (also sometimes called the iterative fixed point). Though the inflationary fixed point extension of first order logic (IFP) is often used interchangeably with LFP, as the two have the same expressive power on finite structures, we show that in the context of modal logic, the inflationary fixed point behaves quite differently from the least fixed point.

Least and Inflationary Inductions. We begin by reviewing the known results on the logics LFP and IFP.

(1) On finite structures, LFP and IFP have the same expressive power [12].
(2) It is conjectured that IFP is strictly more expressive than LFP on infinite structures, but only partial results are known. On many interesting infinite structures, for instance in arithmetic $(\omega, +, \cdot)$, LFP and IFP are known to be equally expressive, but the translation of IFP into LFP can make the formulae much more complicated [6].
(3) On ordered finite structures, LFP and IFP express precisely the properties that are decidable in polynomial time.
(4) Simultaneous least or inflationary inductions do not provide more expressive power than simple inductions.
(5) The complexity of evaluating a formula ψ in LFP or IFP on a given finite structure $\mathfrak{A}$ is polynomial in the size of the structure, but exponential in the length of the formula. For formulae with a bounded number k of variables, the evaluation problem is PSPACE-complete [9], even for $k = 2$ and on fixed (and very small) structures. If, in addition to bounding the number of variables one also forbids parameters in fixed point formulae, the evaluation problem for LFP is computationally equivalent to the model checking problem for L_μ [11,17] which is known to be in NP $\cap$ co-NP, in fact in UP $\cap$ co-UP [14], and hard for PTIME. It is an open problem whether this problem can be solved in polynomial time. The model checking problem for bounded variable IFP does not appear to have been studied previously.

We also note that even though IFP does not provide more expressive power than LFP on finite structures, it is often more convenient to use inflationary inductions in explicit constructions. The advantage of using IFP is that one is not restricted to inductions over positive formulae. A non-trivial case in point is the formula defining an order on the k-variable types in a finite structure, an essential ingredient of the proof of the Abiteboul-Vianu Theorem, saying that least and partial fixed point logics coincide if and only if PTIME = PSPACE (see

[4,8,10]). Furthermore, IFP is more robust, in the sense that inflationary fixed points are well-defined, even when other, non-monotone, operators are added to the language (see, for instance, [7]).

Inflationary Inductions in Modal Logic. Given the close relationship between LFP and IFP on finite structures, and the importance of the μ-calculus, it is natural to study also the properties and expressive power of inflationary fixed points in modal logic. In this paper, we undertake a study of an analogue of IFP for modal logic. We define a modal iteration calculus, MIC, by extending basic multi-modal logic with simultaneous inflationary inductions. While deferring formal definitions until Section 2, we begin with an informal explanation.

In L_μ, we can write formulae $\mu X.\varphi$, which are true in state s of a transition system $\mathcal{K}$ if, and only if, s is in the least set X satisfying $X \leftrightarrow \varphi$ in $\mathcal{K}$. We can do this, provided that the variable X appears only positively in φ. This guarantees that φ defines a monotone operator and has a least fixed point. Moreover, the fixed point can be obtained by an iterative process. Starting with the empty set, if we repeatedly apply the operator defined by φ (possibly through a transfinite series of stages), we obtain an increasing sequence of sets, which converges to the desired least fixed point. If, on the other hand, φ is not positive in X, we can still define an increasing sequence of sets, by starting with the empty set, and iteratively taking the union of the current set X with the set of states satisfying $\varphi(X)$, and this sequence must eventually converge to a fixed point (not necessarily of φ, but of the operator that maps X to $X \vee \varphi(X)$). More generally, we allow formulae $\mathbf{ifp}\ X_i : [X_1 \leftarrow \varphi_1, \dots, X_k \leftarrow \varphi_k]$ that construct sets by a simultaneous inflationary induction. At each stage α, we have a tuple of sets $X_1^\alpha, \dots, X_k^\alpha$. Substituting these into the formulae $\varphi_1, \dots, \varphi_k$ we obtain a new tuple of sets, which we *add* to the existing sets $X_1^\alpha, \dots, X_k^\alpha$, to obtain the next stage.

It is clear that MIC is a modal logic in the sense that it is invariant under bisimulation. In fact, on every class of bounded cardinality, inflationary fixed points can be unwound to obtain equivalent infinitary modal formulae. As a consequence, MIC has the tree model property. It is also clear that MIC is at least as expressive as L_μ. The following natural questions now arise.

(1) Is MIC more expressive than L_μ?
(2) Does MIC have the finite model property?
(3) What are the algorithmic properties of MIC? Is the satisfiability problem decidable? Can model checking be performed efficiently (as efficiently as for L_μ)?
(4) Can we eliminate, as in the μ-calculus and as in IFP, simultaneous inductions without losing expressive power?
(5) What is the relationship of MIC with monadic second-order logic and with finite automata? Or more generally, what are the 'right' automata for MIC?
(6) Is MIC the bisimulation-invariant fragment of any natural logic (as L_μ is the bisimulation-invariant fragment of MSO [13])?

We provide answers to most of these questions. From an algorithmic point of view, most of the answers are negative. From the point of view of expressiveness, we can say that in the context of modal logic, inflationary fixed points provide much more expressive power than least fixed points, and MIC has very different structural properties to L_μ. In particular, we establish the following results:

(1) There exist MIC-definable languages that are not regular. Hence MIC is more expressive than the μ-calculus, and does not translate to monadic second-order logic.
(2) MIC does not have the finite model property.
(3) The satisfiability problem for MIC is undecidable. In fact, it is not even in the arithmetic hierarchy.
(4) The model checking problem for MIC is PSPACE-complete.
(5) Simultaneous inflationary inductions do provide more expressive power than simple inflationary inductions. Nevertheless the algorithmic intractability results for MIC apply also to MIC without simultaneous inductions.
(6) There are bisimulation-invariant polynomial time properties that are not expressible in MIC.
(7) All languages in DTIME$(O(n))$ are MIC-definable.

No doubt, these properties exclude MIC as a candidate logic for hardware verification. On the other hand, the present study is an investigation into the structure of the inflationary fixed point operator and may suggest tractable fragments of the logic MIC, which involve crucial use of an inflationary operator, just as logics like CTL and alternation-free L_μ carve out efficiently tractable fragments of L_μ. In any case, it delineates the differences between inflationary and least fixed point constructs in the context of modal logic

In the rest of this paper, we begin in Section 2 by giving the necessary background on modal logic and fixed points, and giving the definition of MIC, along with an example that illustrates how this calculus has higher expressive power than L_μ. Section 3 establishes that MIC fails to have the finite model property and that the satisfiability problem is highly undecidable. This is established separately for MIC, and 1MIC, its fragment without simultaneous inductions. We also show that MIC is more expressive than 1MIC. In Section 5 we investigate questions of the computational complexity of MIC in the context of finite transitions systems. We show that the model checking problem is PSPACE-complete, that the class of models of any MIC formula is decidable in both polynomial time and linear space, and that there are polynomial time bisimulation-invariant properties that are not expressible in MIC. Finally, Section 6 investigates the expressive power of MIC on finite words, establishing that there are languages definable in MIC that are not context-free, and that every linear time decidable language is expressible in MIC.

Due to space limitations we only sketch the proofs of some results and defer the details to the full version of the paper [5].

2 The Modal Iteration Calculus

Before we define the modal iteration calculus, we briefly recall the definitions of propositional modal logic ML and the μ-calculus L_μ.

2.1 Propositional Modal Logic

Transition Systems. Modal logics are interpreted on transition systems (also called Kripke structures). Fix a set A of actions and a set $\mathcal{V}$ of atomic propositions. A transition system for A and $\mathcal{V}$ is a structure $\mathcal{K}$ with universe V (whose elements are called states) binary relations $E_a \subseteq V \times V$ for each $a \in A$ and monadic relations $p \subseteq V$ for each atomic proposition $p \in \mathcal{V}$ (we do not distinguish notationally between atomic propositions and their interpretations.)

Syntax of ML. For a set A of actions and a set $\mathcal{V}$ of proposition variables, the formulae of ML are built from *false*, *true* and the variables $p \in \mathcal{V}$ by means of Boolean connectives $\wedge$, $\vee$, $\neg$ and modal operators $\langle a \rangle$ and $[a]$. That is, if ψ is a formula of ML and $a \in A$ is an action, then $\langle a \rangle\psi$ and $[a]\psi$ are also formulae of ML. If there is only one action in A, one simply writes $\Box$ and $\Diamond$ for $[a]$ and $\langle a \rangle$, respectively.

Semantics of ML. The formulae of ML are evaluated on transition systems at a particular state. Given a formula ψ and a transition system $\mathcal{K}$ with state v, we write $\mathcal{K}, v \models \psi$ to denote that the formula ψ holds in $\mathcal{K}$ at state v. We also write $[\![\psi]\!]^{\mathcal{K}}$ to denote the set of states v, such that $\mathcal{K}, v \models \psi$. In the case of atomic propositions, $\psi = p$, we have $[\![p]\!]^{\mathcal{K}} = p$. Boolean connectives are treated in the natural way. Finally for the semantics of the modal operators we put

$$[\![\langle a \rangle\psi]\!]^{\mathcal{K}} := \{v : \text{ there exists a state } w \text{ such that } (v,w) \in E_a \text{ and } w \in [\![\psi]\!]^{\mathcal{K}}\}$$
$$[\![[a]\psi]\!]^{\mathcal{K}} := \{v : \text{ for all } w \text{ such that } (v,w) \in E_a, \text{ we have } w \in [\![\psi]\!]^{\mathcal{K}}\}.$$

Hence $\langle a \rangle$ and $[a]$ can be viewed as existential and universal quantifiers 'along a-transitions'.

2.2 The μ-Calculus L_μ

Syntax of L_μ. The μ-calculus extends propositional modal logic ML by the following rule for building fixed point formulae: if ψ is a formula in L_μ and X is a propositional variable that occurs only positively in ψ, then $\mu X.\psi$ and $\nu X.\psi$ are L_μ formulae.

Semantics of L_μ. A formula $\psi(X)$ with a propositional variable X defines on every transition system $\mathcal{K}$ (with state set V, and with interpretations for free variables other than X occurring in ψ) an operator $\psi^{\mathcal{K}} : \mathcal{P}(V) \to \mathcal{P}(V)$ assigning to every set $X \subseteq V$ the set $\psi^{\mathcal{K}}(X) := [\![\psi]\!]^{\mathcal{K},X} = \{v \in V : (\mathcal{K}, X), v \models \psi\}$.

As X occurs only positively in ψ, the operator $\psi^{\mathcal{K}}$ is *monotone* for every $\mathcal{K}$, and therefore, by a well-known theorem due to Knaster and Tarski, has a least fixed point $\mathbf{lfp}(\psi^{\mathcal{K}})$ and a greatest fixed point $\mathbf{gfp}(\psi^{\mathcal{K}})$. Now we put $[\![\mu X.\psi]\!]^{\mathcal{K}} := \mathbf{lfp}(\psi^{\mathcal{K}})$ and $[\![\nu X.\psi]\!]^{\mathcal{K}} := \mathbf{gfp}(\psi^{\mathcal{K}})$.

Least (and greatest) fixed points can also be constructed inductively. Given a formula $\mu X.\psi(X)$, we define for each ordinal α, the stage X^α of the **lfp**-induction of $\psi^{\mathcal{K}}$ by $X^0 := \emptyset$, $X^{\alpha+1} := [\![\psi]\!]^{(\mathcal{K},X^\alpha)}$, and $X^\alpha := \bigcup_{\beta<\alpha} X^\beta$ if α is a limit ordinal.

By monotonicity, the stages of the **lfp**-induction increase until a fixed point is reached. The first ordinal at which this happens is called the *closure ordinal* of the induction. By ordinal induction, one easily proves that this inductively constructed fixed point coincides with the least fixed point. The cardinality of a closure ordinal cannot be larger than the cardinality of $\mathcal{K}$.

For any formula φ, the formula $\nu X.\varphi$ is equivalent to $\neg\mu X.\neg\varphi(\neg X)$, where $\varphi(\neg X)$ denotes the formula obtained from φ by replacing all occurrences of X with $\neg X$.

Simultaneous Fixed Points. There is a variant of L_μ that admits systems of simultaneous fixed points. Here one associates with any tuple $\overline{\psi} = (\psi_1, \dots, \psi_k)$ of formulae $\psi_i(\overline{X}) = \psi_i(X_1, \dots, X_k)$, in which all occurrences of all X_i are positive, a new formula $\varphi = \mu\overline{X}.\overline{\psi}$. The semantics of φ is induced by the least fixed point of the monotone operator $\psi^{\mathcal{K}}$ mapping $\overline{X}$ to $\overline{X}'$ where $X_i' = \{v \in V : (\mathcal{K}, \overline{X}), v \models \psi_i\}$. More precisely, $\mathcal{K}, v \models \varphi$ iff v is an element of the first component of the least fixed point of the above operator. Although these systems are computationally beneficial and sometimes also allow for more straightforward formalisations, they do not increase the expressive power. It is known that simultaneous least fixed points can be eliminated in favour of nested individual fixed points (see e.g. [1, page 27]). Indeed, $\mu XY.[\psi(X,Y), \varphi(X,Y)]$ is equivalent to $\mu X.\psi(X, \mu Y.\varphi(X,Y))$, and this equivalence generalises to larger systems in the obvious way.

Bisimulations and Tree Model Property. Bisimulation is a notion of behavioural equivalence for transition systems. No reasonable modal logic can distinguish between two systems that are bisimulation equivalent. Formally, given two transition systems $\mathcal{K}$ and $\mathcal{K}'$, with distinguished states v and v' respectively, we say that $\mathcal{K}, v$ is bisimulation equivalent to $\mathcal{K}', v'$, written $\mathcal{K}, v \sim \mathcal{K}', v'$ if there is a relation $R \subseteq V \times V'$ between the states of $\mathcal{K}$ and the states of $\mathcal{K}'$ such that: (1) $(v, v') \in R$; (2) for each atomic proposition $p \in \mathcal{V}$ and each $(u, u') \in R$, $u \in [\![p]\!]^{\mathcal{K}}$ if, and only if, $u' \in [\![p]\!]^{\mathcal{K}'}$; (3) for each $(u, u') \in R$, and each $t \in V$ such that $(u, t) \in E_a$, there is a $t' \in V'$ with $(u', t') \in E_a'$ and $(t, t') \in R$; and (4) for each $(u, u') \in R$, and each $t' \in V'$ such that $(u', t') \in E_a'$, there is a $t \in V$ with $(u, t) \in E_a$ and $(t, t) \in R$.

Bisimulation equivalence corresponds to equivalence in an infinitary modal logic ML^∞ [2]. This logic is the closure of ML under disjunctions and conjunctions taken over arbitrary sets of formulae. Thus, if S is any set (possibly infinite) of formulae, then $\bigwedge S$ and $\bigvee S$ are also formulae of ML. It can be shown that for any transition systems $\mathcal{K}$ and $\mathcal{K}'$, $\mathcal{K}, v \sim \mathcal{K}', v'$ if, and only if, $\mathcal{K}, v$ makes true exactly the same formulae of ML^∞ as $\mathcal{K}', v'$.

A transition system is called a *tree*, if for every state v, there is at most one state u, and at most one action a such that $(u,v) \in E_a$ and there is exactly one state r, called the *root* of the tree, for which there is no state having a transition to r, and if every state is reachable from the root. It is known that for every transition system $\mathcal{K}$, and any state v, there is a tree $\mathcal{T}$ with root r such that $\mathcal{K}, v \sim \mathcal{T}, r$. One consequence of this is that any logic that respects bisimulation has the tree model property. For instance, for any formula φ of L_μ, if φ is satisfiable, then there is a tree $\mathcal{T}$ such that $\mathcal{T}, r \models \varphi$.

2.3 The Modal Iteration Calculus

We are now ready to introduce MIC. Informally, MIC is propositional modal logic ML, augmented with simultaneous inflationary fixed points.

Definition 2.1 (Syntax and semantics of MIC**).** MIC *extends propositional multi-modal logic by the following rule: if $\varphi_1, \dots, \varphi_k$ are formulae of* MIC, *and $X_1, \dots, X_k$ are propositional variables, then*

$$S := \begin{cases} X_1 \leftarrow \varphi_1 \\ \quad\vdots \\ X_k \leftarrow \varphi_k \end{cases}$$

is a system *of rules, and* (**ifp** $X_i : S$) *is a formula of* MIC. *If S consists of a single rule $X \leftarrow \varphi$ we simplify the notation and write* (**ifp** $X \leftarrow \varphi$) *instead of* (**ifp** $X : X \leftarrow \varphi$).

Semantics: On every Kripke structure $\mathcal{K}$, the system S defines, for each ordinal α, a tuple $\overline{X}^\alpha = (X_1^\alpha, \dots, X_k^\alpha)$ of sets of states, via the following inflationary induction (for $i = 1, \dots, k$).

$$\begin{aligned} X_i^0 &:= \emptyset, \\ X_i^{\alpha+1} &:= X_i^\alpha \cup [\![\varphi_i]\!]^{(\mathcal{K}, \overline{X}^\alpha)}, \\ X_i^\alpha &:= \bigcup_{\beta<\alpha} X_i^\beta \text{ if } \alpha \text{ is a limit ordinal.} \end{aligned}$$

We call $(X_1^\alpha, \dots, X_k^\alpha)$ the stage α of the inflationary induction of S on $\mathcal{K}$. As the stages are increasing (i.e. $X_i^\alpha \subseteq X_i^\beta$ for any $\alpha < \beta$), this induction reaches a fixed point $(X_1^\infty, \dots, X_k^\infty)$. Now we put $[\![(\mathbf{ifp}\ X_i : S)]\!]^{\mathcal{K}} := X_i^\infty$.

See Section 2.4 and 3 for examples of such formulae.

Lemma 2.2. $L_\mu \subseteq$ MIC. *Further, on every class of structures of bounded cardinality* MIC $\subseteq$ ML$^\infty$.

Proof. Clearly, if X occurs only positively in ψ, then $\mu X.\psi \equiv \mathbf{ifp}\ X \leftarrow \psi$. Hence $L_\mu \subseteq$ MIC.

Now, let S be a system of rules $X_i \leftarrow \varphi_i(X_1, \dots, X_k)$. It is clear that for each ordinal α there exist formulae $\varphi_1^\alpha, \dots, \varphi_k^\alpha \in \mathrm{ML}^\infty$ defining, over any Kripke structure, the stage α of the induction by S. As closure ordinals are bounded on structures of bounded cardinality, the second claim follows. □

Corollary 2.3. MIC *is invariant under bisimulation and has the tree model property.*

Note that on structures of unbounded cardinality, L_μ and MIC are not contained in ML^∞. For instance, well-foundedness is expressed by the L_μ-formula $\mu X.\Box X$, but is known not to be expressible in ML^∞.

2.4 Non-regular Languages

We now demonstrate that MIC is strictly more expressive than L_μ. Recall that every formula of L_μ can be translated into a formula of monadic second order logic (MSO). Moreover, it is known [3] that the only sets of finite words that are expressible in MSO are the regular languages. For our purposes, a finite word is a transition system with only one kind of action, which is a finite tree, and where every state has at most one successor.

Proposition 2.4. *There is a language that is expressible in* MIC *but not in* MSO.

Proof. The language $L := \{a^n b^m : n \leq m\}$ is not regular, hence not definable in monadic second-order logic, but it is definable in MIC. To see this, we consider first the formula $\pi(X) = (\mathbf{ifp}\ Y \leftarrow \Diamond(b \wedge \neg X) \vee \Diamond(a \wedge X \wedge Y))$ which (since the rule is positive in Y) is in fact equivalent to a L_μ-formula. On every word $w = w_0 \cdots w_{n-1} \in \{a,b\}^*$ and $X \subseteq \{0, \dots, n-1\}$, the formula is true if w starts with a (possibly empty) a-sequence inside X followed by a b outside X. Now the formula $(\mathbf{ifp}\ X \leftarrow (a \wedge \pi(X)) \vee (b \wedge \Box X))$ defines (inside a^*b^*) the language L. Note that the language a^*b^* is definable in L_μ, so we can conjoin this definition to the above formula to obtain a definition of L which works on all words in $\{a,b\}^*$ □

The observation in Proposition 2.4 was pointed out to us in discussion by Martin Otto, and was the starting point of the investigation reported here.

3 Interpreting Arithmetic in MIC

In this section we prove that the satisfiability problem of MIC is undecidable in a very strong sense. Given that MIC is invariant under bisimulation, we can restrict attention to trees. In fact we will only consider well-founded trees (i.e. trees satisfying the formula $\mathbf{ifp}\ X \leftarrow \Box X$). The *height* $h(v)$ of a node v in a well-founded tree $\mathcal{T}$ is an ordinal, namely the least upper bound of the heights of its children. For any node v in a tree $\mathcal{T}$, we write $\mathcal{T}(v)$ for the subtree of $\mathcal{T}$ with root v. We first show that the nodes of finite height and the nodes of height ω are definable in MIC.

Lemma 3.1. *Let S be the system*

$$\begin{aligned} X &\leftarrow \Box \mathit{false} \vee (\Box X \wedge \Diamond \neg Y) \\ Y &\leftarrow X. \end{aligned}$$

Then, on every tree $\mathcal{T}$, $[\![\mathbf{ifp}\ X : S]\!]^{\mathcal{T}} = [\![\mathbf{ifp}\ Y : S]\!]^{\mathcal{T}} = \{v : h(v) < \omega\}$.

Proof. By induction we see that for each $i < \omega$, $X^i = \{v : h(v) < i\}$ and $Y^i = X^{i-1} = \{v : h(v) < i - 1\}$. As a consequence $X^\omega = Y^\omega = \{v : h(v) < \omega\}$. One further iteration shows that $X^{\omega+1} = Y^{\omega+1} = X^\omega$. □

With the system S exhibited in Lemma 3.1 we obtain the formulae finite-height := $(\mathbf{ifp}\ X : S)$ and ω-height := $\neg(\mathbf{ifp}\ X : S) \wedge \Box(\mathbf{ifp}\ X : S)$ which define, respectively, the nodes of finite height and the nodes of height ω. Note that ω-height is a satisfiable formula all of whose models are infinite.

Proposition 3.2. MIC *does not have the finite model property.*

We show that the satisfiability problem of MIC is undecidable. In fact MIC interprets full arithmetic on the heights of nodes. To prove this we first define some auxiliary formulae that will be used frequently throughout the paper. We always assume that the underlying structure is a well-founded tree.

- The formula nonempty(φ) := $(\mathbf{ifp}\ X \leftarrow \varphi \vee \Diamond X)$ expresses that φ holds somewhere in the subtree of the current node: $\mathcal{T}, v \models$ nonempty(φ) iff $[\![\varphi]\!]^{\mathcal{T}} \cap \mathcal{T}(v) \neq \emptyset$.
- Dually all(φ) := $(\mathbf{ifp}\ X \leftarrow \varphi \wedge \Box X)$ says that φ holds at all nodes of the subtree $\mathcal{T}(v)$.
- We say that a set X (in a tree $\mathcal{T}$) *encodes* the ordinal α if $X = \{v : h(v) < \alpha\}$. Let ordinal($X$) be the conjunction of the formula all($X \to \Box X$) with

$$\begin{aligned} \neg(\mathbf{ifp}\ Z : Y &\leftarrow \Box Y \\ Z &\leftarrow \text{nonempty}(\neg Y \wedge \Box Y \wedge X) \wedge \text{nonempty}(\neg Y \wedge \Box Y \wedge \neg X)). \end{aligned}$$

  It expresses that X encodes some ordinal. Indeed all($X \to \Box X$) says that with each node $v \in X$, the entire subtree rooted at v is contained in X. The second conjunct performs an inflationary induction incorporating into Y at each stage $\beta + 1$ all nodes of height β (which satisfy $\neg Y \wedge \Box Y$) and incorporates the root of the tree into Z if both X and its complement contain nodes of height β. Hence, at the end of the induction the root of the tree will *not* be contained in Z if, and only if, X does not distinguish between nodes of the same height. Together the two conjuncts imply that X contains all nodes up to some height.
- The formula number(X) = ordinal(X) $\wedge$ nonempty(finite-height $\wedge \neg X$) says that X encodes a natural number n (inside a tree of height $> n$).

Lemma 3.3. *Let $\mathcal{T}$ be a well-founded tree of height ω. There exist formulae* $\text{plus}(S,T)$ *and* $\text{times}(S,T)$ *of* MIC *such that, whenever the sets S and T encode in the tree $\mathcal{T}$ the natural numbers s and t, then* $[\![\text{plus}(S,T)]\!]^{\mathcal{T}}$ *encodes $s+t$, and* $[\![\text{times}(S,T)]\!]^{\mathcal{T}}$ *encodes st.*

Proof. Let

$$\begin{aligned}\text{plus}(S,T) := \textbf{ifp}\ Y : X &\leftarrow \Box X \\ Y &\leftarrow S \vee (\Box Y \wedge \text{nonempty}(X) \wedge \text{all}(X \rightarrow T)).\end{aligned}$$

Obviously at each stage n, we have $X^n = \{v : h(v) < n\}$. We claim that for each n, $Y^{n+1} = \{v : h(v) < s + \min(n,t)\}$. For $n = 0$ this is clear (note that for the case $s = 0$ this is true because the conjunct nonempty(X) prevents the Y-rule from being active at stage 1). For $n > 0$ the inclusion $X^n \subseteq T$ is true iff $n \leq t$. Hence we have $Y^{n+1} = \{v : h(v) < s + n\}$ in the case that $n \leq t$ and $Y^{n+1} = Y^n = \cdots = Y^t$ otherwise. To express multiplication we define

$$\begin{aligned}\text{times}(S,T) := \textbf{ifp}\ Y : X &\leftarrow \Box X \\ Y &\leftarrow \text{plus}(Y,S) \wedge \text{all}(\Box X \rightarrow T).\end{aligned}$$

We claim that $Y^n = \{v : h(v) < s \cdot \min(n,t)\}$. This is trivially true for $n = 0$. If it is true for $n < t$, then $Y^{n+1} = \{v : h(v) < sn + s\} = \{v : h(v) < s(n+1)\}$. Finally for $n \geq t$, the extension of $\Box X^n$ is $\{v : h(v) < n+1\}$ which is not contained in $T = \{v : h(v) < t\}$, hence $Y^{n+1} = Y^n = \cdots = Y^t$. □

Corollary 3.4. *For every polynomial $f(x_1,\dots,x_r)$ with coefficients in the natural numbers there exists a formula $\psi_f(X_1,\dots,X_r) \in$* MIC *such that for every tree $\mathcal{T}$ of height ω and all sets $S_1,\dots,S_r$ encoding numbers $s_1,\dots,s_r \in \omega$*

$$[\![\psi_f(S_1,\dots,S_r)]\!]^{\mathcal{T}} = \{v : h(v) < f(s_1,\dots,s_r)\}.$$

Proof. By induction on f.

- $\psi_0 := \textit{false}$.
- $\psi_1 := \Box \textit{false}$.
- $\psi_X := X$.
- $\psi_{f+g} := \text{plus}[S/\psi_f, T/\psi_g]$, i.e. the formula obtained by replacing in plus(S, T) the variables S and T by, respectively, ψ_f and ψ_g.
- $\psi_{f \cdot g} := \text{times}[S/\psi_f, T/\psi_g]$.

□

Theorem 3.5. *For every first order sentence ψ in the vocabulary $\{+,\cdot,0,1\}$ of arithmetic, there exists a formula $\psi^* \in$* MIC *such that ψ is true in the standard model $(\mathbb{N},+,\cdot,0,1)$ of arithmetic if and only if ψ^* is satisfiable.*

Proof. We have already seen that there exists a MIC-axiom ω-height axiomatising the models that are bisimilar to a tree of height ω. Further, we can express set equalities $X = Y$ by all$(X \leftrightarrow Y)$ and we know how to represent polynomials by MIC-formulae. What remains is to translate quantifiers.

More precisely, we need to show that for each first order formula $\psi(y_1, \ldots, y_r)$ in the language of arithmetic there exists a MIC-formula $\psi^*(Y_1, \ldots, Y_r)$ such that on rooted trees $\mathcal{T}, w$ of height ω and all sets $S_1, \ldots, S_r$ that encode numbers $s_1, \ldots, s_r$ on $\mathcal{T}$ we have that $(\mathbb{N}, +, \cdot, 0, 1) \models \psi(s_1, \ldots, s_r)$ iff $\mathcal{T}, w \models \psi^*(S_1, \ldots, S_r)$.

Only the case of formula of the form $\psi(\overline{y}) := \exists x \varphi(x, \overline{y})$ remains to be considered. By induction hypothesis, we assume that for $\varphi(x, \overline{y})$ the corresponding MIC-formula $\varphi^*(X, \overline{Y})$ has already been constructed. Now let

$$\begin{aligned}\psi^*(\overline{Y}) := \textbf{ifp } Z : \; & X \leftarrow \Box X \\ & Z \leftarrow \varphi^*(X, \overline{Y}) \wedge \text{number}(X).\end{aligned}$$

□

Corollary 3.6. *The satisfiability problem for* MIC *is undecidable. In fact, it is not even in the arithmetical hierarchy.*

The proof given above appears to rely crucially on the use of simultaneous inductions. Indeed, one can show that formulae of MIC involving simultaneous inductions, in particular the formula constructed in the proof of Lemma 3.1, cannot be expressed without simultaneous inductions (see Theorem 4.2). However, it is still the case that first order arithmetic can be reduced to the satisfiability problem for MIC without simultaneous inductions. (See [5] for details.)

4 Simultaneous vs. Non-simultaneous Inductions

It is easy to see that the equivalence $\mu XY.(\psi, \varphi) \equiv \mu X.\psi(X, \mu Y.\varphi(X, Y))$ (sometimes called the Bekic-principle [1]) fails in both directions when we take inflationary instead of least fixed points. However, it still is conceivable that simultaneous inductions could be eliminated by more complicated techniques. It follows from the results below, that this is not the case, i.e. simultaneous inflationary inductions provide more expressive power than simple ones. Let 1MIC denote the fragment of MIC that does not involve simultaneous inductions.

For any ordinal α, let $\mathcal{T}_\alpha$ denote the tree with a root v_α that has a set $\{v_\beta \mid \beta < \alpha\}$ of children indexed by ordinals less than α, where each v_β is the root of a subtree isomorphic to $\mathcal{T}_\beta$.

Lemma 4.1. *Let $\varphi \in$ 1MIC be a formula. If $X_1, \ldots, X_k$ are atomic propositions on $\mathcal{T}_\omega$, closed under bisimulations, such that $v_\omega \notin X_i$ (where v_ω is the root of $\mathcal{T}_\omega$) and $\mathcal{T}_\omega, v_\omega \models \varphi(X_1, \ldots, X_k)$, then there is a finite N such that for all $n > N$ and all nodes v_n of height n, $\mathcal{T}_\omega, v_n \models \varphi(X_1 - \{v_n\}, \ldots, X_k - \{v_n\})$.*

It is a straightforward consequence of this lemma that the formula ω-height defined in Sect. 3 is not equivalent to any formula of 1MIC. We hence have established the following separation result.

Theorem 4.2. MIC *is strictly more powerful than* 1MIC.

5 The Model Checking Problem for MIC

Recall that the model checking problem for the μ-calculus is in UP $\cap$ co-UP, and is conjectured by some to be solvable in polynomial time. We now show that MIC is algorithmically more complicated (unless PSPACE = NP).

We first observe that the naive bottom-up evaluation algorithm for MIC-formulae uses polynomial time with respect to the size of the input structure, and polynomial space (and exponential time) with respect to the length of the formula. Let $\mathcal{K}$ be a transition system with n nodes and m edges. The size $||\mathcal{K}||$ of appropriate encodings of $\mathcal{K}$ as an input for a model checking algorithm is $O(n+m)$. It is well known that the extension $[\![\varphi]\!]^{\mathcal{K}}$ of a basic modal formula φ (without fixed points) on a finite transition system $\mathcal{K}$ can be computed in time $O(|\varphi| \cdot ||\mathcal{K}||)$. Further, any inflationary induction **ifp** $X_i : [X_1 \leftarrow \varphi_1, \dots, X_k \leftarrow \varphi_k]$ reaches a fixed point on $\mathcal{K}$ after at most kn iterations. Hence, the bottom-up evaluation of a MIC-formula ψ with d nested simultaneous inflationary fixed points, each of width k, on $\mathcal{K}$ needs at most $O((kn)^d)$ basic evaluation steps. For each fixed point variable occurring in the formula, $2n$ bits of workspace are needed to record the current value and the last value of the induction. This gives the following complexity results.

Proposition 5.1. *Any* MIC *formula ψ of nesting depth d and simultaneous inductions of width at most k on a transition system $\mathcal{K}$ with n nodes can be evaluated in time $O((kn)^d|\psi| \cdot ||\mathcal{K}||)$ and space $O(|\psi| \cdot n)$.*

In terms of common complexity classes the results can be stated as follows.

Theorem 5.2. *(1) The combined complexity of the model checking problem for* MIC *on finite structures is in* PSPACE.
(2) For any fixed formula $\psi \in$ MIC*, the model checking problem for ψ on finite structures is solvable in polynomial time and linear space.*

We now show that, contrary to the case of the μ-calculus, the complexity results obtained by this naive algorithm cannot be improved essentially.

Theorem 5.3. *There exist transitions systems $\mathcal{K}$, such that the model checking problem for* MIC *on $\mathcal{K}$ is* PSPACE*-complete (even for* 1MIC*).*

The proof is by reduction from QBF (the evaluation problem for quantified Boolean formulae.) We only sketch the argument here.

Let $\mathcal{K}$ be the Kripke-structure consisting of two points $0, 1$, the atomic proposition $p = \{1\}$, and the complete transition relation $\{0,1\} \times \{0,1\}$. Let $\alpha(X) := \neg X \wedge (p \rightarrow \Diamond X)$. Further, let $\varphi[X/\alpha(X)]$ denote the formula obtained from φ by replacing every free occurrence of X by $\alpha(X)$.

We inductively associate with every quantified Boolean formula ψ a MIC-formula ψ^* as follows. For $\psi := X$ we set $\psi^* := (p \wedge X) \vee (\neg p \wedge \Diamond X)$. Further, $(\neg\psi)^* := \neg\psi^*$ and $(\psi \circ \varphi)^* := \psi^* \circ \varphi^*$ for $\circ \in \{\wedge, \vee\}$. Finally, for $\psi := \forall X \varphi$ we put $\psi^* := \Box(\textbf{ifp } X \leftarrow \alpha(X) \wedge \varphi^*[X/\alpha(X)])$.

It can be shown that for any closed QBF-formula ψ, we have $[\![\psi^*]\!]^{\mathcal{K}} = \{0,1\}$ if ψ is true and $[\![\psi^*]\!]^{\mathcal{K}} = \emptyset$ otherwise. The theorem now follows immediately.

In [15], Otto introduced a higher-dimensional μ-calculus, denoted L_μ^ω which extends basic multi-modal logic with an operator for forming least fixed points of arbitrary arity, rather than just sets. He showed that L_μ^ω can express every bisimulation-invariant, polynomial time decidable property of finite structures. Since we know that any collection of finite structures definable in MIC is both bisimulation-invariant and polynomial time decidable, it follows that every formula of MIC can be translated to a formula of L_μ^ω that is equivalent to it on finite structures. We now show that the converse fails. In particular, there are properties of finite trees that are bisimulation-invariant and polynomial time decidable but cannot be expressed in MIC.

Theorem 5.4. *There is a collection $\mathcal{F}$ of finite trees in* PTIME*, closed under bisimulation, which is not expressible in* MIC.

We sketch the proof. Define $\mathcal{F}$ to be the collection of all finite trees $\mathcal{T}$ such that all children of the root of $\mathcal{T}$ are bisimilar. As bisimuation equivalence is decidable in polynomial time, it follows that $\mathcal{F}$ is in PTIME. It is also obvious that $\mathcal{F}$ is closed under bisimulation.

Assume, towards a contradiction, that there is a formula $\varphi \in$ MIC that defines $\mathcal{F}$. We use φ to define an equivalence relation on trees. Informally $\mathcal{T}_1 \sim_\varphi \mathcal{T}_2$ if at all stages of all the **ifp**-inductions in φ, the same subformulae of φ become true in $\mathcal{T}_1$ as in $\mathcal{T}_2$. Now, it can be shown that the index of $\sim_\varphi$ on trees of height n is bounded by $2^{p(n)}$ (for some polynomial $p(n)$ depending only on φ), whereas the bisimulation-index on trees of height n is not bounded by any elementary function. Hence there exist $\mathcal{T}_1 \sim_\varphi \mathcal{T}_2$ with $\mathcal{T}_1 \not\sim \mathcal{T}_2$. It is easy to see that φ cannot distinguish between those trees where every child of the root is the root of a copy of $\mathcal{T}_1$ and those trees where one of these copies is replaced by $\mathcal{T}_2$. But in the first case the tree is in $\mathcal{F}$, and in the second it is not, yielding a contradiction.

6 Languages

In this section we investigate the expressive power of MIC on finite strings. In other words we attempt to determine what languages are definable by formulae of MIC. For our purposes, a word w of length n, in an alphabet Σ is a transition system with n states $v_1, \ldots, v_n$, a single action such that $(v_i, v_j) \in E$ if, and only if, $j = i + 1$ and an atomic proposition s for each $s \in \Sigma$, such that for each v_i, there is a unique s with $v_i \in s$.

We have already seen in Proposition 2.4, that there are non-regular languages that are definable in MIC. We begin this section by strengthening this result and showing that there are languages definable in MIC that are not even context-free.

6.1 Non-CFLs in MIC

Theorem 6.1. *There is a language definable in* MIC *that is not context-free*

Proof. Consider the language $L := \{cwdw \mid w \in \{a,b\}^*\}$ over the alphabet $\{a,b,c,d\}$. It is easily verified that L is not a context-free language. To see that it is definable in MIC, first note that the formula

$$\alpha := c \wedge \Box\text{empty}(c) \wedge \text{nonempty}(d) \wedge \text{all}(d \rightarrow \Box\text{empty}(d))$$

defines the set of strings $\{cxdy \mid x,y \in \{a,b\}^*\}$. Now, the desired formula is the conjunction of α with the negation of the formula

$$\varphi := \textbf{ifp } X \leftarrow [\neg c \wedge (\Box X \vee \Box d)] \vee [c \wedge \text{nonempty}(\psi)]$$

where, ψ is the formula

$$\neg X \wedge \Box X \wedge [(b \wedge \text{nonempty}(a \wedge \Box X \wedge \neg X)) \vee (a \wedge \text{nonempty}(b \wedge \Box X \wedge \neg X))]$$

□

We can also add the observation that the formula constructed in the proof of Theorem 6.1 above does not involve any simultaneous inductions, and therefore there are non-context-free languages definable in 1MIC.

Another measure of the complexity of a language, considered in [16] is *automaticity*. Briefly, the automaticity of a language L is the function A_L which gives for each n the number of states in the smallest deterministic automaton which accepts a language that agrees with L on all strings of length at most n. Clearly, every regular language has constant automaticity. Here, we note that it can be shown that the language used in the proof of Theorem 6.1 has exponential automaticity, which is worst possible.

Finally, to place the expressive power of MIC in the Chomsky hierarchy, we note that every language definable in MIC can be defined by a context-sensitive grammar. This follows from the observation made in Section 5 that any class of finite structures defined by a formula of MIC is decidable in linear space, and the result that all languages decidable by nondeterministic linear space machines are definable by context-sensitive grammars.

6.2 Capturing Linear Time Languages

We have seen in Section 5 that the data complexity of evaluating MIC-formulae is in polynomial time and linear space. It is also clear that MIC can express PTIME-complete properties, as this is already the case for the μ-calculus.

On words the situation is somewhat different. The μ-calculus defines precisely the regular languages and hence is very far away from expressing PTIME-complete properties. On the other side we have already seen that there exist MIC-definable languages that are not even context-free. We will now show that MIC can in fact define all languages that are decidable in linear time (by a Turing machine).

An observation that we will use in the proof, but which may well be of independent interest, is that cardinality comparisons and addition of cardinalities are expressible in MIC on words (recall that none of these are MSO-definable).

Lemma 6.2. *There exists a formula* $\varphi(X, Y)$ *of* MIC *such that on every word* w*, we have* $w, X, Y \models \varphi$ *if and only if* $|X| = |Y|$*. Similarly for* $|X| < |Y|$ *and* $|X| + |Y| = |Z|$.

Theorem 6.3. *Every language* $L \in \text{DTIME}(O(n))$ *is* MIC*-definable.*

Note that we cannot expect to extend the result for linear time to quadratic time or higher. This is because, as we have seen, every language definable in MIC is decidable in linear space, and it is not expected that quadratic time is included in linear space.

References

1. A. Arnold and D. Niwinski. *Rudiments of μ-calculus*, North-Holland, 2001.
2. J. van Benthem. *Modal Logic and Classical Logic.* Bibliopolis, Napoli, 1983.
3. J. Büchi. Weak second-order arithmetic and finite automata. *Zeitschrift für Mathematische Logik und Grundlagen der Mathematik*, 6:66–92, 1960.
4. A. Dawar. *Feasible Computation Through Model Theory.* PhD thesis, University of Pennsylvania, 1993.
5. A. Dawar, E. Grädel, and S. Kreutzer. *Inflationary Fixed Points in Modal Logics.* See "http://www.cl.cam.ac.uk/users/ad260/pubs.html" or "http://www-mgi.informatik.rwth-aachen.de/Publications/".
6. A. Dawar and Y. Gurevich. Fixed point logics. In preparation.
7. A. Dawar and L. Hella. The expressive power of finitely many generalized quantifiers. *Information and Computation*, 123:172–184, 1995.
8. A. Dawar, S. Lindell, and S. Weinstein. Infinitary logic and inductive definability over finite structures. *Information and Computation*, 119:160–175, 1994.
9. S. Dziembowski. Bounded-variable fixpoint queries are PSPACE-complete. In *10th Annual Conference on Computer Science Logic CSL 96. Selected papers*, volume 1258 of *Lecture Notes in Computer Science*, pages 89–105. Springer, 1996.
10. H.-D. Ebbinghaus and J. Flum. *Finite Model Theory.* Springer, 2nd edition, 1999.
11. E. Grädel and M. Otto. On logics with two variables. *Theoretical Computer Science*, 224:73–113, 1999.
12. Y. Gurevich and S. Shelah. Fixed-point extensions of first-order logic. *Annals of Pure and Applied Logic*, 32:265–280, 1986.
13. D. Janin and I. Walukiewicz. On the expressive completeness of the propositional mu-calculus with respect to monadic second order logic. In *Proceedings of 7th International Conference on Concurrency Theory CONCUR '96*, number 1119 in Lecture Notes in Computer Science, pages 263–277. Springer-Verlag, 1996.
14. M. Jurdzinski. Deciding the winner in parity games is in UP ∩ Co-UP. *Information Processing Letters*, 68:119–124, 1998.
15. M. Otto. Bisimulation-invariant Ptime and higher-dimensional mu-calculus. *Theoretical Computer Science*, 224:237–265, 1999.
16. J. Shallit and Y. Breitbart. Automaticity I: Properties of a measure of descriptional complexity. *Journal of Computer and System Sciences*, 53:10–25, 1996.
17. M. Vardi. On the complexity of bounded-variable queries. In *Proc. 14th ACM Symp. on Principles of Database Systems*, pages 266–267, 1995.

Categorical and Kripke Semantics for Constructive S4 Modal Logic

Natasha Alechina[1], Michael Mendler[2], Valeria de Paiva[3], and Eike Ritter[4]

[1] School of Computer Science and IT, Univ. of Nottingham, UK
nza@cs.nott.ac.uk
[2] Department of Computer Science, Univ. of Sheffield, UK
michael@dcs.shef.ac.uk
[3] Xerox Palo Alto Research Center (PARC), USA
paiva@parc.xerox.com
[4] School of Computer Science, Univ. of Birmingham, UK
exr@cs.bham.ac.uk

Abstract. We consider two systems of constructive modal logic which are computationally motivated. Their modalities admit several computational interpretations and are used to capture intensional features such as notions of computation, constraints, concurrency, etc. Both systems have so far been studied mainly from type-theoretic and category-theoretic perspectives, but Kripke models for similar systems were studied independently. Here we bring these threads together and prove duality results which show how to relate Kripke models to algebraic models and these in turn to the appropriate categorical models for these logics.

1 Introduction

This paper is about relating traditional Kripke-style semantics for constructive modal logics to their corresponding categorical semantics. Both forms of semantics have important applications within computer science. Our aim is to persuade traditional modal logicians that categorical semantics is easy, fun and useful; just like Kripke semantics. Additionally we show that categorical semantics generates interesting new constructive modal logics, which differ somewhat from the traditional diet of intuitionistic modal logics[WZ95].

The salient feature of the constructive modal logics considered in this paper is the omission of the axioms $\Diamond(A \vee B) \to \Diamond A \vee \Diamond B$ and $\neg\Diamond\bot$, which are typically assumed for possibility $\Diamond$ not only in classical but also in intuitionistic settings. While in classical (normal) modal logics these principles follow from the properties of necessity $\Box$ there is no a priori reason to adopt them in an intuitionistic setting where the classical duality between $\Box$ and $\Diamond$ breaks down and $\Diamond$ is no longer derivable from $\Box$. In fact, a growing body of work motivated by computer science applications [Wij90,FM97,PD01] rejects these principles from a constructive point of view. In this paper we will study the semantics of two such constructive modal logics, CS4 and PLL, introduced below.

L. Fribourg (Ed.): CSL 2001, LNCS 2142, pp. 292–307, 2001.

We explore three standard types of semantics, Kripke, categorical, and algebraic semantics for CS4 and PLL. The algebraic semantics (CS4-modal algebra, PLL-modal algebra) is concerned only with equivalence of and the relative strength of formulas in terms of abstract semantic values (eg. truth values, proofs, constraints, etc...). It does not explain why a formula is true or why one formula is stronger than another. If one is interested in a more informative presentation and a concrete analysis of semantics, then a Kripke or categorical semantics may be more useful. The former explains 'meaning' in terms of worlds (in models) and validity of assertions at worlds (in models) in a classical Tarski-style interpretation. The 'semantic value' is given by the set of worlds at which a formula is valid. This form of semantics has been very successful for intuitionistic and modal logics alike. More recent and less traditional is the categorical approach. Here, we model not only the 'semantic value' of a formula, but also the 'semantic value' of its derivations/proofs, usually in a given natural deduction calculus. Thus, derivations in the logic are studied as entities in their own right, and have their own semantic objects in the models. Many applications of modal logic to computer science rely on having a term calculus for natural deduction proofs in the logic. Such a term calculus is a suitable variant of the λ-calculus, which is the prototypical functional programming language. From this point of view the semantic value of a formula is given by the collection of normal form programs that witness its assertion. Having a calculus of terms corresponding to derivations in the logic one obtains a direct correspondence between properties of proofs and properties of programs in the functional programming language based on these terms. For a discussion of the necessity modal operator $\Box$ and its interpretation as the 'eval/quote' operator in Lisp the reader is refered to [GL96].

In this sense both Kripke semantics and categorical semantics, presented here for CS4 and PLL, should be seen as two complementary elaborations of the algebraic semantics. They are both intensional refinements of their corresponding modal algebras, and have important applications within computer science. The natural correspondence between the Kripke models and modal algebras will be stated and proved as a *Stone Duality Theorem*. This turns out to require a different approach compared to other more standard intuitionistic modal logics, in particular as regards the $\Diamond$ modality. The other correspondence, between modal algebras and corresponding categorical structures, is essentially that between natural deduction proofs and the appropriate λ-calculus. This is known as the *Extended Curry-Howard Isomorphism*. Whereas the extended Curry-Howard isomorphism between intuitionistic propositional logic and the simply-typed λ-calculus has been known since the late 60s, establishing such isomorphisms for modal logics is a more recent development. In this paper we develop a suitable categorical semantics and associated λ-calculus for CS4 and PLL. It should be mentioned that the results for PLL are not new (see [FM97] for the Kripke and [BBdP98] for categorical semantics for PLL). Our contribution here is to show how PLL is related to CS4 and how these known results for PLL can be derived from those from CS4, or, to put it the other way round, how the known constructions for PLL may be generalised to CS4.

2 The Constructive Modal Systems CS4 and PLL

In this paper we take a fresh look at two prominent constructive modal extensions to intuitionistic propositional logic (IPL), which are particularly interesting because of their various applications in computer science.

To give the reader a taste for these applications, we list a few. Davies and Pfenning [DP96] use the $\Box$-modality to give a λ-calculus for computation in stages. The idea is that a term $\Box t$ represents a delayed computation. Ghani et al. [GdPR98] investigate refinements of this calculus which are suitable for the design of abstract machines. Similar ideas relating $\Box$ with staged evaluation and the distinction between run-time and compile-time semantics have been developed by Moggi et.al. [BMTS99]. Despeyroux and Pfenning [DPS97] use a box modality to encode higher-order abstract syntax in theorem-provers like Elf and Isabelle. Still another use of the $\Box$ modality, to model the `quote` mechanism of Lisp, is proposed by Goubault-Larrecq [GL96]. A $\Diamond$-style modality has been extensively used to distinguish a computation from its result in the λ-calculus: Moggi's [Mog91] influential work on computational monads describes the computational λ-calculus, which corresponds to an intuitionistic modal type theory with a $\Diamond$-like modality (see [BBdP98]). Fairtlough and Mendler [Men93,FMW97,Men00] use the same modality, which they call $\bigcirc$, in their work on lax logic for constraints and hardware verification. The calculus has also been used for denotational semantics of exception handling mechanisms, continuations, etc. On the syntactic side, it has been used, in the monadic-style of functional programming to add a notion of 'encapsulated state' to functional languages.

Despite their relevance for computer science these modal extensions of IPL seem to be less well investigated as modal logics in their own right, perhaps because of the "unusual properties" of their associated modal operators.

2.1 Constructive S4

The first modal system, which we call Constructive S4 (CS4), is a version of the intuitionistic S4 first introduced by Prawitz in his 1965 monograph [Pra65]. The Hilbert-style formulation of CS4 is obtained by extending IPL by a pair $\Box, \Diamond$ of S4-like intuitionistic modalities satisfying the axioms and the necessitation rule listed in Figure 1. The normal basis of CS4, *i.e.*, consisting only of axioms $\Box K$ and $\Diamond K$ plus the axiom $\neg\Diamond\bot$ (which we reject, see below) has been introduced[1] and motivated by Wijesekera [Wij90] as a predecessor to constructive concurrent dynamic logic. The practical importance of CS4 as a type system for functional programming is evident from the literature, e.g. as cited in the beginning of this section, though most applications so far focus on the $\Box$ modality. The formal

[1] Wijesekera considers a first order system, to be precise.

$\Box K : \Box(A \to B) \to (\Box A \to \Box B)$	$\Diamond K : \Box(A \to B) \to (\Diamond A \to \Diamond B)$
$\Box T : \Box A \to A$	$\Diamond T : A \to \Diamond A$
$\Box 4 : \Box A \to \Box\Box A$	$\Diamond 4 : \Diamond\Diamond A \to \Diamond A$
Nec : If A is a theorem then $\Box A$ is a theorem.	

Fig. 1. Hilbert-style system for Constructive S4

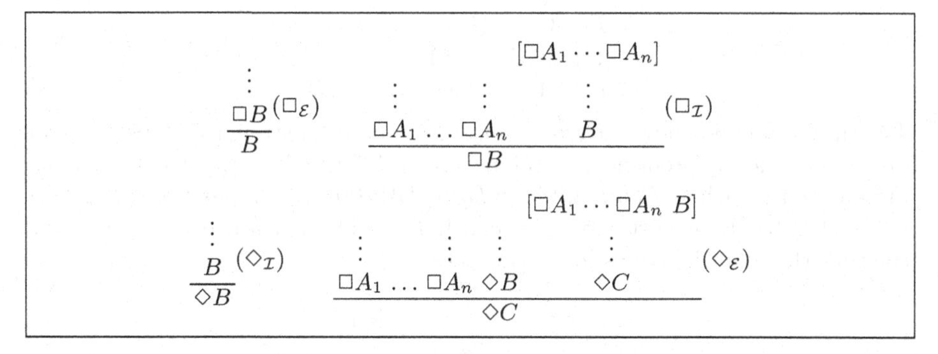

Fig. 2. Natural Deduction rules for Constructive S4

role of $\Diamond$ and its interaction with $\Box$ has recently been studied systematically by Pfenning and Davies [PD01].

The natural deduction formulation of CS4 is subject to some controversy. We recall it in the style of Bierman and de Paiva [BdP96]. The naive introduction rule for $\Box$ (corresponding to the necessitation rule *Nec*) insists that all of the undischarged assumptions at the time of application are modal, i.e. they are all of the form $\Box A_i$. However, the fundamental feature of natural deduction is that it is *closed under substitution* and this naive rule will not be closed under substitution, i.e. substituting a correct derivation in another correct derivation will yield an incorrect one (if this substitution introduces non-modal assumptions). We conclude that $\Box_{\mathcal{I}}$ must be formulated as in Figure 2, where the substitutions are given explicitly. The same sort of problem arises in the rules for $\Diamond_{\mathcal{E}}$ and the same solution (of explicit substitutions) can be used, see the rule $\Diamond_{\mathcal{E}}$ in Figure 2.

Both problems were first observed by Prawitz, who proposed a syntactically more complicated way of solving it [Pra65]. An interesting alternative approach has recently been presented by Pfenning and Davies [PD01], which (essentially) involves two kinds of variables, and two kinds of substitution. Note that in our solution the discharging brackets are used in a slightly different way from traditional natural deduction. In the introduction rule for $\Box$ they mean, discharge *all* assumptions (which must be all boxed in this rule).

The system CS4 is the weakest among the variants of intuitionistic S4 discussed in the literature. In particular, it does not prove the distribution of the possibility operator over disjunction $\Diamond(A \vee B) \to \Diamond A \vee \Diamond B$, nor does it assume $\neg\Diamond\bot$, *i.e.*, that possibly falsum ($\Diamond\bot$) and falsum ($\bot$) are equiprovable (which is the nullary form of the distribution). This version of non-classical S4 without distributivity of $\Diamond$ over $\vee$ is extremely well-behaved. As we will see there is a complete version of the Curry-Howard Isomorphism for it.

2.2 Propositional Lax Logic

The second constructive modal logic we consider is an extension of IPL that features a single modality $\Diamond$ satisfying the axioms

$$\begin{aligned} \Diamond T &: A \to \Diamond A \\ \Diamond 4 &: \Diamond\Diamond A \to \Diamond A \\ \Diamond F &: (A \to B) \to \Diamond A \to \Diamond B. \end{aligned}$$

The third axiom is known (categorically) as 'functorial strength'. This system is discussed under different names and in slightly differing but equivalent axiomatic presentations, such as *Computational Logic* [BBdP98] or *Propositional Lax Logic* (PLL) [FM97]. Henceforth we shall call it PLL. The natural deduction system contains the following rules for $\Diamond$ ([Men93]):

$$\frac{\begin{matrix}\vdots\\ B\end{matrix}}{\Diamond B}(\Diamond_{\mathcal{I}}) \qquad \frac{\begin{matrix}\vdots\\ \Diamond A\end{matrix} \qquad \begin{matrix}[A]\\ \vdots\\ \Diamond B\end{matrix}}{\Diamond B}(\Diamond_{\mathcal{E}})$$

PLL also has a colourful history. As a modal logic it was invented in the forties by Curry [Cur57] (who seems to have dropped it again because of its wild properties) and independently rediscovered in the nineties by Benton et al. and Fairtlough and Mendler, who used the symbol $\bigcirc$ for the modality, as the Curry-Howard isomorphic version of Moggi's computational lambda-calculus. As an algebra the system PLL is well known in abstract topology. The operator $\bigcirc$ arises naturally as a (strong, or multiplicative) closure operator on the lattice of open sets, or more generally as a so-called nucleus in the theory of topoi and sheafification [Joh82]. From this topological perspective, Goldblatt studied a system identical to PLL accommodating Lawvere's suggestion that the $\bigcirc$ modality means "it is locally the case that" by interpreting this in various ways to mean "at all nearby points" [Gol81,Gol93]. The algebraic properties of such operators (on complete Heyting algebras) have been explored by Macnab [Mac81], who calls them "modal operators".

In this paper we show how PLL can be naturally seen as a special CS4 theory or CS4 algebra in the sense that it can be obtained from CS4 by adding the axiom $A \to \Box A$. These results identify $\bigcirc$ as a constructive modality of possibility and provide a satisfactory explanation for why in PLL a modality $\Box$ is missing: it is implicitly built into the semantics already.

3 Kripke Models

Our first step is to develop a suitable Kripke model theory for CS4. While it is easy to agree that a Kripke model of constructive modal logic should consist of a set of worlds W and two accessibility relations, one intuitionistic $\leq$

and the other modal R, it is not so clear how these relations should interact (frame conditions) and just how they should be used to interpret specifically the $\Diamond$ modality. The mainstream approach as exemplified by Ewald [Ewa86], Fischer-Servi [FS80], Plotkin and Stirling [PS86], Simpson [Sim94] is based on the analogy of $\Box$ with $\forall$ and of $\Diamond$ with $\exists$-quantification over the modal accessibility R. Reading these quantifiers intuitionistically, relative to $\leq$, one arrives at the semantic interpretation $w \models \Box A$ iff $\forall v.\ w \leq v \Rightarrow \forall u.\ v\ R\ u \Rightarrow u \models A$ for necessity, and

$$w \models \Diamond A \text{ iff } \exists u.\ w\ R\ u\ \&\ u \models A \tag{1}$$

for possibility. Indeed, as the shown in the literature, this gives a fruitful basis for intuitionistic modal logics. Unfortunately, it is not suitable for CS4, since it forces the axiom $\Diamond(A \vee B) \to (\Diamond A \vee \Diamond B)$ to hold, which we want to avoid. It also requires an extra frame condition to ensure hereditariness of truth, *viz.*, that $w \models \Diamond A$ and $w \leq v$ implies $v \models \Diamond A$. Hereditariness, however, can also be achieved simply by $\forall$-quantifying over all $\leq$-successors in the interpretation of $\Diamond$:

$$w \models \Diamond A \text{ iff } \forall u.\ w \leq u \Rightarrow \exists v.\ u\ R\ v\ \&\ v \models A. \tag{2}$$

Not only does this away with the extra frame condition to force $\Diamond$ hereditary along $\leq$, it also eliminates the unwanted axiom $\Diamond(A \vee B) \to (\Diamond A \vee \Diamond B)$. In fact, as it turns out this works for CS4. This interpretation (2) of $\Diamond$, as far as we are aware, has been introduced by Wijesekera [Wij90] to capture non-deterministic computations and independently in [FM97] as an adequate Kripke interpretation of truth "up to constraints". In both cases the absence of the axioms $\Diamond(A \vee B) \to (\Diamond A \vee \Diamond B)$ is a natural consequence of the semantics.

Wijesekera only considered the normal base $\Box K$, $\Diamond K$ of CS4, yet included the axiom $\neg\Diamond\bot$. To eliminate the axiom $\neg\Diamond\bot$ we follow [FM97] in permitting explicit fallible worlds in our models. What remains, then, is to find suitable frame conditions on $\leq$ and R that are characterised by the CS4 axioms $\Box T$, $\Box 4$, $\Diamond T$, $\Diamond 4$. These are incorporated into the following notion of CS4 model:

Definition 1. *A* Kripke model of CS4 *is a structure $M = (W, \leq, R, \models)$, where W is a non-empty set, $\leq$ and R are reflexive and transitive binary relations on W, and $\models$ a relation between elements $w \in W$ and propositions A, written $w \models A$ ("A satisfied at w in M") such that:*

- *$\leq$ is hereditary with respect to propositional variables, that is, for every variable p and worlds w, w', if $w \leq w'$ and $w \models p$, then $w' \models p$.*
- *R and $\leq$ are related as follows: if wRw' and $w' \leq v$ then there exists v' such that $w \leq v'$ and $v'Rv$. In other words: $(R\, ; \leq) \subseteq (\leq ; R)$.*
- *The relation $\models$ has the following properties:*
  $w \models \top$;
  $w \models A \wedge B$ *iff* $w \models A$ *and* $w \models B$;
  $w \models A \vee B$ *iff* $w \models A$ *or* $w \models B$;

$w \models A \to B$ *iff* $\forall w'.\, w \leq w' \Rightarrow (w' \models A \Rightarrow w' \models B)$
$w \models \Box A$ *iff* $\forall w'.\, w \leq w' \Rightarrow \forall u.\, w'Ru \Rightarrow u \models A$
$w \models \Diamond A$ *iff* $\forall w'.\, w \leq w' \Rightarrow \exists u.\, w'Ru \wedge u \models A$

Notice that we do not have the clause $w \not\models \bot$, *i.e., we allow inconsistent worlds. Instead, we have*

- *if* $w \models \bot$ *and* $w \leq w'$, *then* $w' \models \bot$, *and*
- *if* $w \models \bot$, *then for every propositional variable* p, $w \models p$ *(to make sure that* $\bot \to A$ *is still valid).*

As usual, a formula A is *true* in a model $M = (W, \leq, R, \models)$ if for every $w \in W$, $w \models A$. We sometimes write $M, w \models A$ when we want to make the model explicit. A formula A is *valid* ($\models A$) if it is true in all models; a formula is satisfiable if there is a model and a *consistent* world where it is satisfied. A formula A is a *logical consequence* of a set of formulae Γ if for every M, w if $M, w \models \Gamma$, then $M, w \models A$.

Observe that under the translation of intuitionistic logic into classical S4 which introduces a modality $\Box_I$ corresponding to the intuitionistic accessibility relation $\leq$, our modalities $\Box$ and $\Diamond$ are translated as $\Box_I \Box_M$ and $\Box_I \Diamond_M$, respectively (where $\Box_M$ and $\Diamond_M$ are modalities corresponding to R). This means that our variant of S4 does not fall directly in the scope of Wolter and Zakharyaschev's analysis of intuitionistic modal logics as classical bimodal logics in [WZ97] since they assume $\Diamond$ to be a normal modality. However, analogous techniques could probably be used to give a new proof of decidability and finite modal property of CS4 and PLL.

Theorem 1. CS4 *is sound and strongly complete with respect to the class of models defined above, that is, for every set of formulae* Γ *and formula* A, *we have* $\Gamma \vdash_{\mathsf{CS4}} A \Leftrightarrow \Gamma \models A$.

We can use Theorem 1 to give a new soundness and completeness theorem for PLL. This is based on the observation that PLL models are a sub-class of CS4 models:

Definition 2. *A* Kripke model for PLL *is a Kripke model for* CS4 *where* R *is hereditary, that is, for every formula* A, *if* $w \models A$ *and* wRv, *then* $v \models A$.

The latter requirement corresponds to the strength axiom. It is in fact equivalent to the axiom $A \to \Box A$, so that $\Box$ becomes redundant in Kripke models for PLL. An alternative (slightly stronger) definition to the same effect given by Fairtlough and Mendler requires that R is a subset of $\leq$.

Theorem 2. PLL *is sound and strongly complete with respect to the class of models defined above.*

Proof. Soundness of PLL follows from soundness of CS4 and the fact that PLL-models satisfy the axiom scheme $A \to \Box A$, which renders the strength $\Diamond F$ axiom derivable from $\Diamond K$ of CS4.

For completeness consider an arbitrary set Γ of PLL-formulas, and a PLL-formula B such that $\Gamma \not\vdash_{\mathsf{PLL}} B$. Then, it is not difficult to see that $\Gamma^* \not\vdash_{\mathsf{CS4}} B$ where Γ^* is the theory Γ extended by all instances of the scheme $A \to \Box A$. For otherwise, if $\Gamma^* \vdash_{\mathsf{CS4}} B$, we could transform this derivation into a derivation $\Gamma \vdash_{\mathsf{PLL}} B$ simply by dropping all occurrences of $\Box$ in any formula, which means that every use of a CS4-axiom becomes an application of a PLL-axiom, and any use of an axiom $A \to \Box A$ or rule *Nec* becomes trivial. Note, this holds since if we drop all $\Box$ in a CS4 axiom, we get a PLL-axiom. By strong completeness of CS4 we conclude there exists a CS4-model M such that $M \models \Gamma^*$ but $M \not\models B$. But then not only $M \models \Gamma$ but also M validates all instances of $A \to \Box A$, which means that M is a PLL-model.

4 Modal Algebras and Duality

There is no unique 'right' Kripke semantics for a given system of modal logic. In general, the fit between modal (intuitionistic or classical) logics and Kripke structures is not perfect: apart from several versions of Kripke semantics for the same logic, which already seems suspect to category theorists, there are logics which are not complete for any Kripke semantics ([Fin74,Tho74]). *Modal algebras* have the definite advantage of fitting the logics much better.

One can think of an algebra as a collection of syntactic objects, e.g. formulae of a logic. Representation theorems for algebras show how given an algebra one can build a 'representation' for it - a structure which is a 'concrete' set-theoretic object, e.g. a Kripke model[2].

We define modal algebras corresponding to PLL and CS4 below and show how to construct representations for them. Since the modal algebras can be directly obtained from the respective categorical models, and modal algebras can be shown (see below) to be Stone-dually related to our Kripke models, we obtain an algebraic link (albeit a weak one) between Kripke models and categorical models for the two constructive modal systems considered.

Recall that a *Heyting algebra* H is a structure of the form $\langle A, \leq, \times, +, \Rightarrow, 0\rangle$ where A is a set of objects (one example would be formulae), $\leq$ is a partial order (for formulae, $a \leq b$ means 'a implies b'), $\times$ is a product (which corresponds to $\wedge$ in intuitionistic logic), $+$ a sum (corresponds to $\vee$), $\Rightarrow$ pseudocomplement (corresponds to $\to$) and 0 the least element ($\bot$).

We introduce two additional operators, corresponding to the modalities. Note that $\Box$ distributes over $\times$, but $\Diamond$ does not distribute over $+$.

Definition 3. *A* CS4*-modal algebra* $\mathcal{A} = \langle A, \leq, \times, +, \Rightarrow, 0, \Box, \Diamond\rangle$ *consists of a Heyting algebra* $\langle A, \leq, \times, +, \Rightarrow, 0\rangle$ *with two unary operators* $\Box$ *and* $\Diamond$ *on* A*, such that for every* $a, b \in A$,

$$\begin{array}{lll} \Box(a \times b) = \Box a \times \Box b & \Box a \leq a & a \leq \Diamond a \\ \Diamond a \leq \Diamond(a+b) & \Box a \leq \Box\Box a & \Diamond\Diamond a \leq \Diamond a \\ 1 \leq \Box 1 & \multicolumn{2}{l}{\Box a \times \Diamond b \leq \Diamond(\Box a \times b).} \end{array}$$

[2] More precisely, a general frame; see the discussion below.

Next, we identify the corresponding algebraic structure for PLL, which are also known, in a somewhat different axiomatisation, as "local algebras" [Gol76]:

Definition 4. *A* PLL*-modal algebra* $\mathcal{A} = \langle A, \leq, \times, +, \Rightarrow, 0, \Diamond \rangle$ *consists of a Heyting algebra* $\langle A, \leq, \times, +, \Rightarrow, 0 \rangle$ *with a unary operator* $\Diamond$ *on* A*, such that for every* $a, b \in A$,

$$\Diamond a \leq \Diamond(a+b) \quad a \leq \Diamond a \quad \Diamond\Diamond a \leq \Diamond a \quad a \times \Diamond b \leq \Diamond(a \times b).$$

Obviously, every Kripke model M for CS4 or PLL gives rise to a corresponding modal algebra M^+ (take the set of all definable sets of possible worlds).

Conversely, every modal algebra gives rise to a so-called *general frame.* A general frame is a structure which consists of a set of possible worlds W, two accessibility relations and a collection $\mathcal{W}$ of subsets of W *which can serve as denotations of formulae.* Intuitively, $\mathcal{W}$ should contain $\{w{:}w \models p\}$ for every propositional variable p and be closed under intersection, union and operations which give the set of worlds satisfying $\Box\varphi$ ($\Diamond\varphi$) from the set of worlds satisfying φ. (For more background, see for example [Ben83].)

Here, we will be somewhat sloppy and identify elements of the algebra with logical formulae straightaway. We assume that some subset P of A is arbitrarily designated as a set of propositional variables; $\times$, $+$, $\Rightarrow$ and 0 are interpreted as $\wedge$, $\vee$, $\rightarrow$ and $\perp$. Then we can formulate the representation theorem for models instead of general frames:

Theorem 3 (Representation for CS4). *Let* $\mathcal{A}$ *be a* CS4*-modal algebra. Then the Stone representation of* $\mathcal{A}$, $SR(\mathcal{A}) = (W^*, R^*, \leq^*, \models^*)$ *is a Kripke model for* CS4*, where*

1. W^* *is the set of all pairs* (Γ, Θ) *where* $\Gamma \subseteq \mathcal{A}$ *is a prime filter, and* $\Theta \subseteq \mathcal{A}$ *an arbitrary set of elements such that for all finite, nonempty, choices of elements* $c_1, \ldots, c_n \in \Theta$, $\Diamond(c_1 + \cdots + c_n) \notin \Gamma$.
2. $(\Gamma, \Theta) \leq^* (\Gamma', \Theta')$ *iff* $\Gamma \subseteq \Gamma'$.
3. $(\Gamma, \Theta) R^* (\Gamma', \Theta')$ *iff* $\forall a.\ \Box a \in \Gamma \Rightarrow a \in \Gamma'$ *and* $\Theta \subseteq \Theta'$.
4. *For all* $a \in \mathcal{A}$, $(\Gamma, \Theta) \models^* a$ *iff* $a \in \Gamma$.

Let us call pairs (Γ, Θ) with $\Gamma, \Theta \subseteq A$ *consistent theories* if for any, possibly empty, choice of elements $b_1, \ldots, b_m$ in Γ and any non-empty choice of elements $c_1, \ldots, c_n \in \Theta$, $b_1 \times \ldots \times b_m \not\leq \Diamond(c_1 + \cdots + c_n)$. Then, the worlds of $SR(\mathcal{A})$ are simply the consistent theories (Γ, Θ) where Γ is a prime filter. In the completeness proof we also need a slightly stronger notion of consistency as follows: For $a \in \mathcal{A}$, a theory (Γ, Θ) is *a-consistent* if for any choice of elements $b_1, \ldots, b_m$ in Γ and $c_1, \ldots, c_n \in \Theta$, $b_1 \times \ldots \times b_m \not\leq (a + \Diamond(c_1 + \cdots + c_n))$. This includes the degenerate case $n = 0$ where we simply require $b_1 \times \ldots \times b_m \not\leq a$.

The proof of our Stone Representation Theorem 3 relies on the following lemma.

Lemma 1 (Saturation Lemma). *Let* $a \in \mathcal{A}$ *and* (Γ, Θ) *an a-consistent theory in the* CS4*-algebra* $\mathcal{A}$*. Then* (Γ, Θ) *has a saturated a-consistent extension* (Γ^*, Θ)*, such that* Γ^* *is a prime filter and* $\Gamma \subseteq \Gamma^*$.

We can now extract without extra effort a Stone Representation for PLL algebras from that for CS4 algebras, identical to the one implicit in the completeness proof given in Fairtlough and Mendler [FM97].

Theorem 4 (Representation for PLL). *Let $\mathcal{A}$ be a PLL-modal algebra. Then the Stone representation of $\mathcal{A}$, $SR(\mathcal{A}) = (W^*, R^*, \leq^*, \models^*)$ is a Kripke model for PLL, where W^*, $\leq^*$, $\models^*$ are as above and $(\Gamma, \Theta)R^*(\Gamma', \Theta')$ iff $\Gamma \subseteq \Gamma'$ and $\Theta \subseteq \Theta'$.*

Proof. Observe that every PLL algebra $\mathcal{A}$ is at the same time a CS4 algebra $\mathcal{A}'$ where the operator $\Box$ is taken to be the identity function. Hence, we can construct its CS4 Stone representation $SR(\mathcal{A}')$ as in Theorem 3, which is a CS4 algebra. Now, what properties does the relation R^* have in $SR(\mathcal{A}')$? Well, $(\Gamma_1, \Theta_1)\ R^*\ (\Gamma_2, \Theta_2)$ iff
$\forall a.\ \Box a \in \Gamma_1 \Rightarrow a \in \Gamma_2$ and $\Theta_1 \subseteq \Theta_2$. But since $\Box$ is the identity operator, this is the same as $\Gamma_1 \subseteq \Gamma_2$ and $\Theta_1 \subseteq \Theta_2$ as defined in Theorem 4. Observe further that R^* is a subrelation of $\leq^*$, which means that R^* is hereditary. Thus, $SR(\mathcal{A}')$ is a PLL model.

Section 6 introduces categorical models for CS4 and PLL. Observe that one can view categorical models as modal algebras where the partial order relation $\leq$ is replaced by a collection of morphisms. Intuitively, (again thinking of objects as formulae) while $a \leq b$ in an algebra means that b is implied by a, the category has possibly several morphisms from a to b labelled by encodings of corresponding derivations of b from a.

5 Discussion on Kripke Semantics

Since our Kripke semantics for CS4 is new it deserves some further justification and discussion, which we give in this section.

First, how do our models relate to Wijesekera's? Let us call the class of structures $M = (W, \leq, R, \models)$ with $\leq$ reflexive and transitive but arbitrary R CK-*models* (*i.e.*, drop the requirement that R is reflexive and transitive as well as the frame condition $R;\leq\ \subseteq\ \leq;R$), and further those in which for all worlds $w \not\models \bot$ *infallible* CK models. Then, Wijesekera [Wij90] showed[3] that the theory $\mathsf{IPL} + \Box K + \Diamond K + \neg\Diamond\bot$ with the rules of Modus Ponens and *Nec* is sound and complete for the class of infallible CK models. The proof of Wijesekera can be modified to show that $\mathsf{CK} = \mathsf{IPL} + \Box K + \Diamond K$ is sound and complete for all CK models. Our CS4-models may then be seen as the special class of CK models characterised by the additional axioms $\Diamond T, \Box T, \Diamond 4, \Box 4$.

Following [FM97] we permitted *fallible worlds* to render the formula $\neg\Diamond\bot$ invalid. This makes CS4 different from traditional intuitionistic modal logics which invariably accept this axiom. Fallible worlds were used originally to provide an

[3] Actually, Wijesekera also lists the axiom $\Box A \wedge \Diamond(A \to B) \to \Diamond B$, but this is derivable already.

intuitionistic meta-theory for intuitionistic logic, *e.g.*,[TvD88,Dum77]. For intuitionistic propositional logics, with a classical meta-theory, fallible worlds are redundant. However, this is no longer true for modal logics. There, the presence or absence of fallible worlds is reflected in the absence or presence of the theorem $\neg\Diamond\bot$. In particular note that in the standard classical setting, i.e., without fallible worlds and $w \models \Diamond A$ meaning $\exists v.\ w\ R\ v\ \&\ v \models A$, the axiom $\neg\Diamond\bot$ (as well as $\Diamond(A \vee B) \rightarrow \Diamond A \vee \Diamond B$) is automatically validated.

It is not only the fallible worlds but also the extension by sets Θ, capturing hereditary refutation information, that distinguishes the representation of constructive modal logic, such as CS4, from that for standard intuitionistic modal logics, such as those of [PS86,FS80,Ewa86]. Indeed, if the axioms $\neg\Diamond\bot$ and $\Diamond(\phi \vee \psi) \rightarrow \Diamond\phi \vee \Diamond\psi$ are adopted the sets Θ and fallible worlds become redundant. Without these axioms, however, we also need the "negative" information in Θ to characterise truth at a world fully. It is also worthwhile to note that the model representation of Thm. 3 for CS4 is simpler than the one given by Wijesekera [Wij90] in the completeness proof for CK $+ \neg\Diamond\bot$. There, the Θ are (essentially) *sets of sets* of propositions, in which every element in Θ is a *set* of all possible future worlds for (Γ, Θ) that are accessible through R^*. This too, expresses negative information, though of a second-order nature. A quite different, but still second-order representation of CK models has been proposed by Hilken [Hil96]. As we have shown, however, the representation for CS4 can be done in a first-order fashion.

Our constructive S4 models satisfy the inclusion $R;\leq\ \subseteq\ \leq;R$, a frame condition that is typically assumed in standard intuitionistic modal logic already for system IK. One may wonder about the converse $\leq;R \subseteq R;\leq$ of this inclusion. One can show that in our models it generates the independent axiom scheme $((\Box A \rightarrow \Diamond B) \wedge \Box(A \vee \Diamond B)) \rightarrow \Diamond B$, thus inducing a proper extension of CS4.

As pointed out before, traditional intuitionistic modal logics such as those considered by Fischer-Servi [FS80] or Plotkin and Stirling [PS86] adopt a fundamentally different interpretation of $\Diamond$, defining $w \models \Diamond A$ *iff* $\exists v.\ w\ R\ v\ \&\ v \models A$. This enforces validity of $\Diamond(A \vee B) \rightarrow (\Diamond A \vee \Diamond B)$ but requires a frame condition $\leq^{-1};R \subseteq R;\leq^{-1}$ (confluence of $\leq$ and R) to make $\Diamond$ hereditary along $\leq$. It is not surprising, then, that for our constructive modal models, where hereditariness is built in by the semantic interpretation, this frame condition obtains the axiom scheme $\Diamond(A \vee B) \rightarrow (\Diamond A \vee \Diamond B)$, again inducing a proper extension.

We leave it as an open question if the above-mentioned axioms $((\Box A \rightarrow \Diamond B) \wedge \Box(A \vee \Diamond B)) \rightarrow \Diamond B$ or $\Diamond(A \vee B) \rightarrow (\Diamond A \vee \Diamond B)$ are complete for the frame conditions $\leq;R \subseteq R;\leq$ or $\leq^{-1};R \subseteq R;\leq^{-1}$, respectively. At least for PLL [FM97] it is known that $\leq^{-1};R \subseteq R;\leq^{-1}$ is completely captured by the axiom $\Diamond(A \vee B) \rightarrow (\Diamond A \vee \Diamond B)$, and in [Wij90] this axiom is linked with sequentiality of R.

6 Categorical Models

Categorical models distinguish between different proofs of the same formula. A category consists of objects, which model the propositional variables, and for every two objects A and B each morphism in the category from A to B, corresponds to a proof of B using A as hypothesis.

Cartesian closed categories (with coproducts) are the categorical models for intuitionistic propositional logic. For a proper explanation the reader should consult Lambek and Scott [LS85]; Here we just outline the intuitions. Conjunction is modelled by cartesian products, a suitable generalisation of the products in Heyting algebras. The usual logical relationship between conjunction and implication

$$A \wedge B \longrightarrow C \text{ if and only if } A \longrightarrow (B \rightarrow C)$$

is modelled by an adjunction and this defines categorically the implication connective. Thus we require that for any two objects B and C there is an object $B \rightarrow C$ such that there is a bijection between morphisms from $A \wedge B$ to C and morphisms from A to $B \rightarrow C$. Disjunctions are modelled by coproducts, again a suitable generalisation of the sums of Heyting algebras. True and false are modelled by the empty product (called a terminal object) and co-product (the initial object), respectively. Finally negation, as traditional in constructive logic, is modelled as implication into falsum. A cartesian closed category (with coproducts) is sometimes shortened to a ccc (respectively a bi-ccc). **Set**, the category where the objects are sets and morphisms between sets are functions, is the standard example of a bi-cartesian closed category.

To present a categorical model of constructive S4 we must add to a bi-ccc the structure needed to model the modalities. In previous work [BdP96] it was shown that to model the S4 necessity $\Box$ operator one needs a *monoidal comonad*. Such a monoidal comonad consists of an endofunctor $\Box\colon \mathcal{C} \longrightarrow \mathcal{C}$ together with natural transformations $\delta_A\colon \Box A \longrightarrow \Box\Box A$ and $\epsilon_A\colon \Box A \longrightarrow A$ and $m_{A,B}\colon \Box A \times \Box B \longrightarrow \Box(A \times B)$ and a map $m_1\colon 1 \rightarrow \Box 1$, satisfying some commuting conditions. These natural transformations model the axioms **4** and **T** together with the necessitation rule and the **K** axiom.

Here we assume that the modal operator $\Diamond$ is dually modelled by a *monad* with certain special characteristics: namely we want our monad to be *strong* with respect to the $\Box$ operator, i.e. we assume a natural transformation $st_{A,B}\colon \Box A \times \Diamond B \longrightarrow \Diamond(\Box A \times B)$ satisfying the conditions detailed in [Kob97]. The strength is needed to model the explicit substitution in the $\Diamond_{\mathcal{E}}$-rule.

Definition 5. *A* CS4*-category consists of a cartesian closed category $\mathcal{C}$ with coproducts, a monoidal comonad $(\Box, \delta, \epsilon, m_{-,-}, m_1)$ where $\Box\colon \mathcal{C} \longrightarrow \mathcal{C}$ and a $\Box$-strong monad $(\Diamond, \mu, \eta)$ where $\Diamond\colon \mathcal{C} \longrightarrow \mathcal{C}$.*

The soundness theorem shows in detail how the categorical semantics models the modal logic.

Theorem 5 (Soundness). *Let $\mathcal{C}$ be any* CS4*-category. Then there is a canonical interpretation $[\![\_]\!]$ of* CS4 *in $\mathcal{C}$ such that*

- *a formula A is mapped to an object $[\![A]\!]$ of $\mathcal{C}$;*
- *a natural deduction proof ψ of B using formulae $A_1, \ldots, A_n$ as hypotheses is mapped to a morphism $[\![\psi]\!]$ from $[\![A_1]\!] \times \cdots \times [\![A_n]\!]$ to $[\![B]\!]$;*
- *each two natural deduction proofs ϕ and ψ of B using formulae $A_1, \ldots, A_n$ as hypotheses which are equal (modulo normalisation of proofs) are mapped to the same morphism, in other words $[\![\phi]\!] = [\![\psi]\!]$.*

A trivial degenerate example of an CS4-category consists of taking any bi-ccc, say **Set** for example and considering the identity functor (both as a monoidal comonad and as monad) on it. Less trivial, but still degenerate models are Heyting algebras (the poset version of a bi-ccc) together with a closure and a co-closure operator. Non-degenerate models (but quite complicated ones) can be found in [GL96]. To prove categorical completeness we use a term model construction.

Theorem 6 (Completeness).

(i) There exists a CS4-category such that all morphisms are interpretations of natural deduction proofs.
(ii) If the interpretation of two natural deduction proofs is equal in all CS4-categories, then the two proofs are equal modulo proof-normalisation in natural deduction.

A categorical model of PLL consists of a cartesian closed category with a strong monad. These models were in fact the original semantics for Moggi's computational lambda-calculus and PLL can be seen as reverse engineering from that [BBdP98]. Hence we refrain from stating categorical soundness and completeness for this system, but of course they hold as expected [Kob97].

In the logic, PLL arises as a special case of CS4 when we assume the derivability of $A \to \Box A$. A similar statement holds in category theory. We have an inclusion functor from the category of PLL-categories into the category of CS4-categories: each PLL-category is a CS4-category where the co-monad is the identity functor. Conversely, each CS4-category such that $\Box A$ is isomorphic to A is a CS4-category.

7 Conclusions

This paper shows how traditional Kripke semantics for two systems of intuitionistic modal logic, CS4 and PLL, can be related via duality theory to the categorical semantics of (natural deduction) proofs for these logics. The associated notions of modal algebras serve as an intermediate reference point. From this point of view the results of this paper may be seen as presenting two kinds of representations for these modal algebras.

The first representation explains the semantics of an element in the algebra in terms of sets of worlds and truth within Kripke models. To this end we have developed an appropriate class of Kripke models for CS4 and proved a Stone representation theorem for it. As far as we are aware the model representation for CS4 is new. Its essential first-order character contrasts with the second order representations for the weaker system CK given by Wijesekera and Hilken. We have also shown how the canonical model construction of [FM97] for PLL follows from that for CS4 as a special case. Goldblatt [Gol76] proved a standard representation theorem for PLL algebras in terms of $\mathcal{J}$-frames, that only requires prime filters rather than pairs (Γ, Θ). However, Goldblatt's work explains $\bigcirc$ as a constructive modality of *necessity*, which is an altogether different way to look at $\bigcirc$.

The contribution of this paper regarding PLL lies in showing that the modality $\bigcirc$ of PLL is a constructive modality of *possibility*, in the sense that it can be obtained by adding to CS4 the axiom $A \to \Box A$. This is not the only way to derive PLL from CS4, but probably the most simple one so far proposed. Pfenning and Davies [PD01] give a full and faithful syntactic embedding PLL $\hookrightarrow$ CS4 that reads $\bigcirc A$ as $\Diamond \Box A$ and $A \to B$ as $\Box A \to B$. Both possibilities can be used to generate different semantics for PLL from that of CS4. The embedding discussed in this paper most closely reflects the notion of constraint models for PLL introduced in [FM97].

The second representation given in this paper explains the semantics of an element in the algebra in terms of provability in a natural deduction calculus. The representation theorem establishes a λ-calculus and Curry-Howard correspondence for CS4. In general, modal algebras can be extended to categorical models by adding information about proofs (replacing $\leq$ of the algebra by the collection of morphisms of the category), but this process is not trivial.

This extra information about proofs is crucial in applications of logic to model computational phenomena. While λ terms (encodings of proofs in intuitionistic propositional logic) can be seen as semantic counterparts of functional programs, addition of modalities to intuitionistic propositional logic makes it possible to obtain more sophisticated semantics of programs reflecting such computational phenomena as, for example, non-termination, non-determinism, side effects, etc. [Mog91]. Information about proofs can also be necessary in other applications of logic to computer science, where not just the truth (or falsity) of a formula is important, but also the justification (proof) of the claimed truth (see e.g. [Men93,FMW97,Men00]). One example we are considering is the verification of protocols.

The results in this paper partially depend on having a natural deduction presentation of the logic following the standard Prawitz/Dummett pattern of logical connectives described by introduction and elimination rules. This is true for CS4 and for PLL, but not for weaker logics, for example for a modal logic where $\Box$ satisfies only the K-axiom. Thus, our main challenge is to extend this work on categorical semantics to other modal logics.

Next we would like to apply our techniques to constructive temporal logics. Another direction we would like to pursue is providing concrete mathematical models for CS4. Some such applications might be generated as generalisation of our previous work on constraint verification in PLL. Meanwhile we shall continue our work on applications of constructive modal logics to programming.

Acknowledgements

The second author is supported by EPSRC (grant GR/M99637). We would like to thank Gavin Bierman, Richard Crouch and Matt Fairtlough for their useful comments and suggestions.

References

[BBdP98] N. Benton, G. Bierman, and V. de Paiva. Computational types from a logical perspective. *Journal of Functional Programming*, 8(2), 1998.

[BdP96] G. M. Bierman and V. de Paiva. Intuitionistic necessity revisited. Technical Report CSR-96-10, University of Birmingham, School of Computer Science, June 1996.

[Ben83] J. Benthem, van. *Modal logic and classical logic.* Bibliopolis, Naples, 1983.

[BMTS99] Z. Benaissa, E. Moggi, W. Taha, and T. Sheard. Logical modalities and multi-stage programming. In *Workshop on Intuitionistic Modal Logics and Application (IMLA'99)*, Satellite to FLoC'99, Trento, Italy, 6th July 1999. Proceedings available from `http://www.dcs.shef.ac.uk/~floc99im`.

[Cur57] H. B. Curry. *A Theory of Formal Deducibility*, volume 6 of *Notre Dame Mathematical Lectures.* Notre Dame, Indiana, second edition, 1957.

[DP96] R. Davies and F. Pfenning. A modal analysis of staged computation. In Guy Steele, Jr., editor, *Proc. of 23rd POPL*, pages 258–270. ACM Press, 1996.

[DPS97] J. Despeyroux, F. Pfenning, and C. Schürmann. Primitive recursion for higher-order abstract syntax. In P. de Groote and J. Roger Hindley, editors, *Proc. of TLCA'97*, pages 147–163. LNCS 242, Springer Verlag, 1997.

[Dum77] M. Dummett. *Elements of Intuitionism.* Clarendon Press, Oxford, 1977.

[Ewa86] W. B. Ewald. Intuitionistic tense and modal logic. *Journal of Symbolic Logic*, 51, 1986.

[Fin74] K. Fine. An incomplete logic containing S4. *Theoria*, 39:31 - 42, 1974.

[FM97] M. Fairtlough and M. Mendler. Propositional lax logic. *Information and Computation*, 137, 1997.

[FMW97] M. Fairtlough, M. Mendler, and M. Walton. First-order lax logic as a framework for constraint logic programming. Technical Report MIP-9714, University of Passau, July 1997. Postscript available through `http://www.dcs.shef.ac.uk/~michael`.

[FS80] G. Fischer-Servi. Semantics for a class of intuitionistic modal calculi. In M. L. Dalla Chiara, editor, *Italian Studies in the Philosophy of Science*, pages 59–72. Reidel, 1980.

[GdPR98] Neil Ghani, Valeria de Paiva, and Eike Ritter. Explicit Substitutions for Constructive Necessity. In *Proceedings ICALP'98*, 1998.

[GL96] J. Goubault-Larrecq. Logical foundations of eval/quote mechanisms, and the modal logic S4. Manuscript, 1996.

[Gol76] R. Goldblatt. Metamathematics of modal logic. *Reports on Mathematical Logic*, 6,7:31 - 42, 21 - 52, 1976.

[Gol81] R. Goldblatt. Grothendieck Topology as Geometric Modality. *Zeitschrift für Mathematische Logik und Grundlagen der Mathematik*, 27:495–529, 1981.

[Gol93] R. Goldblatt. *Mathematics of Modality.* CSLI Lecture Notes No. 43. Center for the Study of Language and Information, Stanford University, 1993.

[Hil96] B. P. Hilken. Duality for intuitionistic modal algebras. *Journal of Pure and Applied Algebra*, 148:171 – 189, 2000.

[Joh82] P. T. Johnstone. *Stone Spaces.* Cambridge University Press, 1982.

[Kob97] S. Kobayashi. Monad as modality. *Theoretical Computer Science*, 175:29 – 74, 1997.

[LS85] J. Lambek and Ph. J. Scott. *Introduction to Higher-Order Categorical Logic.* Cambridge University Press, 1985.

[Mac81] D. S. Macnab. Modal operators on Heyting algebras. *Algebra Universalis*, 12:5–29, 1981.

[Men93] M. Mendler. *A Modal Logic for Handling Behavioural Constraints in Formal Hardware Verification.* PhD thesis, Department of Computer Science, University of Edinburgh, ECS-LFCS-93-255, March 1993.

[Men00] M. Mendler. Characterising combinational timing analyses in intuitionistic modal logic. *The Logic Journal of the IGPL*, 8(6):821–852, November 2000.

[Mog91] E. Moggi. Notions of computation and monads. *Information and Computation*, 93(1):55–92, July 1991.

[PD01] F. Pfenning and R. Davies. A judgemental reconstruction of modal logic. *Mathematical Structures in Computer Science*, 2001.

[Pra65] D. Prawitz. *Natural Deduction: A Proof-Theoretic Study.* Almqvist and Wiksell, 1965.

[PS86] G. Plotkin and C. Stirling. A framework for intuitionistic modal logics. In *Theoretical aspects of reasoning about knowledge*, Monterey, 1986.

[Sim94] A.K. Simpson. *The Proof Theory and Semantics of Intuitionistic Modal Logic.* PhD thesis, University of Edinburgh, 1994.

[Tho74] S.K. Thomason. An incompleteness theorem in modal logic. *Theoria*, 40:30 – 34, 1974.

[TvD88] A. S. Troelstra and D. van Dalen. *Constructivism in Mathematics*, volume II. North-Holland, 1988.

[Wij90] D. Wijesekera. Constructive modal logic I. *Annals of Pure and Applied Logic*, 50:271–301, 1990.

[WZ95] F. Wolter and M. Zakharyaschev. Intuitionistic Modal Logics. In *Logic in Florence*, 1995.

[WZ97] F. Wolter and M. Zakharyaschev. Intuitionistic modal logics as fragments of classical bimodal logics. In E. Orlowska, editor, *Logic at Work.* Kluwer, 1997.

Labelled Natural Deduction for Interval Logics

Thomas Marthedal Rasmussen

Informatics and Mathematical Modeling,
Technical University of Denmark, Building 321,
DK-2800 Kgs. Lyngby, Denmark
tmr@imm.dtu.dk

Abstract. We develop a Labelled Natural Deduction framework for a certain class of interval logics. With emphasis on Signed Interval Logic we consider normalization properties and show that normal derivations satisfy a subformula property.
We have encoded our framework in the generic theorem proving system Isabelle. The labelled formalism turns out very convenient for conducting proofs and seems much closer to informal "pen and paper" reasoning than other proof systems. We give an example which supports this claim.
We also sketch how the results are applicable to (non-signed) interval logic and Duration Calculus.

1 Introduction

Interval logics (e.g. [9,17,4,16,12]) are modal logics of temporal intervals. Such logics have proven useful for the specification and verification of real-time and safety-critical systems.

Signed Interval Logic (SIL) [12] was proposed as an extension of Interval Temporal Logic (ITL) [9] with the introduction of the notion of a direction of an interval. SIL includes (as ITL) only one interval modality but SIL is (contrary to ITL) capable of specifying liveness properties. Other interval logics capable of this (such as Neighbourhood Logic [16]) have more than one interval modality.

The interval modality of ITL is the binary *chop*: $\frown$. A formula $\phi\frown\psi$ holds on an interval iff it can be split into two consecutive subintervals where ϕ and ψ holds, respectively. With chop of ITL one can only reach subintervals of the current interval, hence it is only possible to specify safety properties of a system. SIL only has the chop modality too but because of the intervals with directions (the signed intervals), it is possible to specify liveness properties as the two "sub"-intervals can now reach outside the current interval. Signed intervals are represented as pairs of elements from some temporal domain; (b, e) and (e, b) represent the same interval but have opposite directions. Figure 1 illustrates the semantics of $\frown$ in SIL (note how the direction of an interval is indicated by an arrowhead). ITL includes a special symbol ℓ which represents the length of an interval. SIL inherits this symbol from ITL; ℓ now gives the signed length of a signed interval.

If ITL and SIL (and other interval logics alike) are to be used more widely it is very important to develop tools to help automate reasoning in these logics.

L. Fribourg (Ed.): CSL 2001, LNCS 2142, pp. 308–323, 2001.

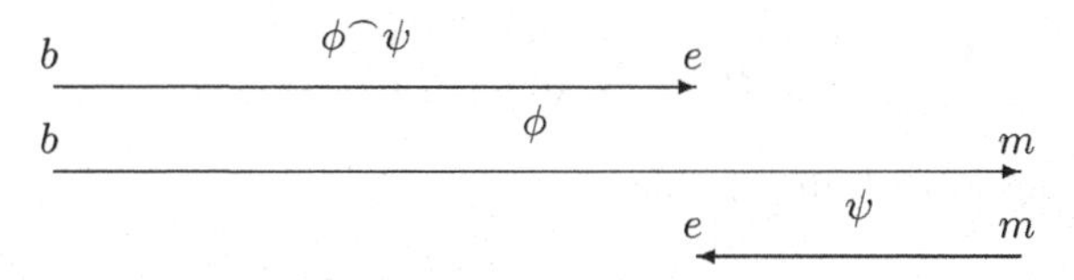

Fig. 1. $\phi^\frown\psi$ holds on (b, e) iff there is m such that ϕ holds on (b, m) and ψ on (m, e)

In [13] a proposal for a sequent calculus proof system for SIL is given. This system has some nice properties and is encoded in Isabelle in a way such that a substantial amount of automation is supported. Despite this, some proofs are still tedious to perform as the way to the informal "pen and paper" level of abstraction seems fairly long. The present paper is an attempt to narrow this gap. It is inspired by work on Labelled Natural Deduction (LND) [15,3] which combines classical natural deduction [11,14] with labelled deductive systems [5]. The LND formalism has shown its worth for traditional modal logics [1,2].

The rest of this paper is organized as follows: In Section 2 we consider propositional logics with a binary modality. These can be seen as the propositional basis for SIL and ITL. We consider LND systems for these logics and discuss normalization properties. Building on these results we then in Section 3 discuss results for the full SIL logic, including a LND system and normalization and subformula properties. In Section 4 we consider an encoding of the LND system for SIL in Isabelle. We give an example which supports our claim that a labelled formalism is very convenient for conducting proofs and seems much closer to informal "pen and paper" reasoning than other proof systems. In Section 5 we briefly sketch how such a labelled formalism is useful for ITL and Duration Calculus (DC) as well before concluding in Section 6.

2 Logics with a Binary Modality

In this section we consider propositional logics with a binary modality. We start by giving the definition of a logic $\mathcal{L}^\frown$ with a binary modality $^\frown$ [7,8]. Thereafter, we give a LND system for this class of logics and discuss normalization properties for such systems.

Formulas $(\alpha, \beta, \gamma, \dots)$ of $\mathcal{L}^\frown$ are constructed from an infinite set of propositional letters $(p, q, r, \dots)$ and $\bot$ (denoting falsity), using the usual Boolean operators ($\rightarrow$, $\vee$, $\wedge$ and $\neg$) and $^\frown$. As we work with classical logic we will often restrict attention to a (functionally complete) set of operators (primarily $\{\rightarrow, \bot\}$, in which case $\neg\alpha$ is $\alpha \rightarrow \bot$) for the propositional logic part. We will collectively refer to operators and modalities as *connectives*. Formulas with no connectives are called *atomic*. We adopt the following precedences: 1) $\neg$, 2) $^\frown$, 3) $\wedge, \vee$ and 4) $\rightarrow, \leftrightarrow$.

For $\mathcal{L}^\frown$ to be a logic with a binary modality it must [7,8] (at least) include the following: 1) All propositional tautologies and modus ponens. 2) Axioms saying that $^\frown$ distributes over $\vee$:

$$\mathrm{K}: \begin{array}{rcl} \alpha^\frown(\beta \vee \gamma) & \to & (\alpha^\frown\beta) \vee (\alpha^\frown\gamma) \\ (\alpha \vee \beta)^\frown\gamma & \to & (\alpha^\frown\gamma) \vee (\beta^\frown\gamma) \end{array} ,$$

and 3) the following monotonicity rules:

$$\mathrm{M}: \quad \frac{\alpha \to \beta}{(\alpha^\frown\gamma) \to (\beta^\frown\gamma)} \quad \frac{\alpha \to \beta}{(\gamma^\frown\alpha) \to (\gamma^\frown\beta)} .$$

The *minimal* logic with a binary modality is the logic with a binary modality consisting only of the above axioms and rules. If we include necessitation rules:

$$\mathrm{N}: \quad \frac{\alpha}{\neg(\neg\alpha^\frown\beta)} \quad \frac{\alpha}{\neg(\beta^\frown\neg\alpha)} ,$$

we call the logic *normal*. We can thus speak of the minimal normal logic with a binary modality. We will denote this specific logic $\mathcal{L}^\frown_{\mathrm{AF}}$.[1] Provability (derivability) in $\mathcal{L}^\frown$ is defined the standard way.

A model $\mathfrak{M}$ for $\mathcal{L}^\frown$ is a triple $(\mathcal{W}, \mathcal{R}, \mathcal{V})$, where $\mathcal{W}$ is a non-empty set of (possible) worlds, $\mathcal{R}$ is a ternary accessibility relation on $\mathcal{W}$ and $\mathcal{V}$ is a function mapping propositional letters to subsets of $\mathcal{W}$. The pair $(\mathcal{W}, \mathcal{R})$ is called the frame of the model and we say that a model $(\mathcal{W}, \mathcal{R}, \mathcal{V})$ is based on the frame $(\mathcal{W}, \mathcal{R})$. We define satisfaction of a formula α in a world $w \in \mathcal{W}$ in a model $\mathfrak{M} = (\mathcal{W}, \mathcal{R}, \mathcal{V})$ (written $\mathfrak{M}, w \models \alpha$) as follows:

$$\begin{array}{lll} \mathfrak{M}, w \not\models \bot , & & \\ \mathfrak{M}, w \models p & \text{iff} & w \in \mathcal{V}(p) , \\ \mathfrak{M}, w \models \alpha \to \beta & \text{iff} & \mathfrak{M}, w \models \alpha \text{ implies } \mathfrak{M}, w \models \beta , \\ \mathfrak{M}, w \models \alpha^\frown\beta & \text{iff} & \mathfrak{M}, v \models \alpha \text{ and } \mathfrak{M}, u \models \beta \text{ and} \\ & & \mathcal{R}(v, u, w) \text{ for some } v, u \in \mathcal{W} . \end{array}$$

We say that a formula α is valid in a class of frames $\mathfrak{F}$ if for all frames F of $\mathfrak{F}$, for all models $\mathfrak{M}$ based on F and for all worlds w of $\mathfrak{M}$, $\mathfrak{M}, w \models \alpha$.

If α is a theorem of $\mathcal{L}^\frown_{\mathrm{AF}}$ we write $\vdash_{\mathrm{AF}} \alpha$. It is easy to check that all axioms of $\mathcal{L}_{\mathrm{AF}}$ are valid and that all inference rules of $\mathcal{L}_{\mathrm{AF}}$ preserve validity in the class of all frames. In fact, we have a much stronger result [7,8]: $\mathcal{L}^\frown_{\mathrm{AF}}$ is characterized exactly by the class of all frames, viz.

Theorem 1. $\models_{\mathrm{AF}} \alpha$ *iff* $\vdash_{\mathrm{AF}} \alpha$,

where $\models_{\mathrm{AF}} \alpha$ denotes validity of α in the class of all frames.

2.1 Labelled Natural Deduction

A *labelled formula* is a pair of a possible world $w \in \mathcal{W}$ and a formula α, written $w : \alpha$. A *relational formula* is a triple of possible worlds v, u, w, written $R(v, u, w)$. We let η denote an arbitrary labelled/relational formula.

[1] AF in $\mathcal{L}^\frown_{\mathrm{AF}}$ is short for "All Frames". The reason for this will be made clear below.

We define satisfaction of η in a model $\mathfrak{M} = (\mathcal{W}, \mathcal{R}, \mathcal{V})$ (written $\mathfrak{M} \Vdash \eta$) as follows:

$$\begin{array}{lll}
\mathfrak{M} \nVdash w : \bot \ , & & \\
\mathfrak{M} \Vdash R(v,u,w) & \text{iff} & \mathcal{R}(v,u,w) \ , \\
\mathfrak{M} \Vdash w : p & \text{iff} & w \in \mathcal{V}(p) \ , \\
\mathfrak{M} \Vdash w : \alpha \to \beta & \text{iff} & \mathfrak{M} \Vdash w : \alpha \text{ implies } \mathfrak{M} \Vdash w : \beta \ , \\
\mathfrak{M} \Vdash w : \alpha \frown \beta & \text{iff} & \mathfrak{M} \Vdash v : \alpha \text{ and } \mathfrak{M} \Vdash u : \beta \text{ and} \\
 & & \mathfrak{M} \Vdash R(v,u,w) \text{ for some } v, u \in \mathcal{W} \ .
\end{array}$$

We say that a labelled formula $w : \alpha$ is valid in a class of frames $\mathfrak{F}$ if for all frames F of $\mathfrak{F}$, for all models $\mathfrak{M}$ based on F, $\mathfrak{M} \Vdash w : \alpha$.

It is clear that $\mathfrak{M} \Vdash w : \alpha$ iff $\mathfrak{M}, w \models \alpha$. Thus, if we let $\Vdash_{\mathrm{AF}} w : \alpha$ denote validity of $w : \alpha$ in the class of all frames we have

Proposition 1. $\models_{\mathrm{AF}} \alpha$ *iff* $\Vdash_{\mathrm{AF}} w : \alpha$.

We define a LND system for $\mathcal{L}_{\mathrm{AF}}^{\frown}$ in the style of [1,3]:

$$\dfrac{\begin{array}{c}[w : \alpha] \\ \vdots \\ w : \beta\end{array}}{w : \alpha \to \beta} \to I \qquad \dfrac{w : \alpha \to \beta \quad w : \alpha}{w : \beta} \to E \qquad \dfrac{\begin{array}{c}[w : \alpha \to \bot] \\ \vdots \\ v : \bot\end{array}}{w : \alpha} \bot E$$

$$\dfrac{v : \alpha \quad u : \beta \quad R(v,u,w)}{w : \alpha \frown \beta} \frown I \qquad \dfrac{w : \alpha \frown \beta \quad \begin{array}{c}[v : \alpha] \ [u : \beta] \ [R(v,u,w)] \\ \vdots \\ w' : \gamma\end{array}}{w' : \gamma} \frown E$$

In $\frown E$, v and u are different from both w, w' and each other, and do not occur in any assumption on which the upper occurrence of $w' : \gamma$ depends except $v : \alpha, u : \beta$ and $R(v,u,w)$. The rule $\bot E$ can be regarded as an E(limination)-rule for $\bot$ (hence the name) but when we henceforth collectively refer to E-rules this will *not* include $\bot E$. The *major premise* of an E-rule is the premise containing the connective being eliminated. A premise which is not major is called *minor*.

When assumptions are *closed* (indicated by $[\cdots]$) by one of the rules $\to I$, $\bot E$ or $\frown E$, we will use a natural number to identify which rules close which assumptions. Assumptions which are not closed by any rule are called *open*.

Definition 1. *A (LND)* derivation *of a labelled formula $w : \alpha$ from a set of labelled formulas Γ and a set of relational formulas Δ in a logic $\mathcal{L}$ is a tree formed using the (LND) rules of $\mathcal{L}$ with $w : \alpha$ as root and where the leafs are either closed assumptions or open assumptions belonging to Γ or Δ. We will use Π to denote derivations. Provability and theoremhood are defined as usual.*

We write $\vdash_{\mathrm{AF}}^{\mathrm{ND}} w : \alpha$ if there is a proof of $w : \alpha$ in the LND system for $\mathcal{L}_{\mathrm{AF}}^{\frown}$.

In [3] a general soundness and completeness result concerning labelled propositional logics with n-ary modalities is proved:

Theorem 2. $\Vdash_{\mathrm{AF}} w : \alpha$ *iff* $\vdash^{\mathrm{ND}}_{\mathrm{AF}} w : \alpha$.

If we are only interested in completeness with respect to the standard semantics we can prove it more directly:

Proposition 2. $\models_{\mathrm{AF}} \alpha$ *iff* $\vdash^{\mathrm{ND}}_{\mathrm{AF}} w : \alpha$.

Proof. Soundness is straightforward; using Proposition 1 we show that the LND rules preserve labelled validity.[2] For the completeness part we utilize Theorem 1: We show that if $\vdash_{\mathrm{AF}} \alpha$ then $\vdash^{\mathrm{ND}}_{\mathrm{AF}} w : \alpha$ for all w. This can be shown by induction on the length of the proof of $\vdash_{\mathrm{AF}} \alpha$ which amounts to showing that all axioms are provable and that the rules preserve provability in the LND system. The propositional part is completely standard, hence we can restrict our attention to K, M and N. We here only show the latter case:

$$\dfrac{\dfrac{[w : (\alpha \to \bot)^\frown \beta]^1 \qquad \dfrac{\dfrac{\dfrac{[v : \alpha \to \bot]^2 \quad v : \alpha}{v : \bot} \to E}{w : \bot} \bot E}{}}{w : \bot} {}^\frown E^2}{w : (\alpha \to \bot)^\frown \beta \to \bot} \to I^1$$

This derivation is valid as $\vdash^{\mathrm{ND}}_{\mathrm{AF}} u : \alpha$ for all u by the induction hypothesis. We can in particular assume $\vdash^{\mathrm{ND}}_{\mathrm{AF}} v : \alpha$. □

2.2 Normalization

We now turn to consider normalization properties of the LND system for $\mathcal{L}^\frown_{\mathrm{AF}}$.

From the semantics we observe that $^\frown$ is an existential $\Diamond$-like binary modality. The normalization proofs get simpler if we instead consider an universal $\Box$-like binary modality $^\smile$. We define $^\smile$ by the following LND rules:

$$\dfrac{\begin{matrix}[v : \alpha][R(v,u,w)] \\ \vdots \\ u : \beta\end{matrix}}{w : \alpha^\smile \beta} {}^\smile I \qquad \dfrac{w : \alpha^\smile \beta \quad v : \alpha \quad R(v,u,w)}{u : \beta} {}^\smile E$$

where in $^\smile I$, v and u are different from both w and each other, and u does not occur in any assumption on which $u : \beta$ depends other than $R(v,u,w)$.

We can show that $\bot, \to, ^\smile$ is a functionally complete fragment for $\mathcal{L}^\frown_{\mathrm{AF}}$ by defining $w : \alpha^\frown\beta$ iff $w : \neg(\alpha^\smile\neg\beta)$ and showing that the rules for $^\frown$ can be derived from those for $^\smile$. Below we give the case for $^\frown I$:

$$\dfrac{\dfrac{\dfrac{\dfrac{[w : \alpha^\smile(\beta \to \bot)]^1 \quad v : \alpha \quad R(v,u,w)}{u : \beta \to \bot} {}^\smile E \qquad u : \beta}{u : \bot} \to E}{w : \bot} \bot E}{w : \alpha^\smile(\beta \to \bot) \to \bot} \to I^1$$

[2] This is (not surprisingly) also the way soundness of Theorem 2 is proved in [3].

We could similarly derive the rules for $\smile$ from those for $\frown$. In the following we will restrict attention to the $\bot, \to, \smile$ fragment. The *size* of a formula is the number of connectives in the formula.

Proposition 3. *For any derivation of $w : \alpha$ in the LND system for $\mathcal{L}^{\frown}_{\mathrm{AF}}$ there is a derivation of $w : \alpha$ with the following restrictions on the $\bot E$ rule: 1) The conclusion is always atomic, and 2) there are no applications immediately following each other.*

Proof. 1) In the original derivation, pick out an application of $\bot E$ where the conclusion has maximal size. If not atomic (in which case we are done), this conclusion will have form $\alpha \to \beta$ or $\alpha^\smile\beta$. Below we only consider the latter case (the former follows analogously). We replace the derivation with one where the conclusion of the affected $\bot E$ has less size by the following transformation (denoted by $\rightsquigarrow$) of part of the derivation tree (the rest of the derivation tree is unchanged):

$$\dfrac{\begin{matrix}[w:\alpha^\smile\beta\to\bot]^1\\ \Pi\\ v':\bot\end{matrix}}{w:\alpha^\smile\beta}\,\bot E^1 \quad\rightsquigarrow\quad \dfrac{\begin{matrix}\dfrac{\dfrac{\dfrac{[u:\beta\to\bot]^2 \qquad \dfrac{[w:\alpha^\smile\beta]^1[v:\alpha]^3[R(v,u,w)]^3}{u:\beta}\,{}^\smile E}{u:\bot}\,\to E}{w:\bot}\,\bot E}{w:\alpha^\smile\beta\to\bot}\,\to I^1\\ \Pi\\ \dfrac{v':\bot}{u:\beta}\,\bot E^2\end{matrix}}{w:\alpha^\smile\beta}\,{}^\smile I^3$$

By induction it is now easy to see that repeated applications of this transformation yield the desired derivation. In the case of 2) we notice that if there is to be two $\bot E$ rules immediately following each other the uppermost has to have conclusion $v : \bot$ (for some v). But then it is clearly superfluous and can thus be removed. □

Definition 2. *A* maximal formula *in a derivation is a labelled formula which is both the conclusion of an introduction rule and the major premise of an elimination rule. A derivation is* normal *if it contains no maximal formulas, all applications of $\bot E$ have atomic consequences and there are no applications of $\bot E$ immediately following each other.*

An introduced labelled formula which is immediately eliminated does clearly not contribute to the derivation, hence a maximal formula can be removed by a transformation called a *contraction step*. Below we show the case of $\smile$:

$$\dfrac{\dfrac{\begin{matrix}[v':\alpha]^1[R(v',u',w)]^1\\ \Pi\\ u':\beta\end{matrix}}{w:\alpha^\smile\beta}\,{}^\smile I^1 \qquad v:\alpha \quad R(v,u,w)}{u:\beta}\,{}^\smile E \quad\rightsquigarrow\quad \begin{matrix}v:\alpha \quad R(v,u,w)\\ \Pi[v/v',u/u']\\ u:\beta\end{matrix}$$

where $\Pi[v/v', u/u']$ is obtained from Π by systematically substituting v for v' and u for u', with a suitable renaming of the variables to avoid clashes.

Repeated applications of such contractions steps, together with Proposition 3, yields the following.

Theorem 3. *Any derivation can be transformed into a normal derivation.*

Definition 3. *A* track *in a derivation Π is a sequence of labelled formulas $w_1 : \alpha_1, w_2 : \alpha_2, \ldots, w_n : \alpha_n$ where $w_1 : \alpha_1$ is a leaf, and for $1 \le i < n$, $w_{i+1} : \alpha_{i+1}$ is immediately below $w_i : \alpha_i$ and $w_i : \alpha_i$ is* not *the minor premise of a $\to E$ or $\smile E$ rule. A track* of order 0 *ends in the root of Π; a track of* order $n+1$ *ends in the minor premise of an E-rule with major premise belonging to a track of order n.*

The above definition of a track is an extension of that of [11] (we use the terminology of [14]) for propositional logic to $\mathcal{L}_{\widehat{\mathrm{AF}}}$. The key observation is that the structure of the rules $\to I$ and $\to E$ is similar to that of $\smile I$ and $\smile E$, respectively (disregarding judgments concerning the accessibility relation R).

Proposition 4. *Let $w_1 : \alpha_1, w_2 : \alpha_2, \ldots, w_n : \alpha_n$ be a track in a normal derivation. There is a* minimal formula *α_i such that 1) $w_j : \alpha_j$ $(j < i)$ is a major premise of an E-rule and α_j is a superformula of α_{j+1}, 2) $w_i : \alpha_i$ $(i \ne n)$ is a premise of an I-rule or $\bot E$, and 3) $w_j : \alpha_j$ $(i < j < n)$ is a premise of an I-rule and α_j is a subformula of α_{j+1}.*

Proof. As the derivation is normal, in the track, an E-rule cannot follow an I-rule (there are no maximal formulas) and it cannot follow a $\bot E$ rule (the consequence is atomic). The $\bot E$ rule cannot follow an I-rule as the premise is $\bot$ and by normality there will thus at most be one $\bot E$ rule. □

Definition 4. *Consider a derivation Π of $w : \alpha$ from Γ and Δ. Let $S = \{\alpha\} \cup \{\gamma \mid u : \gamma \in \Gamma$ for some $u\}$. Π is said to have the* subformula property *if for any labelled formula $v : \beta$ in Π, 1) β is $\bot$, 2) β is a subformula of some formula in S, or 3) β is $\neg\beta'$ and β' is a subformula of some formula in S.*

Theorem 4. *Any normal derivation satisfies the subformula property.*

Proof. We start by observing that any formula in a derivation belongs to some track $w_1 : \alpha_1, w_2 : \alpha_2, \ldots, w_n : \alpha_n$. By Proposition 4, all α_i $(1 \le i \le n)$ are subformulas of either α_1 or α_n. By induction on the order of the track we now conclude that any formula in a normal derivation will be a subformula of either the root or a leaf. Consider a closed assumption: If it is closed by one of the I-rules it will be a subformula of the conclusion of that I-rule. If it is closed by $\bot E$ it will have the form $\neg\beta$ and β will be the conclusion of that $\bot E$ rule. □

3 Labelled Signed Interval Logic

In this section we develop a LND system for SIL. This is done in steps where we first extend $\mathcal{L}_{\widehat{\mathrm{AF}}}$ to first order logic, then to so called S-models and finally to the full system for SIL. We conclude by considering normalization and subformula properties as for $\mathcal{L}_{\widehat{\mathrm{AF}}}$.

3.1 First Order Logic with Equality

In this section we extend the LND system for $\mathcal{L}_{\widehat{\mathrm{AF}}}$ to include first order logic with equality. We therefore, in a standard way, extend the syntax to include predicate and function symbols, variables and quantifiers. We will use $x, y, z, \dots$ for variables, $s, t, u, \dots$ for terms and $\phi, \psi, \varphi, \dots$ for formulas.[3] We only consider constant domains of individuals but we want to distinguish between rigid and flexible symbols; rigid symbols have the same meaning in all worlds whereas the meaning of flexible terms can vary. In the presence of flexible symbols it is necessary to put extra side conditions on quantifier and equality rules to retain soundness. For this we introduce two new judgments, ri(s) and cf(ϕ), stating, respectively, that s is a rigid term and that ϕ is a chop-free formula. A chop-free formula is a formula which does not contain any modalities (in particular, neither $\frown$ nor $\smile$). We will also have to consider rigid formulas; we overload the ri judgment to state this: ri(ϕ). Rigidity and chop-freeness is straightforwardly inductively defined over the structure of terms and formulas. For example:

$$\frac{\mathrm{cf}(\phi) \quad \mathrm{cf}(\psi)}{\mathrm{cf}(\phi \wedge \psi)}\ \mathrm{cf}{\wedge}I \qquad \frac{\mathrm{cf}(\phi \wedge \psi)}{\mathrm{cf}(\phi)}\ \mathrm{cf}{\wedge}E \qquad \frac{}{\mathrm{cf}(\phi)}\ \mathrm{cf}A \qquad \frac{\mathrm{ri(s)} \quad \mathrm{ri(t)}}{\mathrm{ri(s = t)}}\ \mathrm{ri}{=}I \qquad \frac{\begin{matrix}[\mathrm{ri(x)}] \\ \vdots \\ \mathrm{ri}(\phi)\end{matrix}}{\mathrm{ri}((\forall \mathrm{x})\phi)}\ \mathrm{ri}\forall I$$

where cf$A$ has the side condition that ϕ must be atomic. It should be clear how I- and E-rules are defined for the remaining connectives in the case of both rigidity and chop-freeness. We can now give the LND rules for the quantifiers:[4]

$$\frac{\begin{matrix}[\mathrm{ri(x)}] \\ \vdots \\ w : \phi\end{matrix}}{w : (\forall x)\phi}\ \forall I \qquad \frac{w : (\forall x)\phi \quad \mathrm{ri(t)}}{w : \phi[t/x]}\ \forall E_{\mathrm{ri}} \qquad \frac{w : (\forall x)\phi \quad \mathrm{cf}(\phi)}{w : \phi[t/x]}\ \forall E_{\mathrm{cf}}$$

$$\frac{w : \phi[t/x] \quad \mathrm{ri(t)}}{w : (\exists x)\phi}\ \exists I_{\mathrm{ri}} \qquad \frac{w : \phi[t/x] \quad \mathrm{cf}(\phi)}{w : (\exists x)\phi}\ \exists I_{\mathrm{cf}} \qquad \frac{w : (\exists x)\phi \qquad \begin{matrix}[w : \phi][\mathrm{ri(x)}] \\ \vdots \\ v : \psi\end{matrix}}{v : \psi}\ \exists E$$

---

[3] We use $\phi, \psi, \dots$ for formulas (instead of $\alpha, \beta, \dots$) to emphasis the move from propositional to first order logic.

[4] Note that the form of these rules ensure that variables always act as rigid.

In $\forall I$, x is not free in any assumption on which ϕ depends except ri(x). In $\exists E$, x is not free in ψ nor in any assumption on which ψ depends except ϕ and ri(x). As known from classical logic, the rules for $\exists$ are derivable from those for $\forall$ (this still holds for the above modified rules). Thus, we can restrict attention to $\forall$ in the following.

Equality rules and a structural rule for rigid formulas can now be defined.

$$\frac{w : \phi[s/x] \quad w : s = t \quad \text{ri(s)} \quad \text{ri(t)}}{w : \phi[t/x]} \text{Subst}_{\text{ri}} \qquad \frac{}{w : s = s} \text{Refl}$$

$$\frac{w : \phi[s/x] \quad w : s = t \quad \text{cf}(\phi)}{w : \phi[t/x]} \text{Subst}_{\text{cf}} \qquad \frac{v : \phi \quad \text{ri}(\phi)}{w : \phi} R$$

A contraction step for $\forall$ can be defined and Theorem 3 can be modified accordingly. Also Definition 3 can be extended to include the case for $\forall$. The method of the proof of 1) in Proposition 3 can be repeated to prove the following.

Lemma 1. *The rules* Subst_{ri} *and* Subst_{cf} *can can be restricted to applications where* ϕ *is atomic.*

To end this section, we will briefly consider the structure of derivations involving the rules for ri/cf: We observe that the derivation of a labelled formula might depend on the ri/cf rules whereas derivations of ri/cf judgments never depend on labelled formulas or each other; thus, the derivation tree can be seen as "decorated" with independent derivations of ri/cf judgments. Because the rules for ri/cf have a very simple form we can, in these cases, easily define normal derivations which will satisfy a subformula property and furthermore be unique.

3.2 S-Models

The semantics for $\mathcal{L}_{\widehat{\text{AF}}}$ can be extended in the obvious way to include first order models. Here we restrict attention to so called *S-models.* These are first order models which include the flexible symbol ℓ and have a certain uniqueness constraint on worlds. The Hilbert system for $\mathcal{L}_{\widehat{\text{AF}}}$ can be extended to a first order logic version as well, and adding certain axioms gives the following result: $\models_S \phi$ iff $\vdash_S \phi$. We will here not go into further details but refer to [4,12]. Based on the standard semantics we can straightforwardly define a corresponding labelled semantics (as for $\mathcal{L}_{\widehat{\text{AF}}}$) and prove: $\models_S \phi$ iff $\Vdash_S w : \phi$.

We now define a LND system for S. This includes the rules of $\mathcal{L}_{\widehat{\text{AF}}}$, first order logic with equality and the following two rules:

$$\frac{u' : \phi \quad v : \ell = s \quad v' : \ell = s \quad R(v, u, w) \quad R(v', u', w) \quad \text{ri(s)}}{u : \phi} S1$$

$$\frac{v' : \phi \quad u : \ell = s \quad u' : \ell = s \quad R(v, u, w) \quad R(v', u', w) \quad \text{ri(s)}}{v : \phi} S2$$

In the same way as Proposition 2 we can prove:

Proposition 5. $\models_S \phi$ *iff* $\vdash_S^{ND} w : \phi$.

3.3 Signed Interval Logic

We now further restrict S-models so as to obtain SIL-models. The prominent feature of SIL-models is that the worlds of $\mathcal{W}$ are pairs of elements taken from a temporal domain T, and $\mathcal{R} = \{((i,k),(k,j),(i,j)) \mid i,j,k \in T\}$. Note, how this means that the worlds are signed intervals as discussed in the introduction, and how $\mathcal{R}$ gives a formal definition of the semantics of $\frown$ as illustrated in Figure 1. Besides this, ℓ is interpreted by a certain measure and we require the domain of individuals to have the structure of an Abelian group. We further extend the Hilbert system for S with suitable axioms and we get: $\models_{\text{SIL}} \phi$ iff $\vdash_{\text{SIL}} \phi$ [12]. Also as for S, we can define a labelled semantics for SIL and show: $\models_{\text{SIL}} \phi$ iff $\Vdash_{\text{SIL}} w : \phi$.

We now turn to a LND system for SIL. We start out by observing that because of the special structure of worlds and the accessibility relation we do not have to explicitly include the R judgment in the rules but can make it implicit. In the case of $\smile E$ we e.g. have (and similarly for the other rules):

$$\frac{(i,j) : \phi \smile \psi \quad (i,k) : \phi}{(k,j) : \psi} \smile E$$

We can now define the LND system for SIL as that for S with the addition of the following rules:

$$\frac{}{(i,i) : \ell = 0} \ell 0$$

$$\frac{(i,k) : \ell = s \quad (k,j) : \ell = t \quad \text{ri(s)} \quad \text{ri(t)}}{(i,j) : \ell = s + t} \ell{+}I \qquad \frac{(i,j) : \ell = s + t \quad \begin{array}{c}[(k,j) : \ell = t] \; [\text{ri(t)}] \\ [(i,k) : \ell = s] \; [\text{ri(s)}] \\ \vdots \\ (m,n) : \varphi\end{array}}{(m,n) : \varphi} \ell{+}E$$

and four axioms (i.e. rules with no premises) defining the properties of an Abelian group:[5] $(i,j) : s + (t + u) = (s + t) + u$, $(i,j) : s + 0 = s$, $(i,j) : s + -s = 0$ and $(i,j) : s + t = t + s$. In $\ell{+}E$, k is different from i,j,m,n, and does not occur in any assumption on which the upper occurrence of $(m,n) : \varphi$ depends except $(i,k) : \phi$ and $(k,j) : \psi$.

Extending Proposition 5 we can prove:

Theorem 5. $\models_{\text{SIL}} \phi$ *iff* $\vdash_{\text{SIL}}^{ND} w : \phi$.

Lemma 2. *The rules R, S1 and S2 can be restricted to applications where ϕ is atomic.*

This lemma is proven in the same way as Lemma 1 but notice that it does *not* hold for general accessibility relations as $\mathcal{R}$ has to be total (which in particular

[5] Formally, this requires the definition of a derivation to be extended to the cases where leaves can be axioms as well.

is the case for SIL-models). Furthermore, by inspecting the completeness proof for Labelled SIL (Theorem 5) we derive the following lemma:

Lemma 3. *The rule $\ell{+}E$ can be restricted to applications where φ is $\bot$.*

3.4 Normalization

In this section we consider normalization properties of the LND system for SIL.

We extend the definition of a normal derivation (Definition 2) to include the requirements that applications of the rules Subst_{ri}, Subst_{cf}, R, $S1$ and $S2$ are on atomic formulas only, and that applications of $\ell{+}E$ are on $\bot$ only. By the Lemmas 1, 2 and 3 we have that Theorem 3 is valid for the full SIL system too.

Unfortunately, we will not have as nice properties of tracks as those of Proposition 4. But because of the structure of normal derivations, tracks can still be divided in three parts: An elimination part, a part working on atomic formulas and/or $\bot$, and an introduction part. It is possible to go into more detail concerning the structure of the middle part – such as the ordering of the rules, the maximum number of certain rules, etc. – but we will for space reasons not do this here. The important thing is that a normal derivation as defined is enough to achieve a subformula property. First, though, we have to address what we mean by subformula in a first order logic with equality: We say that ϕ is a subformula of $\psi[s/t]$ if ϕ is a subformula of ψ, independently of t and s. In other words, we do not take the term level into account. Given this, if we extend the definition of the subformula property of a derivation (Definition 4) to include the case where a formula in the derivation is allowed to be an arbitrary atomic formula we can show the following:

Theorem 6. *Any normal derivation in the LND system for SIL satisfies the subformula property.*

Note that this result relies on the fact that with the above definition of the subformula property it is not a problem to add axioms to the system as long as they are atomic. This is in particular the case for the SIL system.

4 Isabelle Encoding

In this section we discuss and describe an encoding of Labelled SIL in Isabelle [10]. Below is a dump of (part of) the theory file for Isabelle/LSIL.[6]

```
LSIL = Pure +
types T o D
classes term < logic
arities T,o :: logic, D :: term
default term
consts
  LF            :: "[T, T, o] => prop"        ("(<_,_> : (_))" [6,6,5] 4)
```

[6] We have for space reasons omitted the definitions of some of the constants as well as their defining rules but otherwise the dump includes all necessary definitions.

```
  RI          :: 'a::logic => prop          ("(RI _)")
  CF          ::          o => prop          ("(CF _)")
  True        :: o
  "&"         :: "[o, o] => o"               (infixr 35)
  "^"         :: "[o, o] => o"               (infixr 38)
  "="         :: "[D, D] => o"               (infixl 50)
  Ex          :: "(D => o) => o"             (binder "EX " 10)
  len         :: D
  conv        :: o => o
rules
  conjI       "[| <i,j>:P;  <i,j>:Q |] ==> <i,j>:P&Q"
  chopI       "[| <i,k>:P; <k,j>:Q |] ==> <i,j>:P^Q"
  chopE       "[| <i,j>:P^Q; !!k. [| <i,k>:P; <k,j>:Q |] ==> <l,m>:R |] ==> <l,m>:R"
  lenZero     "<i,i>:len=0"
  conv_def    "conv(P) == (EX x. (len=x) & ((len=0) & (len=x)^P)^True)"
  exIRI       "[| RI s; <i,j>:P(s) |] ==> <i,j>:(EX x. P(x))"
  exE         "[| <i,j>:EX x. P(x); !!x. [| RI x; <i,j>:P(x) |] ==> <k,l>:R |] ==> <k,l>:R"
end
```

Three types are defined: `T` (the temporal domain), `o` (formulas) and `D` (terms). Three judgments are defined (in Isabelle, judgments are coercions from the object level to truth values (the type `prop`) of the meta level): `LF` (labelled formulas), `RI` (note how the definition is polymorphic such that rigidity can be defined for both formulas and terms using the same judgment) and `CF`. To capture that atomic formulas are chop-free we simply introduce an axiom explicitly saying this for each predicate symbol we introduce in the logic. In the case of `=` we e.g. have `CF (s=t)`.

By comparing the rules of the theory file with the LND rules of the previous sections we see a convincing one-to-one correspondence. The only thing worth mentioning is how the side conditions concerning freeness of variables in assumptions (as for $\exists E$) and the non-occurrence of worlds in assumptions (as for $\frown E$) is handled in Isabelle: Both cases are neatly taken care of by meta quantification (`!!`) as illustrated by `exE` and `chopE`.

4.1 The Converse Modality

In this section we will give an example of reasoning in Isabelle/LSIL. The example is concerned with properties of an unary modality $^{-1}$, definable in SIL, which "reverses" the direction of an interval:

$$\phi^{-1} \;\widehat{=}\; (\exists x)(\,(\ell = x) \wedge (\,(\ell = 0) \wedge (\ell = x)^\frown \phi\,)^\frown \mathrm{true}\,)\;.$$

We would like to show the following properties of $^{-1}$:

$$1)\;\;(\phi^{-1})^{-1} \leftrightarrow \phi\;, \qquad 2)\;\;(\phi^\frown\psi)^{-1} \leftrightarrow (\psi^{-1\frown}\phi^{-1})\;.$$

We have proven both theorems in Isabelle/LSIL. For this it was very convenient to first prove two derived rules:

$$\frac{(i,j):\phi}{(j,i):\phi^{-1}}\;{}^{-1}I \qquad\qquad \frac{(i,j):\phi^{-1}}{(j,i):\phi}\;{}^{-1}E$$

We will discuss the proof of ^{-1}I in some detail in the following. We start out by giving an informal "pen and paper" proof: We want to show $(j,i):\phi^{-1}$ (read:

"ϕ^{-1} holds on the signed interval (j,i)") under the assumption $(i,j) : \phi$. First, we notice that any interval has a length; assume the length of (j,i) is a, i.e., $(j,i) : \ell = a$. Thus, after expanding the definition of $^{-1}$ and instantiating the existential quantifier, we are left with proving $(j,i) : ((\ell = 0) \wedge (\ell = a)^\frown\phi)^\frown\text{true}$. This can be illustrated as follows:

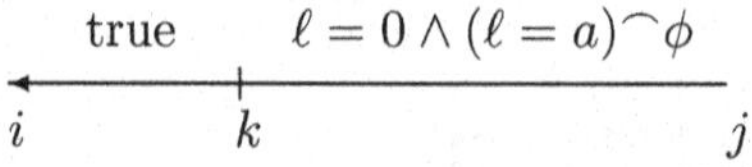

$$\begin{array}{ccc} & \text{true} & \ell = 0 \wedge (\ell = a)^\frown\phi \\ i & k & j \end{array}$$

i.e., we have to find a k such that (k,i) : true and $(j,k) : (\ell = 0) \wedge (\ell = a)^\frown\phi$. As true holds on any interval and $\ell = 0$ only holds on point intervals we take k to be j and thus have to prove that $(j,j) : (\ell = a)^\frown\phi$. We then need to find a m, viz.

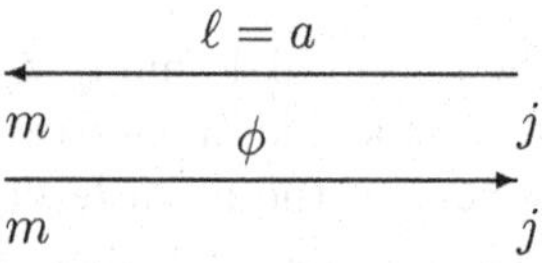

$$\begin{array}{ccc} & \ell = a & \\ m & \phi & j \\ m & & j \end{array}$$

But $(j,i) : \ell = a$ and we are thus done as $(i,j) : \phi$ by assumption.

We will now consider the proof of ^{-1}I in Isabelle/LSIL. For this we will need two simple lemmas (easily derivable):

```
val len_ex = "(<?i,?j> : EX x. len = x ==> <?i,?j> : ?P) ==> <?i,?j> : ?P" : thm
val TrueGlob = "<?i,?j> : True" : thm
```

saying, respectively, that any interval has some length and that true holds on any interval.

The following proof of ^{-1}I is taken verbatim[7] from an Isabelle session. We start by stating the goal (in the proof script, `P` is used for ϕ and `conv(P)` is used for ϕ^{-1}):

```
> Goalw [conv_def] "<i,j>:P ==> <j,i>:conv(P)";
<i,j> : P ==> <j,i> : conv(P)
 1. <i,j> : P ==> <j,i> : EX x. len = x & (len = 0 & len = x ^ P) ^ True
```

Isabelle responds with the goal to be proved and what subgoals are required to establish it. Note how the definition of `conv` is expanded as indicated in the statement of the goal (via `[conv_def]`). We now apply (using `br` or `be`) a series of named rules to suitable subgoals, possibly solving subgoals by assumption (using `ba`):

```
> br len_ex 1;
 1. [| <i,j> : P; <j,i> : EX x. len = x |]
    ==> <j,i> : EX x. len = x & (len = 0 & len = x ^ P) ^ True

> be exE 1;
 1. !!x. [| <i,j> : P; RI x; <j,i> : len = x |]
         ==> <j,i> : EX x. len = x & (len = 0 & len = x ^ P) ^ True

> be exIRI 1;
 1. !!x. [| <i,j> : P; <j,i> : len = x |]
         ==> <j,i> : len = x & (len = 0 & len = x ^ P) ^ True
```

[7] With the omission of some diagnostic output and minor pretty-printing.

```
> br conjI 1;
 1. !!x. [| <i,j> : P; <j,i> : len = x |] ==> <j,i> : len = x
 2. !!x. [| <i,j> : P; <j,i> : len = x |] ==> <j,i> : (len = 0 & len = x ^ P) ^ True

> ba 1;
 1. !!x. [| <i,j> : P; <j,i> : len = x |] ==> <j,i> : (len = 0 & len = x ^ P) ^ True

> br chopI 1;
 1. !!x. [| <i,j> : P; <j,i> : len = x |] ==> <j,?k4(x)> : len = 0 & len = x ^ P
 2. !!x. [| <i,j> : P; <j,i> : len = x |] ==> <?k4(x),i> : True

> br TrueGlob 2;
 1. !!x. [| <i,j> : P; <j,i> : len = x |] ==> <j,?k4(x)> : len = 0 & len = x ^ P

> br conjI 1;
 1. !!x. [| <i,j> : P; <j,i> : len = x |] ==> <j,?k4(x)> : len = 0
 2. !!x. [| <i,j> : P; <j,i> : len = x |] ==> <j,?k4(x)> : len = x ^ P

> br lenZero 1;
 1. !!x. [| <i,j> : P; <j,i> : len = x |] ==> <j,j> : len = x ^ P

> br chopI 1;
 1. !!x. [| <i,j> : P; <j,i> : len = x |] ==> <j,?k8(x)> : len = x
 2. !!x. [| <i,j> : P; <j,i> : len = x |] ==> <?k8(x),j> : P

> ba 1;
 1. !!x. [| <i,j> : P; <j,i> : len = x |] ==> <i,j> : P

> ba 1;
No subgoals!
```

Notice that the two schematic variables `?k4(x)` and `?k8(x)` correspond, respectively, to k and m in the “pen and paper” proof.

We could in a similar way prove ${}^{-1}E$ and the theorems 1) and 2) follow easily. It should be clear by the above example that the proofs in Isabelle/LSIL are very close to the abstraction level of “pen and paper” reasoning.

It is interesting to compare the effort needed to prove 1) and 2) in Isabelle/LSIL with corresponding proofs in the sequent calculus for SIL described in [13]: The above proof is much shorter than the one in the sequent calculus; this despite the fact that some effort has been put into automating reasoning in the sequent calculus and no automation (so far) has been put into Isabelle/LSIL. To partly explain this very noticeable difference we make some observations: In the sequent calculus of [13] we cannot have derived rules such as ${}^{-1}I$ as we cannot explicitly refer to intervals within the proof system (formulas are not labelled). Furthermore, we cannot reason “independently” of subintervals but have to “collapse” them by means of the axiom $\ell = s + t \leftrightarrow (\ell = s)^\frown(\ell = t)$ and related techniques. This means that proofs get more complex in the sequent calculus of [13] as it is more difficult to modularize proofs and separate concerns.

5 Labelled Duration Calculus

So far we have concentrated on investigating a LND system for SIL. In this section we briefly sketch how a similar framework can be developed for ITL and DC.

Semantically, ITL can be regarded as a restriction of SIL as only a subset of the signed intervals are allowed: The intervals where the end point is greater than the beginning point. This intuition motivates the following LND rules for chop in ITL.

$$\frac{(i,k):\phi \quad (k,j):\psi \quad i \sqsubseteq k \quad k \sqsubseteq j}{(i,j):\phi^\frown\psi}\ {}^\frown I \qquad \frac{(i,j):\phi^\frown\psi \qquad \begin{matrix}[(k,j):\psi]\ [k \sqsubseteq j] \\ [(i,k):\phi]\ [i \sqsubseteq k] \\ \vdots \\ (m,n):\varphi\end{matrix}}{(m,n):\varphi}\ {}^\frown E$$

Loosely speaking, by restricting the other rules of SIL in a similar way we will get a LND system for ITL. The relation $\sqsubseteq$ defines an ordering over the temporal domain and to reason with $\sqsubseteq$ we add rules defining its properties (such as transitivity). DC is an extension of ITL [6] and LND rules for DC can be added conservatively to the above system. An encoding of ITL and DC in Isabelle can be carried out along the very same lines as the one for SIL in the previous section.

6 Conclusion

We have developed a LND system for SIL and encoded it in Isabelle.

From a theoretical viewpoint the main result of the paper is the theorem stating that normal derivations in the LND system satisfy a subformula property.

From a pragmatic viewpoint we feel that our example of reasoning in Isabelle/LSIL convincingly conveys the benefits of using a labelled framework.

By sketching how the framework can be modified for ITL and DC we have also indicated how the ideas of the paper have a broader applicability.

References

1. D. Basin, S. Matthews, and L. Viganò. Labelled Propositional Modal Logics: Theory and Practice. *Journal of Logic and Computation*, 7(6):685–717, 1997.
2. D. Basin, S. Matthews, and L. Viganò. Labelled Modal Logics: Quantifiers. *Journal of Logic, Language and Information*, 7(3):237–263, 1998.
3. D. Basin, S. Matthews, and L. Viganò. Natural Deduction for Non-Classical Logics. *Studia Logica*, 60(1):119–160, 1998.
4. B. Dutertre. Complete Proof Systems for First Order Interval Temporal Logic. In *LICS'95*, pages 36–43. IEEE Press, 1995.
5. D.M. Gabbay. *Labelled Deductive Systems.* Oxford, 1996.
6. M.R. Hansen and Zhou Chaochen. Duration Calculus: Logical Foundations. *Formal Aspects of Computing*, 9(3):283–330, 1997.
7. Á. Kurucz, I. Németi, I. Sain, and A. Simon. Decidable and Undecidable Logics with a Binary Modality. *Journal of Logic, Language and Information*, 4:191–206, 1995.
8. M. Marx and Y. Venema. *Multi-Dimensional Modal Logic.* Kluwer, 1997.
9. B. Moszkowski. A Temporal Logic for Multilevel Reasoning about Hardware. *IEEE Computer*, 18(2):10–19, 1985.
10. L.C. Paulson. *Isabelle, A Generic Theorem Prover*, volume 828 of *LNCS*. Springer, 1994.

11. D. Prawitz. *Natural Deduction. A Proof-Theoretical Study.* Almquist & Wiksell, 1965.
12. T.M. Rasmussen. Signed Interval Logic. In *CSL'99*, volume 1683 of *LNCS*, pages 157–171. Springer, 1999.
13. T.M. Rasmussen. A Sequent Calculus for Signed Interval Logic. Technical Report IMM-TR-2001-06, Informatics and Mathematical Modeling, Technical University of Denmark, 2001.
14. A.S. Troelstra and H. Schwichtenberg. *Basic Proof Theory.* Cambridge, 1996.
15. L. Viganò. *Labelled Non-Classical Logics.* Kluwer, 2000.
16. Zhou Chaochen and M.R. Hansen. An Adequate First Order Interval Logic. In *COMPOS'97*, volume 1536 of *LNCS*, pages 584–608. Springer, 1998.
17. Zhou Chaochen, C.A.R. Hoare, and A.P. Ravn. A Calculus of Durations. *Information Processing Letters*, 40(5):269–276, 1991.

Decidable Navigation Logics for Object Structures

Frank S. de Boer and Rogier M. van Eijk

Institute of Information and Computing Sciences
Utrecht University, P.O. Box 80.089
3508 TB Utrecht, The Netherlands
{frankb,rogier}@cs.uu.nl

Abstract. In this paper, we introduce decidable multimodal logics to describe and reason about navigation across object structures. The starting point of these navigation logics is the modelling of object structures as Kripke models that contain a family of deterministic accessibility relations; one for each pointer attribute. These pointer attributes are used in the logics both as first-order terms in equalities and as modal operators. To handle the ambiguities of pointer attributes the logics also cover a mechanism to bind logical variables to objects that are reachable by a pointer. The main result of this paper is a tableau construction for deciding the validity of formulas in the navigation logics.

1 Introduction

In describing structures of objects, we distinguish two main levels of abstraction. First, there is the *modelling level* as specified by the Unified Modelling Language (UML) [10]. At this level, in UML, a structure of objects is described in terms of a diagram consisting of classes that are related to each other via associations. Consider for instance the diagram depicted in Figure 1, which covers a class *Person* that is related to a class *Book* via the association *Author* and a class *Company* that is related to the class *Book* via the association *Publisher*. As indicated in the diagram, the multiplicity of these associations *Author* and *Book* is one-to-many.

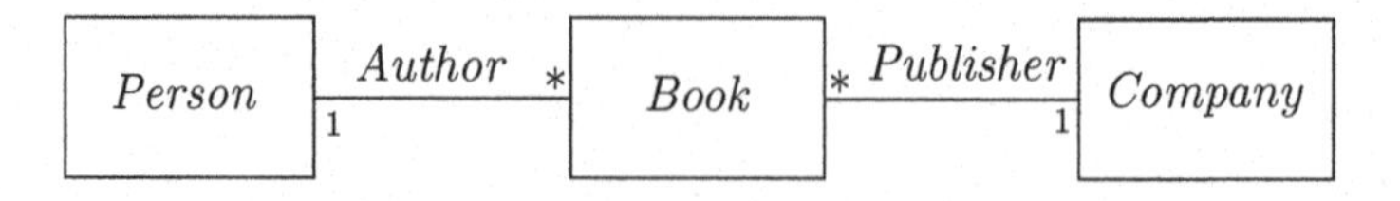

Fig. 1. A class diagram at the modelling level

Second, we distinguish the *implementation level*, which describes the execution of an object-oriented programming language, like for instance JAVA, directly in terms of the class *instances*. At this level, the associations of the high-level

L. Fribourg (Ed.): CSL 2001, LNCS 2142, pp. 324–338, 2001.

class diagram are implemented by means of pointer attributes and *collection objects* like sets, sequences, enumerations and bags that act as *multiplexers*. Consider for example the object structure in Figure 2, which depicts an instance p of the class *Person*, instances b_1, b_2 and b_3 of the class *Book*, instances c_1 and c_2 of the class *Company*, and multiplexers m_1, m_2 and m_3. Persons have a pointer named *AuthorOf* to a multiplexer, multiplexers have pointers named *item*(1), *item*(2), *item*(3) etc. to books, books have a pointer *AuthoredBy* to a person and a pointer *PublishedBy* to a company, and companies have a pointer *PublisherOf* to a multiplexer.

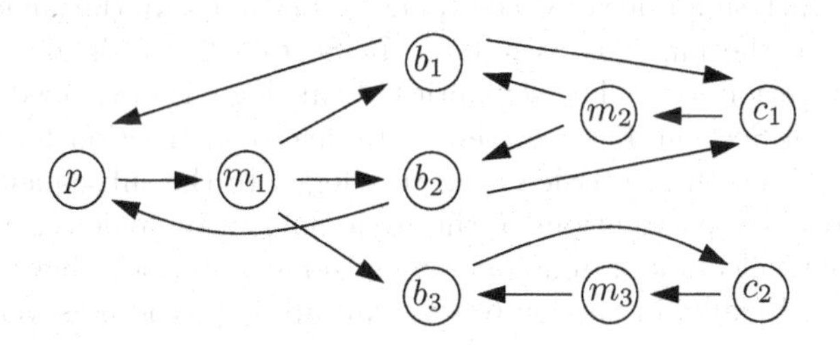

Fig. 2. An object structure at the implementation level

A central concept for the description of object structures at both the modelling and implementation level is *navigation* [7]. The main problem addressed in this paper, is a logical description of navigation at the implementation level, which denotes the operation of successively following pointers across an object structure. A crucial property of navigation is *reachability*; i.e., the question whether starting at an object it is possible to reach another object by navigation. For instance, in the object structure of Figure 2, person p is the author of a book that is published by company c_2. In terms of navigation this means that it is possible to reach c_2 from p by navigating the attributes *AuthorOf*, *item*(3) and *PublishedBy*. Conversely, company c_2 is the publisher of a book that is authored by p. However, it is not possible to reach p from c_2 by navigation because there is no pointer from b_3 to p. This could indicate an *error* in the programming code.

Standard *first-order logic*, which allows only quantification over objects, is not expressive enough to describe reachability in a network of objects. Moreover, the validity problem for first-order logic is undecidable. Standard *modal logics* [9] are suited to navigate through Kripke models, but although they are decidable, they also lack expressive power because they do not distinguish between *bisimilar* Kripke models, such as for instance between a loop and its infinite unfolding.

Over the last decade, a new family of modal logics has come up, which combine modal operators with first-order variable-binding mechanisms. These languages are referred to as the family of *hybrid logics* [2]. In contrast to standard *first-order modal logic* [6], these logics cover mechanisms to bind variables to the *worlds* of a Kripke model. In particular, in [5], a general logical framework is presented that extends the navigation mechanism of standard modal logic with a variable-binding mechanism that allows the binding of variables to worlds that are in the domain of the current world, like for instance the worlds that are accessible.

The starting point of the navigation logics presented in this paper is the modelling of object structures at the implementation level as Kripke models that contain a family of *deterministic* accessibility relations; one for each pointer attribute. Additionally, we introduce a variable-binding mechanism along the lines of [5] that allows the binding of logical variables to objects that are reachable by a pointer. These pointer attributes can be used in the logic both as first-order terms in equalities and as modal operators. The main result of the paper is a tableau construction for deciding the validity of formulas in the navigation logics.

The plan of the paper is as follows. In Section 2, we define the syntax of the basic navigation logic. The semantics of this logic is developed in Section 3. Additionally, in Section 4, we present a tableau construction for deciding the validity of formulas in the basic navigation logic. In the subsequent section, we discuss two decidable extensions of the basic navigation logic; viz. an extension that includes a collection of jump operators and an extension that covers navigation programs. Finally, in Section 6, we wrap up by providing several directions for future research.

2 Basic Navigation Logic

In this section, we define the basic navigation logic. Let $\mathcal{A} = \{A_1, \ldots, A_n\}$ be a finite set of *pointer attributes* and $\mathcal{C} = \{nil,\ self\}$ be the set of *constants*. The constant *nil* is used to denote 'undefined' and the constant *self* to denote an object itself. Additionally, let *Var* be a set of *variables* with typical elements x, y, z. The set $\mathcal{T} = \mathcal{A} \cup \mathcal{C} \cup Var$, with typical element t, denotes the set of *terms*. Finally, let $\mathcal{P}$ be a set of *propositional atoms*, with typical elements p, q and r.

Definition 1. *The* basic navigation language $\mathcal{L}_0$ *consists of formulas* φ *that are generated by the following BNF-grammar:*

$$\varphi ::= p \mid (t_1 = t_2) \mid \neg\varphi \mid \varphi_1 \vee \varphi_2 \mid \langle A \rangle \varphi \mid \exists_{x=t}(\varphi)$$

We have the usual abbreviations $\varphi \rightarrow \psi$ *for* $\neg\varphi \vee \psi$, $\varphi_1 \wedge \varphi_2$ *for* $\neg(\neg\varphi_1 \vee \neg\varphi_2)$, *and* $\varphi \leftrightarrow \psi$ *for* $\varphi \rightarrow \psi \wedge \psi \rightarrow \varphi$.

In the language $\mathcal{L}_0$, pointer attributes can be used both as first-order terms in equalities and as *modal* operators for navigation, which are applied to *formulas*: A formula $\langle A \rangle \varphi$ expresses that φ holds for the object that results from following the pointer A.

Different objects, and in particular different objects from the same class, can have attributes with the same name. To handle the ambiguities of these pointer attributes, i.e., in order to be able to compare the pointer attributes of different objects, the language covers a *variable-binding* mechanism: $\exists_{x=t}(\varphi)$ denotes the formula φ in which the variable x is bound to the object denoted by the term t.

For instance, the formula $\exists_{x=A}\langle B\rangle\neg(x = A)$ expresses that the pointer attribute A of the current object and the pointer attribute A of the object that is reached by following the pointer B from the current object, have different denotations. This variable-binding mechanism is further illustrated in the example below.

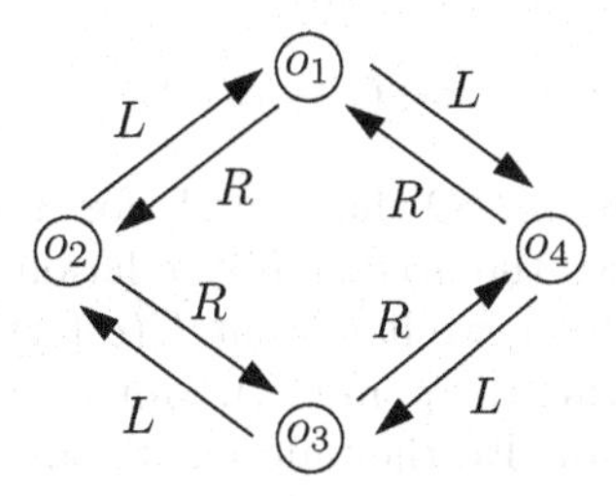

Fig. 3. Objects arranged in a ring structure

Example 1. Consider the structure depicted in Figure 3 in which the objects are arranged in a ring. Each object has two pointers L and R to denote its left and its right neighbour, respectively. As an example, we have that each object is the left neighbour of its right neighbour; formally, for each object in the ring the formula $\exists_{x=self}\langle R\rangle(L = x)$ is true. Second, each object's right neighbour is the same object as the left neighbour of the left neighbour of its left neighbour; that is, for each object the formula $\exists_{x=R}\langle L\rangle\langle L\rangle(L = x)$ holds.

The above example illustrates the difference with the variable-binding mechanism of *first-order logic*, in which this form of binding can be modelled by substitution. That is, in first-order logic, the formula $\exists_{x=t}(\varphi)$ can be modelled by $\varphi[t/x]$, which denotes the substitution of t for x in φ. If we would apply such substitution to the formula $\exists_{x=R}\langle L\rangle\langle L\rangle(L = x)$ we obtain $\langle L\rangle\langle L\rangle(L = R)$, which expresses that for the left neighbour of an object's left neighbour it holds that the left neighbour and the right neighbour are the same. This does not hold for any of the objects in the above figure.

Finally, as our primary concern is the development of a *decidable* navigation logic, we only include the datatype of pointers. We assume a *propositional encoding* of information about the state of an object, which may include other datatypes like integers and booleans; e.g., the fact $x < y$, which expresses that the value of the integer variable x is less than the value of y, can be represented by a particular proposition p.

3 Semantics

In this section, we introduce the semantics of $\mathcal{L}_0$ that is based on a formal description of object structures in terms of Kripke models that contain a deterministic accessibility relation for each pointer attribute.

Definition 2. *A* Kripke model *M is a triple (O, K, π), where O denotes the set of objects, which form the* states *or* worlds *of the Kripke model, K is a total function of type*

$$K : O \to (\mathcal{A} \to O)$$

and π is a valuation function of type

$$\pi : O \to \wp(\mathcal{P}).$$

We assume an element $\perp \in O$ that stands for 'undefined', i.e., the value of *nil*. The function K defines the accessibility relations of the model; that is, for each object o and attribute A, we have that $K(o)(A)$ denotes the object that is accessible from o by following the pointer A. Moreover, if $K(o)(A) = \perp$ then A is called a nil pointer. Additionally, the function π constitutes a valuation function that maps each world o to the set of propositions that are true in it.

Definition 3. *Given a model $M = (O, K, \pi)$, a state $o \in O$ and an assignment function $s : Var \to O$, the* interpretation $[\![t]\!]_{M,o,s}$ of a term $t \in \mathcal{T}$ *is defined by:*

$$\begin{aligned} [\![nil]\!]_{M,o,s} &= \perp \\ [\![self]\!]_{M,o,s} &= o \\ [\![A]\!]_{M,o,s} &= K(o)(A) \\ [\![x]\!]_{M,o,s} &= s(x) \end{aligned}$$

Additionally, the truth definition $M, o, s \models \varphi$ for the language $\mathcal{L}_0$ *is inductively given as follows:*

$$\begin{array}{lcl} M, o, s \models (t_1 = t_2) & \Leftrightarrow & [\![t_1]\!]_{M,o,s} = [\![t_2]\!]_{M,o,s} \\ M, o, s \models p & \Leftrightarrow & p \in \pi(o) \\ M, o, s \models \varphi_1 \vee \varphi_2 & \Leftrightarrow & M, o, s \models \varphi_1 \textit{ or } M, o, s \models \varphi_2 \\ M, o, s \models \neg\varphi & \Leftrightarrow & M, o, s \not\models \varphi \\ M, o, s \models \langle A \rangle \varphi & \Leftrightarrow & M, o', s \models \varphi, \textit{ where } o' = K(o)(A) \neq \perp \\ M, o, s \models \exists_{x=t}(\varphi) & \Leftrightarrow & M, o, s\{x \mapsto o'\} \models \varphi, \textit{ where } o' = [\![t]\!]_{M,o,s} \end{array}$$

where $s\{x \mapsto o'\}$ denotes the function that behaves like s except for the input x for which it yields the output o'.

We define $M, o \models \varphi$ if for all assignments s it holds that $M, o, s \models \varphi$. Additionally, we have $M \models \varphi$ if $M, o \models \varphi$ holds for all objects o. Finally, $\models \varphi$ holds if $M \models \varphi$ for all Kripke models M.

In standard modal logic with deterministic Kripke models, there is only a subtle difference between the interpretation of the possibility and the necessitation operator: The possibility operator requires *exactly one* accessible world to exist, while in the case of the necessitation operator it is required that *at most one* accessible world exists. Here, we have chosen for the *possibility* interpretation of the navigation operator; the *necessitation* reading of the operator would be:

$$M, o, s \models [A]\varphi \;\Leftrightarrow\; \text{if } o' = K(o)(A) \neq \perp \text{ then } M, o', s \models \varphi.$$

Thus, the difference is that in the latter reading the formula is also true if there is no accessible state. Note that $\models [A]\varphi \leftrightarrow (\langle A\rangle\varphi \vee A = nil)$. An alternative to dealing with nil pointers is to use a three-valued logic, which besides the truth values true and false includes a value for undefinedness, but this would unnecessarily complicate the technical treatment. Note that in the evaluation of $\exists_{x=t}(\varphi)$ we do allow t to be a nil pointer.

Finally, we remark that in the evaluation of formulas, the roles of constants and variables are interchanged: The interpretation of variables remains *constant* during the evaluation of a formula, in other words, their interpretation is 'frozen', while the interpretation of constants *varies* with the current state of evaluation.

4 Decidability of Basic Navigation Logic

In this section, we show that the validity of formulas in $\mathcal{L}_0$ can be decided by means of a semantic tableau procedure.

Definition 4. *Given a set O of objects, we construct a* semantic tableau, *which is a tree-like structure with nodes of the form $o : \Gamma$, where $o \in O \setminus \{\bot\}$ and $\Gamma \subseteq \mathcal{L}_0$.*

The construction of a tableau involves three types of branch extension rules, *namely* conjunctive rules, disjunctive rules *and* navigation rules. *They are of the following general format, respectively:*

$$\frac{o : \Delta, \varphi}{o : \Delta, \Gamma} \qquad \frac{o : \Delta, \varphi}{o : \Delta, \Gamma \mid o : \Delta, \Gamma'} \qquad o : \Gamma \xrightarrow{A} o' : \Gamma'$$

The rules are as follows (we omit the context $\Delta \subseteq \mathcal{L}_0$):

$$\frac{o : \neg(\varphi_1 \vee \varphi_2)}{o : \neg\varphi_1,\ \neg\varphi_2} \qquad \frac{o : \exists_{x=t}\varphi}{o : x = t,\ \varphi} \qquad \frac{o : \neg\exists_{x=t}(\varphi)}{o :\ x = t,\ \neg\varphi} \qquad \frac{o : \neg\neg\varphi}{o : \varphi}$$

$$\frac{o : \varphi_1 \vee \varphi_2}{o : \varphi_1 \mid o : \varphi_2} \qquad \frac{o : \neg\langle A\rangle\varphi}{o : \langle A\rangle\neg\varphi \mid o : A = nil}$$

$$o : \Gamma \xrightarrow{A} o' : \Gamma_A$$

where $\Gamma_A = \{\varphi \mid \langle A\rangle\varphi \in \Gamma\}$ and o' is a fresh label.

Given an input formula $\varphi \in \mathcal{L}_0$, the construction of the tableau for φ then proceeds as follows:

(0) *Start with an initial tree consisting of the root node $o : \varphi$.*
(1) *If a conjunctive or disjunctive rule is applicable then apply it and goto* (1), *else goto* (2)
(2) *If the navigation rule is applicable then apply it once for each attribute A that occurs in Γ such that Γ_A is non-empty and goto* (1), *else terminate.*

The construction of the tableau *always* terminates. Formally, this can be shown by a straightforward induction on the length of the input formula φ.

Before considering the soundness and completeness of the tableau construction, let us briefly sketch how a model can be extracted from a tableau. A tableau consists of a number of branches; the branches with a consistent theory can be used in the model construction. The theory of a branch is given by the set of literals that occur on it, but as the interpretation of these literals is relative to the nodes of the branch, we need to replace them by absolute literals whose interpretation is node-independent. A model is then obtained from a consistent branch by identifying the nodes that are expressed to be equal in the theory; in other words, each world of the model represents an equivalence class of identified nodes. Although the branch is a tree structure, the corresponding model may contain cycles due to these identifications. Finally, the accessibility relation and the valuation function of the model can be extracted from the branch in a straightforward manner.

The remainder of this section is devoted to an outline of the soundness and completeness proofs of the above semantic tableau method. From now on, we fix O, with $\perp \in O$, to be the set of objects used in the tableau construction.

First, we give a definition of terms and propositions, so-called *absolute terms* and *absolute propositions*, whose evaluation does not depend on the current object.

Definition 5. *The set* $\mathcal{T}_{abs}$ *of* absolute terms *and the set* $\mathcal{P}_{abs}$ *of* absolute propositions *are given by:*

$$\mathcal{T}_{abs} = Var \cup O \cup \{o.A \mid o \in O,\ A \in \mathcal{A}\}$$

$$\mathcal{P}_{abs} = \{o.p \mid o \in O,\ p \in \mathcal{P}\}$$

An absolute literal *is an (in)equation between terms of* $\mathcal{T}_{abs}$ *or (the negation of) an absolute proposition of* $\mathcal{P}_{abs}$.

Next, we define the interpretation of absolute terms and propositions.

Definition 6. *Given a Kripke model* $M = (O', K, \pi)$*, a corresponding assignment* s*, and a strict (with respect to the respective bottom elements* $\perp \in O$ *and* $\perp' \in O'$*) mapping* $\theta \in O \to O'$*, we define the* interpretation $[\![t]\!]_{M,s,\theta}$ of an absolute term t *as follows:*

$$\begin{aligned} [\![x]\!]_{M,s,\theta} &= s(x) \\ [\![o]\!]_{M,s,\theta} &= \theta(o) \\ [\![o.A]\!]_{M,s,\theta} &= K(\theta(o))(A) \end{aligned}$$

Additionally, for all $t_1, t_2 \in \mathcal{T}_{abs}$ *and* $o.p \in \mathcal{P}_{abs}$ *we define:*

$$\begin{aligned} M, s, \theta \models (t_1 = t_2) &\Leftrightarrow [\![t_1]\!]_{M,s,\theta} = [\![t_2]\!]_{M,s,\theta} \\ M, s, \theta \models o.p &\Leftrightarrow p \in \pi(\theta(o)) \end{aligned}$$

Note that the strictness of θ yields $[\![\bot]\!]_{M,s,\theta} = \bot'$.

The following definition associates with each $o \in O \setminus \{\bot\}$ an operation that transforms a term $t \in \mathcal{T}$ into a corresponding absolute term.

Definition 7. *For each term $t \in \mathcal{T}$ and object $o \in O \setminus \{\bot\}$ we define the term $t^o \in \mathcal{T}_{abs}$ by:*

$$\begin{aligned} nil^o &= \bot \\ self^o &= o \\ A^o &= o.A \\ x^o &= x \end{aligned}$$

The following lemma states some truth-preserving properties of this operation.

Lemma 1. *For every model $M = (O', K, \pi)$, corresponding variable assignment s, and strict mapping $\theta \in O \to O'$ we that have for every term $t \in \mathcal{T}$:*

$$[\![t]\!]_{M,\theta(o),s} = [\![t^o]\!]_{M,s,\theta}$$

Similarly, for every equality $(t_1 = t_2) \in \mathcal{L}_0$ and every proposition $p \in \mathcal{P}$ we have:

$$\begin{aligned} M, \theta(o), s &\models (t_1 = t_2) \;\; \textit{iff} \;\; M, s, \theta \models (t_1^o = t_2^o) \\ M, \theta(o), s &\models p \;\; \textit{iff} \;\; M, s, \theta \models o.p \end{aligned}$$

Next, we define the deductive closure of a set of absolute literals.

Definition 8. *For each set Γ of absolute literals, its* deductive closure *$Clos(\Gamma)$ is defined to be the smallest set that contains Γ and that is closed under the following rules:*

- *$(t = t) \in Clos(\Gamma)$, for every term t occurring in Γ*
- *if $(t_1 = t_2) \in Clos(\Gamma)$ then $(t_2 = t_1) \in Clos(\Gamma)$*
- *if $(t_1 = t_2)$, $(t_2 = t_3) \in Clos(\Gamma)$ then $(t_1 = t_3) \in Clos(\Gamma)$*
- *if $t_1 = t_2$, $\varphi \in Clos(\Gamma)$ then $\varphi[t_2/t_1] \in Clos(\Gamma)$*

It is not difficult to see that if Γ is a finite set of absolute literals then $Clos(\Gamma)$ is also finite.

A *branch* T of a semantic tableau is a substructure that contains the root node and that for each application of a conjunctive rule contains its child, for each application of a disjunctive rule contains precisely one of its children, and for each application of the navigation rule contains all its children. The theory of a tableau branch as a deductively closed set of absolute literals, is introduced in the following definition.

Definition 9. *The* theory of a branch *T is given by $Th(T) = Clos(\Gamma)$,*

$$\begin{aligned} \textit{where } \Gamma \;=\; & \{o.p \mid o : p \in T\} \cup \{\neg o.p \mid o : \neg p \in T\} \cup \\ & \{t_1^o = t_2^o \mid o : (t_1 = t_2) \in T\} \cup \{\neg(t_1^o = t_2^o) \mid o : \neg(t_1 = t_2) \in T\} \cup \\ & \{o.A = o' \mid o \xrightarrow{A} o' \in T\} \cup \\ & \{\neg(o = \bot) \mid o \in T\}. \end{aligned}$$

(When we write $o : \varphi \in T$ we mean that there exists a node $o : \Delta, \varphi$ in T, for some set of formulas Δ. Additionally, by $o \xrightarrow{A} o' \in T$ we mean that in T there exists an A-link from a node labelled by o to one labelled by o'.)

It is worthwhile to observe that $Th(T)$ is a *finite* set of absolute literals. Next, we define when a tableau branch is closed, that is, does not give rise to a model.

Definition 10. *A branch T is called* closed *if $Th(T)$ is inconsistent (i.e., contains both a literal and its negation). If T is not closed it is called* open.

In the following definition, we introduce the notion of a homomorphic mapping of a branch of a semantic tableau into a Kripke model.

Definition 11. *A strict mapping $\theta : O \rightarrow O'$ is a* homomorphic mapping *of a branch T of a tableau into a model $M = (O', K, \pi)$ with respect to a variable assignment s if the following hold:*

- *if $o \xrightarrow{A} o' \in T$ then $K(\theta(o))(A) = \theta(o')$*
- *if $o : \varphi \in T$ then $M, \theta(o), s \models \varphi$*

Finally, we are in a position to state and prove the soundness of the tableau method.

Theorem 1. *For all $\varphi \in \mathcal{L}_0$, if φ is satisfiable then its tableau has an open branch.*

Proof. Without loss of generality we may assume that in φ each variable x is bound at most once and does not occur both free and bound. Let $M = (O', K, \pi)$ be a model and s a variable assignment such that $M, o', s \models \varphi$, for some o'. For an inductive argument, suppose we have constructed a variable assignment s and a corresponding homomorphic mapping $\theta : O \rightarrow O'$ into M of all the nodes of a branch of the tableau of φ which occur at a depth smaller than some $n \geq 0$. We consider the following two main extensions.

- The branch with node $o : \Delta, \exists_{x=t}(\psi)$ at level n is extended with $o : \Delta, x = t, \psi$. From the existence of the homomorphism θ we derive that $M, \theta(o), s \models \exists_{x=t}(\psi)$. Using the truth definition we obtain $M, \theta(o), s\{x \mapsto o'\} \models \psi$ for $o' = [\![t]\!]_{K,\theta(o),s}$. Thus, we have $M, \theta(o), s\{x \mapsto o'\} \models (x = t) \wedge \psi$. It follows by the above assumption on φ that x does not occur (free) in any formula occurring in the tableau other than ψ. So we have that θ is also a homomorphism into M with respect to the variable assignment $s\{x \mapsto o'\}$ and which now includes also the node $o : \Delta, x = t, \psi$.
- A node $o : \Gamma \in T$ on this branch at level n is A-linked to a new node $o' : \Gamma_A$, for some attribute A. From the assumption $M, \theta(o), s \models \Gamma$ we derive $M, o'', s \models \Gamma_A$ for $o'' = K(\theta(o))(A) \neq \perp$. So, we can extend θ to a homomorphism $\theta\{o' \mapsto o''\}$ which includes the new node $o' : \Gamma_A$.

Via Lemma 1 and the construction of the homomorphism θ we derive $M, s, \theta \models Th(T)$, for some branch T. From this it follows that $Th(T)$ is consistent, i.e., that T is an open branch. □

To prove completeness we first show how to construct a model from an open branch.

Definition 12. *Given an open branch T, we define the Kripke model $M_T = (O', K, \pi)$ as follows:*

- *The underlying set O' of worlds consists of the equivalence classes:*

$$[t] = \{t' \mid (t = t') \in Th(T)\},$$

where $t, t' \in \mathcal{T}_{abs} \setminus Var$. The equivalence class $[\bot]$ plays the role of bottom in O'.
- *The accessibility function K is defined by:*

$$\begin{array}{l} K([o])(A) = [o.A] \\ K([t])(A) = [nil], \text{ if there does not exist an } o \in [t]. \end{array}$$

- *The valuation function π is defined by:*

$$\pi([t]) = \{p \mid o : p \in T \text{ for some } o \in [t]\}.$$

It is straightforward to check that M_T is indeed well-defined. Moreover, by construction of M_T, an absolute term denotes in M_T its corresponding equivalence class under the strict mapping θ which assigns to every $o \in O$ its equivalence class $[o]$. This is expressed formally by the following lemma.

Lemma 2. *Let s be an assignment that assigns to each variable an object of M_T, i.e., an equivalence class of absolute terms. Additionally, let $\theta \in O \to O'$ be a strict mapping which assigns to every $o \in O$ its corresponding equivalence class $[o]$ in M_T. For each $t \in \mathcal{T}_{abs}$ let $[t]_s$ denote the equivalence class $[t]$ itself, in case $t \notin Var$, and $[x]_s = s(x)$, otherwise. For each term $t \in \mathcal{T}_{abs}$ we then have:*

$$[\![t]\!]_{M_T,s,\theta} = [t]_s$$

Given the above, we are now in a position to prove the completeness theorem.

Theorem 2. *For all $\varphi \in \mathcal{L}_0$, if the tableau for φ has an open branch then φ is satisfiable.*

Proof. Let T be an open branch of the tableau for φ. We show that for all $o : \psi \in T$ we have $M_T, [o], s \models \psi$ for the assignment s that is defined by: $s(x) = [t]$, where $(x = t) \in Th(T)$ and $t \in \mathcal{T}_{abs} \setminus Var$. Note that the existence of such an absolute term follows from the fact that the initial formula φ is assumed not to contain free variables. The proof proceeds by induction on the length of ψ. We treat the following characteristic cases.

- **case** p. From $o : p \in T$ and $o \in [o]$ we derive $M_T, [o], s \models p$.
- **case** $\neg p$. From $o : \neg p \in T$ and the fact that T is open we derive $o' : p \notin T$ for all $o' \in [o]$. In other words, $M_T, [o], s \models \neg p$.

- **case** $t_1 = t_2$. Suppose $o : (t_1 = t_2) \in T$. By definition of $Th(T)$ we have $t_1^o = t_2^o \in Th(T)$. By construction of the model M_T and the definition of s we derive $[t_1^o]_s = [t_2^o]_s$. From Lemma 2 we subsequently infer $M_T, s, \theta \models (t_1^o = t_2^o)$, where $\theta \in O \rightarrow O'$ is a strict mapping which assigns to every $o \in O$ its corresponding equivalence class $[o]$ in M_T. Finally, from Lemma 1 we conclude $M_T, [o], s \models (t_1 = t_2)$.
- **case** $\neg(t_1 = t_2)$. Suppose $o : \neg(t_1 = t_2) \in T$. By definition of $Th(T)$ we have $\neg(t_1^o = t_2^o) \in Th(T)$. Since $Th(T)$ is consistent we have $t_1^o = t_2^o \notin Th(T)$. By construction of the model M_T and the definition of s it thus follows that $[t_1^o]_s \neq [t_2^o]_s$, from which we derive by Lemma 2 that $M_T, s, \theta \models \neg(t_1^o = t_2^o)$, where $\theta \in O \rightarrow O'$ is a strict mapping which assigns to every $o \in O$ its corresponding equivalence class $[o]$ in M_T. From Lemma 1 we conclude $M_T, [o], s \models \neg(t_1 = t_2)$.
- **case** $\langle A \rangle \varphi$. Suppose $o : \langle A \rangle \varphi \in T$. From the construction of the tableau we derive that there exists a node o' with $o \stackrel{A}{\longrightarrow} o' \in T$ and $o' : \varphi \in T$. The induction hypothesis yields $M_T, [o'], s \models \varphi$. The construction of the model M_T then yields $M_T, [o], s \models \langle A \rangle \varphi$.
- **case** $\exists_{x=t}(\varphi)$. Suppose $o : \exists_{x=t}(\varphi) \in T$. From the construction of the tableau we derive that also $o : x = t, \varphi \in T$. Applying the induction hypothesis we obtain $M_T, [o], s \models (x = t)$ and $M_T, [o], s \models \varphi$. From this we conclude $M_T, [o], s \models \exists_{x=t}(\varphi)$. □

Note that from the construction of the model in the completeness theorem it follows that the language $\mathcal{L}_0$ satisfies the *finite model property*, i.e., every satisfiable formula is satisfiable in a finite model. Moreover, note that the constructed model may contain cycles, which indicates that the language $\mathcal{L}_0$ does not satisfy the *tree model property*, stating that every satisfiable sentence has a model that is a tree of bounded branching [12]. A counterexample is the formula $\exists_{x=self} \langle A \rangle (x = self)$, which is only satisfiable in worlds that are A-linked to themselves.

5 Extensions

In this section, we briefly discuss some interesting decidable extensions of the basic navigation logic presented above.

5.1 Jump Operators

The logic $\mathcal{L}_1$ extends the basic navigation logic $\mathcal{L}_0$ with *jump operators*; i.e., we generalise the syntax of the navigation operator to $\langle t \rangle \varphi$, where t is a term in $\mathcal{T}$. So, if the index is a variable x, the operator can be used to jump back to the state to which x is bound.

Definition 13. *The* truth definition $M, o, s \models \varphi$ for the language $\mathcal{L}_1$ *is the same as for* $\mathcal{L}_0$ *except:*

$$M, o, s \models \langle t \rangle \varphi \Leftrightarrow M, o', s \models \varphi, \text{ where } o' = [\![t]\!]_{M,o,s} \neq \bot.$$

As an example, consider the object structure M in Figure 2. At the world p, the following formula is true:

$$\langle AuthorOf\rangle \exists_{x=self}\langle item(1)\rangle \exists_{y=PublishedBy}\langle x\rangle\langle item(2)\rangle(PublishedBy = y)$$

which means that person p is the author of two books that are published by the *same* company.

The tableau construction $\mathcal{L}_1$ is similar to that of $\mathcal{L}_0$ modulo some minor modifications.

Definition 14. *The* tableau rules for $\mathcal{L}_1$ *are the same as for* $\mathcal{L}_0$ *except for the navigation rule, which is generalised as follows:*

$$o : \Gamma \quad \stackrel{t}{\longrightarrow} \quad o' : \Gamma_t$$

where $\Gamma_t = \{\varphi \mid \langle t\rangle\varphi \in \Gamma\}$ *and* o' *is a fresh label.*

Moreover, it suffices to modify the theory $Th(T)$ of a branch as given in Definition 9 as follows. The support set Γ of absolute literals additionally contains:

$$\{x = o' \mid o \stackrel{x}{\longrightarrow} o' \in T\} \cup$$
$$\{\bot = o' \mid o \stackrel{nil}{\longrightarrow} o' \in T\} \cup$$
$$\{o = o' \mid o \stackrel{self}{\longrightarrow} o' \in T\}.$$

The soundness and completeness proofs are similar to the proofs for $\mathcal{L}_0$.

5.2 Navigation Programs

In the extension $\mathcal{L}_2$, which is inspired by the dynamic logic of [8], we include formulas of the form $\langle \Pi\rangle\varphi$, where Π is a *navigation program* that defines a particular navigation strategy.

Definition 15. *The* navigation language $\mathcal{L}_2$, *consisting of boolean conditions* b, *navigation programs* Π *and formulas* φ, *is generated by the following BNF-grammar:*

$$b \ ::= p \ \mid \ t_1 = t_2 \ \mid \ \neg b \ \mid \ b_1 \vee b_2$$

$$\Pi ::= A \ \mid \ \Pi_1; \Pi_2 \ \mid \ \textsf{if } b \textsf{ then } \Pi_1 \textsf{ else } \Pi_2 \ \mid \ \textsf{while } b \textsf{ do } \Pi$$

$$\varphi ::= b \ \mid \ \neg\varphi \ \mid \ \varphi_1 \vee \varphi_2 \ \mid \ \exists_{x=t}\varphi \ \mid \ \langle \Pi\rangle\varphi$$

To formalise the meaning of a formula in $\mathcal{L}_2$ we first define the meaning of navigation programs.

Definition 16. *Given a model* $M = (O, K, \pi)$ *and an assignment* s, *we have the following (standard)* denotational semantics *which assigns to each program*

Π a strict mapping $\mathcal{M}(\Pi): O \to O$:

$$\begin{array}{ll}
\mathcal{M}(A)(o) & = K(o)(A) \\
\mathcal{M}(\Pi_1;\Pi_2)(o) & = \mathcal{M}(\Pi_2)(\mathcal{M}(\Pi_1)(o)) \\
\mathcal{M}(\mathsf{if}\ b\ \mathsf{then}\ \Pi_1\ \mathsf{else}\ \Pi_2)(o) & = \begin{cases} \mathcal{M}(\Pi_1)(o) & \text{if } M,o,s \models b \\ \mathcal{M}(\Pi_2)(o) & \text{otherwise} \end{cases} \\
\mathcal{M}(\mathsf{while}\ b\ \mathsf{do}\ \Pi)(o) & = \begin{cases} \mathcal{M}(\mathsf{while}\ b\ \mathsf{do}\ \Pi)(\mathcal{M}(\Pi)(o)) & \text{if } M,o,s \models b \\ o & \text{otherwise} \end{cases}
\end{array}$$

The above recursive definition of $\mathcal{M}$ can be justified by means of a (standard) least fixpoint construction defined in terms of the discrete complete partial order $(O, \sqsubseteq)$ that is defined by: $\bot \sqsubseteq o$ and $o \not\sqsubseteq o'$ for all distinct objects o and o'. From this fixpoint construction it follows that $\mathcal{M}(\mathsf{while}\ b\ \mathsf{do}\ \Pi)(o) = \bot$ in case the program $\mathsf{while}\ b\ \mathsf{do}\ \Pi$ does not terminate in o.

We have the following truth definition.

Definition 17. *The* truth definition $M,o,s \models \varphi$ for the language $\mathcal{L}_2$ *is similar to that for $\mathcal{L}_0$ except:*

$$M,o,s \models \langle \Pi \rangle \varphi \;\Leftrightarrow\; M,o',s \models \varphi, \text{ where } o' = \mathcal{M}(\Pi)(o) \neq \bot$$

It is worthwhile to observe that the truth of the formula $\langle \mathsf{while}\ b\ \mathsf{do}\ \Pi \rangle \mathit{true}$ implies that the program $\mathsf{while}\ b\ \mathsf{do}\ \Pi$ terminates. In the light of the Halting Problem, however, this does not imply that we cannot decide the validity of formulas in the language $\mathcal{L}_2$, because our navigation programs are not Turing complete.

As an example of $\mathcal{L}_2$, the formula

$$\langle \mathsf{while}\ \neg(y = \mathit{self})\ \mathsf{do}\ A \rangle \mathit{true}$$

states that the object denoted by y is reachable from the current object by a finite chain of A-links.

Definition 18. *The* tableau rules for $\mathcal{L}_2$ *are given by the rules for $\mathcal{L}_0$ together with the following rules (we omit the context Δ):*

$$\frac{o : \langle \Pi_1;\Pi_2 \rangle \varphi}{o : \langle \Pi_1 \rangle \langle \Pi_2 \rangle \varphi}$$

$$\frac{o : \langle \mathsf{if}\ b\ \mathsf{then}\ \Pi_1\ \mathsf{else}\ \Pi_2 \rangle \varphi}{o : b, \langle \Pi_1 \rangle \varphi \mid o : \neg b, \langle \Pi_2 \rangle \varphi} \qquad \frac{o : \neg\langle \mathsf{if}\ b\ \mathsf{then}\ \Pi_1\ \mathsf{else}\ \Pi_2 \rangle \varphi}{o : b, \neg\langle \Pi_1 \rangle \varphi \mid o : \neg b, \neg\langle \Pi_2 \rangle \varphi}$$

$$\frac{o : \langle \mathsf{while}\ b\ \mathsf{do}\ \Pi \rangle \varphi}{o : \neg b, \varphi \mid o : b, \langle \Pi \rangle \langle \mathsf{while}\ b\ \mathsf{do}\ \Pi \rangle \varphi} \qquad \frac{o : \neg\langle \mathsf{while}\ b\ \mathsf{do}\ \Pi \rangle \varphi}{o : \neg b, \neg\varphi \mid o : b, \neg\langle \Pi \rangle \langle \mathsf{while}\ b\ \mathsf{do}\ \Pi \rangle \varphi}$$

The resulting tableau method may give rise to *non-termination*. However, in the tableau construction, we can stop applying any further rules to a leaf $o : \Gamma$ in case there already exists an ancestor node with the same set of formulas

Γ. With this additional rule the method is guaranteed to terminate, because starting from an initial formula, only a finite number of *different* formulas can be generated.

Additionally, a branch T of a tableau now also closes if one of its leafs $o : \Gamma$ contains a formula of the form $\langle \mathsf{while}\ b\ \mathsf{do}\ \Pi \rangle \psi$. (Note that by definition there also exists an ancestor node of the form $o' : \Gamma$.) The new tableau method is sound and complete because in case its theory does not contain an inconsistency, such a branch T corresponds to a model in which the program $\mathsf{while}\ b\ \mathsf{do}\ \Pi$ 'loops', i.e., $\langle \mathsf{while}\ b\ \mathsf{do}\ \Pi \rangle \psi$ does not hold. For the same reason, a branch T of a tableau with a consistent theory and a leaf which contains a formula of the form $\neg \langle \mathsf{while}\ b\ \mathsf{do}\ \Pi \rangle \psi$ does constitute a model.

6 Conclusions and Future Research

In this paper, we have presented multimodal logics for navigation across object structures. The starting point of these logics is the modelling of object structures at the implementation level as Kripke models that contain a family of deterministic accessibility relations, namely one relation for each pointer attribute. The logics cover a variable-binding mechanism that allows the binding of logical variables to objects that are reachable by a pointer. In this way, pointer attributes can be used in the logic both as first-order terms in equalities and as modal operators. The main result of the paper is a tableau construction for deciding the validity of formulas in these navigation logics.

In [1], it is stated that for hybrid languages, i.e., modal logics that include mechanisms for naming the worlds of a given Kripke model: "it seems unlikely that restricted forms of (label) binding will lead to decidable systems." However, in this paper, we have shown that decidable hybrid languages can be obtained by restricting to particular classes of models. This point is also illustrated in [3], where decidability of a hybrid language is obtained for the class of strict partial orders. In [11], a decision procedure based on a tableau construction is given for hybrid logics that do not involve variable-binding mechanisms. Of interest in this context would be an extension of our tableau construction to hybrid languages that do include variable-binding.

In the graphical modelling language UML, class diagrams can also be annotated with so-called *constraints*, which are formulated in the Object Constraint Language, OCL for short [13]. The language OCL is a textual language for the description of object structures mainly at the modelling level. In contrast to our approach, in OCL, navigation is modelled as a *dereferencing operator* that is applied to first-order terms. For instance, the term $t.A$ denotes the value of the pointer attribute A of the object denoted by t. By means of a formalisation of navigation in terms of a modal logic, however, we are able to identify decidable navigation logics that are still expressive enough to express interesting properties of object structures. Future work concerns an extension of our approach to the modelling level of class diagrams and the development of tools for computer-aided-verification by means of an implementation of the corresponding tableau procedure.

At the implementation level another interesting line of future work concerns an application of our decidable navigation logics to the computer-aided-verification of the correctness of object-oriented programs. Such an application involves the definition of a *weakest precondition calculus* [4] for our navigation logics.

Acknowledgements

The authors would like to thank the anonymous referees for their comments.

References

1. C. Areces, P. Blackburn, and M. Marx. A road-map on the complexity of hybrid logics. In J. Flum and M. Rodríguez-Artalejo, editors, *Computer Science Logic, Proceedings of CSL'99*, volume 1683 of *Lecture Notes in Computer Science*, pages 307–321. Springer-Verlag, Heidelberg, 1999.
2. P. Blackburn and J. Seligman. Hybrid languages. *Journal of Logic, Language and Information*, 4:251–272, 1995.
3. P. Blackburn and J. Seligman. What are hybrid languages? In M. Kracht, M. de Rijke, H. Wansing, and M. Zakharyaschev, editors, *Advances in Modal Logic'96*. CSLI Publications, Stanford, California, 1997.
4. F.S. de Boer. A WP-calculus for OO. In *Proceedings of Foundations of Software Science and Computation Structures (FOSSACS'99)*, volume 1578 of *Lecture Notes in Computer Science*, pages 135–149, 1999.
5. R.M. van Eijk, F.S. de Boer, W. van der Hoek, and J.-J.Ch. Meyer. Modal logic with bounded quantification over worlds. *Journal of Logic and Computation*, 2001. To appear.
6. M. Fitting and R.L. Mendelsohn. *First-Order Modal Logic*. Kluwer Academic Publishers, Dordrecht, The Netherlands, 1998.
7. A. Hamie, J. Howse, and S. Kent. Navigation expressions in object-oriented modelling. In *Fundamental Approaches to Software Engineering, Proceedings of FASE'98*, volume 1382 of *Lecture Notes in Computer Science*, pages 123–137. Springer-Verlag, Heidelberg, 1998.
8. D. Harel. *First-Order Dynamic Logic*, volume 68 of *Lecture Notes in Computer Science*. Springer-Verlag, Heidelberg, 1979.
9. G.E. Hughes and M.J. Cresswell. *An Introduction to Modal Logic*. Methuen and Co. Ltd, London, 1968.
10. J. Rumbauch, I. Jacobson, and G. Booch. *The Unified Modeling Language Reference Manual*. Addison-Wesley, Reading, Massachusetts, 1999.
11. M. Tzakova. Tableau calculi for hybrid logics. In N.V. Murray, editor, *Automated Reasoning with Analytic Tableaux and Related Methods, Proceedings of TABLEAUX'99*, volume 1617 of *Lecture Notes in Artificial Intelligence*, pages 278–292. Springer-Verlag, Heidelberg, 1999.
12. M.Y. Vardi. Why is modal logic so robustly decidable? In N. Immerman and P. Kolaitis, editors, *Descriptive Complexity and Finite Models*, volume 31 of *DIMACS Series in Discrete Mathematics and Theoretical Computer Science*, pages 149–184. AMS, 1997.
13. J.B. Warmer and A.G. Kleppe. *The Object Constraint Language: Precise Modeling with UML*. Addison-Wesley, Reading, Massachusetts, 1998.

The Decidability of Model Checking Mobile Ambients

Witold Charatonik[1,2] and Jean-Marc Talbot[3]

[1] Max-Planck-Institut für Informatik, Germany
[2] University of Wrocław, Poland
[3] Laboratoire d'Informatique Fondamentale de Lille, France

Abstract. The ambient calculus is a formalism for describing the mobility of both software and hardware. The ambient logic is a modal logic designed to specify properties of distributed and mobile computations programmed in the ambient calculus. In this paper we investigate the border between decidable and undecidable cases of model checking mobile ambients for some fragments of the ambient calculus and the ambient logic.
Recently, Cardelli and Gordon presented a model-checking algorithm for a fragment of the calculus (without name restriction and without replication) against a fragment of the logic (without composition adjunct) and asked the question, whether this algorithm could be extended to include either replication in the calculus or composition adjunct in the logic. Here we answer this question negatively: it is not possible to extend the algorithm, because each of these extensions leads to undecidability of the problem. On the other hand, we extend the algorithm to the calculus with name restriction and logic with new constructs for reasoning about restricted names.

1 Introduction

The ambient calculus [6,4,2] is a process calculus for modeling mobile computations and mobile devices; one can describe the mobility of both software and hardware in this formalism. An ambient is a named cluster of running processes and nested sub-ambients. Each computation state has a spatial structure, the tree induced by the nesting of ambients. Mobility is abstractly represented by re-arrangement of this tree: an ambient may move inside or outside other ambients.

The ambient logic [5] is a modal logic designed to specify properties of distributed and mobile computations programmed in the ambient calculus. As well as standard temporal modalities for describing the evolution of ambient processes over the time, the logic includes spatial modalities for describing the tree structure of processes. In a recent paper, Cardelli and Gordon extend the logic with the constructs for describing private names [7]. Other work on the ambient logic includes a study of the process equivalence induced by the satisfaction relation [11] and the use of spatial modalities to describe the tree structure of semistructured databases [1].

L. Fribourg (Ed.): CSL 2001, LNCS 2142, pp. 339–354, 2001.

The *model-checking* problem is to decide whether a given object (in our case, an ambient process) satisfies (that is, is a model of) a given formula. Cardelli and Gordon [5] give a model-checking algorithm for the fragment of the calculus in which the processes contain no replications and no dynamic name generation against a fragment of the logic in which formulas contain no composition adjunct. It was then proved in [8] that model checking of this fragment of the calculus against this fragment of the logic is PSPACE-complete.

Our results. Cardelli and Gordon raised in [5] the question whether their algorithm for model-checking could be extended to include either replication in the calculus or composition adjunct in the logic. Here we answer this question negatively: it is not possible to extend the algorithm, because each of these extensions leads to undecidability. On the other hand, we show that the restriction to public names was not necessary: model checking remains decidable for the replication-free fragment with dynamic name generation, even if we extend the logic with constructs from [7] for reasoning about restricted names. Moreover, this extension does not increase the complexity of the problem: one can obtain PSPACE algorithm by combining the abstract algorithm presented in Section 5 with the representation of processes proposed in [8].

We start by recalling the calculus and the logic in the next section. Then we prove the undecidability of the problem for the two mentioned extensions: in Section 3 for the case of the calculus with replication and in Section 4 for the case of the logic with composition adjunct. Finally, in Section 5 we present the model-checking algorithm extended to the case of the calculus with name restriction and the logic with constructs for reasoning about these names.

2 Review of the Ambient Calculus and Logic

In this section we recall the ambient calculus and logic from [6,5,7].

2.1 The Ambient Calculus

The following table describes the expressions and processes of our calculus.

Processes and Capabilities:

$P ::=$	processes	$M ::=$	capabilities
$\mathbf{0}$	inactivity	n	name
$M[P]$	ambient	$\mathtt{in}\ M$	can enter M
$P \mid Q$	composition	$\mathtt{out}\ M$	can exit M
$M.P$	capability action	$\mathtt{open}\ M$	can open M
$(n).P$	input action	ϵ	null
$\langle M \rangle$	output action	$M.M'$	path
$!\,P$	replication		
$(\nu n)P$	name restriction		

The sets $bn(P)$ and $fn(P)$ of bound and free names of a given process P are defined in a usual way keeping in mind that (νn) and (n) are name binders. We identify processes up to renaming of bound names. We write $P\{n{\leftarrow}M\}$ for the substitution of the expression M for the name n in the process P.

The semantics of the calculus is given by the relations $P \equiv Q$ and $P \to Q$. The *reduction* relation, $P \to Q$, defines the evolution of processes over time. The *structural congruence* relation, $P \equiv Q$, is an auxiliary relation used in the definition of reduction. When we define the satisfaction relation of the modal logic in the next section, we use an auxiliary relation, the *sublocation* relation, $P \downarrow Q$, which holds when Q is the whole interior of a top-level ambient in P. We write $\to^*$ and $\downarrow^*$ for the reflexive and transitive closure of $\to$ and $\downarrow$, respectively.

Structural Congruence $P \equiv Q$:

$P \equiv P$	(Str Refl)	$(\nu n)(\nu m)P \equiv (\nu m)(\nu n)P$	(Str Res Res)
$P \equiv Q \Rightarrow Q \equiv P$	(Str Symm)	$(\nu n)\mathbf{0} \equiv \mathbf{0}$	(Str Res Zero)
$P \equiv Q, Q \equiv R \Rightarrow P \equiv R$	(Str Trans)	$(\nu n)(P \mid Q) \equiv P \mid (\nu n)Q$ if $n \notin fn(P)$	(Str Res Par)
$P \equiv Q \Rightarrow (\nu n)P \equiv (\nu n)Q$	(Str Res)	$(\nu n)(m[P]) \equiv m[(\nu n)P]$ if $n \neq m$	(Str Res Amb)
$P \equiv Q \Rightarrow P \mid R \equiv Q \mid R$	(Str Par)	$P \mid \mathbf{0} \equiv P$	(Str Zero Par)
$P \equiv Q \Rightarrow !P \equiv !Q$	(Str Repl)	$P \mid Q \equiv Q \mid P$	(Str Par Comm)
$P \equiv Q \Rightarrow n[P] \equiv n[Q]$	(Str Amb)	$(P \mid Q) \mid R \equiv P \mid (Q \mid R)$	(Str Par Assoc)
$P \equiv Q \Rightarrow M.P \equiv M.Q$	(Str Action)		
$P \equiv Q \Rightarrow (n).P \equiv (n).Q$	(Str Input)	$!\mathbf{0} \equiv \mathbf{0}$	(Str Repl Zero)
		$!(P \mid Q) \equiv !P \mid !Q$	(Str Repl Par)
$\epsilon.P \equiv P$	(Str ϵ)	$!P \equiv P \mid !P$	(Str Repl Copy)
$(M.M').P \equiv M.M'.P$	(Str .)	$!P \equiv !!P$	(Str Repl Repl)

Reduction $P \to Q$ and Sublocation $P \downarrow Q$:

$n[\mathsf{in}\ m.P \mid Q] \mid m[R] \to m[n[P \mid Q] \mid R]$	(Red In)
$m[n[\mathsf{out}\ m.P \mid Q] \mid R] \to n[P \mid Q] \mid m[R]$	(Red Out)
$\mathsf{open}\ n.P \mid n[Q] \to P \mid Q$	(Red Open)
$\langle M \rangle \mid (n).P \to P\{n{\leftarrow}M\}$	(Red I/O)
$P \to Q \Rightarrow (\nu n)P \to (\nu n)Q$	(Red Res)
$P \to Q \Rightarrow P \mid R \to Q \mid R$	(Red Par)
$P \to Q \Rightarrow n[P] \to n[Q]$	(Red Amb)
$P' \equiv P, P \to Q, Q \equiv Q' \Rightarrow P' \to Q'$	(Red $\equiv$)
$P \equiv n[Q] \mid P' \Rightarrow P \downarrow Q$	(Loc)

When no confusion is possible, we omit the inactive process $\mathbf{0}$; for instance, we shorten $\mathsf{open}\ n.\mathbf{0}$ and $m[\mathbf{0}]$ as $\mathsf{open}\ n$ and $m[]$, respectively.

2.2 The Ambient Logic

We describe the formulas and satisfaction relation of the logic. This is the logic defined in [5] extended with the two revelation constructs for handling name restriction [7].

Logical Formulas:

η	a name n or a variable x		
$\mathcal{A}, \mathcal{B} ::=$	formula	$\eta[\mathcal{A}]$	location
$\mathbf{T}$	true	$\mathcal{A}@\eta$	location adjunct
$\neg\mathcal{A}$	negation	$\eta®\mathcal{A}$	revelation
$\mathcal{A} \vee \mathcal{B}$	disjunction	$\mathcal{A}\oslash\eta$	revelation adjunct
$\mathbf{0}$	void	$\Diamond\mathcal{A}$	sometime modality
$\mathcal{A} \mid \mathcal{B}$	composition match	$\diamond\!\!\!\!+\mathcal{A}$	somewhere modality
$\mathcal{A} \triangleright \mathcal{B}$	composition adjunct	$\exists x.\mathcal{A}$	existential quantification

We assume that names and variables belong to two disjoint vocabularies. We write $\mathcal{A}\{x \leftarrow m\}$ for the outcome of substituting each free occurrence of the variable x in the formula $\mathcal{A}$ with the name m. The sets of bound and free variables of a given formula are defined in a usual way keeping in mind that $\exists$ is the only variable binder. There are no name binders for formulas, so all names occurring in a formula are free. We say a formula $\mathcal{A}$ is closed if and only if it has no free variables (though it may contain free names).

Intuitively, we interpret closed formulas as follows. The formulas $\mathbf{T}$, $\neg\mathcal{A}$, and $\mathcal{A} \vee \mathcal{B}$ embed propositional logic. The formulas $\mathbf{0}$, $\eta[\mathcal{A}]$, and $\mathcal{A} \mid \mathcal{B}$ are spatial modalities. A process satisfies $\mathbf{0}$ if it is structurally congruent to the inactive process. It satisfies $n[\mathcal{A}]$ if it is structurally congruent to an ambient $n[P]$ where P satisfies $\mathcal{A}$. A process P satisfies $\mathcal{A} \mid \mathcal{B}$ if it can be decomposed into two subprocesses, $P \equiv Q \mid R$, where Q satisfies $\mathcal{A}$, and R satisfies $\mathcal{B}$. The formula $\exists x.\mathcal{A}$ is an existential quantification over names. The formulas $\Diamond\mathcal{A}$ (sometime) and $\diamond\!\!\!\!+\mathcal{A}$ (somewhere) quantify over time and space, respectively. A process satisfies $\Diamond\mathcal{A}$ if it has a temporal successor, that is, a process into which it evolves, that satisfies $\mathcal{A}$. A process satisfies $\diamond\!\!\!\!+\mathcal{A}$ if it has a spatial successor, that is, a sublocation, that satisfies $\mathcal{A}$. A process P satisfies the formula $\mathcal{A}@n$ if the ambient $n[P]$ satisfies $\mathcal{A}$. A process P satisfies $\mathcal{A} \triangleright \mathcal{B}$ if for all P', the process $P \mid P'$ guarantees $\mathcal{B}$ assuming that P' satisfies $\mathcal{A}$. Finally, we discuss shortly the two logical constructs for reasoning about restricted names that were not present in [5,8]. Intuitively, a process P satisfies the formula $n®\mathcal{A}$ (read "reveal n then $\mathcal{A}$") if it is possible to pull a restricted name from P to the top and rename it n and then strip off the restriction to leave a residual process that satisfies $\mathcal{A}$. The inverse of revelation is hiding: a process P satisfies $\mathcal{A}\oslash n$ (read "hide n then $\mathcal{A}$") if it is possible to hide n in P and then satisfy $\mathcal{A}$.

The satisfaction relation $P \models \mathcal{A}$ provides the semantics of our logic.

Satisfaction $P \models \mathcal{A}$ (for $\mathcal{A}$ Closed):

$P \models \mathbf{T}$

$P \models \neg\mathcal{A} \triangleq \neg(P \models \mathcal{A})$

$P \models \mathcal{A} \vee \mathcal{B} \triangleq P \models \mathcal{A} \vee P \models \mathcal{B}$

$P \models \mathbf{0} \triangleq P \equiv \mathbf{0}$

$P \models \mathcal{A} \mid \mathcal{B} \triangleq \exists P', P''.\ P \equiv P' \mid P'' \wedge P' \models \mathcal{A} \wedge P'' \models \mathcal{B}$

$P \models \mathcal{A} \triangleright \mathcal{B} \triangleq \forall P'.P' \models \mathcal{A} \Rightarrow P \mid P' \models \mathcal{B}$

$P \models n[\mathcal{A}] \triangleq \exists P'.P \equiv n[P'] \wedge P' \models \mathcal{A}$

$P \models \mathcal{A}@n \triangleq n[P] \models \mathcal{A}$

$P \models n®\mathcal{A} \triangleq \exists P'.P \equiv (\nu n)P' \wedge P' \models \mathcal{A}$

$P \models \mathcal{A}\oslash n \triangleq (\nu n)P \models \mathcal{A}$

$P \models \Diamond\mathcal{A} \triangleq \exists P'.P \rightarrow^* P' \wedge P' \models \mathcal{A}$

$P \models \diamond\!\!\!\!+\mathcal{A} \triangleq \exists P'.P \downarrow^* P' \wedge P' \models \mathcal{A}$

$P \models \exists x.\mathcal{A} \triangleq \exists m.P \models \mathcal{A}\{x \leftarrow m\}$

We use $\Box\mathcal{A}$ (everytime modality), $❑\mathcal{A}$ (everywhere modality) and $\forall x.\mathcal{A}$ (universal quantification) as abbreviations for $\neg(\Diamond\neg\mathcal{A})$, $\neg(\diamondsuit\neg\mathcal{A})$ and $\neg(\exists x.\neg\mathcal{A})$, respectively.

3 Calculus with Replication

In this section we show that model checking for the ambient calculus with replication (but still without name restriction, and in fact without communication) against the ambient logic (without composition adjunct) is undecidable. We use here α, β, γ for words in $\{a,b\}^*$, σ for letters in $\{a,b\}$ and ϵ for the empty word. Lower-case strings (possibly with subscripts) like $c_i, n_i, w_i, start_i, word_i, compare$ denote ambient names, while upper-case strings like *Concatenate*, *Compare*, $Word_i$ denote processes.

Encoding of PCP. The undecidability proof is done by a reduction of the Post correspondence problem (PCP). An instance of the problem is a set of pairs of words $\{\langle\alpha_1,\beta_1\rangle,\dots,\langle\alpha_n,\beta_n\rangle\}$ over the two-letter alphabet $\{a,b\}$ (that is, $\alpha_i,\beta_i \in \{a,b\}^*$). The question is whether there exists a sequence of numbers $1 \le i_0, i_1, \dots, i_k \le n$ such that $\alpha_{i_0}\cdot\ldots\cdot\alpha_{i_k} = \beta_{i_0}\cdot\ldots\cdot\beta_{i_k}$, where $\cdot$ denotes word concatenation. It is well-known that Post correspondence problem is undecidable [10].

The idea of the reduction is to construct for a given instance of PCP a process P whose reduction simulates all possible concatenations of pairs of words in the instance. Then we have to only check if a process representing two equal words is reachable.

The process P is defined as the parallel composition

$$P \triangleq start_1[] \mid start_2[] \mid Word_1(\epsilon) \mid Word_2(\epsilon) \mid Concatenate \mid Compare,$$

where $start_1$ and $start_2$ are two different ambient names (we write ambient names with lower-case letters and (meta-)names of processes in upper case) and $Word_i(w)$ is a process representing the word w (we start with the empty word). Before we give the precise definition of the processes $Word_i(w)$, *Concatenate*, *Compare*, we briefly describe the intuition behind them. *Concatenate* is a process responsible for concatenating pairs of words from the given instance of PCP: it chooses nondeterministically a pair $\langle\alpha_i,\beta_i\rangle$ and rewrites $Word_1(\alpha) \mid Word_2(\beta)$ to $Word_1(\alpha_i\cdot\alpha) \mid Word_2(\beta_i\cdot\beta)$; this is done again and again. At some nondeterministically chosen point of time the process *Compare* activates — it stops *Concatenate* and starts comparing the two words represented by $Word_1$ and $Word_2$ by nondeterministically choosing the letter a or b and trying to delete it simultaneously from both words; this is repeated until both words are empty or they start with a different letter. Clearly, the instance of PCP has a solution if and only if there exists a (nonempty) execution of the process that ends with the representation of two empty words.

$$\begin{aligned} \mathit{Concatenate} \triangleq\ & !\,(\textbf{open}\ \mathit{start}_1.\textbf{open}\ \mathit{start}_2.\textbf{open}\ \mathit{pair})\ \mid \\ & !\,\mathit{pair}[\mathit{Concatenate}_1(\alpha_1) \mid \mathit{Concatenate}_2(\beta_1)]\ \mid \\ & \ldots \mid \\ & !\,\mathit{pair}[\mathit{Concatenate}_1(\alpha_n) \mid \mathit{Concatenate}_2(\beta_n)] \end{aligned}$$

The two ambients $\mathit{start}_1[]$ and $\mathit{start}_2[]$ are used for synchronization — the only possible reduction is to open $\mathit{start}_1[]$ and then $\mathit{start}_2[]$; after this the two ambients disappear and they will appear again only after $\mathit{Concatenate}_1$ and $\mathit{Concatenate}_2$ finish their jobs. In this way we avoid processing two different pairs at the same time (and thus confusing different pairs during computation).

Thus, in every iteration of *Concatenate*, we rewrite in several steps a process of the form $\mathit{start}_1[] \mid \mathit{start}_2[] \mid \mathit{Word}_1(\alpha) \mid \mathit{Word}_2(\beta) \mid \mathit{Concatenate} \mid \mathit{Compare}$ to

$$\begin{aligned} \mathit{Word}_1(\alpha) \mid \mathit{Word}_2(\beta) \mid\ & \mathit{Concatenate}_1(\alpha') \mid \mathit{Concatenate}_2(\beta') \\ \mid\ & \mathit{Concatenate} \mid \mathit{Compare} \end{aligned}$$

for some words α, β and some pair $\langle \alpha', \beta' \rangle$ from the instance of PCP. Intuitively, two words $\gamma = \sigma_1 \ldots \sigma_k$ and $\gamma' = \sigma'_1 \ldots \sigma'_{k'}$ in $\{a, b\}^*$ are represented by ambients $\sigma_1[\sigma_2[\ldots\sigma_k[]]\ldots]$ and $\sigma'_1[\sigma'_2[\ldots\sigma'_{k'}[]]\ldots]$ and the process $\mathit{Concatenate}_i(\gamma')$ leads the process $\mathit{Word}_i(\gamma)$ inside $\sigma'_{k'}[]$ and generates an ambient $\mathit{start}_i[]$ so that

$$\mathit{Word}_i(\gamma) \mid \mathit{Concatenate}_i(\gamma') \quad \rightarrow^* \quad \mathit{Word}_i(\gamma' \cdot \gamma) \mid \mathit{start}_i[].$$

The details are quite technical and are presented in the appendix. Then the initial process rewrites to

$$\mathit{start}_1[] \mid \mathit{start}_2[] \mid \mathit{Word}_1(\alpha' \cdot \alpha) \mid \mathit{Word}_2(\beta' \cdot \beta) \mid \mathit{Concatenate} \mid \mathit{Compare}$$

The process *Compare* works in a similar way.

$$\begin{aligned} \mathit{Compare} \ \triangleq\ \ \mathit{compare}[] \mid \mathit{Initialize}.(&!\,(\textbf{open}\ \mathit{compare}.\mathit{Consume}(a))\ \mid \\ &!\,(\textbf{open}\ \mathit{compare}.\mathit{Consume}(b))) \end{aligned}$$

The initialization essentially opens start_1 and start_2 so that *Concatenate* is blocked. The process $\mathit{Consume}(a)$ replaces the representation of the two words α, β by α', β' if $\alpha = a\alpha'$ and $\beta = a\beta'$ by simply opening the leading ambients $a[\ldots]$ in the representation of both words, similarly $\mathit{Consume}(b)$ opens the leading $b[\ldots]$ if both words start with b. The ambient $\mathit{compare}[]$ is used for synchronization to avoid deleting different letters from the two words. The details are presented in the appendix.

We have the following theorem.

Theorem 3.1. *The model checking problem for the ambient calculus with replication against the ambient logic is undecidable.*

Proof. Let P be the process defined above (note that the definition of P depends on the instance of PCP). We have already seen that the instance has a solution if and only if there exists an execution of P starting with the concatenation of at least one pair and ending in a configuration representing the pair of empty words. This can be expressed by the formula

$$\mathcal{A} \triangleq \Diamond(\mathit{nonempty}(w_1) \wedge \Diamond(\mathit{empty}(w_1) \wedge \mathit{empty}(w_2)))$$

where

$$\mathit{nonempty}(w_i) \triangleq \diamondsuit w_i[(a[\mathbf{T}] \vee b[\mathbf{T}]) \mid \mathbf{T}]$$
$$\mathit{empty}(w_i) \triangleq \neg \mathit{nonempty}(w_i).$$

Here, w_i is an ambient name used in the encoding of the process $\mathit{Word}_i(\gamma)$ (see the appendix for details), and the formula $a[\mathbf{T}] \vee b[\mathbf{T}]$ is matched by (the encoding of) the first letter in the word γ. Then $P \models \mathcal{A}$ if and only if the instance of PCP has a solution. □

It should be noticed that our proof of undecidability of model-checking the ambient calculus with replication but without private names implies that reachability via reduction for ambient processes with public names and with replication is undecidable.

4 Logic with Composition Adjunct

We investigate in this section the problem of model checking ambients against formulas that may contain composition adjunct. Let us first show that the model checking problem of formulas with composition adjunct subsumes the satisfiability problem of formulas without composition adjunct.

Proposition 4.1. *The process* $\mathbf{0}$ *satisfies the formula* $\neg(\mathbf{T} \triangleright \neg\mathcal{A})$ *if and only if the formula* $\mathcal{A}$ *is satisfiable.*

Proof. By definition, $\mathbf{0} \models \mathbf{T} \triangleright \neg\mathcal{A}$ if and only if for all processes P that satisfy $\mathbf{T}$, the process $P \mid \mathbf{0}$ satisfies $\neg\mathcal{A}$. Since all processes satisfy $\mathbf{T}$ and $P \mid \mathbf{0}$ is equivalent to P, by the definition of satisfaction for negation we have that $\mathbf{0} \models \neg(\mathbf{T} \triangleright \neg\mathcal{A})$ if and only if there exists P that satisfies $\mathcal{A}$. □

We show now that the satisfiability problem for ambient formulas (even without composition adjunct) is an undecidable problem [1]. Thus, it implies

Theorem 4.1. *The model checking problem of ambient processes without replication and name restriction against formulas with composition adjunct is undecidable.*

Let us consider the set $\mathcal{F}$ of first-order formulas defined over a countable set of variables $x, y, z, \ldots$ and some relational symbols $\{R_1, \ldots, R_k\}$, each of those symbols having strictly positive arity. The set of formulas $\mathcal{F}$ is the least set such

[1] Actually we consider a very small fragment of the logic, in particular without temporal modalities.

that (i) for any R_i with arity l, $\mathcal{F}$ contains $R_i(x_1, \ldots, x_l)$, and (ii) for all φ and φ' in $\mathcal{F}$, $\varphi \wedge \varphi'$, $\neg\varphi$ and $\exists x\varphi$ belong to $\mathcal{F}$.

Formulas from $\mathcal{F}$ are interpreted over structures; a structure $\mathcal{S}$ over some domain $\mathcal{D}$ is simply a set of objects of the form $R_i(a_1, \ldots, a_l)$ where R_i is an l-ary relational symbol and $a_1, \ldots, a_l$ are elements of $\mathcal{D}$. We say that a structure S is finite whenever its domain $\mathcal{D}$ is finite.

A formula is said to be closed if it has no free variables. We assume wlog. that in formulas bound variables are pairwise distinct. For a formula φ and a structure S with domain $\mathcal{D}$, a valuation σ is a mapping from the free variables of φ to $\mathcal{D}$. A structure S is a model of a formula φ under a valuation σ (written $S, \sigma \models \varphi$) if

- $R_i(x_1, \ldots, x_l)\sigma \in S$ for $\varphi = R_i(x_1, \ldots, x_l)$,
- $S, \sigma \models \varphi'$ and $S, \sigma \models \varphi''$ for $\varphi = \varphi' \wedge \varphi''$,
- $S, \sigma \not\models \varphi'$ for $\varphi = \neg\varphi'$,
- there exists a in $\mathcal{D}$ such that $S, \sigma\{x \leftarrow a\} \models \varphi'$ for $\varphi = \exists x\varphi'$.

Theorem 4.2 (Trakhtenbrot [12]). *Given a closed first-order formula φ, it is undecidable to know whether φ admits a finite model.*

With a formula φ from $\mathcal{F}$ we associate a formula $[\![\varphi]\!]$ from the ambient logic inductively defined as follows:

- $[\![R_i(x_1, \ldots, x_l)]\!] = r_i[x_1[x_2[\ldots[x_l[\mathbf{0}]]\ldots]]] \mid \mathbf{T}$,
- $[\![\varphi \wedge \varphi']\!] = [\![\varphi]\!] \wedge [\![\varphi']\!]$
- $[\![\neg\varphi]\!] = \neg[\![\varphi]\!]$,
- $[\![\exists x\varphi]\!] = \exists x((d[x[\mathbf{0}]] \mid \mathbf{T}) \wedge [\![\varphi]\!])$.

Note that we identify first-order variables in formulas from $\mathcal{F}$ with variables of the ambient logic. Therefore, free variables of φ and $[\![\varphi]\!]$ coincide.

The key idea of this encoding is to consider the parallel operator of the ambient calculus as a (multi-)set constructor. Then, the finite domain $\mathcal{D}$ as well as the structure $\mathcal{S}$ are encoded in a straightforward way using simply ambient name d for elements from $\mathcal{D}$ and ambient names r_i for the relational symbols R_i in $\mathcal{S}$.

Lemma 4.1. *A closed formula φ from $\mathcal{F}$ admits a finite model iff there exists an ambient process P without replication and without name restriction such that $P \models [\![\varphi]\!]$.*

The proof of Lemma 4.1 can be found in the appendix. It is straightforward that Lemma 4.1 and Theorem 4.2 yield the undecidability of the satisfiability problem of the logic without composition adjunct over ambient processes without replication and name restriction. Hence, Theorem 4.1 follows.

5 Calculus with Name Restriction

In this section we show that the model-checking algorithm from [5] can be extended to the calculus with name restriction. Moreover, the additional logical operators introduced in [7] do not influence the decidability. We recall that, as in [5], the logic does not contain the composition adjunct and the calculus does not contain replication.

First, we fix the representation of the processes: Using α-renaming of restricted names and the rules (Str Res Par) and (Str Res Amb) of the congruence relation, we group together all name-restriction operators by transforming every process to one of the form $(\nu n_1)\ldots(\nu n_k)P$ where every occurrence of name restriction is guarded by an action, that is every name restriction occurring in P occurs in a subprocess of the form $M.P'$ or $(n).P'$ with $M \not\equiv \epsilon$. Formally, a process is guarded if it is of the form

$$\mathtt{in}\, M.P,\ \mathtt{out}\, M.P,\ \mathtt{open}\, M.P,\ n.P,\ (n).P,\ \text{or}\ \langle M\rangle$$

for some process P and expression M. Note that neither a guarded process nor any of its subprocesses can be reduced (a guarded process can be reduced only if it occurs in a parallel composition with other processes) and that the sublocation relation $\downarrow^*$ does not look inside a guarded process. We separate bounded names from the unguarded part of a process using the following function *separate*. Recall that we assume that all bounded names are renamed apart so that they are different.

Separating Bounded Names from a Process

$$\begin{array}{ll}
\mathit{separate}(P) \triangleq \langle \varnothing, P\rangle & \text{if } P \text{ is guarded}\\
\mathit{separate}((\nu n)P) \triangleq \langle N\cup\{n\}, P'\rangle & \text{if } \mathit{separate}(P) = \langle N, P'\rangle\\
\mathit{separate}(n[P]) \triangleq \langle N, n[P']\rangle & \text{if } \mathit{separate}(P) = \langle N, P'\rangle\\
\mathit{separate}(P\,|\,Q) \triangleq \langle N\cup N', P'\,|\,Q'\rangle & \text{if } \mathit{separate}(P) = \langle N, P'\rangle \text{ and}\\
& \quad \mathit{separate}(Q) = \langle N', Q'\rangle
\end{array}$$

The definition of the $\rightarrow$ and $\downarrow$ relations extends in a straightforward way to the representation with bounded names separated from the process. We say that $\langle N, P\rangle \rightarrow \langle N', P'\rangle$ if there exists P'' such that $P \rightarrow P''$, $\mathit{separate}(P'') = \langle N'', P'\rangle$ and $N' = N \cup N''$. Similarly, $\langle N, P\rangle \downarrow \langle N, P'\rangle$ if there exist n, P'' such that $P \equiv n[P'] \mid P''$ and $n \notin N$. As usually, $\rightarrow^*$ and $\downarrow^*$ are reflexive and transitive closures of $\rightarrow$ and $\downarrow$ respectively. We define

$$\mathit{Reachable}(N,P) \triangleq \{\langle N', P'\rangle \mid \langle N,P\rangle \rightarrow^* \langle N', P'\rangle\}$$
$$\mathit{Sublocations}(N,P) \triangleq \{\langle N, P'\rangle \mid \langle N,P\rangle \downarrow^* \langle N, P'\rangle\}$$

Example 5.1. Consider the processes $P_1 = (m).(\nu n)(m[\mathtt{in}\, n] \mid n[]) \mid \langle a\rangle$, $P_2 = (\nu n)(m).(m[\mathtt{in}\, n] \mid n[]) \mid \langle a\rangle$ and the formulas $\mathcal{A}_1 = \Diamond(n®n[a[\mathbf{T}]])$ and $\mathcal{A}_2 = n®(\Diamond n[a[\mathbf{T}]])$.

The process P_1 is guarded, so $separate(P_1) = \langle \varnothing, P_1 \rangle$; the process P_2 is not guarded and $separate(P_2) = \langle \{n\}, (m).(m[\mathtt{in}\, n] \mid n[]) \mid \langle a \rangle \rangle$. In both cases, that is for $i = 1, 2$, using the reduction (Red I/O) followed by (Red In), we reduce $separate(P_i) \to \langle \{n\}, a[\mathtt{in}\, n] \mid n[] \rangle \to \langle \{n\}, \mid n[a[]] \rangle$.

The reader can check that $P_1 \models \mathcal{A}_1$ and $P_2 \models \mathcal{A}_2$. Moreover, $P_2 \models \mathcal{A}_1$, but $P_1 \not\models \mathcal{A}_2$.

Now we are ready to define our model-checking algorithm. It is an extension of the algorithms from [5,8] to the calculus and logic with name restriction. We write $\dot{\cup}$ for disjoint union, that is, $A = B \dot{\cup} C$ if $A = B \cup C \wedge B \cap C = \varnothing$. We recall the assumption that all bound names in the process are renamed apart so that they are all different from each other and different from all free names occurring in the process and the formula.

Model-Checking Algorithm

$Check(N, P, \mathbf{T}) \triangleq \mathbf{T}$

$Check(N, P, \neg\mathcal{A}) \triangleq \neg Check(N, P, \mathcal{A})$

$Check(N, P, \mathcal{A} \vee \mathcal{B}) \triangleq Check(N, P, \mathcal{A}) \vee Check(N, P, \mathcal{B})$

$Check(N, P, \mathbf{0}) \triangleq \begin{cases} \mathbf{T} \text{ if } P \equiv \mathbf{0} \\ \mathbf{F} \text{ otherwise} \end{cases}$

$Check(N, P, \mathcal{A} \mid \mathcal{B}) \triangleq \bigvee_{N_1 \dot{\cup} N_2 = N} \bigvee_{P_1 \mid P_2 \equiv P} Check(N_1, P_1, \mathcal{A}) \wedge Check(N_2, P_2, \mathcal{B}) \wedge$
$\qquad fn(P_1) \cap N_2 = \varnothing \wedge fn(P_2) \cap N_1 = \varnothing$

$Check(N, P, n[\mathcal{A}]) \triangleq P \equiv n[Q] \wedge n \notin N \wedge Check(N, Q, \mathcal{A})$

$Check(N, P, \mathcal{A}@n) \triangleq Check(N, n[P], \mathcal{A})$

$Check(N, P, n®\mathcal{A}) \triangleq \bigvee_{m \in N} Check(N - \{m\}, P\{m \leftarrow n\}, \mathcal{A})$
$\qquad \vee (n \notin fn(P) \wedge Check(N, P, \mathcal{A}))$

$Check(N, P, \mathcal{A} \oslash n) \triangleq Check(N \cup \{n\}, P, \mathcal{A})$

$Check(N, P, \Diamond\mathcal{A}) \triangleq \bigvee_{\langle N', P' \rangle \in Reachable(N, P)} Check(N', P', \mathcal{A})$

$Check(N, P, \diamondsuit\mathcal{A}) \triangleq \bigvee_{\langle N, P' \rangle \in Sublocations(N, P)} Check(N, P', \mathcal{A})$

$Check(N, P, \exists x.\mathcal{A}) \triangleq$ let $n_0 \notin N \cup fn(P) \cup bn(P)$ be a fresh name in
$\qquad \bigvee_{n \in fn(N, P) \cup fn(\mathcal{A})} Check(N, P, \mathcal{A}\{x \leftarrow n\})$
$\qquad \vee Check(N, P, \mathcal{A}\{x \leftarrow n_0\})$

The correctness of the algorithm (Theorem 5.1) is based on the following lemmas and propositions.

Proposition 5.1.

(1) $(\nu n)P \equiv \mathbf{0}$ *if and only if* $P \equiv \mathbf{0}$.
(2) *If* n *and* m *are different names, then* $(\nu n)P \equiv m[Q]$ *if and only if there exists a process* R *such that* $P \equiv m[R]$ *and* $Q \equiv (\nu n)R$.
(3) $(\nu n)P \equiv Q \mid Q'$ *if and only if there exist processes* R, R' *such that either* $Q \equiv (\nu n)R$ *and* $Q' \equiv R'$ *and* $n \notin fn(Q')$ *or* $Q \equiv R$ *and* $Q' \equiv (\nu n)R'$ *and* $n \notin fn(Q)$.

Sketch of proof. The "if" implications follow from (Str Res Zero), (Str Res Amb) and (Str Res Par) congruence rules. The first two of the equivalences above are stated as Inversion Lemmas (Lemmas 2-2) in [7] with a reference to [9] for a proof. However, the proof from [9] cannot be applied directly, since spatial congruence from [9] is not the same as structural congruence. In particular, it does not distinguish between the processes P_1 and P_2 from Example 5.1. On the other hand, the methods of [9] can be easily extended to the case of structural congruence.

The third equivalence is a stronger version of the third Inversion Lemma (the Inversion Lemma does not mention the $n \notin fn(Q)$ condition), again proved in the case of spatial congruence in [9]. Again, the same methods can be used to show our stronger version in the case of structural congruence (in the general case of processes with replication the proof is not very easy, but in the absence of replication, as we have it here, it is enough to use the Inversion Lemma together with the observation that equivalent processes have the same numbers of occurrences of free names). □

Lemma 5.1. *If $n \notin fn(P)$ then $(\nu n)P \equiv P$.*

Proof. By the rules (Str Res) and (Str Zero Par), $(\nu n)P$ is equivalent to $(\nu n)(P \mid \mathbf{0})$, which by (Str Res Par), (Str Res Zero) and (Str Zero Par) is equivalent to P. □

Proposition 5.2. *Consider any process P and a closed $\triangleright$-free formula $\mathcal{A}$. Let $fn(P) \cup fn(\mathcal{A}) = \{n_1, \dots, n_k\}$, and suppose $n_0 \notin \{n_1, \dots, n_k\}$. Then $P \models \exists x.\mathcal{A}$ if and only if $P \models \mathcal{A}\{x \leftarrow n_i\}$ for some $i \in \{0, \dots, k\}$.*

The proof follows the lines of the proof of Proposition 4.11 in [3]. □

Lemma 5.2. *For all processes P, P'*

(1) *$P \downarrow P'$ if and only if $separate(P) \downarrow separate(P')$*
(2) *$P \to P'$ if and only if $separate(P) \to separate(P')$*
(3) *the sets $Reachable(separate(P))$ and $Sublocations(separate(P))$ are finite and effectively computable.*

Sketch of proof. The first two equivalences follow from Proposition 5.1 and the definitions of $\downarrow$ and $\to$ relations. The third one follows from the finiteness of the analogous sets for the replication-free fragment of the ambient calculus with public names [5,8] and a simple observation that $P \downarrow P'$ and $P \to P'$ imply $public(P) \downarrow public(P')$ and $public(P) \to public(P')$, respectively, where $public(P)$ is the process obtained from P by removing all ν quantifiers. □

Theorem 5.1. *For all replication-free processes P and closed $\triangleright$-free formulas $\mathcal{A}$, we have $P \models \mathcal{A}$ if and only if $Check(separate(P), \mathcal{A}) = \mathbf{T}$.*

Sketch of proof. The proof goes by induction on the formula $\mathcal{A}$. In the cases of $\mathbf{T}$, $\neg\mathcal{A}$, $\mathcal{A} \vee \mathcal{B}$, $\mathcal{A}@n$, $\mathcal{A}\oslash n$ it follows directly from the definition of the satisfaction relation (the side condition $n_0 \notin N \cup fn(P)$ in the case of $\mathcal{A}\oslash n$ reflects only our convention that all bounded names are renamed apart). In the case of $\mathbf{0}$ and $n[\mathcal{A}]$ it follows from Proposition 5.1(1) and (2), in the case of $\mathcal{A} \mid \mathcal{B}$ from Proposition 5.1(3) by induction on the number of names in N. The cases of $\diamond\!\!\!\cdot\,\mathcal{A}$ and $\Diamond\mathcal{A}$ follow from Lemma 5.2 and the definitions of *Reachable* and *Sublocations*; $\exists x.\mathcal{A}$ from Proposition 5.2. Finally, the case of $n®\mathcal{A}$ reflects the two possibilities that either n is one of the bounded names occurring in the process or it does not occur there (in the latter case we appeal to Lemma 5.1). □

Theorem 5.2. *The model-checking problem for replication-free processes against* $\triangleright$*-free ambient logic is decidable. Moreover, it is PSPACE-complete.*

Sketch of proof. Decidability follows from Theorem 5.1. One obtains the PSPACE upper bound by combining the above algorithm with the polynomial-size representation of processes from [8] and implementing disjunction in polynomial space, as it is done in [8]. The PSPACE lower bound is proved in [8]. □

6 Conclusion

We investigate in this paper borders of the decidability for model-checking mobile ambient against the ambient logic. We have started from the fragments of mobile calculus without name restriction and without replication against the logic without composition adjunct for which decidability of model-checking has been showed in [5]. We have showed that adding either replication in the calculus or composition adjunct in the logic leads to undecidability for the model-checking problem. On the other hand, we have considered the extension of the calculus with private names and the adequate operators in the logic to manipulate those names. We have proved this extension to preserve decidability of model-checking as well as the complexity of the original fragments.

A Encoding of PCP: Concatenation and Comparison of Words

A.1 Concatenation

Here we show how to rewrite $\mathit{Word}_i(\gamma) \mid \mathit{Concatenate}_i(\gamma')$ to $\mathit{Word}_i(\gamma' \cdot \gamma) \mid \mathit{start}_i[]$. For this, we need precise definition of Word_i and $\mathit{Concatenate}_i$. For $i = 1, 2$ we introduce fresh ambient names $\mathit{word}_i, c_i, n_i, v_i, w_i$; similarly, we introduce fresh names a, b corresponding to the two letters of the alphabet. Let $\gamma = \sigma_1 \ldots \sigma_k$ and $\gamma' = \sigma'_1 \ldots \sigma'_{k'}$ be two words in $\{a, b\}^*$. We define

$$\mathit{Word}_i(\gamma) \triangleq \mathit{word}_i[!\,\mathtt{open}\,c_i \mid w_i[\mathtt{open}\,n_i \mid \mathit{String}(\gamma)]]$$

$$\mathit{Concatenate}_i(\gamma') \triangleq c_i[\mathtt{in}\,\mathit{word}_i.\mathit{MvIn}_i(\gamma') \mid \mathit{String}'(\gamma', \mathit{Continue}_i)]$$

where

$$String(\sigma_1 \ldots \sigma_k) \triangleq \sigma_1[\sigma_2[\ldots \sigma_k[] \ldots]]$$
$$String'(\sigma'_1 \ldots \sigma'_{k'}, P) \triangleq s_i[\mathtt{in}\, v_i.\mathtt{in}\, w_i \mid \sigma'_1[\ldots \sigma'_k[P] \ldots]]$$
$$MvIn_i(\sigma'_1 \ldots \sigma'_k) \triangleq n_i[\mathtt{in}\, w_i.\mathtt{in}\, s_i.\mathtt{in}\, \sigma'_1 \ldots \mathtt{in}\, \sigma'_k]$$
$$MvOut(\sigma'_1 \ldots \sigma'_k) \triangleq \mathtt{out}\, \sigma'_k \ldots \mathtt{out}\, \sigma'_1$$
$$Continue_i \triangleq \mathtt{open}\, w_i.v_i[MvOut(\gamma').Tinue_i]$$
$$Tinue_i \triangleq \mathtt{in}\, w_i \mid w_i[\mathtt{open}\, n_i \mid \mathtt{open}\, s.\mathtt{out}\, v_i.Nue_i]$$
$$Nue_i \triangleq \mathtt{open}\, v_i.start[\mathtt{out}\, w_i.\mathtt{out}\, word_i].$$

Then $Word_i(\gamma) \mid Concatenate_i(\gamma')$ reduces by moving the ambient $c_i[\ldots]$ inside $word_i[\ldots]$ and opening it to

$$word_i[!\,\mathtt{open}\, c_i \mid MvIn_i(\gamma') \mid String'(\gamma', Continue_i)$$
$$w_i[\mathtt{open}\, n_i \mid String(\gamma)]],$$

the ambient $n_i[\ldots]$ goes inside $w_i[\ldots]$ and gets opened there

$$word_i[!\,\mathtt{open}\, c_i \mid String'(\gamma', Continue_i)$$
$$w_i[\mathtt{in}\, s_i.\mathtt{in}\, \sigma'_1 \ldots \mathtt{in}\, \sigma'_k \mid String(\gamma)]],$$

$w_i[\ldots]$ goes inside $String'(\ldots)$

$$word_i[!\,\mathtt{open}\, c_i \mid$$
$$String'(\gamma', (w_i[String(\gamma)] \mid Continue_i))],$$

$Continue_i$ opens w_i and v_i moves out of γ'

$$word_i[!\,\mathtt{open}\, c_i \mid$$
$$String'(\gamma' w, \mathbf{0}) \mid v_i[Tinue_i]],$$

$s_i[\ldots]$ goes inside v_i, then inside w_i and gets opened there

$$word_i[!\,\mathtt{open}\, c_i \mid$$
$$v_i[\mathtt{in}\, w_i \mid w_i[\mathtt{open}\, n_i \mid \mathtt{out}\, v_i.Nue_i \mid String(\gamma'\gamma)]]],$$

w_i gets out of v_i and v_i gets into w_i

$$word_i[!\,\mathtt{open}\, c_i \mid$$
$$w_i[\mathtt{open}\, n_i \mid v_i[] \mid Nue_i \mid String(\gamma'\gamma)]],$$

Nue_i opens v_i; $start_i$ goes out of w_i and out of $word_i$

$$Word_i(\gamma'\gamma) \mid start_i[]$$

which is the desired process. Note that since guarded processes cannot be reduced, this was the only possible execution of the process $Word_i(\gamma) \mid Concatenate_i(\gamma')$.

A.2 Comparing the Words

First we define the two missing processes.

$$Initialize \triangleq \texttt{open}\ start_1.\texttt{open}\ start_2.\texttt{open}\ word_1.\texttt{open}\ word_2$$
$$Consume(\sigma) \triangleq n_1[\texttt{in}\ w_1.\texttt{open}\ \sigma.(\texttt{open}\ n_1 \mid Nsume(\sigma))]$$

where

$$Nsume(\sigma) \triangleq n_2[\texttt{out}\ w_1.\texttt{in}\ w_2.\texttt{open}\ \sigma.(\texttt{open}\ n_2 \mid compare[\texttt{out}\ w_2])].$$

Then

$$start_1[] \mid start_2[] \mid Word_1(\alpha) \mid Word_2(\beta) \mid Compare$$

reduces to

$$\begin{array}{l} !\,\texttt{open}\ c_1 \mid\ !\,\texttt{open}\ c_2 \mid \\ w_1[\texttt{open}\ n_1 \mid String(\alpha)] \mid w_2[\texttt{open}\ n_2 \mid String(\beta)] \mid \\ compare[] \mid\ !\,(\texttt{open}\ compare.Consume(a)) \mid\ !\,(\texttt{open}\ compare.Consume(b))). \end{array}$$

The two processes $!\,\texttt{open}\ c_i$ remain inactive, since the names c_i will never occur again. The only possibility of executing the process is to choose one of the two subprocesses consuming a or b; each of them opens $compare$, so the desired property now is that

$$w_1[\texttt{open}\ n_1 \mid String(\sigma\alpha)] \mid w_2[\texttt{open}\ n_2 \mid String(\sigma\beta)] \mid Consume(\sigma)$$

reduces to

$$w_1[\texttt{open}\ n_1 \mid String(\alpha)] \mid w_2[\texttt{open}\ n_2 \mid String(\beta)] \mid compare[].$$

This can be easily checked by the reader: The process $Consume(\sigma)$ is an ambient named n_1, it goes inside w_1, gets opened there, opens σ thus deleting the leading letter from σu, leaves the capability $\texttt{open}\ n_1$ for the next iteration, and as $Nsume(\sigma)$ goes out of w_1; then it repeats the same thing with w_2 and leaves the ambient $compare[]$ at the top level. Note that if the two words u and v start with two different letters a and b then the process $w_1[\texttt{open}\ n_1 \mid String(\alpha)] \mid w_2[\texttt{open}\ n_2 \mid String(\beta)] \mid Consume(\sigma)$ deadlocks after reaching a configuration where it tries to open σ but there is no ambient named σ at the respective place. If this happens, no further reduction of the whole process is possible.

B Satisfiability of the Ambient Logic

We give here the proof of Lemma 4.1. We consider the relation $\rightsquigarrow_{Proc}$ between finite structures and ambient processes without replication and name restriction. For a process P and a structure $\mathcal{S}$ whose domain is $\mathcal{D}$, we have $\mathcal{S} \rightsquigarrow_{Proc} P$ if:

- there exists P' such that $P \equiv d[a[\mathbf{0}]] \mid P'$ iff a belongs to $\mathcal{D}$,
- whenever $a_1, \ldots, a_l$ belong to $\mathcal{D}$, there exists P'' such that P is structurally congruent to $r_i[a_1[a_2[\ldots[a_l[\mathbf{0}]]\ldots]]] \mid P''$ iff $R_i(a_1, \ldots, a_l)$ belongs to $\mathcal{S}$.

We denote $\rightsquigarrow_{Struct}$ the symmetric relation of $\rightsquigarrow_{Proc}$. Notice that $\rightsquigarrow_{Proc} \circ \rightsquigarrow_{Struct}$ is the identity relation over structures and $\rightsquigarrow_{Struct} \circ \rightsquigarrow_{Proc}$ simply contains the identity relation.

We prove the following proposition which implies Lemma 4.1 in case where the formula φ is closed.

Proposition B.1. *Let φ be a formula from $\mathcal{F}$. Then,*

(i) let $\mathcal{S}$ be a finite structure over a domain $\mathcal{D}$ and α be a valuation for the free variables of φ. If $\mathcal{S}, \alpha \models \varphi$ and $\mathcal{S} \rightsquigarrow_{Proc} P$ then $P \models [\![\varphi]\!]\alpha$,

(ii) let P be an ambient process without replication and name restriction and α be a mapping from variables of φ to names. If $P \models [\![\varphi]\!]\alpha$ and $P \rightsquigarrow_{Struct} \mathcal{S}$ then $\mathcal{S}, \alpha \models \varphi$.

Proof. The proof goes by induction over the structure of φ.

- for $\varphi = R_i(x_1, \ldots, x_l)$:
  Case (i): $\mathcal{S}, \alpha \models R_i(x_1, \ldots, x_l)$. Therefore, for α equal to $\{x_1 \leftarrow a_1, \ldots, x_l \leftarrow a_l\}$, $R_i(a_1, \ldots, a_l)$ belongs to $\mathcal{S}$. Therefore, by definition of $\mathcal{S} \rightsquigarrow_{Proc} P$, P is structurally congruent to $r_i[a_1[\ldots[a_l[\mathbf{0}]]\ldots]] \mid P'$ for some P'. So, P is a model of $(r_i[x_1[\ldots[x_l[\mathbf{0}]]\ldots]] \mid \mathbf{T})\alpha$, that is $P \models [\![R_i(x_1, \ldots, x_l)]\!]\alpha$.
  Case (ii): $P \models [\![R_i(x_1, \ldots, x_l)]\!]\alpha$. Therefore, for $\alpha = \{x_1 \leftarrow a_1, \ldots, x_l \leftarrow a_l\}$, $P \models r_i[a_1[\ldots[a_l[\mathbf{0}]]\ldots]] \mid \mathbf{T}$. Hence, by definition of the satisfaction relation, there exists P' such that $P \equiv r_i[a_1[\ldots[a_l[\mathbf{0}]]\ldots]] \mid P'$. Since $P \rightsquigarrow_{Struct} \mathcal{S}$, we have that $a_1, \ldots, a_l$ belong to $\mathcal{D}$ and $R_i(a_1, \ldots, a_l)$ belongs to $\mathcal{S}$. Thus, $\mathcal{S}, \alpha \models R_i(x_1, \ldots, x_l)$.
- for $\varphi = \varphi' \wedge \varphi''$:
  Case (i): as $\mathcal{S}, \alpha \models \varphi$, $\mathcal{S}, \alpha \models \varphi'$ and $\mathcal{S}, \alpha \models \varphi''$. So, by induction hypothesis, $P \models \varphi'\alpha$ and $P \models \varphi''\alpha$. Thus, $P \models \varphi$.
  Case (ii): dual from the previous case.
- for $\varphi = \neg\varphi'$:
  Case (i): $\mathcal{S}, \alpha \models \neg\varphi'$. So, $\mathcal{S}, \alpha \not\models \varphi'$. Hence, by induction hypothesis for Case (ii), either $P \not\models [\![\varphi']\!]\alpha$ or $P \not\rightsquigarrow_{Struct} \mathcal{S}$. Furthermore, we know by assumption that $\mathcal{S} \rightsquigarrow_{Proc} P$, and so, $P \rightsquigarrow_{Struct} \mathcal{S}$. Hence, $P \not\models [\![\varphi']\!]\alpha$ holds. Therefore, $P \models \neg([\![\varphi']\!]\alpha)$. Finally, $P \models [\![\varphi]\!]\alpha$.
  Case (ii): $P \models [\![\neg\varphi']\!]\alpha$. So, $P \not\models [\![\varphi']\!]\alpha$. Hence, by induction hypothesis for Case (i), either $\mathcal{S}, \alpha \not\models \varphi'$ or $\mathcal{S} \not\rightsquigarrow_{Proc} P$. As by assumption, $P \rightsquigarrow_{Struct} \mathcal{S}$, we have $\mathcal{S} \rightsquigarrow_{Proc} P$. So, $\mathcal{S}, \alpha \not\models \varphi'$ holds. Hence, $\mathcal{S}, \alpha \models \neg\varphi'$.
- $\varphi = \exists x \varphi'$:
  Case (i): $\mathcal{S}, \alpha \models \exists x \varphi'$. By definition, there exists $a \in \mathcal{D}$ such that $\mathcal{S}, \alpha\{x \leftarrow a\} \models \varphi'$. By assumption $\mathcal{S} \rightsquigarrow_{Proc} P$, so there exists P' such that $P \equiv d[a[\mathbf{0}]] \mid P'$. Moreover, by induction hypothesis, $P \models [\![\varphi']\!]\alpha\{x \leftarrow a\}$. Hence, $P \models (d[a[\mathbf{0}]] \mid \mathbf{T}) \wedge ([\![\varphi']\!]\alpha\{x \leftarrow a\})$. Hence, $P \models ((d[x[\mathbf{0}]] \mid \mathbf{T}) \wedge [\![\varphi']\!])\alpha\{x \leftarrow a\}$. So, $P \models \exists x.((d[x[\mathbf{0}]] \mid \mathbf{T}) \wedge [\![\varphi']\!])\alpha$.

Case (*ii*): $P \models [\![\exists x \varphi']\!]\alpha$, that is by definition $P \models (\exists x.((d[x[\mathbf{0}]] \mid \mathbf{T}) \wedge [\![\varphi']\!]))\alpha$. Therefore, by definition of satisfiability, there exists a name a such that $P \models ((d[x[\mathbf{0}]] \mid \mathbf{T}) \wedge [\![\varphi']\!])\alpha\{x \leftarrow a\}$. This implies that

- there exists P' such that $P \equiv d[a[\mathbf{0}]] \mid P'$,
- $P \models [\![\varphi']\!]\alpha\{x \leftarrow a\}$.

As $P \rightsquigarrow_{Struct} \mathcal{S}$, the first point implies that $a \in \mathcal{D}$. This latter together with the second point and the induction hypothesis implies that $\mathcal{S}, \alpha\{x \leftarrow a\} \models \varphi'$. So, $\mathcal{S}, \alpha \models \exists x \varphi'$.

References

1. L. Cardelli and G. Ghelli. A query language based on the ambient logic. In *Proceedings of the 9th European Symposium on Programming ESOP'01*, volume 2028 of *LNCS*, pages 1–22. Springer, 2001.
2. L. Cardelli and A. D. Gordon. Equational properties of mobile ambients. In *Proceedings FoSSaCS'99*, volume 1578 of *LNCS*, pages 212–226. Springer, 1999. An extended version appears as Microsoft Research Technical Report MSR–TR–99–11, April 1999.
3. L. Cardelli and A. D. Gordon. Modal logics for mobile ambients: Semantic reasoning. Unpublished annex to [5], 1999.
4. L. Cardelli and A. D. Gordon. Types for mobile ambients. In *Proceedings POPL'99*, pages 79–92. ACM, Jan. 1999.
5. L. Cardelli and A. D. Gordon. Anytime, anywhere: Modal logics for mobile ambients. In *Proceedings POPL'00*, pages 365–377. ACM, Jan. 2000.
6. L. Cardelli and A. D. Gordon. Mobile ambients. *Theoretical Computer Science*, 240(1):177–213, 2000.
7. L. Cardelli and A. D. Gordon. Logical properties of name restriction. In *Proceedings of the 5th International Conference on Typed Lambda Calculi and Applications (TLCA'01)*, volume 2044 of *LNCS*, pages 46–60. Springer, 2001.
8. W. Charatonik, S. Dal Zilio, A. D. Gordon, S. Mukhopadhyay, and J.-M. Talbot. The complexity of model checking mobile ambients. In *Proceedings FoSSaCS'01*, volume 2030 of *LNCS*, pages 152–167. Springer, 2001.
9. S. Dal Zilio. Spatial congruence for ambients is decidable. In *Proceedings of the 6th Asian Computing Science Conference (ASIAN'00)*, volume 1961 of *LNCS*, pages 88–103. Springer, 2000. A full version *Technical Report MSR-TR-2000-57, Microsoft Research.*
10. E. L. Post. Recursively Enumerable Sets of Positive Integers and their Decision Problems. *Bulletion of the American Mathematical Society*, 50:284–316, 1944.
11. D. Sangiorgi. Extensionality and intensionality of the ambient logics. In *Proceedings POPL'01*, pages 4–13. ACM, Jan. 2001.
12. B. A. Trakhtenbrot. The impossibility of an algorithm for the decision problem for finite models. *Doklady Akademii Nauk SSR*, 70:569–572, 1950.

A Generalization of the Büchi-Elgot-Trakhtenbrot Theorem

Matthias Galota and Heribert Vollmer

Theoretische Informatik, Universität Würzburg,
Am Hubland, 97074 Würzburg, Germany

Abstract. We consider the power of nondeterministic finite automata with generalized acceptance criteria and the corresponding logics. In particular, we examine the expressive power of monadic second-order logic enriched with monadic second-order generalized quantifiers for algebraic word-problems. Extending a well-known result by Büchi, Elgot, and Trakhtenbrot, we show that considering monoidal quantifiers, the obtained logic captures the class of regular languages. We also consider monadic second-order groupoidal quantifiers and show that these are powerful enough to define every language in LOGCFL.

1 Introduction

Nondeterministic finite automata with generalized acceptance criteria were introduced in [PV01]. Usually, an NFA M is said to accept its input word w if it has at least one accepting computation on w, or, in other words, if in the computation tree of $M(w)$ there is at least one accepting path. One might ask what class of languages one obtains when an input is defined to be accepted if another condition holds, e.g., the number of accepting paths is divisible by some prime number p. Even more generally, let us suppose that every path in the computation tree produces an output symbol (which can be 0, 1 or something else), and say that the *leafstring* of M on w is the sequence of these symbols over all paths read from left to right (in an order defined formally below in Sect. 2). A generalized acceptance criterion in the sense of [PV01] now is a leaf language A, i.e., a set of leafstrings. Automaton M with leaf language A *by definition* accepts an input w if the leafstring of $M(w)$ is an element of A. Clearly, the leaf language for usual nondeterministic acceptance is the language $L_\exists$ of all binary words with at least one occurrence of the letter "1", while that for modulo-p acceptance is the language $L_{\mathrm{mod}\ p}$ of all words with a number of "1"s that is a multiple of p. Leaf languages drawn from various complexity and formal language classes were examined in [PV01], and the power of NFAs using these to define acceptance was clarified.

In this paper, we address the question, to which logical framework finite automata with leaf languages correspond. We show that a suitable way to characterize NFAs with leaf languages in a logical way is to consider monadic second-order logic enhanced with monadic second-order Lindström quantifiers

L. Fribourg (Ed.): CSL 2001, LNCS 2142, pp. 355–368, 2001.

(a.k.a. generalized quantifiers, see, e.g., [EF95, Chap. 10.1]). We prove the following general theorem: The languages accepted by finite automata with leaf language A are exactly those definable by formulas with one second-order Lindström quantifier given by A and no further second-order quantifiers. For example, the above condition "divisible by p" corresponds to a monadic modular counting quantifier.

Our result gives a comparison between automata and logic which we hope will prove useful in many contexts. Here, we apply our general theorem to obtain consequences in two directions.

First we consider monadic *monoidal* quantifiers. These are defined by monoid word-problems, i.e., regular languages. Thus, by our general theorem, these correspond to regular leaf languages. Using a result of Peichl and Vollmer [PV01] about the power of finite automata with regular leaf languages, we conclude that formulas with monadic second-order monoidal quantifiers cannot define non-regular languages.

Second, we consider monadic *groupoidal* quantifiers, that is, quantifiers given by groupoid word-problems. (A groupoid is a finite multiplication table with an identity, i.e., informally, a monoid without the associativity requirement.) Such word-problems are context-free languages; hence, by the above, groupoid quantifiers correspond to context-free leaf languages. Using a result of Lautemann et al. [LMSV01] giving a model-theoretic characterization of LOGCFL, we conclude that finite automata with context-free leaf language can accept every language in LOGCFL (the class of languages logspace-reducible to context-free languages; it is known that these are exactly those languages accepted by auxiliary pushdown-automata operating simultaneously in logarithmic space and polynomial time [Coo71, Sud78]). Hence, in this case the situation is different from the monoidal case: There, "regular quantifiers" could not define non-regular languages, while now "context-free quantifiers" can define a strict superclass of CFL.

2 Finite Automata and Leaf Languages

In this section we will define *finite automata with generalized acceptance criterion*, as introduced in [PV01].

The basic model we use is that of nondeterministic finite automata. On an input word w, such a device defines a tree of possible computations. We want to consider this tree, but with a natural order on the leaves. Therefore we make the following definition:

A *finite leaf automaton* is a tuple $M = (Q, \Sigma, \delta, s, \Gamma, \beta)$ where Q is the finite set of *states*, Σ is an alphabet, the *input alphabet*, $\delta\colon Q \times \Sigma \to Q^+$ is the *transition function*, $s \in Q$ is the *initial state*, Γ is an alphabet, the *leaf alphabet*, and $\beta\colon Q \to \Gamma$ is a function that associates a state q with its *value* $\beta(q)$. The sequence $\delta(q, a)$, for $q \in Q$ and $a \in \Sigma$, contains all possible successor states of M when reading letter a while in state q, and the order of letters in that sequence defines a *total order on these successor states.* This definition allows the same state to appear more than once as a successor in $\delta(q, a)$.

Let M be as above. The computation tree $T_M(w)$ of M on input w is a labeled directed rooted tree defined as follows:

- The root of $T_M(w)$ is labeled (s, w).
- Let v be a node in $T_M(w)$ labeled by (q, x), where $x \neq \epsilon$ (the empty word), $x = ay$ for $a \in \Sigma$, $y \in \Sigma^*$. Let $\delta(q, a) = q_1 q_2 \cdots q_k$. Then v has k children in $T_M(w)$, and these are labeled by $(q_1, y), (q_2, y), \ldots, (q_k, y)$ in this order.

If we look at the tree $T_M(w)$ and attach the symbol $\beta(q)$ to a leaf in this tree with label (q, ε), then $\text{leafstring}^M(w)$ is defined to be the string of symbols attached to the leaves, read from left to right in the order induced by δ.

Definition 1. For $A \subseteq \Gamma^*$, the class $\text{Leaf}^{\text{FA}}(A)$ consists of all languages $B \subseteq \Sigma^*$, for which there is a leaf automaton M as just defined, with input alphabet Σ and leaf alphabet Γ such that for all $w \in \Sigma^*$, $w \in B$ iff $\text{leafstring}^M(w) \in A$.

The following result from [PV01] about the power of regular leaf languages for finite automata will be central for one of our results below:

Proposition 2. $\text{Leaf}^{\text{FA}}(\text{REG}) = \text{REG}$.

3 The Logical Framework

We follow standard notation for monadic second-order logic with linear order, see, e.g., [Str94]. We restrict our attention to *string signatures*, i.e., signatures of the form $\langle P_{a_1}, \ldots, P_{a_s} \rangle$, where all the predicates P_{a_i} are unary, and in every structure $\mathcal{A}$, $\mathcal{A} \models P_{a_i}(j)$ iff the jth symbol in the input is the letter a_i. Such structures are thus words over the alphabet $\{a_1, \ldots, a_s\}$; first-order variables range over positions within such a word, i.e., from 1 to the word length n; second-order variables range over subsets of $\{1, \ldots, n\}$. The logic's linear order symbol refers to numerical order on $\{1, \ldots, n\}$. For technical reasons to be motivated shortly, we also assume that every alphabet has a built-in linear order, and we write alphabets as sequences of symbols to indicate that order, e.g., in the above case we write $(a_1, \ldots, a_s)$.

Our basic formulas are built from first- and second-order variables in the usual way, using the Boolean connectives $\{\wedge, \vee, \neg\}$, the relevant predicates P_{a_i} together with $\{=, <\}$, the constants min and max, the first- and second-order quantifiers $\{\exists, \forall\}$, and parentheses. For a formula φ, L_φ denotes the language defined by φ, i.e., the set of all models of ϕ; that is: L_φ is a set of words. We use $\triangleq$ to name formulas, e.g.: $\varphi \triangleq \exists x P_1(x)$.

SOM is the class of all languages definable using formulas as just described. (The letters SOM stand for second order monadic logic; in the literature, this logic is sometimes denoted by MSO.) FO is the subclass of SOM restricted to languages definable by first-order formulas. It is known [MP71] that FO is equal to the class of star-free regular languages. In this paper we are mainly interested in the following earlier result (see [BE58, Büc62, Tra61]):

Proposition 3 (Büchi-Elgot-Trakhtenbrot Theorem). *The class* SOM *is equal to the class* REG *of regular languages.*

Next, we extend the logical language allowing generalized quantifiers.

Definition 4. Consider a language L over an alphabet $\Sigma = (a_1, a_2, \ldots, a_s)$. Such a language gives rise to a Lindström quantifier Q_L, that may be applied to any sequence of $s-1$ formulas as follows:

Let $\overline{x}$ be a k-tuple of variables (each of which ranges from 1 to the "input length" n, as we have seen). We assume the lexical ordering on $\{1, 2, \ldots, n\}^k$, and we write $\overline{x}^{(1)} < \overline{x}^{(2)} < \cdots < \overline{x}^{(n^k)}$ for the sequence of potential values taken on by $\overline{x}$. The k-ary *Lindström quantifier* Q_L binding $\overline{x}$ takes a meaning if $s-1$ formulas, each having as free variables the variables in $\overline{x}$ (and possibly others), are available. Let $\varphi_1(\overline{x})$, $\varphi_2(\overline{x})$, ..., $\varphi_{s-1}(\overline{x})$ be these $s-1$ formulas. Then $Q_L\overline{x}\big[\varphi_1(\overline{x}), \varphi_2(\overline{x}), \ldots, \varphi_{s-1}(\overline{x})\big]$ holds on a string $w = w_1 \cdots w_n$, iff the word of length n^k whose ith letter, $1 \leq i \leq n^k$, is

$$\begin{cases} a_1 \text{ if } w \models \varphi_1(\overline{x}^{(i)}), \\ a_2 \text{ if } w \models \neg\varphi_1(\overline{x}^{(i)}) \wedge \varphi_2(\overline{x}^{(i)}), \\ \vdots \\ a_s \text{ if } w \models \neg\varphi_1(\overline{x}^{(i)}) \wedge \neg\varphi_2(\overline{x}^{(i)}) \wedge \cdots \wedge \neg\varphi_{s-1}(\overline{x}^{(i)}), \end{cases}$$

belongs to L. We denote this word by $f^{\overline{x}}_{\varphi_1 \cdots \varphi_{s-1}}(w)$.

As an example, take $s = 2$ and consider $L_\exists =_{\text{def}} 0^*1(0+1)^*$; then $Q_{L_\exists}$ is the usual first-order existential quantifier. Similarly, the universal quantifier can be expressed using the language $L_\forall =_{\text{def}} 1^*$. The quantifiers $Q_{L_{\text{mod } p}}$ for $p > 1$ are known as modular counting quantifiers [Str94].

The Lindström quantifiers of Def. 4 are precisely what has been referred to as "Lindström quantifiers on strings" [BV98]. The original more general definition [Lin66] uses transformations to arbitrary structures, not necessarily of string signature. However, in the context of this paper, only reductions to (mostly regular or context-free) languages or algebraic word-problems will be important, and hence the above definition seems to be the most natural here.

Fix a finite monoid M. Each $S \subseteq M$ defines an M-word-problem, i.e., a language $\mathcal{W}(S, M)$ composed of all words w, over the alphabet M, that "multiply out" to an element of S.

The following definition is due to Barrington, Immerman, and Straubing [BIS90].

Definition 5. A *monoidal quantifier* is a Lindström quantifier Q_L where L is a word-problem of some finite monoid.

It is well-known that any regular language is a homomorphic pre-image of (i.e., reduces via a homomorphism to) a word-problem over some monoid, and, vice-versa, every word-problem of a finite monoid is regular. Hence, a monoidal quantifier is nothing other than a Lindström quantifier Q_L where L is a regular language. Coming back for a moment to the classical definition of Lindström quantifiers, we thus see that a Lindström quantifier on strings defined by a regular language is nothing else than a Lindström quantifier (in the classical sense) defined by a structure that is a finite monoid multiplication table.

The class $Q_{\text{Mon}}\text{FO}$ is the class of all languages definable by applying a single monoidal quantifier to an appropriate tuple of FO formulas. The class $\text{FO}(Q_{\text{Mon}})$ is defined analogously, but allowing monoidal quantifiers to be used as any other quantifier would (i.e., allowing arbitrary nesting).

It was known [BIS90] that first-order logic with unnested unary monoidal quantifiers characterizes the class of regular languages. Recently this has been extended in [LMSV01] as follows:

Theorem 6. $\text{FO}(Q_{\text{Mon}}) = Q_{\text{Mon}}\text{FO} = \text{REG}$.

One of our results below will be that replacing in this statement the first-order quantifiers by monadic second-order quantifiers does not enlarge the expressive power.

A *groupoid* is a finite multiplication table with an identity element. For a fixed groupoid G, each $S \subseteq G$ defines a G-word-problem, i.e., a language $\mathcal{W}(S, G)$ composed of all words w, over the alphabet G, that can be bracketed in such a way that w multiplies out to an element of S. Groupoid word-problems relate to context-free languages in the same way as monoid word-problems relate to regular languages: Every such word-problem is context-free, and every context-free language is a homomorphic pre-image of a groupoid word-problem (this result is credited to Valiant in [BLM93]).

The following definition is due to Bédard, Lemieux, and McKenzie [BLM93]:

Definition 7. A *groupoidal quantifier* is a Lindström quantifier Q_L where L is a word-problem of some finite groupoid.

Usage of groupoidal quantifiers in our logical language is signalled by Q_{Grp}, used in the same way as described for Q_{Mon} above.

Second-order Lindström quantifiers on strings were introduced in [BV98]. Here, we are mainly interested in those binding only set variables, so called *monadic quantifiers*.

Definition 8. Consider a language L over an alphabet $\Sigma = (a_1, a_2, \ldots, a_s)$. Let $\overline{X} = (X_1, \ldots, X_k)$ be a k-tuple of unary second-order variables, i.e., set variables. There are 2^{nk} different instances (assignments) of $\overline{X}$. We assume the following ordering on those instances: Let each instance of a single X_i be encoded by a bit string $s_1^i \cdots s_n^i$ with the meaning $s_j^i = 1 \iff j \in X_i$. Then we encode an instance of $\overline{X}$ by the bit string $s_1^1 s_1^2 \cdots s_1^k s_2^1 s_2^2 \cdots s_2^k \cdots s_n^1 s_n^2 \cdots s_n^k$ and order the instances lexicographically by their codes. We write $\overline{X}^{(1)} < \overline{X}^{(2)} < \cdots < \overline{X}^{(2^{nk})}$ for the sequence of all instances in that order. The *monadic second-order Lindström quantifier* Q_L binding $\overline{X}$ takes a meaning if $s-1$ formulas, each having free variables $\overline{X}$, are available. Let $\varphi_1(\overline{X})$, $\varphi_2(\overline{X})$, ..., $\varphi_{s-1}(\overline{X})$ be these

$s-1$ formulas. Then $\varphi = Q_L\overline{X}\big[\varphi_1(\overline{X}), \varphi_2(\overline{X}), \ldots, \varphi_{s-1}(\overline{X})\big]$ holds on a string $w = w_1 \cdots w_n$, iff the word of length 2^{nk} whose ith letter, $1 \leq i \leq 2^{nk}$, is

$$\begin{cases} a_1 \text{ if } w \models \varphi_1(\overline{X}^{(i)}), \\ a_2 \text{ if } w \models \neg\varphi_1(\overline{X}^{(i)}) \wedge \varphi_2(\overline{X}^{(i)}), \\ \quad\vdots \\ a_s \text{ if } w \models \neg\varphi_1(\overline{X}^{(i)}) \wedge \neg\varphi_2(\overline{X}^{(i)}) \wedge \cdots \wedge \neg\varphi_{s-1}(\overline{X}^{(i)}), \end{cases}$$

belongs to L. We denote this word by $f^{\overline{X}}_{\varphi_1 \cdots \varphi_{s-1}}(w)$.

Again, taking as examples the languages $L_\exists$ and $L_\forall$, we obtain the usual second-order existential and universal quantifiers.

If L above is a monoid word-problem, then we say that Q_L is a monadic second-order monoidal quantifier; if L is a groupoid word-problem then Q_L is called monadic second-order groupoidal quantifier.

The class mon-Q^1_LFO is the class of all languages describable by applying a specific monadic second-order monoidal quantifier Q_L to an appropriate tuple of formulas without further occurrences of second-order quantifiers. The class mon-Q^1_{Mon}FO is defined analogously using arbitrary monadic second-order monoidal quantifiers. The class mon-Q^1_{Mon}SOM is defined analogously using tuples of SOM formulas. The class SOM(mon-Q^1_{Mon}) is defined analogously, but allowing monoidal quantifiers to be used as any other quantifier would (i. e., allowing arbitrary nesting). Analogous notation with mon-Q^1_{Grp} will be used.

4 Leaf Languages vs. Generalized Quantifiers

We start by proving a technical lemma which we will need later on. We follow the treatment of formulas with free variables developed in detail in [Str94, pp. 14ff], i.e., such formulas define languages of words with (first- and second-order) variables attached to its letters, so called $(\mathcal{V}_1, \mathcal{V}_2)$-*structures* for a set $\mathcal{V}_1$ of first-order and a set $\mathcal{V}_2$ of second-order variables; formally a $(\mathcal{V}_1, \mathcal{V}_2)$-structure over Σ is a word over the alphabet $\Sigma \times 2^{\mathcal{V}_1} \times 2^{\mathcal{V}_2}$ where Σ is the original (letter) alphabet.

Lemma 9. *Let Σ be an alphabet and $\mathcal{V}_1$ and $\mathcal{V}_2$ be sets of first- and second-order variables. Let $\overline{X} = (X_1, \ldots, X_k)$ be a vector of monadic second-order variables. Let $\varphi_1(\overline{X}), \varphi_2(\overline{X}), \ldots, \varphi_{s-1}(\overline{X})$ be formulas (from any calculus) that define regular sets of $(\mathcal{V}_1, \mathcal{V}_2 \cup \{X_1, \ldots, X_k\})$-structures over Σ.*

Let $\Gamma = (a_1, a_2, \ldots, a_s)$ be an ordered alphabet, and let $L \subseteq \Gamma^*$. Define $\varphi \triangleq Q_L\overline{X}\big[\varphi_1(\overline{X}), \varphi_2(\overline{X}), \ldots, \varphi_{s-1}(\overline{X})\big]$; hence φ is a formula which defines a set of $(\mathcal{V}_1, \mathcal{V}_2)$-structures over Σ. Then there exists a leaf automaton M that accepts the language L_φ when working with leaf language L.

Proof. Let $M_1, \ldots, M_{s-1}$ be the deterministic automata that accept $L_{\varphi_1}, \ldots, L_{\varphi_{s-1}}$. For $1 \leq i < s$, let $M_i = (Q_i, \Sigma \times 2^{\mathcal{V}_1} \times 2^{\mathcal{V}_2 \cup \{X_1, \ldots, X_k\}}, \delta_i, s_i, A_i)$. We consider the usual parallel product of the M_i, i.e., $M_\times =_{\mathrm{def}} \left(\times_{i=1}^{s-1} Q_i, \Sigma \times 2^{\mathcal{V}_1} \times 2^{\mathcal{V}_2 \cup \{X_1,\ldots,X_k\}}, \delta_\times, (s_1, s_2, \ldots, s_{s-1}), \times_{i=1}^{s-1} A_i\right)$ with a transition function given by $\delta_\times((q_1, q_2, \ldots, q_{s-1}), (v, S, T)) = (\delta_1(q_1, (v, S, T)), \ldots, \delta_{s-1}(q_{s-1}, (v, S, T)))$.

We define our leaf automaton

$$M =_{\mathrm{def}} \left(\times_{i=1}^{s-1} Q_i, \Sigma \times 2^{\mathcal{V}_1} \times 2^{\mathcal{V}_2}, \delta, (s_1, s_2, \ldots, s_{s-1}), \Gamma, \beta\right)$$

as follows:

We want M to accept only $(\mathcal{V}_1, \mathcal{V}_2)$-structures as input. Certainly we have to simulate the behavior of $M_\times$ on every possible instance of $\overline{X} = (X_1, \ldots, X_k)$. Each position of an input of $M_\times$ may or may not be an element of each (set) variable X_i—in other words: If w is an input of M, then for each read letter $(w_j, \mathcal{V}_1^j, \mathcal{V}_2^j)$ of w, where $w_j \in \Sigma$, $\mathcal{V}_1^j \subseteq \mathcal{V}_1$, $\mathcal{V}_2^j \subseteq \mathcal{V}_2$, we have to simulate $M_\times$ for the cases $X_i \in \mathcal{V}_2^j$ and $X_i \notin \mathcal{V}_2^j$. So there are 2^k possibilities for simulated inputs for each letter of w. We encode the members of $\mathcal{P}(\{X_1, \ldots, X_k\})$ by $S_i, 1 \le i \le 2^k$ where each set S_i is defined by $X_j \in S_i$ iff the jth bit of $\mathrm{bin}_k(i-1)$ (the length k binary encoding of the natural number $i-1$) is 1.

The transition function of M is now defined as (recall that the transition function for leaf automata maps into *sequences* of states)

$$\begin{aligned} \delta_\varphi((q_1, q_2, \ldots, q_{s-1}), (x, S, T)) = {} & \delta_\times((q_1, q_2, \ldots, q_{s-1}), (x, S, T \cup \{S_1\})) \\ & \delta_\times((q_1, q_2, \ldots, q_{s-1}), (x, S, T \cup \{S_2\})) \\ & \vdots \\ & \delta_\times((q_1, q_2, \ldots, q_{s-1}), (x, S, T \cup \{S_{2^k}\})) \end{aligned}$$

Finally we define

$$\beta(q_1, q_2, \ldots, q_{s-1}) =_{\mathrm{def}} \begin{cases} a_1 \text{ if } q_1 \in A_1, \\ a_2 \text{ if } q_1 \notin A_1 \wedge q_2 \in A_2, \\ \vdots \\ a_s \text{ if } q_1 \notin A_1 \wedge q_2 \notin A_2 \wedge \cdots \wedge q_{s-1} \notin A_{s-1} \end{cases}$$

Now M on input $w = w_1 \ldots w_n$ spans a computation tree $T_M(w)$ of depth n and branching width 2^k. Such a tree has $(2^k)^n = 2^{nk}$ leaves. We will now prove that the leaf string on $T_M(w)$ is identical to the word $f^{\overline{X}}_{\varphi_1 \cdots \varphi_{s-1}}(w)$:

Each of the 2^k branches on an input letter corresponds to a set $S_i, 1 \le i \le 2^k$. Thus we can encode the branch with the bit string $\mathrm{bin}_k(i-1)$ defining S_i as above. When ordering the leaves on $T_M(w)$, the branching on earlier input letters is of higher importance than the branching on later input letters. So we can encode a path in $T_M(w)$ by the bit string $\mathrm{bin}_k(i_1 - 1)\mathrm{bin}_k(i_2 - 1) \cdots \mathrm{bin}_k(i_n - 1)$ of length nk, meaning: on input letter j the set S_{i_j} was chosen. If path l has a lexicographically smaller encoding than path l', then the leaf of l has a position in the leaf string of $T_M(w)$ to the left of the leaf on path l'. Since there are exactly 2^k different bit strings $\mathrm{bin}_k(i)$, each of those strings corresponds to one set S_i, and as such it corresponds to the branch i in a forking of the computation tree. Thus the space of all 2^{nk} binary strings of length nk is completely filled with encodings of computation paths. From this follows, that path number l in the computation tree is encoded by the bit string of length nk which is the binary representation of $l-1$.

Now we want to show that on path l the tuple of sets $\overline{X}^{(l)}$ (using the notation from Def. 8) is simulated. The encoding of path l is the binary string z representing the number $l-1$. We can split z in n parts of length k: $z = z_1 z_2 \ldots z_n$. As said before, the bit string $z_j = z_j^1 z_j^2 \ldots z_j^k$ encodes the simulation of sets X_i in this way: $z_j^i = 1$ iff the set X_i was chosen on input letter j. So the value of the variable X_i depends on the values of $z_1^i, z_2^i, \ldots, z_n^i$ of the string $z = z_1^1 z_1^2 \ldots z_1^k z_2^1 z_2^2 \ldots z_2^k \ldots z_n^1 z_n^2 \ldots z_n^k$. But since z represents the number $l-1$, this is exactly the definition of the tuple $\overline{X}^{(l)}$.

Thus we conclude that on path l of the computation tree we simulate the behavior of the machines $M_1, \ldots, M_n$ on input w supplemented by $\overline{X}^{(l)}$. Each M_k simulates the formula φ_k. So, if on reading the last input letter the state q_k is reached on machine M_k then it holds $q_k \in A_k$ iff $w \models \varphi_k(\overline{X}^{(l)})$.

From the definition of the function β we see that the leaf letter on path l is

$$\begin{cases} a_1 \text{ if } w \models \varphi_1(\overline{X}^{(l)}), \\ a_2 \text{ if } w \models \neg\varphi_1(\overline{X}^{(l)}) \wedge \varphi_2(\overline{X}^{(l)}), \\ \quad \vdots \\ a_s \text{ if } w \models \neg\varphi_1(\overline{X}^{(l)}) \wedge \neg\varphi_2(\overline{X}^{(l)}) \wedge \cdots \wedge \neg\varphi_{s-1}(\overline{X}^{(l)}). \end{cases}$$

But this is the letter l of the word $f^{\overline{X}}_{\varphi_1 \cdots \varphi_{s-1}}(w)$ (see Def. 8). This completes the proof that the leaf string on computation tree $T_M(w)$ is identical to the word $f^{\overline{X}}_{\varphi_1 \cdots \varphi_{s-1}}(w)$.

This implies $w \models Q_L \overline{X}\left[\varphi_1(\overline{X}), \varphi_2(\overline{X}), \ldots, \varphi_{s-1}(\overline{X})\right]$ iff $f^{\overline{X}}_{\varphi_1 \cdots \varphi_{s-1}}(w) \in L$ iff $\text{leafstring}^M(w) \in L$. Thus M with leaf language L accepts the language L_φ. □

Let $\mathcal{N}$ be the class of all languages that have a neutral letter; i.e., $L \subseteq \Gamma^*$ is in $\mathcal{N}$ if there is a letter $\epsilon \in \Gamma$ such that, for all $u, v \in \Gamma^*$, we have $uv \in L \iff u\epsilon v \in L$.

Our main result in this section states that, for all languages $L \in \mathcal{N}$, finite automata with leaf language L accept exactly those languages definable with a monadic Q_L^1 quantifier.

Theorem 10. *For any $L \in \mathcal{N}$, $\text{Leaf}^{\text{FA}}(L) = \text{mon-}Q_L^1\text{FO}$.*

Proof. We first consider the inclusion $\text{Leaf}^{\text{FA}}(L) \supseteq \text{mon-}Q_L^1\text{FO}$.

Let φ be a mon-Q_L^1FO-formula which defines a language $L_\varphi \subseteq \Sigma^*$. Then φ will have the form $Q_L \overline{X}\left[\varphi_1(\overline{X}), \varphi_2(\overline{X}), \ldots, \varphi_{s-1}(\overline{X})\right]$ for some formulas $\varphi_i(\overline{X})$, $1 \le i \le s-1$, which are first order except of the occurrence of the monadic variables in $\overline{X}$. Hence we conclude by Lemma 9 that there must exist a leaf automaton M_φ which accepts the language L_φ.

It remains to prove the opposite inclusion $\text{Leaf}^{\text{FA}}(L) \subseteq \text{mon-}Q_L^1\text{FO}$.

Let $\Gamma = \{a_1, a_2, \ldots, a_{s-1}, \epsilon\}$, let $L \in \mathcal{N} \cap \mathcal{P}(\Gamma^*)$ where ϵ is neutral, and let $M = (Q, \Sigma, \delta, q_1, \Gamma, \beta)$ be a leaf automaton. Suppose that M working with leaf language L accepts a certain language $A \subseteq \Sigma^*$. We have to construct a mon-Q_L^1FO-formula φ such that $L_\varphi = A$.

First, order the leaf alphabet in such a way that ϵ is the last symbol in Γ, e.g., $\Gamma = (a_1, a_2, \ldots, a_{s-1}, \epsilon)$. Let $Q = \{q_1, \ldots, q_l\}$ and let $m =_{\text{def}} \max\{\, |\delta(q, a)| \mid q \in Q, a \in \Sigma \,\}$. Formula φ will be of the form $Q_L \overline{X}\left[\varphi_1(\overline{X}), \varphi_2(\overline{X}), \ldots, \varphi_{s-1}(\overline{X})\right]$ where $\overline{X} =_{\text{def}} (X_m, \ldots, X_1, Y_l, \ldots, Y_1)$.

We want to encode the number of a computation path of M on input w in the variables X_i. This will be done in the following way: $j \in X_i$ iff M reading the j-th symbol of w chooses the i-th alternative of the successor function.

In the variables Y_i we want to encode the sequence of states of M on input w on a certain computation path. We shall do this in the following form: $j \in Y_i$ iff M (on that computation path whose number is given by variables X_l) while reading the j-th symbol of the input is in state q_i.

The formulas φ_k on input w and an instance of $\overline{X}$ thus have to express the following:

1. All φ_k have to test whether the X_i encode a computation path of M on w. If this is not the case (e.g. if there exists a j and $i_1 \neq i_2$ with $j \in X_{i_1}$ and $j \in X_{i_2}$) then $w \not\models \varphi_k(\overline{X})$ for all $k \in \{1, \ldots, s-1\}$. (A look at the definition of $f^{\overline{X}}_{\varphi_1 \cdots \varphi_{s-1}}(w)$ shows that the letter of $f^{\overline{X}}_{\varphi_1 \cdots \varphi_{s-1}}(w)$ yielded by $\overline{X}$ will then be ϵ.)
2. All φ_k have to test whether the Y_i encode the correct sequence of states of M for the computation path encoded by the X_i. If this is not the case then $w \not\models \varphi_k(\overline{X})$ for all $k \in \{1, \ldots, s-1\}$. (The letter of $f^{\overline{X}}_{\varphi_1 \cdots \varphi_{s-1}}(w)$ yielded by $\overline{X}$ thus again will be ϵ.)
3. If $\overline{X}$ encodes a computation of M—i. e., if the X_i encode the number of a path of M and the Y_i encode the correct sequence of states of M on that path—then $w \models \varphi_k(\overline{X})$ iff M on the computation path encoded by the X_i produces the leaf letter a_k. (The letter of $f^{\overline{X}}_{\varphi_1 \cdots \varphi_{s-1}}(w)$ yielded by $\overline{X}$ will then be identical to the leaf letter produced by M on the computation path encoded by the X_i.)

It is easy to construct FO-formulas with this behavior.

We now show that the word $f^{\overline{X}}_{\varphi_1 \cdots \varphi_{s-1}}(w)$ is essentially equal to the word leafstring$^M(w)$, more precisely: that both words are identical once all occurrences of the letter ϵ are deleted from them; hence they are equivalent w.r.t. membership in L.

Since each computation of M is encoded by a certain instance of $\overline{X}$, each letter in leafstring$^M(w)$ is contained in $f^{\overline{X}}_{\varphi_1 \cdots \varphi_{s-1}}(w)$, also. And since each instance of $\overline{X}$ not encoding a sensible computation of M yields the letter ϵ in $f^{\overline{X}}_{\varphi_1 \cdots \varphi_{s-1}}(w)$, we know that the amount of each letter different from ϵ in both words is identical. All there is left to do is showing that their sequences are identical in both words.

Let p_1 and p_2 be computation paths in M, u and v their respective leaf letters and $\overline{X_1}$ and $\overline{X_2}$ the encoding of the sensible computations associated with these paths (it was shown above that then u and v are also the letters of $f^{\overline{X}}_{\varphi_1\cdots\varphi_{s-1}}(w)$ produced by $\overline{X_1}$ and $\overline{X_2}$). We show that p_1 is positioned before p_2 iff the encoding of $\overline{X_1}$ is of lower order than the encoding of $\overline{X_2}$.

The encoding of an instance $\overline{X_k}$, according to the definition of $f^{\overline{X}}_{\varphi_1\cdots\varphi_{s-1}}(w)$, can be separated into n bit strings of length $m+l$. Each of those strings is of the form $x_m \ldots, x_1, y_l, \ldots, y_1$. The j-th such string corresponds to the input letter j with the meaning $x_i = 1 \leftrightarrow j \in X_i$ (there will always be only one $x_i = 1$, since the $\overline{X_k}$ are correct encodings) and $y_i = 1 \leftrightarrow j \in Y_i$ (there will also always be only one $y_i = 1$).

Two computation paths that differ have a word position on which the first difference occurs. We call it d. Up to position d not only the sequences of successor function choices are identical for both paths but also the sequence of states M assumes while reading letters of w; this holds because the computations so far are identical.

But from this it follows that the encodings of the $\overline{X_k}$ are identical up to the bit string number d. So, this string decides which encoding is of higher order. Let the d-th string of $\overline{X_1}$ be $x^1_m \ldots x^1_1 y^1_l \ldots y^1_1$ and of $\overline{X_2}$ be $x^2_m \ldots x^2_1 y^2_l \ldots y^2_1$.

The unique $x^1_a = 1$ is the choice for the successor function on computation path p_1 and input letter d while the $x^2_b = 1$ is the corresponding choice for path p_2. Then $a < b$ iff p_1 is ordered before p_2 in $T_M(w)$ (this is due to the ordering which the successor function imposes on the computation tree) and the string $x^1_m \ldots x^1_b \ldots x^1_a \ldots x^1_1 y^1_l \ldots y^1_1 = 0 \ldots 0 \ldots 1 \ldots 0 y^1_l \ldots y^1_1$ is smaller than $x^2_m \ldots x^2_b \ldots x^2_a \ldots x^2_1 y^2_l \ldots y^2_1 = 0 \ldots 1 \ldots 0 \ldots 0 y^2_l \ldots y^2_1$. Thus the ordering of $\overline{X_1}$ and $\overline{X_2}$ is identical to the ordering of p_1 and p_2.

This concludes the proof that the words $f^{\overline{X}}_{\varphi_1\cdots\varphi_{s-1}}(w)$ and leafstring$^M(w)$ are identical if the letters ϵ are not considered. Thus $w \in A$ iff leafstring$^M(w) \in L$ iff $f^{\overline{X}}_{\varphi_1\cdots\varphi_{s-1}}(w) \in L$ iff $w \in L_\varphi$, hence we conclude that φ defines the language A accepted by M. □

5 Monoidal Quantifiers

We can now give an extension of Proposition 3: if we extend the monadic second-order formalism by monoidal quantifiers, we do not gain expressive power.

Theorem 11.
$$\begin{aligned} \text{mon-}Q^1_{\text{Mon}}\text{FO} &= \text{FO}(\text{mon-}Q^1_{\text{Mon}}) = \\ \text{mon-}Q^1_{\text{Mon}}\text{SOM} &= \text{SOM}(\text{mon-}Q^1_{\text{Mon}}) = \text{REG}. \end{aligned}$$

Proof. The inclusions mon-Q^1_{Mon}FO $\subseteq$ mon-Q^1_{Mon}SOM $\subseteq$ SOM(mon-Q^1_{Mon}) and mon-Q^1_{Mon}FO $\subseteq$ FO(mon-Q^1_{Mon}) $\subseteq$ SOM(mon-Q^1_{Mon}) are trivial.

For the inclusion REG $\subseteq$ mon-Q^1_{Mon}FO we use the fact that every regular language can be defined by an SOM formula with only one second-order existential quantifier preceding a first-order formula [Tho82], and an existential quantifier is one particular monoidal quantifier—as already mentioned, it is equivalent to $Q_{L_\exists}$ with $L_\exists = 0^*1(0+1)^*$.

Now all there is left to do is proving that $\text{SOM}(\text{mon-}Q^1_{\text{Mon}}) \subseteq \text{REG}$. We do this closely following the proof of Theorem III.1.1 in [Str94]. There, SOM $\subseteq$ REG is proven proceeding by induction on the construction of second order monadic formulas. Since here we have SOM-formulas with one additional constructive element—the Lindström quantifier—we have to extend the inductive proof by one step:

Let Σ be an alphabet and $\mathcal{V}_1$ and $\mathcal{V}_2$ be sets of first- and second-order variables. Then a formula $\varphi \in \text{SOM}(\text{mon-}Q^1_{\text{Mon}})$ with free variables $\mathcal{V}_1 \cup \mathcal{V}_2$ accepts a language $L_\varphi \in (\Sigma \times 2^{\mathcal{V}_1} \times 2^{\mathcal{V}_2})^*$. We have to prove that L_φ is a regular language. As mentioned, the proof proceeds by induction, see [Str94, pp. 21ff]. The additional case in the induction step for us here is that of $\varphi \triangleq Q_L\overline{X}\big[\varphi_1(\overline{X}), \varphi_2(\overline{X}), \ldots, \varphi_{s-1}(\overline{X})\big]$: By induction hypothesis there exist deterministic automata $M_1, M_2, \ldots, M_{s-1}$ that accept the $(\mathcal{V}_1, \mathcal{V}_2 \cup \{X_1, \ldots, X_k\})$-structures that model $\varphi_1, \varphi_2, \ldots, \varphi_{s-1}$. Hence we can use Lemma 9 to conclude that there is a leaf automaton M_φ that accepts L_φ when working w.r.t. leaf language L.

But since L is regular, we can use Proposition 2 to see that then L_φ is regular, too. This concludes the induction and the proof of the theorem. □

This result, together with the result of Thomas [Tho82] mentioned in the above proof, leads immediately to the following *normal form for monadic second-order monoidal logic*:

Corollary 12. *Every* $\text{SOM}(\text{mon-}Q^1_{\text{Mon}})$*-formula is equivalent to a formula of the form* $\exists X\varphi(X)$*, where* X *is a set variable and* φ *is a formula without second-order quantifiers.*

6 Groupoidal Quantifiers

Finally, we examine the power of monadic second-order groupoidal quantifiers. These are powerful enough to define every language in LOGCFL, as we prove next.

Theorem 13. $\text{LOGCFL} \subseteq \text{mon-}Q^1_{\text{Grp}}\text{FO}$.

Proof. Lautemann et al. in [LMSV01] showed that $\text{LOGCFL} = Q_{\text{Grp}}\text{FO}$. This means that for every language $L \in$ LOGCFL there is a sentence φ in first-order logic prefixed with a generalized quantifier for a context-free language such that $L = L_\varphi$.

However, such a CFL-quantifier can easily be simulated by a monadic second-order Lindström quantifier defined by another context-free language as follows: Consider the formula $\varphi \triangleq Q_L x_1 \cdots x_k\big[\psi(x_1 \cdots x_k)\big]$ for first-order formula ψ and context-free language L over $(0,1)$ (the case of larger alphabets proceeds completely analogously). Let L' be obtained from L by adding a neutral letter, i.e., L' is defined over alphabet $(0, 1, \epsilon)$ where ϵ is neutral. Define

$$\varphi' \triangleq Q_{L'} X_1 \cdots X_k \big[\psi_1(X_1 \cdots X_k), \psi_2(X_1 \cdots X_k)\big],$$

where

$$\psi_1(X_1 \cdots X_k) \triangleq \exists x_1 \cdots x_k \Big(\bigwedge_{i=1}^{k} (x_i \in X_i \wedge \forall y (y \in X_i \to x_i = y)) \wedge \psi(x) \Big),$$
$$\psi_2(X_1 \cdots X_k) \triangleq \exists x_1 \cdots x_k \Big(\bigwedge_{i=1}^{k} (x_i \in X_i \wedge \forall y (y \in X_i \to x_i = y)) \wedge \neg\psi(x) \Big).$$

Then the only assignments of a set variable X_i that do not lead to output symbol ϵ in $f^{(X_1,\ldots,X_k)}_{\psi_1\psi_2}$ are the singleton sets, and these correspond (in the same relative order) to assignments of x_i in ψ. Hence, for every input w, $f^{(X_1,\ldots,X_k)}_{\psi_1\psi_2}(w) \in L' \iff f^{(x_1,\ldots,x_k)}_{\psi}(w) \in L$, and thus, $L_\varphi = L_{\varphi'}$.

This proves LOGCFL $= Q_{\text{Grp}}\text{FO} \subseteq \text{mon-}Q^1_{\text{Grp}}\text{FO}$. □

Peichl and Vollmer, after introducing finite leaf automata in [PV01], examined the power of various classes of leaf languages and thus obtained a number of new characterizations of complexity classes and formal language classes. One particular case, left open in that paper, is the following: What is the power of finite leaf automata with context-free leaf languages, i.e., how does the class $\text{Leaf}^{\text{FA}}(\text{CFL})$ relate to other well-known classes? The only upper and lower bounds known were $\text{CFL} \subsetneq \text{Leaf}^{\text{FA}}(\text{CFL}) \subseteq \text{DSPACE}(n^2) \cap \text{DTIME}(2^{O(n)})$. Here, we improve the lower bound.

Corollary 14. $\text{LOGCFL} \subseteq \text{Leaf}^{\text{FA}}(\text{CFL})$.

Proof. From Theorem 13, we obtain $\text{LOGCFL} \subseteq \text{mon-}Q^1_{\text{Grp}}\text{FO} = \text{mon-}Q^1_{\text{CFL}}\text{FO}$. By Theorem 10, however, the latter class is equal to $\text{Leaf}^{\text{FA}}(\text{CFL})$, yielding $\text{LOGCFL} \subseteq \text{Leaf}^{\text{FA}}(\text{CFL})$. □

7 Conclusion

We considered monadic second-order logic extended with generalized quantifiers. We contributed to the study of the expressive power of such logics as well as to the study of the power of finite automata with generalized acceptance criteria. So far, we did not exploit the close connection between logical reductions and generalized quantifiers. This might lead to further interesting consequences.

One of the results of this paper is that by adding *arbitrary* monadic second-order monoidal quantifiers to SOM does not give us more power. On the other hand one might ask what specific regular languages L besides $L_\exists$ (for nondeterministic acceptance, see Theorem 11 and Corollary 12) there are, for which $Q^1_L\text{FO} = \text{REG}$. Even for a very "easy" language obtained from deterministic acceptance of regular languages, namely $L_{\text{det}} = 1(0+1)^*$, we have $Q^1_{L_{\text{det}}}\text{FO} =$

REG. (This implies easily that also $Q^1_{L_{\mathrm{mod}\ p}}\mathrm{FO} = \mathrm{REG}$.) Can interesting subclasses of REG be defined with restricted quantifiers?

Still open remains a complete clarification of the power of $\mathrm{Leaf}^{\mathrm{FA}}(\mathrm{CFL})$. We proved $\mathrm{LOGCFL} \subseteq \mathrm{Leaf}^{\mathrm{FA}}(\mathrm{CFL})$. It is relatively easy to see that every language in $\mathrm{Leaf}^{\mathrm{FA}}(\mathrm{CFL})$ can be accepted by auxiliary pushdown automata with linear space and simultaneous exponential time bound; hence $\mathrm{LOGCFL} = \mathrm{NAuxPDA\text{-}SPACE\text{-}TIME}(\log n, n^{O(1)}) \subseteq \mathrm{Leaf}^{\mathrm{FA}}(\mathrm{CFL}) \subseteq \mathrm{NAuxPDA\text{-}SPACE\text{-}TIME}(n, 2^{O(n)})$. Is one of these inclusions actually an equality? And how do the logics in the range from mon-$Q^1_{\mathrm{Grp}}\mathrm{FO}$ to $\mathrm{SOM}(\text{mon-}Q^1_{\mathrm{Grp}})$ relate to these classes?

Acknowledgment

We thank Gerhard Buntrock (Lübeck), Klaus-Jörn Lange (Tübingen), Klaus Reinhard (Tübingen), and, in particular, Sven Kosub (Würzburg) for helpful discussions. We also thank the anonymous referees for hints that helped to improve the presentation of the paper.

References

[BE58] J. R. Büchi and C. C. Elgot. Decision problems of weak second order arithmetics and finite automata, Part I. *Notices of the American Mathematical Society*, 5:834, 1958.

[BIS90] D. A. Mix Barrington, N. Immerman, and H. Straubing. On uniformity within NC^1. *Journal of Computer and System Sciences*, 41:274–306, 1990.

[BLM93] F. Bédard, F. Lemieux, and P. McKenzie. Extensions to Barrington's M-program model. *Theoretical Computer Science*, 107:31–61, 1993.

[Büc62] J. R. Büchi. On a decision method in restricted second-order arithmetic. In *Proceedings Logic, Methodology and Philosophy of Sciences 1960*, Stanford, CA, 1962. Stanford University Press.

[BV98] H.-J. Burtschick and H. Vollmer. Lindström quantifiers and leaf language definability. *International Journal of Foundations of Computer Science*, 9:277–294, 1998.

[Coo71] S. A. Cook. Characterizations of pushdown machines in terms of time-bounded computers. *Journal of the Association for Computing Machinery*, 18:4–18, 1971.

[EF95] H.-D. Ebbinghaus and J. Flum. *Finite Model Theory.* Perspectives in Mathematical Logic. Springer Verlag, Berlin Heidelberg, 1995.

[Lin66] P. Lindström. First order predicate logic with generalized quantifiers. *Theoria*, 32:186–195, 1966.

[LMSV01] C. Lautemann, P. McKenzie, T. Schwentick, and H. Vollmer. The descriptive complexity approach to LOGCFL. *Journal of Computer and Systems Sciences*, 2001. To appear. A preliminary version appeared in the *Proceedings of the 16th Symposium on Theoretical Aspects of Computer Science*, Lecture Notes in Computer Science Vol. 1563, pp. 444–454, Springer Verlag, 1999.

[MP71] R. McNaughton and S. Papert. *Counter-Free Automata.* MIT Press, 1971.

[PV01] T. Peichl and H. Vollmer. Finite automata with generalized acceptance criteria. *Discrete Mathematics and Theoretical Computer Science*, 2001. To appear. A preliminary version appeared in the *Proceedings of the 26th International Colloqium on Automata, Languages, and Programming,* Lecture Notes in Computer Science Vol. 1644, pp. 605–614, Springer Verlag, 1999.

[Str94] H. Straubing. *Finite Automata, Formal Logic, and Circuit Complexity.* Birkhäuser, Boston, 1994.

[Sud78] I. H. Sudborough. On the tape complexity of deterministic context-free languages. *Journal of the Association for Computing Machinery*, 25:405–414, 1978.

[Tho82] W. Thomas. Classifying regular events in symbolic logic. *Journal of Computer and Systems Sciences*, 25:360–376, 1982.

[Tra61] B. A. Trakhtenbrot. Finite automata and logic of monadic predicates. *Doklady Akademii Nauk SSSR*, 140:326–329, 1961. In Russian.

An Effective Extension of the Wagner Hierarchy to Blind Counter Automata

Olivier Finkel

Equipe de Logique Mathématique
U.F.R. de Mathématiques, Université Paris 7
2 Place Jussieu 75251 Paris cedex 05, France
finkel@logique.jussieu.fr

Abstract. The extension of the Wagner hierarchy to blind counter automata accepting infinite words with a Muller acceptance condition is effective. We determine precisely this hierarchy.

Keywords: ω-languages; blind counter automata; effective extension of the Wagner hierarchy; topological properties; Wadge hierarchy; Wadge games.

1 Introduction

Regular ω-languages are accepted by (deterministic) Muller automata. Finite machines having a stronger expressive power when reading infinite words have also been investigated [Sta97a]. Recently Engelfriet and Hoogeboom studied **X**-automata, i.e. automata equipped with a storage type **X**, including the cases of pushdown automata, Turing machines, Petri nets [EH93]. A way to investigate the expressive power of such machines is to study the topological complexity of the ω-languages they accept. For deterministic machines, it is shown in [EH93] that every **X**-automaton accepts boolean combinations of $\mathbf{\Pi}_2^0$-sets. Hence in order to distinguish the different storage types it turned out that the study of the Wadge hierarchy is suitable. The Wadge hierarchy is a great refinement of the Borel hierarchy, recently studied by Duparc [Dup99a]. The Wadge hierarchy of ω-regular languages has been determined in an efective way by Wagner [Wag79]. Several extensions of this hierarchy have been recently determined as the extension to deterministic pushdown automata, to k-blind counter automata, [DFR01] [Dup99b] [Fin00b]. We present here the extension to (one) blind counter automata, which is the first known **effective** extension. We study Muller blind counter automata (MBCA), and define chains and superchains as Wagner did for Muller automata. The essential difference between the two hierarchies relies on the existence of superchains of transfinite length $\alpha < \omega^2$ for MBCA. The hierarchy is effective and leads to effective winning strategies in Wadge games between MBCA. The hierarchy of Muller automata equipped with several blind counters is presented in a non effective way in [Fin00b][DFR01].

L. Fribourg (Ed.): CSL 2001, LNCS 2142, pp. 369–383, 2001.

2 Regular and Blind Counter ω-Languages

We assume the reader to be familiar with the theory of formal languages and of ω-regular languages, see for example [HU69], [Tho90]. We first recall some definitions and results concerning ω-regular languages and omega pushdown automata and introduce blind counter automata as a special case of pushdown automata [Tho90] [Sta97a].

When Σ is a finite alphabet, a finite string (word) over Σ is any sequence $x = x_1 \ldots x_k$, where $x_i \in \Sigma$ for $i = 1, \ldots, k$, and k is an integer ≥ 1. The length of x is k, denoted by $|x|$. If $|x| = 0$, x is the empty word denoted by λ.

We write $x(i) = x_i$ and $x[i] = x(1) \ldots x(i)$ for $i \leq k$ and $x[0] = \lambda$. $\Sigma^\star$ is the set of finite words over Σ. The first infinite ordinal is ω. An ω-word over Σ is an ω -sequence $a_1 \ldots a_n \ldots$, where $a_i \in \Sigma, \forall i \geq 1$. When σ is an ω-word over Σ, we write $\sigma = \sigma(1)\sigma(2)\ldots\sigma(n)\ldots$ and $\sigma[n] = \sigma(1)\sigma(2)\ldots\sigma(n)$ the finite word of length n, prefix of σ. The set of ω-words over the alphabet Σ is denoted by Σ^ω. An ω-language over an alphabet Σ is a subset of Σ^ω.

The usual concatenation product of two finite words u and v is denoted $u.v$ (and sometimes just uv). This product is extended to the product $u.v$ of a finite word u and an ω-word v.

For $V \subseteq \Sigma^\star$, $V^\omega = \{\sigma = u_1 \ldots u_n \ldots \in \Sigma^\omega / u_i \in V, \forall i \geq 1\}$ is the ω-power of V.

R. Mc Naughton established that the expressive power of deterministic Muller automata (DMA) is equal to the expressive power of non deterministic Muller automata (MA) [Tho90]. An ω-language is regular iff it is accepted by a Muller automaton. The class REG_ω of ω-regular languages is the ω-Kleene closure of the class REG of (finitary) regular languages where the ω-Kleene closure of a family L of finitary languages is:

$$\omega - KC(L) = \{\cup_{i=1}^{n} U_i.V_i^\omega / U_i, V_i \in L, \forall i \in [1, n]\}$$

We now define the (blind) one counter machines which we assume here to be realtime and deterministic, and the corresponding classes of blind counter ω-languages.

Definition 1. *A (realtime deterministic) pushdown machine (PDM) is a 6-tuple $M = (K, \Sigma, \Gamma, \delta, q_0, Z_0)$, where K is a finite set of states, Σ is a finite input alphabet, Γ is the finite pushdown alphabet, $q_0 \in K$ is the initial state, $Z_0 \in \Gamma$ is the start symbol, and δ is a mapping from $K \times \Sigma \times \Gamma$ into $K \times \Gamma^\star$.*

If $\gamma \in \Gamma^+$ describes the pushdown store content, the leftmost symbol will be assumed to be on "top" of the store. A configuration of a PDM is a pair (q, γ) where $q \in K$ and $\gamma \in \Gamma^\star$.

For $a \in \Sigma$, $\gamma, \beta \in \Gamma^\star$ and $Z \in \Gamma$, if (p, β) is in $\delta(q, a, Z)$, then we write $a : (q, Z\gamma) \mapsto_M (p, \beta\gamma)$.

$\mapsto_M^\star$ is the transitive and reflexive closure of $\mapsto_M$. (The subscript M will be omitted whenever the meaning remains clear).

Let $\sigma = a_1 a_2 \ldots a_n \ldots$ be an ω-word over Σ. An infinite sequence of configurations $r = (q_i, \gamma_i)_{i \geq 1}$ is called a run of M on σ, starting in configuration (p, γ), iff:

1. $(q_1, \gamma_1) = (p, \gamma)$
2. for each $i \geq 1$, $a_i : (q_i, \gamma_i) \mapsto_M (q_{i+1}, \gamma_{i+1})$

For every such run, $In(r)$ *is the set of all states entered infinitely often during run* r.

A run r *of* M *on* σ, *starting in configuration* (q_0, Z_0), *will be simply called "a run of* M *on* σ".

A one counter machine is a PDM such that $\Gamma = \{Z_0, I\}$ *where* Z_0 *is the bottom symbol and always remains at the bottom of the store. So the pushdown store is used like a counter whose value is the integer* n *if the content of the pushdown store is* $I^n Z_0$.

A one blind counter machine is a one counter machine such that every transition which is enabled at zero level is also enabled at non zero level, i.e. if $\delta(q, a, Z_0) = (p, I^n Z_0)$, *for some* $p, q \in K$, $a \in \Sigma$ *and* $n \geq 0$, *then* $\delta(q, a, I) = (p, I^{n+1})$. *But the converse may not be true, i.e. some transition may be enabled at non zero level but not at zero level.*

Definition 2. *A Muller (realtime deterministic) blind counter automaton* (**MBCA**) *is a 7-tuple* $\mathcal{A} = (K, \Sigma, \Gamma, \delta, q_0, Z_0, \mathcal{F})$ *where* $\mathcal{A}' = (K, \Sigma, \Gamma, \delta, q_0, Z_0)$ *is a (realtime deterministic) one blind counter machine and* $\mathcal{F} \subseteq 2^K$ *is the collection of designated state sets.*

The ω*-language accepted by* M *is* $L(\mathcal{A}) = \{\sigma \in \Sigma^\omega$ / *there exists a run* r *of* $\mathcal{A}$ *on* σ *such that* $In(r) \in \mathcal{F}\}$.

The class of ω*-languages accepted by* **MBCA** *will be denoted* **BC**.

Remark 3. *Machines we call here one blind counter machines are sometimes called one partially blind counter machines as in [Gre78].*

Remark 4. *If* M *is a deterministic pushdown machine, then for every* $\sigma \in \Sigma^\omega$, *there exists at most one run* r *of* M *on* σ *determined by the starting configuration. Each* ω*-language accepted by a Muller deterministic pushdown automaton (DMPDA) can be accepted by a DMPDA such that for every* $\sigma \in \Sigma^\omega$, *there exists such a run of* M *on* σ.

But this is not true for **MBCA** *because some words* x *may be rejected by an MBCA* $\mathcal{A}$ *because the machine* $\mathcal{A}$ *blocks at zero level when reading* x. *This is connected with the fact that the class* **BC** *is not closed under complementation as it is shown by the following example.*

Example 5. *It is easy to see that the* ω*-language* $L = \{a^n b^p c^\omega \ / \ p \leq n\}$ *is accepted by a deterministic MBCA, but its complement is not accepted by any deterministic MBCA because* $L' = \{a^n b^p c^\omega \ / \ p > n\}$ *is not accepted by any deterministic MBCA.*

3 Topology

We assume the reader to be familiar with basic notions of topology which may be found in [Kur66][LT94] [Sta97a] [PP98].

Topology is an important tool for the study of ω-languages, and leads to characterization of several classes of ω-languages.

For a finite alphabet X, we consider X^ω as a topological space with the Cantor topology (see [LT94] [Sta97a] [PP98]). The open sets of X^ω are the sets in the form $W.X^\omega$, where $W \subseteq X^\star$. A set $L \subseteq X^\omega$ is a closed set iff its complement $X^\omega - L$ is an open set. The class of open sets of X^ω will be denoted by $\mathbf{G}$ or by $\mathbf{\Sigma^0_1}$. The class of closed sets will be denoted by $\mathbf{F}$ or by $\mathbf{\Pi^0_1}$. Closed sets are characterized by the following:

Proposition 6. *A set $L \subseteq X^\omega$ is a closed set of X^ω iff for every $\sigma \in X^\omega$,*
$[\forall n \geq 1, \exists u \in X^\omega$ such that $\sigma(1)\ldots\sigma(n).u \in L]$ implies that $\sigma \in L$.

Define now the next classes of the Hierarchy of Borel sets of finite rank:

Definition 7. *The classes $\mathbf{\Sigma^0_n}$ and $\mathbf{\Pi^0_n}$ of the Borel Hierarchy on the topological space X^ω are defined as follows:*
$\mathbf{\Sigma^0_1}$ is the class of open sets of X^ω.
$\mathbf{\Pi^0_1}$ is the class of closed sets of X^ω.
$\mathbf{\Pi^0_2}$ or $\mathbf{G}_\delta$ is the class of countable intersections of open sets of X^ω.
$\mathbf{\Sigma^0_2}$ or $\mathbf{F}_\sigma$ is the class of countable unions of closed sets of X^ω.
And for any integer $n \geq 1$:
$\mathbf{\Sigma^0_{n+1}}$ is the class of countable unions of $\mathbf{\Pi^0_n}$-subsets of X^ω.
$\mathbf{\Pi^0_{n+1}}$ is the class of countable intersections of $\mathbf{\Sigma^0_n}$-subsets of X^ω.

There is a nice characterization of $\mathbf{\Pi^0_2}$-subsets of X^ω. First define the notion of W^δ:

Definition 8. *For $W \subseteq X^\star$, let:*
$W^\delta = \{\sigma \in X^\omega / \exists^\omega i$ such that $\sigma[i] \in W\}$.
($\sigma \in W^\delta$ iff σ has infinitely many prefixes in W).

Then we can state the following Proposition:

Proposition 9. *A subset L of X^ω is a $\mathbf{\Pi^0_2}$-subset of X^ω iff there exists a set $W \subseteq X^\star$ such that $L = W^\delta$.*

Mc Naughton's Theorem implies that every ω-regular language is a boolean combination of G_δ-sets, hence a $\mathbf{\Delta^0_3} = (\mathbf{\Pi^0_3} \cap \mathbf{\Sigma^0_3})$-set. This result holds in fact for every ω-language accepted by a deterministic $\mathbf{X}$-automaton in the sense of [EH93], i.e. an automaton equipped with a storage type $\mathbf{X}$, including the case of the Turing machine. A way to distinguish the expressive power of finite machines reading ω-words is the Wadge hierarchy which we now introduce.

Definition 10. *For $E \subseteq X^\omega$ and $F \subseteq Y^\omega$, E is said to be Wadge reducible to F ($E \leq_W F$) iff there exists a continuous function $f : X^\omega \to Y^\omega$, such that $E = f^{-1}(F)$.*

E and F are Wadge equivalent iff $E \leq_W F$ and $F \leq_W E$. This will be denoted by $E \equiv_W F$. And we shall say that $E <_W F$ iff $E \leq_W F$ but not $F \leq_W E$.

A set $E \subseteq X^\omega$ is said to be self dual iff $E \equiv_W (X^\omega - E)$, and otherwise it is said to be non self dual.

The relation $\leq_W$ is reflexive and transitive, and $\equiv_W$ is an equivalence relation.

The equivalence classes of $\equiv_W$ are called wadge degrees.

WH is the class of Borel subsets of finite rank of a set X^ω, where X is a finite set, equipped with $\leq_W$ and with $\equiv_W$.

For $E \subseteq X^\omega$ and $F \subseteq Y^\omega$, if $E \leq_W F$ and $E = f^{-1}(F)$ where f is a continuous function from X^ω into Y^ω, then f is called a continuous reduction of E to F. Intuitively it means that E is less complicated than F because to check whether $x \in E$ it suffices to check whether $f(x) \in F$ where f is a continuous function. Hence the Wadge degree of an ω-language is a measure of its topological complexity.

Remark 11. *In the above definition, we consider that a subset $E \subseteq X^\omega$ is given together with the alphabet X. This is necessary as it is shown by the following example.*

Let $E = \{0,1\}^\omega$ considered as an ω-language over the alphabet $X = \{0,1\}$ and let $F = \{0,1\}^\omega$ be the same ω-language considered as an ω-language over the alphabet $Y = \{0,1,2\}$. Then E is an open and closed subset of $\{0,1\}^\omega$ but F is a closed and non open subset of $\{0,1,2\}^\omega$. It is easy to check that $E <_W F$ hence E and F are not Wadge equivalent.

Then we can define the Wadge class of a set F:

Definition 12. *Let F be a subset of X^ω. The wadge class of F is $[F]$ defined by: $[F] = \{E/E \subseteq Y^\omega$ for a finite alphabet Y and $E \leq_W F\}$.*

Recall that each Borel class $\mathbf{\Sigma}^0_\mathbf{n}$ and $\mathbf{\Pi}^0_\mathbf{n}$ is a Wadge class.

There is a close relationship between Wadge reducibility and games which we now introduce. Define first the Wadge game $W(A, B)$ for $A \subseteq X_A^\omega$ and $B \subseteq X_B^\omega$:

Definition 13. *The Wadge game $W(A, B)$ is a game with perfect information between two players, player 1 who is in charge of A and player 2 who is in charge of B.*
Player 1 first writes a letter $a_1 \in X_A$, then player 2 writes a letter $b_1 \in X_B$, then player 1 writes a letter $a_2 \in X_A$, and so on ...
The two players alternatively write letters a_n of X_A for player 1 and b_n of X_B for player 2.
After ω steps, the player 1 has written an ω-word $a \in X_A^\omega$ and the player 2 has written an ω-word $b \in X_B^\omega$.
The player 2 is allowed to skip, even infinitely often, provided he really write an ω-word in ω steps.

The player 2 wins the play iff $[a \in A \leftrightarrow b \in B]$, *i.e. iff*
$[(a \in A \text{ and } b \in B) \text{ or } (a \notin A \text{ and } b \notin B \text{ and } b \text{ is infinite})]$.

Recall that a strategy for player 1 is a function $\sigma : (X_B \cup \{s\})^\star \to X_A$. And a strategy for player 2 is a function $f : X_A^+ \to X_B \cup \{s\}$.

σ is a winning stategy (w.s.) for player 1 iff he always wins a play when he uses the strategy σ, i.e. when the nth letter he writes is given by $a_n = \sigma(b_1 \ldots b_{n-1})$, where b_i is the letter written by player 2 at step i and $b_i = s$ if player 2 skips at step i.

A winning strategy for player 2 is defined in a similar manner.

Martin's Theorem states that every Gale-Stewart Game $G(X)$ (see [Tho90] [PP98] for more details), with X a borel set, is determined and this implies the following:

Theorem 14 (Wadge). *Let* $A \subseteq X_A^\omega$ *and* $B \subseteq X_B^\omega$ *be two Borel sets, where* X_A *and* X_B *are finite alphabets. Then the Wadge game* $W(A, B)$ *is determined: one of the two players has a winning strategy. And* $A \leq_W B$ *iff the player 2 has a winning strategy in the game* $W(A, B)$.

Recall that a set X is well ordered by a binary relation $<$ iff $<$ is a linear order on X and there is not any strictly decreasing (for $<$) infinite sequence of elements in X.

Theorem 15 (Wadge). *Up to the complement and* $\equiv_W$, *the class of Borel subsets of finite rank of* X^ω, *for* X *a finite alphabet, is a well ordered hierarchy. There is an ordinal* $|WH|$, *called the length of the hierarchy, and a map* d_W^0 *from* WH *onto* $|WH|$, *such that for all* $A, B \in WH$:
$d_W^0 A < d_W^0 B \leftrightarrow A <_W B$ *and*
$d_W^0 A = d_W^0 B \leftrightarrow [A \equiv_W B \text{ or } A \equiv_W B^-]$.

Remark 16. *We do not give here the ordinal* $|WH|$. *Details may be found in [Dup99a].*

4 Wagner Hierarchy and Its Extension to Blind Counter Automata

Consider now ω-regular languages. Landweber studied first the topological properties of ω-regular languages. He characterized the ω-regular languages in each of the Borel classes $\mathbf{F}, \mathbf{G}, \mathbf{F}_\sigma, \mathbf{G}_\delta$, and showed that one can decide, for an effectively given ω-regular language L, whether L is in $\mathbf{F}, \mathbf{G}, \mathbf{F}_\sigma$, or $\mathbf{G}_\delta$.

It turned out that an ω-regular language is in the class $\mathbf{G}_\delta$ iff it is accepted by a deterministic Büchi automaton. These results were refined by K. Wagner who studied the Wadge Hierarchy of ω-regular languages. In fact there is an effective version of the Wadge Hierarchy restricted to ω-regular languages:

Theorem 17 (Corollary of Büchi-Landweber's Theorem [BL69]). *For* A *and* B *some* ω*-regular sets, one can effectively decide which player has a w.s. in the game* $W(A, B)$ *and the winner has a w.s. given by a transducer.*

The hierarchy obtained on ω-regular languages is now called the Wagner hierarchy and has length ω^ω. Wagner [Wag79] gave an automata structure characterization, based on notion of chain and superchain, for an automaton to be in a given class and showed that the Wadge degree of an ω-regular language is computable. Wilke and Yoo proved in [WY95] that this can be done in polynomial time. Wagner's hierarchy has been recently studied by Carton and Perrin in connection with the theory of ω-semigroups [CP97] [CP98] [PP98] and by Selivanov in [Sel98].

We present in this paper an extension of the Wagner hierarchy to the class of blind counter ω-languages, using analogous notions of chains and superchains. We shall first define positive and negative loops, next chains and superchains. A crucial fact which allows this definition is the following lemma:

Lemma 18. *Let $\mathcal{A} = (K, \Sigma, \Gamma, \delta, q_0, Z_0, \mathcal{F})$ be a MBCA and $x \in \Sigma^\omega$ such that there exists an infinite run $r = (q_i, I^{n_i} Z_0)_{i \geq 1}$ of $\mathcal{A}$ over x such that $Inf(r) = F \subseteq K$. Then there exist infinitely many integers i such that for all $j \geq i$, $n_j \geq n_i$. Among these integers there exist infinitely many integers i_k, $k \geq 1$, and a state $q \in K$ such that for all $k \geq 1$, $q_{i_k} = q$. Then there exist two integers s, s' such that between steps i_s and $i_{s'}$ of the run r, $\mathcal{A}$ enters in every state of F and in not any other state of K, because $Inf(r) = F$.*

Proof. With the hypotheses of the lemma, assume that $r = (q_i, I^{n_i} Z_0)_{i \geq 1}$ is an infinite run of M over x. If there exist only finitely many integers i such that for all $j \geq i$, $n_j \geq n_i$, then there exists a largest one l. But then if j_0 is an integer $> l$ there exists an integer $j_1 > j_0$ such that $n_{j_1} < n_{j_0}$. By induction one could construct a sequence of integers $(j_k)_{k \geq 0}$ such that for all k, $n_{j_{k+1}} < n_{j_k}$. This would lead to a contradiction because every integer n_i is positive.

Then there exist infinitely many integers i such that $\forall j \geq i$, $n_j \geq n_i$. The set of states is finite, hence there exists a state $q \in K$ and infinitely many such integers i_k, $k \geq 1$, such that for all $k \geq 1$, $q_{i_k} = q$ and $n_{i_k} > 0$ or for all $k \geq 1$, $q_{i_k} = q$ and $n_{i_k} = 0$. Now if $Inf(r) = F$, the states not in F occur only finitely many times during run r thus there exist two integers $s < s'$ such that the set of states $\mathcal{A}$ enters between steps i_s and $i_{s'}$ of the run r is exactly F.

Remark 19. *The proof of Lemma 18 relies on a simple property of local minima of functions mapping natural numbers to themselves. A similar argument is due to Linna [Lin77].*

Then we shall write

(a) $(q, I) \overset{F^\star}{\mapsto} (q, I^+)$ if $n_{i_s} > 0$ and $n_{i_{s'}} > n_{i_s}$

(b) $(q, I) \overset{F^\star}{\mapsto} (q, I^=)$ if $n_{i_s} > 0$ and $n_{i_{s'}} = n_{i_s}$

(c) $(q, Z_0) \overset{F^\star}{\mapsto} (q, Z_0)$ if $n_{i_s} = 0$ and $n_{i_{s'}} = 0$

The set F is said to be an essential set (of states) and we shall say that in the case (a) there exists a loop $L(q, I, F, +)$, in the case (b) there exists a loop

$L(q, I, F, =)$, in the case (c) there exists a loop $L(q, Z_0, F, =)$. Such a loop is positive if $F \in \mathcal{F}$ and it is negative if $F \notin \mathcal{F}$. We then denote the loop $L(q, I, F, =)$ by $L^+(q, I, F, =)$ or $L^-(q, I, F, =)$ and similarly in the other cases.

Lemma 20. *The set of essential sets and the set of positive and negative loops of a MBCA is effectively computable.*

This follows from the decidability of the emptiness problem for context free languages accepted by pushdown automata.

We assume now some familiarity with the Wagner hierarchy as presented in [Wag79] [Sta97a]. The next step is to define, following Wagner's study, the (alternating) chains. Let E^+ (respectively E^-) be the set of essential sets in $\mathcal{F}$ (respectively not in $\mathcal{F}$). An alternating chain of length n is in the form

$$F_1 \subset F_2 \subset F_3 \subset \ldots F_n$$

where $F_i \in E^+$ iff $F_{i+1} \in E^-$ for $1 \leq i < n$. It is a positive chain if $F_1 \in E^+$ and a negative chain if $F_1 \in E^-$.

As in the case of Muller automata [Sta97a], one can see that if F is a maximal essential set then all (alternating) chains of maximal length contained in F have the same sign (positive or negative) because in every chain of maximal length contained in F one can replace the last essential set by F itself. Let then $l(F)$ be the maximal length of chains contained in F and $s(F)$ be the sign of these chains.

We now define the first invariant of the MBCA $\mathcal{A}$ as $\mathrm{m}(\mathcal{A})$ being the maximal length of chains of essential sets. Lemma 18 is crucial because it makes every essential set F_i of a chain $F_1 \subset F_2 \subset F_3 \subset \ldots F_n$ to be indefinitely reachable from (q, I) (respectively (q, Z_0)) if there exists a loop $L(q, I, F_n, + \textit{ or } =)$, (respectively $L(q, Z_0, F_n, =)$).

The great difference between the case of Muller automata and the case of MBCA comes with the notion of superchain. Briefly speaking in a MA $\mathcal{A}$ a superchain of length n is a sequence $S_1, \ldots, S_n$ of chains of length $\mathrm{m}(\mathcal{A})$ such that for every integer i, $1 \leq i < n$, S_{i+1} is reachable from S_i and S_{i+1} is positive iff S_i is negative. In the case of MA, S_i cannot be reachable from S_{i+1} otherwise there would exist a chain of length $>\mathrm{m}(\mathcal{A})$.

But in the case of MBCA, in such a superchain, S_i may be reachable from S_{i+1} but **with a reachability which is limited by the counter**. This leads to the notion of superchains of length ω, where ω is the first infinite ordinal, and next of length α where α is an ordinal $< \omega^2$.

An example of a MBCA $\mathcal{A}$ with $\mathrm{m}(\mathcal{A}) = m$ and a superchain of length ω is obtained from two MA $\mathcal{B}$ and $\mathcal{B}'$ such that the graph of $\mathcal{B}$ is just constituted by a positive chain of length m with a maximal essential set $F_m = \{q_1, \ldots q_m\}$ and the graph of $\mathcal{B}'$ is just constituted by a negative chain of length m with a maximal essential set $F'_m = \{q'_1, \ldots q'_m\}$. The behaviour of the MBCA $\mathcal{A}$ is as follows: at the beginning of an infinite run, the counter may be increased up to a counter value N; then there exist transitions from state q_1 to q'_1 and conversely from state q'_1 to q_1 but these transitions make the counter value decrease. Moreover

$\mathcal{A}$ has also the transitions of the two MA $\mathcal{B}$ and $\mathcal{B}'$ but these transitions do not change the counter value. Then one can see thet after a first transition from state q_1 to q_1' or from q_1' to q_1 the number of such transitions is bounded by the counter value N, but this initial value may be chosen $> n_0$ where n_0 is any given integer.

Let then $\mathcal{A}$ be a MBCA such that m($\mathcal{A}$)= m and such that $\mathcal{A}$ has positive and negative chains of length m. A superchain of length ω is formed by two maximal loops $L^+(q, I, F_m, + \textit{ or } =)$ and $L^-(q', I, F'_m, + \textit{ or } =)$ of such chains, i.e. F_m is the last element of a positive chain of length m and F'_m is the last element of a negative chain of length m; moreover, for all $p_0 > 1$, configurations $(q, I^p Z_0)$ are reachable for integers $p > p_0$, and there exist transitions implying that

$$(q, I^p Z_0) \mapsto^\star (q', I^{p'} Z_0) \mapsto^\star (q, I^{p''} Z_0)$$

for some integers p, p', p''. the MBCA $\mathcal{A}$ having not any chain of length $> m$, it holds that $p'' < p$, because otherwise there would exist an essential set $F \supseteq F_m \cup F'_m$ and then there would exist a chain of length $> m$. And the loop $L^+(q, I, F_m, + \textit{ or } =)$ is in fact $L^+(q, I, F_m, =)$ and similarly $L^-(q', I, F'_m, + \textit{ or } =)$ is $L^-(q', I, F'_m, =)$

One can informally say that F_m is reachable from F'_m and conversely but after such transitions the counter value has decreased hence there is a limitation to this reachability.

Lemma 21. *The set of superchains of length ω of a MBCA is effectively computable.*

Now one can define superchains of length $\omega.p$ for an integer $p \geq 1$. Informally speaking a superchain of length $\omega.p$ is a sequence $\Omega_1, \ldots, \Omega_p$ of superchains of length ω such that any state q of an essential set of Ω_{i+1} is reachable with unbounded values of the counter from any state of an essential set of Ω_i. It is now easy to define superchains of length $\omega.p + s \geq 1$, (with p, s some integers ≥ 0), which are a sequence of a superchain of length s followed by a superchain of length $\omega.p$.

In the case $s > 0$, the superchain is said to be positive if it begins with a positive chain and it is said to be negative if it begins with a negative chain.

In the case $s = 0$, we consider now that a superchain: $\Omega_1, \ldots, \Omega_p$, of length $\omega.p$, is given with a loop L. Then it is said to be positive (respectively, negative) if Ω_1 is formed by two maximal loops $L^+(q, I, F_m, =)$ and $L^-(q', I, F'_m, =)$ of chains of length m($\mathcal{A}$)= m and configurations $(q, I^p Z_0)$ are reachable for unbounded values of $p \geq 1$ from the positive loop L (respectively, from the negative loop L).

We define now the second invariant of the MBCA $\mathcal{A}$ as n($\mathcal{A}$) being the maximal length of superchains (n($\mathcal{A}$) $< \omega^2$). The MBCA is said to be prime if all superchains of length n($\mathcal{A}$) have the same sign, i.e. all are positive or all are negative. Denote s($\mathcal{A}$)= 0 if $\mathcal{A}$ is not prime, s($\mathcal{A}$)= 1 if all longest superchains are positive, and s($\mathcal{A}$)= -1 if all longest superchains are negative.

Lemma 22. *Let $\mathcal{A}$ be a MBCA. Then n($\mathcal{A}$) and s($\mathcal{A}$) are computable. Moreover the set of superchains of length n($\mathcal{A}$) is computable.*

We can now follow Wagner's study and define for α an ordinal $< \omega^2$ and m an integer ≥ 1:
$C_m^\alpha = \{L(\mathcal{A})$ / s($\mathcal{A}$)= 1 and m($\mathcal{A}$)= m and n($\mathcal{A}$)=$\alpha\}$
$D_m^\alpha = \{L(\mathcal{A})$ / s($\mathcal{A}$)= -1 and m($\mathcal{A}$)= m and n($\mathcal{A}$)=$\alpha\}$
$E_m^\alpha = \{L(\mathcal{A})$ / s($\mathcal{A}$)= 0 and m($\mathcal{A}$)= m and n($\mathcal{A}$)=$\alpha\}$

Using the Wadge game, one can now show that each class C_m^α or D_m^α defines a Wadge degree, i.e. all ω-languages in the same class C_m^α or D_m^α are Wadge equivalent. In other words C_m^α and D_m^α are the restrictions to the class **BC** of some Wadge degrees.

Moreover when $\alpha = n$ is an integer, this degree corresponds to the degree obtained in the Wagner hierarchy for the classes C_m^n or D_m^n.

The classes C_m^α, D_m^α, and E_m^α, for m an integer ≥ 1 and α a non null ordinal $< \omega^2$, form the coarse structure of the Wadge hierarchy of **BC**. It is a strict extension of the coarse structure of the Wagner hierarchy studied in [Wag79] and it satisfies the following Theorem.

Theorem 23. *Let $\mathcal{A}$ and $\mathcal{B}$ be two MBCA accepting the ω-languages $L(\mathcal{A})$ and $L(\mathcal{B})$. Then it holds that:*

1. *If m($\mathcal{A}$) < m($\mathcal{B}$), then $L(\mathcal{A}) <_W L(\mathcal{B})$.*
2. *If m($\mathcal{A}$) = m($\mathcal{B}$), and n($\mathcal{A}$) < n($\mathcal{B}$), then $L(\mathcal{A}) <_W L(\mathcal{B})$.*
3. *If m($\mathcal{A}$) = m($\mathcal{B}$), n($\mathcal{A}$) = n($\mathcal{B}$), s($\mathcal{A}$) = 1 or s($\mathcal{A}$) = -1, and s($\mathcal{B}$) = 0, then $L(\mathcal{A}) <_W L(\mathcal{B})$.*
4. *If m($\mathcal{A}$) = m($\mathcal{B}$), n($\mathcal{A}$) = n($\mathcal{B}$), s($\mathcal{A}$) = 1 and s($\mathcal{B}$) = -1, then $L(\mathcal{A})$ and $L(\mathcal{B})$ are non self dual and $L(\mathcal{A}) \equiv_W L(\mathcal{B})^-$.*

From this Theorem one can easily infer that the integer m($\mathcal{A}$), the ordinal n($\mathcal{A}$), and s($\mathcal{A}$) $\in \{-1, 0, 1\}$, are invariants of the ω-language $L(\mathcal{A})$ and not only of the MBCA $\mathcal{A}$:

Corollary 24. *Let $\mathcal{A}$ and $\mathcal{B}$ be two MBCA accepting the same ω-language $L(\mathcal{A}) = L(\mathcal{B})$. Then m($\mathcal{A}$) = m($\mathcal{B}$), n($\mathcal{A}$) = n($\mathcal{B}$), and s($\mathcal{A}$) = s($\mathcal{B}$).*

One can give a canonical member in each of the classes C_m^α, D_m^α, and E_m^α, for m an integer ≥ 1 and α a non null ordinal $< \omega^2$. And one can easily deduce that the length of the coarse structure of the Wadge hierarchy of blind counter ω-languages is the ordinal ω^3, while the length of the coarse structure of the Wagner hierarchy was the ordinal ω^2.

The coarse structure of the class **BC** is effective but it is not exactly the Wadge hierarchy of **BC**, because each class E_m^α is the union of countably many (restrictions of) Wadge degrees. We can next define a sort of derivation as Wagner did for Muller automata.

Two MBCA $\mathcal{A}$ and $\mathcal{B}$ in the same class E_m^α have essentially the same "most difficult parts" because they have positive and negative superchains of length

$n(\mathcal{A}) = n(\mathcal{B})$. Hence, in the case of Muller automata (then α is an integer), Wagner's idea was to cut off the superchains of length $n(\mathcal{A}) = n(\mathcal{B})$ of $\mathcal{A}$ and $\mathcal{B}$; this way one get some new automata $\partial\mathcal{A}$ and $\partial\mathcal{B}$ which are called the derivations of $\mathcal{A}$ and $\mathcal{B}$ and the comparison of $\mathcal{A}$ and $\mathcal{B}$ with regard to $\leq_W$ is reduced to the comparison of their derivations $\partial\mathcal{A}$ and $\partial\mathcal{B}$.

In the case of MBCA one do as in the case of MA but with some modification. We first define the derivation $\partial\mathcal{A}$ of a MBCA in E_m^α: $\mathcal{A} = (K, \Sigma, \Gamma, \delta, q_0, Z_0, \mathcal{F})$ as follows.

Let ∂K be the set of states in K from which some positive **and** some negative superchains of length $n(\mathcal{A})$ are reachable. In fact for each such $q \in \partial K$, it may exist an integer n_q such that positive **and** negative superchains of length $n(\mathcal{A})$ are reachable only from configurations $(q, I^n Z_0)$ with $n \geq n_q$. And these integers n_q are effectively computable. Let us define now

$$\partial\mathcal{A} = (\partial K, \Sigma, \Gamma = \{I, Z_0\}, \partial\delta, q_0, Z_0, \partial\mathcal{F})$$

where $\partial\delta$ is defined by:

for each $q \in \partial K$, $a \in \Sigma$, $Z \in \Gamma$:
$\partial\delta(q, a, Z) = \delta(q, a, Z)$ if $\delta(q, a, Z) = (p, \gamma)$ for some $\gamma \in \Gamma^\star$ and $p \in \partial K$.

Otherwise $\partial\delta(q, a, Z)$ is undefined.

And $\partial\mathcal{F} = \{F \ / \ F \subseteq \partial K \text{ and } F \in \mathcal{F}\}$

We consider now the MBCA $\partial\mathcal{A}$ given with the integers n_q, for $q \in \partial K$. Then we study the loops of $\partial\mathcal{A}$ as above but **we keep only loops in the form** $L(q, Z_0$ **or** $I, F, +or-)$ **such that state** q **is reachable with a counter value** $n \geq n_q$. We can next define chains and superchains for $\partial'\mathcal{A}$=$(\partial\mathcal{A}, (n_q)_{q\in\partial K})$. We define $m(\partial'\mathcal{A})$, $n(\partial'\mathcal{A})$, and $s(\partial'\mathcal{A})$, and it holds that $m(\partial'\mathcal{A}) < m(\mathcal{A})$. We then attribute a class $C^{n(\partial'\mathcal{A})}_{m(\partial'\mathcal{A})}$, $D^{n(\partial'\mathcal{A})}_{m(\partial'\mathcal{A})}$, or $E^{n(\partial'\mathcal{A})}_{m(\partial'\mathcal{A})}$, to $\partial'\mathcal{A}$ as we did for $\mathcal{A}$. It may happen that there does not exist any loop for $\partial'\mathcal{A}$=$(\partial\mathcal{A}, (n_q)_{q\in\partial K})$; in that case we associate the class E to $\partial'\mathcal{A}$. Now we can iterate this process and associate to the MBCA $\mathcal{A}$ a name $N(\mathcal{A})$ which is inductively defined by:

1. If $\mathcal{A}$ is prime and $s(\mathcal{A}) = 1$, then $N(\mathcal{A}) = C^{n(\mathcal{A})}_{m(\mathcal{A})}$.
2. If $\mathcal{A}$ is prime and $s(\mathcal{A}) = -1$, then $N(\mathcal{A}) = D^{n(\mathcal{A})}_{m(\mathcal{A})}$.
3. If $\mathcal{A}$ is not prime then $N(\mathcal{A}) = E^{n(\mathcal{A})}_{m(\mathcal{A})} N(\partial'\mathcal{A})$.

This name depends only on the ω-language $L(\mathcal{A})$ accepted by the MBCA $\mathcal{A}$ and is effectively computable. We can write it in a similar fashion as in Wagner's study: we associate with each blind counter ω-language $L(\mathcal{A})$ in **BC** a name in the form:

$$N(\mathcal{A}) = E^{\alpha_1}_{m_1} \ldots E^{\alpha_k}_{m_k} H^{\alpha_{k+1}}_{m_{k+1}}$$

where $m_1 > m_2 > \ldots > m_k > m_{k+1}$ are integers; each α_i is an ordinal $< \omega^2$; and $H \in \{C, D\}$, or in the form:

$$N(\mathcal{A}) = E^{\alpha_1}_{m_1} \ldots E^{\alpha_k}_{m_k} E$$

which we shall simply denote by

$$N(\mathcal{A}\ ) = E_{m_1}^{\alpha_1} \ldots E_{m_k}^{\alpha_k}$$

where $m_1 > m_2 > \ldots > m_k$ are integers and each α_i is an ordinal $< \omega^2$.

One can show that each such name is really the name of an ω-language in **BC**. And the Wadge relation $\leq_W$ is now computable because of the following result.

Theorem 25. *Let $\mathcal{A}$ and $\mathcal{B}$ be two MBCA accepting the ω-languages $L(\mathcal{A})$ and $L(\mathcal{B})$. Assume that the names associated with the MBCA $\mathcal{A}$ and $\mathcal{B}$ are:*

$$N(\mathcal{A}\ ) = E_{m_1}^{\alpha_1} \ldots E_{m_k}^{\alpha_k} H_{m_{k+1}}^{\alpha_{k+1}}$$

$$N(\mathcal{B}) = E_{m'_1}^{\alpha'_1} \ldots E_{m'_l}^{\alpha'_l} H'^{\alpha'_{l+1}}_{m'_{l+1}}$$

where ($H = E$ or $H = C$ or $H = D$), and ($H' = E$ or $H' = C$ or $H' = D$).

Then $L(\mathcal{A}) \leq_W L(\mathcal{B})$ if there exists an integer $j \leq min(k+1, l+1)$ such that $m_i = m'_i$ and $n_i = n'_i$ for $1 \leq i \leq j$ and one of the two following properties holds.

1. *$j = k+1 \leq l+1$ and $H' = E$ or $H = H'$.*
2. *$j < min(k+1, l+1)$ and $m_{j+1} < m'_{j+1}$ or ($m_{j+1} = m'_{j+1}$ and $\alpha_{j+1} < \alpha'_{j+1}$).*

Then the structure of the Wadge hierarchy of ω-languages in **BC** is completely determined. One can show that a blind counter ω-language $L(\mathcal{A})$, where $\mathcal{A}$ is a MBCA, is in the class $\mathbf{\Delta}_2^0$ iff m($\mathcal{A}$) < 2, i.e. iff the name of $\mathcal{A}$ is in the form C_1^α, D_1^α, or E_1^α, for $\alpha < \omega^2$. Thus the Wadge hierarchy restricted to the class $\mathbf{BC} \cap \mathbf{\Delta}_2^0$ has length ω^2, while the Wadge hierarchy restricted to $REG_\omega \cap \mathbf{\Delta}_2^0$ has length ω. The Wadge hierarchy of $\mathbf{BC} \cap \mathbf{\Delta}_2^0$ is then a great extension of the Wagner hierarchy restricted to the class $\mathbf{\Delta}_2^0$. This phenomenon is still true for larger Wadge degrees and non $\mathbf{\Delta}_2^0$-sets. Considering the length of the whole hierarchy of **BC** we get the following:

Corollary 26. *(a) The length of the Wadge hierarchy of blind counter ω-languages in $\mathbf{\Delta}_2^0$ is ω^2.*

(b) The length of the Wadge hierarchy of blind counter ω-languages is the ordinal ω^ω (hence it is equal to the length of the Wagner hierarchy).

Once the structures of two MBCA $\mathcal{A}$ and $\mathcal{B}$ are determined as well as their names $N(\mathcal{A}\ )$ and $N(\mathcal{B})$ are effectively computed, one can construct winning strategies in Wadge games $W(L(\mathcal{A}), L(\mathcal{B}))$ and $W(L(\mathcal{B}), L(\mathcal{A}))$. These strategies may be defined by blind counter transducers, and this extends Wagner's result to blind counter automata.

5 Concluding Remarks

This extended abstract is still a very summarized presentation of our results, which will need exposition of many other details we could not include in this paper [Fin00a].

We have considered above deterministic real time blind counter automata, which form a subclass of the class of deterministic pushdown automata and of the class of deterministic k-blind counter automata. The Wadge hierarchies of ω-languages in each of these classes have been determined in a non effective way, by other methods, in [Dup99b] [Fin99b] [Fin00b], and these results had been announced in the survey [DFR01]. The Wadge degrees in these hierarchies may be described with similar names

$$N(\mathcal{A}) = E_{m_1}^{\alpha_1} \ldots E_{m_k}^{\alpha_k} H_{m_{k+1}}^{\alpha_{k+1}}$$

where $m_1 > m_2 > \ldots > m_k > m_{k+1}$ are integers ≥ 1 and $H \in \{C, D, E\}$, and

1. each α_i is an ordinal $< \omega^{k+1}$, in the case of **k-blind counter automata.**
2. each α_i is an ordinal $< \omega^{\omega}$, in the case of **deterministic pushdown automata.**

We will further extend the results of the present paper in both directions to get decidability results and effective winning strategies in Wadge games. The above case of (one) blind counter automata already introduces some of the fundamental ideas which we will apply in further cases.

Another problem is to study the complexity of the problem: "determine the Wadge degree of a blind counter ω-language", extending this way the results of Wilke and Yoo to blind counter ω-languages.

Further study would be the investigation of links between the problems of simulation and bisimulation [Jan00] [JKM00] [JMS99] [Kuc00] and the problem of finding winning strategies in Wadge games.

A Wadge game between two blind counter ω-languages, whose complements are also blind counter ω-languages, can easily be reduced to a Gale-stewart game, (see [Tho95] [PP98]), with a winning set accepted by a deterministic 2-blind-counter automaton. This suggests that Walukiewicz's result, the proof of the existence of effective winning strategies in a Gale-stewart game with a winning set accepted by a deterministic pushdown automaton, [Wal96], could be extended to the case of a winning set accepted by a deterministic multi blind counter automata, giving additional results as asked by Thomas in [Tho95].

Acknowledgements

Thanks to Jean-Pierre Ressayre and Jacques Duparc for many helpful discussions about Wadge and Wagner Hierarchies.

Thanks also to the anonymous referees for useful comments on the preliminary version of this paper. In particular the remark 19 is due to one of them.

References

BL69. J.R. Büchi and L. H. Landweber, Solving sequential conditions by finite state strategies. Trans. Amer. Math. Soc. 138 (1969).

CP97. O. Carton and D. Perrin, Chains and Superchains for ω-Rational sets, Automata and semigroups, International Journal of Algebra and Computation Vol. 7, No. 7(1997) p. 673-695.

CP98. O. Carton and D. Perrin, The Wagner Hierarchy of ω-Rational sets, International Journal of Algebra and Computation, vol. 9, no. 5, pp. 597-620, 1999.

Dup99a. J. Duparc, Wadge Hierarchy and Veblen hierarchy: part 1: Borel sets of finite rank, Journal of Symbolic Logic, March 2001.

Dup99b. J. Duparc, A Hierarchy of Context Free Omega Languages, Theoretical Computer Science, to appear. Available from http://www.logigue.jussieu.fr/www.duparc

DFR01. J. Duparc, O. Finkel and J-P. Ressayre, Computer Science and the Fine Structure of Borel Sets, Theoretical Computer Science, Volume 257 (1-2), April 2001, p.85-105.

EH93. J. Engelfriet and H. J. Hoogeboom, X-automata on ω-words, Theoretical Computer Science 110 (1993) 1, 1-51.

Fin99a. O. Finkel, Wadge Hierarchy of Omega Context Free Languages, Theoretical Computer Science, to appear.

Fin99b. O. Finkel, Wadge Hierarchy of Deterministic Omega Context Free Languages, in preparation.

Fin00a. O. Finkel, An Effective Extension of the Wagner Hierarchy to Blind Counter Automata, full version, in preparation.

Fin00b. O. Finkel, Wadge Hierarchy of Petri Net Omega Languages, in preparation.

Gre78. S.A. Greibach, Remarks on Blind and Partially Blind One Way Multicounter Machines, Theoretical Computer Science 7 (1978) p. 311-324.

HU69. J.E. Hopcroft and J.D. Ullman, Formal Languages and their Relation to Automata, Addison-Wesley Publishing Company, Reading, Massachussetts, 1969.

Jan00. P. Jancar, Decidability of bisimilarity for one-counter processes, Information and Computation 158, 2000, pp. 1-17 (Academic Press) (A preliminary version appeared at ICALP'97.)

JKM00. P. Jancar, A. Kucera, and F. Moller, Simulation and Bisimulation over One-Counter Processes, In Proceedings of 17th International Symposium on Theoretical Aspects of Computer Science (STACS 2000), pages 334-345, volume 1770 of LNCS, Springer-Verlag, 2000.

JMS99. P. Jancar, F. Moller and Z. Sawa, Simulation Problems for One-Counter Machines, in Proc. SOFSEM'99 (Milovy, Czech Rep., November 1999), Lecture Notes in Computer Science, Vol. 1725, Springer 1999, pp. 404-413.

Kuc00. A. Kucera, Efficient Verification Algorithms for One-Counter Processes, In Proceedings of 27th International Colloquium on Automata, Languages, and Programming (ICALP 2000), pages 317-328, volume 1853 of LNCS, Springer-Verlag, 2000.

Kur66. K. Kuratowski, Topology, Academic Press, New York 1966.

Lan69. L. H. Landweber, Decision problems for ω-automata, Math. Syst. Theory 3 (1969) 4,376-384.

Lin77. M. Linna, A decidability result for deterministic ω-context-free languages, Theoretical Computer Science 4 (1977), 83-98.

LT94. H. Lescow and W. Thomas, Logical specifications of infinite computations, In: "A Decade of Concurrency" (J. W. de Bakker et al., eds), Springer LNCS 803 (1994), 583-621.

PP98. D. Perrin and J.-E. Pin, Infinite Words, Book in preparation, available from http://www.liafa.jussieu.fr/~jep/InfiniteWords.html

Sel98. V. Selivanov, Fine hierarchy of regular ω-languages, Theoretical Computer Science 191(1998) p.37-59.

Sta86. L. Staiger, Hierarchies of Recursive ω-Languages, Jour. Inform. Process. Cybernetics EIK 22 (1986) 5/6, 219-241.

Sta97a. L. Staiger, ω-languages, Chapter of the Handbook of Formal Languages, Vol 3, edited by G. Rozenberg and A. Salomaa, Springer-Verlag, Berlin, 1997.

SW74. L. Staiger and K. Wagner, Automatentheoretische und Automatenfreie Charakterisierungen Topologischer Klassen Regulärer Folgenmengen. Elektron. Informationsverarb. Kybernetik EIK 10 (1974) 7, 379-392.

Tho90. W. Thomas, Automata on Infinite Objects, in: J. Van Leeuwen, ed., Handbook of Theorical Computer Science, Vol. B (Elsevier, Amsterdam, 1990), p. 133-191.

Tho95. W. Thomas, On the synthesis of strategies in infinite games, in STACS'95, Volume 900 of LNCS, p.1-13, 1995.

Wad84. W.W. Wadge, Ph. D. Thesis, Berkeley, 1984.

Wag79. K. Wagner, On Omega Regular Sets, Inform. and Control 43 (1979) p. 123-177.

Wal96. I. Walukiewicz, Pushdown Processes: Games and Model Checking, Information and Computation 164 (2) p. 234-263, 2001.

WY95. Th. Wilke and H. Yoo, Computing the Wadge Degree, the Lifschitz Degree and the Rabin Index of a Regular Language of Infinite Words in Polynomial Time, in: TAPSOFT' 95: Theory and Practice of Software Development (eds. P.D. Mosses, M. Nielsen and M.I. Schwartzbach), L.N.C.S. 915, p. 288-302, 1995.

Decision Procedure for an Extension of WS1S

Felix Klaedtke

Institut für Informatik, Albert-Ludwigs-Universität
Georges-Köhler-Allee 52
79085 Freiburg i. Br., Germany
Phone: +49 (761) 203-8245
Fax: +49 (761) 203-8241
klaedtke@informatik.uni-freiburg.de

Abstract. We define an extension of the weak monadic second-order logic of one successor (WS1S) with an infinite family of relations and show its decidability. Analogously to the decision procedure for WS1S, automata are used. But instead of using word automata, we use tree automata that accept or reject words. In particular, we encode a word in a complete leaf labeled tree and restrict the acceptance condition for tree automata to trees that encode words. As applications, we show how this extension can be applied to reason automatically about parameterized families of combinational tree-structured circuits and used to solve certain decision problems.

Keywords: tree automata, word languages, weak monadic second-order logic of one successor, WS1S

1 Introduction

The tight relation between automata and logics has been used to show the decidability of several decision problems for monadic second-order logics. For instance, the decidability of the *weak monadic second-order logic of one successor, WS1S*, was shown using word automata [5,11,24], and the decidability of the *weak monadic second-order logic of two successors, WS2S*, was shown using tree automata [22,8]. Although the theory of WS1S is non-elementary decidable [20], the *Mona system* [16,9] – decision procedures for WS1S and WS2S – has proved useful and efficient in practice, e.g. [3,1,10].

The decidability result presented here, exploits – analogously to the decision procedures for WS1S and WS2S – the relation between automata and monadic second-order logics. But instead of using word automata that read words and tree automata that read trees, we use *tree* automata to describe *word* languages. In particular, we encode a word in a complete leaf labeled tree and restrict the acceptance condition of tree automata to this subset of trees. The set of word languages that can be described by tree automata is a proper superset of the regular word languages and has useful properties, e.g., it is effectively closed under the Boolean operations and the emptiness problem is decidable. We show that, as a result, there is a decidable extension of WS1S with an infinite family of relations.

L. Fribourg (Ed.): CSL 2001, LNCS 2142, pp. 384–398, 2001.

We present two applications of this extension. First, we show how the extension can be used to reason automatically about parameterized families of combinational tree-structured circuits. Second, we show how results that where previously established using other techniques, naturally fall out of our formalism.

For example, the subset $\mathbb{P}$ of the natural numbers that are a power of two, is a predicate that is included in the family of relations. The decidability of the theory of the weak monadic second-order logic over the structure $(\mathbb{N}, \text{succ}, \mathbb{P})$ was already shown by Elgot and Rabin in [12]. They construct an ω-word automaton $\mathcal{A}$, for a sentence φ of the monadic second-order logic over of $(\mathbb{N}, \text{succ}, \mathbb{P})$, treating the predicate symbol $\mathbb{P}$ as a free second-order variable. They show that it is decidable if $\mathcal{A}$ accepts the characteristic ω-word $x_{\mathbb{P}}$ (the ith letter of $x_{\mathbb{P}}$ is 1 iff $i \in \mathbb{P}$).[1] The decidability of the theory of the weak monadic over the structure $(\mathbb{N}, \text{succ}, \mathbb{P})$ follows immediately from the fact that finiteness of a subset of natural numbers can be expressed in the *monadic second-order logic of one successor, S1S.* This decision procedure only holds for closed formulas and is based on the decidability of S1S, i.e., it uses ω-word automata.

Using tree automata to decide the theory of the weak monadic second-order logic of $(\mathbb{N}, \text{succ}, \mathbb{P})$ has advantages in implementing a decision procedure. First, we avoid the algorithmically difficult constructions of ω-word automata, like complementing Büchi word automata. Second, tree automata can be efficiently minimized (the author is unaware of any implementation that minimizes ω-automata efficiently in general). Third, we can use an existing implementation of tree automata, such as in the Mona system [4]. Further, we can implement a decision procedure that returns a counter-example and a fulfilling substitution for the free variables of a formula.

We proceed as follows. In §2 we provide background material. In §3 we define how tree automata can be used to describe word languages and prove properties about the set of word languages characterized by tree automata. In §4 we show the decidability of our extension of WS1S. We also apply the extension to reason about parameterized families of combinational tree-structured circuits. Finally, in §5 we discuss future work.

2 Background

2.1 Words and Trees

Σ^* is the set of all words over the alphabet Σ. We write λ for the *empty word.* For $u, v \in \Sigma^*$, the *concatenation* of u and v is written as uv, and $|u|$ denotes u's *length.* Let $\mathbb{B}$ be the alphabet $\{0, 1\}$.

A *(full binary) Σ-labeled tree* is a function t where the range of t is Σ and the domain of t, $\text{dom}(t)$, is a finite nonempty subset of $\mathbb{B}^*$ where (i) $\text{dom}(t)$ is prefix-closed, and (ii) $u0 \in \text{dom}(t)$ iff $u1 \in \text{dom}(t)$. The elements of $\text{dom}(t)$ are called *nodes*, and $\lambda \in \text{dom}(t)$ is called the *root.* The node $ub \in \text{dom}(t)$, with

[1] More recently, in [6] it was shown that for any *morphic predicate* $P \subseteq \mathbb{N}$ the problem if an ω-word automaton accepts the ω-word x_P is decidable.

$b \in \mathbb{B}$, is a *successor* of u. A node is an *inner node* if it has successors and is a *leaf* otherwise. A node $u \in \mathrm{dom}(t)$ is *Γ-labeled*, for $\Gamma \subseteq \Sigma$, if $t(u) \in \Gamma$.

The *height* of t is $|t| = \max\{|u| \mid u \in \mathrm{dom}(t)\}$. The *subtree* of t with the root $u \in \mathrm{dom}(t)$ is denoted by t_u; it is $\mathrm{dom}(t_u) = \{v \mid uv \in \mathrm{dom}(t)\}$ and $t_u(v) = t(uv)$. The *frontier* of t is the word in Σ^* where the ith letter is the label of the ith leaf in t (from the left). A tree is *complete* if all its leaves have the same length.

2.2 Word and Tree Automata

We briefly recall the definitions of nondeterministic word and tree automata.

A *(nondeterministic) word automaton* $\mathcal{A}$ is a tuple $(\Sigma, Q, Q_0, F, \delta)$, where Σ is a nonempty finite alphabet, Q is a nonempty finite set of states, $Q_0 \subseteq Q$ is a set of initial states, $F \subseteq Q$ is a set of final states, and $\delta : Q \times \Sigma \to \mathcal{P}(Q)$ is a transition function. A *run* of $\mathcal{A}$ on a word $w = a_1 \ldots a_n \in \Sigma^*$ is a nonempty word $\pi = s_1 \ldots s_{n+1} \in Q^*$ with $s_1 \in Q_0$ and $s_{i+1} \in \delta(s_i, a_i)$ for $1 \leq i \leq n$. π is *accepting* if $s_{n+1} \in F$. $\mathcal{A}$ *accepts* w if there is an accepting run of $\mathcal{A}$ on w; the *accepted language* of $\mathcal{A}$, $L(\mathcal{A})$, is the set of words over Σ that $\mathcal{A}$ accepts.

A *(nondeterministic binary top-down) tree automaton* is defined analogously: $\mathcal{A}$ is a tuple $(\Sigma, Q, Q_0, F, \delta)$, where Σ, Q, Q_0, and F are as above. The transition function is $\delta : Q \times \Sigma \to \mathcal{P}(Q \times Q)$. A *run* of $\mathcal{A}$ on a Σ-labeled tree t is a Q-labeled tree π, where $\mathrm{dom}(\pi) = \{\lambda\} \cup \{ub \mid u \in \mathrm{dom}(t) \text{ and } b \in \mathbb{B}\}$. Moreover, $\pi(\lambda) \in Q_0$ and for $u \in \mathrm{dom}(t)$, $(\pi(u0), \pi(u1)) \in \delta(\pi(u), t(u))$. The run π is *accepting* if $\pi(u) \in F$ for any leaf $u \in \mathrm{dom}(\pi)$. $\mathcal{A}$ *accepts* t if there is an accepting run of $\mathcal{A}$ on t; $\mathcal{A}$ *accepts t from $q \in Q$* if $(\Sigma, Q, \{q\}, F, \delta)$ accepts t. The *accepted language* of $\mathcal{A}$ is the set of Σ-labeled trees that $\mathcal{A}$ accepts.

Word automata and tree automata recognize the regular word and tree languages, and are effectively closed under intersection, union, complement and projection. For a detailed account of regular word and tree languages see [17] and [13,7], respectively.

3 Tree Automata over Words

A natural way to define an acceptance condition for a tree automaton $\mathcal{A}$ for words is the following: $\mathcal{A}$ *accepts* the word w if there is a tree t with frontier w and $\mathcal{A}$ accepts t. It is known that the set of word languages that are described in this way are exactly the context-free languages (see [13]). The context-free languages are not closed under intersection and complement. In the following, we characterize a set of word languages which is effectively closed under the Boolean operations, contains the regular word languages, and for which the emptiness problem is decidable. For doing so, we restrict the input of tree automata to complete leaf labeled trees.

Let Σ be a nonempty finite alphabet, and for the remainder of the text let $\Box$ and $\#$ be two new symbols, i.e. $\Box, \# \notin \Sigma$. We use the notation $\Sigma_\Box$ for $\Sigma \cup \{\Box\}$, $\Sigma_\#$ for $\Sigma \cup \{\#\}$, and $\Sigma_{\Box,\#}$ for $\Sigma \cup \{\Box, \#\}$. A *Σ-leaf labeled tree t* is a $\Sigma_\#$-labeled tree where all inner nodes are labeled with $\#$ and all leaves are Σ-labeled. The complete $\Sigma_\Box$-leaf labeled tree t where the frontier is $w\Box^k$ (for some $k \geq 0$) is an *input tree* of the word $w \in \Sigma^*$. t is *minimal* if $k = \min\{m \mid |w| + m \text{ is a power of } 2\}$. Note that the minimal input tree of a word in Σ^* is unique.

3.1 Acceptance Conditions

We define two different conditions for tree automata for accepting words. We will show in Lemma 3 that both are related, i.e., describe the same set of word languages. We will switch between the two acceptance conditions, for proving properties (e.g. Lemma 6 and Theorem 9) about this set of word languages.

Definition 1. *Let $\mathcal{A} = (\Sigma_{\Box,\#}, Q, Q_0, F, \delta)$ be a tree automaton.*

i) $\mathcal{A}$ star-accepts *the word $w \in \Sigma^*$ if $\mathcal{A}$ accepts an input tree of w. $W_*(\mathcal{A}) \subseteq \Sigma^*$ denotes the set of words that $\mathcal{A}$ star-accepts.*
ii) $\mathcal{A}$ minimal-accepts *the word $w \in \Sigma^*$ if $\mathcal{A}$ accepts the minimal input tree of w. $W(\mathcal{A}) \subseteq \Sigma^*$ denotes the set of words that $\mathcal{A}$ minimal-accepts.*

Example 2. Let $\mathcal{A}$ be the tree automaton $(\mathbb{B}_{\Box,\#}, \{q_1, q_0, q_{ok}\}, \{q_1\}, \{q_{ok}\}, \delta)$ with

$$\begin{aligned} &\delta(q_1, 0) = \emptyset && \delta(q_0, 0) = \{(q_{ok}, q_{ok})\} \\ &\delta(q_1, 1) = \{(q_{ok}, q_{ok})\} && \delta(q_0, 1) = \emptyset \\ &\delta(q_1, \Box) = \emptyset && \delta(q_0, \Box) = \emptyset \\ &\delta(q_1, \#) = \{(q_0, q_1)\} && \delta(q_0, \#) = \{(q_0, q_0)\} \end{aligned}$$

and $\delta(q_{ok}, a) = \emptyset$, for $a \in \mathbb{B}_{\Box,\#}$. $\mathcal{A}$ minimal-accepts $w \in \mathbb{B}^*$ iff $w = 0\ldots01$ and $|w|$ is a power of 2. Note that in this example, $W(\mathcal{A}) = W_*(\mathcal{A})$.

The two different acceptance conditions are related as follows:

Lemma 3. *Let $\mathcal{A} = (\Sigma_{\Box,\#}, Q, Q_0, F, \delta)$ be a tree automaton.*

a) We can construct a tree automaton $\mathcal{B}$ with $W(\mathcal{A}) = W_*(\mathcal{B})$.
b) We can construct a tree automaton $\mathcal{C}$ with $W_*(\mathcal{A}) = W(\mathcal{C})$.

Proof (sketch). a) We construct $\mathcal{B}$ such that it accepts the intersection of the set of trees that $\mathcal{A}$ accepts and the set of trees, where the right subtree of the root contains a leaf that is not labeled with $\Box$.

The set of states of the tree automaton $\mathcal{B}$ is $Q \cup \{\overset{*}{q} \mid q \in Q\} \cup \{p_0\}$ where p_0 is new, $\{p_0\}$ is the set of initial states and the set of final states is $F \cup \{\overset{*}{q} \mid q \in F\}$. To define the transition function ρ of $\mathcal{B}$, let $q \in Q$ and $a \in \Sigma_{\Box,\#}$. Then ρ is

$$\rho(q, a) = \delta(q, a)\,,$$

$$\rho(\overset{*}{q}, a) = \begin{cases} \{(\overset{*}{q}_L, q_R) \mid (q_L, q_R) \in \delta(q, a)\} \cup \\ \qquad \{(q_L, \overset{*}{q}_R) \mid (q_L, q_R) \in \delta(q, a)\} & \text{if } a = \#, \\ \emptyset & \text{if } a = \Box, \\ \delta(q, a) & \text{otherwise,} \end{cases}$$

and

$$\rho(p_0, a) = \begin{cases} \{(q_L, \overset{*}{q}_R) \,|\, (q_L, q_R) \in \bigcup_{q \in Q_0} \delta(q, \#)\} & \text{if } a = \#, \\ \bigcup_{q \in Q_0} \delta(q, a) & \text{otherwise.} \end{cases}$$

A simple induction over the height of an input tree t shows that if a node u in an accepting run π of $\mathcal{B}$ on t is labeled with a state $\overset{*}{q}$, then there exists a leaf in the subtree t_u that is Σ-labeled. From the definition of $\delta(p_0, a)$ it follows that if $\mathcal{B}$ accepts an input tree then there is a Σ-labeled leaf in the right subtree of the root. Thus, t is minimal. From the definition of ρ it follows that $W(\mathcal{A}) = W_*(\mathcal{B})$.

b) The idea of the construction of $\mathcal{C}$ is that $\mathcal{C}$ guesses the state from which to minimal-accept the tree that $\mathcal{A}$ accepts, while taking care that this state is reachable from the initial state such that only $\square$s are generated to the right of the word.

The set of states of the tree automaton $\mathcal{C}$ is $P = Q \cup \{(q, T) \,|\, q \in Q \text{ and } T \subseteq Q\}$ and the set of final states is $E = F \cup \{(q, T) \,|\, q \in F \text{ and } T \subseteq F\}$. The transition function ρ of $\mathcal{B}$ is as follows: ρ restricted to $Q \times \Sigma_{\square,\#}$ is identical to δ, and

$$\rho((q,T),\#) = \{(q', (q'', T')) \mid (q', q'') \in \delta(q, \#) \text{ and for every } t \in T, \\ \delta(t, \#) \cap T' \times T' \neq \emptyset\},$$

and for $a \in \Sigma_\square$

$$\rho((q,T),a) = \{(q', (q'', T')) \mid (q', q'') \in \delta(q, a) \text{ and for every } t \in T, \\ \delta(t, \square) \cap T' \times T' \neq \emptyset\}.$$

We say that $\mathcal{C}$ *reaches* the state (q, T) in $h \geq 0$ steps if there exists a run π of $\mathcal{A}$ on the complete $\{\#\}$-labeled tree t of height h with $\pi(\lambda) \in Q_0$, $\pi(0^h) = q$, and $\{\pi(u) \,|\, |u| = h \text{ and } u \neq 0^h\} = T$. The set of initial states is $P_0 = \{(q, T) \,|\, \mathcal{C} \text{ reaches } (q, T) \text{ in } h \geq 0 \text{ steps}\}$. A pigeon-hole argument proves that $\mathcal{C}$ reaches (q, T) in $h \geq 0$ steps iff $\mathcal{C}$ reaches (q, T) in less than $|Q|2^{|Q|} + 1$ steps. Thus, P_0 is computable.

It can be shown by induction over the height of an input tree that $W_*(\mathcal{A}) = W(\mathcal{C})$. We omit it. □

3.2 Relation to Regular Languages

A word language $L \subseteq \Sigma^*$ is *tree-accepted* if $L = W(\mathcal{A})$ for some tree automaton $\mathcal{A}$.

Theorem 4. *The set of regular word languages is a proper subset of the set of word languages that are tree-accepted.*

Proof (sketch). The language in Example 2 is not regular (it is even not context-free). It remains to show that for every word automaton $\mathcal{A} = (\Sigma, Q, Q_0, F, \delta)$

there exists a tree automaton $\mathcal{B}$ with the alphabet $\Sigma_{\Box,\#}$ that tree-accepts the word language of $\mathcal{A}$.

The set of states of $\mathcal{B}$ is $P = (Q \times Q) \cup \{q_{ok}\}$, the set of initial states is $P_0 = \{(q, q') \mid q \in Q_0 \text{ and } q' \in F\}$, the set of final states is $E = \{q_{ok}\}$, and the transition function ρ is as follows: $\rho(q_{ok}, a) = \emptyset$, for $a \in \Sigma_{\#,\Box}$, and

$$\rho((q,q'),a) = \begin{cases} \{((q,p),(p,q')) \mid p \in Q\} & \text{if } a = \#, \\ \{(q_{ok},q_{ok})\} & \text{if } a \in \Sigma \text{ and } q' \in \delta(q,a), \\ \{(q_{ok},q_{ok})\} & \text{if } a = \Box \text{ and } q = q', \\ \emptyset & \text{otherwise.} \end{cases}$$

Intuitively, $\mathcal{B}$ guesses with a state in P_0 the initial and the final state of an accepting run of $\mathcal{A}$ on $w \in \Sigma^*$. From level to level of the minimal input tree of w, $\mathcal{B}$ guesses the missing states of the run of $\mathcal{A}$. Finally, $\mathcal{B}$ checks at the leaves of the minimal input tree if the guesses really correspond to a run of $\mathcal{A}$ on w. □

3.3 Closure Properties

The *projection* of $L \subseteq \Sigma^*$ w.r.t. the equivalence relation $\sim \subseteq \Sigma \times \Sigma$ is the set $L/\!\sim = \{a_1 \dots a_n \in \Sigma^* \mid \text{it exists } b_1 \dots b_n \in L \text{ with } a_i \sim b_i \text{ for all } 1 \leq i \leq n\}$. We can use the standard constructions for tree automata to show the closure under the Boolean operations. The correctness of the construction for complementation follows from the fact that the minimal input tree of a word is unique.

Theorem 5. *The set of tree-accepted word languages is effectively closed under complement, union, and projection.*

Closure under complement and union is used in §4 to handle the connectives $\neg$ and $\vee$ in the weak monadic second-order logic. To handle the existential quantification over finite subsets of $\mathbb{N}$, we need – additionally to the closure under projection – the notion of a right quotient and the closure property stated in the next lemma. The *right quotient* of $L \subseteq \Sigma^*$ w.r.t. $L' \subseteq \Sigma^*$ is $L/L' = \{w \in \Sigma^* \mid \text{there exists a } w' \in L' \text{ with } ww' \in L\}$.

Lemma 6. *Let $\mathcal{A} = (\Sigma_{\Box,\#}, Q, Q_0, F, \delta)$ be a tree automaton and let $\Gamma \subseteq \Sigma$. We can construct a tree automaton $\mathcal{B}$ with $W(\mathcal{B}) = W(\mathcal{A})/\Gamma^*$.*

Proof (sketch). By Lemma 3b) it is sufficient to construct a tree automaton $\mathcal{B}$ with $W_*(\mathcal{B}) = W(\mathcal{A})/\Gamma^*$. We first give the intuition how $\mathcal{B}$ simulates runs of $\mathcal{A}$. Let t be the minimal input tree of the word ww' with $w \in \Sigma^*$ and $w' \in \Gamma^*$. $\mathcal{B}$ guesses the input tree t' of w with $|t| = |t'|$. Note that $\text{dom}(t) = \text{dom}(t')$ and that $t(u) = t'(u) = \#$ for all inner nodes u. $\mathcal{B}$ makes on an inner node u the same transitions as $\mathcal{A}$ on u. Additionally, $\mathcal{B}$ guesses if the subtree t_u is a $\{\Box\}$-leaf labeled tree. For a leaf u with $t'(u) = \Box$, $\mathcal{B}$ guesses the labeling $t(u)$.

Let $\mathcal{B} = (\Sigma_{\Box,\#}, P, Q_0, E, \rho)$ with $P = Q \cup \{\overset{*}{q} \mid q \in Q\}$, $E = F \cup \{\overset{*}{q} \mid q \in F\}$, and for $a \in \Sigma_{\Box,\#}$ and $q \in Q$, let

$$\rho(q,a) = \begin{cases} \bigcup_{b \in \Gamma} \delta(q,b) & \text{if } a = \Box, \\ \delta(q,\#) \cup \{(q_L, \overset{*}{q}_R) \mid (q_L,q_R) \in \delta(q,\#)\} \cup \\ \qquad \{(\overset{*}{q}_L, \overset{*}{q}_R) \mid (q_L,q_R) \in \delta(q,\#)\} & \text{if } a = \#, \\ \delta(q,a) & \text{otherwise,} \end{cases}$$

and

$$\rho(\overset{*}{q},a) = \begin{cases} \emptyset & \text{if } a \in \Sigma, \\ \{(\overset{*}{q}_L, \overset{*}{q}_R) \mid (q_L,q_R) \in \delta(q,a)\} & \text{otherwise.} \end{cases}$$

$W(\mathcal{A})/\Gamma^* \subseteq W_*(\mathcal{B})$: Assume that $ww' \in W(\mathcal{A})$ with $w' \in \Gamma^*$. We have to show that $\mathcal{B}$ star-accepts w. Let t be the minimal input tree of ww', and let t' be the input tree of w with $|t'| = |t|$. $\mathcal{B}$ makes the same transitions on t' as $\mathcal{A}$ on t with the difference that if all the leaves of the subtree t_u for $u \in \text{dom}(t)$ are labeled with $\Box$, then $\mathcal{B}$ guesses the state $\overset{*}{q}$ instead of $q \in Q$.

$W_*(\mathcal{B}) \subseteq W(\mathcal{A})/\Gamma^*$: Assume that $\mathcal{B}$ star-accepts $w \in \Sigma^*$. We have to show that there is a $w' \in \Gamma^*$ such that $\mathcal{A}$ minimal-accepts ww'. Let t be an input tree of w and u a node of t labeled with $\Box$. Note that $\mathcal{B}$ can only guess in u a letter in Γ, if it is in a state $q \in Q$. But by definition of $\delta(q,\#)$, $\mathcal{B}$ must also guess a letter in Γ for the nodes u' that are labeled with $\Box$ and are left of u. Thus, there exists a $w' \in \Gamma^*$ with $ww' \in W(\mathcal{A})$. $\Box$

3.4 Emptiness Problem

The *emptiness problem (for tree automata over words)* is to decide if a tree automaton minimal-accepts some word. We give an algorithm that runs in linear space in the size of the given tree automaton deciding if it star-accepts some word. The emptiness problem can then be decided with Lemma 3a).

For the remainder of this section, let $\mathcal{A} = (\Sigma_{\Box,\#}, Q, Q_0, F, \delta)$ be a tree automaton. To determine if $W_*(\mathcal{A}) = \emptyset$, we define the sets $P^h_\Sigma, P^h_\Box, P^h_{\Sigma\Box} \subseteq Q$ for $h \geq 0$. Intuitively, $q \in P^h_\Sigma$ iff $\mathcal{A}$ accepts from q a complete Σ-leaf labeled tree of height h, $q \in P^h_\Box$ iff $\mathcal{A}$ accepts from q a complete $\{\Box\}$-leaf labeled tree of height h, and $q \in P^h_{\Sigma\Box}$ iff $\mathcal{A}$ accepts from q an input tree of height h. Formally,

$$P^h_\Sigma = \begin{cases} \{q \in Q \mid \delta(q,a) \cap F \times F \neq \emptyset \text{ with } a \in \Sigma\} & \text{if } h = 0, \\ \{q \in Q \mid \delta(q,\#) \cap P^{h-1}_\Sigma \times P^{h-1}_\Sigma \neq \emptyset\} & \text{otherwise,} \end{cases}$$

$$P^h_\Box = \begin{cases} \{q \in Q \mid \delta(q,\Box) \cap F \times F \neq \emptyset\} & \text{if } h = 0, \\ \{q \in Q \mid \delta(q,\#) \cap P^{h-1}_\Box \times P^{h-1}_\Box \neq \emptyset\} & \text{otherwise,} \end{cases}$$

and

$$P^h_{\Sigma\Box} = \begin{cases} P^0_\Sigma \cup P^0_\Box & \text{if } h = 0, \\ P^h_\Sigma \cup \{q \in Q \mid \delta(q,\#) \cap P^{h-1}_{\Sigma\Box} \times P^{h-1}_\Box \neq \emptyset\} & \text{otherwise.} \end{cases}$$

The following lemma can be proven by induction over the height of an input tree.

Lemma 7. *Let q be a state of $\mathcal{A}$. Then*

$$\mathcal{A} \text{ accepts some input tree } t \text{ of height } h \text{ from } q \qquad \text{iff} \qquad q \in P^h_{\Sigma\Box}\,.$$

Lemma 8. *For $q \in Q$ and $h \geq 0$ with $q \in P^h_{\Sigma\Box}$ there exists an h' with $h' \leq |Q|2^{2|Q|}$ and $q \in P^{h'}_{\Sigma\Box}$.*

Proof (sketch). Assume that $h > |Q|2^{2|Q|}$ and h is minimal, i.e., for all $h' < h$, $q \notin P^{h'}_{\Sigma\Box}$. By Lemma 7, there is an input tree t such that $|t| = h$ and $\mathcal{A}$ accepts t from q. We show by an application of the pigeon hole argument that $\mathcal{A}$ accepts an input tree t' from q with $|t'| \leq |Q|2^{2|Q|}$. By Lemma 7, $q \in P^{|t'|}_{\Sigma\Box}$. This contradicts the assumption that h is minimal.

Let π be an accepting run on t. Note that π is a complete Q-labeled tree with $|\pi| = |t|+1$ and each leaf of π is F-labeled. We extend π to subsets of nodes, i.e., for $U \subseteq \text{dom}(\pi)$, $\pi(U) = \{\pi(u) \mid u \in U\}$. The nodes in a level $0 \leq k \leq |t|$ of π can be partitioned into a triple (U_k, w_k, V_k) such that $\pi(U_k) \subseteq P^k_{\Sigma}$, $\pi(w_k) \in P^k_{\Sigma\Box}$ and $\pi(V_k) \subseteq P^k_{\Box}$. This can be shown by induction over k.

Since there are $2^{|Q|}$ many subsets of Q, there are $m, n \leq |Q|2^{2|Q|}$ with $m < n$ such that

$$\pi(U_m) = \pi(U_n)\,, \qquad \pi(w_m) = \pi(w_n)\,, \qquad \text{and} \qquad \pi(V_m) = \pi(V_n)\,.$$

Note that for each node $u \in U_m$ there is a node $u' \in U_n$ with $\pi(u) = \pi(u')$, and similarly for each node $u \in V_m$. We can construct an accepting run π' for an input tree t' with $|t'| < |t|$: Each subtree π_u for $u \in U_m$ can be replaced by a subtree $\pi_{u'}$ with $u' \in U_n$ and $\pi(u) = \pi(u')$, the subtree π_{w_m} can be replaced by the subtree π_{w_n} and each subtree π_u for $u \in V_m$ can be replaced by a subtree $\pi_{u'}$ with $u' \in V_n$ and $\pi(u) = \pi(u')$. □

The correctness of the algorithm in Figure 1 follows from the two lemmas above. The algorithm runs in $O(|Q|)$ space.

Theorem 9. *The emptiness problem for tree automata over words is* PSPACE-*complete.*

Proof. The emptiness problem is in PSPACE since the construction in Lemma 3a) can be done in polynomial time and the algorithm in Figure 1 runs in polynomial space.

We show the hardness by reducing the *reachability problem for* 2-*set systems*[2] to the emptiness problem. In [21] it was shown that the reachability problem for 2-set systems is PSPACE-hard.

---

[2] A *set system* over a finite set S is a functional relation $\longrightarrow \subseteq \mathcal{P}(S) \times \mathcal{P}(S)$. We write $X \longrightarrow Y$, for $(X, Y) \in \longrightarrow$. $N \subseteq S$ is the *successor* of $M \subseteq S$, $M \Longrightarrow N$, if $N = \bigcup_{X \longrightarrow Y, X \subseteq M} Y$. $\Longrightarrow^*$ denotes the reflexive, transitive closure of $\Longrightarrow$. The functional relation $\longrightarrow$ is a 2-*set system* if $|X| = 2$ and $1 \leq |Y| \leq 2$ for each $X \longrightarrow Y$. The *reachability problems for* 2-*set systems* is to decide if for a 2-set system $\longrightarrow$ over S and $S_0, S_1 \subseteq S$, there exists a set $S' \subseteq S$ such that $S_0 \Longrightarrow^* S'$ and $S' \cap S_1 \neq \emptyset$.

INPUT: tree automaton $\mathcal{A} = (\Sigma_{\Box,\#}, Q, Q_0, F, \delta)$
OUTPUT: returns true iff $W_*(\mathcal{A}) \neq \emptyset$

$h := 0$;
$P_\Sigma := \{q \in Q \mid \delta(q,a) \cap F \times F \neq \emptyset \text{ for } a \in \Sigma\}$;
$P_\Box := \{q \in Q \mid \delta(q,\Box) \cap F \times F \neq \emptyset\}$;
$P_{\Sigma\Box} := P_\Sigma \cup P_\Box$;
while $h \leq |Q|2^{2|Q|}$ **and** $Q_0 \cap P_{\Sigma\Box} = \emptyset$ **do begin**
  $h := h+1$;
  $P_\Sigma := \{q \in Q \mid \delta(q,\#) \cap P_\Sigma \times P_\Sigma \neq \emptyset\}$;
  $P_{\Sigma\Box} := P_\Sigma \cup \{q \in Q \mid \delta(q,\#) \cap P_{\Sigma\Box} \times P_\Box \neq \emptyset\}$;
  $P_\Box := \{q \in Q \mid \delta(q,\#) \cap P_\Box \times P_\Box \neq \emptyset\}$;
end;
if $h \leq |Q|2^{2|Q|}$ **then return** true **else return** false;

Fig. 1. Algorithm for deciding if a tree automaton star-accepts some word.

For a 2-set system $\longrightarrow$ over S and $S_0, S_1 \subseteq S$, let $\mathcal{A} = (\{0,\Box,\#\}, S, S_1, S_0, \delta)$ with

$$\delta(y,a) = \begin{cases} \{(x_1,x_2) \mid \{x_1,x_2\} \longrightarrow Y \in P \text{ and } y \in Y\} & \text{if } a \neq \Box, \\ \emptyset & \text{otherwise.} \end{cases}$$

We show that $\mathcal{A}$ minimal-accepts some word iff there is an S' with $S_0 \Longrightarrow^* S'$ and $S' \cap S_1 \neq \emptyset$.

($\Rightarrow$) Let π be an accepting run on the minimal input tree of some word in $\{0\}^*$. For $0 \leq k \leq |\pi|$, let $T_k = \{\pi(u) \mid |u| = k\}$, and let

$$M_0 = S_0 \qquad \text{and} \qquad M_{k+1} = \bigcup_{X \longrightarrow Y, X \subseteq M_k} Y,$$

for $0 \leq k < |\pi|$. By the definition of the successor, $M_k \Longrightarrow M_{k+1}$. We show by induction over k that $T_{|\pi|-k} \subseteq M_k$, for all $0 \leq k \leq |\pi|$. The base case for $k=0$ follows from the fact that $T_{|\pi|} \subseteq S_0$ since π is accepting. The step case for $k>0$ follows directly from the definition of δ. Thus, $M_\pi \cap S_1 \neq \emptyset$ since $T_0 \cap S_1 \neq \emptyset$.

($\Leftarrow$) Let $M_0, \ldots, M_h \subseteq S$ be a solution for the reachability problem, i.e., $M_0 = S_0$, $M_k \Longrightarrow M_{k+1}$, for $0 \leq k < h$, and $M_h \cap S_1 \neq \emptyset$. It is straightforward to construct an accepting run of $\mathcal{A}$ on the minimal input tree of the word 0^{2^h}. $\Box$

4 Decidable Extension of WS1S

In this section, we define an extension of WS1S with an infinite family of relations and show its decidability. A relation in this family is defined by a tree automaton with an alphabet of the form $\mathbb{B}^r_{\Box,\#}$ for $r > 0$. Hence, we assume in the following that a tree automaton has an alphabet of this form.

We encode a tuple of finite subsets of $\mathbb{N}$ in words. Let Λ denote the set of all finite subsets of $\mathbb{N}$. A word $b_1 \dots b_n \in \mathbb{B}^*$ *represents* $X \in \Lambda$ if $n \geq \max(\{0\} \cup X)$ and $b_k = 1$ iff $k \in X$, for all $1 \leq k \leq n$. Let $w \in (\mathbb{B}^r)^*$ where the ith letter of w has the form $(b_1^i, \dots, b_r^i) \in \mathbb{B}^r$. The word w *represents* the tuple $(X_1, \dots, X_r) \in \Lambda^r$ if each word $b_j^1 \dots b_j^{|w|} \in \mathbb{B}^*$ represents X_j, for $1 \leq j \leq r$.

Let $(R_\mathcal{A})_{\mathcal{A} \text{ t.a.}}$ stand for the infinite family of relations, where each $R_\mathcal{A} \subseteq \Lambda^r$ in this family for the tree automata $\mathcal{A}$ (with alphabet $\mathbb{B}^r_{\square,\#}$) is the relation defined by

$$(X_1, \dots, X_r) \in R_\mathcal{A} \quad \text{iff} \quad \mathcal{A} \text{ minimal-accepts a word that represents } (X_1, \dots, X_r).$$

The next lemma shows the closure under cylindrification. The construction is technical but straightforward, we omit it.

Lemma 10. *For a tree automaton $\mathcal{A}$ with alphabet $\mathbb{B}^r_{\square,\#}$ and $r' \geq 0$ we can construct a tree automaton $\mathcal{A}'$ with alphabet $\mathbb{B}^{r+r'}_{\square,\#}$ such that*

$$(X_1, \dots, X_r) \in R_\mathcal{A} \quad \textit{iff} \quad (X_1, \dots, X_r, Y_1, \dots, Y_{r'}) \in R_{\mathcal{A}'}$$

for $X_1, \dots, X_r, Y_1, \dots, Y_{r'} \in \Lambda$.

We show that the weak monadic second-order logic over the structure $\mathfrak{N} = (\mathbb{N}, \text{succ}, (R_\mathcal{A})_{\mathcal{A} \text{ t.a}})$, *WS1S*$^+$ for short, is decidable. Analogously to the idea in [5,11,24] we use automata to prove decidability of WS1S$^+$: Roughly speaking, we construct recursively for each sub-formula of a formula, a tree automaton that minimal-accepts the representatives of the satisfying interpretations.

We briefly sketch syntax and semantic of WS1S$^+$. A *formula* is built from the *atomic* formulas as in WS1S, namely[3] $x = y$, $X(x)$, $\text{succ}(x, y)$, and additionally, we have for each tree automaton $\mathcal{A}$ with alphabet $\mathbb{B}^r_{\square,\#}$ the atomic formula $R_\mathcal{A}(X_1, \dots, X_r)$. Further, we have the connectives $\neg$ and $\vee$, and the existential quantifier $\exists$ for first-order and monadic second-order variables.[4]

A *(weak) interpretation* I of φ is a function mapping first-order variables occurring in φ to elements of $\mathbb{N}$ and second-order variables occurring in φ to *finite* subsets of $\mathbb{N}$. The *truth value* of φ in $\mathfrak{N}$ w.r.t. I, $\mathfrak{N}, I \models \varphi$, is defined as usual. Note that the existential quantifier for second-order variables only ranges over *finite* subsets of $\mathbb{N}$. If φ contains no free variables, i.e., φ is a *sentence*, we write $\mathfrak{N} \models \varphi$. The *theory of WS1S*$^+$ is the set $\{\varphi \text{ sentence} \mid \mathfrak{N} \models \varphi\}$.

A formula φ with the variables $x_1, \dots, x_r$ and $X_1, \dots, X_s$ defines the word language

$$L(\varphi) = \{w \in (\mathbb{B}^{r+s})^* \mid \text{there is an interpretation } I \text{ with } \mathfrak{N}, I \models \varphi \text{ and } w \text{ represents } (\{I(x_1)\}, \dots, \{I(x_r)\}, I(X_1), \dots, I(X_s))\}.$$

---

[3] Lower case letters, like $x, y, \dots$, denote first-order variables and upper case letters, like $X, Y, \dots$, denote second-order variables.

[4] We also use the connectives $\wedge$, $\rightarrow$ and $\leftrightarrow$, and the universal quantifiers for first-order and second-order variables. We use the standard conventions for omitting parenthesis.

Lemma 11. *For a formula φ, $L(\varphi)$ is a tree-accepted word language. Moreover, we can construct a tree automaton $\mathcal{A}$ such that $W(\mathcal{A}) = L(\varphi)$.*

Proof (sketch). We use a restricted logical system, which we call *MSO_0-logic* (see e.g. [19,23]). It has a simpler syntax, in which the first-order variables are canceled. Elements of $\mathbb{N}$ are simulated by singletons and quantification over elements is simulated by quantification over singletons. Formulas in MSO_0-logic are constructed from the atomic formulas $R_{\mathcal{A}}(X_1, \dots, X_r)$ where $\mathcal{A}$ is a tree automaton with alphabet $\mathbb{B}^r_{\Box,\#}$. Formulas in MSO_0-logic are interpreted over the structure $(\mathbb{N}, (R_{\mathcal{A}})_{\mathcal{A}\text{ t.a.}})$.

The translation $\ulcorner . \urcorner$ of atomic formulas of $WS1S^+$ to formulas in MSO_0-logic is defined as

$$\begin{aligned}
\ulcorner x = y \urcorner &= R_{\mathcal{A}_{\text{Single}}}(\tilde{x}) \wedge R_{\mathcal{A}_{\text{Single}}}(\tilde{y}) \wedge R_{\mathcal{A}_{\subseteq}}(\tilde{x}, \tilde{y}) \wedge R_{\mathcal{A}_{\subseteq}}(\tilde{y}, \tilde{x}) \\
\ulcorner X(x) \urcorner &= R_{\mathcal{A}_{\text{Single}}}(\tilde{x}) \wedge R_{\mathcal{A}_{\subseteq}}(X) \\
\ulcorner \text{succ}(x, y) \urcorner &= R_{\mathcal{A}_{\text{Single}}}(\tilde{x}) \wedge R_{\mathcal{A}_{\text{Single}}}(\tilde{y}) \wedge R_{\mathcal{A}_{\text{Succ}}}(\tilde{x}, \tilde{y})
\end{aligned}$$

where $\tilde{x}, \tilde{y}$ denote new second-order variables for the first-order variables x and y, respectively. $\mathcal{A}_{\text{Single}}$ is the tree-automaton that minimal-accepts the words $0 \dots 01$, $\mathcal{A}_{\subseteq}$ tree-accepts the word language $(\mathbb{B}^2 \setminus \{(1,0)\})^*$ and $\mathcal{A}_{\text{Succ}}$ minimal-accepts a word $b_1 \dots b_n$ with $b_i = (b_i^1, b_i^2)$ iff $b_n^1 = 0$, and $b_{i-1}^1 = 1$ iff $b_i^2 = 1$, for $0 < i \leq n$. Since these languages are regular, the tree automata $\mathcal{A}_{\text{Single}}$, $\mathcal{A}_{\subseteq}$, and $\mathcal{A}_{\text{Succ}}$ exist by Theorem 4. The definition of $\ulcorner . \urcorner$ for the step cases is straightforward. It can be easily checked that $L(\varphi) = L(\ulcorner \varphi \urcorner)$.

Let φ be a formula in MSO_0-logic, where $X_1, \dots, X_s$ are the variables occurring in φ. We construct for each sub-formula ψ of φ a tree automaton $\mathcal{A}_\psi$ with the alphabet $\mathbb{B}^s_{\Box,\#}$ such that for all interpretations I of φ

$$(\mathbb{N}, (R_{\mathcal{A}})_{\mathcal{A}\text{ t.a.}}), I \models \psi \quad \text{iff} \quad \mathcal{A}_\psi \text{ minimal-accepts a word that represents } (I(X_1), \dots, I(X_s)).$$

We omit the details, as they are analogue to the details of the proof of decidability of WS1S in [19].

The base case for $\psi = R_{\mathcal{A}}(X_1, \dots, X_r)$ follows from Lemma 6. Note that we can assume that all variables $X_1, \dots, X_r$ are pairwise distinct, and by Lemma 10 that $r = s$. The step cases $\psi = \neg\chi$ and $\psi = \chi_1 \vee \chi_2$ follow from Theorem 5. The step case $\psi = \exists X\, \chi$ follows from Lemma 6 and Theorem 5. □

Corollary 12. *The theory of $WS1S^+$ is decidable.*

Proof. Let φ be a sentence. By Lemma 11, we can construct a tree automaton $\mathcal{A}$ such that $W(\mathcal{A}) = L(\varphi)$. φ is false in $\mathfrak{N}$ iff $L(\varphi) = \emptyset$, which can be checked by exploiting Theorem 9. □

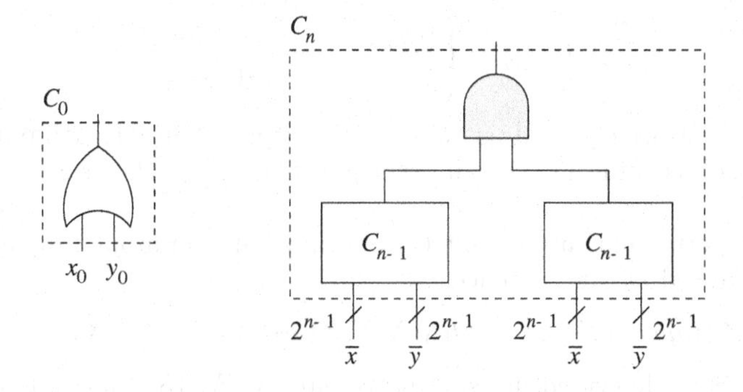

Fig. 2. Example of a parameterized combinational tree-structured circuit.

4.1 Applications

The first application shows that the predicate of natural numbers that are a power of 2 can be defined in WS1S$^+$. Let $\mathcal{A}$ be the tree automaton of Example 2. $\mathbb{P}(x)$ stands for the formula $\exists X\,\exists x'\,(X(x') \wedge \mathrm{succ}(x', x) \wedge R_{\mathcal{A}}(X))$. $\mathbb{P}(x)$ is true iff x is interpreted with a power of 2. The decidability of the weak monadic theory over $(\mathbb{N}, \mathrm{succ}, P)$ with $P = \{c^n \mid n \in \mathbb{N}\}$, for some $c \geq 2$, can be shown by using tree automata over trees over full $c$ary trees instead of using tree automata over full binary trees.

We introduce more syntactic sugar. $\mathbb{P}(\mathrm{succ}(x))$ is an abbreviation for the formula $\exists z\,(\mathrm{succ}(x, z) \wedge \mathbb{P}(x))$. The formula $\forall Z\,(Z(y) \wedge \forall z\,(\exists z'\,(\mathrm{succ}(z, z') \wedge Z(z')) \to Z(z)) \to Z(x))$ is true iff the interpretation of x is less than or equal to the interpretation of y; we write $x \leq y$.

The second application illustrates how WS1S$^+$ can be used to reason about parameterized combinational tree-structured circuits. For doing so, we use the (toy) family of circuits C_n, $n \geq 0$, depicted in Figure 2. We verify that the circuit C_n outputs 1 iff the input propagates a bit, i.e., for all $0 \leq i < 2^n$ either input pin x_i or input pin y_i has value 1.

An input of the circuit C_n can be encoded by three finite subsets of $\mathbb{N}$, namely X, Y, Z where $i \in X$ iff the input pin x_i has value 1, $i \in Y$ iff the input pin y_i has value 1, and $Z = \{0, \ldots, 2^n - 1\}$, i.e., Z determines the size of the circuit. More formally,

$$\begin{aligned} \mathit{input}(X, Y, Z) = \exists m\,\big(&\forall u\,(X(u) \vee Y(u) \vee Z(u) \to u \leq m) \wedge \\ &\forall u\,(u \leq m \to Z(u)) \wedge \mathbb{P}(\mathrm{succ}(m))\big)\,. \end{aligned}$$

The output value of the circuit C_n for any $n \geq 0$, can be described by the tree automaton $\mathcal{A} = (\mathbb{B}^3_{\Box,\#}, \{q, q_{ok}\}, \{q\}, \{q_{\mathrm{ok}}\}, \delta)$ where the step case of the recursive defined circuit family is reflected in the transition $\delta(q, \#) = \{(q, q)\}$. This can be intuitively read as "circuit C_n outputs 1 iff the left circuit C_{n-1} *and* the right circuit C_{n-1} outputs 1". The base case is reflected in the transitions

$$\delta(q,(b,b',1)) = \begin{cases} \{(q_{ok},q_{ok})\} & b=1 \text{ or } b'=1, \\ \emptyset & \text{otherwise.} \end{cases}$$

This can be intuitively read as "circuit C_0 outputs 1 iff either input pin x_0 *or* input pin y_0 has value 1". For all other letters in $\mathbb{B}^3_{\square,\#}$, the transitions of δ are arbitrary.

The property that any circuit C_n outputs 1 iff the input propagates a bit, can be expressed by the sentence

$$\forall X \forall Y \forall Z \left(\mathit{input}(X,Y,Z) \to \left(R_{\mathcal{A}}(X,Y,Z) \leftrightarrow \forall x \left(Z(x) \to X(x) \vee Y(x) \right) \right) \right) .$$

With the above described decision procedure for WS1S$^+$, we can check if this sentence holds in the structure $\mathfrak{N}$.

5 Future Work

We have shown how to encode words as complete leaf labeled trees and restricted tree automata to operate on such encodings. This leads to a characterization of a set of word languages, including the regular languages, that has the needed properties to define a decidable extension of WS1S. As applications, we have shown how this extension can be used to solve certain decision problems and how it can be applied to reason automatically about parameterized families of combinational tree-structured circuits.

In [15,14,2,18] exponentially inductive functions (EIFs) were used to describe parameterized families of combinational tree-structured circuits. It is possible to represent EIFs in WS1S$^+$ since in [2,18] it was shown that they are equivalent to alternating tree automata restricted on complete leaf labeled trees.

In [3,1] WS1S and WS2S were used to reason about serial and tree-structured parameterized circuits. Since the network topology of a tree-structured circuit is normally a complete tree where the leaves are the inputs of the circuit, WS1S$^+$ appears to be more natural than WS2S for modeling tree-structured circuits. Confirming this however, requires larger case studies.

Moreover, it seems that WS1S$^+$ cannot be embedded in WS2S. This conjecture is based on the following observation. Since the set of all complete trees is not a regular tree language, it is unlikely that the weak monadic second-order logic over $(\mathbb{N}, \mathrm{succ}, \mathbb{P})$ can be translated to WS2S. A rigorous comparison of WS2S and WS1S$^+$, and a more "natural" definition of the family of relations of WS1S$^+$ remains as future work. This includes a more complete investigation of the set of tree-accepted languages.

Implementing a decision procedure for WS1S$^+$ is also future work. To this end, we plan to use the existing implementation of tree automata in the Mona system [4].

Acknowledgments

I want to thank Abdelwaheb Ayari, David Basin, and the anonymous referees for their helpful comments.

References

1. A. Ayari, D. Basin, and S. Friedrich. Structural and behavioral modeling with monadic logics. In *29th IEEE International Symposium on Multiple-Valued Logic*, pages 142–151, 1999.
2. A. Ayari, D. Basin, and F. Klaedtke. Decision procedures for inductive boolean functions based on alternating automata. In *12th International Conference on Computer-Aided Verification, CAV'00*, volume 1855 of *LNCS*, pages 170–186, 2000.
3. D. Basin and N. Klarlund. Automata based symbolic reasoning in hardware verification. *The Journal of Formal Methods in Systems Design*, 13(3):255–288, 1998.
4. M. Biehl, N. Klarlund, and T. Rauhe. Algorithms for guided tree automata. In *1st International Workshop on Implementing Automata, WIA '96*, volume 1260 of *LNCS*, pages 6–25, 1997.
5. J. Büchi. Weak second-order arithmetic and finite automata. *Zeitschrift der mathematischen Logik und Grundlagen der Mathematik*, 6:66–92, 1960.
6. O. Carton and W. Thomas. The monadic theory of morphic infinite words and generalizations. In *25th International Symposium on Mathematical Foundations of Computer Science, MFCS 2000*, volume 1893 of *LNCS*, pages 275–284, 2000.
7. H. Comon, M. Dauchet, R. Gilleron, F. Jacquemard, D. Lugiez, S. Tison, and M. Tommasi. Tree automata techniques and applications. Available on `http://www.grappa.univ-lille3.fr/tata`, 1997.
8. J. Doner. Tree acceptors and some of their applications. *Journal of Computer and System Sciences*, 4:406–451, 1970.
9. J. Elgaard, N. Klarlund, and A. Møller. Mona 1.x: New techniques for WS1S and WS2S. In *10th International Conference on Computer Aided Verification, CAV'98*, volume 1427 of *LNCS*, pages 516–520, 1998.
10. J. Elgaard, A. Møller, and M. Schwartzbach. Compile-time debugging of C programs working on trees. In *9th European Symposium on Programming, ESOP'00*, volume 1782 of *LNCS*, pages 119–134, 2000.
11. C. Elgot. Decision problems of finite automata design and related arithmetics. *Transactions of the AMS*, 98:21–51, 1961.
12. C. Elgot and M. Rabin. Decidability and undecidability of extensions of second (first) order theory of (generalized) successor. *Journal of Symbolic Logic*, 31(2):169–181, 1966.
13. F. Gécseg and M. Steinby. *Tree Automata.* Akadémiai Kiadó, Budapest, 1984.
14. A. Gupta. *Inductive Boolean Function Manipulation: A Hardware Verification Methodology for Automatic Induction.* PhD thesis, School of Computer Science, Carnegie Mellon University, Pittsburgh, USA, 1994.
15. A. Gupta and A. Fisher. Parametric circuit representation using inductive boolean functions. In *5th International Conference on Computer-Aided Verification, CAV'93*, volume 697 of *LNCS*, pages 15–28, 1993.
16. J. Henriksen, J. Jensen, M. Jørgensen, N. Klarlund, B. Paige, T. Rauhe, and A. Sandholm. Mona: Monadic second-order logic in practice. In *1st International Workshop on Tools and Algorithms for the Construction and Analysis of Systems, TACAS'95*, volume 1019 of *LNCS*, pages 89–110, 1996.
17. J. Hopcroft and J. Ullman. *Formal Languages and their Relation to Automata.* Addison-Wesley, 1969.
18. F. Klaedtke. Induktive boolesche Funktionen, endliche Automaten und monadische Logik zweiter Stufe. Diplomarbeit, Institut für Informatik, Albert-Ludwigs-Universität, Freiburg i. Br., 2000. In German.

19. N. Klarlund and A. Møller. *MONA Version 1.3 User Manual.* BRICS Notes Series NS-98-3 (2.revision), Department of Computer Science, University of Aarhus, 1998.
20. A. Meyer. Weak monadic second-order theory of successor is not elementary-recursive. In *Logic Colloquium*, volume 453 of *Lecture Notes in Mathematics*, pages 132–154, 1975.
21. A. Monti and A. Roncato. Completeness results concerning systolic tree automata and E0L languages. *Information Processing Letters*, 53:11–16, 1995.
22. J. Thatcher and J. Wright. Generalized finite automata theory with an application to a decision problem in second-order logic. *Mathematical Systems Theory*, 2:57–81, 1968.
23. W. Thomas. Languages, automata, and logic. In A. Salomaa and G. Rozenberg, editors, *Handbook of Formal Languages*, volume 3, chapter 7, pages 389–455. Springer-Verlag, 1997.
24. B. Trakhtenbrot. Finite automata and the logic of one-place predicates. *AMS, Transl., II. Ser.*, 59:23–55, 1966.

Limiting Partial Combinatory Algebras towards Infinitary Lambda-Calculi and Classical Logic*

Yohji Akama

Mathematical Institute, Tohoku University Sendai Miyagi Japan, 980-8578
akama@math.tohoku.ac.jp
Tel. +81-22-214-0364 Fax. +81-22-214-0364

Abstract. We will construct from every partial combinatory algebra (PCA, for short) $\mathcal{A}$ a PCA a-lim($\mathcal{A}$) s.t. (1) every representable numeric function $\varphi(n)$ of a-lim($\mathcal{A}$) is exactly of the form $\lim_t \xi(t,n)$ with $\xi(t,n)$ being a representable numeric function of $\mathcal{A}$, and (2) $\mathcal{A}$ can be embedded into a-lim($\mathcal{A}$) which has a synchronous application operator. Here, a-lim($\mathcal{A}$) is $\mathcal{A}$ equipped with a limit structure in the sense that each element of a-lim($\mathcal{A}$) is the limit of a countable sequence of $\mathcal{A}$-elements. We will discuss limit structures for $\mathcal{A}$ in terms of Barendregt's range property. Moreover, we will repeat the construction lim(−) transfinite times to interpret infinitary λ-calculi. Finally, we will interpret affine type-free $\lambda\mu$-calculus by introducing another partial applicative structure which has an asynchronous application operator and allows a parallel limit operation. keywords: partial combinatory algebra, limiting recursive functions, realizability interpretation, discontinuity, infinitary lambda-calculi, $\lambda\mu$-calculus. In the interpretation, μ-variables(=continuations) are interpreted as streams of λ-terms.

1 Introduction

Partial combinatory algebras (PCA, for short) are partial applicative structures axiomatized by the same axioms as combinatory algebras, except that the application operators can be partial operators. The PCAs are important in connection with the realizability interpretations of intuitionistic logics. The realizability interpretations extract the computational content from intuitionistic logic's proofs as programs. Using PCAs to carry out the realizability interpretations, we can obtain 'realizability' models of typed term calculi and constructive set theories in which we can do mathematics.

Recently, a new realizability interpretation was introduced by Nakata and Hayashi [9] to extract the computational content of semi-classical logic's proofs as approximation algorithms.

They first noticed that Gold's limiting recursive functions [5], which was originally introduced to formulate the learning processes of machines, is useful

* The author acknowledges Susumu Hayashi, Mariko Yasugi, Stefano Berardi, and Ken-etsu Fujita. The comment by anonymous referees was useful to partly improve the presentation.

L. Fribourg (Ed.): CSL 2001, LNCS 2142, pp. 399–414, 2001.

for what they call animation of proofs [7]. Gold's limiting recursive function is of the form $f(x)$ s.t.

$$f(x) = y \iff \exists t_0 \forall t > t_0 .\, g(t, x) = y \iff \lim_t g(t, x) = y,$$

where $g(t, x)$ is called a guessing function, and t is a limit variable. Then, they proved that some limiting recursive functions approximate a realizer of a semi-classical principle $\neg\neg\exists y \forall x .\, g(x, y) = 0 \to \exists y \forall x .\, g(x, y) = 0$. Also, they showed that the semi-classical principle is sufficient for usual mathematics and for software synthesis ([6]).

In this way, Nakata and Hayashi opened up the possibility that limiting operations provide realizability interpretation of semi-classical logical systems.

They formulated the set of the limiting recursive functions as a *Basic Recursive Function Theory*(BRFT, for short. Wagner [17] and Strong [14]). Then Nakata and Hayashi carried out their realizability interpretation using the BRFT.

If we can formulate the set of limiting algorithms as a PCA $\mathcal{L}$, then by carrying out Nakata and Hayashi's realizability interpretation using $\mathcal{L}$, we may be able to construct 'realizability' models of

1. semi-classical typed term calculi. For example, typed Parigot's $\lambda\mu$-calculus, typed term calculi with control operators.
2. semi-classical constructive set theory.

Motivated by the above, we will introduce a construction from a given PCA $\mathcal{A}$ another PCA a-lim($\mathcal{A}$). Our idea is (1) limit variable t is the clock of the processing element(MPU). (2) guessing function $g(t, x)$ is a generator of a stream $\langle g(0, x), g(1, x), \ldots \rangle$. (3) limiting recursive function $\lim_t g(t, x)$ is the stream modulo a symmetric, transitive relation $\sim$. Here, two streams are related with $\sim$ if for all natural number t except finite numbers, the two t-th elements of the two streams have the same value.

In a-lim($\mathcal{A}$), every allowed limit is exactly of the form $\lim_t a_t$ s.t. the infinite sequence $\langle\, a_t \in \mathcal{A} \,\rangle_t$ is (1) indexed by $\mathbb{N}$; and (2) generated inside the PCA $\mathcal{A}$. We will call this $\lim_t a_t$ an autonomous limit. Owing to the above (1,2), we will be able to prove that every representable partial function of a-lim($\mathcal{A}$) is exactly a limiting recursive function guessed by some representable partial functions of $\mathcal{A}$ (see Section 3).

Based on this result, we have only to take $\mathcal{L}$ as a-lim($\mathbb{N}$), in order to find realizability models of both semi-classical typed term calculi and a semi-classical constructive set theory.

In order to construct from the $\mathcal{A}$ a PCA with stronger computation power, we will first consider the following two forms of limits $\lim_\lambda a_\lambda$.

(R) $\langle\, a_\lambda \in \mathcal{A} \,\rangle_\lambda$ is indexed by the whole $\mathcal{A}$.
(N) $\langle\, a_\lambda \in \mathcal{A} \,\rangle_\lambda$ is any countable sequence of $\mathcal{A}$-elements.

We will prove that (R) does not always strengthen the computation power of $\mathcal{A}$. However, (N) has an extreme effect on the strength of the $\mathcal{A}$. We will introduce another construction from any PCA $\mathcal{A}$ to another PCA n-lim($\mathcal{A}$) where only allowed limit is of the form (N). Then, the set of representable functions in n-lim($\mathcal{A}$) is the set of all partial numeric functions. Moreover, it can compute a discontinuous function from $\mathbb{R}$ to $\mathbb{N}$. The construction n-lim($-$) may be interesting itself since it applies for all signature of partial algebras. See Section 4.

By using our results on limits over PCAs, we aim to interpret the following infinite λ-calculi. Infinite λ-calculi have been studied in proof- and recursion-theoretic contexts and are now being studied in the analysis of infinite streams (for input/output) and non-terminating recursive calls of functional programming languages.

1. Tait's typed calculus of infinitely long terms [15]: An infinite sequence of type A terms is again a type A term. His motivation was proof-theoretic. He wanted to define a large class of calculable functionals of finite types.
2. S. Feferman's typed calculus T_0 of infinitely long terms (For details, see Schwichtenberg and Wainer [11]): An infinite sequence $\langle P_1, P_2, \ldots \rangle$ of type A terms is again a type A term, if there is a term that calculates for each i the code of P_i. After Feferman developed Tait's typed λ-calculus in a proof-theoretic context, he introduced T_0 and studied T_0 in a recursion-theoretic context.
3. two systems of infinite type-free λ-calculi by Kennaway-Klop-Sleep-de Vries [8] and Berarducci [3]: Both have terms representing infinite Böhm-, Lévy-Longo-, or Berarducci-trees.
4. The type-free $\lambda\mu$-calculus (Parigot [10]): The typed version corresponds to the classical logic, and a typed/type-free version relates to a typed/type-free functional programming language with control operators such as `call/cc`. Type-free $\lambda\mu$-calculus has μ-variables to represent continuations. By regarding μ-variables as infinite sequences of usual variables, we can regard $\lambda\mu$-calculus as an infinite λ-calculus.

The relationship between Tait's infinite typed λ-calculus and Feferman's typed λ-calculus is comparable to the relationship between n-lim($-$) and a-lim($-$).

The infinite λ-calculi (1,2,3) have an infinite term consisting of infinite terms, while our constructions a-lim($\mathcal{A}$) and n-lim($\mathcal{A}$) introduce an element infinitely depending on elements which are "finite"(i.e., in $\mathcal{A}$). In order to interpret the infinite λ-calculi, we will repeat our constructions a-lim($-$) and/or n-lim($-$). The resulting PCA allows repeated limit $\lim_{t_1} \cdots \lim_{t_n} f(t_1, \ldots, t_n)$. See Section 5.

However, Parigot's type-free $\lambda\mu$-calculus is difficult to interpret with a-lim and/or n-lim. We will introduce another construction n-LIM($-$) which extends a given PCA $\mathcal{A}$ to a partial applicative structure n-LIM($\mathcal{A}$) s.t.

- every allowed limit in n-LIM($\mathcal{A}$) is a parallel limit $\lim_{t_1,\ldots,t_n} f(t_1, \ldots, t_n)$.
- the application operator is asynchronous.

With the help of concurrency theory, we will try to clarify the parallelism hidden in the parallel limits of n-LIM($-$). Then, we will interpret the affine type-free $\lambda\mu$-calculus in Section 6.

From now on, the symbol "$\simeq$" means "if one-side is defined, then the other side is defined as having the same value," while the symbol "$=$" means that "both sides are defined as having the same value." The symbol "$\downarrow$" means "is defined." So, $t\downarrow$ is equivalent to an equation $t = t$. For any operation f, we assume that $f(a_1, \dots, a_n)\downarrow$ implies $a_1\downarrow, \dots$ and $a_n\downarrow$. The set of partial functions from A to B is denoted by $A \rightharpoonup B$. We write $\overset{\infty}{\forall}\, x \in A.\, \varphi(x)$ to express that "$\{x \in A \mid \neg\varphi(x)\}$ is a finite set."

2 Limiting Recursion

Definition 1. We say a partial numeric function $\varphi(n_1, \dots, n_k)$ is *guessed* by a partial numeric function $\xi(t, n_1, \dots, n_k)$ as t goes to infinity, if $\forall n_1, \dots, n_k \exists t_0 \forall t > t_0.\, \varphi(n_1, \dots, n_k) \simeq \xi(t, n_1, \dots, n_k)$. We write $\varphi(n_1, \dots, n_k) \simeq \lim_t \xi(t, n_1, \dots, n_k)$. For every class $\mathcal{F}$ of partial numeric functions, $\lim(\mathcal{F})$ denotes the set of partial numeric functions guessed by a partial numeric function in $\mathcal{F}$.

In calibrating the computational power of $\lim(\mathcal{F})$, we recall the *limiting recursive functions* introduced by Gold [5]. We assume the knowledge about the arithmetical hierarchy of sets and complete sets w.r.t. many-one reducibility. The standard reference is Soare's book [12].

Proposition 1 (Gold [5]).

1. A total function guessed by a partial recursive function can be guessed by a primitive recursive function.
2. A (partial) function guessed by a total recursive function is exactly a (partial) recursive function in the halting set $\mathcal{K}$ (called a *limiting recursive function.*)

The set **PRF** of partial recursive functions is contained in $\lim(\mathbf{PRF})$, because every $\varphi \in \mathbf{PRF}$ is guessed by φ itself composed with projections. We will show that $\lim(\mathbf{PRF})$ strictly includes the set of limiting recursive functions.

Proposition 2 (Takeshi Yamazaki, or folklore).

1. For every total function, it is guessed by a total recursive function if and only if the graph of the function is in a class Δ^0_2 of the arithmetical hierarchy.
2. For every partial function, if it is guessed by a partial recursive function, then the graph and the domain are both in a class Σ^0_3 of the arithmetical hierarchy. Moreover, there is a partial function f s.t. f is guessed by a partial recursive function and the graph of f and the domain of f are both complete w.r.t. many-one reducibility.

Proof. (2)Let $\mathcal{W}_n$ be the domain of the unary partial recursive function of index n. It is well-known $\mathbf{Cof} = \{n \mid \mathcal{W}_n \text{ is cofinite}\} = \{n \mid \exists s \forall t > s.\, t \in \mathcal{W}_n\}$ is Σ^0_3 complete. Define $\xi(t, n) = 1$, if $t \in \mathcal{W}_n$, and undefined, otherwise. Then $x \in \mathbf{Cof} = \mathrm{dom}(\lim_t \xi(t, n))$ iff $(x, 1) \in \mathrm{graph}(\lim_t \xi(t, n))$. □

The operation $\lim(-)$ preserves an abstract structure for basic recursion theory, as Nakata and Hayashi showed in [9]. We recall *ω-basic recursive function theory with a successor function* of Strong [14] and Wagner [17]. If $\mathcal{F}$ is an ω-BRFT with suc, so is $\lim(\mathcal{F})$.

3 Autonomous Limit

A *partial combinatory algebra*(PCA, *for short*) is $\mathcal{A} = \langle |\mathcal{A}|, \cdot, s, k\rangle$ s.t. $\cdot$ is a binary partial operator(*application operator*) on a set $|\mathcal{A}|$, and $s, k \in |\mathcal{A}|$ are distinct elements subject to $(k \cdot a) \cdot b = a, (s \cdot a) \cdot b \downarrow, ((s \cdot a) \cdot b) \cdot c \simeq (a \cdot c) \cdot (b \cdot c)$. Examples of PCA are the set $\mathbb{N}$ of natural numbers and the set of values of the call-by-value λ-calculus. $\mathcal{A}, \mathcal{B}, \ldots$ range over PCAs.

Given a PCA, we can simulate a λ-abstraction; for a "polynomial" $t[x]$, there is a "polynomial" $\lambda x.\, t[x]$ s.t. $(\lambda x.\, t[x]) \cdot a \simeq t[a]$. The *Church numeral* of a natural number n, denoted by $\overline{n}$, is a polynomial $\lambda y \lambda x.\, y \cdot (\cdots (y \cdot x) \cdots)$ with the y successively applied n-times to x.

We say a partial numerical function $\psi(t, n_1, \ldots, n_k)$ is *represented* by an element $a \in \mathcal{A}$, whenever $(\cdots((a \cdot \overline{t}) \cdot \overline{n_1}) \cdots) \cdot \overline{n_k} = \overline{m}$ iff $\psi(t, n_1, \ldots, n_k) = m$. The set of partial numerical functions representable in $\mathcal{A}$ is denoted by $\mathrm{RpFn}(\mathcal{A})$. It is well-known that $\mathrm{RpFn}(\mathcal{A})$ contains **PRF**.

Given a PCA $\mathcal{A}$, we will construct a PCA a-lim$(\mathcal{A})$ s.t. RpFn(a-lim$(\mathcal{A})$) = lim(RpFn$(\mathcal{A})$).

Definition 2 (Autonomous Limit of PCA). Given a PCA $\mathcal{A} = \langle |\mathcal{A}|, \cdot, s, k\rangle$. *The extension* a-lim$(\mathcal{A})$ *of* $\mathcal{A}$ *by the autonomous limits* is a PCA a-lim$(\mathcal{A})$ = $\langle |\text{a-lim}(\mathcal{A})|, *, \mathbf{s}, \mathbf{k}\rangle$, where

- $|\text{a-lim}(\mathcal{A})|$ is a quotient set $\{a \in |\mathcal{A}| \mid a \sim a\}/\sim$, where $a \sim b$ is defined as $\overset{\infty}{\forall}\, t \in \mathbb{N}.\, (a \cdot \overline{t} = b \cdot \overline{t})$.
- $\mathbf{s} = [k \cdot s]_\sim, \mathbf{k} = [k \cdot k]_\sim$ ("s, k for any value t", t being "time");
- $[a]_\sim * [b]_\sim = [(s \cdot a) \cdot b]_\sim$ ("synchronous application").

As $\sim$ is a symmetric, transitive relation on $|\mathcal{A}|$, the quotient set $|\text{a-lim}(\mathcal{A})|$ is well-defined. The element is an equivalence class $[a]_\sim$ with $a \in \mathcal{A}$. When a is undefined, so will $[a]_\sim$ be undefined. The operation $*$ is well-defined; Suppose $a \sim a'$ and $b \sim b'$ and $s \cdot a \cdot b \cdot \overline{n} \simeq (a \cdot \overline{n}) \cdot (b \cdot \overline{n})$ is defined for values of n which are large enough. Then, because $a \cdot \overline{n} = a' \cdot \overline{n}$ and $b \cdot \overline{n} = b' \cdot \overline{n}$, we have $s \cdot a \cdot b \cdot \overline{n} \simeq (a \cdot \overline{n}) \cdot (b \cdot \overline{n}) \simeq (a' \cdot \overline{n}) \cdot (b' \cdot \overline{n}) \simeq s \cdot a' \cdot b' \cdot \overline{n}$. Therefore, $[a] * [b] \simeq [a'] * [b']$.

Theorem 1. a-lim$(\mathcal{A})$ is a PCA s.t. RpFn(a-lim$(\mathcal{A})$) = lim(RpFn$(\mathcal{A})$).

Proof. We will first prove that it is a PCA. Let $t \in \mathbb{N}$ be sufficiently large. (1) $\mathbf{s} * [a] * [b] * [c] \simeq [s(s(s(ks)a)b)c]$ while $[a] * [c] * ([b] * [c]) \simeq [s(sac)(sbc)]$. An axiom for s implies $s(s(s(ks)a)b)c\overline{t} \simeq (ks\overline{t})(a\overline{t})(b\overline{t})(c\overline{t})$. By using the axiom for k with $\overline{t}\downarrow$, the last is $\simeq s(a\overline{t})(b\overline{t})(c\overline{t})$. It is, by an axiom for s, $\simeq s(sac)(sbc)\overline{t}$. So, $\mathbf{s}*[a]*[b]*[c] \simeq [a]*[c]*([b]*[c])$. (2) $\mathbf{s}*[a]*[b] \simeq [s(s(ks)a)b]$. An axiom of s implies $s(s(ks)a)b\overline{t} \simeq ks\overline{t}(a\overline{t})(b\overline{t})$. By using the axiom of k with $\overline{t}\downarrow$, it is $\simeq s(a\overline{t})(b\overline{t})$. It is always defined because $a\overline{t}\downarrow$, $b\overline{t}\downarrow$ and an axiom of s. So, $\mathbf{s} * [a] * [b] \downarrow$. (3) $\mathbf{k} * [a] * [b] = [s(s(kk)a)b]$ and $s(s(kk)a)b\overline{t} \simeq (kk\overline{t})(a\overline{t})(b\overline{t})$. By using the axiom of k with $\overline{t}\downarrow$, it is $\simeq k(a\overline{t})(b\overline{t})$. It is $\simeq a\overline{t}$ as $b\overline{t}\downarrow$. So, $\mathbf{k} * [a] * [b] = [a]$. Therefore, a-lim$(\mathcal{A})$ is indeed a PCA.

Both of RpFn(a-lim$(\mathcal{A})$) and lim(RpFn$(\mathcal{A})$) have the nowhere defined partial functions of any arity. Let φ be a unary partial function somewhere defined. $\varphi \in$

$\lim(\mathrm{RpFn}(\mathcal{A}))$ is equivalent to $\exists \xi \in \mathrm{RpFn}(\mathcal{A}) \forall n, m(\lim_t \xi(t,n) = m$ iff $\varphi(n) = m)$ As φ is somewhere defined, we have $\varphi(l)\downarrow$ for some l and so, $\overset{\infty}{\forall} t.\, \xi(t,l)\downarrow$. Thus, it is equivalent to $\exists a \in \mathcal{A}(\overset{\infty}{\forall} t.\, a\bar{t}\downarrow \;\&\; \forall n, m(\overset{\infty}{\forall} t \in \mathbb{N}.\, a\bar{t}\bar{n} = \overline{m})$ iff $\varphi(n) = m$). Because a Church numeral $\overline{\mathbf{n}}$ of a-lim($\mathcal{A}$) is $[k\overline{n}]_\sim$ with the latter Church numeral $\overline{n}$ being in $\mathcal{A}$, it is equivalent to $\exists a \in \mathcal{A}(a \sim a \;\&\; \forall n, m([a]_\sim * \overline{\mathbf{n}} = \overline{\mathbf{m}})$ iff $\varphi(n) = m$). Therefore, $\varphi \in \lim(\mathrm{RpFn}(\mathcal{A}))$ iff $\varphi \in \mathrm{RpFn}(\text{a-lim}(\mathcal{A}))$. For φ with the arity being greater than 1, it is similarly proved. □

Definition 3 (homomorphism). A function f from a PCA $\mathcal{A}$ to a PCA $\mathcal{B}$ is a *homomorphism*, if f preserves the operators as relations. That is, $f(s) = \mathbf{s}$, $f(k) = \mathbf{k}$, and $ab = c$ implies $f(a)f(b) = f(c)$. We denote by $\mathbb{PCA}$ the category of PCAs and homomorphisms. An injective, surjective homomorphism is called an *isomorphism*. A homomorphism is abbreviated as homo.

Theorem 2. The function $\iota_{\mathcal{A}} : \mathcal{A} \to \text{a-lim}(\mathcal{A});\; a \mapsto [k \cdot a]_\sim$ is an injective, non-surjective homo. such that $a \cdot a' = b \iff \iota_{\mathcal{A}}(a) * \iota_{\mathcal{A}}(a') = \iota_{\mathcal{A}}(b)$. Moreover,

$$[a]_\sim = \iota_{\mathcal{A}}(b) \iff \overset{\infty}{\forall} t \in \mathbb{N}.\; a \cdot \bar{t} = b \quad (\iff \text{“}\lim_t\; a \cdot \bar{t} = b\text{”})$$

4 Possible Limit Structures

Each element of the PCA a-lim($\mathcal{A}$) is of the form $\lim_t a_t$, where the

1. parameter t can be any natural number; and
2. the sequence $\langle a_t \rangle_t$ is of the form $\langle a \cdot \bar{t} \rangle_t$ for some $a \in \mathcal{A}$. In this case, the sequence is “autonomously tracked” by an $\mathcal{A}$-element a.

To justify the necessity of the two conditions, we will discuss following alternative $\lim_t a_t$ for $\mathcal{A}$: (**R**) t of $\lim_t a_t$ can be any element of $\mathcal{A}$ (See Subsection 4.1); or (**N**) the sequence $\langle a_t \rangle_t$ is any countable sequence.(See Subsection 4.2)

4.1 Range Property and Limit

We consider another lim operation: $\lim_a f(\boldsymbol{x}, a) = y \iff \overset{\infty}{\forall} a \in |\mathcal{A}|.\; y = f(\boldsymbol{x}, a)$. Although this limit is useful in a PCA $\mathbb{N}$, it is not the case for the set $\mathcal{M}$ of closed λ-terms modulo β-equality. $\mathcal{M}$ is called a closed term model of $\lambda\beta$-calculus. Even though we can construct from $\mathcal{M}$ another PCA $A(\mathcal{M})$ with the limit being above, we have $\mathbf{PRF} = \mathrm{RpFn}(A(\mathcal{M})) = \mathrm{RpFn}(\mathcal{M}) \neq \lim(\mathrm{RpFn}(\mathcal{M})) = \lim(\mathbf{PRF})$, because of the range property of Barendregt([1, p.517]): In $\mathcal{M}$, the range of any combinator is either a singleton or an infinite set. Indeed, if $f : \mathcal{M} \times \mathcal{M} \to \mathcal{M}$ is representable, so is $F_x(a) = f(x, a)$. When $\lim_a f(x, a)$ in the above sense has a value, F_x has the finite range $\{F_x(a) \mid a \in \mathcal{M}\}$. However, it must be a singleton because of the range property. Therefore, the lim of above will be useless.

4.2 Limit along All the Countable Sequences

Theorem 3. Let $\mathcal{M}$ be a partial algebra of a signature $\langle f_1, \ldots;\ u_1 = u_1', \ldots;\ v_1 \simeq v_1', \ldots \rangle$. Here $f_1, \ldots$ are (partial) operators, and $u_1 = u_1', \ldots; v_1 \simeq v_1', \ldots$ are axioms. Each u_i, u_i', v_i, v_i' is built up from the variables and the operators $f_1, \ldots$.

The extension n-lim$(\mathcal{M})$ *of* $\mathcal{M}$ *by the non-constructive limits* is $\langle |\text{n-lim}(\mathcal{M})|,\ \mathbf{f}, \ldots \rangle$ s.t.

1. $|\text{n-lim}(\mathcal{M})| = \{a : \mathbb{N} \rightharpoonup \mathcal{M} \mid a \sim a\}/\sim$, where $a \sim b$ is defined as $\overset{\infty}{\forall}\, k \in \mathbb{N}.\, a(k) = b(k)$.
2. For $\mathbf{c}^{(k)} = [\ c^{(k)}\ ]_\sim \in |\text{n-lim}(\mathcal{M})|$, define $\mathbf{f}(\mathbf{c}^{(1)},\ \mathbf{c}^{(1)}, \ldots) \simeq [\ \varphi\ ]_\sim$ with $\varphi(i) \simeq f(\ c^{(1)}(i),\ c^{(2)}(i), \ldots)$.

Then n-lim$(\mathcal{M})$ is a partial algebra of the same signature.

As we defined homomorphisms for PCAs, we will define a homomorphism for partial algebras as a function which preserves the operators as relations.

Theorem 4. The function $\iota_\mathcal{M} : \mathcal{M} \to \text{n-lim}(\mathcal{M});\ b \mapsto [a]_\sim,$ where $a(t) = b$. is an injective homo. such that $\mathbf{f}(b, \ldots) = b'$ in $\mathcal{M} \iff \iota_\mathcal{M}(\mathbf{f})(\iota_\mathcal{M}(b), \ldots) = \iota_\mathcal{M}(b')$ in n-lim$(\mathcal{M})$. If $\mathcal{M}$ has at least two elements, it is non-surjective. Moreover,

$$[a]_\sim = \iota_\mathcal{M}(b) \iff (\overset{\infty}{\forall}\, t \in \mathbb{N}.\ a(t) = b) \qquad (\iff \text{“}\lim_t\ a(t) = b\text{”}).$$

Proof. Let $a : \mathbb{N} \rightharpoonup \mathcal{M}$ s. t. $a(t) = b_{t \bmod 2}$ with distinct $b_0, b_1 \in \mathcal{M}$. If $\iota_\mathcal{M}$ is surjective, $\exists b\ \overset{\infty}{\forall}\, t \in \mathbb{N}\, (a(t) = b)$. Contradiction. The other claims are clear. □

Partial algebras appear as algebraic specification of software. For instance, stacks. It is indeed partial, because `pop(nil)` can be undefined. The signature of stacks is $\langle \mathtt{pop}, \mathtt{push}, \mathtt{nil}, 1;\ \mathtt{pop}(\mathtt{push}(\mathtt{x}, \mathtt{y})) = \mathtt{x} \rangle$. By taking $\mathcal{M}$ of above theorems as PCAs, we will have the following:

Theorem 5. If $\mathcal{A}$ is a PCA, n-lim$(\mathcal{A})$ is a PCA s. t. RpFn(n-lim$(\mathcal{A})$) is the set of all partial numeric functions.

Proof. We will prove only the second part. Since any partial numeric function of finite domain is representable in any PCA, every partial numeric m-ary function φ is represented by $[f]_\sim$ s. t. $f(t)$ represents $\varphi \restriction \{0, 1, \ldots, t-1, t\}^m$ in $\mathcal{A}$. □

Remark 1. There are other PCAs where every partial numeric function is representable. For instance, D_∞ introduced by Scott and the partial continuous operations (PCO, for short. See [16, Ch.9, Sect.4]) introduced by Kleene. However, no n-lim$(\mathcal{A})$ is isomorphic to them. This fact is explained by comparing the three extensional collapses [16] of n-lim$(\mathcal{A})$, D_∞ and PCO.

On the one hand, the extensional collapses of both D_∞ and PCO are the same; a type structure consisting of all total Kleene-Kreisel continuous functionals over

$\mathbb{N}$. By a Kleene-Kreisel continuous functional, we mean that the output of the functional depends only on finite subinformation of the input functionals.

On the other hand, the extensional collapse of n-lim($\mathcal{A}$) contains a discontinuous functional over $\mathbb{N}$. One such example is the Gauss function, a function given a real number x returning the smallest integer not greater than x. The Gauss function is actually a functional of type $(\mathbb{N} \to \mathbb{N}) \to \mathbb{N}$ by using a suitable coding of $\mathbb{Q}$ to $\mathbb{N}$. The Gauss function is not Kleene-Kreisel continuous, because the output needs infinite precision on the input real number to determine whether the input is an integer or not. As Gauss function is rewritten as the limit according to Yasugi-Brattka-Washihara [18], it is in the extensional collapse of n-lim($\mathcal{A}$). Hence, no n-lim($\mathcal{A}$) is isomorphic to D_∞ or PCO.

5 Repeating Limits

In functional programming, we use infinite lists as streams(for input/output) and non-terminating recursive calls. These objects are usually the unwinding("limit") of finite objects. To analyze such infinitary objects, of interest is the *infinitary* λ-calculus which was introduced by Kennaway-Klop-Sleep-de Vries [8] and that which was introduced by Berarducci [3]. Both calculi admit terms like $\lambda x.y^\omega x$, $\lambda x.\, y^\omega(x^\omega)$, and have terms with the limit operation $(-)^\omega$ being nested. So, to interpret all of the infinitary λ-terms, it is necessary to repeat the limit constructions ω-times.

Let $\mathcal{A}$ be a combinatory algebra. Then the interpretations of ordinary λ-terms are in $\mathcal{A}$. If an infinitary λ-term M has k-nested $(-)^\omega$, then the interpretation is in a-lim$^k(\mathcal{A})$ or n-lim$^k(\mathcal{A})$.

Definition 4 (α-times repeated limits). For every PCA $\mathcal{A}$ and every ordinal number α, let us define a PCA $\lim^\alpha(\mathcal{A})$ and the canonical injective homo. ι^α_β : $\lim^\beta(\mathcal{A}) \to \lim^\alpha(\mathcal{A})$ for each $0 < \beta \le \alpha$.

- a-lim$^0(\mathcal{A}) = \mathcal{A}$, and ι^β_β is the identity for each ordinal number β.
- a-lim$^{\beta+1}(\mathcal{A})$ = a-lim $\left(\text{a-lim}^\beta(\mathcal{A})\right)$, and $\iota^{\beta+1}_\gamma = \iota_{\text{a-lim}^\beta(\mathcal{A})} \circ \iota^\beta_\gamma$.
- For α being a limit ordinal number, a-lim$^\alpha(\mathcal{A})$ is an inductive limit of $\langle \iota^\delta_\gamma \rangle_{0<\gamma<\delta<\alpha}$, and each ι^α_δ is a natural injection of the inductive limit.

Theorem 6. For each ordinal number $\alpha > \beta \ge 0$,

1. a-lim$^\alpha(\mathcal{A})$ is indeed a PCA, and ι^α_β is an injective, non-surjective homo.;
2. RpFn(a-lim$^\alpha(\mathcal{A})$) is $\lim^\alpha$(RpFn($\mathcal{A}$)) for α finite, and is $\bigcup_{\beta\in\mathbb{N}} \lim^\beta$ (RpFn($\mathcal{A}$)) otherwise.

Proof. (1) By transfinite induction on α. (2) Note that RpFn(a-lim$^\alpha(\mathcal{A})$) is $\bigcup_{\beta<\alpha}$ RpFn(a-lim$^\beta(\mathcal{A})$). Theorem 1 proves the case for finite α. Let's first consider the case $\alpha = \omega+1$. RpFn(a-lim$^{\omega+1}(\mathcal{A})$) = lim $\left(\bigcup_{\beta\in\mathbf{N}} \text{RpFn(a-lim}^\beta(\mathcal{A}))\right)$ =

$\lim\left(\bigcup_{\beta\in\mathbf{N}}\lim^{\beta}(\mathrm{RpFn}(\mathcal{A}))\right)$. Therefore, $\varphi(-)\in\mathrm{RpFn}(\lim^{\omega+1}(\mathcal{A}))$ iff $\exists\beta<\omega\exists\xi\in\lim^{\beta}(\mathrm{RpFn}(\mathcal{A}))\forall n.\,\exists t_0'\forall t_0>t_0'\varphi(n)\simeq\xi(t_0,n)$, iff $\varphi(-)\in\mathrm{RpFn}(\lim^{\omega+1}(\mathcal{A}))$ iff $\exists\beta<\omega\exists\xi'\in\mathrm{RpFn}(\mathcal{A})\,\exists t_0'\forall t_0>t_0'\cdots\exists t_\beta'\forall t_\beta>t_\beta'\varphi(n)\simeq\xi'(t_0,\cdots,t_\beta,n)$. So, $\mathrm{RpFn}(\lim^{\omega+1}(\mathcal{A}))$ is $\mathrm{RpFn}(\lim^{\omega}(\mathcal{A}))$. In this way, we can prove the second part of the theorem. □

We similarly define n-$\lim^{\alpha}(\mathcal{A})$. Theorem similar to Theorem 2 also holds. We have above Theorem with a-lim$(-)$ replaced by n-lim$(-)$, but RpFn(n-$\lim^{\alpha}(\mathcal{A})$) is the set of all partial numeric functions for $\alpha>0$.

We do not know whether there is an ordinal number α such that a-$\lim^{\alpha}(\mathcal{A})$ is isomorphic to a-lim (a-$\lim^{\alpha}(\mathcal{A})$). $\iota_{\alpha}^{\alpha+1}$ cannot be such an isomorphism, according to above Theorem. There is no $\mathcal{A}$ isomorphic to n-lim$(\mathcal{A})$, because of the cardinality argument. Although the construction a-lim$(-)$ is an endofunctor of a category $\mathbb{PCA}$, it is difficult to employ a category-theoretic version of Tarski's fixpoint theorem arguments; because two homo's a-lim$(\iota_{\beta}^{\beta+1})$ and $\iota_{\beta+1}^{\beta+2}$ are not equal but their equalizer can be proved to be $\iota_{\beta}^{\beta+1}$.

An infinitely long term $\langle P_1,P_2,\ldots\rangle$ of Tait's and Feferman's λ-calculi (resp.) can be interpreted as $\langle a_1,a_2,a_3,\ldots,\rangle\simeq[t\mapsto\langle a_1,a_2,a_3,\ldots,a_t\rangle_t]_\sim$ in n-$\lim^{\alpha+1}(\mathcal{A})$ and a-$\lim^{\alpha+1}(\mathcal{A})$ (resp.) whenever each P_i is interpreted as $a_i\in$ a-$\lim^{\alpha}(\mathcal{A})$. In particular, if P_i is an ordinary λ-term, then $\alpha=0$. Here, $\langle a_1,a_2,a_3,\ldots,a_t\rangle$ is the abbreviation of $\mathbf{cons}\ a_1\ (\cdots(\mathbf{cons}\ a_t\ \mathbf{i})\cdots)$, for a pairing $\mathbf{car}\ (\mathbf{cons}\ a\ b)=a$, and $\mathbf{cdr}\ (\mathbf{cons}\ a\ b)=b$. For example, we can define $\mathbf{cons}=\lambda xyz.\,xyz$, $\mathbf{car}=\lambda x.\,x\mathbf{k}$, $\mathbf{cdr}=\lambda x.\,x(\mathbf{ki})$.

6 An Interpretation of Type-Free $\lambda\mu$-Calculus

In [10], Parigot introduced the typed $\lambda\mu$-calculus which corresponds to classical propositional logic via Curry-Howard isomorphism. By forgetting the types in the $\lambda\mu$-calculus, we obtain a *type-free $\lambda\mu$-calculus.* Both calculi are related to type-free/typed programming languages with control operators (call/cc, $\mathcal{C}$ introduced by M. Felleisen and D. Friedman, raise-handle, ...).

Type-free $\lambda\mu$-calculus is specified by defining $\lambda\mu$-terms, and the reduction rules (the β-reduction rule and the μ-reduction rule).

$\lambda\mu$-Terms are generated by $M::=c\mid x\mid MM\mid\lambda x.\,M\mid[\alpha]M\mid\mu\alpha.\,M$. An occurrence of α in $\mu\alpha.\,\cdots$ will be called a *bound* occurrence of α. An occurrence of α which is not bound is called *free.* A $\lambda\mu$-term $[\alpha]M$ is regarded as application of M to the continuation bound to α. $\mu\alpha.\,M$ is regarded as functional abstraction over the continuation variable α on the level of continuation semantics. Here, $x,y,z,\ldots$ range over term-variables. $\alpha,\beta,\gamma,\ldots$ range over *μ-variables* which are distinguished from the term-variables.

Mixed substitution. A *context* is generated by the above grammar with a special constant ( ). By replacing the occurrences of ( ) of a context $C(\ )$ with a $\lambda\mu$-term M, we obtain a $\lambda\mu$-term $C(M)$. For any context $C(\ )$ and a $\lambda\mu$-term N, we define a *mixed substitution* $\vartheta = \big[[\alpha](\ ) := C(\ )\big]$ as follows; If N is a $\lambda\mu$-term, so is a $N\vartheta$. If N does not have a free μ-variable α, then $N\vartheta$ is N. So, if N is a variable or $\mu\alpha.\,M$, then $N\vartheta \equiv N$. The mixed substitution commutes with a λ-abstraction and an application. The mixed substitution $\vartheta = \big[[\alpha](\ ) := C(\ )\big]$ satisfies $([\beta]M)\vartheta \equiv [\beta](M\vartheta)$ and $(\mu\beta.\,M)\vartheta \equiv \mu\beta.\,(M\vartheta)$, provided $\beta \neq \alpha$. Finally, $([\alpha]M)\big[[\alpha](\ ) := C(\ )\big] \equiv C\big(M\big[\,[\alpha](\ ) := C(\ )\,\big]\big)$.

μ-reduction is specified by the rule: $(\mu\alpha.\,M)N \to_\mu \mu\alpha.\Big(M\big[[\alpha](\ ) := [\alpha]((\ )N)\big]\Big)$. In graphical notation, the rule is

$$(\mu\alpha.(\cdots([\alpha]P)\cdots))N \to_\mu \mu\alpha.(\cdots([\alpha](P'N))\cdots). \tag{1}$$

For instance, $\big(\mu\alpha.\,[\alpha]\,(y[\alpha]x)\big)z \to_\mu \mu\alpha.\,[\alpha]\,(y\,([\alpha](xz))\,z)$. Of course, the type-free $\lambda\mu$-calculus has the usual β-reduction.

6.1 Informal Semantics of $\lambda\mu$

μ-application(-abstraction, -reduction) = infinite applications(abstractions, β-reduction). Consider an informal translation from the type-free $\lambda\mu$-calculus to (infinitely long) type-free λ-calculus:

$$[\alpha]P \mapsto \tilde{P}\alpha_0 \ldots \alpha_m \cdots \qquad \text{and} \qquad \mu\alpha.\,M \mapsto \lambda\alpha_0 \ldots \alpha_m \ldots .\,\tilde{M}\ . \tag{2}$$

Then, the above rewriting rule (1) is translated to

$$(\lambda\alpha_0\alpha_1 \ldots (\cdots(\tilde{P}\alpha_0\alpha_1 \ldots)\cdots))\tilde{N} \to_\beta \lambda\alpha_0\alpha_1 \ldots (\cdots(\tilde{P}'\tilde{N}\alpha_0\alpha_1 \ldots)\cdots), \tag{3}$$

which turns out to be a β-reduction between infinite terms, by renaming bound variables infinite times $\lambda\alpha_1 \ldots (\cdot\ \cdot\ \cdot(\tilde{P}'\tilde{N}\alpha_1 \ldots)\ \cdot\ \cdot) \equiv \lambda\alpha_0\alpha_1 \ldots (\cdots(\tilde{P}'\tilde{N}\alpha_0\alpha_1 \ldots)\cdots)$. The η-like reduction on continuation $\mu\alpha.\,[\alpha]P \to P$ (if α is not free in P.) turns out to be just *infinite* η-reductions(see Theorem 9). This idea of Parigot is being studied by Fujita [4]; with type-regime, the translation above precisely corresponds to Gödel translation and the length of $\alpha_0, \alpha_1, \ldots$ is finite.

μ-variable = infinite stream. The idea of (2) leads us to interpret $[\alpha]P \simeq \lim_t \tilde{P}\alpha_0\alpha_1 \ldots \alpha_t \simeq [t \mapsto \tilde{P}\alpha_0\alpha_1 \ldots \alpha_t]_\sim$ $\alpha_i \simeq \mathbf{car} * (\mathbf{cdr} * \cdots (\mathbf{cdr} * \alpha)\cdots)$, where **cdr** is successively applied i-times to α. Then, we have a *Swap rule* of Streicher-Reus's version of type-free $\lambda\mu$-calculus [13]: $[\mathbf{cons}*\ M * N]P \simeq [N](P * M)$, if we allow more general *continuation terms* than mere μ-(continuation) variables namely pure λ-terms stacked to a μ-variable $\mathbf{cons} * M_1 * (\mathbf{cons} * M_2 * \ldots * (\mathbf{cons} * M_n * \alpha)\ldots)$.

μ-reduction causes delay in a stream. The translation result of the rewriting rule (1) is $f \to g$ s. t.

$$f(t) \simeq (\lambda a_0 \ldots a_t.\,(\cdots(\tilde{P}a_0 \ldots a_t)\cdots))N \simeq \lambda a_1 \ldots a_t.\,(\cdots(\tilde{P}\tilde{N}a_2 \ldots a_t)\cdots),$$
$$g(t) \simeq \lambda a_0 \ldots a_t.(\cdots(\tilde{P}\tilde{N}a_0 \ldots a_t)\cdots)$$

But, $f(t+1) \simeq g(t)$. Because it takes 1 'clock time' to compute(by β-reduction) g from f, a delay will occur because of the extra computational time required. Anyway, $[f]_\sim \neq [g]_\sim$ in n-lim($\mathcal{A}$). So, n-lim($\mathcal{A}$) does not interpret the calculus. Neither does a-lim($\mathcal{A}$).

To equate f and g above, let us replace the symmetric, transitive relation $\sim$ with the smallest symmetric, transitive relation $\approx$ containing $\sim$ and the 'delay' rule $f \approx (t \mapsto f(t+1))$. Unfortunately, a quotient set $(\mathbb{N} \rightharpoonup |\mathcal{A}|) \mathbin{/} \approx$ cannot have the synchronous application operator $[f]_\approx * [g]_\approx \simeq [t \mapsto f(t) \cdot g(t)]_\approx$ well-defined.

6.2 Asynchronous Applicative Structure and Parallel Limit

Given a PCA $\mathcal{A}$, we introduce another partial algebra n-LIM($\mathcal{A}$) where an appropriate application operator can be defined. The carrier set of n-LIM($\mathcal{A}$) is $\{f \mid \exists n \geq 0.\, f : \mathbb{N}^n \rightharpoonup |\mathcal{A}| \text{ and } f \sim f\}/\sim$, where $\sim$ is a symmetric, transitive relation over $\bigcup_{n\geq 0}(\mathbb{N}^n \rightharpoonup |\mathcal{A}|)$ defined by the symmetricity rule, the transitivity rule plus the following two rules. Let $f : \mathbb{N}^n \rightharpoonup \mathcal{A}$. Then

1. The 'delay' rule. $f \sim \mathbb{N}^n \xrightarrow{id\times\cdots\times id\times suc\times id\cdots\times id} \mathbb{N}^n \xrightarrow{f} \mathcal{A}$ where id is the identity function on $\mathbb{N}$ and suc is the successor function.
   As for $\times$, it is an associative operator and for all $f_i : A_i \rightharpoonup B_i$ we have $f_1 \times f_2 :\ A_1 \times A_2 \rightharpoonup B_1 \times B_2$ s.t. $(f_1 \times f_2)(a_1, a_2)$ is $(f_1(a_1), f_2(a_2))$ if each $f_i(a_i)$ is defined, and it is undefined otherwise.
   The rule is necessary to have $\lim_t f(t) \simeq \lim_t f(t+1)$.
2. The 'exchange' and 'weakening' rule. $f \sim \mathbb{N}^m \xrightarrow{(\pi^m_{\sigma(1)},\pi^m_{\sigma(2)},\ldots,\pi^m_{\sigma(n)})} \mathbb{N}^n \xrightarrow{f} \mathcal{A}$, where $m \geq n$, σ is a permutation on $\{1,\ldots,n\}$, and for each $k = 1,\ldots,m$ the function π^m_k returns the k-th argument.
   For all $f_i : A \rightharpoonup B_i$ we have $(f_1,\ldots,f_n) :\ A \rightharpoonup B_1 \times \cdots \times B_n$ s.t. $(f_1,\ldots,f_n)(a)$ is $(f_1(a),\ldots,f_n(a))$ if each $f_i(a)$ is defined, and it is undefined otherwise. The rule will make the following application operator well-defined.

Lemma 1. Let $i, j, k \geq 0$. If $\overset{\infty}{\forall}\, (u_1,\ldots,u_k) \in \mathbb{N}^k.\ \forall s_1,\ldots,s_i,t_1,\ldots,t_j \in \mathbb{N}.$ $f_1(u_1,\ldots,u_k,s_1,\ldots,s_i) \simeq f_2(u_1,\ldots,u_k,t_1,\ldots,t_j)$, then $f_1 \sim f_2$,

Proof. Assume that $\boldsymbol{t} > t_0$ implies $f_1(\boldsymbol{t},\boldsymbol{t_1}) \simeq f_2(\boldsymbol{t},\boldsymbol{t_2})$. Let $g_i(\boldsymbol{t},\boldsymbol{t_i}) \simeq f_i(t_1 + t_0,\ldots,t_k+t_0,\boldsymbol{t_i})$. Then, $f_i \sim g_i$ by kt_0 times repeated applications of the 'delay' rule. Because $g_1(\boldsymbol{t},\boldsymbol{t_1}) \simeq g_2(\boldsymbol{t},\boldsymbol{t_2})$, by using the 'exchange' and 'weakening' rule from we have $g_1 \sim g_2$. By the transitivity of $\sim$, we have $f_1 \sim f_2$. □

Remark 2. Consistent with Lemma 1, for each X = a-lim($\mathcal{A}$), n-lim($\mathcal{A}$), or n-LIM($\mathcal{A}$), we mention the relationship between the symmetric, transitive relation $\sim$ for X and the limit structure of X.

- In a-lim($\mathcal{A}$), we have always $i = j = 0$ and $k = 1$ and for an autonomous sequence of $\mathcal{A}$-elements the sequential limit $\lim_t a \cdot \overline{t}$ corresponds to $[a]_\sim \in$ a-lim($\mathcal{A}$).

- In n-lim($\mathcal{A}$), we have always $i = j = 0$ and $k = 1$ and for a $f : \mathbb{N} \rightharpoonup \mathcal{A}$ the sequential limit $\lim_t f(t)$ corresponds to $[f]_\sim \in$ n-lim($\mathcal{A}$).
- In n-LIM($\mathcal{A}$), for a $f : \mathbb{N}^{k+i} \rightharpoonup \mathcal{A}$ the parallel limit $\lim_{t_1,\ldots,t_{k+i}} f(t_1, \ldots, t_{k+i})$ corresponds to $[f]_\sim \in$ n-LIM($\mathcal{A}$).

An asynchronous application operator. For $f : \mathbb{N}^n \rightharpoonup \mathcal{A}$ and $g : \mathbb{N}^m \rightharpoonup \mathcal{A}$, define

$$[f]_\sim * [g]_\sim \simeq [h]_\sim \ , \text{ with } h = \mathbb{N}^n \times \mathbb{N}^m \xrightarrow{f \times g} \mathcal{A} \times \mathcal{A} \xrightarrow{(-)\cdot(-)} \mathcal{A}$$

where $(-) \cdot (-)$ is the application operator of a given PCA $\mathcal{A}$. The operator $*$ is 'asynchronous' in the sense that $f \times g$ is involved. The operator, as well as the relation $\sim$, permits 'delay' in the arguments (as streams). Therefore,

Lemma 2. $(-) * (-)$ is well-defined.

We say n-LIM($\mathcal{A}$) is *the extension of a* PCA $\mathcal{A}$ *by the non-constructive parallel limits and the asynchronous application.*

Remark 3. We explain the application operator with the vocabulary of the concurrency theory:

1. f and g (resp.) is a process having at most n and m (resp.) independent clocks. For all time slices[1] except for finite numbers, we can observe $\mathcal{A}$-elements.
2. h is a process having the clocks of both f and g. Given a time slice, let a be the observation of f at a given time slice and let b be the observation of g at a given time slice. Then the observation of h at a given time slice is $a \cdot b$.

As is common, the contraction rule is seen as a communication(synchronization). In defining $\sim$, we cannot replace the 'exchange' and 'weakening' rule with the 'exchange,' 'weakening' and 'contraction' rule $\mathbb{N}^n \xrightarrow{f} \mathcal{A} \sim \mathbb{N}^m \xrightarrow{(\pi^m_{\sigma(1)},\pi^m_{\sigma(2)},\ldots,\pi^m_{\sigma(n)})} \mathbb{N}^n \xrightarrow{f} \mathcal{A}$ where $m \geq n$, σ is a function on $\{1, \ldots, n\}$.

Lemma 3. In n-LIM($\mathcal{A}$), we have $[k]_\sim * x * y = x$. It is not the case that $[s]_\sim * x * y * z \simeq x * z * (y * z)$. But, it is the case if $x = [h]_\sim$ with some $h \in \mathcal{A}$.

Proof. Let $x = [f]_\sim, y = [g]_\sim, z = [h]_\sim$, and the arities of f, g, h be n, m, l. $[k]_\sim * x * y$ is $[p]_\sim$ s.t. $p = f \circ (\pi_1^{n+m} \times \cdots \times \pi_n^{n+m}) \sim f$ by the exchange and weakening rule. Therefore $[k]_\sim * x * y = x$. On the other hand, $[s]_\sim * x * y * z$ is $[u]_\sim$ with u being naturally an $(n+m+l)$-ary partial function, while $x * z * (y * z)$ is $[v]_\sim$ with v being naturally $(n + l + m + l)$-ary. So, it is difficult to have the equation unless $l = 0$ (i.e., $h \in \mathcal{A}$). □

Definition 5. For every polynomial $t[x]$, define the polynomial $\lambda x.\, t[x]$ as follows. Let x not occur in u. $\lambda x.\, a \simeq [k]_\sim * a$. $\lambda x.\, x \simeq [\lambda x.\, x]_\sim$, $\lambda x.\, u * x \simeq u$.

[1] $(t_1, \ldots, t_n)$ where each t_i is the value of the clock.

– $\lambda x. t[x] * u \simeq [\lambda xyz. xzy]_\sim * (\lambda x. t[x]) * u$.
– $\lambda x. u * t[x] \simeq [\lambda xyz. x(yz)]_\sim * u * (\lambda x. t[x])$.
– $\lambda x. t[x] * t'[x] \simeq [s]_\sim * (\lambda x. t[x]) * (\lambda x. t'[x])$.

Lemma 4. For every $a \in$ n-LIM($\mathcal{A}$) we have $(\lambda x. t[x])a \simeq t[a]$, if x occurs in $t[x]$ at most once, or if $a = [h]_\sim$ with some $h \in \mathcal{A}$.

By using the λ-abstraction of Definition 5, we re-define the *Church numeral* $\bar{t}$, and the pairing by the equations at the end of last section. Let $\mathbf{nth} \simeq \lambda xy. \mathbf{car} * (y * \mathbf{cdr} * x)$.

Lemma 5. n-LIM($\mathcal{A}$) has a pairing **cons**, **car**, **cdr**, and **nth** s.t.

$$\mathbf{nth} * \langle x_0, \ldots, x_n \rangle * \bar{t} \simeq x_t.$$

We can define similarly *the extension* a-LIM($\mathcal{A}$) *of a* PCA $\mathcal{A}$ *by the autonomous parallel limits and the asynchronous application.*

6.3 The Interpretation

Convention 1 In every $\lambda\mu$-term M, every bound μ-variable is distinct[2] and different from any free μ-variables.

Definition 6. Given a type-free $\lambda\mu$-term M. Let $\{\alpha, \beta, \ldots\}$ be the set of μ-variables in M, then a set $\{t_\alpha, t_\beta, \ldots\}$ of numeric variables is denoted by M^p.

We will define the partial function M^g returns at most one $\mathcal{A}$-element, when the values of M^p is determined; $x^g \simeq x$ ranges over the elements of $\mathcal{A}$, $(MN)^g \simeq M^g N^g$ and $(\lambda x. M)^g \simeq \lambda x. M^g$.

$$([\alpha]M)^g \simeq M^g(\mathbf{nth}\ \alpha\ \overline{0}) \ldots (\mathbf{nth}\ \alpha\ \overline{t_\alpha});$$
$$(\mu\alpha. M)^g \simeq \lambda\alpha_0 \ldots \alpha_{t_\alpha}. M^g\big[\alpha := \langle \alpha_0, \ldots, \alpha_{t_\alpha} \rangle\big].$$

The interpretation $[\![M]\!]$ of M in n-LIM($\mathcal{A}$) is $[M^g]_\sim$.

We will prove that this interpretation works for the affine type-free $\lambda\mu$-calculus.

Lemma 6. $(P[x := Q])^g \simeq P^g[x := Q^g]$.

Proof. By induction on P. We abbreviate $[x := Q]$ as θ, $[x := Q^g]$ as θ^g. Case 1. P is $[\alpha]M$. Then L.H.S. $([\alpha](M\theta))^g$ is $\simeq (M\theta)^g\ (\mathbf{nth}\ \alpha\ \overline{0})\ \ldots\ (\mathbf{nth}\ \alpha\ \overline{t_\alpha})$. By I.H., it is $M^g\theta^g\ (\mathbf{nth}\ \alpha\ \overline{0})\ \ldots\ (\mathbf{nth}\ \alpha\ \overline{t_\alpha}) \simeq \big(M^g\ (\mathbf{nth}\ \alpha\ \overline{0})\ \ldots\ \big)\theta^g$, which is the R.H.S.. Case 2. P is $\mu\beta. M$. Then the L.H.S. is $(\mu\beta. (M\theta))^g \simeq \lambda\beta_0 \ldots \beta_{t_\beta}. (M\theta)^g\ [\beta := \langle \beta_0, \ldots, \beta_{t_\beta} \rangle]$. By I.H., it is $\lambda\beta_0 \ldots \beta_{t_\beta}. M^g\theta^g\ [\beta := \langle \beta_0, \ldots, \beta_{t_\beta} \rangle]$. Because we can assume that β is not in Q without loss of generality, the last is $\big(\lambda\beta_0 \ldots \beta_{t_\beta}. M^g\ [\beta := \langle \beta_0, \ldots, \beta_{t_\beta} \rangle]\big)\ \theta^g$, which is the R.H.S.. When P is of the other forms, then it is trivial. □

[2] That is, $\alpha \neq \beta$, if M is $\cdots(\mu\alpha. \cdots\alpha\cdots)\cdots(\mu\beta. \cdots\beta\cdots)\cdots$.

Lemma 7. $\left(P[[\alpha](\,) := (\,)x_0 \dots x_{t_\alpha}] \right)^g \simeq P^g[\,\alpha := \langle x_0, \dots, x_{t_\alpha} \rangle\,]$.

Proof. By induction on P. ϑ stands for $[\,[\alpha](\,) := (\,)x_0 \dots x_{t_\alpha}\,]$ and θ for $[\,\alpha := \langle x_0, \dots, x_{t_\alpha} \rangle\,]$. Case 1. P is $[\alpha]M$. Then the L.H.S. is $(M\vartheta\; x_0 \dots x_{t_\alpha})^g$. By I.H., it is $M^g\theta x_0 \dots x_{t_\alpha}$, which is the R.H.S.. Case 2. P is $\mu\beta.\, M$ with $\beta \neq \alpha$. Then the L.H.S. is $(\mu\beta.\, M\vartheta)^g$, which is $\lambda\beta_0 \dots \beta_{t_\beta}.\,(M\vartheta)^g\;[\beta := \langle \beta_0,\; \dots\; \beta_{t_\beta} \rangle]$. By I.H., it is $\lambda\beta_0 \dots \beta_{t_\beta}.\, M^g\theta\;[\beta := \langle \beta_0,\; \dots\; \beta_{t_\beta} \rangle]$. It is $\lambda\beta_0 \dots \beta_{t_\beta}.\, M^g[\beta := \langle \beta_0,\; \dots\; \beta_{t_\beta} \rangle]\theta$, which is the R.H.S.. When P is of the other forms, then it is trivial. □

Theorem 7. If x occurs free at most once in M, or if no μ-variable occurs in N, then $((\lambda x.\, M)N)^g \simeq (M[x := N])^g$. Hence, $[\![(\lambda x.\, M)N]\!] \simeq [\![M[x := N]]\!]$.

Proof. The L.H.S. is $(\lambda x.\, M^g)N^g$, where x occurs free at most once in M^g, or $N \simeq [h]_\sim$ for some $h \in \mathcal{A}$. So, the L.H.S. is $M^g[x := N^g]$, by Lemma 4. It is the R.H.S. by Lemma 6. □

Theorem 8. Let α occur free at most once in P, or let no μ-variable occur in Q. Then we have

μ-reduction.	$[\![(\mu\alpha.\, P)Q]\!] \simeq [\![\mu\alpha.\, P[\,[\alpha](\,) := [\alpha]((\,)Q)\,]]\!]$
Parigot's S3 rule.	$[\![\mu\alpha.\, M]\!] \quad \simeq [\![\lambda x.\, \mu\alpha.\, M[\,[\alpha](-) \;:=\; [\alpha]((-)x)\,]]\!]$.

Proof. In order to prove the two statements, we will claim resp.

1. if $f(t_\alpha, t_\beta) \simeq ((\mu\alpha.\, P)Q)^g$, then $f(t_\alpha{+}1, t_\beta) \simeq \left(\mu\alpha.\, P[[\alpha](\,) := [\alpha]((\,)Q)]\right)^g$; and
2. if $f(t_\alpha, t_\beta) \simeq (\mu\alpha.\, P)^g$, then $f(t_\alpha{+}1, t_\beta) \simeq \left(\lambda x.\, \mu\alpha.\, P[\,[\alpha](\,) := [\alpha]((\,)x)\,]\right)^g$.

(1) The L.H.S. $f(t_\alpha + 1, t_\beta)$ is $\left(\lambda\alpha_0 \dots \alpha_{t_\alpha+1}.\, P^g[\alpha := \langle \alpha_0, \dots, \alpha_{t_\alpha+1} \rangle\,]\right)Q^g$, which is, by Lemma 7, $(\lambda\alpha_0 \dots\; (\,P[\,[\alpha](\,) := (\,)\alpha_0 \dots \alpha_{t_\alpha+1}\,]\,)^g)Q^g$, which is, by Lemma 6 and the premise, $\lambda\alpha_1 \dots \alpha_{t_\alpha+1}.\, \left(\,P[\,[\alpha](\,) := (\,)Q\alpha_1 \dots \alpha_{t_\alpha+1}\,]\,\right)^g$, which is, by renaming bound variables, $\lambda\alpha_0 \dots \alpha_{t_\alpha}.\, \left(P[[\alpha](\,) := (\,)Q\alpha_0 \dots \alpha_{t_\alpha}]\right)^g$. Because of Convention 1, the mixed substitution above is a composition of two mixed substitutions: $\left[\,[\alpha](\,) := [\alpha]((\,)Q)\,\right]\left[\,[\alpha](\,) := (\,)\alpha_0 \dots \alpha_{t_\alpha}\,\right]$. By Lemma 7, $f(t_\alpha + 1, t_\beta) \simeq \lambda\alpha_0 \dots \alpha_{t_\alpha}.\left(P[[\alpha](\;) := [\alpha]((\;)Q)]\right)^g\;[\alpha := \langle \alpha_0, \dots, \alpha_{t_\alpha} \rangle]$, which is the R.H.S. $\left(\mu\alpha.\, P[[\alpha](\,) := [\alpha]((\,)Q)]\right)^g$. (2) is similarly proved. □

Theorem 9 (η_{cont})**.** If α is not free in M, then $(\;\mu\alpha.\,[\alpha]M\;)^g \simeq M^g$, hence $[\![\,\mu\alpha.\,[\alpha]M\,]\!] \simeq [\![M]\!]$.

Proof. $(-)^g$ unwinds each occurrence of α to the same sequence $\alpha_0, \dots, \alpha_{t_\alpha}$. □

When we validate the μ-reduction but not η_{cont}-reduction, we can replace $(-)^g$ with another translation $(-)^G$ satisfying the following two conditions .

(1) for the same μ-variable α, we will distinguish the different occurrences by $\alpha^1, \alpha^2, \ldots$. let $(-)^G$ unwind α^i to $\alpha_0, \alpha_1, \ldots, \alpha_{t_{\alpha^i}}$. (2) For the μ-reduction with the following graphical notation

$$(\mu\alpha^1.(\cdots([\alpha^2]P)\cdots([\alpha^3]Q)\cdots))N \to_\mu \mu\alpha^1.(\cdots([\alpha^2](P'N))\cdots([\alpha^3]Q')\cdots),$$

we have $t_1 \geq t_2, t_3, \ldots$.

7 Final Remarks

In Berardi [2], to obtain an interpretation of the intuitionistic logic + $QxA(x) \vee \neg QxA(x)$ ($A(x)$ is quantifier-free), he uses a completion idea, similar to the topological completion producing $\mathbb{R}$ out of $\mathbb{Q}$. He is concerned with the processes of computing the limit values. Based on those processes, he directly interprets his semi-formal system of Δ^0_2 maps. In constructing his model, he uses intuitionistic reasoning, and, consequently, the cofinally true conditions supersedes the definite true condition (classically, they are the same for converging limits). For $l : D \to \mathbb{N}$ being a converging limit, he defines $P(l)$ is cofinally true iff $P(l(d))$ is true cofinally on D. So, he can effectively find $d \in D$ such that $P(l(d))$. But the only way we have of finding this is to go through all possible values for d.

We conjecture that our non-constructive limit construction n-lim$(-)$ in some constructive set theory is the same as our autonomous limit operation a-lim$(-)$.

References

1. H.P. Barendregt. *The Lambda Calculus.* 1984.
2. S. Berardi. Classical logic as limit completion I&II, submitted, 2001.
3. A. Berarducci. Infinite λ-calculus and non-sensible models. In *Logic and algebra (Pontignano, 1994)*, pp. 339–377. Dekker, New York, 1996.
4. K. Fujita. On proof terms and embeddings of classical substructural logics. *Studia Logica*, 61(2):199–221, 1998.
5. E. Mark Gold. Limiting recursion. *JSL*, 30:28–48, 1965.
6. S. Hayashi and M. Nakata. Towards limit computable mathematics. In *TYPES2000*, LNCS. Springer, 2001.
7. S. Hayashi, R. Sumitomo, and K. Shii. Towards animation of proofs – testing proofs by examples –. *TCS*, 2001. to appear.
8. J.R. Kennaway, J.W. Klop, M.R. Sleep, and F.J. de Vries. Infinitary lambda calculus. *TCS*, 175(1):93–125, 1997.
9. M. Nakata and S. Hayashi. Limiting first order realizability interpretation. *Sci. Math. Japonicae*, 2001. to appear.
10. M. Parigot. λμ-calculus: an algorithmic interpretation of classical natural deduction. In *Logic programming and automated reasoning.* Springer, 1992.
11. H. Schwichtenberg and S. S. Wainer. Infinite terms and recursion in higher types. In J. Diller and G. H. Müller, eds., ⊨ *ISILC Proof Theory Symposion* , pp. 341–364. LNM., Vol. 500, 1975. Springer.
12. R. I. Soare. *Recursively Enumerable Sets and Degrees.* Springer, 1987.

13. Th. Streicher and B. Reus. Classical logic, continuation semantics and abstract machines. *J. Funct. Programming*, 8(6):543–572, 1998.
14. H.R. Strong. Algebraically generalized recursive function theory. *IBM Journal of Research and Development*, 12:465–475, 1968.
15. W.W. Tait. Infinitely long terms of transfinite type. In Crossley and Dummet, eds., *Formal Systems and Recursive Functions*, pp. 176–185. North-Holland, 1965.
16. A. S. Troelstra and D. van Dalen. *Constructivism in Mathematics*, 1988.
17. E. G. Wagner. Uniformly reflexive structures: On the nature of Gödelizations and relative computability. *Trans. of the Amer. Math. Soc.*, 144:1–41, 1969.
18. M. Yasugi, V. Brattka, and M. Washihara. Computability aspects of some discontinuous functions. http://www.kyoto-su.ac.jp/~yasugi/Recent/gaussnew.ps, 2001.

Intersection Logic

Simona Ronchi Della Rocca and Luca Roversi

Dipartimento di Informatica
C.so Svizzera, n. 185
10149 Torino, Italy
{ronchi,rover}@di.unito.it
http://www.di.unito.it/~{ronchi,rover}

Abstract. The intersection type assignment system IT uses the formulas of the negative fragment of the predicate calculus (LJ) as types for the λ-terms. However, the deductions of IT only correspond to the proper sub-set of the derivations of LJ, obtained by imposing a meta-theoretic condition about the use of the conjunction of LJ. This paper proposes a logical foundation for IT. This is done by introducing a logic IL. Intuitively, a derivation of IL is a set of derivations in LJ such that the derivations in the set can be thought of as writable in parallel. This way of looking at LJ, by means of IL, allows to transform the meta-theoretic condition, mentioned above, into a purely structural property of IL. The relation between IL and LJ surely has a first main benefit: the strong normalization of LJ directly implies the same property on IL, which translates in a very simple proof of the strong normalizability of the λ-terms typable with IT.

1 Introduction

The intersection type assignment system (IT) is a set of rules for assigning types to terms of the untyped λ-calculus. The types of IT are formulas of the predicate logic, built from the two connectives "implication" ($\rightarrow$) and "conjunction" ($\wedge$).

IT was introduced in the early Eighty by Mariangiola Dezani and Mario Coppo [6], in order to enhance the tipability power of the Curry type assignment system. The system characterizes important properties of the λ-terms, like the normalization and the strong normalization. Indeed, it has be proved that IT assigns types to all and only the strong normalizing terms [19]. Moreover, if the set of types is extended to contain a "universal type" ω, that can be given to all the λ-terms, then the normalizing terms are exactly those that can be given a type free of occurrences of ω [4].

Intersection types have been particularly useful in studying the semantics of various kinds of λ-calculi. This can be done by extending the system with suitable sub-typing relations. In this way the type assignment is a finitary tool to reason about the interpretation of the terms in the topological models of λ-calculus, like Scott domains, DI-domains, coherence spaces [1,4,7,10,14,15,22].

Unlike other type assignment systems (like Curry type assignment [8], or the type assignment version of the II order λ-calculus [12,17]), IT has not been

L. Fribourg (Ed.): CSL 2001, LNCS 2142, pp. 414–429, 2001.

designed starting from a logic, and up to now, the relationship between IL and the implicational and conjunctive fragment of the predicate calculus (LJ) are far from being clear. This was firstly pointed out by Hindley [13]. The problem is the logical interpretation of the rule introducing the conjunction:

$$(\wedge I_{IT})\frac{\Gamma \vdash_{\wedge} M : \sigma \quad \Gamma \vdash_{\wedge} M : \tau}{\Gamma \vdash_{\wedge} M : \sigma \wedge \tau} \tag{1}$$

If we reason according to the Curry-Howard isomorphism, which is the standard relationship between logic and λ-calculus, then we can observe that in the rule (1) here above the λ-term M denotes two proofs. In particular, (1) says that an intersection type $\sigma \wedge \tau$ can be built from the two components σ and τ only when they can be proved by two "isomorphic" proofs, according to a notion of isomorphism that relates proofs encoded by the same λ-term. Our point is that the use of a λ-term to express the isomorphism of two derivations is a meta-theoretical restriction on the introduction of the conjunction, and for this reason LJ is not the logic which IT is based on.

In this paper we establish a logical foundation for IT, and clarify the relationship between IT and LJ. More precisely, we define the new logic IL, such that every of its deductions corresponds to a set of deductions in LJ, sharing some structural properties. IL is the desired "bridge" between LJ and IT, since a deduction in IT can be obtained as a partial decoration of a deduction in IL. Moreover, IL has all the good properties we ask for a logic. In particular it enjoys the strong normalization property, whose proof directly derives from the analogous property of LJ [20]. Moreover, thanks to the relation between IL and IT, we obtain for free that if a term is typable by an intersection type, then it is also strongly normalizable. As a side result, a typed version of λ-terms with intersection type can be obtained through a full decoration of deductions in IL. But this is subject of a forthcoming paper.

The literature presents other proof-theoretical investigations of the intersection type assignment. Barbanera and Alessi [2], refining a previous attempt of Mints [18], proved that IT, equipped with both β and η-reduction, gives a complete realizability semantics of the predicate logic with implication and "strong conjunction". This result has been further extended to other connectives in [3].

The first attempt to give a logical foundation to intersection types was by Venneri [23]. She proposed an Hilbert style logic corresponding to a system which assigns intersection types to the terms of Combinatory Logic. A further extension with union types is in [9]. Our feeling is that the approach in [23,9] is not suitable for the λ-calculus.

Moreover, in [5] there is the proof that a typed version of IT can be obtained, through the Curry-Howard isomorphism, from a logic, where formulas are sequence of formulas of LJ. The untyped version of such a language is quite similar to the language in [16], which has been defined to highlight the intrinsic parallelism of the β-reduction. Other typed versions of IT have been defined by Reynolds [21] and Wells [24], but they did not follow a logical approach, so are unrelated with the topic of the present paper.

To conclude, we believe that our approach to a logical foundation of IT could evolve by adopting the principles at the base of Ludics, the project initiated by J.-Y. Girard to re-found logic [11].

The paper is organized as follows. In Section 2 and 3 the systems LJ and IT are briefly recalled. In Section 4 the intersection logic IL is defined, and its relation with LJ is stated. In Section 5 the strong normalization of IL is proved. Section 6 contains the main theorem, which formalizes the relation between IL and IT.

2 Intuitionistic Logic: Implicative and Conjunctive Fragment

We start by recalling the natural deduction of the implicative and conjunctive fragment of Intuitionistic Logic that, somewhat abusing the name, we call LJ.

Definition 1. *i) The* formulas *of LJ belong to the language generated by the grammar:*

$$\sigma ::= \alpha \mid (\sigma \to \sigma) \mid (\sigma \wedge \sigma)$$

where $\alpha \in V$ (a denumerable set of constants). The formulas are denoted by ρ, σ, τ. We assume the associativity to the right for $\to$ and the one to the left for $\wedge$.

ii) A context *is a* finite sequence $\sigma_1, \ldots, \sigma_m$ *of formulas, denoted by Γ, Δ, possibly indexed.*

iii) The natural *deduction system LJ proves statements $\Gamma \vdash_{LJ} \sigma$, where Γ is a context and σ a formula. It consists of the following rules:*

$$(A_{LJ})\frac{}{\sigma \vdash_{LJ} \sigma} \qquad (W_{LJ})\frac{\Gamma \vdash_{LJ} \sigma}{\Gamma, \tau \vdash_{LJ} \sigma}$$

$$(X_{LJ})\frac{\Gamma, \sigma, \tau, \Delta \vdash_{LJ} \rho}{\Gamma, \tau, \sigma, \Delta \vdash_{LJ} \rho} \qquad (\wedge I_{LJ})\frac{\Gamma \vdash_{LJ} \sigma \quad \Gamma \vdash_{LJ} \tau}{\Gamma \vdash_{LJ} \sigma \wedge \tau}$$

$$(\wedge E^l_{LJ})\frac{\Gamma \vdash_{LJ} \sigma \wedge \tau}{\Gamma \vdash_{LJ} \sigma} \qquad (\wedge E^r_{LJ})\frac{\Gamma \vdash_{LJ} \sigma \wedge \tau}{\Gamma \vdash_{LJ} \tau}$$

$$(\to I_{LJ})\frac{\Gamma, \sigma \vdash_{LJ} \tau}{\Gamma \vdash_{LJ} \sigma \to \tau} \qquad (\to E_{LJ})\frac{\Gamma \vdash_{LJ} \sigma \to \tau \quad \Gamma \vdash_{LJ} \sigma}{\Gamma \vdash_{LJ} \tau}$$

The deductions of LJ are denoted by $\Pi, \Pi_1, \ldots$. Moreover, $\Pi : \Gamma \vdash_{LJ} \sigma$ means that Π concludes, proving $\Gamma \vdash_{LJ} \sigma$.

Example 1. Let σ denote the formula $(\alpha \wedge \beta) \wedge \gamma$. Define the following three deductions:

$$\Pi_0 : (\wedge E^l_{LJ})\frac{(\wedge E^l_{LJ})\dfrac{(A_{LJ})\dfrac{}{\sigma \vdash_{LJ} (\alpha \wedge \beta) \wedge \gamma}}{\sigma \vdash_{LJ} \alpha \wedge \beta}}{\sigma \vdash_{LJ} \alpha}$$

$$\Pi_1 : (\wedge E^r_{LJ})\frac{(A_{LJ})\dfrac{}{\sigma \vdash_{LJ} (\alpha \wedge \beta) \wedge \gamma}}{\sigma \vdash_{LJ} \gamma}$$

$$\Pi_2 : (\to I_{LJ}) \frac{(A_{LJ}) \overline{\alpha \vdash_{LJ} \alpha}}{\vdash_{LJ} \alpha \to \alpha}$$

Then, Π_3 is:

$$(\wedge I_{LJ}) \frac{(\to I_{LJ}) \dfrac{(\wedge I_{LJ}) \dfrac{\Pi_0 : \sigma \vdash_{LJ} \alpha \quad \Pi_1 : \sigma \vdash_{LJ} \gamma}{\sigma \vdash_{LJ} \alpha \wedge \gamma}}{\vdash_{LJ} (\alpha \wedge \beta) \wedge \gamma \to (\alpha \wedge \gamma)} \qquad \Pi_2 : \vdash_{LJ} \alpha \to \alpha}{\vdash_{LJ} ((\alpha \wedge \beta) \wedge \gamma \to (\alpha \wedge \gamma)) \wedge (\alpha \to \alpha)}$$

Let us recall the strong normalization property of the deductions in LJ.

Definition 2. *Let Π be a deduction of LJ.*

i) *A $\wedge$-LJ-*redex *of Π is a sequence of* two *rules, formed by an instance of $(\wedge I_{LJ})$, followed by an instance of either $(\wedge E^l_{LJ})$ or $(\wedge E^r_{LJ})$;*
ii) *A $\to$-LJ-*redex *of Π is a sequence of* two *rules, formed by an instance of $(\to I_{LJ})$, followed by an instance of $(\to E_{LJ})$;*
iii) *Π is* normal *if it does not contain neither $\wedge$-LJ-redexes nor $\to$-LJ-redexes.*

Lemma 1. *Consider two derivations $\Pi : \Gamma \vdash_{LJ} \sigma$ and $\Pi' : \Gamma, \sigma \vdash_{LJ} \tau$. Call $S(\Pi, \Pi')$ the deductive structure obtained by replacing the conclusion of Π for every occurrence (A_{LJ}) deriving $\sigma \vdash_{LJ} \sigma$, and such that σ to left of $\vdash_{LJ}$ is free in the conclusion of Π'. Then, $S(\Pi, \Pi') : \Gamma \vdash_{LJ} \tau$.*

Definition 3. *The rewriting relation $\rightsquigarrow_{LJ}$ between derivations is defined as follows:*

– $$(\wedge E^s_{LJ}) \frac{(\wedge I_{LJ}) \dfrac{\Pi_l : \Gamma \vdash_{LJ} \sigma_l \quad \Pi_r : \Gamma \vdash_{LJ} \sigma_r}{\Gamma \vdash_{LJ} \sigma_l \wedge \sigma_r}}{\Gamma \vdash_{LJ} \sigma_s} \rightsquigarrow_{LJ} \Pi_s : \Gamma \vdash_{LJ} \sigma_s,$$
where $s \in \{l, r\}$.

– $$(\to E_{LJ}) \frac{(\to I_{LJ}) \dfrac{\Pi_0 : \Gamma, \tau \vdash_{LJ} \sigma}{\Gamma \vdash_{LJ} \tau \to \sigma} \quad \Pi_1 : \Gamma \vdash_{LJ} \tau}{\Gamma \vdash_{LJ} \sigma} \rightsquigarrow_{LJ} S(\Pi_1, \Pi_0) : \Gamma \vdash_{LJ} \sigma$$

Theorem 1 (Prawitz). *The rewriting relation $\rightsquigarrow_{LJ}$ is strongly normalizing.*

3 Intersection Types

We briefly recall the system of Intersection Types (IT), which works as a type assignment system for the untyped λ-calculus.

Definition 4. *i)* *The set of* types *of IT coincides to the formulas of LJ.*
ii) *An IT-context is a finite set of pairs $\{x_1 : \sigma_1, \ldots, x_n : \sigma_n\}$ that assigns types to λ-variables so that $i \neq j$ implies $x_i \not\equiv x_j$. By abusing the notation, Γ and Δ denote IT-contexts.*
iii) *IT is a deductive system that derives judgments $\Gamma \vdash_{IT} M : \sigma$ where M is a λ-term, Γ is an IT-context, and σ is a type. IT consists of the following rules:*

$$(A_{IT})\frac{x:\sigma\in\Gamma}{\Gamma\vdash_{IT}x:\sigma}\qquad(\wedge I_{IT})\frac{\Gamma\vdash_{IT}M:\sigma\quad\Gamma\vdash_{IT}M:\tau}{\Gamma\vdash_{IT}M:\sigma\wedge\tau}$$

$$(\wedge E^{l}_{IT})\frac{\Gamma\vdash_{IT}M:\sigma\wedge\tau}{\Gamma\vdash_{IT}M:\sigma}\qquad(\wedge E^{r}_{IT})\frac{\Gamma\vdash_{IT}M:\sigma\wedge\tau}{\Gamma\vdash_{IT}M:\tau}$$

$$(\rightarrow I_{IT})\frac{\Gamma\cup\{x:\sigma\}\vdash_{IT}M:\tau}{\Gamma\vdash_{IT}\lambda x.M:\sigma\rightarrow\tau}\qquad(\rightarrow E_{IT})\frac{\Gamma\vdash_{IT}M:\sigma\rightarrow\tau\quad\Gamma\vdash_{IT}N:\sigma}{\Gamma\vdash_{IT}MN:\tau}$$

We keep using $\Pi,\Pi_1,\ldots$ to denote the deductions of IT. Moreover, the meaning of $\Pi:\Gamma\vdash_{IT}M:\sigma$ is analogous to the one of $\Pi:\Gamma\vdash_{LJ}\sigma$.

Example 2. Let σ denote the type $(\alpha\wedge\beta)\wedge\gamma$ and let Π_4 be the following deduction:

$$(\rightarrow I_{IT})\frac{(\wedge I_{IT})\dfrac{(\wedge E^{l}_{IT})\dfrac{(\wedge E^{l}_{IT})\dfrac{(A_{IT})\overline{x:\sigma\vdash_{IT}x:\sigma}}{x:\sigma\vdash_{IT}x:\alpha\wedge\beta}}{x:\sigma\vdash_{IT}x:\alpha}\quad(\wedge E^{r}_{IT})\dfrac{(A_{IT})\overline{x:\sigma\vdash_{IT}x:\sigma}}{x:\sigma\vdash_{IT}x:\gamma}}{x:\sigma\vdash_{IT}x:\alpha\wedge\gamma}}{\vdash_{IT}\lambda x.x:(\alpha\wedge\beta)\wedge\gamma\rightarrow(\alpha\wedge\gamma)}$$

Let Π_5 be the following deduction:

$$(\rightarrow I_{IT})\frac{(A_{IT})\overline{x:\alpha\vdash_{IT}x:\alpha}}{\vdash_{IT}\lambda x.x:\alpha\rightarrow\alpha}$$

And finally let Π_6 be:

$$(\wedge I_{IT})\frac{\Pi_4:\vdash_{IT}\lambda x.x:(\alpha\wedge\beta)\wedge\gamma\rightarrow(\alpha\wedge\gamma)\quad\Pi_2:\vdash_{IT}\lambda x.x:\alpha\rightarrow\alpha}{\vdash_{IT}\lambda x.x:((\alpha\wedge\beta)\wedge\gamma\rightarrow(\alpha\wedge\gamma))\wedge(\alpha\rightarrow\alpha)}$$

4 Intersection Logic

In this section we introduce *Intersection Logic* (IL), whose derivations correspond to sets of derivations in LJ, sharing some similarity in the structure. The formulas of IL are binary trees, whose leaves are labeled by formulas of LJ. The relation between IL and LJ can be informally described as follows: a derivation Π of IL groups a set, say $LJ(\Pi)$, of derivations of LJ. Every derivation $\Pi'\in LJ(\Pi)$ can be obtained by taking the leaf of a given path in every tree of Π. In particular, the elements of $LJ(\Pi)$ share both the number of instances and the order of application of the rules introducing and eliminating the connective $\rightarrow$.

We need to introduce some preliminary notions.

Definition 5. *i) A* kit *is a binary tree in the language generated by the following grammar: $K ::= \sigma \mid [K,K]$. The leaves of any kit, which we call also* atoms, *are* formulas of LJ. *The kits are denoted by H,K.*

ii) Two kits overlap *if they are two trees with exactly the same structure, but which may differ only on the name of their atoms; $H\simeq K$ denotes two overlapping kits H and K. For example, $\sigma\simeq\tau$.*

iii) *Two overlapping kits map to a kit with only arrows as its leaves, by the function* $(\ )^+$, *defined as follows:*

$$[\sigma, \tau]^+ \mapsto \sigma \to \tau$$
$$[[H_1, H_2], [K_1, K_2]]^+ \mapsto [[H_1, K_1]^+, [H_2, K_2]^+]$$

$H \to K$ denotes the result of $[H, K]^+$, with $H \simeq K$. Otherwise, $H \to K$ is undefined.

iv) *A* path *is a string built over the set $\{l, r\}$; p, q, possibly indexed, denote paths, and ϵ denotes the empty path. The subtree of a kit H at the path p is inductively defined as follows:*

$$H^\epsilon = H, \qquad [H_1, H_2]^{lp} = H_1^p, \qquad [H_1, H_2]^{rp} = H_2^p$$

Otherwise, σ^p is undefined.
A path p is defined *in H if H^p is defined, and it is* terminal *in H if H^p is an atom; the set of terminal paths of a kit H is denoted by $P_T(H)$.*
$H[p := K]$ denotes the kit resulting from the replacement of K for H^p in H.

v) *Let $s \in \{l, r\}$, and let ps be a path, defined in H. The* pruning *of H at path ps, is defined as follows: $H \backslash^{ps} = H[p := H^{ps}]$.*

vi) *$\equiv$ is the syntactical identity of both atoms, kits and paths.*

By definition, we have:

Fact 1 *1.* $P_T(H \to K) = P_T(H) = P_T(K)$;
2. $H\backslash^p \to K\backslash^p \equiv (H \to K)\backslash^p$;
3. If p and q are two different paths, then $(H\backslash^p)\backslash^q \equiv (H\backslash^q)\backslash^p$.

The definition here below is about the deductive system $\vdash_{pIL}$ on which we shall define IL. The key feature of $\vdash_{pIL}$ is that every of its judgments exclusively contains overlapping kits. Intuitively, this invariance on the judgment form formalizes that every derivation of $\vdash_{pIL}$, being introduced, stands for a set of deductions of LJ, which share structural properties.

Definition 6 (Natural deduction $\vdash_{pIL}$). *The natural deduction system $\vdash_{pIL}$, that we call* pre Intersection Logic, *derives judgments $\Gamma \vdash_{pIL} K$, where Γ is a sequence of kits and K is a kit. It consists of the following rules:*

$$(A_{pIL})\frac{}{H \vdash_{pIL} H} \qquad (X_{pIL})\frac{\Gamma, H, H', \Delta \vdash_{pIL} K}{\Gamma, H', H, \Delta \vdash_{pIL} K}$$

$$(W_{pIL})\frac{H_1, \ldots, H_n \vdash_{pIL} H \quad H_i \simeq H' \quad (1 \leq i \leq n)}{H_1, \ldots, H_n, H' \vdash_{pIL} H}$$

$$(P_{pIL})\frac{\Gamma \vdash_{pIL} K \quad s \in \{l, r\} \quad ps \in P_T(K)}{\Gamma\backslash^{ps} \vdash_{pIL} K\backslash^{ps}}$$

$$(\to E_{pIL})\frac{\Gamma \vdash_{pIL} H \to K \quad \Gamma \vdash_{pIL} H}{\Gamma \vdash_{pIL} K} \qquad (\to I_{pIL})\frac{\Gamma, H \vdash_{pIL} K}{\Gamma \vdash_{pIL} H \to K}$$

$$(\wedge E^l_{pIL})\frac{\Gamma \vdash_{pIL} K[p := \sigma \wedge \tau]}{\Gamma \vdash_{pIL} K[p := \sigma]} \qquad (\wedge E^r_{pIL})\frac{\Gamma \vdash_{pIL} K[p := \sigma \wedge \tau]}{\Gamma \vdash_{pIL} K[p := \tau]}$$

$$(\wedge I_{pIL})\frac{H_1[p := [\sigma_1, \sigma_1]], ..., H_n[p := [\sigma_n, \sigma_n]] \vdash_{pIL} K[p := [\sigma, \tau]]}{H_1[p := \sigma_1], ..., H_n[p := \sigma_n] \vdash_{pIL} K[p := \sigma \wedge \tau]}$$

where, in rule (P_{pIL})*, the notation* $\Gamma\backslash^{ps}$ *stands for the distribution of the pruning to the components of* Γ.

The judgments of $\vdash_{pIL}$ enjoy an invariant:

Lemma 2. *If* $H_1, \ldots, H_n \vdash_{pIL} K$*, then* $H_i \simeq K$ $(1 \leq i \leq n)$.

Proof. By structural induction on the deduction of $H_1, \ldots, H_n \vdash_{pIL} K$.

In the following definition we introduce a "decoration" of all the systems previously defined, inspired to the so called "Curry-Howard isomorphism": every deduction Π is associated to a λ-term to keep track of some structural properties of Π. Note that this decoration is not standard: the λ-term associated to Π is *untyped*, and does not encode *the whole* structure of Π, but just the order of the occurrences of the rules which introduce and eliminate $\rightarrow$. Moreover, the decoration is a partial function when applied to a derivation of LJ. Indeed, a decoration of a proof whose last rule is $(\wedge I_{LJ})$ is defined only if the derivations of its two premises are decorated by the same term.

Definition 7. *Let* $\vdash_* \in \{LJ, pIL\}$*, and* $A, B, C, ...$ *be meta-variables for denoting either atoms or kits. Also, let* Δ *denote a sequence built over* $\{A, B, C, ...\}$.

i) Every Π *proving* $\Delta \vdash_* A$ *can be decorated by a* λ*-term* $T_{dom(\Delta^*)}(\Pi)$*, where* Δ^* *is a decoration of* Δ*, and, if* $\Delta^* \equiv x_1 : A_1, \ldots, x_n : A_n$*, then* $dom(\Delta^*)$ *is the sequence* $x_1, \ldots, x_n$*. The decoration of* $\vdash_*$ *is denoted by* $\vdash^+_*$ *and is inductively defined as:*

- $\Pi : (A_*)\frac{}{A \vdash_* A} \Rightarrow (A^+_*)\frac{}{x : A \vdash^+_* x : A}$
  and $T_x(\Pi) \equiv x$;
- $\Pi \ : \ (W_*)\frac{\Pi_1 : \Delta \vdash_* A}{\Delta, B \vdash_* A} \ \Rightarrow \ (W^+_*)\frac{\Delta^* \vdash^+_* T_{dom(\Delta^*)}(\Pi_1) : A}{\Delta, x : B \vdash^+_* T_{dom(\Delta^*),x}(\Pi_1) : A}$
  where Δ^* *is the decoration of* Δ*,* x *is fresh, and* $T_{dom(\Delta^*),x}(\Pi) \equiv T_{dom(\Delta^*)}(\Pi_1)$;
- $\Pi : (\rightarrow I_*)\frac{\Pi_1 : \Delta, A \vdash_* B}{\Delta \vdash_* A \rightarrow B} \Rightarrow$
  $$(\rightarrow I^+_*)\frac{\Delta^*, x : A \vdash^+_* T_{dom(\Delta^*),x}(\Pi_1) : B}{\Delta^* \vdash^+_* \lambda x.T_{dom(\Delta^*),x}(\Pi_1) : A \rightarrow B}$$
  and $T_{dom(\Delta^*)}(\Pi) \equiv \lambda x.T_{dom(\Delta^*),x}(\Pi_1)$;

- $\Pi : (\to E_*)\dfrac{\Pi_1 : \Delta \vdash_* A \to B \quad \Pi_2 : \Delta \vdash_* A}{\Delta \vdash_* B} \Rightarrow$

$$(\to E_*^+)\frac{\Delta^* \vdash_*^+ T_{dom(\Delta^*)}(\Pi_1) : A \to B \quad \Delta^* \vdash_*^+ T_{dom(\Delta^*)}(\Pi_2) : A}{\Delta^* \vdash_*^+ T_{dom(\Delta^*)}(\Pi_1)T_{dom(\Delta^*)}(\Pi_2) : B}$$

and $T_{dom(\Delta^*)}(\Pi) \equiv T_{dom(\Delta^*)}(\Pi_1)T_{dom(\Delta^*)}(\Pi_2)$*;*

- $\Pi : (P_{pIL})\dfrac{\Pi_1 : K_1, \ldots, K_n \vdash_{pIL} H}{K_1\backslash^{ps}, \ldots, K_n\backslash^{ps} \vdash_{pIL} H\backslash^{ps}} \Rightarrow$

$$(P_{pIL}^+)\frac{x_1 : K_1, \ldots, x_n : K_n \vdash_{pIL}^+ T_{x_1,\ldots,x_n}(\Pi_1) : H}{x_1 : K_1\backslash^{ps}, \ldots, x_n : K_n\backslash^{ps} \vdash_{pIL}^+ T_{x_1,\ldots,x_n}(\Pi) : H\backslash^{ps}}$$

and $T_{x_1,\ldots,x_n}(\Pi) \equiv T_{x_1,\ldots,x_n}(\Pi_1)$*;*

- $\Pi : (\wedge I_{pIL})\dfrac{\Pi_1 : H_1[p := [\sigma_1, \sigma_1]], ..., H_n[p := [\sigma_n, \sigma_n]] \vdash_{pIL} K[p := [\sigma, \tau]]}{H_1[p := \sigma_1], ..., H_n[p := \sigma_n] \vdash_{pIL} K[p := \sigma \wedge \tau]}$

$\Rightarrow$

$$(\wedge I_{pIL}^+)\frac{x_1 : H_1[p := [\sigma_1, \sigma_1]], ..., x_n : H_n[p := [\sigma_n, \sigma_n]] \vdash_{pIL}^+ T_{x_1,\ldots,x_n}(\Pi_1) : K[p := [\sigma, \tau]]}{x_1 : H_1[p := \sigma_1], ..., x_n : H_n[p := \sigma_n] \vdash_{pIL}^+ T_{x_1,\ldots,x_n}(\Pi_1) : K[p := \sigma \wedge \tau]}$$

and $T_{x_1,\ldots,x_n}(\Pi) \equiv T_{x_1,\ldots,x_n}(\Pi_1)$*;*

- *with*

$$\Pi : (\wedge I_{LJ})\frac{\Pi_1 : \Delta \vdash_{LJ} \sigma \quad \Pi_2 : \Delta \vdash_{LJ} \tau}{\Delta \vdash_{LJ} \sigma \wedge \tau}$$

let $\Pi_1^+ : \Delta^* \vdash_{LJ}^+ T_{dom(\Delta^*)}(\Pi_1) : \sigma$ *and* $\Pi_2^+ : \Delta^* \vdash_{LJ}^+ T_{dom(\Delta^*)}(\Pi_2) : \tau$. *If* $T_{dom(\Delta^*)}(\Pi_1) \equiv T_{dom(\Delta^*)}(\Pi_2)$*, then the decoration is:*

$$(\wedge I_{LJ}^+)\frac{\Delta^* \vdash_{LJ} T_{dom(\Delta^*)}(\Pi_1) : \sigma \quad \Delta^* \vdash_{LJ} T_{dom(\Delta^*)}(\Pi_2) : \tau}{\Delta^* \vdash_{LJ} T_{dom(\Delta^*)}(\Pi) : \sigma \wedge \tau}$$

where $T_{dom(\Delta^*)}(\Pi_1) \equiv T_{dom(\Delta^*)}(\Pi)$.
Otherwise, if $T_{dom(\Delta^*)}(\Pi_1) \not\equiv T_{dom(\Delta^*)}(\Pi_2)$*, then* $T_{dom(\Delta^*)}(\Pi)$ *does not exist.*

- $\Pi : (\circ)\dfrac{\Pi_1 : \Delta \vdash_* A}{\Delta \vdash_* B} \Rightarrow (\circ^+)\dfrac{\Delta^* \vdash_*^+ T_{dom(\Delta^*)}(\Pi_1) : A}{\Delta^* \vdash_*^+ T_{dom(\Delta^*)}(\Pi_1) : B}$

and $T_{dom(\Delta^*)}(\Pi) \equiv T_{dom(\Delta^*)}(\Pi_1)$*, for all rules* $(\circ)$ *not in the set:* $\{(\to I_*), (\to E_*), (A_*), (W_*), (P_{pIL}), (\wedge I_*)\}$.

ii) Let Π *be a deduction in the system* $\vdash_*$*. Then,* $U(\Pi)$ *is the set:*

$$\{T_{dom(\Delta^*)}(\Pi) \mid \Pi : \Delta \vdash_* A, T_{dom(\Delta^*)}(\Pi) \text{ exists and } \Delta^* \text{ is a decoration of } \Delta\}$$

Notice that, by construction, $M, N \in U(\Pi)$ implies M and N can differ each other by renaming of both free and bound variables. We will call $U(\Pi)$ the *form* of Π.

The next theorem shows that every derivation of $\vdash_{pIL}$ corresponds to a set of derivations in LJ with the same form.

Theorem 2 (From $\vdash_{pIL}$ to LJ). *Let $\Pi : K_1,\ldots,K_n \vdash_{pIL} H$. For all path p terminal in H, it is $\Pi^p : K_1^p,\ldots,K_n^p \vdash_{LJ} H^p$, and $U(\Pi^p)$ is defined, and $U(\Pi^p) = U(\Pi)$.*

Proof. By induction on Π.

We conclude this section by a definition that eliminates unnecessary differentiations among the deductions of $\vdash_{pIL}$. In particular, it allows to consider the deductions of $\vdash_{pIL}$ up to the order of applications of the rules, involving the manipulations of the kits. Equivalently, the definition here below introduces a set of commuting equivalences. We could get rid of them, simply by introducing new versions of the rules $(\wedge E^l_{pIL}), (\wedge E^r_{pIL}), (\wedge I_{pIL})$, working in parallel on disjoint paths of the same kit. Our opinion is that such a solution would have obscured the clarity of the logical system $\vdash_{pIL}$.

Definition 8 (Intersection Logic). Intersection Logic, *abbreviated as IL, is the set* ${}^{\vdash_{pIL}}/_{\sim}$ *of all the deductions of* $\vdash_{pIL}$, *quotiented by the congruence* $\sim$, *defined as:*

$$(\wedge E^{s'}_{pIL})\dfrac{(\wedge E^{s}_{pIL})\dfrac{\Gamma \vdash_{pIL} K[p := \sigma_l \wedge \sigma_r][q := \tau_l \wedge \tau_r]}{\Gamma \vdash_{pIL} K[p := \sigma_s][q := \tau_l \wedge \tau_r]}}{\Gamma \vdash_{pIL} K[p := \sigma_s][q := \tau_{s'}]} \sim$$

$$(\wedge E^{s}_{pIL})\dfrac{(\wedge E^{s'}_{pIL})\dfrac{\Gamma \vdash_{pIL} K[p := \sigma_l \wedge \sigma_r][q := \tau_l \wedge \tau_r]}{\Gamma \vdash_{pIL} K[p := \sigma_l \wedge \sigma_r][q := \tau_{s'}]}}{\Gamma \vdash_{pIL} K[p := \sigma_s][q := \tau_{s'}]}$$

$$(\wedge E^{s}_{pIL})\dfrac{(\wedge I_{pIL})\dfrac{\Gamma \vdash_{pIL} K[p := \sigma_l \wedge \sigma_r][q := [\tau_l, \tau_r]}{\Gamma\backslash^{qs'} \vdash_{pIL} K[p := \sigma_l \wedge \sigma_r][q := \tau_l \wedge \tau_r]}}{\Gamma\backslash^{qs'} \vdash_{pIL} K[p := \sigma_s][q := \tau_l \wedge \tau_r]} \sim$$

$$(\wedge I_{pIL})\dfrac{(\wedge E^{s}_{pIL})\dfrac{\Gamma \vdash_{pIL} K[p := \sigma_l \wedge \sigma_r][q := [\tau_l, \tau_r]]}{\Gamma \vdash_{pIL} K[p := \sigma_s][q := \tau_l \wedge \tau_r]}}{\Gamma\backslash^{qs'} \vdash_{pIL} K[p := \sigma_s][q := \tau_l \wedge \tau_r]}$$

being $s, s' \in \{l, r\}$, *and* p, q *two different paths.*

An equivalence class in IL whose deductions prove $H_1,\ldots,H_n \vdash_{pIL} K$, *is denoted by* $H_1,\ldots,H_n \vdash_{IL} K$, *or* $\pi : H_1,\ldots,H_n \vdash_{IL} K$.

Definition 8 also assures that the term decorating the conclusion of two deductions of the same equivalence class of IL are the same:

Fact 2 $\pi : H_1,\ldots,H_n \vdash_{IL} K$ *implies* $T_{x_1,\ldots,x_n}(\Pi) \equiv T_{x_1,\ldots,x_n}(\Pi')$, *for every* $\Pi, \Pi' \in \pi$.

So, we can safely identify $T_{x_1,\ldots,x_n}(\pi)$ and $T_{x_1,\ldots,x_n}(\Pi)$, if $\Pi \in \pi$. Moreover, we can extend our terminology to to say that $U(\Pi)$ is the form of π as well.

Example 3. Let σ denote the formula $(\alpha\wedge\beta)\wedge\gamma$. Let also the following deductions be given:

$$\Pi_7 : (\to I_{pIL})\dfrac{(\wedge I_{pIL})\dfrac{(\wedge E^r_{pIL})\dfrac{(\wedge E^l_{pIL})\dfrac{(\wedge E^l_{pIL})\dfrac{(A_{pIL})\dfrac{}{[[\sigma,\sigma],\alpha] \vdash_{pIL} [[\sigma,\sigma],\alpha]}}{[[\sigma,\sigma],\alpha] \vdash_{pIL} [[\alpha\wedge\beta,\sigma],\alpha]}}{[[\sigma,\sigma],\alpha] \vdash_{pIL} [[\alpha,\sigma],\alpha]}}{[[\sigma,\sigma],\alpha] \vdash_{pIL} [[\alpha,\gamma],\alpha]}}{[\sigma,\alpha] \vdash_{pIL} [\alpha\wedge\gamma,\alpha]}}{\vdash_{pIL} [\sigma\to(\alpha\wedge\gamma),\alpha\to\alpha]}$$

$$\Pi_8 : (\wedge I_{pIL})\dfrac{\Pi_7 : \vdash_{pIL} [\sigma\to(\alpha\wedge\gamma),\alpha\to\alpha]}{\vdash_{pIL} (\sigma\to(\alpha\wedge\gamma))\wedge(\alpha\to\alpha)}$$

$$\Pi_9 : (\to I_{pIL})\dfrac{(\wedge I_{pIL})\dfrac{(\wedge E^l_{pIL})\dfrac{(\wedge E^r_{pIL})\dfrac{(\wedge E^l_{pIL})\dfrac{(A_{pIL})\dfrac{}{[[\sigma,\sigma],\alpha] \vdash_{pIL} [[\sigma,\sigma],\alpha]}}{[[\sigma,\sigma],\alpha] \vdash_{pIL} [[\alpha\wedge\beta,\sigma],\alpha]}}{[[\sigma,\sigma],\alpha] \vdash_{pIL} [[\alpha\wedge\beta,\gamma],\alpha]}}{[[\sigma,\sigma],\alpha] \vdash_{pIL} [[\alpha,\gamma],\alpha]}}{[\sigma,\alpha] \vdash_{pIL} [\alpha\wedge\gamma,\alpha]}}{\vdash_{pIL} [\sigma\to(\alpha\wedge\gamma),\alpha\to\alpha]}$$

$$\Pi_{10} : (\wedge I_{pIL})\dfrac{\Pi_9 : \vdash_{pIL} [\sigma\to(\alpha\wedge\gamma),\alpha\to\alpha]}{\vdash_{pIL} (\sigma\to(\alpha\wedge\gamma))\wedge(\alpha\to\alpha)}$$

As an exercise, verify that: i) $\Pi_7 \sim \Pi_9$, ii) if π is the equivalence class of Π_7, then $T_x[\pi] \equiv \lambda x.x$, and iii) the deductions corresponding to Π_7 and Π_{10} in the system LJ, according to Theorem 2, are respectively: $(\Pi_7)^l \equiv \Pi_1$, $(\Pi_7)^r \equiv \Pi_2$, $(\Pi_{10})^\epsilon \equiv \Pi_3$, being Π_1, Π_2 and Π_3 defined in Example 1.

5 Strong Normalization

In this section we shall prove that IL is strongly normalizable. This property follows from the strong normalizability of $\vdash_{pIL}$, proved by reducing to the analogous property of LJ.

Definition 9. *Let $s \in \{l, r\}$.*

i) The P-commuting conversions *on the sequent calculus $\vdash_{pIL}$ are the following rewriting rules:*

$$(P_{pIL})\dfrac{(A_{pIL})\dfrac{}{\Gamma, H \vdash_{pIL} H}}{\Gamma\backslash^p, H\backslash^p \vdash_{pIL} H\backslash^p} \leadsto_P (A_{pIL})\dfrac{}{\Gamma\backslash^p, H\backslash^p \vdash_{pIL} H\backslash^p}$$

$$(P_{pIL})\dfrac{(X_{pIL})\dfrac{\Gamma, H, H', \Delta \vdash_{pIL} K}{\Gamma, H', H, \Delta \vdash_{pIL} K}}{\Gamma\backslash^p, H'\backslash^p, H\backslash^p, \Delta\backslash^p \vdash_{pIL} K\backslash^p} \leadsto_P$$

$$(X_{pIL})\dfrac{(P_{pIL})\dfrac{\Gamma, H, H', \Delta \vdash_{pIL} K}{\Gamma\backslash^p, H\backslash^p, H'\backslash^p, \Delta\backslash^p \vdash_{pIL} K\backslash^p}}{\Gamma\backslash^p, H'\backslash^p, H\backslash^p, \Delta\backslash^p \vdash_{pIL} K\backslash^p}$$

$$(P_{pIL})\dfrac{(W_{pIL})\dfrac{\Gamma \vdash_{pIL} K}{\Gamma, H \vdash_{pIL} K}}{\Gamma\backslash^p, H\backslash^p \vdash_{pIL} K\backslash^p} \leadsto_P (W_{pIL})\dfrac{(P_{pIL})\dfrac{\Gamma \vdash_{pIL} K}{\Gamma\backslash^p \vdash_{pIL} K\backslash^p}}{\Gamma\backslash^p, H\backslash^p \vdash_{pIL} K\backslash^p}$$

$$(P_{pIL})\dfrac{(\to I_{pIL})\dfrac{\Gamma, H \vdash_{pIL} K}{\Gamma \vdash_{pIL} H \to K}}{\Gamma\backslash^p \vdash_{pIL} (H \to K)\backslash^p} \leadsto_P (\to I_{pIL})\dfrac{(P_{pIL})\dfrac{\Gamma, H \vdash_{pIL} K}{\Gamma\backslash^p, H\backslash^p \vdash_{pIL} K\backslash^p}}{\Gamma\backslash^p \vdash_{pIL} (H \to K)\backslash^p}$$

$$(P_{pIL})\dfrac{(\to E_{pIL})\dfrac{\Gamma \vdash_{pIL} H \to K \quad \Gamma \vdash_{pIL} H}{\Gamma \vdash_{pIL} K}}{\Gamma\backslash^p \vdash_{pIL} K\backslash^p} \leadsto_P$$

$$(\to E_{pIL})\dfrac{(P_{pIL})\dfrac{\Gamma \vdash_{pIL} H \to K}{\Gamma\backslash^p \vdash_{pIL} (H \to K)\backslash^p} \qquad (P_{pIL})\dfrac{\Gamma \vdash_{pIL} H}{\Gamma\backslash^p \vdash_{pIL} H\backslash^p}}{\Gamma\backslash^p \vdash_{pIL} K\backslash^p}$$

$$(P_{pIL})\dfrac{(\wedge I_{pIL})\dfrac{\Gamma \vdash_{pIL} K[q := [\sigma, \tau]]}{\Gamma\backslash^q \vdash_{pIL} K[q := \sigma \wedge \tau]}}{(\Gamma\backslash^q)\backslash^p \vdash_{pIL} (K[q := \sigma \wedge \tau])\backslash^p} \leadsto_P$$

$$(\wedge I_{pIL})\dfrac{(P_{pIL})\dfrac{\Gamma \vdash_{pIL} K[q := [\sigma, \tau]]}{\Gamma\backslash^p \vdash_{pIL} (K[q := [\sigma, \tau]])\backslash^p}}{(\Gamma\backslash^p)\backslash^q \vdash_{pIL} ((K[q := \sigma \wedge \tau])\backslash^p)\backslash^q}$$

$$(P_{pIL})\dfrac{(\wedge E^s_{pIL})\dfrac{\Gamma \vdash_{pIL} K[q := \sigma_l \wedge \sigma_r]}{\Gamma \vdash_{pIL} K[q := \sigma_s]}}{\Gamma\backslash^p \vdash_{pIL} (K[q := \sigma_s])\backslash^p} \leadsto_P$$

$$(\wedge E^s_{pIL})\dfrac{(P_{pIL})\dfrac{\Gamma \vdash_{pIL} K[q := \sigma_l \wedge \sigma_r]}{\Gamma\backslash^p \vdash_{pIL} (K[q := \sigma_l \wedge \sigma_r])\backslash^p}}{\Gamma\backslash^p \vdash_{pIL} (K[q := \sigma_s])\backslash^p}$$

Under the standard terminology, every sequence of rules to the left of $\leadsto$ *is a P-redex, while that one to its right is a P-reduct.*

ii) A derivation free of occurrences of (P_{pIL}) *is P-normal.*

iii) We say that a class π *of IL reduces to another class* π' *of IL, under* $\leadsto_P$*, and we write* $\pi \leadsto_P \pi'$*, if there are* $\Pi \in \pi$ *and* $\Pi' \in \pi'$ *such that* $\Pi \leadsto_P \Pi'$*.*

Remark that the fourth and sixth P-commuting conversions exploits Fact 1.

Lemma 3 (P-strong normalization). *Every* $\Pi : \Gamma \vdash_{pIL} H$ *can be reduced to a P-normal* $\Pi' : \Gamma \vdash_{pIL} H$*, under any strategy.*

Proof. Observe that the commuting conversions shift every occurrence of (P_{pIL}) upwards, which, eventually, gets erased.

Definition 10. *Let* $s \in \{l, r\}$ *and* $\Pi \in \pi$*.*

i) A $\wedge$-IL-redex *of* Π *is the sequence:*

$$(\wedge E^s_{pIL})\dfrac{(\wedge I_{pIL})\dfrac{H_1[p := [\sigma_1, \sigma_1]], \ldots, H_n[p := [\sigma_n, \sigma_n]] \vdash_{pIL} K[p := [\sigma_l, \sigma_r]]}{H_1[p := \sigma_1], \ldots, H_n[p := \sigma_n] \vdash_{pIL} K[p := \sigma_l \wedge \sigma_r]}}{H_1[p := \sigma_1], \ldots, H_n[p := \sigma_n] \vdash_{pIL} K[p := \sigma_s]}$$

ii) A ∧-IL-rewriting step *on* Π *is:*

$$(\wedge E^s_{pIL})\dfrac{(\wedge I_{pIL})\dfrac{H_1[p:=[\sigma_1,\sigma_1]],\ldots,H_n[p:=[\sigma_n,\sigma_n]]\vdash_{pIL} K[p:=[\sigma_l,\sigma_r]]}{H_1[p:=\sigma_1],\ldots,H_n[p:=\sigma_n]\vdash_{pIL} K[p:=\sigma_l\wedge\sigma_r]}}{H_1[p:=\sigma_1],\ldots,H_n[p:=\sigma_n]\vdash_{pIL} K[p:=\sigma_s]}\ \rightsquigarrow_\wedge$$

$$(P_{pIL})\dfrac{H_1[p:=[\sigma_1,\sigma_1]],\ldots,H_n[p:=[\sigma_n,\sigma_n]]\vdash_{pIL} K[p:=[\sigma_l,\sigma_r]]}{(H_1[p:=[\sigma_1,\sigma_1]])\backslash^{ps},\ldots,(H_n[p:=[\sigma_n,\sigma_n]])\backslash^{ps}\vdash_{pIL} K\backslash^{ps}}$$

where $(H_i[p:=[\sigma_i,\sigma_i]])\backslash^{ps}\equiv H_i[p:=\sigma_i]$, *with* $1\leq i\leq n$, *and* $K\backslash^{ps}\equiv K[p:=\sigma_s]$.

iii) We say that a class π *of IL reduces to another class* π' *of IL, under a* ∧*-IL-rewriting step, and we write* $\pi\rightsquigarrow_\wedge\pi'$*, if there are* $\Pi\in\pi$ *and* $\Pi'\in\pi'$ *such that* $\Pi\rightsquigarrow_\wedge\Pi'$.

Lemma 4. *Consider* $\Pi:\Gamma\vdash_{pIL} H$ *and* $\Pi':\Gamma,H\vdash_{pIL} K$. *Call* $S(\Pi,\Pi')$ *the deductive structure, obtained by replacing the conclusion of* Π *for every occurrence* (A_{pIL}) *deriving* $H\vdash_{pIL} H$*, and such that* H *to the left of* $\vdash_{pIL}$ *is free in* Π'. *Then,* $S(\Pi,\Pi'):\Gamma\vdash_{pIL} K$.

Proof. By structural induction on the deduction of $\Gamma,H\vdash_{pIL} K$.

Definition 11. *Let* $\Pi\in\pi$.

i) A →-IL-redex *of* Π *is the sequence:*

$$(\rightarrow E_{pIL})\dfrac{(\rightarrow I_{pIL})\dfrac{\Gamma,H\vdash_{pIL} K}{\Gamma\vdash_{pIL} H\rightarrow K}\qquad \Gamma\vdash_{pIL} H}{\Gamma\vdash_{pIL} K}$$

ii) A →-IL-rewriting *step on* Π *is:*

$$(\rightarrow E_{pIL})\dfrac{(\rightarrow I_{pIL})\dfrac{\Pi_0:\Gamma,H\vdash_{pIL} K}{\Gamma\vdash_{pIL} H\rightarrow K}\qquad \Pi_1:\Gamma\vdash_{pIL} H}{\Gamma\vdash_{pIL} K}\ \rightsquigarrow_\rightarrow$$

$$S(\Pi_1,\Pi_0):\Gamma\vdash K\ .$$

iii) We say that a class π *of IL reduces to another class* π' *of IL, under a* →*-IL-rewriting step, and we write* $\pi\rightsquigarrow_\rightarrow\pi'$*, if there are* $\Pi\in\pi$ *and* $\Pi'\in\pi'$ *such that* $\Pi\rightsquigarrow_\rightarrow\Pi'$.

Definition 12. *A deduction of IL is* normal *if it is free of* (P), ∧, *and* →*-IL-redexes.*

Definition 13. $\rightsquigarrow$ *is the smallest contextual, reflexive and transitive closure of* $\rightsquigarrow_P\cup\rightsquigarrow_\wedge\cup\rightsquigarrow_\rightarrow$.

Lemma 5. $\rightsquigarrow$ *is strongly normalizing.*

Proof. The proof proceeds by exploiting the embedding of IL into LJ, which allows to show the number of both the →-IL-redexes and the ∧-IL-redexes, of any derivation Π of IL, can be bound by the number of the analogous redexes of any projection of Π into LJ. The existence of the P-commuting conversions in $\rightsquigarrow$, which are strongly normalizable, is completely transparent to the embedding.

Theorem 3. *IL is strongly normalizable.*

Proof. From Definition 13 and Lemma 5.

6 Intersection Types and Intersection Logic

In this section, we trace the relationship between IL and the intersection type system IT. On one side, every deduction $\pi : \Gamma \vdash_{IL} H$ corresponds to a set of type assignments. Every of such type assignments gives a type to the form of π, which, we recall, is a λ-term. The type is one of the leaves of H. On the other side, every type assignment $\{x_1 : \sigma_1, ..., x_n : \sigma_n\} \vdash_{\wedge} M : \sigma$ corresponds to a deduction $\pi : \sigma_1, ..., x : \sigma_n \vdash_{IL} \sigma$ such that M is the form of π.

Lemma 6. *$\pi : H_1, \ldots, H_n \vdash_{IL} K$ implies $\{x_1 : H_1^p, .., x_n : H_n^p\} \vdash_{\wedge} T_{x_1 \ldots x_n}(\pi) : K^p$, for every $p \in P_T(K)$.*

Proof. The proof is reduced to proving the statement: "$\Pi : H_1, \ldots, H_n \vdash_{pIL} K$ implies $\{x_1 : H_1^p, .., x_n : H_n^p\} \vdash_{\wedge} T_{x_1 \ldots x_n}(\Pi) : K^p$, for every $p \in P_T(K)$, and $\Pi \in \pi$" by induction on Π. Then, the final statement is a corollary of Fact 2.

In order to study the opposite direction of the correspondence between $\vdash_{IL}$ and $\vdash_{\wedge}$, we need an auxiliary lemma.

Lemma 7. *Let assume the notations:*

$$\Delta_1 \equiv H_1, \ldots, H_n, \quad \Delta_2 \equiv H'_1, \ldots, H'_n, \quad \Delta \equiv [H_1, H'_1], \ldots, [H_n, H'_n] \ .$$

Take any $\Pi_1 : \Delta_1 \vdash_{pIL} K_1$, and $\Pi_2 : \Delta_2 \vdash_{pIL} K_2$ such that $T_{dom(\Delta_1^*)}(\Pi_1) \equiv T_{dom(\Delta_2^*)}(\Pi_2)$, for every decoration Δ_1^*, Δ_2^*, such that $dom(\Delta_1^*) \equiv dom(\Delta_2^*)$. There is $\Pi : \Delta \vdash_{pIL} [K_1, K_2]$ such that $T_{dom(\Delta^*)}(\Pi) \equiv T_{dom(\Delta_1^*)}(\Pi_1)$, whenever $dom(\Delta^*) \equiv dom(\Delta_1^*)$.

Proof. The proof can proceed by structural induction on Π_1. As an example, we show the details about an instance of one of the more interesting cases.

Let:

$$(\rightarrow I_{pIL}) \frac{\Delta_1, K'_1 \vdash_{pIL} K''_1}{\Delta_1 \vdash_{pIL} K'_1 \rightarrow K''_1}$$

be the last rule of Π_1. For any decorations Δ_1^* and Δ_2^*, such that $dom(\Delta_1^*) \equiv dom(\Delta_2^*)$, the assumption $T_{dom(\Delta_1^*)}(\Pi_1) \equiv T_{dom(\Delta_2^*)}(\Pi_2)$, assures that the last rules of Π_2 can only be an instance of $(\rightarrow I_{pIL})$, followed by a, possibly empty, sequence $R_1, \ldots, R_m$ of rules, each belonging to $\{(P_{pIL}), (\wedge I_{pIL}), (\wedge E_{pIL}), (X_{pIL}), (W_{pIL})\}$, and such that they apply to the paths $p_1, \ldots, p_m$ of the conclusion of Π_2. Assume:

$$(\rightarrow I_{pIL}) \frac{\hat{\Delta}_2, K'_2 \vdash_{pIL} K''_2}{\hat{\Delta}_2 \vdash_{pIL} K'_2 \rightarrow K''_2}$$

be the last $(\rightarrow I_{pIL})$ instance of Π_2, where $\hat{\Delta}_2 \equiv \hat{H}'_1, \ldots, \hat{H}'_n$. Tanks to the α-equivalence on the λ-terms, we can always end up with decorations such that $dom(\Delta_1^*, x : K'_1) \equiv dom(\hat{\Delta}_2^*, x : K'_2)$, for some suitable x. So, by induction, there exists a deduction $\bar{\Pi} : \bar{\Delta}, [K'_1, K'_2] \vdash_{pIL} [K''_1, K''_2]$, such that $\bar{\Delta} \equiv [H_1, \hat{H}'_1], \ldots, [H_n, \hat{H}'_n]$, and whose decoration is

$$T_{dom(\bar{\Delta}^*, x:K'_1)}(\bar{\Pi}) \equiv T_{dom(\Delta_1^*, x:K'_1)}(\Pi_1).$$

Now, we can firstly extend $\bar{\Pi}$ to $\breve{\Pi}$ by a $(\rightarrow I_{pIL})$, as follows:

$$\check{\Pi} : (\rightarrow I)\frac{\bar{\Pi} : \bar{\Delta}, [K_1', K_2'] \vdash [K_1'', K_2'']}{\bar{\Delta} \vdash_{pIL} [K_1' \rightarrow K_1'', K_2' \rightarrow K_2'']}$$

By Definition 7, it must be $\lambda x.T_{dom(\bar{\Delta}^*, x:K_1')}(\bar{\Pi}) \equiv \lambda x.T_{dom(\Delta_1^*, x:K_1')}(\Pi_1)$. To conclude, it is enough to apply $R_1, \ldots, R_m$ to the paths $rp_1, \ldots, rp_m$ of $\check{\Pi}$.

The other case, which requires some work to be proved, has $(\rightarrow E_{pIL})$ as last rule of Π_1.

All the remaining cases, instead, exploit the induction in the simplest way.

Lemma 8. *$\Pi : \{x_1 : \sigma_1, \ldots, x_n : \sigma_n\} \vdash_\wedge M : \tau$ implies $\pi : \sigma_1, \ldots, \sigma_n \vdash_{IL} \tau$ such that $M \equiv T_{x_1 \ldots x_n}(\pi)$.*

Proof. The proof is by induction on Π. Here we limit to sketch only the not obvious case. Assume to prove $\Pi : \{x_1 : \sigma_1, \ldots, x_n : \sigma_n\} \vdash_\wedge M : \sigma \wedge \tau$ from the assumptions $\{x_1 : \sigma_1, \ldots, x_n : \sigma_n\} \vdash_\wedge M : \sigma$ and $\{x_1 : \sigma_1, \ldots, x_n : \sigma_n\} \vdash_\wedge M : \tau$. By induction, we get both $\sigma_1, \ldots, \sigma_n \vdash_{IL} \sigma$ and $\sigma_1, \ldots, \sigma_n \vdash_{IL} \tau$. Then, Lemma 7 implies $[\sigma_1, \sigma_1], \ldots, [\sigma_n, \sigma_n] \vdash_{IL} [\sigma, \tau]$, to which we can apply $(\wedge I_{IL})$ to conclude.

Definition 14. *A judgment $K_1, \ldots, K_n \vdash_{IL} H$ is* proper *if, and only if, H and every K_i is an atom.*

We are finally in the position to relate IL and IT:

Theorem 4. *$\pi : \sigma_1, \ldots, \sigma_n \vdash_{IL} \tau$ if, and only if, $x_1 : \sigma_1, \ldots, x_n : \sigma_n \vdash_\wedge T_{x_1 \ldots x_n}(\pi) : \tau$, for every $\pi : \sigma_1, \ldots, \sigma_n \vdash_{IL} \tau$ proper.*

Proof. Directly from Lemma 6 and Lemma 8.

Example 4. Let π be the equivalence class which Π_8 (or Π_{10}) in Example 3 belongs to. The corresponding deduction of $\vdash_\wedge$ is Π_6 of Example 2.

The correspondence between IL and IT allows to derive for free the property of strong normalization of the λ-terms, typable in IT, with respect to the β-reduction. This property has been first proved in [19].

Theorem 5. *Let $\Gamma \vdash_\wedge M : \sigma$. Then, M is strongly normalizable.*

Proof. The proof proceeds in two steps. Firstly, we embed the derivation of $\Gamma \vdash_\wedge M : \sigma$ into LJ, getting a derivation Π. Secondly, we assume the existence of a redex of M not present in the normal form of Π, getting a contradiction.

References

1. Samson Abramsky. Domain theory in logical form. *Ann. Pure Appl. Logic*, 51(1-2):1–77, 1991.
2. F. Alessi and F. Barbanera. Strong conjunction and intersection types. In *16h International Symposium on Mathematical Foundation of Computer Science (MFCS91)*, volume Lecture Notes in Computer Science 520. Springer-Verlag, 1991.

3. F. Barbanera and S. Martini. Proof-functional connectives and realizability. *Archive for Mathematical Logic*, 33:189–211, 1994.
4. Henk Barendregt, Mario Coppo, and Mariangiola Dezani-Ciancaglini. A filter lambda model and the completeness of type assignment. *J. Symbolic Logic*, 48(4):931–940 (1984), 1983.
5. B. Capitani, M. Loreti, and Venneri B. Hyperformulae, parallel deductions and intersection types. To appear in "Workshop on Bohm Theorem", ICALP 2001, Creta (Greece), 2001.
6. Mario Coppo and Mariangiola Dezani-Ciancaglini. An extension of the basic functionality theory for the λ-calculus. *Notre Dame J. Formal Logic*, 21(4):685–693, 1980.
7. Mario Coppo, Mariangiola Dezani-Ciancaglini, Furio Honsell, and Giuseppe Longo. Extended type structures and filter lambda models. In *Logic colloquium '82 (Florence, 1982)*, pages 241–262. North-Holland, Amsterdam, 1984.
8. H. Curry, R. Feys, and W. Craig. *Combinatory Logic*, volume 1. North Holland, 1958.
9. M. Dezani-Ciancaglini, S. Ghilezan, and B. Venneri. The "relevance" of intersection and union types. *Notre Dame J. Formal Logic*, 38(2):246–269, 1997.
10. Lavinia Egidi, Furio Honsell, and Simona Ronchi Della Rocca. Operational, denotational and logical descriptions: a case study. *Fund. Inform.*, 16(2):149–169, 1992. Mathematical foundations of computer science '91 (Kazimierz Dolny, 1991).
11. Jean-Yves Girard. Locus solum: From the rules of logic to the logic of rules. Internal Report, IML, Marseille, 2001.
12. J.Y. Girard. *Interpretation Fonctionelle et Elimination des Coupures de l'Arithmetique d'Ordre Superieur.* PhD thesis, Université Paris VII, 1972.
13. J. Roger Hindley. Coppo Dezani types do not correspond to propositional logic. *Theoret. Comput. Sci.*, 28(1-2):235–236, 1984.
14. Furio Honsell and Simona Ronchi Della Rocca. Reasoning about interpretations in qualitative lambda-models. In *Programming Concepts and Methods*, pages 505–522. North Holland, 1990.
15. Furio Honsell and Simona Ronchi Della Rocca. An approximation theorem for topological lambda models and the topological incompleteness of lambda calculus. *J. Comput. System Sci.*, 45(1):49–75, 1992.
16. Assaf Kfoury. Beta-reduction as unification. In *Logic Algebra and Computer Science*, pages 241–262. Polish Academy of Science, Warsaw, 1999.
17. Daniel Leivant. Polymorphic type inference. *Symposium on Principles of Programming Languages*, 1983.
18. G. E. Mints. The completeness of provable realizability. *Notre Dame J. Formal Logic*, 30(3):420–441, 1989.
19. Garrel Pottinger. A type assignment for the strongly normalizable λ-terms. In *To H. B. Curry: essays on combinatory logic, lambda calculus and formalism*, pages 561–577. Academic Press, London, 1980.
20. Dag Prawitz. *Natural Deduction.* Almquist & Wiksell.
21. J. C. Reynolds. Design of the programming language Forsythe. In P. O'Hearn and R.D. Tennent, editors, *Algol-like Languages.* Birkhauser, 1996.
22. L. Roversi. a Type-Free Resource-Aware λ-Calculus. In *Fifth Annual Conference of the EACSL (CSL'96)*, volume 1258 of *Lecture Notes in Computer Science*, pages 399 – 413, Utrecht (The Nederland), September 1996. Springer-Verlag.
23. Betti Venneri. Intersection types as logical formulae. *J. Logic Comput.*, 4(2):109–124, 1994.

24. J. B. Wells, Allyn Dimock, Robert Muller, and Franklyn Turbak. A typed intermediate language for flow-directed compilation. In *7th International Joint Conference on Theory and Practice of Software Development (TAPSOFT97)*, pages 757–771.

Life without the Terminal Type

Lutz Schröder

BISS, Department of Computer Science, Bremen University

Abstract. We introduce a method of extending arbitrary categories by a terminal object and apply this method in various type theoretic settings. In particular, we show that categories that are cartesian closed except for the lack of a terminal object have a universal full extension to a cartesian closed category, and we characterize categories for which the latter category is a topos. Both the basic construction and its correctness proof are extremely simple. This is quite surprising in view of the fact that the corresponding results for the simply typed λ-calculus with surjective pairing, in particular concerning the decision problem for equality of terms in the presence of a terminal type, are comparatively involved.

Introduction

Cartesian closed categories have attracted considerable interest in theoretical computer science due to their close relation to the λ-calculus [14,5,17]. Indeed, up to a certain point, cartesian closed categories and simply typed λ-calculi with surjective pairing and terminal type ($\lambda\pi*$) can be regarded as being essentially the same [14].

It is this terminal or unit type that we are concerned with here. Terminal types are a standard feature in many type systems [4]. In connection with object-oriented subtyping paradigms, the terminal object has been regarded as playing the role of a maximal type [6].

Now it has turned out that, rather unexpectedly, the presence of a terminal type in a λ-calculus leads to severe complications concerning confluence and hence decidability of equality for terms [6,12]. The equivalence of λ-calculi and cartesian closed categories offers a way around these difficulties: calling a category *almost cartesian closed* if it is cartesian closed except that it may lack a terminal object, one has an equivalence between λ-calculi with surjective pairing ($\lambda\pi$) on the one hand and almost cartesian closed categories on the other hand. Now the question whether extending a given $\lambda\pi$-calculus by a terminal type is conservative translates into the question whether each almost cartesian closed category can be (fully) extended by a terminal object to yield a cartesian closed category. One of the results presented here answers the latter question in the positive, thus generalizing known solutions [14,18] for the case of almost cartesian closed categories with only one object, commonly called *C-monoids.* (*C*-monoids are essentially the same as untyped $\lambda\pi$-calculi.)

L. Fribourg (Ed.): CSL 2001, LNCS 2142, pp. 429–442, 2001.

The most striking aspect of this extension result is that it does not really depend on cartesian closedness at all; rather, it is an instance of a corresponding result for entirely arbitrary categories which, given the complexity that the problem assumes on the purely syntactical side, is surprisingly easy to state and prove. Roughly speaking, this result says that an 'invisible' terminal object, along with the 'elements' or 'constant operations' as which one may interpret morphisms with terminal domain, are 'nearly' unambiguously encoded in any category. This 'nearly' disappears gradually as one ascends through the hierarchy of type systems, up to the point that, for almost cartesian closed categories, the terminal object can be regarded as entirely implicit (in fact, we observe on the side that the terminal object does not really belong to the structure of a cartesian closed category at all, but can rather be viewed as a 'derived operation').

At the top of the said hierarchy of type systems, we consider the question of when the relevant extension produces a topos (i.e. an intuitionistic type theory [14]). The result is a characterization of 'toposes without terminal objects' which shows that the classification of 'singleton subobjects' as well as of subobjects of the terminal object is hidden in the remaining structure of a topos.

For unexplained categorical terminology see [1,15]; all categories are assumed to be *locally small* (i.e. $hom(A, B) = \{f | f : A \to B\}$ is a set for all objects A, B).

1 Adding a Terminal Object

To start off, we are going to show, in a nutshell, that arbitrary categories admit rather few full extensions that add a terminal object. We fix some notation: 1 always refers to a selected terminal object. The unique morphism $A \to 1$ is denoted by $!_A$.

Now, given any category $\mathbf{A}$, there is always a trivial solution to the problem of extending $\mathbf{A}$ by a terminal object: just add a new object 1 and a single new morphism $A \to 1$ for each object A. However, this will rarely produce the desired result. In most situations, one will expect there to be morphisms $1 \to A$ as well; such morphisms are often called *elements* of A. Now such a morphism $f : 1 \to A$ gives rise to a family of morphisms $f_C = f \circ !_C : C \to A$, where C ranges over all objects. This family has the property that

$$f_C \circ g = f_B$$

for each morphism $g : B \to C$ — and this rather trivial observation is really all we need in order to solve the problem.

Definition 1. Let A be an object in a category $\mathbf{A}$. A *structural element of* A is a family $(f_C : C \to A)$ of morphisms, indexed over all objects C of $\mathbf{A}$, such that $f_C \circ g = f_B$ for each $g : B \to C$ in $\mathbf{A}$. A is called *structurally nonempty* if it has a structural element.

The class of structural elements of an object A is small, since it injects into $hom(A, A)$. A structural element may be regarded as a cocone for $id_{\mathbf{A}}$. If $\mathbf{A}$ has a terminal object, then structural elements are essentially the same as elements. In particular, in that case an object A is *nonempty*, i.e. has an element $1 \to A$, iff it is structurally nonempty.

Definition 2. A functor $F : \mathbf{A} \to \mathbf{B}$ *preserves* a structural element (f_C) in $\mathbf{A}$ if the family $(Ff_C)_{C \in \mathrm{Ob}\,\mathbf{A}}$ *extends* to a structural element (g_D), i.e. $g_{FC} = Ff_C$ for each object C of $\mathbf{A}$. If this is the case for *all* (f_C), then F *preserves structural elements.*

Note that the extension (g_D) is unique: any other extension (h_D) must satisfy $h_D = h_{FA} \circ h_D = Ff_A \circ h_D = g_{FA} \circ h_D = g_D$ for each D. If $\mathbf{B}$ has a terminal object, then the structural element associated to a morphism $g : 1 \to FA$ extends (Ff_C) iff $g \circ !_{FC} = Ff_C$ for each C; in this case, we say that g *extends* (Ff_C).

In the presence of a terminal object, preservation of structural elements reduces to a more familiar property:

Proposition 3. *Let $F : \mathbf{A} \to \mathbf{B}$ be a functor, and let $\mathbf{A}$ have a terminal object. Then F preserves structural elements iff it preserves the terminal object.*

Proof. If F preserves structural elements, then $F!_1 = id_{F1}$ is part of a structural element of $F1$, which implies that $F1$ is terminal. The converse implication is trivial.

It suffices to check preservation of a single structural element:

Lemma 1. *Let $F : \mathbf{A} \to \mathbf{B}$ be a functor, and let (f_C) be a structural element of A in $\mathbf{A}$. If F preserves (f_C), then F preserves structural elements.*

Proof. Let (g_D) be a structural element of FA that extends (Ff_C), and let (h_C) be a structural element of B in $\mathbf{A}$. Put $k_D = Fh_A \circ g_D$ for each object D in $\mathbf{B}$. Then (k_D) is a structural element of FB, and $k_{FC} = Fh_A \circ g_{FC} = Fh_A \circ Ff_C = F(h_A \circ f_C) = Fh_C$ for each object C in $\mathbf{A}$.

Corollary 4. *If $E : \mathbf{A} \hookrightarrow \mathbf{B}$ is a full embedding and $\mathbf{A}$ contains a structurally nonempty object of $\mathbf{B}$, then E preserves structural elements.*

Proof. Let A be an object of $\mathbf{A}$ that has a structural element (f_C) in $\mathbf{B}$. Then the restriction $(f_C)_{C \in \mathrm{Ob}\,\mathbf{A}}$ is a structural element of A in $\mathbf{A}$ which is preserved by E. By Lemma 1, this implies the claim.

Since full embeddings also *reflect* structural elements in the obvious sense, this has a quite striking consequence: any full extension of a category $\mathbf{A}$ will either leave the set of structural elements of each object of $\mathbf{A}$ essentially unchanged or make all objects of $\mathbf{A}$ structurally empty. In particular, this means that there are at most two ways of fully extending $\mathbf{A}$ by a terminal object: one where all objects of $\mathbf{A}$ become empty, and one where the elements of A are essentially the previous structural elements. We have seen above how to construct extensions of the first type; extensions of the (rather more interesting) second type always exist as well:

Theorem 5. *Given a category* $\mathbf{A}$*, there is an essentially unique way of fully extending* $\mathbf{A}$ *by a terminal object such that the resulting extension*

$$E : \mathbf{A} \to \mathbf{A}[1]$$

preserves structural elements. E *is universal among the functors* $F : \mathbf{A} \to \mathbf{C}$ *such that* $\mathbf{C}$ *has a terminal object and* F *preserves structural elements. Moreover,* E *preserves limits.*

Proof. We begin by constructing the desired extension: let $\mathbf{A}[1]$ denote the full subcategory of $\mathbf{Set}^{\mathbf{A}^{op}}$ spanned by the hom-functors $H_A = hom(\_, A)$ and a terminal object 1, i.e. a functor that maps all objects to singleton sets. Let $E : \mathbf{A} \to \mathbf{A}[1]$ denote the codomain restriction of the Yoneda embedding $\mathbf{A} \to [\mathbf{A}^{op}, \mathbf{Set}]$; as such, E preserves limits ([3], 2.15.5). 1 is a terminal object of $\mathbf{A}[1]$, and natural transformations $1 \to H_A$ are essentially the same as structural elements of A (in particular, they form a set). We regard $\mathbf{A}$ as an actual subcategory of $\mathbf{A}[1]$, i.e. we identify H_A with A.

Uniqueness has been discussed above. To prove the universal property, let $F : \mathbf{A} \to \mathbf{C}$ be as in the statement. In order to extend F to a functor $F^\# : \mathbf{A}[1] \to \mathbf{C}$ that preserves the terminal object, put $F^\#1 = 1$ and $F^\#!_A = !_{FA}$. Given $f : 1 \to A$ in $\mathbf{A}[1]$, i.e. a structural element (f_C) of A, we have to define $F^\# f$ as the unique element $1 \to FA$ that extends (Ff_C). All that remains to be shown is that $F^\#$ preserves composition. This is clear for composites of the type $\bullet \to \bullet \to 1$. Preservation of composites $h \circ f$, where f is as above and $h : A \to B$ in $\mathbf{A}$, follows from the fact that $Fh \circ F^\# f$ extends $(F(h \circ f_C))$. Finally, $\bar{F}(f \circ !_C) = Ff_C = F^\# f \circ !_{FC} = F^\# f \circ F^\# !_C$, where the second equality follows from the fact that $F^\# f$ extends (Ff_C).

Remark 6. Of course, the factorization $F^\#$ constructed in the above proof is unique only up to a (unique) natural isomorphism, since the terminal object of $\mathbf{C}$ is unique up to isomorphism. Therefore, the universal property determines $\mathbf{A}[1]$ uniquely up to equivalence. Corresponding remarks hold for all similar universality statements below.

It is helpful to restate the construction of $\mathbf{A}[1]$ more explicitly: morphisms $1 \to A$ in $\mathbf{A}[1]$ are structural elements $f = (f_C)$ of A. Given a morphism $h : A \to B$ in $\mathbf{A}$, $h \circ f$ is the structural element (hf_C); the composite $f \circ !_B$ is just f_B.

Remark 7. If $\mathbf{A}$ has a structurally nonempty object U, then it is easily seen that structural elements of A in $\mathbf{A}$ are essentially the same as constant morphisms $U \to A$. Indeed, there is an alternative construction of $\mathbf{A}[1]$ in this case: we can fix a structural element (u_C) of U, take 1 as a copy of U, $!_A = u_A$ for each object A, and the constant morphisms $U \to A$ as morphisms $1 \to A$. This is more or less the description which drops out of the observation that the Karoubi envelope of a category $\mathbf{A}$ (i.e. the category of idempotents; see e.g. [13,14]) has a terminal object iff $\mathbf{A}$ has a structurally nonempty object U — in that case, the terminal object is the idempotent u_U, where (u_C) is as above.

Remark 8. As formulated in [19], categories without further structure can be regarded as 'a rather bland theory of types' - namely, a type theory that admits only unary functions (and constants, if we add a terminal object). For comparison with more complex type theories, we give an example of how preservation of structural elements affects the notion of 'model' here: consider the category $\mathbf{A}$ with a single object A and a single nontrivial morphism $f : A \to A$ such that $f \circ f = f$. A 'model' of $\mathbf{A}$, i.e. a functor from $\mathbf{A}$ into, say, $\mathbf{Set}$, will consist of a set X equipped with an idempotent unary operation α. The class of models becomes a lot smaller if we additionally require preservation of structural elements: f 'is' a structural element of A, the preservation of which amounts to requiring that $X \neq \emptyset$ and α is constant — i.e. the restricted models are essentially pointed sets.

A notion related to preservation of structural elements is preservation of structurally nonempty objects, for which there is an all-or-nothing statement similar to Lemma 1:

Lemma 2. *Let $F : \mathbf{A} \to \mathbf{B}$ be a functor. If $F[\mathbf{A}]$ contains a structurally nonempty object, then F preserves structurally nonempty objects.*

Proof. Let (g_B) be a structural element of FA, where $A \in \operatorname{Ob} \mathbf{A}$, and let (f_C) be a structural element of D, where $D \in \operatorname{Ob} \mathbf{A}$. Then $(Ff_A \circ g_B)_{B \in \operatorname{Ob} \mathbf{B}}$ is a structural element of FD.

As seen in Remark 8, there are functors that preserve structurally nonempty objects, but not structural elements.

Lemma 3. *A functor preserves structural elements iff it preserves structurally nonempty objects and constant morphisms with structurally nonempty domain.*

Proof. If all objects of the domain of the functor are structurally empty, there is nothing to show. Otherwise, the statement is a corollary of the observation made in Remark 7 that structural elements can be represented by constant morphisms with a fixed structurally nonempty domain.

(Functors that preserve structural elements need not preserve all constant morphisms: consider, e.g., the inclusion $\mathbf{A} \hookrightarrow \mathbf{Set}$, where $\mathbf{A}$ 'consists' of the map $\{0,1\} \to \{0\}$.)

2 Almost Cartesian Categories

We now briefly discuss how the notions introduced above relate to cartesian categories (categories with finite products), i.e. to algebraic type theory [4]. The main upshot is that the associated functors already behave rather civil with respect to structural elements.

Definition 9. A category with products of pairs is called *almost cartesian*. A functor between such categories is called almost cartesian if it preserves products of pairs.

The projections for products $A \times B$ are written π_i^{AB}, $i = 1, 2$; the factorization of a pair $(f : C \to A, g : C \to B)$ through $A \times B$ is denoted $\langle f, g \rangle$.

Concerning structural elements, the world is a lot simpler for almost cartesian functors than for arbitrary ones:

Proposition 10. *Almost cartesian functors preserve constant morphisms.*

Proof. In an almost cartesian category, a morphism $f : A \to B$ is constant iff $f \circ \pi_1^{AA} = f \circ \pi_2^{AA}$.

Corollary 11. *An almost cartesian functor preserves structural elements iff it preserves structurally nonempty objects.*

Thus, the choice that an almost cartesian functor is facing is even more extreme than that for an arbitrary functor (cf. Lemma 1): it can either preserve structural elements or make all objects empty. (Note that this implies in particular that an almost cartesian functor between *cartesian* categories will either make all objects empty or preserve the terminal object.)

Theorem 12. *If* $\mathbf{A}$ *is almost cartesian, then* $\mathbf{A} \hookrightarrow \mathbf{A}[1]$ *is universal among all cartesian functors* $F : \mathbf{A} \to \mathbf{B}$ *such that* $\mathbf{B}$ *is cartesian and* F *preserves structurally nonempty objects.*

Proof. All that remains to be shown is that $\mathbf{A}[1]$ is cartesian. By Theorem 5, $\mathbf{A}[1]$ has a terminal object and products of pairs in $\mathbf{A}$. In any category with terminal object, each object A is a product of A and 1 (and such products are preserved by any functor that preserves the terminal object).

Remark 13. This translates nicely into the language of algebraic theories: given an almost cartesian category $\mathbf{A}$, $\mathbf{A}[1]$ is a (multisorted) algebraic theory in the standard sense. The models of $\mathbf{A}[1]$ in, say, **Set**, i.e. the cartesian functors $\mathbf{A}[1] \to \mathbf{Set}$, are essentially the same as the almost cartesian functors $\mathbf{A} \to \mathbf{Set}$ that preserve structurally nonempty objects. If we drop the latter condition, respectively if we admit almost cartesian functors as models of $\mathbf{A}[1]$, the only additional model that crops up is the one where all carriers are empty (of course, target categories other than **Set** will, in general, have more than one empty object).

The introduction of the terminal type can be made explicit on the type theoretical side by introducing additional rules for term formation and equality in context as shown in Figure 1. It is assumed that a set of rules for an equational theory with pairing in a fictitious, though rather standard notation is given and is extended by the new rules. The fact that there are (still) two possible ways of adding a new terminal type 1 is reflected by the dotted line, which separates the common (trivial) part from the particularities of the construction of $\mathbf{A}[1]$ captured by the two rules at the bottom.

The first of these is a new term formation rule which introduces new closed terms that correspond to previous structural elements. In the premises of this

$$\frac{}{\vdash * : 1} \qquad \frac{}{x : 1 \vdash x = * : 1}$$

$$\frac{x : B \vdash s_B : A \text{ for each type } B \neq 1 \quad x, y : A \vdash t[x] = t[y] : A}{\vdash t[*] : A} \qquad \frac{\vdash t[*] : A}{x : A \vdash t[x] = t[*] : A}$$

Fig. 1. Terminal type rules in equational logic

rule, use is made of the fact that a constant term of type A with a single free variable $x : A$ belongs to a (uniquely determined) structural element of A iff, for each 'old' type B, there exists a term s_B of type A with a single free variable $x : B$. We use the notational trick of 'overloading' such constant terms, allowing to substitute the unit constant $*$ for $x : A$ in order to obtain the required closed terms. The second rule asserts that these new constants actually represent the given structural element.

3 Almost Cartesian Closed Categories

The next level upwards in the hierarchy of type systems is that of λ-calculi or, in the terminology of [4], functional type theories. As indicated in the introduction, such type theories correspond to cartesian closed categories.

Definition 14. A category $\mathbf{A}$ is called *almost cartesian closed* if it is almost cartesian and all functors $\_ \times A$ have right adjoints A^-. As usual, the co-universal arrows for $\_ \times A$ are called *evaluation maps* and are denoted $ev_{AB} : B^A \times A \to B$; B^A is called a *function space.* A functor between almost cartesian closed categories is called almost cartesian closed if it preserves this structure (i.e. products of pairs and function spaces) up to isomorphism.

For example, C-monoids as defined in [14] are one-object almost cartesian closed categories.

Just as cartesian closed categories are essentially the same as $\lambda\pi*$-calculi [14,4], almost cartesian closed categories are essentially the same as $\lambda\pi$-calculi: given a $\lambda\pi$-calculus $\mathcal{L}$, one can build an almost cartesian closed classifying category $\mathsf{Cl}_0(\mathcal{L})$ from syntactic material, taking types (not contexts!) as objects and typed terms modulo provable equality as morphisms. Conversely, an almost cartesian closed category gives rise to a $\lambda\pi$-calculus in the shape of an internal language [4]. (Note that the construction of the latter is necessarily slightly different from the one given in [14]; in particular, one has an operation symbol $f : A_1, \ldots, A_n \to B$ for each morphism $f : A_1 \times \cdots \times A_n \to B$, $n \geq 1$, rather than only constant symbols.)

As laid out in [4], one of the benefits of the internal language is that it can be used to define morphims and prove their equality in a somewhat shorter notation than via (otherwise equally easy) categorical arguments. We shall use this facility to define a structural element of B^A that corresponds to a given morphism $f : A \to B$ in an almost cartesian closed category: for each object C, let $\nu_C^f : C \to B^A$ be the morphism represented by the term $f^* = \lambda y : B.\, f(y)$ in the variable $x : C$ (one is tempted to write $f^* = f$, but this is impossible, since f is not a term). Then (ν_C^f) is a structural element: if $g : D \to C$ is a morphism, then $\nu_C^f \circ g$ is represented by $\nu_C^f(g(z)) = f^*$, where $z : D$, hence equal to ν_D^f.

We have already noticed that almost cartesian functors come quite close to preserving structural elements. Rather more radically,

Proposition 15. *Almost cartesian closed functors preserve structural elements.*

Proof. By Lemma 3 and Proposition 10, all that remains to be shown is that almost cartesian closed functors preserve structurally nonempty objects. This follows from Lemma 2 and the fact that objects of the form A^A are structurally nonempty in any almost cartesian closed category, since they have the structural element $(\nu_C^{id_A})$.

This implies that almost cartesian closed functors between cartesian closed categories are cartesian closed, i.e. preserve the terminal object as well (this is, of course, easily seen directly). Strangely enough, this fact seems to have gone unnoticed up to now.

At any rate, thanks to this observation, cartesian closed categories and cartesian closed functors form a full subcategory of the category of almost cartesian closed categories and almost cartesian closed functors. Thus, the universality statement in this context is

Theorem 16. *If* $\mathbf{A}$ *is almost cartesian closed, then* $\mathbf{A}[1]$ *is the cartesian closed reflection of* $\mathbf{A}$*.*

Proof. The 'cartesian part' follows from Theorem 12. In any cartesian category, the functor $\_ \times 1$ is co-adjoint (being naturally isomorphic to the identity), with the projection $A \times 1 \to A$ as evaluation map (i.e. $A^1 \cong A$). Moreover, $1 \times A \to 1$ is an evaluation map (i.e. $1^A \cong 1$).

In order to see that $\mathbf{A}[1]$ is cartesian closed and that $\mathbf{A} \hookrightarrow \mathbf{A}[1]$ is almost cartesian closed, it remains to be shown that $\mathbf{A} \hookrightarrow \mathbf{A}[1]$ preserves function spaces. This follows from the known fact that the Yoneda embedding has this property [19]. However, since the latter statement requires dealing with the somewhat intricate construction of exponentials in the 'overly large' category $[\mathbf{A}^{op}, \mathbf{Set}]$, and since the recurring theme here is 'simplicity', we give a short direct proof: Taking $1 \times A = A$, we have to show that the co-universal property of $ev_{AB} : B^A \times A \to B$ holds also for morphisms $f : 1 \times A = A \to B$. Of course, the associated morphism $1 \to B^A$ we are looking for is the structural element $\nu^f = (\nu_C^f)$. It is easy to see that $\nu^f \times A : A \to B^A \times A$ is really $\langle \nu_A^f, id_A \rangle$, so that $ev_{AB} \circ (\nu^f \times A) = f$ follows by a calculation in the internal language: $ev_{AB}(\langle \nu_A^f, id_A \rangle(x)) = \nu_A^f(x)x = f^*x = f(x)$, where x is a variable of type A.

Lastly, we have to verify that the extension $\mathbf{A}[1] \to \mathbf{B}$ of an almost cartesian functor $\mathbf{A} \to \mathbf{B}$, where $\mathbf{B}$ is cartesian closed, preserves function spaces. This is clear, since all 'new' function spaces are trivial.

In model theoretic terms, Theorem 16 implies the slogan '$\mathbf{A}[1]$ and $\mathbf{A}$ have essentially the same models'. Syntactically, the situation is, thanks to Proposition 15, a lot simpler than for equational theories: it suffices to add a new ground type 1, a new symbol $*$, and the well-known rules

$$\frac{}{\vdash * : 1} \qquad \frac{}{x : 1 \vdash x = * : 1}$$

to a given $\lambda\pi$-calculus $\mathcal{L}$. Let $\mathcal{L}_1$ denote the resulting $\lambda\pi*$-calculus. Explicit introduction of closed terms as in Figure 1 is unnecessary: a structural element (f_C) of A is represented by the (preexisting!) closed term $f_{A^A}(\lambda x : A.\, x)$.

The precise relationship between $\mathcal{L}_1$ and the result of the categorical construction, namely, $\mathsf{Cl}_0(\mathcal{L})[1]$, requires some clarification: the typed terms in context of $\mathcal{L}_1$ are the morphisms of a classifying category $\mathsf{Cl}(\mathcal{L}_1)$, constructed as e.g. in [4] (unlike $\mathsf{Cl}_0(\mathcal{L}_1)$, $\mathsf{Cl}(\mathcal{L}_1)$ has contexts as objects). Now the obvious functor

$$\mathsf{Cl}_0(\mathcal{L}) \to \mathsf{Cl}(\mathcal{L}_1)$$

has the same universal property as $\mathsf{Cl}_0(\mathcal{L}) \to \mathsf{Cl}_0(\mathcal{L})[1]$: given an almost cartesian functor $F : \mathsf{Cl}_0(\mathcal{L}) \to \mathbf{A}$, where $\mathbf{A}$ is cartesian closed, a factorization $F^\# : \mathsf{Cl}(\mathcal{L}_1) \to \mathbf{A}$ is constructed by recursion over the structure of types and terms, respectively, taking 1 to a terminal object. (In other words, models of $\mathcal{L}$ in $\mathbf{A}$ can be extended to models of $\mathcal{L}_1$ in the obvious way.) In particular, we have a recursively defined equivalence (cf. Remark 6)

$$\mathsf{Cl}(\mathcal{L}_1) \to \mathsf{Cl}_0(\mathcal{L})[1].$$

This leads to the following decision procedure for equality of terms in $\mathcal{L}_1$: apply the above equivalence functor, i.e. recursively build morphisms in $\mathsf{Cl}_0(\mathcal{L})[1]$ (this is the essence of the construction of 'top-free' terms outlined in the conclusion of [6]). If both the domain and the codomain of the result are nonterminal, then these morphisms are terms in $\mathcal{L}$ with a single free variable; thus, the problem is reduced to deciding equality in $\mathcal{L}$. The other cases are either trivial or reducible to the first one by composition with $!_{A^A}$, where $A \neq 1$.

By itself, the statement that any almost cartesian category extends to a cartesian closed category (alternatively: any typed λ-calculus with surjective pairing extends to one that has a terminal type) has hardly any claim to originality. Indeed, this problem has, despite its apparent triviality, received such an amount of previous attention that the existing solutions deserve to be listed:

- As outlined in Remark 7, an alternative to using the Yoneda embedding for the construction of $\mathbf{A}[1]$ consists in forming a corresponding subcategory of the Karoubi envelope (note that, by the above, a nonempty almost cartesian closed category always has structurally nonempty objects.) For the case of C-monoids, it has been noticed by Scott [19] that the Karoubi envelope is cartesian closed (see [13] for a detailed exposition), and in [14], it is pointed out that the terminal object and the 'original' reflexive object suffice. It is surprising that the more general observation that the same process works for arbitrary almost cartesian closed categories does not seem to have been explicitly made.

- A rather more complicated syntactic construction which seems to work only in the case of C-monoids is exhibited in [18].
- On the side of the λ-calculus, the problem is discharged in [14], Ch. 12, by a sketch of a method for eliminating the terminal object for purposes of deciding existence and equality of terms. This method is made slightly more explicit in the conclusion of [6].
- It is pointed out in [14], Ch. 13, that confluence fails for the rewrite system obtained by making the usual equations of $\lambda\pi*$ directed. It is comparatively easy to perform a 'manual Knuth-Bendix completion' on this system, obtaining a system that is, by construction, weakly confluent. This prodecure is laid out in [6], where it is also shown that the resulting system is indeed confluent (and thus provides a decision procedure for equality). Since the standard methods of establishing normalization fail, the proof of this statement is rather involved.
- In the conclusion of [6], several other methods of obtaining the decidability and conservativity results proved there via the mentioned confluent reduction system are discussed, including [10,16].
- As an alternative to the approach of [6], it has been suggested to replace η-contraction by (restricted) η-expansion, thus obtaining a confluent reduction to long $\beta\eta$-normal forms for the $\lambda\pi*$-calculus [12]. This method is extendible to polymorphism [8] and even the calculus of constructions [9]. Moreover, it lends itself to a certain amount of modularization [7], thus allowing simpler proofs than η-contraction.
- In [11], it is more or less shown that the proper categorical models for the simply typed λ-calculus (without terminal object or products) are closed subcategories of cartesian closed categories, which translates back into the statement that products and a terminal object can be conservatively added to the simply typed λ-calculus. Less mysteriously put: the classifiying category of a simply typed λ-theory [11], consisting of *contexts* as objects and tuples of typed terms in context modulo provable equality as morphisms, is cartesian closed. The internal language of that category in the sense of [14] is a $\lambda\pi*$-calculus which conservatively extends the originally given λ-calculus. (Similar considerations are outlined at the end of [6].)

What we believe is new here is the insight that the extension in question is unique, that this fact is not particular to cartesian closed categories, but rather an instance of a (very simple) statement about categories in general, and that the extension is in fact universal.

4 Subobject Classifiers

The most complex type of structured category we are going to consider here is that of a topos; on the type theoretical side, the notion of topos corresponds to full intuitionistic type theory [14].

The most economical definition of topos for our purposes is the one suggested in [1]: a cartesian closed category $\mathbf{A}$ is a topos iff it has a *subobject classifier* $\top : 1 \to \Omega$ in the sense that each diagram of the form

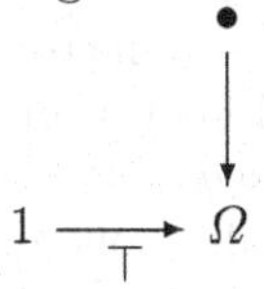

has a pullback, and for each subobject (monomorphism) $m : A \to B$, there exists a unique morphism $m^* : B \to \Omega$ such that

$$\begin{array}{ccc} A & \xrightarrow{m} & B \\ \downarrow & & \downarrow{\scriptstyle m^*} \\ 1 & \xrightarrow[\top]{} & \Omega \end{array}$$

is a pullback. It follows that $\mathbf{A}$ has all finite limits.

It is clear that, in the absence of a terminal object, the element $\top$ has to be replaced by a structural element $(\top_C)$ of Ω. The diagrams above can then be interpreted as equalizers of pairs $(f, \top_B)$, where $f : B \to \Omega$. The problem that arises is that morphisms $1 \to B$ are also 'singleton subobjects' and therefore are encoded as morphisms $B \to \Omega$. Hence the following, not entirely pleasing

Definition 17. A *structural subobject classifier* is a structural element $\top = (\top_C)$ of an object Ω such that

(i) whenever $f : B \to \Omega$, then the pair $(f, \top_B)$ has an equalizer, or for each object C, there exists a unique morphism $h : C \to B$ such that $f \circ h = \top_C$;
(ii) for each subobject $m : A \to B$, there exists a unique morphism $m^* : B \to \Omega$ such that m is an equalizer of $(m^*, \top_B)$.

Thus, the pullback existence condition for subobject classifiers has been coded in a more or less obvious way in the definition of a structural subobject classifier. However, there are still a few subtleties attached to establishing that subobjects $m : A \to B$, where one of A and B is 1, are properly 'classified' in $\mathbf{A}[1]$ (recall that, in a topos, 1 may have rather a lot of subobjects!):

Theorem 18. *If Ω is a structural subobject classifier in an almost cartesian category $\mathbf{A}$, then Ω is a subobject classifier in $\mathbf{A}[1]$.*

Proof. To begin, let $!_P : P \to 1$ be a subobject in $\mathbf{A}[1]$ (w.l.o.g., $P \in \mathrm{Ob}\,\mathbf{A}$). Then the projection

$$\pi_1^{CP} : C \times P \to C$$

is a monomorphism for each object C in $\mathbf{A}$. Let p_C be the unique morphism $C \to \Omega$ such that π_1^{CP} is an equalizer of $(p_C, \top_C)$. Then the family $p = (p_C)$ is a structural element of Ω: by the uniqueness requirement in Condition (ii) of Definition 17, it suffices to show that, given a morphism $g : C \to B$ in $\mathbf{A}$, π_1^{CP} is an equalizer of $(p_B g, \top_C)$. To this end, let $h : A \to C$ be a morphism such that $p_B \circ g \circ h = \top_C \circ h$. Since $\top_C = \top_B \circ g$, the definition of p_B implies that there exists $k : A \to B \times P$; the required (necessarily unique) factorizing morphism $A \to C \times P$ for h is $\langle h, \pi_2^{BP} \circ k \rangle$.

Generally, a morphism $f : A \to C$ factors through π_1^{CP} iff $!_A$ factors through $!_P$, i.e. iff there exists $A \to P$. Moreover, given a structural element $\bar{p} = (\bar{p}_B)$ of Ω, g equalizes $(\bar{p}_C, \top_C)$ (i.e. $\bar{p}_C \circ g = \top_C \circ g$) iff $!_A$ equalizes $(\bar{p}, \top)$ (namely, iff $\bar{p}_A = \top_A$). Thus, for C structurally nonempty (say, $C = \Omega$), $!_P$ is an equalizer of $(\bar{p}, \top)$ iff π_1^{CP} is an equalizer of $(\bar{p}_C, \top_C)$, i.e. iff $\bar{p} = p$.

Secondly, let $f : 1 \to A$ be a subobject in $\mathbf{A}[1]$ associated to a structural element (f_C) of A. Then

$$\langle id_A, f_A \rangle : A \to A \times A$$

is a monomorphism, hence an equalizer of $(\top_{A \times A}, \tilde{f})$ for some $\tilde{f} : A \times A \to \Omega$. Put

$$f^* = \tilde{f} \circ \langle f_A, id_A \rangle : A \to \Omega.$$

f is an equalizer of $(f^*, \top_A)$: let $g : C \to A$ be a morphism such that $f^* \circ g = \top_A \circ g$. Then $\tilde{f} \circ \langle f_A, id_A \rangle \circ g = \top_{A \times A} \circ \langle f_A, id_A \rangle \circ g$; hence there exists $\bar{g} : C \to A$ such that $\langle id_A, f_A \rangle \circ \bar{g} = \langle f_A, id_A \rangle \circ g$. It follows that $g = f_A \circ \bar{g} = f_C = f \circ !_C$, i.e. $!_C$ is the (necessarily unique) factorizing morphism for g.

It remains to be shown that f^* is unique. Let f be an equalizer of $(f^+, \top_A)$ for some $f^+ : A \to \Omega$. Then $\langle id_A, f_A \rangle$ is an equalizer of $(f^+ \circ \pi_2^{AA}, \top_{A \times A})$: if $f^+ \circ \pi_2^{AA} \circ g = \top_{A \times A} \circ g$ for some $g : C \to A \times A$,

$$\begin{array}{ccccc}
1 & \xrightarrow{f} & A & \xrightarrow{\langle id_A, f_A \rangle} & A \times A \\
 & & f^+ \Big\downarrow\Big\downarrow \top_A & & f^+ \circ \pi_2^{AA} \Big\downarrow\Big\downarrow \top_{A \times A} \\
 & & \Omega & = = = = & \Omega
\end{array}$$

then $f^+ \circ \pi_2^{AA} \circ g = \top_A \circ \pi_2^{AA} \circ g$, so that the assumption on f^+ implies $\pi_2^{AA} \circ g = f_C$. Thus,

$$\begin{aligned}
g = \langle \pi_1^{AA} \circ g, \pi_2^{AA} \circ g \rangle &= \langle \pi_1^{AA} \circ g, f_C \rangle \\
&= \langle \pi_1^{AA} \circ g, f_A \circ \pi_1^{AA} \circ g \rangle = \langle id_A, f_A \rangle \circ \pi_1^{AA} \circ g,
\end{aligned}$$

i.e. $\pi_1^{AA} \circ g$ is the (necessarily unique) factorizing morphism for g. Now by uniqueness of $\tilde{f}$, $f^+ \circ \pi_2^{AA} = \tilde{f}$, hence $f^+ = f^+ \circ \pi_2^{AA} \circ \langle f_A, id_A \rangle = f^*$.

Corollary 19. $\mathbf{A}[1]$ *is a topos iff* $\mathbf{A}$ *is an almost cartesian closed category and has a structural subobject classifier.*

Remark 20. There is an important intermediate step missing here: as shown in [20], Martin-Löf style dependent type theories are, on the categorical side, equivalent to locally cartesian closed categories, i.e. categories $\mathbf{A}$ with a terminal object such that all slice categories $\mathbf{A}/A$ are cartesian closed. Locally cartesian closed categories are cartesian closed and finitely complete; every topos is locally cartesian closed. It is an open problem to find a reasonable characterization of categories $\mathbf{A}$ for which $\mathbf{A}[1]$ is locally cartesian closed; the difficulty is that the introduction of a terminal object (in the base category, not in the slices, which automatically have terminal objects) adds nontrivial new objects to the slices.

Conclusion

We have shown that, for arbitrary categories, there is a single way of adding a terminal object apart from the trivial one, and we have discussed in some depth how this observation relates to various type theories. In particular, we have demonstrated that the terminal type, although, of course, convenient, can be entirely ignored in the typed λ-calculus whenever it causes theoretical difficulties such as the ones treated in [6,12]. Moreover, we have characterized the notion of 'topos without terminal object', showing that not only the terminal object itself, but also the classification of subobjects that involve the terminal object can be reconstructed.

Two main points have been stressed: the first is the unexpected rigidity which, even in simple contexts, governs the elements that a type may or may not have in possible extensions. The other is the extreme simplicity of the underlying constructions and arguments made possible by a categorical treatment.

Future work will focus on obtaining similar results for dependent type theories, i.e. locally cartesian closed categories, and polymorphic type theories (covered in [6]), which, on the categorical side, correspond to a suitable class of hyperdoctrines [21]. Moreover, it should be investigated how this work relates to singleton types as featured e.g. in [2].

Acknowledgements

The author wishes to thank Christoph Lüth and Till Mossakowski for useful comments and discussions.

References

1. J. Adámek, H. Herrlich, and G. E. Strecker, *Abstract and concrete categories*, Wiley, New York, 1990.
2. D. Aspinall, *Subtyping with singleton types*, Computer Science Logic, LNCS, vol. 933, Springer, 1995, pp. 1–15.
3. F. Borceux, *Handbook of categorical algebra 1*, Cambridge, 1994.
4. R. L. Crole, *Categories for types*, Cambridge, 1994.

5. P.-L. Curien, *Categorical combinators, sequential algorithms, and functional programming*, 2nd ed., Birkhäuser, Boston, 1993.
6. P.-L. Curien and R. Di Cosmo, *A confluent reduction for the λ-calculus with surjective pairing and terminal object*, J. Funct. Programming **6** (1996), 299–327.
7. R. Di Cosmo, *On the power of simple diagrams*, Rewriting Techniques and Applications, LNCS, vol. 1103, Springer, 1996, pp. 200–214.
8. R. DiCosmo and D. Kesner, *Rewriting with polymorphic extensional λ-calculus*, Computer Science Logic, LNCS, vol. 1092, Springer, 1996, pp. 215–232.
9. N. Ghani, *Eta-expansions in dependent type theory — the calculus of constructions*, Typed Lambda Calculus and Applications, LNCS, vol. 1210, Springer, 1997, pp. 164–180.
10. T. Hardin, *Confluence results for the pure strong categorical logic CCL; lambda-calculi as subsystems of CCL*, Theoret. Comput. Sci. **65** (1989), 291–342.
11. B. Jacobs, *Categorical logic and type theory*, Elsevier, Amsterdam, 1999.
12. C. B. Jay and N. Ghani, *The virtues of eta-expansion*, J. Funct. Programming **5** (1995), 135–154.
13. C. P. J. Koymans, *Models of the lambda calculus*, CWI, Amsterdam, 1984.
14. J. Lambek and P. J. Scott, *Introduction to higher order categorical logic*, Cambridge, 1986.
15. S. Mac Lane, *Categories for the working mathematician*, Springer, 1997.
16. Adam Obtułowicz, *Algebra of constructions I. The word problem for partial algebras*, Inform. and Comput. **73** (1987), 129–173.
17. Axel Poigné, *Cartesian closure - higher types in categories*, Category Theory and Computer Programming, LNCS, vol. 240, Springer, 1985, pp. 58–75.
18. P. H. Rodenberg and F. J. van der Linden, *Manufacturing a cartesian closed category with exactly two objects out of a C-monoid*, Stud. Log. **48** (1989), 279–283.
19. D. S. Scott, *Relating theories of the λ-calculus*, To H.B. Curry: Essays in Combinatory Logic, Lambda Calculus and Formalisms, Academic Press, 1980, pp. 403–450.
20. R. A. G. Seely, *Locally cartesian closed categories and type theory*, Math. Proc. Cambridge Philos. Soc. **95** (1984), 33–48.
21. R. A. G. Seely, *Categorical semantics for higher order polymorphic lambda calclus*, J. Symbolic Logic **52** (1987), 969–989.

Fully Complete Minimal PER Models for the Simply Typed λ-Calculus*

Samson Abramsky[1] and Marina Lenisa[2]

[1] Oxford University Computing Laboratory
Wolfson Building, Parks Road, OX1 3QD, England
Samson.Abramsky@comlab.ox.ac.uk
[2] Dipartimento di Matematica e Informatica, Università di Udine,
Via delle Scienze 206, 33100 Udine, Italy
lenisa@dimi.uniud.it

Abstract. We show how to build a *fully complete* model for the *maximal theory* of the simply typed λ-calculus with k ground constants, λ_k. This is obtained by *linear realizability* over an *affine combinatory algebra* of *partial involutions* from natural numbers into natural numbers. For simplicitly, we give the details of the construction of a fully complete model for λ_k extended with ground *permutations*. The fully complete minimal model for λ_k can be obtained by carrying out the previous construction over a suitable subalgebra of partial involutions. The full completeness result is then put to use in order to prove some simple results on the maximal theory.

Introduction

A *categorical* model of a type theory (or logic) is said to be *fully-complete* ([AJ94a]) if, for all *types* (*formulae*) A, B, all morphisms $f : [\![A]\!] \to [\![B]\!]$, from the interpretation of A into the interpretation of B, are denotations of a *proof-term* of the entailment $A \vdash B$, i.e. if the interpretation function from the category of syntactical objects to the category of denotations is *full*. The notion of full-completeness is a counterpart to the notion of *full abstraction* for programming languages. A fully complete model indicates that there is a very tight connection between syntax and semantics. Equivalently, one can say that the term model has been made into a mathematically respectable structure.

Over the past decade, Game Semantics has been used successfully by various people to define fully-complete models for various fragments of Linear Logic, and to give fully-abstract models for many programming languages, including PCF, and other functional and non-functional languages. Recently, a new technique, called *linear realizability* (see [AL99,AL00]), has been proposed as a valid and less complex alternative to Game Semantics in providing fully complete and fully abstract models. In particular, this technique has been used in [AL99,AL00] to define a model fully complete w.r.t. the fragment of system F consisting of

* Work partially supported by TMR Linear FMRX-CT98-0170.

L. Fribourg (Ed.): CSL 2001, LNCS 2142, pp. 443–457, 2001.

ML polymorphic types, and in [AL99a] to provide a fully complete model for PCF. The linear (linear affine) realizability technique amounts to constructing a category of *Partial Equivalence Relations (PERs)* over a *Linear Combinatory Algebra, LCA*, (Affine Combinatory Algebra, ACA). This category turns out to be *linear* (affine), and to form an *adjoint model* with its co-Kleisli category. The notion of Linear (Affine) Combinatory Algebra introduced by the first author ([Abr97]) refines the standard notion of Combinatory Algebra, in the same way in which intuitionistic linear (affine) logic refines intuitionistic logic. The construction of PER models from LCA's (ACA's) of [AL99,AL00] is quite simple and clear, and it yields models with *extensionality* properties. Many examples of linear combinatory algebras arise in the context of Abramsky's categorical version of Girard's *Geometry of Interaction* ([AJ94,Abr97,Abr96,AHS98]).

In this paper, we define a *fully complete PER model* for the *maximal theory* $\approx$ on λ_k. λ_k is the simply typed λ-calculus with finitely many ground constants in the ground type o. The theory $\approx$ equates two closed λ-terms M, N of type $T_1 \to \ldots \to T_n \to o$ if and only if, for all $P_1 \approx Q_1$ of type T_1, ..., $P_n \approx Q_n$ of type T_n, $MP_1 \ldots P_n =_\beta NQ_1 \ldots Q_n$. To our knowledge, our model is the first model of $\approx$ different from the term model.

For simplicitly, we show first how to build a fully complete minimal PER model $\mathcal{M}_{\mathcal{O}_k}$ for the simply typed λ-calculus with k ground constants extended with *permutations* of ground type. The fully complete minimal model for λ_k can then be obtained by cutting down the combinatory algebra. The model $\mathcal{M}_{\mathcal{O}_k}$ for the extended language arises from the special affine combinatory algebra of *partial involutions* used in [AL99,AL00] for modeling System F. It consists, essentially, of the hierarchy of simple PERs over a PER having exactly k distinct equivalence classes (for any $k \geq 2$). The proof of full completeness carried out in this paper is based on the *linear affine* analysis of the *intuitionistic* arrow, which is possible in our PER category. Our proof uses a *Decomposition Theorem*, which is by now a standard tool in discussing full completeness. In the present case, given a partial involution which inhabits a PER interpreting a simple type, the Decomposition Theorem allows to recover the *top-level structure* (up-to permutations) of the (possibly infinite) *typed Böhm tree* corresponding to the given partial involution. Once we have the Decomposition Theorem, in order to prove λ-definability, we still need to rule out possibly infinite typed Böhm trees from the model. In order to do this, we prove an *Approximation Theorem*, and we study an intermediate PER model $\mathcal{M}_{\mathcal{O}_k^\perp}$ for λ_k extended with a new ground constant $\perp$, intended to denote the undefined constant. A variant of this model, $\mathcal{M}_{\mathcal{O}}$, for the special case of two gound constants, $\top$ and $\perp$, has been used implicitly as an intermediate construction in the work of [AL99] on system F.

In order to get a fully complete minimal model for λ_k, we only need to cut down the algebra of involutions by putting an extra constraint, which allows us to rule out permutations from the model $\mathcal{M}_{\mathcal{O}_k}$.

The full completeness result is then put to use to prove (or re-prove) some simple facts on the maximal λ-theory. In particular, the *Context Lemma* follows immediately from the full completeness of the model. Moreover, we are able to prove that this holds also for the simply typed (possibly infinite) Böhm trees, in the special case of exactly two ground constants. We give also some decidability results, such as whether a *finite* partial involution belongs to a given type. These give a *semantical* procedure to decide, in some special cases, whether two terms are equivalent in the maximal theory, alternative to the *syntactical* procedures by Padovani [Pad95] and Loader [Loa97].

The paper is organized as follows. In Section 1, we recall some definitions and notations concerning the simply typed λ-calculus with finitely many base constants, the construction of a PER category over an ACA, and we present the special ACA of partial involutions. In Section 2 we study the PER model $\mathcal{M}_{\mathcal{O}_k}$ for λ_k and we prove that it is fully complete and minimal w.r.t. the extended calculus. In Section 3, we present the construction of the fully complete and minimal model for the simply typed λ-calculus. In Section 4, we illustrate some uses of the full completeness result. Conclusions and directions for future work appear in Section 5.

Notation. Vectors are written in bold. For $f, f' : \mathrm{Nat} \rightharpoonup \mathrm{Nat}$ partial functions from Nat to Nat, and $n, n' \in \mathrm{Nat}$, we denote, respectively, by $f(n) \downarrow$ and by $f(n) \uparrow$ the fact that f is defined on n and the fact that f is not defined (diverges) on n. We denote by $f(n) \updownarrow f'(n')$ the equiconvergence predicate to be read as: $f(n) \downarrow \Leftrightarrow f'(n') \downarrow$. Let X, Y be sets, then (l, x) and (r, y), where $x \in X$ and $y \in Y$, denote elements of $X + Y$. Let $X_1, \ldots, X_n$ be sets, often we will omit the parentheses in denoting $((X_1 + \ldots) + X_{n-1}) + X_n$. We will use the abbreviated notation (i, x), for denoting the element of $X_1 + \ldots + X_n$ coming from an element $x \in X_i$.

1 Preliminaries

This section consists of two parts. In the first part, we recall some definitions and notations concerning the simply typed λ-calculus λ_k with k base constants, the maximal theory, and its models. In the second part, we recall the notion of *affine combinatory algebra* (ACA) [Abr97], the construction of a PER category over an ACA [AL99,AL00], and we present the special ACA of *partial involutions*.

1.1 Simply Typed λ-Calculus, Maximal Theories, Models

Definition 1 (λ_k). *The class* SimType *of simple types over a ground type o is defined by:*

$$(\mathit{SimType} \ni)\ T \ ::=\ o \mid T \to T\ .$$

Raw Terms are defined as follows:

$$\Lambda \ni M \ ::= c_i \mid x \mid \lambda x : T.M \mid MM\ ,$$

where $c_i \in \mathit{Const}_k = \{c_1, \ldots, c_k\}$, is the set of ground constants, $x \in \mathit{Var}$. We denote by Λ^0 the set of closed λ-terms.

Well-typed terms. *We introduce a* proof system *for deriving* typing judgements *of the form* $\Delta \vdash M : T$*, where* Δ *is a* type assignment, *i.e. a finite list* $x_1 : T_1, \ldots, x_n : T_n$. *The rules of the proof system are the following:*

$$\frac{}{\Delta \vdash c_i : o} \qquad \frac{}{\Delta, x : T, \Delta' \vdash x : T}$$

$$\frac{\Delta, x : T \vdash M : S}{\Delta \vdash \lambda x : T.M : T \to S} \qquad \frac{\Delta \vdash M : T \to S \quad \Delta \vdash N : T}{\Delta \vdash MN : S}$$

β-conversion. *β-conversion between well-typed terms is the least relation generated by the following rule and the rules for congruence closure (which we omit):* $\Delta \vdash (\lambda x : T.M)N = M[N/x] \ : \ S$*, where* $\Delta, x : T \vdash M : S$*, and* $\Delta \vdash N : T$.

It is well-known that the *maximal theory* over λ_k can be characterized as follows:

Definition 2 (Maximal Theory on λ_k). *Let* $M, N \in \Lambda$*. We define the equivalence* $\approx \subseteq \Lambda^0 \times \Lambda^0$ *by induction on types as follows:*
$M \approx_{T_1 \to \ldots T_n \to o} N$ *iff* $\vdash M : T_1 \to \ldots T_n \to o,\ \ \vdash N : T_1 \to \ldots T_n \to o,$ *and* $\forall P_1 \approx_{T_1} Q_1 \ldots P_n \approx_{T_n} Q_n.\ MP_1 \ldots P_n =_\beta NQ_1 \ldots Q_n$.

In this paper, we will focus on categorical models for the simply typed λ-calculus λ_k. As usual, categorical models of λ_k are cartesian closed categorie, in which types are interpreted by objects and terms in contexts are interpreted by morphisms.

Definition 3 (Fully Complete Model). *A categorical model of* λ_k *is* fully complete *if, for all simple types* T *and for all* $h : 1 \to [\![T]\!]$*, there exists* $M \in \Lambda^0$ *such that* $h = [\![\vdash M : T]\!]$*, where* $[\![\ ]\!]$ *is the interpretation function in the model.*

1.2 PERs over Affine Combinatory Algebras

In this section, we briefly recall the construction of a PER category from an affine combinatory algebra (ACA) (see [AL99,AL00] for more details on this construction). This category turns out to be *affine*, while its co-Kleisli category turns out to be cartesian closed. In particular, we focus on a combinatory algebra of partial involutions $\mathcal{A}_{\mathsf{PInv}}$, and we consider the model of λ_k induced by the cartesian closed subcategory of PERs over $\mathcal{A}_{\mathsf{PInv}}$ generated by the special PER $\mathcal{O}_k$. This latter PER is intended to denote the ground type o.

We start by giving the notion of *affine combinatory algebra* (ACA):

Definition 4 (Affine Combinatory Algebra, [Abr97]). *An* affine combinatory algebra $\mathcal{A} = (A, \bullet, !)$ *is an applicative structure* $(A, \bullet)$ *with a unary (injective) operation* $!$*, and distinguished elements (combinators)* $\boldsymbol{B}, \boldsymbol{C}, \boldsymbol{I}, \boldsymbol{K}, \boldsymbol{W}, \boldsymbol{D}, \delta, \boldsymbol{F}$ *satisfying the following equations:*

$\boldsymbol{I}x = x$	$\boldsymbol{B}xyz = x(yz)$	$\boldsymbol{C}xyz = (xz)y$	$\boldsymbol{K}xy = x$
$\boldsymbol{W}x!y = x!y!y$	$\boldsymbol{D}!x = x$	$\delta !x = !!x$	$\boldsymbol{F}!x!y = !(xy)$.

Proposition 1 ([AL99,AL00]). *Let $\mathcal{A} = (A, \bullet)$ be an ACA. We define the category $PER_{\mathcal{A}}$ as follows.*
Objects*: PERs $\mathcal{R} \subseteq A \times A$, i.e. symmetric and transitive relations.*
Morphisms*: a morphism f from $\mathcal{R}$ to $\mathcal{S}$ is an equivalence class of the PER $\mathcal{R} \multimap \mathcal{S}$, where $\mathcal{R} \multimap \mathcal{S}$ is defined by: $\alpha(\mathcal{R} \multimap \mathcal{S})\beta$ iff $\forall \gamma \, \mathcal{R} \, \gamma' . \; \alpha \bullet \gamma \, \mathcal{S} \, \beta \bullet \gamma'$.*
Let P be the pairing combinator, *i.e. (using λ-notation) $P = \lambda xyz.zxy$. Then, for all PERs $\mathcal{R}, \mathcal{S}$, let $\mathcal{R} \otimes \mathcal{S}$ be the PER defined by:*

$$\mathcal{R} \otimes \mathcal{S} = \{(P\alpha\beta, P\alpha'\beta') \mid \alpha \, \mathcal{R} \, \alpha' \wedge \beta \, \mathcal{S} \, \beta'\} .$$

$\otimes$ gives rise to a tensor product on $PER_{\mathcal{A}}$. We define the PER $\mathcal{I} = \{(\boldsymbol{I}, \boldsymbol{I})\}$ to be the tensor identity.
For all $\mathcal{R}$, we define $! \, \mathcal{R} = \{(!\alpha, !\beta) \mid \alpha \, \mathcal{R} \, \beta\}$. $!$ gives rise to a symmetric monoidal comonad $(!, der, \delta)$ on $PER_{\mathcal{A}}$, where $der = [\boldsymbol{D}]$, and δ denotes, by abuse of notation, the equivalence class of the combinator δ.
Summarizing, we have:
- *The category $PER_{\mathcal{A}}$ is affine.*
- *The co-Kleisli category $(PER_{\mathcal{A}})_!$, induced by the comonad $!$ on the category $PER_{\mathcal{A}}$, is cartesian closed.*
- *The categories $PER_{\mathcal{A}}$ and $(PER_{\mathcal{A}})_!$ form an adjoint model.*

The first author introduced in [Abr97] a basic important example of an ACA on the space [Nat ⇀ Nat] of partial functions from natural numbers into natural numbers. Here we briefly recall the definition of this ACA (see [AL99,AL00] for more details). The ACA of partial involutions, which we will consider in the next section, arises as subalgebra of this.

Let us consider the space [Nat ⇀ Nat] of partial functions from natural numbers to natural numbers. For any $\alpha \in$ [Nat ⇀ Nat] injective, we denote by α^{-1} the inverse of α. We start by fixing two injective *coding* functions t and p:

$$t : \mathrm{Nat} + \mathrm{Nat} \rightharpoonup \mathrm{Nat} \quad , \quad p : \mathrm{Nat} \times \mathrm{Nat} \rightharpoonup \mathrm{Nat} .$$

The first is used in order to define application, and it allows to transform a one-input/one-output function into a two-input/two-output function. The latter is used for creating *infinitely* many copies of a one-input/one-output function α, i.e. for defining $!\alpha$. Application can be explained *geometrically*, using the language of "boxes and wires" which arises in the general setting of traced symmetric monoidal categories (see [JSV96] for an abstract treatment). Let us represent a one-input/one-output function $\alpha \in$ [Nat ⇀ Nat] by the one-input-port/one-output-port box in Fig. 1(i) below. In order to define the application $\alpha \bullet \beta$, for $\alpha, \beta \in$ [Nat ⇀ Nat], we regard α as a two-input/two-output function via the coding t. In particular, $t; \alpha; t^{-1} : \mathrm{Nat} + \mathrm{Nat} \rightharpoonup \mathrm{Nat} + \mathrm{Nat}$ can be described as a matrix of 4 one-input/one-output functions, where each entry $\alpha_{ij} : \mathrm{Nat} \rightharpoonup \mathrm{Nat}$, $\alpha_{ij} = in_i; t; \alpha; t^{-1}; in_j^{-1}$ accounts for the contribution from the i-th input wire into the j-th output wire (see Fig. 1(ii)).

The result of the application $\alpha \bullet \beta$ is the following one-input/one-output function (see Fig. 1(iii)): $\alpha \bullet \beta = \alpha_{22} \cup \alpha_{21}; (\beta; \alpha_{11})^{\star}; \beta; \alpha_{12}$,
where $\cup$ denotes union of graph relations, and $(\beta; \alpha_{11})^{\star}$ denotes $\bigcup_{n \geq 0} (\beta; \alpha_{11})^n$.

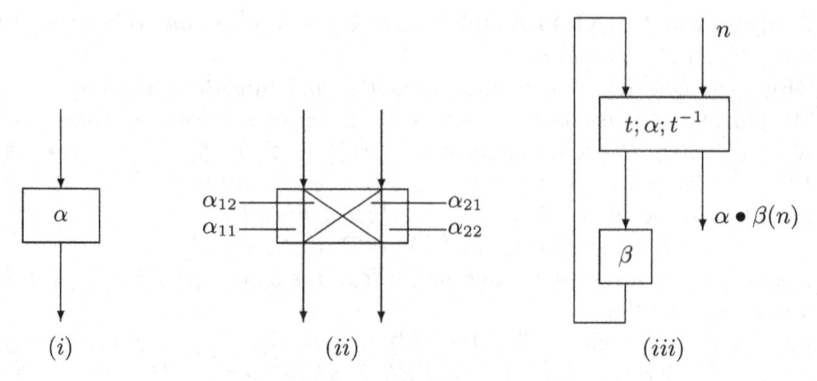

Fig. 1. Geometrical description of linear application.

The above formula for computing the application is essentially the *Execution Formula* from Girard's Geometry of Interaction ([Gir89]).

The definition of the !-operation on our applicative structure is quite simple. The operation ! is intended to produce, from a single copy of α, *infinitely* many copies of α. These are obtained by simply tagging each of these copies with a natural number, i.e. we define: $!\alpha = p^{-1}; (id_{\mathrm{Nat}} \times \alpha); p$.

For the definition of the affine combinators on $([\mathrm{Nat} \rightharpoonup \mathrm{Nat}], \bullet, !)$ see [AL00].

There are many possible conditions that can be imposed on partial functions in order to cut down the space $[\mathrm{Nat} \rightharpoonup \mathrm{Nat}]$, still maintaining closure under the application, !, and all the affine combinators. The subalgebra which we are interested in is obtained by considering *partial involutions*:

Proposition 2. *Let $f : Nat \rightharpoonup Nat$. f is a* partial involution *iff its graph is a symmetric relation. Let us denote by $[Nat \rightharpoonup_{\mathsf{Inv}} Nat]$ the space of partial involutions from Nat to Nat. Then $\mathcal{A}_{\mathsf{PInv}} = ([Nat \rightharpoonup_{\mathsf{Inv}} Nat], \bullet, !)$ is an affine combinatory algebra.*

$\mathcal{A}_{\mathsf{PInv}}$ is a highly constrained algebra. Partial involutions are reminiscent of the *copy-cat* strategies of game categories. Notice that the only computational effect that the combinators have is that of *copying* information from input to output wires. Partial involutions f on a set S correspond biuniquely to pair-wise disjoint families of subsets $\{x, y\}$ of S, where $\{x, y\}$ is in the family if and only if $f(x) = y$ (and hence also $f(y) = x$). We can think of these as abstract families of "axiom links" as in the proof-nets of Linear Logic.

2 A Fully Complete Minimal Model for an Extended Calculus

In this section, we define a model for λ_k in a suitable subcategory of the co-Kleisli category of $\mathrm{PER}_{\mathcal{A}_{\mathsf{PInv}}}$. We prove that this model is fully complete w.r.t.

λ_k extended with constants for all *transpositions* [1] of type $o \to o$. The proof of full completeness will use an *Approximation Theorem* and an intermediate model, which allows for partial elements.

We start by introducing the PER $\mathcal{O}_k$ on $\mathcal{A}_{\mathsf{PInv}}$, with k distinct equivalence classes. Our model is defined in the co-Klesli category of the affine category freely generated by $\mathcal{O}_k$.

Definition 5 (The PER $\mathcal{O}_k$). *Fix distinct natural numbers ("moves") $*_1, \ldots *_k$, $a_1, \ldots, a_k$. Let $\mathcal{O}_k$ be the PER on the combinatory algebra $\mathcal{A}_{\mathsf{PInv}}$ consisting of k equivalence classes defined by:*
- $c_1 = \{f : Nat \rightharpoonup_{\mathsf{Inv}} Nat \mid f(*_1) = a_1 \ \wedge\ \forall m \neq 1.\ *_m, a_m \notin dom(f)\}$
- $\ldots$
- $c_k = \{f : Nat \rightharpoonup_{\mathsf{Inv}} Nat \mid f(*_k) = a_k \ \wedge\ \forall m \neq k.\ *_m, a_m \notin dom(f)\}$.

Definition 6 ($\mathcal{M}_{\mathcal{O}_k}$). *Let $\mathcal{M}_{\mathcal{O}_k} = (CCPER_{\mathcal{O}_k}, \bullet_{\mathcal{O}_k}, [\![\ ]\!]^{\mathcal{O}_k})$ be the model of λ_k, where $CCPER_{\mathcal{O}_k}$ is the co-Klesli category of the (linear) affine category freely generated by $\mathcal{O}_k$.*

First of all notice that, by the fact that the PER $\mathcal{O}_k$ has only a finite number of equivalence classes and by extensionality of the PER model, each PER in $\mathrm{CCPER}_{\mathcal{O}_k}$ has only *finitely many* equivalence classes. Moreover, it is easy to check that the equivalence classes of the PER $\mathcal{O}_k \multimap \mathcal{O}_k$ correspond to the permutations from $\mathcal{O}_k$ to $\mathcal{O}_k$. I.e. an involution f belongs to $\mathrm{dom}(\mathcal{O}_k \multimap \mathcal{O}_k)$ if and only if $\forall c_i \exists c_j.\ f \bullet c_i = c_j$. Different permutations are in different equivalence classes of $\mathcal{O}_k \multimap \mathcal{O}_k$. It is standard that all permutations can be obtained by suitably composing elementary permutations, i.e. *transpositions.* Permutations (transpositions) of ground type are sufficient to λ-define all the elements of $\mathcal{M}_{\mathcal{O}_k}$, i.e. $\mathcal{M}_{\mathcal{O}_k}$ is fully complete. The proof of this fact is based on a *Decomposition Theorem* for the partial involutions in the domains of the PERs interpreting a simple type:

Theorem 1 (Decomposition). *Let $T = T_1 \to \ldots \to T_n \to o \in SimType$, $n \geq 0$, where, for all $i = 1, \ldots, n$, $T_i = U_{i1} \to \ldots \to U_{iq_i} \to o$. If $f \in dom(\otimes_{i=1}^n ![\![T_i]\!]^{\mathcal{O}_k} \multimap \mathcal{O}_k)$, then*

- *either $f \in dom([\![\boldsymbol{x} : \boldsymbol{T} \vdash c_i : o]\!]^{\mathcal{O}_k})$, for some c_i*
- *or $\exists i \in \{1, \ldots, n\}$, $\exists p_{\mathcal{O}_k} : \mathcal{O}_k \multimap \mathcal{O}_k$ permutation, and $\exists g_1, \ldots, g_{q_i}$, where $\forall j \in \{1, \ldots q_i\}.\ g_j \in dom(\otimes_{i=1}^n ![\![T_i]\!]^{\mathcal{O}_k} \multimap [\![U_{ij}]\!]^{\mathcal{O}_k})$, such that*

$$f\ (\underset{i=1}{\overset{n}{\times}} ![\![T_i]\!]^{\mathcal{O}_k} \multimap \mathcal{O}_k)\ (con_{\otimes_{i=1}^n ![\![T_i]\!]^{\mathcal{O}_k}}; (\pi_i^n \otimes \langle g_1, \ldots, g_{q_i} \rangle^\dagger); Ap); p_{\mathcal{O}_k}\ ,$$

where, by abuse of notation, we denote representatives of equivalence classes of some canonical morphisms in the affine category freely generated by $\mathcal{O}_k$ by the canonical morphisms themselves.

---

[1] I.e. permutations which exchange exactly two elements.

Since the g's appearing in the Decomposition Thereom still live (up-to uncurrying) in a PER interpreting a simple type, we could keep on iterating the decomposition, expanding in turn these g's, thus getting (up-to permutations) a possible infinite tree from f:

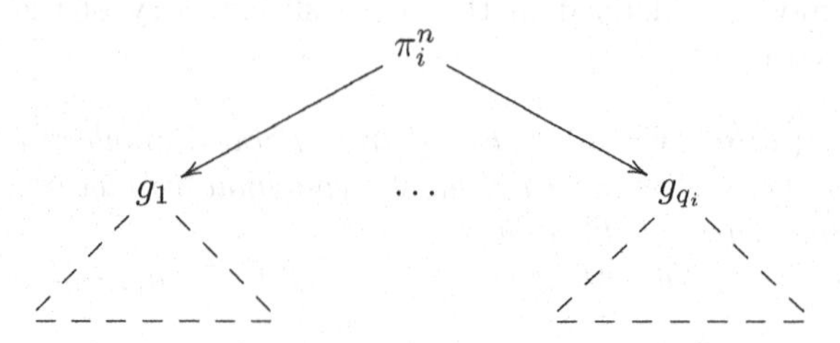

Partial involutions which generate, under Decomposition a *finite* tree are easily proved to be λ-*definable*. Therefore, if the Decomposition Theorem holds, in order to get the full completeness result, we are only left to deal with partial involutions generating trees whose height is infinite, which would correspond to *infinite* typed Böhm trees. The proof of Theorem 1 follows a standard pattern, and it is carried out in detail in Section 2.1. The proof of λ-definability uses the Decomposition Theorem, and two furher ingredients: an *Approximation Theorem*, and an intermediate model. These are discussed in Section 2.2. From the λ-definability result, we have immediately:

Theorem 2 (Full Completeness for an Extended Language). *The model $\mathcal{M}_{\mathcal{O}_k}$ is fully complete and minimal w.r.t. λ_k, enriched with constants for all transpositions* [2] *of type $o \to o$.*

Notice that we use the full completeness result in order to show that $\mathcal{M}_{\mathcal{O}_k}$ is minimal, ie. it realizes the maximal theory on the extended calculus.

2.1 Proof of the Decomposition Theorem

Let T $\in$SimType, and $\mathcal{R}= [\![T]\!]^{\mathcal{O}_k}$, where $\mathcal{R}= \otimes_{i=1}^n ! \ \mathcal{R}_i \multimap \mathcal{O}_k$, and, for all $i = 1,\ldots,n$, $\mathcal{R}_i= \otimes_{j=1}^{q_i} !S_{ij} \multimap \mathcal{O}_k$. Let $f \in \mathrm{dom}(\otimes_{i=1}^n ! \ \mathcal{R}_i \multimap \mathcal{O}_k)$. We analyze the behaviour of f as operator in an application to arguments $!g_1 \in ! \ \mathcal{R}_1,\ldots,!g_n \in ! \ \mathcal{R}_n$. I.e., let us apply the coding functions t, p of Section 1.2, in order to get $\overline{f} : ((\mathrm{Nat} \times \mathrm{Nat}) + \ldots + (\mathrm{Nat} \times \mathrm{Nat})) + \mathrm{Nat} \rightharpoonup$
$((\mathrm{Nat} \times \mathrm{Nat}) + \ldots + (\mathrm{Nat} \times \mathrm{Nat})) + \mathrm{Nat}$,
where in the domain (codomain) of $\overline{f}$ there are n occurrences of Nat $\times$ Nat, each one corresponding to one of the n arguments to which f applies. In the interaction with the i-th argument $!g_i$ only the i-th occurrence of Nat $\times$ Nat in the domain (codomain) of $\overline{f}$ is involved. In particular, the lefthand occurrence of Nat in Nat $\times$ Nat refers to the copy of the argument $!g_i$ used, while the righthand occurence of Nat carries the values from (to) g_i.

We have two possibilities, according to the behaviour of $\overline{f}$ on the inputs $*_1,\ldots,*_k$ (other input values are not relevant, by definition of the PER $\mathcal{O}_k$):

[2] I.e. permutations which exchange exactly two elements.

Lemma 1. *Let* $f \in dom(\boldsymbol{\mathcal{R}} \multimap \mathcal{O}_k)$, *where* $\mathcal{R} = \otimes_{i=1}^{n} ! \, \mathcal{R}_i \multimap \mathcal{O}_k$, *and, for all* $i = 1, \ldots, n$, $\mathcal{R}_i = \otimes_{j=1}^{q_i} !S_{ij} \multimap \mathcal{O}_k$. *Then*

1. *either* $\exists *_j \, . \, \overline{f}(r, *_j) = (r, a_j) \ \wedge \ \forall *_k \neq q_j. \ ((r, *_k), (r, a_k)) \notin dom(\overline{f})$.
   I.e. $f \in [\![\boldsymbol{x} : \boldsymbol{T} \vdash c_j : o]\!]^{\mathcal{O}_k}$.
2. *or* $\exists i \in \{1, \ldots, n\}$ *such that*
   (a) $\forall *_j \ \ \exists i_0. \ \overline{f}(r, *_j) = (l, (i, (i_0, m))) \ \wedge \ m = (r, y_k)$ *where* $y_k \in \{*_k, a_k\}$ *and*
   (b) $\forall i, j \ (\overline{f}(r, *_j) = (l, (i, (i_0, (r, *_k)))) \ \Rightarrow \ \overline{f}(l, (i, (i_0, (r, a_k)))) = (r, a_j) \ \wedge$
   $\overline{f}(r, *_j) = (l, (i, (i_0, (r, a_k)))) \ \Rightarrow \ \overline{f}(l, (i, (i_0, (r, *_k)))) = (r, a_j)$.

Proof. Item 1 is easy to prove. We focus on the proof of item 2. As far as item 2a, one can easily show that, for any given $*_j$, $\exists i, i_0$ s.t. $\overline{f}(r, *_j) = (l, (i, (i_0, m)))$ and $m = (r, y_k)$. The proof of the fact that, for all $*_j$, the argument i interrogated by f is the same is based on a "counting argument". If we "split" the responses to initial questions $*_j$ among different arguments, then we lose totality, because of the constraints of being a partial involution. Namely, assume by contradiction that it is not the case that for all initial questions $*_j$ the responses are in the same argument. Then, for all $i = 1, \ldots, n$ $\exists p_i$ s.t.

$$\forall *_j \ \forall i_0 \in \text{Nat}. \ \overline{f}(r, *_j) \neq (l, (i, (i_0, (r, x_{p_i})))) \ , \text{ for } x \in \{*, a\}.$$

Then consider constants $(\mathbf{K}c_{p_1}), \ldots, (\mathbf{K}c_{p_n})$ in $\mathcal{R}_1, \ldots, \mathcal{R}_n$. Then we have $\forall *_j \, . \, f \bullet (\mathbf{K}c_{p_1}) \ldots (\mathbf{K}c_{p_n})(*_j) \uparrow$, i.e. $f \notin \text{dom}(\boldsymbol{\mathcal{R}} \multimap \mathcal{O}_k)$. Contradiction.

Finally, in order to prove item 2b, one can proceed by contradiction, and by case analysis. This concludes the proof of Lemma 1. □

Using Lemma 1 above, one can easily prove the following two lemmata: *Linearization of Head Occurrence* and *Linear Function Extensionality.* The factorization of the proof of the Decomposition Theorem in these two lemmata is standard, but notice the special form of *Linear Function Extensionality*, where permutations come into play.

Lemma 2 (Linearization of Head Occurrence). *Let* $\mathcal{R} = \otimes_{i=1}^{n} ! \, \mathcal{R}_i \multimap \mathcal{O}_k$, *where, for all* $i = 1, \ldots, n$, $\mathcal{R}_i = \otimes_{j=1}^{q_i} !S_{ij} \multimap \mathcal{O}_k$. *Then, for all* $f \in dom(\boldsymbol{\mathcal{R}} \multimap \mathcal{O}_k)$, *where* $\boldsymbol{\mathcal{R}}$ *is an abbreviation for* $\otimes_{i=1}^{n} ! \, \mathcal{R}_i$, *there exist* $i \in \{1, \ldots, n\}$ *and* $f' \in dom(R_i \multimap \boldsymbol{\mathcal{R}} \multimap \mathcal{O}_k)$ strict, *i.e.* $t + t; t; f'; t^{-1}; t^{-1} + t^{-1}(r, (r, *)) = (l, (r, *))$, *such that* $\quad f \ \mathcal{R} \ (con_{\boldsymbol{\mathcal{R}}}; \pi_i^n \otimes id_{\boldsymbol{\mathcal{R}}}; \Lambda^{-1}(f'))$. □

Now we examine the structure of f'. One can show that:

Lemma 3 (Linear Function Extensionality). *Let* $\mathcal{S}, \mathcal{R}$ *be PERs. Then, for all* $f \in dom((\mathcal{S} \multimap \mathcal{O}_k) \multimap (\mathcal{R} \multimap \mathcal{O}_k))$ *strict, there exists* $f' \in dom(\mathcal{R} \multimap \mathcal{S})$ *such that* $\quad f \ \ ((\mathcal{S} \multimap \mathcal{O}_k) \multimap (\mathcal{R} \multimap \mathcal{O}_k)) \ \ (\Lambda(((id_{\mathcal{S}} \multimap p_{\mathcal{O}_k}) \otimes f'); Ap))$,
where $p_{\mathcal{O}_k}$ *is a permutation in the PER* $\mathcal{O}_k \multimap \mathcal{O}_k$.

The last technical lemma that we need in order to prove the Decomposition Theorem amounts to a general fact which follows by construction of the category of PERs over an ACA (see Section 1.2):

Lemma 4 (Uniformity of Threads). *The following isomorphism holds, for all PERs* $\mathcal{R}, \mathcal{S}$,

$(id_{!\mathcal{R}} \multimap ders) \;:\; (!\mathcal{R} \multimap !\mathcal{S} \simeq\, !\mathcal{R} \multimap \mathcal{S}) \;:\; (\lambda[f] \in\, !\mathcal{R} \multimap \mathcal{S}\,.\,([f])^{\dagger}_{\mathcal{R},\mathcal{S}})$,

where $(\ )^{\dagger}_{\mathcal{R},\mathcal{S}}$ *is the canonical morphism induced by the comonad !.*

Finally, we have:

Proof of the Decomposition Theorem 1. If case 1 of Lemma 1 applies, then we are done. If case 2 applies, then, by Lemma 2, there exist $i \in \{1,\dots,n\}$ and $f' \in \mathrm{dom}(\mathcal{R}_i \multimap (\boldsymbol{\mathcal{R}} \multimap \mathcal{O}_k))$ s.t. $f\ (\boldsymbol{\mathcal{R}} \multimap \mathcal{O}_k)\ (\mathrm{con}_{\boldsymbol{\mathcal{R}}}; \pi^n_i \otimes \mathrm{id}_{\boldsymbol{\mathcal{R}}}; \Lambda^{-1}(f'))$. By Lemma 3, $\exists g, p_{\mathcal{O}_k}$ s.t. $f'\ ((\boldsymbol{\mathcal{S}_i} \multimap \mathcal{O}_k) \multimap (\boldsymbol{\mathcal{R}} \multimap \mathcal{O}_k))\ \Lambda(((\mathrm{id}_{\boldsymbol{\mathcal{S}_i}} \multimap p_{\mathcal{O}_k}) \otimes g); \mathrm{Ap})$. Then $f\ (\boldsymbol{\mathcal{R}} \multimap \mathcal{O}_k)\ \mathrm{con}_{\boldsymbol{\mathcal{R}}}; (\mathrm{id}_{\boldsymbol{\mathcal{S}_i}} \multimap p_{\mathcal{O}_k}); \pi^n_i \otimes g; \mathrm{Ap}$. Finally, by Lemma 4, by definition of the product of PERs and by the universality property of the product, we obtain $g\ (\boldsymbol{\mathcal{R}} \multimap \boldsymbol{\mathcal{S}_i})\ \langle g_1, \dots, g_{q_i}\rangle^{\dagger}$, for some $g_1 \in \mathcal{S}_{i1}, \dots, g_{q_i} \in \mathcal{S}_{iq_i}$. □

2.2 Proof of λ-Definability

The proof of the λ-definability property of $\mathcal{M}_{\mathcal{O}_k}$ is quite involved and it uses an *Approximation Theorem* and an intermediate PER model $\mathcal{M}_{\mathcal{O}_k^{\perp}}$ for the simply typed λ-calculus λ_k with k ground constants plus an extra *undefined* ground constant ⊥. This model allows for *partial* elements (*approximants*).

The Approximation Theorem. We start by introducing the notion of *approximant* of a partial involution. By repeatedly applying the Decomposition Theorem to a partial involution f in the model $\mathcal{M}_{\mathcal{O}_k}$, we obtain a (possibly) infinite typed Böhm tree (up to permutations). The j-th approximant of f is a partial involution obtained by truncating at level j this tree, and by substituting the *empty partial involution* for each possibly erased subtree. Notice that approximants are not in the model, in general. Formally:

Definition 7 (Approximants).
Let $f \in \mathrm{dom}(\mathcal{R})$, *where* $\mathcal{R} = [\![T]\!]^{\mathcal{O}_k}$, *for some simple type* T.

- *We define the* j-th tree, $t_j(f)$, *of height at most* j, *generated from* f *after iterated applications of the Decomposition Theorem by induction on* j *as follows:*
  - $t_0(f)$ *is the tree of height* 0 *with only a root labeled by* f;
  - *given the tree* $t_j(f)$ *of height at most* j, *the tree* $t_{j+1}(f)$ *is obtained from the tree* $t_j(f)$ *by expanding the possible leaves at level* j *via the Decomposition Theorem.*
- *We define the* j-th approximant *of* f, $p_j(f)$, *as the partial involution obtained from the tree* $t_j(f)$ *by substituting any partial involution at level* j *by the* empty *partial involution.*

By monotonicity of • we have immediately:

Lemma 5. *Let* $T_1 \to \dots \to T_n \to o \in SimType$, *let* $f \in \mathrm{dom}(\bigotimes_{i=1}^{n} ![\![T_i]\!]^{\mathcal{O}_k} \multimap \mathcal{O}_k)$. *Then* $\forall j \geq 0.\ p_j(f) \subseteq p_{j+1}(f)$.

Theorem 3 (Approximation). *Let $T_1 \to \ldots \to T_n \to o$ be a simple type, and let $f \in dom(\bigotimes_{i=1}^{n} ![\![T_i]\!]^{\mathcal{O}_k} \multimap \mathcal{O}_k)$. Then*

$$f \; (\bigotimes_{i=1}^{n} ![\![T_i]\!]^{\mathcal{O}_k} \multimap \mathcal{O}_k) \; (\bigcup_{j\in\omega} p_j(f)) \; .$$

Proof. We have to show that, $\forall \boldsymbol{g} \in \bigotimes_{i=1}^{n} ![\![T_i]\!]^{\mathcal{O}_k}$, $f \bullet \boldsymbol{g} \; \mathcal{O}_k \; (\bigcup_{j\in\omega} p_j(f)) \bullet \boldsymbol{g}$. I.e.:
i) $\forall *_i \, . \, f \bullet \boldsymbol{g}(*_i) \uparrow \iff (\bigcup_{j\in\omega} p_j(f)) \bullet \boldsymbol{g}(*_i) \uparrow$ and
ii) $\forall *_i, a_i . \, f \bullet \boldsymbol{g}(*_i) = a_i \iff (\bigcup_{j\in\omega} p_j(f)) \bullet \boldsymbol{g}(*_i) = a_i$.
The implications ($\Rightarrow$) in i) and ($\Leftarrow$) in ii) follow from monotonicity of $\bullet$ and from the fact that, for all j the graph of $p_j(f)$ is contained in the graph of f. In order to show i)($\Leftarrow$) and ii)($\Rightarrow$), one can check that, if $f \bullet \boldsymbol{g}(*_i) = a_i$ and the result is obtained with a thread of length at most $2j$, then $p_j(f) \bullet \boldsymbol{g}(*_i) = a_i$. □

The PER Model $\mathcal{M}_{\mathcal{O}_k^\perp}$.

Definition 8. *Let $\mathcal{M}_{\mathcal{O}_k^\perp}$ be the PER model induced by the ground PER $\mathcal{O}_k^\perp$ defined as follows. Fix distinct natural numbers $*_1, \ldots *_k$, $a_1, \ldots, a_k$. Let $\mathcal{O}_k^\perp$ be the PER on the ACA $\mathcal{A}_{\mathsf{PInv}}$ consisting of $k+1$ equivalence classes defined by:*
- $\perp = \{f : Nat \rightharpoonup_{\mathsf{Inv}} Nat \mid \forall i \in \{1, \ldots, k\}. \, *_i, a_i \not\in dom(f)\}$
- $c_1 = \{f : Nat \rightharpoonup_{\mathsf{Inv}} Nat \mid f(*_1) = a_1 \; \wedge \; \forall m \neq 1. \, *_m, a_m \not\in dom(f)\}$
- ...
- $c_k = \{f : Nat \rightharpoonup_{\mathsf{Inv}} Nat \mid f(*_k) = a_k \; \wedge \; \forall m \neq k. \, *_m, a_m \not\in dom(f)\}$.

$\mathcal{M}_{\mathcal{O}_k^\perp}$ is a model of $\lambda_k^\perp$, i.e. the simply typed λ-calculus with k ground constants plus the extra ground constant $\perp$. In particular, approximants are in $\mathcal{M}_{\mathcal{O}_k^\perp}$. The relationship with the model $\mathcal{M}_{\mathcal{O}_k}$ is given by the following lemma:

Lemma 6. *Let $f \in dom([\![\boldsymbol{T} \multimap o]\!]^{\mathcal{O}_k})$. Then*
i) $f \in dom([\![\boldsymbol{T} \multimap o]\!]^{\mathcal{O}_k^\perp})$.
ii) $\exists J \geq 0. \; f \; [\![\boldsymbol{T} \multimap o]\!]^{\mathcal{O}_k^\perp} \; p_J(f)$.

Proof. i) Assume that $f \in \mathrm{dom}([\![\boldsymbol{T} \multimap o]\!]^{\mathcal{O}_k})$. By the Approximation Theorem 3, $f \sim \bigcup_{j\in\omega} p_j(f)$ in $\mathcal{M}_{\mathcal{O}_k}$. Moreover, for all j, $p_j(f) \in \mathrm{dom}([\![\boldsymbol{T} \multimap o]\!]^{\mathcal{O}_k^\perp})$, and hence also $\bigcup_{j\in\omega} p_j(f) \in \mathrm{dom}([\![\boldsymbol{T} \multimap o]\!]^{\mathcal{O}_k^\perp})$. Then, using an argument similar to the one used in the proof of the Approximation Theorem, one can check that, $\forall \boldsymbol{g} \in [\![\boldsymbol{T}]\!]^{\mathcal{O}_k^\perp} . \; \forall *_i \, . \, f \bullet \boldsymbol{g}(*_i) \simeq \bigcup_{j\in\omega} p_j(f) \bullet \boldsymbol{g}(*_i)$. Hence, in particular, $f \in \mathrm{dom}([\![\boldsymbol{T} \multimap o]\!]^{\mathcal{O}_k^\perp})$.
ii) By the proof of item i) of this lemma, $f \sim \bigcup_{j\in\omega} p_j(f)$ in $\mathcal{M}_{\mathcal{O}_k^\perp}$, and $\forall j. \, p_j(f) \in \mathrm{dom}([\![\boldsymbol{T} \multimap o]\!]^{\mathcal{O}_k^\perp})$. Moreover, since $[\![\boldsymbol{T} \multimap o]\!]^{\mathcal{O}_k^\perp}$ has only finitely many equivalence classes, by Lemma 5, there exists $J \geq 0$ such that, for all $m_1, m_2 \geq J$, $p_{m_1}(f) [\![\boldsymbol{T} \multimap o]\!]^{\mathcal{O}_k^\perp} p_{m_2}(f)$. Hence $f \sim p_J(f)$ in $\mathcal{M}_{\mathcal{O}_k^\perp}$. □

λ-Definability of $\mathcal{M}_{\mathcal{O}_k}$. Finally, we are in the position of proving that all partial involutions $f \in \mathrm{dom}([\![\boldsymbol{T}\multimap o]\!]^{\mathcal{O}_k})$ are λ-definable. We proceed by induction on types. The base case is easy. Let us consider the induction step.

By the Approximation Theorem 3, $f[\![\boldsymbol{T}\multimap o]\!]^{\mathcal{O}_k}(\bigcup_{j\in\omega} p_j(f))$. By Lemma 6ii), there exists J such that $f \sim p_J(f)$ in the model $\mathcal{M}_{\mathcal{O}_k^\perp}$. Hence, using Lemma 6i), we have, in particular, that: $\forall \boldsymbol{g} \in [\![T]\!]^{\mathcal{O}_k}.\ \forall *_i\ .\ f \bullet \boldsymbol{g}(*_i) \simeq p_J(f) \bullet \boldsymbol{g}(*_i)$. Therefore, $p_J(f) \sim f$ in $\mathcal{M}_{\mathcal{O}_k}$. Let us call P_J the λ-term whose interpretation in $\mathcal{M}_{\mathcal{O}_k^\perp}$ is $p_J(f)$. Two cases can arise:
1) $\exists \boldsymbol{M} : \boldsymbol{T}$, $\boldsymbol{M}$ $\perp$-free, such that $P_J\boldsymbol{M} =_\beta \perp$.
2) $\forall \boldsymbol{M} : \boldsymbol{T}$, $\boldsymbol{M}$ $\perp$-free, such that $P_J\boldsymbol{M} \neq_\beta \perp$.
If case 1 applies, then we have a contradiction, since $\perp \notin \mathcal{O}_k$. If case 2 applies, then one can check that P_J is equivalent in the maximal theory to any λ-term P' obtained from P_J by substituting any constant $c \neq \perp$ for each possible occurrence of $\perp$ in P_J. But then, since, by induction hypothesis, $\forall \boldsymbol{g} \in [\![T]\!]^{\mathcal{O}_k}$, $\boldsymbol{g}$ is λ-definable, $[\![P']\!]^{\mathcal{O}_k} \bullet \boldsymbol{g} \sim [\![P_J]\!]^{\mathcal{O}_k} \bullet \boldsymbol{g}$ in $\mathcal{M}_{\mathcal{O}_k}$, and hence $[\![P']\!]^{\mathcal{O}_k} \sim p_J(f) \sim f$, i.e. f is λ-definable.

This concludes the proof of Theorem 2.

3 A Fully Complete Minimal Model for the Simply Typed λ-Calculus

One can build a fully complete model for the simply typed λ-calculus λ_k, by getting rid of permutations in the models $\mathcal{M}_{\mathcal{O}_k}$. There are two ways of doing this. The first is "low-level", and it amounts to cutting down the affine combinatory algebra of partial involutions, by placing additional constraints on the partial involutions, similarly to what one does in [AL99a] for getting a fully abstract model for PCF. Alternatively, one can define a suitable *logical relation* and use it to cut down the PER model. We briefly sketch the first technique.

For the sake of simplicitly, let us consider, in place of the set of natural numbers, the following set of inductively defined moves:

Definition 9. *Let $k \in Nat$. We define*

$$(\boldsymbol{M}_k \ni)\ m\ ::=\ *_i\ \mid\ a_i\ \mid\ (l,m)\ \mid\ (r,m)\ \mid\ \langle j,m\rangle\ ,$$

where $i = 1,\ldots,k$ and $j \in Nat$.
We regard $\boldsymbol{M}_k$ as equipped with the intrinsic coding functions $[l,r] : \boldsymbol{M}_k + \boldsymbol{M}_k \to \boldsymbol{M}_k$, and $\langle\ ,\ \rangle : Nat \times \boldsymbol{M}_k \to \boldsymbol{M}_k$.

One could equivalently take the moves to be natural numbers (under suitable assumptions on coding functions), but the set $\mathbf{M}_k$ simplifies the argument. We can immediately define a function v on moves, which, for any move m, provides the index i of the basic move $*_i$ or a_i which the move m is made up. I.e.:

Definition 10. *Let $v : \boldsymbol{M}_k \to \boldsymbol{M}_k$ be defined as follows. For all $i \in Nat$, for all $m \in \boldsymbol{M}_k$,*
$v(*_i) = v(a_i) = i$
$v((l,m)) = v((r,m)) = v(\langle i,m\rangle) = v(m)$.

Partial involutions which preserves the function v still form an affine combinatory algebra.

Proposition 3 (Full Completeness and Minimality). *Let $\mathcal{A}_{\mathtt{vPInv}}$ be the affine combinatory algebra whose carrier is the set of partial involutions $f : \boldsymbol{M}_k \to \boldsymbol{M}_k$ such that, for all $m \in dom(f)$, $v(f(m)) = v(m)$. Then the model induced by the PER $\mathcal{O}_k$ over $\mathcal{A}_{\mathtt{vPInv}}$ is fully complete and minimal w.r.t. λ_k.*

4 Some Applications of the Full Completeness Results

Our fully complete model $\mathcal{M}_{\mathcal{O}_k}$ provides immediately *semantical* proofs of some interesting facts concerning the maximal theory $\approx$ over the simply typed λ-calculus with or without permutations.

Context Lemma.

Definition 11 (Applicative Equivalence). *Let $\approx^{\mathsf{app}} \subseteq \Lambda^0 \times \Lambda^0$ be defined by*

$$M \approx^{\mathsf{app}} N \;\Leftrightarrow\; \forall P_1, \ldots, P_n \in \Lambda^0.\; MP_1 \ldots P_n =_\beta NP_1 \ldots P_n\ .$$

Lemma 7 (Context Lemma). *The theory $\approx$ admits an* applicative *characterization, i.e. $M \approx N \iff M \approx^{\mathsf{app}} N$.*

The Theory on Infinitary Böhm Trees is Conservative. Let us call $\mathcal{M}_{\mathcal{O}}$ the the variant of the model $\mathcal{M}_{\mathcal{O}_k^\perp}$ obtained by considering two ground constants, $\perp$ and $\top$, and by defining the ground Sierpinski PER as follows: fix $* \in$ Nat, $\perp = \{f : \mathrm{Nat} \rightharpoonup_{\mathsf{Inv}} \mathrm{Nat} \mid * \notin \mathrm{dom}(f)\}$, $\top = \{f : \mathrm{Nat} \rightharpoonup_{\mathsf{Inv}} \mathrm{Nat} \mid f(*) = *\}$.

Then one can check that $\mathcal{M}_{\mathcal{O}}$ is fully complete and minimal directly for the simply typed λ-calculus with two ground constants. Actually $\mathcal{M}_{\mathcal{O}}$ is a fully complete model for the maximal theory on the *infinitary calculus*, i.e. the simply typed, possibly infinite, Böhm trees:

Definition 12 (Infinitary Typed Böhm Trees). *We define the* infinitary *typed Böhm trees as the trees obtained as supremums of typed Böhm trees corresponding to approximants.*

Using the model $\mathcal{M}_{\mathcal{O}}$, we can show that, in the case of two ground constants, the theory over infinitary typed Böhm trees $\approx^\infty$ is a conservative extension of $\approx$ w.r.t. terms in $\lambda_{\top}^{\perp}$, i.e.:

Proposition 4. $\approx^\infty_{|\mathsf{TBT}} = \approx_{|\mathsf{TBT}}$,
where $\approx^\infty_{|\mathsf{TBT}}$, $\approx_{|\mathsf{TBT}}$, denote the theory $\approx^\infty$ and the theory $\approx$ restricted to the finite typed Böhm trees, respectively, in the case of two ground constants.

Decidability Results. The following decidability result can be proved in $\mathcal{M}_{\mathcal{O}_k}$:

Theorem 4. *Let $\mathcal{R}$ be a simple PER, i.e. $\mathcal{R}=\mathcal{R}_1 \to \ldots \to \mathcal{R}_n \to \mathcal{O}_k$. For all $f : Nat \to Nat$ whose graph is finite, it is decidable whether $f \in dom(\mathcal{R})$.*

Proof. In order to decide whether $f \in \mathrm{dom}(\mathcal{R})$, it is sufficient to check the behaviour of f when applied to the "relevant" $\boldsymbol{g} \in \boldsymbol{\mathcal{R}}$, i.e. to the g's whose

domains (and codomains), roughly, are contained in a suitable subset of the domains of h, h'. More precisely, let
$\overline{h}, \overline{h}' : ((\mathrm{Nat} \times \mathrm{Nat}) + \ldots + (\mathrm{Nat} \times \mathrm{Nat})) + \mathrm{Nat} \rightharpoonup$
$((\mathrm{Nat} \times \mathrm{Nat}) + \ldots + (\mathrm{Nat} \times \mathrm{Nat})) + \mathrm{Nat}$
be the partial involutions obtained from h, h' using the coding functions t, p. Then g_i is "relevant" to h, h' if $\mathrm{dom}(g_i) \subseteq \{n \mid \exists k.\ (l,(i,(k,n))) \in \mathrm{dom}(\overline{h}) \cup \mathrm{dom}(\overline{h}')\}$. These $\boldsymbol{g}$'s are the only "relevant" ones for h, h' in the sense that, for any other "non-relevant" $\boldsymbol{g}$, there exists $\boldsymbol{g}'$ "relevant" such that, for all $*_i$, $h \bullet \boldsymbol{g}(*_i) \simeq h \bullet \boldsymbol{g}'(*_i)$ and $h' \bullet \boldsymbol{g}(*_i) \simeq h' \bullet \boldsymbol{g}'(*_i)$. Since h, h' have finite graphs, then there are only finitely many $g_1, \ldots, g_n$ whose graphs are finite. We can easily generate all these "relevant" partial involutions $g_1, \ldots, g_n$. At this point, we have to eliminate the relevant $\boldsymbol{g}$'s which are not in $\mathrm{dom}(\boldsymbol{\mathcal{R}})$. Moreover, we need to know, for all relevant g_i, g'_i, whether g_i is equivalent to g'_i. In order to decide this, we compute, in turn, the partial involutions which are relevant for g_i and g'_i. And, recursively, we have to compute the relevant partial involutions of the relevant partial involutions, until we reach the ground PER $\mathcal{R}$. Once we have eliminated those $\boldsymbol{g}$ which are not in $\mathrm{dom}(\boldsymbol{\mathcal{R}})$, and we have divided the set of relevant $\boldsymbol{g}$'s in equivalence classes, we can check, finally, the applicative behaviour of f. The computations $h \bullet \boldsymbol{g}(*_i)$ and $h' \bullet \boldsymbol{g}(*_i)$ always terminate, since, by definition of partial involution, and by the fact that the graphs of $h, h', g_1, \ldots, g_n$ are all finite, there cannot be an infinite (possibly cyclic) computation. Namely, the computation $h \bullet \boldsymbol{g}(*_i)$ either converges to a_i or diverges because h or $\boldsymbol{g}$ are not defined on some element. This concludes the proof.

□

Using a similar argument, we can prove:

Proposition 5. *Let* $\vdash M : T$, $\vdash N : T$ *be such that* $[\![\vdash M : T]\!]^{\mathcal{O}_k}$, $[\![\vdash N : T]\!]^{\mathcal{O}_k}$ *have representatives with finite graphs. Then it is decidable whether* $M \approx N$.

However, there are very few λ-terms whose interpretation is finite in the sense of Proposition 5 above. E.g. the identity of type $((o \to o) \to o) \to ((o \to o) \to o)$ has no representatives with finite graph. Intuitively, this depends on the fact that it can ask for *any* number of copies of its argument.

5 Conclusions and Future Work

In this paper, we have studied *fully complete* PER models for the *maximal theory* of the simply typed λ-calculus with finitely many base constants, and we have seen some applications of this construction. Here we summarize a list of remarks and interesting issues which remain to be addressed:
– Following [Abr00], we could abstract axioms for full completeness from the lemmata in our proof. However, these would not imply faithfulness w.r.t. the maximal theory, i.e. that the theory of models would be maximal.
– One could define fully complete Game Models for λ_k by considering strategies in the style of [AJM00]. But, by the intensionality of the Game Semantics, these models would not realize the maximal theory, but rather the $\beta\eta$-theory.

– In [Lai00], a fully abstract translation of "Finitary μ-PCF" into the simply typed lambda calculus with constants is given. An interesting consequence of this result is that our model is also fully abstract for this finitary μ-PCF.
– We feel that the fully complete models based on partial involutions defined in this paper should provide a semantical proof of the decidability of the maximal theory, alternative to those of Padovani and Loader. We should capitalize on the possibility of checking equivalence of involutions by evaluating them on *finite* sets of inputs (moves).
– We conjecture that the model $\mathcal{M}_{\mathcal{O}_k^{\perp}}$, for $k = 1$, is fully complete for the decidable fragment of PCF called "Unary PCF".
– We end this paper with avery speculative comment. Partial involutions provide models for *reversible* computations ([Abr01]). Is this related to the decidability of the various theories which can be modeled by partial involutions?

References

[Abr96] S.Abramsky. Retracing some paths in Process Algebra, *Concur'96*, U. Montanari and V. Sassone eds., 1996, 1–17.

[Abr97] S.Abramsky. Interaction, Combinators, and Complexity, Notes, Siena (Italy), 1997.

[Abr00] S.Abramsky. Axioms for Definability and Full Completeness, in *Proof, Language and Interaction: Essays in Honour of Robin Milner*, G. Plotkin, C. Stirling and M. Tofte, eds., MIT Press, 2000, 55–75.

[Abr01] S.Abramsky. A Structural Approach to Reversible Computation, May 2001.

[AHS98] S.Abramsky, E.Haghverdi, P.Scott. Geometry of Interaction and Models of Combinatory Logic, 1998, to appear.

[AJ94] S.Abramsky, R.Jagadeesan. New foundations for the Geometry of Interaction, *Inf. and Comp.* **111**(1), 1994, 53–119.

[AJ94a] S.Abramsky, R.Jagadeesan. Games and Full Completeness for Multiplicative Linear Logic, *J. of Symbolic Logic* **59**(2), 1994, 543–574.

[AJM00] S.Abramsky, R.Jagadeesan, P.Malacaria. Full Abstraction for PCF, *Inf. and Comp.* **163**, 2000, 409–470.

[AL99] S.Abramsky, M.Lenisa. Fully Complete Models for ML Polymorphic Types, Technical Report **ECS-LFCS-99-414**, LFCS, 1999 (available at http://www.dimi.uniud.it/~lenisa/Papers/Soft-copy-ps/lfcs99.ps.gz).

[AL00] S.Abramsky, M.Lenisa. A Fully-complete PER Model for ML Polymorphic Types, *CSL'00*, LNCS **1862**, 2000, 140–155.

[AL99a] S.Abramsky, J.Longley. Realizability models based on history-free strategies, 1999.

[Gir89] J.Y.Girard. Towards a Geometry of Interaction, *Contemporary Mathematics* **92**, 1989, 69–108.

[KNO99] A.Ker, H.Nickau, L.Ong. More Universal Game Models of Untyped λ-Calculus: The Böhm Tree Strikes Back, *CSL'99*, LNCS, 1999.

[JSV96] A.Joyal, R.Street, D.Verity. Traced monoidal categories, *Math. Proc. Comb. Phil. Soc.* **119**, 1996, 447–468.

[Lai00] J.Laird. Games, control and full abstraction, February 2000.

[Loa97] An Algorithm for the Minimal Model, Note, 1997.

[Pad95] V.Padovani. Decidability of all Minimal Models, *TYPES'95*, LNCS **1185**, 201–215.

A Principle of Induction

Keye Martin

Oxford University Computing Laboratory
Wolfson Building, Parks Road, Oxford OX1 3QD
kmartin@comlab.ox.ac.uk
http://web.comlab.ox.ac.uk/oucl/work/keye.martin

Abstract. We introduce an induction principle on complete partial orders and consider its applications to program verification and analysis on the real line. The highlight of this technique is that it allows one to make inductive arguments over continuous as well as discrete forms of data without ever having to distinguish between the two.

1 Introduction

This paper introduces an induction principle based on complete partial orders and certain selfmaps called ideal mappings. We first study the ideal mappings, including the important issue of how one proves in practice that a map is in fact ideal. We then proceed to the induction principle and apply it to establish the correctness of mergesort. One of the interesting aspects about this application to program verification is the ease with which this form of induction may be applied. Whereas in domain theory, where the usual fixed point induction [1] would require us to undertake the nontrivial task of realizing the algorithm as the least fixed point of a higher order operator, the induction principle presented here requires no transitional step between theory and practice: An actual ML program is *already* in the form that the theory requires. We will also see that it captures the usual fixed point induction as a trivial consequence. However, because the principle presented here is based on mappings that in general are not monotone, it admits applications outside the scope of domain theory, such as inductive proofs of the compactness and connectedness of the unit interval.

2 Background

A *poset* is a partially ordered set [1], that is, a set P together with a binary relation $\sqsubseteq\, \subseteq P^2$ which is reflexive, transitive and antisymmetric.

Definition 1. Let $(P, \sqsubseteq)$ be a partially ordered set. An *upper bound* of a subset $S \subseteq P$ is an element $u \in P$ such that $s \sqsubseteq u$, for all $s \in S$. If $S \subseteq P$ has an upper bound u such that $u \sqsubseteq v$, for any upper bound v of S, then we call u the *supremum* of S and write $u = \bigsqcup S$.

L. Fribourg (Ed.): CSL 2001, LNCS 2142, pp. 458–468, 2001.

Definition 2. Let $(P, \sqsubseteq)$ be a partially ordered set. A sequence (x_n) in P is *increasing* if $x_n \sqsubseteq x_{n+1}$, for all $n \geq 1$. A *cpo* is a poset in which every increasing sequence has a supremum.

Definition 3. A function $f : D \to E$ between cpo's is continuous if

(i) f is monotone: $x \sqsubseteq y \Rightarrow f(x) \sqsubseteq f(y)$, and
(ii) f preserves suprema of increasing sequences: If (x_n) is an increasing sequence in D, then

$$f(\bigsqcup_{n \in \mathbb{N}} x_n) = \bigsqcup_{n \in \mathbb{N}} f(x_n).$$

We consider a few examples.

Example 1. The set of nonnegative reals $[0, \infty)^*$ in their opposite order,

$$x \sqsubseteq y \Leftrightarrow x \geq y,$$

is a cpo. If (x_n) is increasing in $[0, \infty)^*$, $\bigsqcup x_n = \lim x_n = \inf\{x_n : n \in \mathbb{N}\}$.

Example 2. The interval cpo. The collection of compact intervals of the real line

$$\mathbf{I}\mathbb{R} = \{[a, b] : a, b \in \mathbb{R} \;\&\; a \leq b\}$$

ordered under reverse inclusion

$$[a, b] \sqsubseteq [c, d] \Leftrightarrow [c, d] \subseteq [a, b]$$

is a cpo. The supremum of an increasing sequence (x_n) in $\mathbf{I}\mathbb{R}$ is $\bigcap x_n$.

Our final example is the cpo $[S]$ of finite lists [6] over a set $(S, \leq)$.

Definition 4. A *list* over S is a function $x : \{1, ..., n\} \to S$, for $n \geq 0$. The *length* of a list x is $|\mathrm{dom}\, x|$. The set of all (finite) lists over S is $[S]$.

A list x can be written as $[x(1), ..., x(n)]$, where the *empty list* (the list of length 0) is written $[\,]$. We also write lists as $a :: x$, where $a \in S$ is the *first element* of the list $a :: x$, and $x \in [S]$ is the *rest* of the list $a :: x$. For example, the list $[1, 2, 3]$ is written $1 :: [2, 3]$.

Definition 5. A set $K \subseteq \mathbb{N}$ is *convex* if $a, b \in K \;\&\; a \leq x \leq b \Rightarrow x \in K$. Given a finite convex set $K \subseteq \mathbb{N}$, the map $\mathrm{scale}(K) : \{1, ..., |K|\} \to K$ given by

$$\mathrm{scale}(K)(i) = \min K + i - 1$$

relabels the elements of K so that they begin with one.

Definition 6. For $x, y \in [S]$, x is a *sublist* of y iff there is a convex subset $K \subseteq \{1, \ldots, \mathrm{length}\ y\}$ such that $y \circ \mathrm{scale}\, K = x$.

Example 3. If $L = [1,2,3,4,5,6]$, then $[1,2,3], [4,5,6], [3,4,5], [2,3,4], [3,4], [5]$ and $[\,]$ are all sublists of L, while $[1,4,5,6], [1,3]$ and $[2,4]$ are *not* sublists of L.

Lemma 1. *The finite lists $[S]$ over a set S, ordered under reverse convex containment,*

$$x \sqsubseteq y \Leftrightarrow y \text{ is a sublist of } x,$$

form a cpo. In fact, every increasing sequence in $[S]$ is finite.

The order on $[S]$ is based on computational progress: Intuitively, it is easier to solve a problem on input $[\,]$ than for any other input x, hence $x \sqsubseteq [\,]$.

3 Ideal Mappings

The induction principle of the next section makes use of ideal splittings.

Definition 7. A *splitting* on a poset is a selfmap $s : P \to P$ such that $x \sqsubseteq s(x)$, for all $x \in P$.

The fixed points of a splitting s are denoted by $\mathrm{fix}(s) = \{x \in P : s(x) = x\}$. It is easy to see that every splitting on a poset P has a fixed point iff P has at least one *maximal element,* that is, an element $x \in P$ such that

$$\uparrow\!x := \{y \in P : x \sqsubseteq y\} = \{x\}.$$

However, the cpo's used in computation always have maximal elements, so existence of fixed points for splittings is not a concern. What is important, however, is finding a class of splittings whose fixed points may be calculated naturally.

Definition 8. A splitting $s : D \to D$ on a cpo D is *ideal* if for all sequences (a_n) in D with $s(a_n) \sqsubseteq a_{n+1}$ for all n, we have $s(\bigsqcup a_n) = \bigsqcup s(a_n)$.

Lemma 2 (Martin [6]). *Let $s : D \to D$ be a splitting on a cpo D. Then*

(i) *The map s is ideal iff for all sequences (a_n) with $s(a_n) \sqsubseteq a_{n+1}$, we have $s(\bigsqcup a_n) = \bigsqcup a_n$.*
(ii) *If s is ideal, then $\bigsqcup s^n(x) \in \mathrm{fix}(s)$, for all $x \in D$.*

Now we turn to some basic techniques which enable us to *prove* that splittings are ideal.

Lemma 3. *A continuous splitting is ideal.*

In particular, if $f : D \to D$ is continuous, then its restriction to the cpo $I(f) = \{x \in D : x \sqsubseteq f(x)\}$ is ideal.

Lemma 4. *For a cpo D, the following are equivalent:*

(i) *Every splitting on D is ideal.*

(ii) *The supremum of every strictly increasing sequence in D is maximal.*

Proof. (i) $\Rightarrow$ (ii): Let (a_n) be an increasing sequence with $a_n \neq a_{n+1}$ for all n. Write $a = \bigsqcup a_n$ and let $m \in D$ be any element with $a \sqsubseteq m$. Define a splitting

$$s : D \to D$$

$$s(x) = \begin{cases} a_{n+1} & \text{if } x = a_n; \\ m & \text{if } x = a; \\ x & \text{otherwise.} \end{cases}$$

Then $\bigsqcup a_n = \bigsqcup s^n(a_1) = a$. However, the splitting s is ideal, so $a \in \text{fix}(s)$. This gives $s(a) = a = m$ which proves that $\bigsqcup a_n$ is maximal. (ii) $\Rightarrow$ (i): Let $s : D \to D$ be a splitting and $s(a_n) \sqsubseteq a_{n+1}$. If this sequence is eventually constant, its supremum is a fixed point of s, and so obviously one preserved by s. On the other hand, if it is not eventually constant, its supremum must be maximal by (ii). But

$$\bigsqcup_{n \geq 0} a_n \sqsubseteq s(\bigsqcup_{n \geq 0} a_n) = \bigsqcup_{n \geq 0} s(a_n),$$

so the inequality on the left is actually an equality. □

Thus, by Lemma 4, every splitting on the cpo of lists $[S]$ is ideal. However, the most useful technique for establishing that a splitting is ideal is the next result.

Proposition 1. *Let D be a cpo with a continuous map $\mu : D \to [0, \infty)^*$ which is strictly monotone, that is, for all $x, y \in D$,*

$$x \sqsubseteq y \ \& \ \mu x = \mu y \Rightarrow x = y.$$

If $s : D \to D$ is a splitting such that

(i) *There is $0 \leq r < 1$ such that $\mu s(x) \leq r \cdot \mu x$, for all $x \in D$, or*
(ii) *The map $\mu \circ s : D \to [0, \infty)^*$ is continuous,*

then s is ideal.

Proof. (i) It is easy to see that any element x with $\mu x = 0$ is maximal. If (a_n) is a sequence with $a_n \sqsubseteq s(a_n) \sqsubseteq a_{n+1}$, then $(\forall n > 0)\ \mu a_{n+1} \leq r^n \mu a_1$. Thus,

$$\mu(\bigsqcup a_n) \leq \mu a_{n+1} \leq r^n \mu a_1,$$

for all $n > 0$. Then $\bigsqcup a_n$ is maximal in D and so must be a fixed point. This proves s is ideal by Lemma 2(i). (ii) If (a_n) is a sequence with $a_n \sqsubseteq s(a_n) \sqsubseteq a_{n+1}$, then

$$\mu s(\bigsqcup a_n) = \lim_{n \to \infty} \mu s(a_n) = \mu(\bigsqcup s(a_n)),$$

and since $\bigsqcup a_n \sqsubseteq \bigsqcup s(a_n)$, the strict monotonicity implies that these two are equal. □

The following example of a nonmonotonic splitting will give the reader a better feel for the usefulness of the last proposition. Notice too the ease with which nonmonotonic splittings arise in practice.

Example 4. The splitting left : $\mathbf{I}\mathbb{R} \to \mathbf{I}\mathbb{R}$ given by

$$\text{left}[a,b] = [a,(a+b)/2]$$

is *not* monotone. In particular, it is not continuous. However, it is easy to see that

$$\mu\,\text{left}[a,b] = \frac{\mu[a,b]}{2},$$

where $\mu : \mathbf{I}\mathbb{R} \to [0,\infty)^*$ is the length function $\mu[a,b] = b - a$. By Prop. 1, left is ideal.

For more on strictly monotone mappings like μ above, and how it is that they arise naturally in the study of computation on cpo's, see [6] and [7].

4 Induction

A subset P of a cpo D is called a *subcpo* if for every increasing sequence (x_n) in P, we have $\bigsqcup x_n \in P$, where the supremum is taken in D. A simple example of a subcpo is provided by the set of fixed points fix(s) of an ideal map $s : D \to D$. Another is given by that of an *inductive property.*

Definition 9. An *inductive property* on a cpo D is a subcpo $P \subseteq D$ and two ideal maps $l : D \to D$ and $r : D \to D$ such that

$$x \in P \Rightarrow lx \in P \text{ or } rx \in P$$

for all $x \in D$. We can write an inductive property as a triple (P, l, r).

The complement of an inductive property is a deductive property.

Definition 10. A *deductive property* on a cpo D is a subset $P \subseteq D$ and two ideal maps $l : D \to D$ and $r : D \to D$ such that the triple $(D \backslash P, l, r)$ forms an inductive property. We write a deductive property as a triple (P, l, r).

Theorem 1 (Induction). *If P is an inductive property on a cpo D, then*

$$x \in P \Rightarrow P \cap \uparrow x \cap (\text{fix}(l) \cup \text{fix}(r)) \neq \emptyset,$$

for all $x \in D$.

Proof. Define a splitting $s : P \to P$ by

$$s(x) = \begin{cases} l(x) & \text{if } l(x) \in P; \\ r(x) & \text{otherwise.} \end{cases}$$

Let $(a_n)_{n\geq 0}$ be a sequence in P with $a_n \sqsubseteq s(a_n) \sqsubseteq a_{n+1}$ for all n. Then there is an infinite subsequence of (a_n) named (b_i), which has the same supremum as (a_n), and for which we also have that either

$$(\forall i)\, l(b_i) \sqsubseteq b_{i+1} \;\&\; l(b_i) \in P$$

or

$$(\forall i)\, r(b_i) \sqsubseteq b_{i+1} \;\&\; l(b_i) \notin P.$$

In the first case, the idealness of l, combined with the fact that P is a cpo, gives

$$l(\bigsqcup a_n) = l(\bigsqcup b_i) = \bigsqcup l(b_i) = \bigsqcup b_i = \bigsqcup a_n \in P.$$

Thus, $a_0 \sqsubseteq l(\bigsqcup a_n) = \bigsqcup a_n$. In the second case, we must have $r(b_i) \in P$, and so the same argument gives $a_0 \sqsubseteq r(\bigsqcup a_n) = \bigsqcup a_n$. Finally, given a point $x \in P$, we set $a_n = s^n x$, for $n \geq 0$, and see that

$$\bigsqcup s^n(x) \in P \cap \uparrow x \cap (\mathrm{fix}(l) \cup \mathrm{fix}(r)),$$

which finishes the proof. □

Corollary 1 (Deduction). *If P is a deductive property on a cpo D, then*

$$\uparrow x \cap (\mathrm{fix}(l) \cup \mathrm{fix}(r)) \subseteq P \Rightarrow x \in P,$$

for all $x \in D$.

It is interesting that induction on the naturals has the form of Theorem 1.

Example 5. Let $p : \mathbb{N} \to \{\bot, \top\}$ be a function. The set

$$P = \{n \in \mathbb{N} \cup \{\infty\} : (\forall k < n)\, p(k) = \top\}$$

is a subcpo of $\mathbb{N} \cup \{\infty\}$. The successor function

$$\mathrm{succ}\, n = \begin{cases} n+1 & \text{if } n \in \mathbb{N} \\ \infty & \text{if } n = \infty \end{cases}$$

is ideal. If p has the property that for all $n \in \mathbb{N}$,

$$p(n) = \top \Rightarrow p(n+1) = \top,$$

then $(P, \mathrm{succ}, \mathrm{succ})$ is an inductive property on $\mathbb{N} \cup \{\infty\}$. In this case, Theorem 1 says that $p(0) = \top \Rightarrow p(n) = \top$ for all $n \in \mathbb{N}$.

The connectedness of $\mathbb{R}$ may now be proven by induction.

Example 6. Let $f : \mathbb{R} \to \mathbb{R}$ be a continuous map on the real line. The set

$$C(f) = \{[a,b] \in \mathbf{I}\mathbb{R} : f(a) \cdot f(b) \leq 0\}$$

is subcpo of $\mathbf{I}\mathbb{R}$ by the continuity of f. The mappings left : $\mathbf{I}\mathbb{R} \to \mathbf{I}\mathbb{R}$, left$[a,b] = [a,(a+b)/2]$, and right : $\mathbf{I}\mathbb{R} \to \mathbf{I}\mathbb{R}$, right$[a,b] = [(a+b)/2,b]$, are ideal. In addition,

$$x \in C(f) \Rightarrow \text{left}\, x \in C(f) \text{ or right}\, x \in C(f),$$

for all $x \in \mathbf{I}\mathbb{R}$. Thus, $(C(f), \text{left}, \text{right})$ is an inductive property on $\mathbf{I}\mathbb{R}$. By induction, if f changes sign on $[a,b]$, it must have at least one zero on $[a,b]$. This implies that the real line is connected.

The proof of the induction principle reveals more than is actually stated in Theorem 1. Not only does it show that $(\exists\alpha)\, \alpha \in P \cap \uparrow x \cap (\text{fix}(l) \cup \text{fix}(r))$, it also reveals *how to obtain* such an α. At times, this is worth remembering: In the last example, this fixed point is a zero of the function f, and the process by which it is obtained is the *bisection method* [6]. It is also important in the next example.

Example 7. Let D be a cpo with *least element* $\perp$ (meaning $(\forall x) \perp \sqsubseteq x$). If $f : D \to D$ is continuous and P is a subcpo such that

(i) $\perp \in P$, and
(ii) $(\forall x)\, x \in P \Rightarrow f(x) \in P$,

then the *least fixed point* $\text{fix}(f) = \bigsqcup f^n(\perp) \in P$.

This is the basic fixed point induction principle of domain theory [1]. But what is worth pointing out is that it is a special case of Theorem 1. To see this, set $l = r = f|_{I(f)}$, where $I(f) = \{x \in D : x \sqsubseteq f(x)\}$, and notice that $(P \cap I(f), l, r)$ is an inductive property over $I(f)$. Because $\perp \in P \cap I(f)$, $\bigsqcup s^n(\perp) \in P$, where s is as defined in the proof of Theorem 1. But s is nothing more than $f|_{I(f)}$ in this case, which finishes the demonstration.

Deduction (Corollary 1) is simply a special case of the induction principle since (P, l, r) is deductive iff $(D \setminus P, l, r)$ is inductive. However, it is an instance worthy of distinction. Very much in the spirit of classical induction on the naturals, it works as follows: To prove that x has property P, we need only establish the *base case* that the fixed points of l and r above x have property P.

For instance, in the next example, we will use deduction to establish the compactness of the unit interval $[0,1]$. The essence of the argument is this: Because the points of $[0,1]$ are all compact (base case), the unit interval itself is compact. We should point out before proceeding that only the completeness of $\mathbb{R}$ is required to prove that $\mathbf{I}\mathbb{R}$ is a cpo.

Example 8. Let $\{U_\alpha\}$ be an open cover of $[0,1]$. Consider the set

$$P = \{x \in \mathbf{I}\mathbb{R} : x \textit{ can be finitely covered by } \{U_\alpha\}\}.$$

First, if $[a,b] \in P$, then write $[a,b] \subseteq \bigcup_{i=1}^{n} U_i = V$. Because V is an open subset of $\mathbb{R}$,

$$(\exists\, \varepsilon > 0\,)\; a \in (a-\varepsilon, a+\varepsilon) \subseteq V \;\&\; b \in (b-\varepsilon, b+\varepsilon) \subseteq V,$$

which means $\{x \in \mathbf{IR} : [a,b] \subseteq x \subseteq (a-\varepsilon, b+\varepsilon)\} \subseteq P$. This implies that $\mathbf{IR} \setminus P$ is a subcpo of $\mathbf{IR}$. Now observe that for all $x \in \mathbf{IR}$,

$$\text{left}\, x \in P \;\&\; \text{right}\, x \in P \Rightarrow x \in P.$$

Thus, $(P, \text{left}, \text{right})$ is a deductive property. Finally,

$$\uparrow[0,1] \cap (\text{fix(left)} \cup \text{fix(right)}) = \{[t] : t \in [0,1]\} \subseteq P,$$

and so by deduction $[0,1] \in P$.

Another application of the induction principle is to program correctness.

5 Program Verification

We derive the following method for verifying an algorithm:

(i) For an algorithm a, let $\Diamond a$ be its domain, the set of all *possible* inputs, and let $\Box a := \{x \in \Diamond a : a \text{ works correctly on } x\}$. Trivially, then, $\Box a \subseteq \Diamond a$.
(ii) Show that $\Box a$ is a deductive property over $\Diamond a$.
(iii) Use deduction to show $\Diamond a \subseteq \Box a$.
(iv) Conclude that a works correctly on all inputs since $\Box a = \Diamond a$.

We now apply this idea to list processing algorithms on $[S]$. Notice that Lemma 4 implies that all splittings on $[S]$ and $[S]^2$ are ideal. In addition, all subsets of $[S]^2$ and $[S]$ are subcpo's. Before proceeding, we need to formalize some basic computational aspects.

Definition 11. The *concatenation* of two lists x and y is written $x \cdot y$. Formally, it is the list $x \cdot y : \{1, \ldots, |x| + |y|\} \to S$ given by $(x \cdot y)|_{\{1,\ldots,|x|\}} = x$ and $(x \cdot y)|_{\{|x|+1,\ldots,|x|+|y|\}} = y$.

Definition 12. If S is a poset, then we say that a list $x \in [S]$ is *sorted* if x is monotone as a map between posets. The set of all sorted lists is denoted $[S]_{\leq}$.

Definition 13. A *permutation* of a list $x \in [S]$ is a list $y \in [S]$ for which there is a bijection $\phi : \{1, \ldots, |x|\} \to \{1, \ldots, |y|\}$ such that $x = y \circ \phi$. The set of all permutations of x is denoted by $x!$.

We now use the induction principle to give a proof of the correctness of mergesort. There are two stages: First the verification of the merge operation, and then the sorting algorithm itself. Notice that we are able to work with each program in its *natural* state.

Example 9. Consider the the following ML program to merge two sorted lists of integers:

$$\begin{aligned}
\text{fun merge}(\ [\,], ys\ ) &= ys : \text{int list}\\
\mid \text{merge}(\ xs, [\,]\ ) &= xs\\
\mid \text{merge}(\ x :: xs, y :: ys\ ) &= \text{if } x \le y \text{ then}\\
&\qquad x :: \text{merge}(\ xs, y :: ys\ )\\
&\quad \text{else}\\
&\qquad y :: \text{merge}(\ x :: xs, ys\ );
\end{aligned}$$

(1) *Identify the domain of the algorithm and define correctness.*

$$\Diamond\text{merge} := \{\ (x, y) \in [\text{int}]^2 : x, y \in [\text{int}]_{\le}\}$$
$$\Box\text{merge} := \{\ (x, y) \in \Diamond\text{merge} : \text{merge}(x, y) \in (x \cdot y)! \cap [\text{int}]_{\le}\}$$

(2) *Show that* $\Box$merge *is a deductive property on* $\Diamond$merge.

Consider the splitting

$$\pi : \Diamond\text{merge} \to \Diamond\text{merge}$$

defined by

$$\begin{aligned}
\pi(\ [\,], ys\ ) &= (\ [\,], ys\ )\\
\pi(\ xs, [\,]\ ) &= (\ xs, [\,]\ )\\
\pi(\ x :: xs, y :: ys\ ) &= (\ xs, y :: ys\ ) \text{ if } x \le y\\
&= (\ x :: xs, ys\ ) \text{ otherwise.}
\end{aligned}$$

If $(x, y) \in \Diamond$merge, then

$$\pi(x, y) \in \Box\text{merge} \Rightarrow (x, y) \in \Box\text{merge}.$$

Thus, ($\Box$merge, π, π) is a deductive property over $\Diamond$merge.

(3) *Use deduction to show that* $\Diamond$merge $\subseteq$ $\Box$merge.

The fixed points of π are $\text{fix}(\pi) = \{([\,], y) : y \text{ sorted}\} \cup \{(x, [\,]) : x \text{ sorted}\}$.

Given $(x, y) \in \Diamond$merge,

$$\uparrow\!(x, y) \cap \text{fix}(\pi) \subseteq \text{fix}(\pi) \subseteq \Box\text{merge}.$$

Then, by deduction, $(x, y) \in \Box$merge.

(4) Consequently, merge is a correct algorithm.

In the next example, take(n, xs) returns the first n elements of the list xs, while drop(n, xs) removes the first n elements of xs and returns the remainder.

Example 10. Consider the ML implementation of mergesort for lists of integers:

```
fun sort [ ] = [ ]
  | sort [x]= [x]
  | sort xs = let val n = length xs div 2
              in merge( sort( take(n,xs) ),
                        sort( drop(n,xs) ) )
              end;
```

The domain of sort is

$\Diamond$sort = [int] and $\Box$sort := $\{ x \in [\text{int}] : \text{sort}(x) \in x! \cap [\text{int}]_{\leq} \}$.

Now consider the splittings

left : [int] $\to$ [int]	right : [int] $\to$ [int]
left xs = take(length xs div 2, xs)	right xs = drop(length xs div 2, xs)

By the correctness of merge,

($\Box$sort, left, right) is a deductive property over [int].

Finally,

fix(left) = $\{[\,]\} \subseteq \Box$sort

and

fix(right) = $\{[x] : x \in \text{int}\} \cup \{[\,]\} \subseteq \Box$sort,

which by deduction proves that $\Box$sort = [int], i.e., sort is correct.

6 Related Work

As we have already mentioned implicitly numerous times, the idea most closely related to the one given here is fixed point induction [1]. Other versions of induction tend to be rather specific in nature, as more often than not they focus on a specific data type [5], or on a specific topological space([3],[4]). The principle given here emphasizes the form that inductive arguments take when looked at from the informatic viewpoint: A single technique enables us to reason about the continuous (the real line) as well as the discrete (ML programs) without ever drawing a distinction between the two.

7 Presentation

Great care has been taken to keep the mathematics as simple as possible as a means of enhancing applicability. However, an application like Example 8 is probably better dealt with using the approximation relation and the μ *topology* on $\mathbb{IR}$. For more on this, see the third chapter of [6].

8 Conclusion

An induction principle based on complete partial orders and splittings has been introduced. It has been shown to capture the form of the usual induction on the naturals, to imply the basic fixed point induction of domain theory, to admit proofs of the compactness of $[0,1]$ and the connectedness of $\mathbb{R}$, and to be useful for establishing the correctness of algorithms in their *natural* state. The reason for this diversity of application is that it is based on nonmonotonic mappings. For more on splittings, their applications to numerical analysis, and the sense in which the "recursive part" of an algorithm may always be modeled by a splitting, see the author's Ph.D. thesis [6].

References

1. S. Abramsky and A. Jung, *Domain theory.* In S. Abramsky, D. M. Gabbay, T. S. E. Maibaum, editors, Handbook of Logic in Computer Science, vol. III. Oxford University Press, 1994.
2. P. Aczel, *An introduction to inductive definitions.* In Handbook of Mathematical Logic, J. Barwise, Editor, North-Holland, p.739-782.
3. T. Coquand, *Constructive topology and combinatorics.* Lecture Notes in Computer Science, vol. 613, p.159–164.
4. T. Coquand, *A note on the open induction principle.* `http://www.cs.chalmers.se/ coquand/open.ps.Z`
5. M. Escardo and T. Streicher, *Induction and recursion on the partial real line with applications to Real PCF.* Theoretical Computer Science, volume 210, number 1, p. 121–157, 1999.
6. K. Martin, *A foundation for computation.* Ph.D. Thesis, Department of Mathematics, Tulane University, May 2000. `http://web.comlab.ox.ac.uk/oucl/work/keye.martin`
7. K. Martin, *The measurement process in domain theory.* Proceedings of the 27th International Colloquium on Automata, Languages and Programming (ICALP), Lecture Notes in Computer Science, vol. 1853, Springer-Verlag, 2000.
8. L.C. Paulson, *ML for the Working Programmer.* Cambridge University Press, 1991.

On a Generalisation of Herbrand's Theorem

Matthias Baaz and Georg Moser

Vienna University of Technology, Institut für Algebra und Computermathematik,
E118.2, Wiedner Hauptstrasse 8–10
{baaz,moser}@logic.at

Abstract. In this paper we investigate the purely logical rule of *term induction*, i.e. induction deriving numerals instead of arbitrary terms. In this system it is not possible to bound the length of Herbrand disjunctions in terms of proof length and logical complexity of the end-formula as usual. The main result is that we can bound the length of the reduct of Herbrand disjunctions in this way. (Reducts are defined by omitting numerals.)

1 Introduction

Let $\exists \bar{x} F(\bar{x})$ be an existential formula which is provable in a usual Hilbert or Gentzen type system of pure logic. We assert the existence of a proof Π of length k.[1] Using Herbrand's Theorem we find a valid disjunction (called *Herbrand disjunction*)

$$C_1 \vee \cdots \vee C_c \tag{1}$$

such that the C_i are instances of $F(\bar{a})$. It is a well-known fact that c, the number of disjuncts, can be bounded by a primitive recursive function which depends only on k and the logical complexity of $\exists \bar{x} F(\bar{x})$.[2]

In this paper we take a close look on a formal first order system $\mathbf{T}^{(tind)}$ in the standard language $\mathcal{L}$, of **LK**, see [20]. $\mathbf{T}^{(tind)}$ extends **LK** by the following valid first-order inference rule (A is quantifier-free).

$$\frac{\Gamma, A(a), \Lambda \rightarrow \Delta, A(s(a)), \Theta}{\Gamma, A(0), \Lambda \rightarrow \Delta, A(s^n(0)), \Theta}\,(tind) \tag{2}$$

This rule is called *term induction*, it derives a *restricted* term built from successor s and the constant 0. We call such terms *numerals*. (This inference is in stark contrast to the usual induction principles, as these derive *arbitrary* terms. $\mathbf{T}^{(tind)}$ is *conservative* over pure logic.

On first sight this investigation of a *dependent* rule seems to be a strange task. The reader may wonder what could possibly be gained from such an analysis, as the inference rule (tind) can be immediately replaced by a sequence of

[1] We conceive proofs as rooted trees whose vertices are sequents. The *length*, $|\Pi|$, of a proof Π is the number of vertices.

[2] In fact it is possible to bound c by a function that depends only on k. This follows from the first ϵ-elimination theorem, cf. [12] pages 27–33, compare also [1].

L. Fribourg (Ed.): CSL 2001, LNCS 2142, pp. 469–483, 2001.

implications. Hence our analysis may appears to be nothing else than another analysis of predicate logic. This is obviously true, but beside the point. The difference between $\mathbf{T}^{(tind)}$ and **LK**, is not a question of *logical* transformability, but of *uniform* transformability: In $\mathbf{T}^{(tind)}$, we can no longer prove a *uniform* bound on the number of disjunctions in Herbrand disjunctions, similar to the one above. (We will give a proof of this fact in the next section.)

The main result of this paper can be stated as follows. A *general Herbrand disjunction* H (for $\exists\bar{x}F(\bar{x})$) is a *valid* disjunction

$$\bigvee_{i_1}^{N}\cdots\bigvee_{i_l}^{N} M_1(s^{i_1}(0),\ldots,s^{i_l}(0)) \vee \cdots \vee M_m(s^{i_1}(0),\ldots,s^{i_l}(0)) \tag{3}$$

where the M_i are instances of $F(\bar{a})$ and $N \in \mathbb{N}$. By removing all indicated numerals from the general Herbrand disjunction a disjunction is obtained whose length is independent on the numerals present in H.

$$M_1(a_{i_1},\ldots,a_{i_l}) \vee \cdots \vee M_m(a_{i_1},\ldots,a_{i_l}) \tag{4}$$

where the a_i denote arbitrary distinct free variables. We call this disjunction an $\{s,0\}$*-matrix*, or simply *matrix*.

It is now possible to bound the length m by a primitive recursive function ψ that depends only on k and the logical complexity $\mathrm{ld}(\exists\bar{x}F(\bar{x}))$ of the end-formula. In particular it doesn't depend on the height (and number) of the numerals contained in $\exists\bar{x}F(\bar{x})$.

This implies that we can distinguish two separate parts, namely an *arithmetical* and a *logical* part. The arithmetical part expresses the impact of term induction in Π, i.e. the possibility to derive uniformly formulas that contain arbitrary numerals. On the other hand the logical part is concerned with the aspects of usual logical rules, only. In this respect our results are similar to standard results for Hilbert type systems over pure logic cf. [13,1].

By an extension of the argument that renders the bound on the length, we find a generalisation of the $\{s,0\}$-matrix such that the depth of terms built from function symbols different from s is uniformly bounded. From this we obtain as a corollary that it is not possible in $\mathbf{T}^{(tind)}$ (contrary to a 'full' induction system, cf. [23]) to derive $0 + (\cdots + 0) = 0$ uniformly in a fixed number of steps from $\forall x\; 0 + x = x$ and axioms of identity.

Moreover l – the number of 'big' disjunctions in H – is bounded by a primitive recursive function depending only on the maximal *iteration* of (tind)-inferences in Π and $\mathrm{ld}(\exists\bar{x}F(\bar{x}))$. Note that if we restrict Π to derivations admitting propositional cuts only, l is already bound by the maximal iteration of (tind)-inferences in Π. Furthermore this bound is sharp.

2 Herbrand Disjunctions in $\mathbf{T}^{(tind)}$ Have No Uniform Bounds

We show that $\mathbf{T}^{(tind)}$ does not admit uniform bounds for the length of Herbrand disjunctions in the length of the proofs and the logical complexities of the end formulas. Instead of showing the claim directly for Herbrand disjunctions, we show it w.l.o.g with respect to *Herbrand sequents*, a slightly more general concept.

Let $Qx\, B(x)$, $Q \in \{\forall, \exists\}$ be a sub-formula of F. If $Qx\, B(x)$ occurs positively (negatively) in F, then the occurrence of Q is called *strong* (*weak*) if $Q \equiv \forall$ and *weak* (*strong*), otherwise.

Definition 1. *Let S be a provable sequent, containing only weak quantifiers.*

$$\forall \bar{y}_1 A_1(\bar{y}_1), \ldots, \forall \bar{y}_a A_a(\bar{y}_a) \to \exists \bar{x}_1 B_1(\bar{x}_1), \ldots, \exists \bar{x}_b B_b(\bar{x}_b)$$

Then a valid sequent of the form

$$A_1(s_1^1, \ldots, s_{n1}^1), \ldots, A_a(s_1^{va}, \ldots, s_{na}^{va}) \to B_1(t_1^1, \ldots, t_{m1}^1), \ldots, B_b(t_1^{wb}, \ldots, t_{mb}^{wb})$$

(n_i;m_j are chosen according to the length of $\bar{y}_i, \bar{x}_j$) is called Herbrand sequent

Proposition 1. $\mathbf{T}^{(tind)}$ *does not admit uniform bounds for the length of Herbrand sequents.*

Proof. To prove the claim we proceed indirectly: Assume the existence of a function ϕ, such that for every proof Π of $\exists \bar{x} F(\bar{x})$ in $\mathbf{T}^{(tind)}$, ϕ uniformly bounds the number of sequent formulas (i.e. the *length*) in the corresponding Herbrand sequent. Hence ϕ is independent of the occurring parameters in $\exists \bar{x} F(\bar{x})$. Consider the trivial proof Π (P is atomic) given in Table 1

Table 1.

$$\dfrac{\dfrac{\dfrac{P(a) \to P(a) \quad P(s(a)) \to P(s(a))}{P(a) \supset P(s(a)), P(a) \to P(s(a))}}{\forall x(P(x) \supset P(s(x))), P(a) \to P(s(a))}}{\forall x(P(x) \supset P(s(x))), P(0) \to P(s^n(0))}\ (tind)$$

Obviously any Herbrand sequent of $\forall x(P(x) \supset P(s(x))), P(0) \to P(s^n(0))$ has to contain – we apply the pigeon-hole principle – all n implications $P(s^i(0)) \supset P(s^{i+1}(0))$ in the antecedent. □

An immediate consequence of the example in Table 1 is that the following property $\exists k \forall n \vdash_k \Gamma \to \Delta, A(s^n(0))$ iff $\vdash \Gamma \to \Delta, \forall x A(x)$, which is sometimes called Kreisel's conjecture, does not hold in $\mathbf{T}^{(tind)}$.

Remark 1. Using *structural skolemisation* cf. [2] one can transform any proof Π of a sequent S to a proof Π' with skolemised end-sequent such that $|\Pi'| \leq |\Pi|$. Furthermore one can transform an arbitrary sequent into a sequent in prenex normal form by addition suitable cuts. This is possible by an increase in proof-length at most quadratic in the logical complexity, cf. [3].

3 Extraction of Herbrand Sequent from $\mathbf{T}^{(tind)}$-Proofs

We aim at a characterization of the Herbrand sequents of theorems in $\mathbf{T}^{(tind)}$. Assume Π to be a proof of S in $\mathbf{T}^{(tind)}$ with length $|\Pi| \leq k$.

A well-known result by Parikh [15] shows that the logical complexity of the formulas in Π can be bounded by a primitive recursive function ϕ, depending only on k and $\mathrm{ld}(S)$. Parikh's argument is sufficiently general to be applied to $\mathbf{T}^{(tind)}$ as well. (For a modern presentation of the argument used in [15], see also [10,6,16].) We obtain

Proposition 2. *If a sequent S has a proof (in $\mathbf{T}^{(tind)}$) of length k then there exists a proof Π', $|\Pi'| = k$, of S so that the maximal logical depth of the formulas in Π' is bounded by $2^{2k}\mathrm{ld}(S)$.*

Using the proposition we find a proof Π^2 such that for all formulas A in Π^2, $\mathrm{ld}(A) \leq 2^{2k}\mathrm{ld}(S)$. We introduce a notation that counts the maximal number of iterations of (tind)-rules in $\mathbf{T}^{(tind)}$-proofs.

Definition 2. $\mathrm{it}(\Pi)$ *is defined inductively on the length of Π. Assume $|\Pi| = 0$, then* $\mathrm{it}(\Pi) \stackrel{\mathrm{def}}{=} 0$. *Otherwise assume* $\mathrm{it}(\Pi)$ *has already been defined for proofs Π', $|\Pi'| \leq k$ and let $|\Pi| = k+1$.*

We proceed by case-analysis on the last inference rule Q. Assume Q is a (tind)-rule, then $\mathrm{it}(\Pi) \stackrel{\mathrm{def}}{=} \mathrm{it}(\Pi') + 1$. *Now assume otherwise Q is a binary inference rule with subproof-proofs Π_1, Π_2. Then* $\mathrm{it}(\Pi) = \max\{\mathrm{it}(\Pi_1, \Pi_2)\}$. *In the case Q is an unary rule, simple set* $\mathrm{it}(\Pi)$ *equal to* $\mathrm{it}(\Pi')$.

We perform a last transformation on the given proof Π^2. A formula is called *propositional* if A does not contain any bounded variables. We define

$$2^y_0 \stackrel{\mathrm{def}}{=} y \qquad 2^y_{x+1} \stackrel{\mathrm{def}}{=} 2^{2^y_x}$$

The *cut-degree* $\rho(\Pi)$ of a proof Π is defined by induction. Let $\Pi_i, i = 1,2$ be direct subproofs of Π. Assume the last inference rule in Π is a Cut with cut-formula A. Let $\rho(\Pi) \stackrel{\mathrm{def}}{=} \max(\mathrm{ld}(A), \rho(\Pi_1), \rho(\Pi_2))$. Otherwise let $\rho(\Pi) \stackrel{\mathrm{def}}{=} \max(\rho(\Pi_1), \rho(\Pi_2))$.

Theorem 1. *Let Π be a proof in $\mathbf{T}^{(tind)}$. Then we can transform Π to a proof Π' of the same end-sequent S such that Π' admits propositional cuts only. Moreover*

$$|\Pi'| \leq 2^k_{2(\rho(\Pi)+1)} \qquad \mathrm{it}(\Pi') \leq 2^{\mathrm{it}(\Pi)}_{\rho(\Pi)+1}$$

The proof of the admissibility of cut-reduction is well-known, see e.g. [18]. Hence, it remains to prove the stated result on the iteration of (tind)-inferences. For our purposes it is best to follow Buss' proof of cut-elimination [9]. Note that our notion of proof-length is slightly different. The first step is to transform Π^2 to a proof in which initial sequents are atomic. One can bound the length of the transformed proof Π^3 in terms of k and $\mathrm{ld}(S)$. More precisely $|\Pi_3| \leq 5 \cdot k \cdot 2^{2k}\mathrm{ld}(S)$ $(= l)$, cf. [2].

To show the admissibility of cut-reduction it suffices to investigate a variant of the Reduction Lemma. The theorem then follows by induction on the cut-degree.

Lemma 1. *Let Π_1, Π_2 be derivations of $\Gamma_1 \to \Delta_1, A, \Theta_1$; $\Gamma_2, A, \Lambda_2 \to \Delta_2$, respectively such that $\rho(\Pi_i) < \mathrm{ld}(A)$. Then we can find a proof Π of $\Gamma_1, \Gamma_2, \Lambda_2 \to \Delta_1, \Theta_1, \Delta_2$ and $|\Pi| \leq (|\Pi_1| + |\Pi_2| + 1)^2$ such that $\rho(\Pi) < \mathrm{ld}(A)$. Moreover $\mathrm{it}(\Pi) \leq \mathrm{it}(\Pi_1) + \mathrm{it}(\Pi_2)$.*

Proof. The argument centers around a case analysis of the form of the cut formulas. (We assume acquaintance with [9].) For brevity we concentrate on the case where the cut formula of form $\exists x B(x)$. W.l.o.g we may assume that both cut-formulas are not derived by weakenings. Note that no ancestor of the cut-formula A can be a principal formula of a (tind)-inference. Assume there exist k $\exists$: right introductions for ancestors of A in Π_1. Consider

$$\frac{\Xi \to \Upsilon, B(t), \Omega}{\Xi \to \Upsilon, \exists x B(x), \Omega}$$

We assert an enumeration of all such t: $t_1, \ldots, t_k$. For each t_i a proof $\widehat{\Pi}_2^i$ of $\Gamma_2, B(t_i), \Lambda_2 \to \Delta_2$ is obtained from Π_2. The procedure is standard. Now replace all sequents $\Xi \to \Upsilon$ in Π_1 by $\Xi \to \Upsilon^-$, where Υ^- denotes Υ after removal of all predecessors of A. This step transforms $\exists$: right inferences to

$$\frac{\Xi \to \Upsilon^-, B(t_i), \Omega^-}{\Xi \to \Upsilon^-, \Omega^-}$$

Then the lower sequent follows from the upper sequent by a cut-inference with $\widehat{\Pi}_2^i$. The final proof Π is obtained by induction on k. The bound on the length of Π follows as in [9]. We only need in addition the following observation: Suppose $|\Pi| = k$. Then it is easy to see that $|\Gamma \to \Delta|$ for some sequent $\Gamma \to \Delta$ is bounded by $k + 1$.

It remains to prove the bound on $\mathrm{it}(\Pi)$. Let $Q_1, \ldots, Q_k$ be an enumeration of all (tind)-inferences in Π_1. In Π the same (tind) applications occur, but most likely at different places. We denote them by $\overline{Q}_1, \ldots, \overline{Q}_k$, respectively. It follows by definition that $\mathrm{it}(\bar{Q}_i) \leq \mathrm{it}(Q_i) + \mathrm{it}(\Pi_2)$. Now the bound on $\mathrm{it}(\Pi')$ follows easily. □

Using Theorem 1 we transform Π^3 to a proof $\Pi^\star$ admitting only propositional cuts. We conclude that $|\Pi^\star| \leq 2^l_{\rho(\Pi_1)}$ where $\rho(\Pi_1) \leq 2^{2k}\mathrm{ld}(S)$. To abbreviate the bound for $|\Pi^\star|$ we introduce the primitive recursive function $\varphi(k, \mathrm{ld}(S)) \stackrel{\text{def}}{=} 2^l_{\rho(\Pi_1)}$ (recall that $l = 5 \cdot k \cdot 2^{2k}\mathrm{ld}(S)$). For simplicity we assume that $|\Pi^\star| = \varphi(k, \mathrm{ld}(S)) = r$. We denote the sequence of formulas occurring in the succedent (antecedent)

$$D_k(0, \ldots, 0), \ldots, D_k(s^{N_1}(0), \ldots, s^{N_{qk}}(0))$$

for arbitrary natural numbers $N_1, \ldots, N_{qk}$, by $\bigvee_{(j_1,\ldots,j_{qk})} D_k(s^{j_k}(0), \ldots, s^{j_{qk}}(0))$ $(\bigwedge_{(i_1,\ldots,i_{pk})} D_k(s^{i_k}(0), \ldots, s^{i_{pk}}(0)))$.

Definition 3. *Let T be a Herbrand sequent written as*

$$\bigwedge_{(i_1,\ldots,i_{p1})} C_1(s^{i_1}(0),\ldots,s^{i_{p1}}(0)),\ldots,\bigwedge_{(i_1,\ldots,i_{pc})} C_c(s^{i_1}(0),\ldots,s^{i_{pc}}(0)) \to \\ \to \bigvee_{(j_1,\ldots,j_{q1})} D_1(s^{j_1}(0),\ldots,s^{j_{q1}}(0)),\ldots,\bigvee_{(j_1,\ldots,j_{qd})} D_d(s^{j_1}(0),\ldots,s^{j_{qd}}(0))$$

Then the sequent

$$C_1(a_1,\ldots,a_{p1}),\ldots,C_c(a_1,\ldots,a_{pc}) \to D_1(b_1,\ldots,b_{q1}),\ldots,D_d(b_1,\ldots,b_{qd})$$

is called an $\{s,0\}$-matrix *of T.*

Definition 4. *The $\{s,0\}$-complexity of S, denoted as* $\text{complex}^{\{s,0\}}(S)$, *is the minimal length of the $\{s,0\}$-matrices of Herbrand sequents of S.*

We denote the sequence of m equal sequent formulas $F^1,\ldots,F^m$ by $(F)^m$. The number m is called the *multiplicity* of F.

Lemma 2. *Let $\Pi^\star$, $|\Pi^\star| = r$, be a proof – admitting only propositional cuts – of the end-sequent S (all quantifiers are indicated)*

$$\forall\bar{y}_1 A_1(\bar{y}_1),\ldots,\forall\bar{y}_a A_a(\bar{y}_a) \to \exists\bar{x}_1 B(\bar{x}_1),\ldots,\exists\bar{x}_b B(\bar{x}_b)$$

Then there is a proof Π' of

$$(\forall\bar{y}_1 A_1(\bar{y}_1))^{k_1},\ldots,(\forall\bar{y}_a A_a(\bar{y}_a))^{k_a} \to (\exists\bar{x}_1 B(\bar{x}_1))^{l_1},\ldots,(\exists\bar{x}_b B(\bar{x}_b))^{l_b}$$

such that only quantifier-free formulas are subject to contractions. Moreover $|\Pi'| \leq r^2$ and $\Sigma_i^a k_i + \Sigma_j^b l_j \leq r$.

Proof. By our assumptions on the form of S the initial sequents cannot contain quantifiers. Moreover (tind) and propositional inferences are applied to quantifier-free formulas only. By omitting all contractions on quantified formulas we obtain a proof Π' from $\Pi^\star$ fulfilling the conditions of the lemma. Obviously for any of these formulas the multiplicity is bounded by the number of inferences in Π. Using induction on k it is easy to verify the claim on the proof-length of Π'. □

Theorem 2. *Let $\Pi^\star$, $|\Pi^\star| = r$, be a proof – admitting only propositional cuts – of the end-sequent S with* $\text{it}(\Pi^\star) = l$. *Then we find a Herbrand sequent T of S*

$$\bigwedge_{\bar{p}_1} C_1(s^{i_1}(0),\ldots,s^{i_{p1}}(0)),\ldots,\bigwedge_{\bar{p}_c} C_c(s^{i_1}(0),\ldots,s^{i_{pc}}(0)) \to \\ \to \bigvee_{\bar{q}_1} D_1(s^{j_1}(0),\ldots,s^{j_{q1}}(0)),\ldots,\bigvee_{\bar{q}_d} D_d(s^{j_1}(0),\ldots,s^{j_{qd}}(0))$$

such that the tuples $\bar{p}_k, \bar{q}_l$; $k = 1,\ldots,c; l = 1,\ldots,d$ denote tuples of indices $(i_1,\ldots,i_{pk})$ and $(j_1,\ldots,j_{pl})$ chosen accordingly.

Moreover $\max\{p_1,\ldots,p_c,q_1,\ldots,q_d\} \leq \text{it}(\Pi^\star)$ *(note that* $\text{len}(\bar{p}_k) = p_k$ *and* $\text{len}(\bar{q}_l) = q_l$*) and $c + d \leq r$*

Proof. We apply the lemma 2 to $\Pi^\star$ to obtain a proof Π' with the denoted properties. We proceed by an induction on $|\Pi'|$ by a case-distinction on the last inference rule Q. We concentrate on the cases where Q is either a quantifier rule or a (tind) application. In the other cases the result follows almost trivially by application of the induction hypothesis. With respect to weakening rules we may assume w.l.o.g. that only quantifier-free formulas are introduced.

Assume Q is a $\forall$-*left* introduction. Then, applying the induction hypothesis we assume the existence of a Herbrand sequent T' of the form above. Note that T' is already in the requested form and the rule can be omitted.

Now assume Q is a (tind)-inference, written as (2). W.l.o.g. we conclude the existence of a Herbrand sequent T':

$$\bigwedge_{\bar{p}} C_1(\overline{s^{p'}(0)}, a_{p'}, \ldots, a_p), A(a), \bigwedge_{\bar{q}} C_2(\overline{s^{q'}(0)}, a_{q'}, \ldots, a_q) \rightarrow \\ \bigvee_{\bar{u}} D_1(\overline{s^{u'}(0)}, b_{u'}, \ldots, b_u), A(s(a), \bigvee_{\bar{v}} D_2(\overline{s^{v'}(0)}, b_{v'}, \ldots, b_v)$$

where $\overline{s^{p'}(0)}$, $\overline{s^{q'}(0)}$,$\overline{s^{u'}(0)}$, and $\overline{s^{v'}(0)}$ denote suitable tuples of numerals; the variable a is the eigenvariable of Q. (The other variables indicated previously (i.e. in Π') bound variables, freed by the removal of quantifier rules in the previous construction of T'.)

We obtain n different sequents $T(i)$ by replacing the eigenvariable a in T' by 0, $s^1(0)$, $s^2(0), \ldots, s^{n-1}(0)$, respectively. By iterated cuts and subsequent contractions we derive a sequent T according to the conditions of the theorem. Note that the eigenvariable a may occur in the side-formulas in $T(i)$ as in the construction of T' quantifier introductions above Q are omitted. Furthermore observe, that a can occur only at places previously bounded by quantifiers.

By definition $\mathrm{it}(\Pi^\star)$ is $\mathrm{it}(\Pi_0) + 1$. By the step above the length of 'big' conjunctions and 'big' disjunction is increased by at most one. Hence the bound on $\mathrm{it}(\Pi^\star)$ follows almost trivially. □

Example 1. Consider the sequent $S(l)$.

$$\forall x_1, \ldots, \forall x_n \bigvee_i^n P_i(x_i) \supset P_i(s(x_i)), P_1(0), \ldots, P_n(0) \rightarrow P_1(s^l(0)), \ldots, P_n(s^l(0))$$

$S(l)$ can be uniformly derived cut free, using n iterated (tind) inferences, see Table 2. □

We now argue, that $S(l)$ cannot be derived uniformly, admitting only propositional cuts, with less than n iterated inductions. Every Herbrand sequent of $S(l)$ has the form

$$\ldots \bigvee_i^n P_i(s^{j_i}(0)) \supset P_i(s(s^{j_i}(0))), \ldots, P_1(0), \ldots, P_n(0) \rightarrow P_1(s^l(0)), \ldots, P_n(s^l(0))$$

where $j_1, \ldots, j_n \in \mathbb{N}$ are arbitrary. Assume otherwise that $\bigvee_i^n P_i(s^{r_i}(0)) \supset P_i(s(s^{r_i}(0)))$ for some combination $r_1 \leq l, \ldots, r_n \leq l$ is not present. Define an evaluation v as follows

Table 2. Uniform cut-free derivation of $S(l)$

$$\dfrac{\dfrac{\dfrac{\dfrac{\Pi_0}{\bigvee_i^n P_i(a_i) \supset P_i(s(a_i)), P_1(a_1), \ldots, P_n(a_n) \to P_1(s(a_1)), \ldots, P_n(s(a_n))}}{\forall x_1, \ldots, x_n \bigvee_i^n P_i(x_i) \supset P_i(s(x_i)), P_1(a_1), \ldots, P_n(a_n) \to P_1(s(a_1)), \ldots, P_n(s(a_n))}}{\forall x_1, \ldots, x_n \bigvee_i^n P_i(x_i) \supset P_i(s(x_i)), P_1(0), \ldots, P_n(a_n) \to P_1(s^l(0)), \ldots, P_n(s(a_n))}}{\forall x_1, \ldots, x_n \bigvee_i^n P_i(x_i) \supset P_i(s(x_i)), P_1(0), \ldots, P_n(0) \to P_1(s^l(0)), \ldots, P_n(s^l(0))}$$

Table 3. Proof fragment Π_i

$$\dfrac{\dfrac{\dfrac{P(a_i) \to P(a_i) \quad P(s(a_i)) \to P(s(a_i))}{P(a_i) \supset P(s(a_i)), P(a_i) \to P(s(a_i))}}{\forall x_i P(x_i) \supset P(s(x_i)), P(a_i) \to P(s(a_i))}}{\forall x_i P(x_i) \supset P(s(x_i)), P(0) \to P(s^l(0))}$$

$$v(P_i(s^k(0))) \stackrel{\text{def}}{=} \begin{cases} \textbf{true} & k \leq r_i \\ \textbf{false} & k > r_i \end{cases}$$

Then v falsifies the sequent and shows that every general Herbrand disjunction for $S(l)$ has to contain at least n 'big' disjunctions.

Furthermore the given example shows that our arguments crucially depend on the fact that $S(l)$ is in prefix normal form.

Example 1 (continued). Consider the sequents $S'(l)$

$$\bigvee_i^n \forall x_i P_(x_i) \supset P_i(s(x_i)), P_1(0), \ldots, P_n(0) \to P_1(s^l(0)), \ldots, P_n(s^l(0))$$

These sequents are uniformly derivable from n instances of the proof-fragment Π_i given in Table 3 and $n-1$ $\vee$-left inferences. Note that the quantifiers in $S'(l)$ can be shifted outward by a single cut-inference with the sequent

$$\forall x_1, \ldots, \forall x_n \bigvee_i^n P_i(x_i) \supset P_i(s(x_i)) \to \bigvee_i^n \forall x_i P(x_i) \supset P_i(s(x_i))$$

Hence $S'(l)$ can be uniformly transformed into $S(l)$. This demonstrates that in some cases the cut-reduction necessarily increases the iterations of (tind)-rules. □

The following example shows that for the results above the restriction of (tind) to quantifier-free formulas is necessary. We assume an extension $\mathbf{T}'$ of the system $\mathbf{T}^{(tind)}$ such that in $\mathbf{T}'$ term-induction for Σ_1-formulas is admissible. We introduce the notation $\phi^0(t) = t$; $\phi^{k+1}(t) = \phi(\phi^k(t))$.

Example 2. In $\mathbf{T}'$ the sequent $S(l)$ becomes provable.

$$P(0,0), \forall x, y\ (P(x,y) \supset P(s(x), \phi(y)) \to \exists z\ P(s^l(0), z)$$

Every Herbrand sequent $H(l)$ for fixed l, has the form

$$\ldots, P(s^i(0), \phi^i(0)) \supset P(s^{i+1}(0), \phi^{i+1}(0)), \ldots, P(0,0) \to \ldots, P(s^l(0), \phi^l(0))$$

where $0 \leq i \leq l-1$ are arbitrary. I.e. $\text{complex}^{\{s,0\}}(S(l))$ cannot be bounded uniformly. □

4 Unification and Substitution

To be able to prove the mentioned bound on the term-depth of non-numerals, we apply results from unification theory, especially the theory of *sorted unification.* In this section we introduce the relevant concepts and notations. We assume general familiarity with unification theory. (For general background information on unification theory see e.g. [19,14].)

Let $U \equiv s_1 = t_1, \ldots, s_k = t_k$ be a unification problem. Then we have the following well-known fact [6,14]. (By $\tau(E)$, E some expression, we denote the term-depth of E.) Let $\tau(U) = \max\{\tau(s_i = t_i) : 1 \leq i \leq k\}$; assume the number of variables x in U $\tau(x, U) > 0$ (i.e. occurring at depth at least 1 in U) is n. If σ is the m.g.u. of U, then $\tau(U\sigma) \leq 2^n \tau(U)$. Syntactic unification can be generalised to unification over a *sorted language*, see e.g. [22] for a survey on sorted unification.

We extend $\mathcal{L}$ with a finite set $\mathcal{S}$ of monadic atoms, the *base sorts.* A sort T is a finite set of base sorts which denotes the intersection of the base sorts. For the sort $\emptyset$ we write $\top$, the top sort. We assert a mapping *sort*: $\mathcal{V} \to 2^{\mathcal{S}}$. $T(t)$ is called a *declaration* if $T \in \mathcal{S}$.

The set of *well-sorted* terms $\mathcal{T}_S(\mathcal{K}, \mathcal{V})$, S a sort, are inductively defined as (i) $x \in \mathcal{T}_S(\mathcal{K}, \mathcal{V})$ if $S \subset \text{sort}(x)$. (ii) $t \in \mathcal{T}_S(\mathcal{K}, \mathcal{V})$ if $S(t)$ and S is a base sort. (iii) $t\sigma \in \mathcal{T}_S(\mathcal{K}, \mathcal{V})$ if $t \in \mathcal{T}_S(\mathcal{K}, \mathcal{V})$ and $x\sigma \in \mathcal{T}_{\text{sort}(x)}(\mathcal{K}, \mathcal{V})$, for $x \in \text{var}(t)$. (iv) $t \in \mathcal{T}_{S_1,\ldots,S_n}(\mathcal{K}, \mathcal{V})$ if $t \in \mathcal{T}_{S_i}(\mathcal{K}, \mathcal{V})$ for all i. A unification problem $s_1 = t_1, \ldots, s_k = t_k$ is called *sorted solved* if it is solved syntactically by a unifier σ and for each $x \in \text{dom}(\sigma)$: $x\sigma \in \mathcal{T}_{\text{sort}(x)}(\mathcal{K}, \mathcal{V})$. σ is then called *sorted unifier.*

Note that in general sorted unification is undecidable [22]. However, we are only interested in particular simple sort theories, called *order sorted* sort theories: All declarations are either of the form $S(x_S)$ or $S(f(x_{1,S_1}, \ldots, x_{n,S_n}))$, where all variables are distinct. Sorted unification in order sorted sort theories is of unification type finitary, for a proof see [22].

Lemma 3. *Let $U \equiv s_1 = t_1, \ldots, s_k = t_k$ be a sorted unification problem such that the used sort theory is order sorted, and assume W to be the minimal and complete set of well-sorted unifiers of U. Then, if $\sigma \in W$,*

$$\tau(U\sigma) \leq 2^n \tau(U)$$

where n is the number of variables x in U, such that $\tau(x, U) > 0$.

Proof. The proof is by induction on the employed unification rules. Similar to the standard unification rule-set a correct and complete set of rules for sorted unification can be defined. The key rule is *Weakening*:

$$x = t \wedge U' \longrightarrow x = v \wedge t = v \wedge U$$

where $x \in \mathcal{V}$ and $t \notin \mathcal{V}$ and there exists a declaration $S(v')$ such that $S \in \text{sort}(x)$, v being a renaming of v'. For each stage of the unification procedure we can define a partial substitution σ' that consists of those pairs $x \mapsto t$ where $x = t$ is already sorted solved.

The interesting part is to show that although the number of variables x, $\tau(x, U\sigma') > 0$ can increase through Weakening steps, these newly introduced variables can be ignored. Let $\{x_1, \dots, x_n\}$ denote the set of variables $\{x\colon \tau(x, U) > 0\}$. We need not have that the variable-names in this set is fixed trough out the construction (contrary to 'ordinary' unification). Moreover, Weakening steps may add variables $y_1, \dots, y_k$ distinct from $x_1, \dots, x_n$ to $U\sigma'$ such that $\tau(y_i, U)$ However, only Weakening steps can introduce additional variables y with $\tau(y, U) > 0$ but these variable can occur at most once in $U\sigma'$, according to the definition of the sort theory. Hence we can prove that although the actual set $\{x\colon \tau(x, U) > 0\}$ changes in size, we need only consider renamings of $\{x_1, \dots, x_n\}$. The elimination of any of these variables may double the depth of $U\sigma'$. However this can happen only n times. □

The basic language $\mathcal{L}$ is extended with variables of sort S. S is a base sort, and $\mathcal{T}_S$ is defined through the declarations $S(0)$, $S(s(x_s))$. In the sequel we assume the existence of an infinite set of sorted variables $a_{1,S}, a_{2,S}, \dots$.

5 Term Complexity of Generalisations of Herbrand Sequents

In this section we show that one can define generalisations of Herbrand sequents relative to the complexity of the underlying $\{s, 0\}$-matrices. Let T be a Herbrand sequent of S. We construct a Herbrand sequent T' of S such that the term-depths of non-numerals in T' is bound in terms of the $\{s, 0\}$-complexity of S. To simplify our argumentation we revise the notation for the end-sequent S slightly.

$$\begin{aligned} &\forall \bar{y}_1 A_1(\bar{y}_1; \lambda_1, \dots, \lambda_n), \dots, \forall \bar{y}_a A_a(\bar{y}_a; \lambda_1, \dots, \lambda_n) \to \\ &\qquad \exists \bar{x}_1 B(\bar{x}_1; \lambda_1, \dots, \lambda_n), \dots, \exists \bar{x}_b B(\bar{x}_b; \lambda_1, \dots, \lambda_n) \end{aligned}$$

where the $\bar{y}_1, \dots \bar{y}_a; \bar{x}_1, \dots \bar{x}_b$ denote variables bound by (weak) quantifiers in S. The parameters occurring in S are indicated by $\bar{\lambda} \equiv \lambda_1, \dots, \lambda_n$. Let T be a Herbrand sequent of S written as

$$\begin{aligned} &A_1(s_1^1, \dots, s_{n1}^1; \bar{\lambda}), \dots, A_a(s_1^{va}, \dots, s_{na}^{va}; \bar{\lambda}) \to \\ &\qquad \to B_1(t_1^1, \dots, t_{m1}^1; \bar{\lambda}), \dots, B_b(t_1^{wb}, \dots, t_{mb}^{wb}; \bar{\lambda}) \end{aligned}$$

One can define a *generalisation* T' of T such that the depth of terms in T' is bound in the length of T and the term complexity of the sequent S. We define an *abstraction* H *of* T

$$A_1(a_1^1,\dots,a_{n1}^1;\bar{c}),\dots,A_a(a_1^{va},\dots,a_{na}^{va};\bar{c}) \rightarrow \\ \rightarrow B_1(b_1^1,\dots,b_{m1}^1;\bar{c}),\dots,B_b(b_1^{wb},\dots,b_{mb}^{wb};\bar{c})$$

using free variables $a_j^i, b_{j'}^{i'}, c_i$, respectively. This abstraction naturally induces an unification problem U: Let $P(u_1,\dots,u_n)$, $P(u_1',\dots,u_n')$ be atomic formulas in H. Assume instances $R^1 \equiv P(u_1,\dots,u_n)\delta$, $R^2 \equiv P(u_1',\dots,u_n')\delta$ in T such that $R^1 \equiv R^2$. Then add the equations

$$u_1 = u_1',\dots,u_n = u_n'$$

to U. Clearly this unification problem is solvable with a m.g.u. σ sucht that $H\sigma\delta \equiv T$ for some substitution δ. It follows by definition of U that $H\sigma$ is valid iff T is valid. Moreover $\tau(H\sigma) \leq 2^l\tau(H)$, where l denotes the number of variables in U.

Note that in our case we cannot directly apply this reasoning as the number of variables in U depends on the length of T. However, this length cannot be bound uniformly, cf. Section 2.

Analoguously to the abstraction of T we define the *abstraction of a* $\{s,0\}$-*matrix*. Let N be a $\{s,0\}$-matrix of T. It is useful to denote N as

$$A_1(v_1^1,\dots,v_n^1;\bar{\lambda}),\dots,A_a(v_1^r,\dots,v_n^r;\bar{\lambda}) \rightarrow \\ \rightarrow B_1(w_1^1,\dots,w_m^1;\bar{\lambda}),\dots,B_b(w_1^{r'},\dots,w_m^{r'};\bar{\lambda})$$

W.l.o.g. we may assume that the variables abstracting numerals in N are $\bar{a} \equiv a_1,\dots,a_p$, and $\bar{b} \equiv b_1,\dots,b_q$, respectively. In each s-formula of N we replace occurrences of variables $a_i(b_j)$ that occur as a maximal terms by a new sorted variables x_S. If thus all variables $a_i(b_i')$ are replaced, replace all remaining terms $v_{ij}^z(w_{i'j'}^{z'})$ by free variables $a_{ijz}(b_{i'j'z'})$ of sort $\top$. Otherwise at least one of the $a_i(b_{i'})$ occurs in some $v_{ij}^z(w_{i'j'}^{z'})$. The number of function symbols in this term, e.g. v_{ij}^z, is finite. The construction is by induction on $\tau(v_{ij}^z)$; we concentrate on the step case. Assume $v_{ij}^z = f(t_1,\dots,t_n)$. We assume sort declarations $A_1,\dots,A_n$ for the argument terms t_i including one of the $\bar{a}$. Let $\langle i_1,\dots,i_n\rangle$ denote their indices. We add two declarations $A(f(x_{\top,1},x_{A^\star,i_1},\dots,x_{A^\star,i_n},x_{\top,n}))$; $A(f(x_{\top,1},x_{S,i_1},\dots,x_{S,i_n},x_{\top,n}))$ where for each i_j $A^\star$ denotes the appropriate sort declarations. This is repeated till all occurrences of maximal terms in N containing sorted variables are replaced. All remaining occurrences of maximal terms v_{ij}^k are replaced by variables a_{ijk} of sort $\top$.

Let $\bar{c}$ be tuples of free variables, and let $S^\star$ denote the sequent S where the parameters are replaced by $\bar{c}$, respectively. The *size* of a sequent S, size(S), is the number of symbols in S.

Theorem 3. *Let T be a Herbrand sequent of S written as*

$$\bigwedge_{\bar{p}} C_1(s^{i_1}(0),\dots,s^{i_p}(0);\bar{\lambda}),\dots,\bigwedge_{\bar{p}} C_c(s^{i_1}(0),\dots,s^{i_p}(0);\bar{\lambda}) \rightarrow \\ \rightarrow \bigvee_{\bar{q}} D_1(s^{j_1}(0),\dots,s^{j_q}(0);\bar{\lambda}),\dots,\bigvee_{\bar{q}} D_d(s^{j_1}(0),\dots,s^{j_q}(0);\bar{\lambda})$$

where $\overline{p} \equiv (i_1, \ldots, i_p)$; $\overline{q}_k \equiv (j_1, \ldots, j_q)$. *Then there exists a generalisation* T' *of* T *with* $\{s, 0\}$*-matrix* G $(p' \leq p;\ q' \leq q)$

$$C_1(a_{1,S}, \ldots, a_{p',S}; \lambda'_1, \ldots, \lambda'_n), \ldots, C_c(a_{1,S}, \ldots, a_{p',S}; \lambda'_1, \ldots, \lambda'_n) \rightarrow$$
$$\rightarrow D_1(b_{1,S}, \ldots, b_{q',S}; \mu'_1, \ldots, \mu'_m), \ldots, D_d(b_{1,S}, \ldots, b_{q',S}; \mu'_1, \ldots, \mu'_m)$$

such that

$$\tau(G) \leq 2^{\text{complex}^{\{s,0\}}(T) \cdot \text{size}(S^\star)} \cdot \tau(S^\star)$$

Proof. We construct an abstraction H of the Herbrand sequent T together with the induced unification problem U. Let N be an $\{s, 0\}$-matrix of T. As above we construct an abstraction M of N.

We transform U to a sorted unification problem U_S: For each equation $u = u'$ in U there exists a pair $\langle P(u_1, \ldots, u_n), P(u'_1, \ldots, u'_n) \rangle$ in H such that $u \equiv u_i$ and $u' \equiv u'_i$ for some i and there exists equal instances R^1, R^2 in T. By definition we find abstractions $P(v_1, \ldots, v_n), P(v'_1, \ldots, v'_n)$ of R^1, R^2 in N. By construction, in the abstraction M there exists term tuples

$$w_1, \ldots, w_n \quad w'_1, \ldots, w'_n$$

abstracting v_j and v'_j, $j = 1, \ldots, n$ respectively. Hence the equation $u_i = u'_i$ in U can be represented by $w_i = w'_i$ in U_S.

Clearly the number of variables in U_S depends only on $|\,N\,| \leq \text{complex}^{\{s,0\}}(T)$ and the number of argument positions in S ($\leq \text{size}(S^\star)$). Now we can apply sorted unification. However we must not substitute for variables of sort S, hence we alter the unification procedure slightly. We change the definition of a (partial) substitution σ' wrt. U_S: Assume $U_S \equiv x_s = s^l(y_s) \wedge U'_S$ for some l. Then the equation $x_s = s^l(y_s)$ is ignored in the definition of σ'. (It is straight-forward to check that this does not affect the validity of Lemma 3.)

The unification problem U_S has a solution σ, as T codifies a solution of U. Applying Lemma 3, the maximal term depth in $M\sigma$ is bound by $2^{(c \cdot s)}\tau(S^\star)$, where $c = \text{complex}^{\{s,0\}}(T)$, $s = \text{size}(S^\star)$. The theorem follows by setting $G \stackrel{\text{def}}{=} M\sigma$. □

6 Generalisation of Terms in $\mathbf{T}^{(tind)}$

We obtain a generalisation of large terms in short $\mathbf{T}^{(tind)}$-proofs as consequence of the last sections if these terms are large wrt. to other function symbols than the successor s.

Let t be term, the we define the *reduct* of t informally as the term t° obtained from t by replacing all numerals in t by fresh free variables.

Definition 5. *Let* $S(a)$ *be a sequent in prenex normal form, containing weak quantifiers only; let* N *be a* $\{s, 0\}$*-matrix. A* term basis $T(S(a), N)$ *is a set of terms such that*

1. *For $t \in T(S(a), N)$ there exists a Herbrand sequent T of $S(t)$ with $\{s, 0\}$-matrix N.*
2. *If there exists a Herbrand sequent T of $S(t')$ with $\{s, 0\}$-matrix N then $t' \equiv t\rho$ for some $t \in T(S(a), N)$.*
3. *There exists a h such that if $t \in T(S(a), N)$ then $\tau(t^\circ) \leq h$.*

Theorem 4. *$T(S(a), N)$ exists for any sequent in prenex normal form S with parameter a and any $\{s, 0\}$-matrix N.*

Proof. Collect the generalisations λ' of the parameters λ in the construction of the generalisation of a Herbrand sequent of S, cf. Theorem 3.

Theorem 4 allows some insight into the proof complexity relation between term induction to 'full' induction w.r.t. quantifier-free formulas, where both are conservative over elementary arithmetic.

Theorem 5 (cf. [23]). *Using two instances of the following restricted scheme of identity*

$$t = 0 \supset g(t) = g(0) \tag{5}$$

we can uniformly derive $0^k \stackrel{\text{def}}{=} \overbrace{0 + (0 + \cdots (0 + 0))}^{k \text{ times}} = 0$, from (i) $0 + 0 = 0$, (ii) $\forall x, y, z\ x = y \wedge y = z \supset x = z$, and (iii) $\forall x, y, z\ x + y = y \supset x = 0$.

Proof. Let $r_1(0 + 0) \equiv 0^n + (0^{n-1} + \cdots + (0^2 + 0) \ldots)$ where $0 + 0$ is fully indicated. Let $r_2[0 + 0] \equiv 0^{n-1} + (0^{n-2} + \cdots + (0^2 + (0 + 0)) \ldots)$, where $0 + 0$ in $r_2[0 + 0]$ refers only to the innermost occurring term $0 + 0$. If we employ the the instances of (5) $0 + 0 = 0 \supset r_1(0 + 0) = r_1(0)$ and $0 + 0 = 0 \supset r_2[0 + 0] = r_2[0]$, together with an instance of the transitivity axiom (ii) and instances of axiom (i), then $r_1(0 + 0) = r_2[0]$ is easily derivable. However, $r_1(0 + 0) \equiv 0^n + r_2[0]$, and therefore $r_1(0 + 0) = r_2[0]$ is nothing else than $0^n + r_2[0] = r_2[0]$. Hence one final application of axiom (iii) renders the result of the theorem. □

Proposition 3. *Using full induction we can derive the implication (5) uniformly from (i) $\forall x\ s(x) \neq 0$ and $\forall x\ x = x$.*

This follows from the formal proof given in Table 4 together with a cut with the sequent $\rightarrow 0 = 0 \supset g(0) = g(0)$, which is easily derivable by axiom (ii).

However, in $\mathbf{T}^{(tind)}$ we obtain the following:

Proposition 4. *Assume $\forall x\ 0 + x = x$ and axioms of identity present in Γ then*

$$\exists k \forall n \mathbf{T}^{(tind)} \vdash_k \Gamma \rightarrow \Delta, A(0^n) \textit{ iff } \mathbf{T}^{(tind)} \vdash \Gamma \rightarrow \Delta, \forall x A(x)$$

Proof. Let $S(n)$ denote $\Gamma \rightarrow \Delta, A(0^n)$. Using Theorem 2 there exists a Herbrand disjunction H of S with matrix N. Given short proofs of $S(n)$ for all n, we can apply Theorem 4 to conclude the existence of a term $0^h + b$; b some free variable, in $T(S(a), N)$ such that there exists a Herbrand sequent $T(0^h + b)$ of $S(0^h + b)$. Using the validity of $T(0^h + b)$ we find a proof of $S(0^h + b)$ in **LK**, hence $\mathbf{T}^{(tind)} \vdash \Gamma \rightarrow \Delta, \forall x A(0^h + x)$. Applying $\forall x\ 0 + x = x$ the result follows easily □

Table 4. The restricted identity scheme

$$\dfrac{\to s(a) \neq 0 \quad \dfrac{s(a) \neq 0 \to s(a) \neq 0}{s(a) \neq 0, a = 0 \supset g(a) = g(0) \to s(a) = 0 \supset g(s(a)) = g(0)}}{\dfrac{a = 0 \supset g(a) = g(0) \to s(a) = 0 \supset g(s(a)) = g(0)}{0 = 0 \supset g(0) = g(0) \to t = 0 \supset g(t) = g(0)}}$$

Other properties of full induction, e.g. fast addition see [17] prevail for $\mathbf{T}^{(tind)}$.

Proposition 5. *Assume* $\forall x, y\ x = y \supset s(x) = s(y), \forall x\ 0 + x = x, \forall x, y\ s(x + y) = x + s(y) \subset \Gamma$. *Then there exists* $k \in \mathbb{N}$ *so that*

$$\mathbf{T}^{(tind)} \vdash_k \Gamma \to s^n(0) + s^m(0) = s^{n+m}(0)$$

for arbitrary natural numbers n, m.

Kreisel's conjecture, which fails for $\mathbf{T}^{(tind)}$ holds if sufficiently reformulated. We abbreviate $s(0)$ by 1 and define $1^k \stackrel{\text{def}}{=} 1 + (1 \cdots (1+1))^{k \text{ times}}$. The proposition follows similarly as Proposition 4.

Proposition 6. *Assume that* Γ *contains axioms sufficiently strong to derive* $\mathbf{T}^{(tind)} \vdash \Gamma \to \forall x \bigvee_{i=0}^{n} 1^i = x \lor \exists y\ y + 1^{n+1} = x$. *Then*

$$\exists k \forall n \mathbf{T}^{(tind)} \vdash_k \Gamma \to \Delta, A(1^n) \text{ iff } \mathbf{T}^{(tind)} \vdash \Gamma \to \Delta, \forall x A(x)$$

7 Conclusion

We conceive *mathematical induction* as a principle that states two different things. Firstly (i) a specific generation of terms is stipulated, secondly (ii) a – so-called *closed world* – assumption is imposed that these terms refer to *all* natural numbers.

Term induction formulates the first aspect, and therefore an analysis allows a separate assessment of the different characteristics of mathematical induction. In particular note that (variants of) term induction were already employed in mathematical proofs long before the concept of *mathematical induction* was introduced, c.f [11]. On the other hand note that term induction need not be restricted to (a restriction of) successor induction, it can be generalised to arbitrary *constructor-style* [8] term induction. Furthermore, the main results presented in this paper remain valid wrt. to this extension.

Term induction might therefore provide additional insight in the foundation of mathematical induction and its application to computer science. In particular, further work will be dedicated towards applications in the area of *inductive theorem proving*, see e.g. [7,21,8] and in the area of *automated analysis of proofs*, cf. [3,4,5].

References

1. M. Baaz. Über den allgemeinen Gehalt von Beweisen. In *Contributions to General Algebra 6*, pages 21–29. Hölder-Pichler-Tempsky, Teubner, 1988.
2. M. Baaz and A. Leitsch. On skolemization and proof complexity. *Fund. Informaticae*, 20(4):353–379, 1994.
3. M. Baaz and A. Leitsch. Cut normal forms and proof complexity. *Annals of Pure and Applied Logic*, pages 127–177, 1997.
4. M. Baaz and A. Leitsch. Cut elimination by resolution. *J. Symbolic Computation*, 1999.
5. M. Baaz, A. Leitsch, and G. Moser. System Description: `CutRes 0.1`: Cut Elimination by Resolution. 1999.
6. M .Baaz and R. Zach. Generalizing theorems in real closed fields. *Ann. of Pure and Applied Logics*, 75:3–23, 1995.
7. R.S. Boyer and J.S. Moore. *A Computational Logic Handbook*. Academia Press, 1988.
8. A. Bundy. *The Automation of Proof By Mathematical Induction*. Elsevier Science Publisher, 2001. To appear.
9. S. R. Buss. An introduction to proof theory. In S. R. Buss, editor, *Handbook of Proof Theory*, pages 1–79. Elsevier Science, 1998.
10. W.M. Farmer. A unification-theoretic method for investigating the k-provability problem. *ANAP*, pages 173–214, 1991.
11. H. Gericke. *Mathematik in Antike und Orient*. Springer Verlag, 1984.
12. D. Hilbert and P. Bernays. *Grundlagen der Mathematik 2*. Spinger Verlag, 1970.
13. J. Krajíček and P. Pudlák. The number of proof lines and the size of proofs in first-order logic. *Arch. Math. Logic*, 27:69–84, 1988.
14. A. Leitsch. *The Resolution Calculus*. EATCS - Texts in Theoretical Computer Science. Springer, 1997.
15. R.J. Parikh. Some results on the length of proofs. *Trans. Amer. Math. Soc.*, pages 29–36, 1973.
16. P. Pudlak. The lengths of proofs. In S. Buss, editor, *Handbook of Proof Theory*, pages 547–639. Elsevier, 1998.
17. D. Richardson. Sets of theorems with short proofs. *J. Symbolic Logic*, 39(2):235–242, 1974.
18. H. Schwichtenberg. *Some applications of cut-elimination*, pages 867–897. North Holland, 5th edition, 1989.
19. J. H. Siekmann. *Unification Theory*, pages 1–68. Academic Press, 1990.
20. G. Takeuti. *Proof Theory*. North-Holland, Amsterdam, 2nd edition, 1980.
21. C. Walther. *Mathematical induction*, volume 12, pages 122–227. Oxford University Press, 1994.
22. C. Weidenbach. Sorted unification and tree automata. In Wolfgang Bibel and Peter H. Schmitt, editors, *Automated Deduction - A Basis for Applications*, volume 1 of *Applied Logic*, chapter 9, pages 291–320. Kluwer, 1998.
23. T. Yukami. Some results on speed-up. *Ann. Japan Assoc. Philos. Sci.*, 6:195–205, 1984.

Well-Founded Recursive Relations

Jean Goubault-Larrecq

LSV, ENS Cachan, 61 av. du président-Wilson, F-94235 Cachan Cedex, France
Phone: +33-1 47 40 24 30   Fax: +33-1 47 40 24 64
goubault@lsv.ens-cachan.fr

Abstract. We give a short constructive proof of the fact that certain binary relations $>$ are well-founded, given a lifting $\gg$ à la Ferreira-Zantema and a well-founded relation $\rhd$. This construction generalizes several variants of the recursive path ordering on terms and of the Knuth-Bendix ordering. It also applies to other domains, of graphs, of infinite terms, of word and tree automata notably. We then extend this construction further; the resulting family of well-founded relations generalizes Jouannaud and Rubio's higher-order recursive path orderings.

Keywords: Termination, well-foundedness, path orderings, Knuth-Bendix orderings, λ-calculus, higher-order path orderings, graphs, automata.

1 Introduction

The use of well-founded orderings is a well-established technique to show that term rewrite systems terminate [4]. On the other hand, the tradition in λ-calculus circles, exemplified by the Tait-Girard technique [10], is to show termination by structural induction on terms, backed by auxiliary well-founded inductions. In fact, the recursive path ordering can be proved terminating by structural induction on terms, as noticed in [14].

Prompted by [12], we wrote a direct inductive proof of the termination of the recursive path ordering $\succ_{rpo}$ based on a well-founded precedence $\succ$ [4], which turned out to be surprisingly short. Our point is that this proof generalizes considerably, while remaining short and constructive, and still using only elementary principles of logic. Chasing generalizations and simplifications, we arrived at Theorem 1, which is the core of this paper. We state it and prove it in Section 2. Here the use of the Coq proof assistant [1] helped us in reassuring ourselves that the proof was indeed correct, but more importantly helped us delineate useful generalizations and useless assumptions.

The rest of the paper examines applications and extensions of Theorem 1. First, we shall see in Section 3 that it subsumes Ferreira and Zantema's Theorem [8], and therefore most path orderings of the literature.

Theorem 1 is more general still, if only because it does not depend on term structure. That is, this construction works equally well on other kinds of algebras. We illustrate this by sketching a few well-founded relations resembling the recursive path ordering on graphs, on infinite terms, and on word and tree automata in Section 4.

L. Fribourg (Ed.): CSL 2001, LNCS 2142, pp. 484–498, 2001.

Theorem 1 seems however helpless in establishing that simply-typed λ-terms terminate. We extend Theorem 1 by augmenting its proof with two ingredients that are crucial in the classical Tait-Girard proof of strong normalization [10], and come up with a suitable generalization of Jouannaud and Rubio's higher-order recursive path ordering (horpo) [13] in Section 5, Theorem 2. This provides for variants of the horpo that also generalize semantic path orderings and the general path ordering, just like Theorem 1 did in the first-order case. (Again, we checked the proof in Coq.)

We conclude in Section 6 by pondering over yet unexploited features of Theorem 1 and Theorem 2, and possible generalizations.

Related Work. Proving the termination of term rewrite systems by structural induction on terms is not new. This is classical in the λ-calculus. Similar techniques are extensively used in [7], where several proofs of termination consist in showing that for every substitution σ mapping variables to terminating terms, the term $t\sigma$ terminates by induction on some well-founded measure on t and σ. Jouannaud and Rubio [14] also notice that recursive path orderings can be shown well-founded by the same technique. However, our proof of Theorem 1 is in fact simpler: it does not consider substitutions, and proceeds directly on t. Naturally, Theorem 1 does not consider the higher-order case. Theorem 2 does, and this requires both the use of substitutions as above and replacing strong normalization by the stronger, and more complex notion of reducibility [10], a.k.a. computabiliy [13]. We shall demystify the latter notion in Section 5: reducibility is just ordinary strong normalization, albeit of a richer reduction relation.

On first-order term algebras, the closest result to our first theorem (Theorem 1) is Ferreira and Zantema's Theorem [8], which is almost Theorem 1 specialized to the case of terms. We shall indeed rederive the latter from ours. Theorem 2 can then be seen as the higher-order, abstract version of Ferreira and Zantema's Theorem.

We stress the fact that our results are in no way tied to term structure, and that this allows to design syntactic well-founded path orderings on graphs, infinite terms and automata. We are not aware of any previous termination method resembling recursive path orderings on algebras other than terms.

Finally, Paul-André Melliès [16] suggested a deep connection between Kruskal's Theorem and the termination of recursive path orderings, including Ferreira and Zantema's Theorem. We have not managed to reverse the duality and deduce Kruskal's Theorem from Theorem 1: the problem is that Melliès' definition of a well-founded relation R is different from ours, namely that in any sequence $(s_i)_{i \in \mathbb{N}}$ there is $i < j$ such that not $s_i \; R \; s_j$. Furthermore, this difference seems essential.

2 The First Termination Theorem and Its Proof

Let T be any set, and $\rhd$, $>$ and $\gg$ be three binary relations on T. For short, we write $u \lhd t$ for $t \rhd u$, and $\geq$ for the reflexive closure of $>$. Assume that $>$ has the following property:

Property 1. For every $s, t \in T$, if $s > t$ then either:

(*i*) for some $u \in T$, $s \rhd u$ and $u \geq t$;
(*ii*) or $s \gg t$ and for every $u \lhd t$, $s > u$.

Remark 1. Condition (*ii*) resembles Ferreira and Zantema's [8] condition that $\gg$ be a *term lifting*. We shall see the precise connection in Section 3, Corollary 1.

Remark 2. Although the notation suggests it, none of $\rhd$, $>$, $\gg$ are required to be transitive. Taking them to be orderings, as is standard in the literature, is an orthogonal issue to termination.

Remark 3. A more general definition would be to take $\geq$ as primitive and define $>$ by $s > t$ if and only if $s \geq t$ and $t \not\geq^* s$, where $\geq^*$ is the reflexive transitive closure of $\geq$. We feel that this would only make the presentation more clumsy, while the additional generality can be obtained by reasoning over $T/(\geq^* \cap \leq^*)$.

Remark 4. A general way of obtaining $>$ satisfying Property 1, which we shall use in the sequel, is as follows. Let $\rhd$ be given, and let $R \mapsto \gg_R$ be any monotonic function from binary relations to binary relations. By *monotonic*, we mean that $R \subseteq R'$ implies $\gg_R \subseteq \gg_{R'}$. Let $>$ be the greatest binary relation on T such that Property 1 holds, where $\gg$ abbreviates $\gg_>$. This is well-defined by Tarski's Fixpoint Theorem on the complete lattice of binary relations over T. (In fact, $>$ is then the greatest binary relation such that $s > t$ *if and only if* (*i*) or (*ii*).)

Let us talk about termination. An element $s \in T$ is *accessible* in the binary relation R, a.k.a., s is in the *well-founded part* of R, if and only if every decreasing sequence $s_0 \hat{=} s \; R \; s_1 \; R \; s_2 \; R \; \ldots \; R \; s_k \; R \; \ldots$ starting from s is finite. The relation R is *well-founded*, or *terminating* (over T) if and only if every $s \in T$ is accessible in R.

We shall use a slightly more general notion than accessibility. Let S be any subset of T. Say that S *bars* s in R if and only if every infinite sequence $s_0 \hat{=} s \; R \; s_1 \; R \; s_2 \; R \; \ldots \; R \; s_k \; R \; \ldots$ starting from s meets S, that is, $s_k \in S$ for some $k \geq 0$. The proper, constructive characterization of bars is the following principle, due to Brouwer:

Proposition 1 (Bar induction). *For every property P on T, if:*

1. *every $s \in S$ satisfies $P(s)$,*
2. *and for every $s \in T$, if $P(t)$ for every t such that $s \; R \; t$, then $P(s)$,*

then every $s \in T$ barred by S in R satisfies $P(s)$.

In fact, the set B of all terms s barred by S in R is *defined* as the smallest such that $P \hat{=} \lambda s \cdot s \in B$ satisfies 1 and 2 above.

Remark 5. An element s is accessible in R if and only if the empty set bars s in R.

Our first theorem is the following:

Theorem 1. *Let SN be the set of all $s \in T$ that are accessible in $>$, and $\underline{SN}$ be the set of all s such that, if every $u \lhd s$ is in SN, then s is in SN. Assume that the following conditions hold:*

(iii) $\rhd$ is well-founded on T;
(iv) for every $s \in T$, if every $u \lhd s$ is in SN, then $\underline{SN}$ bars s in $\gg$.

Then $>$ is well-founded on T. Equivalently, $SN = T$.

Proof. We are going to prove this in excruciating detail: the proof is short but subtle.

First observe that SN satisfies the following properties:

$$\text{for every } s \in SN \text{, if } s > t \text{ then } t \in SN \tag{1}$$
$$\text{for every } s \in T \text{, if every } t \text{ such that } s > t \text{ is in } SN \text{, then } s \in SN \tag{2}$$

We shall show that every $s \in T$ is in SN, by well-founded induction on $\rhd$, using (*iii*). I.e., let our first induction hypothesis be:

$$\text{For every } u \lhd s,\ u \in SN \tag{3}$$

and let us show that $s \in SN$. To show the latter, it is enough to show $s \in \underline{SN}$, that is, if every $u \lhd s$ is in SN then $s \in SN$. Indeed $s \in SN$ will follow easily, using (3).

Now let us show $s \in \underline{SN}$ under assumption (3). By (3) and (*iv*), $\underline{SN}$ bars s in $\gg$, so we may prove $s \in \underline{SN}$ by bar induction. Following Proposition 1, we have:

- (Base case) $s \in \underline{SN}$: obvious.
- (Inductive step) Assume that for every t such that $s \gg t$, $t \in \underline{SN}$ holds, and prove $s \in \underline{SN}$. Expanding the definition of $\underline{SN}$, we must show that $s \in SN$ under the assumptions:

$$\text{For every } t \text{ such that } s \gg t \text{ and such that every } u \lhd t \text{ is in } SN,\ t \in SN \tag{4}$$
$$\text{For every } u \lhd s,\ u \in SN \tag{5}$$

Using (2), it is enough to show that whenever $s > t$, $t \in SN$. We show this by well-founded induction on t ordered by $\rhd$ – which is well-founded by (*iii*). Our new induction hypothesis is therefore:

$$\text{For every } u \lhd t \text{, if } s > u \text{ then } u \in SN \tag{6}$$

Now since $s > t$, use Property 1, which yields two cases:
(*i*) For some $u \lhd s$, $u \geq t$. By (5), $u \in SN$. By (1), $t \in SN$.
(*ii*) Or $s \gg t$ and for every $u \lhd t$, $s > u$. By (6) every such u is in SN. To sum up, $s \gg t$ and for every $u \lhd t$, $u \in SN$. By (4) $t \in SN$. □

Remark 6. Condition (*iv*) is a bit hard to apply. Defining $\overline{SN}$ as the set of all $s \in T$ such that every $u \lhd s$ is in SN, a less general but simpler condition is:

(*v*) for every $s \in T$, if every $u \lhd s$ is in SN, then s is accessible in $\gg_{|\overline{SN}}$.

where $R_{|A}$ denotes R restricted to A. Indeed (*v*) implies (*iv*): assume that s is such that every $u \lhd s$ is in SN, and show that s is barred by $\underline{SN}$ in $\gg$ by induction on $\gg_{|\overline{SN}}$, which is legal by (*v*). It is enough to show that every t such that $s \gg t$ is barred by $\underline{SN}$ in $\gg$. If $t \in \overline{SN}$, apply the induction hypothesis. Otherwise $t \notin \overline{SN}$ clearly implies $t \in \underline{SN}$. (Note that this argument is *not* constructive.)

3 Path Orderings

In this section, let T be the set of all first-order terms on a given signature and a given set of variables. Let $\rhd$ denote the *immediate superterm relation*, defined as the smallest such that $f(s_1, \ldots, s_m) \rhd s_i$ for each n-ary function symbol f and every i, $1 \leq i \leq m$.

Path Orderings. Path orderings are obtained by setting $\lhd \hat{=} \lhd$. Condition (*iii*) is then satisfied. The relation $>$ obtained by Remark 4 is then such that $s > t$ if and only if either:

(*i*) $s = f(s_1, \ldots, s_m)$ and $s_i \geq t$ for some i, $1 \leq i \leq m$;
(*ii*) or $s \gg_> t$ and either t is a variable, or $t = g(t_1, \ldots, t_n)$ and $s > t_j$ for all j, $1 \leq j \leq n$.

In this form, $>$ starts looking more like the recursive path ordering and its variants. And indeed, Theorem 1 has the following corollaries:

- Dershowitz' original *recursive path ordering* [3] on the well-founded precedence $\succ$ is well-founded : define $s \gg_R t$ if and only if s is of the form $f(s_1, \ldots, s_m)$, t is of the form $g(t_1, \ldots, t_n)$, and either $f \succ g$ or $f = g$ and $\{s_1, \ldots, s_m\}\ R_{mul}\ \{t_1, \ldots, t_n\}$;
- similarly Kamin and Lévy's *lexicographic path ordering* [15] : $s \gg_R t$ if and only if $s = f(s_1, \ldots, s_m)$, $t = g(t_1, \ldots, t_n)$, and either $f \succ g$, or $f = g$, $m = n$, and $(s_1, \ldots, s_n)\ R_{lex}\ (t_1, \ldots, t_n)$ (meaning $s_1 = t_1$, ..., $s_{k-1} = t_{k-1}$ and $s_k\ R\ t_k$ for some k, $1 \leq k \leq n$);
- The recursive path ordering with status (easy exercise);
- Plaisted's *semantic path ordering* [4] : given a well-founded quasi-ordering $\succeq$ on terms with strict part $\succ$ and equivalence $\approx$, $s \gg_R t$ if and only if either $s \succ t$, or $s \approx t$ and $\{s_1, \ldots, s_m\}\ R_{mul}\ \{t_1, \ldots, t_n\}$ (recall that the *strict part* of a preordering $\succeq$ is $\succ \hat{=} \succeq \setminus \preceq$, while its *equivalence* is $\approx \hat{=} \succeq \cap \preceq$);
- Dershowitz and Hoot's *general path ordering* [5] (easy justification omitted).

One way to show at once that these orderings are well-founded is to show that Ferreira and Zantema's result [8] is a consequence of Theorem 1. We silently assume here that all terms are ground, each variable x being considered as a nullary function symbol.

Corollary 1 (Ferreira-Zantema). *Let $>$ be a partial order on a set of first-order terms. A* term lifting $>^{\Lambda}$ *is a strict ordering on terms such that for each set A of terms, if $>_{|A}$ is well-founded, then $>^{\Lambda}_{|\overline{A}}$ is well-founded, where $\overline{A}$ is the set of terms $f(s_1, \ldots, s_m)$ where $s_i \in A$ for all i, $1 \leq i \leq m$.*

Assume that $>$ has the subterm property (i.e., $s \rhd t$ implies $s > t$). Also, assume that $s \hat{=} f(s_1, \ldots, s_m) > t$ implies either that $s_i \geq t$ for some i, $1 \leq i \leq m$, or $s >^{\Lambda} t$.

Then $>$ is well-founded.

Proof. Define $\gg$ as $>^{\Lambda}$, $>$ and $\rhd$ are already defined.

We must first show that $s > t$ implies (i) or (ii) (see Property 1). So assume $s > t$. By assumption, there are two cases. Case 1: $s_i \geq t$ for some i; then (i) holds with $u \hat{=} s_i$. Case 2: $s >^{\Lambda} t$, so $s \gg t$. In this case every $u \lhd t$ is of the form t_j, $1 \leq j \leq n$; since $>$ has the subterm property, $t > t_j$; since $>$ is transitive and $s > t$, it obtains $s > t_j$, therefore (ii) holds.

Condition (iii) is trivial. Let us show (iv), or rather (v) (Remark 6): let s be such that every $u \lhd s$ is in SN, i.e., $s \in \overline{SN}$. Let A be SN. Then $\overline{A}$ is $\overline{SN}$: by assumption $>^{\Lambda}_{|\overline{A}}$, i.e., $\gg_{|\overline{SN}}$ is well-founded. So s is accessible in $\gg_{|\overline{SN}}$. Then apply Theorem 1. □

Knuth-Bendix Orderings. Theorem 1 also generalizes Dershowitz' version of the Knuth-Bendix ordering [4]. Let $\succeq_{\mathrm{T}}$ and $\succeq_{\mathrm{F}}$ be two preorderings, respectively on terms and on function symbols, with strict parts $\succ_{\mathrm{T}}$ and $\succ_{\mathrm{F}}$, and equivalences $\approx_{\mathrm{T}}$ and $\approx_{\mathrm{F}}$. Choose $\rhd \hat{=} \emptyset$ and let $s \gg_R t$ if and only if (a) $s \succ_{\mathrm{T}} t$, or (b) $s \approx_{\mathrm{T}} t$, $s = f(s_1, \ldots, s_m)$, $t = g(t_1, \ldots, t_n)$, and either $f \succ_{\mathrm{F}} g$ or $f \approx_{\mathrm{F}} g$, $m = n$ and $(s_1, \ldots, s_m)\ R_{lex}\ (t_1, \ldots, t_n)$. Remark 4 builds a binary relation $>$: this is the Knuth-Bendix ordering $\succ_{kbo}$ of [4]. (Note that since $\rhd = \emptyset$, $s > t$ if and only if $s \gg_{>} t$.) Theorem 1 then allows us to retrieve:

Corollary 2. *If $\rhd \subseteq \succ_{\mathrm{T}}$, and $\succ_{\mathrm{T}}$ and $\succ_{\mathrm{F}}$ are well-founded, then $\succ_{kbo}$ is well-founded.*

Proof. Property (iii) is trivial. Let us show (iv), and consider any chain $s = s_0 \gg_{\succ_{kbo}} s_1 \gg_{\succ_{kbo}} \ldots \gg_{\succ_{kbo}} s_k \gg_{\succ_{kbo}} \ldots$ For every $k \geq 0$, for every immediate subterm u of s_k, $u \prec_{\mathrm{T}} s_k$, since $\rhd \subseteq \succ_{\mathrm{T}}$. Since $\gg_{\succ_{kbo}} \subseteq \succeq_{\mathrm{T}}$ and $\succeq_{\mathrm{T}}$ is transitive, $u \prec_{\mathrm{T}} s$, so $u \in SN$ by assumption. In particular, the whole chain is inside $\overline{SN}^{\rhd}$, the set of terms whose immediate subterms all are in SN. Clearly $\gg_{\succ_{kbo}}$ restricted to $\overline{SN}^{\rhd}$ is well-founded, since it is a lexicographic product of $\succeq_{\mathrm{T}}$, $\succeq_{\mathrm{F}}$, and for each function symbol f, of a lexicographic extension of $\succ_{kbo}$ restricted to SN. So the chain is finite, hence s is accessible in $\gg$. This proves (v) (Remark 6). □

Monotonicity, Stability. To show that a rewrite system terminates, it is important that the relation $>$ be *monotonic*, i.e., $s > t$ implies $f(\ldots, s, \ldots) > f(\ldots, t, \ldots)$ – this notation meaning $f(s_1, \ldots, s_{i-1}, s, s_{i+1}, \ldots, s_n) > f(s_1, \ldots, s_{i-1}, t, s_{i+1}, \ldots, s_n)$; and that it be *stable*, i.e., $s > t$ implies $s\sigma > t\sigma$ for every substitution σ. We state conditions under which these properties hold.

Lemma 1. *Let $>$ be defined as in Remark 4. Assume that:*

(vi) whenever $s\,R\,t$, $f(\ldots,s,\ldots) \gg_R f(\ldots,t,\ldots)$;
(vii) whenever $f(\ldots,t,\ldots) \rhd u$, *then* $u = t$ *or* $f(\ldots,s,\ldots) \rhd u$ *for every* $s \in T$.

Then $>$ is monotonic.

Proof. Assume that $s > t$. By (*vi*) $f(\ldots,s,\ldots) \gg_> f(\ldots,t,\ldots)$. It remains to show that for every $u \lhd f(\ldots,t,\ldots)$, $f(\ldots,s,\ldots) > u$. By (*vii*) every such u is such that $f(\ldots,s,\ldots) \rhd u$ – therefore $f(\ldots,s,\ldots) > u$ –, or $t = u$ – then $f(\ldots,s,\ldots) \rhd s > t = u$, so $f(\ldots,s,\ldots) > u$. □

Lemma 2. *Let $>$ be defined as in Remark 4. Recall in particular that $\gg_R$ is monotonic in R. Assume that:*

(viii) $\rhd$ *is stable;*
(ix) $\gg_R$ *is stable whenever R is;*
(x) whenever $s \gg_R x$, *where x is a variable, then* $s \rhd^* x$;
(xi) if $u \lhd t'\sigma$ *and t' is not a variable, then for some* u', $u = u'\sigma$ *and* $u' \lhd t'$.

Then $>$ is stable.

Proof. Let $>'$ be defined by $s >' t$ if and only if for some terms s', t' and for some substitution σ, $s = s'\sigma$, $t = t'\sigma$, and $s' > t'$. We claim that $s >' t$ implies either:

(*i'*) for some $u \lhd s$, $u \geq' t$;
(*ii'*) or $s \gg_{>'} t$ and for every $u \lhd t$, $s >' u$.

Since $> \subseteq >'$ and $>$ is the greatest relation satisfying (*i*) or (*ii*), it will follow that $> = >'$, therefore that $>$ is indeed stable.

So assume $s >' t$, i.e., $s' > t'$ and $s = s'\sigma, t = t'\sigma$. Then either (*i*) $s' \rhd u' \geq t'$, in which case $s'\sigma \rhd u'\sigma$ by (*viii*) and $u'\sigma \geq' t'\sigma$ by definition, so (*i'*) holds; or (*ii*) $s' \gg_> t'$ and for every $u' \lhd t'$, $s' > u'$. In the latter case, since $R \mapsto \gg_R$ is monotonic and $> \subseteq >'$, $s' \gg_{>'} t'$; since $>'$ is stable, by (*ix*) $\gg_{>'}$ is, too, so $s \gg_{>'} t$. On the other hand, we claim that for every $u \lhd t$, $s >' u$; (*ii'*) will follow. Distinguish two cases:

1. If t' is not a variable, by (*xi*) there is a term u' such that $u = u'\sigma$ and $u' \lhd t'$. But by assumption $u' \lhd t'$ implies $s' > u'$, so $s >' u$.
2. If t' is a variable x, by (*x*) $s' \rhd^* x$. By (*viii*) $s = s'\sigma \rhd^* x\sigma = t \rhd u$. So $s \rhd^+ u$. It follows that $s > u$, so trivially $s >' u$. □

Remark 7. Conditions (*vii*), (*viii*) and (*xi*) are automatically verified when $\rhd$ is the empty relation or $\blacktriangleright$, as above. In the recursive and lexicographic cases, as well as the recursive path ordering with status, (*vi*) holds: we retrieve the fact that these orderings are monotonic. In the same cases, (*ix*) and (*x*) hold – the latter because $s \gg_R x$ is always false –, so these orderings are also stable, as is already known.

4 Well-Founded Orderings on Graphs, Infinite Terms, Automata

Theorem 1 does not need to operate on terms. As an application of this remark, let T be the set of all finite edge-labeled rooted graphs – *graphs* for short. Recall that a *graph* G is a 6-tuple $(V, E, \partial_0, \partial_1, F, v_0)$, where V is a finite set of so-called *vertices*, E is a finite set of so-called *edges*, ∂_0 and ∂_1 are functions from E to V – $\partial_0 e$ is the *source* vertex of edge e, $\partial_1 e$ is its *target* –, F is a map from E to *labels* in some fixed set Σ, $v_0 \in V$ is called the *root* of G. We abbreviate the fact that e is an edge with source v_0 and target v_1, and label f, by the notation $e : v_0 \xrightarrow{f} v_1$. We let $root(G)$ be the root of G, and if v is any vertex in V, we let G/v be the graph G with new root v, i.e., $(V, E, \partial_0, \partial_1, F, v)$.

Here are a few natural candidates for the $\rhd$ relation, resembling the immediate superterm relation on terms:

- the *post-edge-erasure* relation $\rhd_+$: if $G = (V, E, \partial_0, \partial_1, F, v_0)$, $G \rhd_+ G'$ if and only if there is an edge $e : v_0 \xrightarrow{f} v_1$ in E such that G' is isomorphic to $(V, E \setminus \{e\}, \partial_0, \partial_1, F, v_1)$ – i.e., we erase e and change the root to the target of e;
- the *pre-edge-erasure* relation $\rhd_-$: with the same notations, $G \rhd_- G'$ if and only if G' is isomorphic to $(V, E \setminus \{e\}, \partial_0, \partial_1, F, v_0)$ – i.e., we erase e and leave the root on the source of e;
- the *edge-collapsing* relation $\rhd_0$: with the same notations again, $G \rhd_0 G'$ if and only if G' is isomorphic to $(V/\sim, E \setminus \{e\}, e \mapsto |\partial_0 e|_\sim, e \mapsto |\partial_1 e|_\sim, F, |v_0|_\sim)$, where $\sim$ is the equivalence relation $v_0 \sim v_1$, and $|v|_\sim$ is the equivalence class of v under $\sim$ – i.e., we equate v_0 and v_1 and remove e;
- the *garbage-collection* relation $\rhd_{gc}$: $G \rhd_{gc} G'$ if and only if G' is obtained from G by removing vertices and edges that are unreachable from $root(G)$.

Any union of any of these relations is well-founded, since all of them decrease the size of graphs, measured as the number of vertices plus the number of edges. Define the reflexive transitive closure $\sqsupseteq_{minor}$ of $\rhd_+ \cup \rhd_- \cup \rhd_0 \cup \rhd_{gc}$. It is natural to say that G' is a *minor* of G whenever $G \sqsupseteq_{minor} G'$ [18], as then G' is obtained by taking a subgraph of some graph obtained from G by collapsing edges.

For every graph G, let $top(G)$ be the multiset of all pairs $(f, G/v_1)$, where $e : root(G) \xrightarrow{f} v_1$ ranges over edges in G with source $root(G)$. We may then define an analogue of the recursive path ordering by the construction of Remark 4 again. Take any precedence $\succ$ on Σ, let $\rhd$ be any union of relations $\rhd_+$, $\rhd_-$, $\rhd_0$, $\rhd_{gc}$, and define $\gg_R$ by, say, $G \gg_R G'$ if and only if $top(G)((\succ, R)_{lex})_{mul} top(G')$. Condition (*iii*) is trivial. Condition (*iv*), in fact (*v*) (Remark 6) also holds. So Theorem 1 allows us to conclude that $>$ is well-founded.

When $\rhd$ is exactly the minor relation $\sqsupseteq_{minor}$, another plausible line of proof to show that $>$ is well-founded would have been by adapting Dershowitz' proof of the termination of the recursive path ordering [3], replacing any use of Kruskal's Theorem by a suitable variant of Robertson and Seymour's Theorem ([18], p. 305). The latter indeed states that (non-rooted, non-labeled) graphs under the embedding-by-minors ordering are well-quasi-ordered. However, the proof of Robertson and Seymour's Theorem is considerably more involved than that of Theorem 1. Moreover, there is no hope of using it to establish that $>$ is well-founded when $\rhd$ is not $\sqsupseteq_{minor}$.

Remark 8. Here Tarski's Fixpoint Theorem is needed in Remark 4. Contrarily to Section 3, there is no unique relation $>$ such that $s > t$ if and only if (i) or (ii) holds, since graphs may contain loops. Choosing the greatest allows us to get a comparison predicate $>$ that makes the most pairs of graphs comparable. We conjecture that simple loop-checking mechanisms will provide an algorithm for deciding $>$.

Remark 9. It is immediate to adapt the above definitions to non-oriented graphs, where edges are just unordered pairs $\{v_0, v_1\}$ of distinct edges.

Remark 10. It is easy to extend the above definitions to ordered *multigraphs*, where labeled edges are rewrite rules of the form $f(v_1, \ldots, v_n) \leftarrow v_0$, where v_0, v_1, $\ldots$, v_n are vertices and the label f is an n-ary function symbol. (This generalizes edges $v_0 \xrightarrow{f} v_1$ when f is unary.) We may let *top*(G), for any multigraph G, be the multiset of all $f(v_1, \ldots, v_n)$ such that $f(v_1, \ldots, v_n) \leftarrow v_0$ is an edge in G with $v_0 = \mathit{root}(G)$, and define $\gg_R$ by comparing such multisets by the multiset extension of the lexicographic product of a precedence $\succ$ on function symbols and R, n times, as in the lexicographic path ordering. (It is of course possible to have function symbols with multiset status, where arguments v_1, $\ldots$, v_n are compared by the multiset extension of R, and in general to use any kind of extraction and termination function as in [5].)

Such multigraphs are exactly non-deterministic finite tree automata with one final state (the root) [9]; in this context the edge $f(v_1, \ldots, v_n) \leftarrow v_0$ is usually written $f(v_1, \ldots, v_n) \rightarrow v_0$. The $>$ relations might have applications in showing that certain sequences of tree automata are ultimately stationary, as needed in building widenings [2] in tree automata-based abstract interpretations such as [17,11].

Remark 11. Multigraphs where, for each vertex v_0, there is exactly one edge of the form $f(v_1, \ldots, v_n) \leftarrow v_0$ are exactly regular infinite terms, as used in Prolog for example. The construction of this section therefore provides well-founded relations that may be of use in extending completion and narrowing-based automated deduction tools to the case of regular infinite terms.

5 Higher-Order Path Orderings

Theorem 1 seems to be insufficient to show that every simply-typed λ-term terminates. While the proof of Theorem 1 proceeds by showing that every $s \in T$ is in SN directly, in the classical strong normalization proof of the simply-typed λ-calculus [10] one shows that for every $s \in T$, $s\sigma$ is *reducible* of its type, where σ is any substitution mapping variables to reducible terms of the correct types, and reducibility is a new property that implies termination. Trying to integrate these notions into Theorem 1, we obtain the following. (For the sake of comparison, we have taken similar numbering conventions as in Section 2, with a $\circ$ superscript; e.g., condition (i) becomes $(i)^\circ$.)

Let T be any set, and $\rhd$, $\blacktriangleright$, $>$ and $\gg$ be four binary relations on T. Assume that:

Property 2. For every $s, t \in T$, if $s > t$ then either:

$(i)^\circ$ for some $u \in T$, $s \blacktriangleright u$ and $u \geq t$;
$(ii)^\circ$ or $s \gg t$ and for every $u \lhd t$, either $s > u$ or for some $v \lhd s$, $v \geq u$.

Remark 12. Compared with Property 1, the main difference is the use of $\blacktriangleright$ instead of $\rhd$ in the first alternative. In general we will have $\blacktriangleright \supseteq \rhd$; in the typed λ-calculus for example, $\blacktriangleright$ will be the union of the immediate superterm relation and of β-contraction at the top. The added complication in $(ii)^\circ$ is inspired from [13].

Theorem 2. *Let $\rhd^\circ$ be any binary relation on T. Let SN° be the set of all $s \in T$ that are accessible in $> \cup \rhd^\circ$, $\overline{SN}^\circ \hat{=} \{s \in T | \forall u (\lhd \cup \lhd^\circ) s \cdot u \in SN^\circ\}$, $\underline{SN}^\circ \hat{=} \{s \in T | if\ s \in \overline{SN}^\circ\ then\ s \in SN^\circ\}$.*

Let S be some set, $\sigma_0 \in S$, and $\_ \cdot \_$ be an (infix) map from $T \times S$ to T. Assume:

$(iii)^\circ$ $\rhd$ is well-founded on T;
$(iv)^\circ$ for every $s \in T$, if for every $u \lhd s$, for every $\sigma \in S$, $u \cdot \sigma$ is in SN°, then for every $\sigma \in S$, $\underline{SN}^\circ$ bars $s \cdot \sigma$ in $\gg$;
(xii) for every $s \in T$, $s \cdot \sigma_0 = s$;
$(xiii)$ for every $s \in T$, $\sigma \in S$, either $s \cdot \sigma \in SN^\circ$ or for every $u (\lhd \cup \lhd^\circ) s \cdot \sigma$, $u \in SN^\circ$ or there is a $v \lhd s$ and $\sigma' \in S$ such that $v \cdot \sigma' (\geq \cup \rhd^\circ) u$;
(xiv) for every $s', t, u \in T$, if $s' \gg t \rhd^\circ u$, then $t \rhd u$ or $s' \rhd^\circ v \geq t$ for some $v \in T$;
(xv) for every $s', u \in T$, if $s' \blacktriangleright u$ and for every $v (\lhd \cup \lhd^\circ) s'$, v is in SN°, then u is in SN°.

Then $T = SN^\circ$. In particular, $>$ is well-founded.

We prove Theorem 2 shortly. The proof is very similar to that of Theorem 1, and the reader is invited to proceed directly to applications following the proof. Meanwhile, observe that Theorem 1 is the special case of Theorem 2 where $\blacktriangleright \hat{=} \rhd$, S is any one-element set $\{\sigma_0\}$, $s \cdot \sigma_0 \hat{=} s$, and $\rhd^\circ$ is the empty relation.

Proof. We show that for every $s \in T$, for every $\sigma \in S$, $s \cdot \sigma$ is in SN°. Using (xii) and $\sigma \hat{=} \sigma_0$, it will follow that $s \in SN^\circ$. This is by induction on $\rhd$, which is legal by $(iii)^\circ$. So the following induction hypothesis is available:

$$\text{For every } u \lhd s \text{, for every } \sigma \in S,\ u \cdot \sigma \in SN^\circ \qquad (3)^\circ$$

Fix σ: we wish to show $s \cdot \sigma \in SN^\circ$. We claim that this is entailed by: $(*)$ $s \cdot \sigma \in \underline{SN}^\circ$. Indeed, by $(xiii)$, either $s \cdot \sigma \in SN^\circ$ and we are done, or for every

$u(\lhd \cup \lhd^\circ) s \cdot \sigma$, $u \in SN^\circ$, or there is a $v \lhd s$ and $\sigma' \in S$ such that $v \cdot \sigma' (\geq \cup \rhd^\circ) u$; in the latter case $v \cdot \sigma' \in SN^\circ$ by $(3)^\circ$, so $u \in SN^\circ$ again: since every $u(\lhd \cup \lhd^\circ) s \cdot \sigma$ is in SN°, $(*)$ indeed implies $s \cdot \sigma \in SN^\circ$, by the definition of $\underline{SN}^\circ$.

It remains to show $(*)$. By $(3)^\circ$ and $(iv)^\circ$, $\underline{SN}^\circ$ bars $s \cdot \sigma$ in $\gg$. We show that $s \cdot \sigma \in \underline{SN}^\circ$ by showing that any s' barred by $\underline{SN}^\circ$ in $\gg$ is in fact in $\underline{SN}^\circ$, by bar induction. The base case is obvious. In the inductive case, assume that for every t such that $s' \gg t$, $t \in \underline{SN}^\circ$, and prove $s \in \underline{SN}^\circ$. Expanding the definition of $\underline{SN}^\circ$, we must show that $s' \in SN^\circ$ under the assumptions:

$$\text{For every } t, \text{ if } s' \gg t \text{ and every } u(\lhd \cup \lhd^\circ) t \text{ is in } SN^\circ, \text{ then } t \in SN^\circ \qquad (4)^\circ$$

$$\text{For every } u(\lhd \cup \lhd^\circ) s', \ u \in SN^\circ \qquad (5)^\circ$$

To show that $s' \in SN^\circ$, it suffices to show that every t such that $s'(> \cup \rhd^\circ) t$ is in SN°, which we show by induction on $\rhd$. So the following induction hypothesis is available:

$$\text{For every } u \lhd t, \text{ if } s'(> \cup \rhd^\circ) u \text{ then } u \in SN^\circ \qquad (6)^\circ$$

Since $s'(> \cup \rhd^\circ) t$, we distinguish three cases, two when $s' > t$, one when $s' \rhd^\circ t$:

- $(i)^\circ$ For some $u \blacktriangleleft s'$, $u \geq t$. By (xv) and $(5)^\circ$, $u \in SN^\circ$, so $t \in SN^\circ$.
- $(ii)^\circ$ Or $s' \gg t$ and for every $u \lhd t$, either $s' > u$ or $v \geq u$ for some $v \lhd s'$. For each $u \lhd t$, if $s' > u$ then by $(6)^\circ$ $u \in SN^\circ$; and if $s' \rhd v \geq u$, then by $(5)^\circ$ $v \in SN^\circ$ so $u \in SN^\circ$ again. To sum up: (†) for every $u \lhd t$, $u \in SN^\circ$. Moreover, for each $u \lhd^\circ t$, by (xiv) either $t \rhd u$ or $s' \rhd^\circ v \geq t$ for some v; if $t \rhd u$, by (†) $u \in SN^\circ$, and if $s' \rhd^\circ v \geq t$, then $v \in SN^\circ$ by $(5)^\circ$, so $t \in SN^\circ$, so $u \in SN^\circ$ since $t \rhd^\circ u$. Summing up and combining with (†), we get: for every $u(\lhd \cup \lhd^\circ) t$, $u \in SN^\circ$. By $(4)^\circ$ it obtains $t \in SN^\circ$.
- Or $s' \rhd^\circ t$. Then $t \in SN^\circ$ by $(5)^\circ$. □

Higher-Order Recursive Path Orderings. Theorem 2 entails that Jouannaud and Rubio's *higher-order path ordering* [13,14] is well-founded. We also take the opportunity to generalize it.

Let $\mathcal{T}$ be an algebra of so-called *types*, including a $\rightarrow$ binary infix operator. That is, whenever τ_1 and τ_2 are types, so is $\tau_1 \rightarrow \tau_2$. Let $>_\mathcal{T}$ be a well-founded ordering on types such that $(\tau_1 \rightarrow \tau_2) >_\mathcal{T} \tau_1$ and $(\tau_1 \rightarrow \tau_2) >_\mathcal{T} \tau_2$. A *signature* Σ is any map from so-called *function symbols* f, g, ..., to *arities* $\tau_1, \ldots, \tau_n \Rightarrow \tau$, where τ_1, ..., τ_n, τ are types in $\mathcal{T}$. For each type $\tau \in \mathcal{T}$, let $\mathcal{X}_\tau$ be pairwise disjoint countably infinite sets of so-called *variables of type* τ, written as x_τ, y_τ, ... We write x, y, ..., instead of the latter when the types are clear from context.

Recall that the language of *algebraic* λ*-terms* over Σ is the smallest collection of sets Λ^Σ_τ when τ ranges over types in $\mathcal{T}$, such that $\mathcal{X}_\tau \subseteq \Lambda^\Sigma_\tau$, such that $MN \in \Lambda^\Sigma_{\tau_2}$ whenever $M \in \Lambda^\Sigma_{\tau_1 \rightarrow \tau_2}$ and $N \in \Lambda^\Sigma_{\tau_1}$, such that $\lambda x_{\tau_1} \cdot M \in \Lambda^\Sigma_{\tau_1 \rightarrow \tau_2}$ whenever $M \in \Lambda^\Sigma_{\tau_2}$, and such that $f(M_1, \ldots, M_n) \in \Lambda^\Sigma_\tau$ whenever $f \in \operatorname{dom} \Sigma$, $\Sigma(f) = \tau_1, \ldots, \tau_n \Rightarrow \tau$, and $M_1 \in \Lambda^\Sigma_{\tau_1}$, ..., $M_n \in \Lambda^\Sigma_{\tau_n}$. We drop type subscripts when irrelevant or clear from context. We also consider that any two α-equivalent terms are equated, i.e., we reason on the set T of equivalence classes of terms in $\Lambda^{\Sigma} \hat{=} \bigcup_{\tau \in \mathcal{T}} \Lambda^{\Sigma}_{\tau}$ modulo α-renaming. By abuse of language, we shall say that $M >_{\mathcal{T}} N$ when $M \in \Lambda^{\Sigma}_{\tau}$, $N \in \Lambda^{\Sigma}_{\tau'}$ and $\tau >_{\mathcal{T}} \tau'$.

Let us build a relation $>$ on typed λ-terms by mimicking Remark 4:

Definition 1. *Let $\triangleright$ be the immediate subterm relation (in particular $\lambda x_{\tau_1} \cdot M \triangleright M[x := y_{\tau_1}]$ for every variable y_{τ_1}, where $M[x := N]$ denotes the capture-avoiding substitution of x for N in M). Let β be the smallest relation such that $(\lambda x \cdot M)N \; \beta \; M[x := N]$, and $\blacktriangleright$ be $\triangleright \cup \beta$. Let $R \mapsto \gg_R$ be given and monotonic, and define $>$ as the largest binary relation such that $s > t$ implies $s \geq_{\mathcal{T}} t$ and also either (i)° or (ii)°, where $\gg$ denotes $\gg_>$.*

Clearly $>$ satisfies Property 2. Moreover $s > t$ implies $s \geq_{\mathcal{T}} t$, by construction.

Remark 13. We get Jouannaud and Rubio's horpo [14] by using the following definition for $\gg_R$. Split function symbols in $\operatorname{dom} \Sigma$ into symbols with *multiset status* and symbols with *lexicographic status*. Let $\succ$ be any strict ordering on $\operatorname{dom} \Sigma$. Let $M \gg_R N$ if and only if either:

1. $M = f(M_1, \ldots, M_m)$, $N = g(N_1, \ldots, N_n)$ $(f, g \in \operatorname{dom} \Sigma)$ and $f \succ g$, or $f = g$ has multiset status and $\{M_1, \ldots, M_m\} \; R_{mul} \; \{N_1, \ldots, N_n\}$, or $f = g$ has lexicographic status, $m = n$, and $(M_1, \ldots, M_n) \; R_{lex} \; (N_1, \ldots, N_n)$;
2. or $M = f(M_1, \ldots, M_m)$, $N = N_1 N_2$ $(f \in \operatorname{dom} \Sigma)$;
3. or $M = M_1 M_2$, $N = N_1 N_2$ and $\{M_1, M_2\} \; R_{mul} \; \{N_1, N_2\}$;
4. or $M = \lambda x \cdot M_1$, $N = \lambda x \cdot N_1$ (where x is the same on both sides, up to α-renaming) and $M_1 \; R \; N_1$.

To apply Theorem 2, first define $M \triangleright^{\circ} N$ by induction on the type of M, ordered by $>_{\mathcal{T}}$, as follows. $M \triangleright^{\circ} N$ if only if M is an *abstraction* $\lambda x_{\tau_1} \cdot M_1$, of type $\tau_1 \to \tau_2$, and $N = M_1[x := N_1]$ for some N_1 in SN° of type τ_1. Note that since $> \cup \triangleright^{\circ} \subseteq \geq_{\mathcal{T}}$, any $(> \cup \triangleright^{\circ})$-reduction starting from N_1 only involves λ-terms of types $\leq_{\mathcal{T}} \tau_1$, on which $\triangleright^{\circ}$ is already defined by induction hypothesis: therefore "N_1 in SN° of type τ_1" is well-defined by induction hypothesis in the definition of $\triangleright^{\circ}$ for terms M of type $\tau_1 \to \tau_2$.

Remark 14. This construction implies that SN°, the set of accessible terms in $> \cup \triangleright^{\circ}$, is also the set of terms M that are in SN (the set of terms that are strongly normalizing for $>$), and such that, if M has type $\tau_1 \to \tau_2$, then whenever $M >^* \lambda x_{\tau_1} \cdot M_1$, for every $N_1 \in SN^{\circ}$ of type τ_1, $M_1[x := N_1]$ is in SN° of type τ_2. This is one of the classical definitions of reducibility (a.k.a., computability) [6]. In this sense, Theorem 2 is a reducibility argument, just like [13].

Second, let S be the set of all substitutions mapping variables x_{τ} to λ-terms in SN° of type τ, let σ_0 be the empty substitution, and let $M \cdot \sigma$ denote application of substitution σ to M. Conditions (*iii*)° and (*xii*) are obvious.

Condition (*xiii*) is justified as follows. Let s be a typed λ-term, $\sigma \in S$. If s is a variable by construction $s \cdot \sigma \in SN^{\circ}$. Otherwise, consider any $u (\triangleleft \cup \triangleleft^{\circ}) s \cdot \sigma$. If $u \triangleleft s \cdot \sigma$, then since s is not a variable, u is written $v \cdot \sigma$ for some immediate subterm v of s; i.e., (*xiii*) is proved with $\sigma' \hat{=} \sigma$, $v \cdot \sigma' = u$. If $u \triangleleft^{\circ} s \cdot \sigma$, then since

s is not a variable, s must be written $\lambda x \cdot M_1$, with $M_1 \cdot \sigma[x := N_1] = u$ for some $N_1 \in SN^\circ$: take $v \hat{=} M_1$, and σ' be the substitution mapping x to N_1 and every other variable y to $y \cdot \sigma$ (taking x outside the domain of σ, by α-renaming).

Condition (xv) is justified as follows. Assume that: $(*)$ every $v(\triangleleft \cup \triangleleft^\circ)s'$ is in SN°. If $s' \blacktriangleright u$, then either $s' \triangleright u$ and (xv) is clear; or $s' \; \beta \; u$. If $s' \; \beta \; u$, then s' is of the form $(\lambda x \cdot M)N$ and $u = M[x := N]$. By $(*)$ N is in SN° and $\lambda x \cdot M$ is in SN° too. So $M[x := N] \triangleleft^\circ \lambda x \cdot M$ is in SN°.

So only conditions $(iv)^\circ$ and (xiv) are not automatically verified. We get the following corollary to Theorem 2:

Lemma 3. *Let $>$ be as in Definition 1, and $\gg$ be $\gg_>$. Assume that:*

1. *for every λ-term M, if every immediate subterm of M is in SN°, then M is accessible in $\gg_{|\overline{SN}^\circ}$;*
2. *if $M \gg \lambda x \cdot M_1$, then M is of the form $\lambda x \cdot M_0$ and $M_0 \cdot \sigma > M_1 \cdot \sigma$ for every $\sigma \in S$.*

Then $>$ is well-founded.

Proof. Condition $(iv)^\circ$ follows from 1 by the same argument as in Remark 6. Let us prove (xiv). Assume $s' \gg t \triangleright^\circ u$. By definition t must be of the form $\lambda x \cdot M_1$ and $u = M_1[x := N_1]$ with $N_1 \in SN^\circ$. By 2, $s' = \lambda x \cdot M_0$ and $M_0[x := N_1] > M_1[x := N_1]$. Take $v \hat{=} M_0[x := N_1]$, then $s' \triangleright^\circ v > u$. □

Lemma 1 still holds in the higher-order case, provided the following condition is added: whenever $s \; R \; t$ and $f(\ldots, s, \ldots)$ and $f(\ldots, t, \ldots)$ are well-typed, then $f(\ldots, s, \ldots) \geq_{\mathcal{T}} f(\ldots, t, \ldots)$; and provided in the notation $f(\ldots, s, \ldots)$, f is taken to be any function symbol, or application, or λ-abstraction. Lemma 2 holds without modification. It follows that Jouannaud and Rubio's horpo (Remark 13) is well-founded, monotonic and stable. By construction, it also includes β-reduction.

Remark 15. Lemma 3 can be seen as a suitable generalization of Ferreira and Zantema's result to the case of higher-order rewrite relations. Note that condition 2, which basically says that any term that is $\gg$ an abstraction must also be an abstraction, is fundamental. Jouannaud and Rubio's horpo in addition requires applications to be $\ll$ any term of the form $f(\ldots)$ with $f \in \Sigma$, and abstraction cannot be $\gg$ any non-abstraction term; these conditions are simply not needed.

6 Conclusion

We hope to have demonstrated that Theorem 1 and Theorem 2 have very general scopes. In term algebras, we retrieve all variants of path orderings, as well as the Knuth-Bendix orderings. The scope of Theorem 1, and naturally of Theorem 2 as well, exceeds term algebras, and we have sketched a few path ordering-like constructions on graphs, automata, and infinite terms. Theorem 2 then provides a general extension of Jouannaud and Rubio's higher-order path ordering.

There are at least two points that deserve further study. First, the proofs of Theorem 1 and Theorem 2 are intuitionistic, and therefore give rise to algorithms that might be used to implement reduction machines and study the complexity of reductions. Second, because Theorem 1 and Theorem 2 deal with general binary relations, we believe that a deeper understanding of the duality of [16] and of possible connections with Theorem 1 would be profitable. It has been pointed out by an anonymous referee that F. Veltman has produced a constructive version of Kruskal's Theorem, which may be useful in this endeavor.

References

1. B. Barras, S. Boutin, C. Cornes, J. Courant, Y. Coscoy, D. Delahaye, D. de Rauglaudre, J.-C. Filliâtre, E. Giménez, H. Herbelin, G. Huet, H. Laulhère, C. Muñoz, C. Murthy, C. Parent-Vigouroux, P. Loiseleur, C. Paulin-Mohring, A. Saïbi, and B. Werner. The Coq proof assistant – reference manual. Available at `http://coq.inria.fr/doc/main.html.`, Dec. 1999. Version 6.3.1.
2. P. Cousot. Semantic foundations of program analysis. In S. Muchnick and N. Jones, editors, *Program Flow Analysis: Theory and Applications*, chapter 10, pages 303–342. Prentice-Hall, Inc., Englewood Cliffs, New Jersey, 1981.
3. N. Dershowitz. Orderings for term rewriting systems. *Theoretical Computer Science*, 17(3):279–301, 1982.
4. N. Dershowitz. Termination of rewriting. *Journal of Symbolic Computation*, 3:69–116, 1987.
5. N. Dershowitz and C. Hoot. Topics in termination. In C. Kirchner, editor, *5th RTA*, pages 198–212. Springer Verlag LNCS 690, 1983.
6. G. Dowek and B. Werner. Proof normalization modulo. Research Report RR-3542, INRIA, Nov. 1998.
7. M. Fernández and J.-P. Jouannaud. Modular termination of term rewriting systems revisited. In E. Astesiano, G. Reggio, and A. Tarlecki, editors, *Recent Trends in Data Type Specification.* Springer-Verlag LNCS 906, 906.
8. M. C. F. Ferreira and H. Zantema. Well-foundedness of term orderings. In N. Dershowitz, editor, *4th International Workshop on Conditional Term Rewriting Systems (CTRS'94)*, pages 106–123. Springer Verlag LNCS 968, 1995.
9. F. Gécseg and M. Steinby. Tree languages. In G. Rozenberg and A. Salomaa, editors, *Handbook of Formal Languages*, volume 3, chapter 1, pages 1–68. Springer-Verlag, 1997. Beyond Words.
10. J.-Y. Girard, Y. Lafont, and P. Taylor. *Proofs and Types*, volume 7 of *Cambridge Tracts in Theoretical Computer Science.* Cambridge University Press, 1989.
11. J. Goubault-Larrecq. A method for automatic cryptographic protocol verification (extended abstract). In *FMPPTA'2000*, pages 977–984. Springer Verlag LNCS 1800, 2000.
12. H. Herbelin, Dec. 1996. Private communication.
13. J.-P. Jouannaud and A. Rubio. The higher-order recursive path ordering. In G. Longo, editor, *14th LICS*, pages 402–411, Trento, Italy, July 1999.
14. J.-P. Jouannaud and A. Rubio. Higher-order recursive path orderings à la carte. Draft available at `ftp://ftp.lri.fr/LRI/articles/jouannaud/horpo-full.ps.gz`, 2001.
15. S. Kamin and J.-J. Lévy. Two generalizations of the recursive path ordering. Technical report, University of Illinois, 1980.

16. P.-A. Melliès. On a duality between Kruskal and Dershowitz theorems. In *25th ICALP*, pages 518–529. Springer Verlag LNCS 1443, 1998.
17. D. Monniaux. Abstracting cryptographic protocols with tree automata. In *6th International Static Analysis Symposium (SAS'99)*. Springer-Verlag LNCS 1694, 1999.
18. C. Thomassen. Embeddings and minors. In R. Graham, M. Grötschel, and L. Lovász, editors, *Handbook of Combinatorics*, chapter 5. Elsevier Science B.V., 1995.

Stratified Context Unification Is in PSPACE

Manfred Schmidt-Schauß

Fachbereich Informatik
Johann Wolfgang Goethe-Universität
Postfach 11 19 32, D-60054 Frankfurt
Germany
schauss@ki.informatik.uni-frankfurt.de

Abstract. [illegible]

1 Introduction

[illegible]

L. Fribourg (Ed.): CSL 2001, LNCS 2142, [illegible], 2001.

Stratified Context Unification Is in PSPACE

Manfred Schmidt-Schauß

Fachbereich Informatik
Johann Wolfgang Goethe-Universität
PoBox 11 19 32, D-60054 Frankfurt
Germany
schauss@cs.uni-frankfurt.de
www.ki.informatik.uni-frankfurt.de
Tel: (+49)69 798 28597 Fax: (+49)69 798 28919

Abstract. Context unification is a variant of second order unification and also a generalization of string unification. Currently it is not known whether context unification is decidable. A decidable specialization of context unification is stratified context unification, which is equivalent to satisfiability of one-step rewrite constraints.
This paper contains an optimization of the decision algorithm, which shows that stratified context unification can be done in polynomial space.

1 Introduction

Context unification is a variant of second order unification and also a generalization of string unification. There are unification procedures for the more general problem of higher-order unification (see e.g. [Pie73,Hue75,SG89]). It is well-known that higher-order unification and second-order unification are undecidable [Gol81,Far91,LV00a] and that string unification is decidable [Mak77]. Recent upper complexity estimations are that it is in EXPSPACE [Gut98], in NEXPTIME [Pla99a] and even in PSPACE [Pla99b].

Context unification problems are restricted second-order unification problems. Context variables represent terms with exactly one hole in contrast to a term with an arbitrary number of holes in the general second-order case. The name *contexts* was coined in [Com93], see also [Com98a,Com98b]. Currently, it is not known whether context unification is decidable. It is known that it is $\mathcal{NP}$-hard (cf. [SSS98]), and that satisfiability of formulas in a logical theory of context unification is undecidable [NPR97a,Vor98].
There are some decidable fragments: If for every context variable X, all occurrences of X have the same argument [Com98a,Com98b]; if the number of occurrences of every first order variable and every context variable is at most two [Lev96], if there are at most two context variables [SSS99], or if the context unification problems are stratified [SS99b]. There is also some work on generalizations of context unification [LV00b] aiming at decidability. A decidable restriction of second order unification similar in spirit to context unification is bounded second order unification [SS99a], where second order variables represent terms with a number of holes that is bounded by a given number.

L. Fribourg (Ed.): CSL 2001, LNCS 2142, pp. 498–513, 2001.

Applications of context unification and stratified context unification are for example in computational linguistics [NPR97a,EN00], in particular as a uniform framework for semantic underspecification of natural language [NPR97b]. The fragment of stratified context unification is expressive enough for applying it in computational linguistics. It was also used in equational unification as an important step in showing decidability of distributive unification [SS98]. It was proved that one-step rewrite constraints and stratified context unification can be interreduced [NTT00]. As a lower bound for the complexity it is known that stratified context unification is $\mathcal{NP}$-hard [SSS98]. The method in [SS99c] estimates the complexity of the unification algorithm on the basis of the main depth of cycles. This appears to be insufficient for upper estimations, so the current paper gives an upper complexity bound based rather on the size of the representation of cycles. The main depth of cycles may be exponential. Basing an upper complexity estimation on the translation to string unification described in [SS94] and then using the PSPACE upper bound in [Pla99b] does not work, since the translation is not polynomial.

This paper analyses the complexity of the decision algorithm for stratified context unification in [SS99b] if it is enhanced with compression techniques like sharing and power-expressions. A key technique is to use nested powers of contexts. The following new result is proved in this paper:

Theorem: Stratified context unification is in PSPACE .

A corollary following from [NTT00] is:

Theorem: Satisfiability of one-step rewrite constraints is in PSPACE.

The upper polynomial space bound for stratified context unification may have an impact on context unification algorithms used in computational linguistics. The paper is structured as follows. First we describe a decision algorithm CSCU for stratified context unification as a translation from [SS99b]. This is done by first describing the data structure, then the adapted rules, and at last a space estimation is given showing that the decision algorithm CSCU is in PSPACE. Note that this paper relies on the description and proofs in [SS99b].

2 Preliminaries

Let Σ be a signature of function symbols. Every function symbol comes with an arity, denoted $ar(f)$, which is a nonnegative integer. Function symbols with $ar(f) = 0$ are also called *constant symbols*. We assume that the signature contains at least one constant symbol and at least one non-constant function symbol, in particular we also allow that the signature may be infinite or monadic. Let $\mathcal{V}_1$ be the set of first-order variables $x, y, z, \ldots$, $\mathcal{V}_2$ be the set of context variables $X, Y, Z, \ldots$, and $\mathcal{V} := \mathcal{V}_1 \cup \mathcal{V}_2$. *Terms* t are formed using the grammar $t ::= x \mid f(t_1, \ldots, t_{ar(f)}) \mid X(t_0)$, where x is a first-order variable, f is a function symbol, X is a context variable, and t_i are terms. For a constant a, we write a

instead of $a()$. We denote terms using the letters s, t. Syntactic equality of terms s, t is denoted as $s \equiv t$. The set of variables occurring in the term s is denoted as $Var(s)$. A term s is called a *ground term* if s has no occurrences of variables. *Contexts* are formed by the grammar $C[\cdot] ::= [\cdot] \mid X(C[\cdot]) \mid f(t_1, \ldots, C[\cdot], \ldots, t_{ar(f)})$, where $[\cdot]$ is called the *hole* (also *trivial context*, *Id*), f is a function symbol, X is a context variable, C is a context, and t_i are terms. Contexts must contain exactly one occurrence of the hole. We denote contexts as $C[\cdot]$, or as C, if it is not ambiguous, and the subterm $X([\cdot])$ is abbreviated as $X(\cdot)$. The notation $C[t]$ means the term where the term t is plugged into the hole of $C[\cdot]$. We denote syntactic equality of contexts by $\equiv$. A *ground context* is a context without occurrences of variables, i.e., it can be seen as a ground term with a single hole, where a signature with the additional constant $[\cdot]$ is used. The length of the position of the hole in the ground context C is called *main depth* and denoted as $|C|$. The size of terms is the number of occurrences of symbols, and the size of contexts is the number of occurrences of symbols not counting the hole. This may be denoted as $size(.)$.

A *(ground) substitution* is a mapping from terms to ground terms with the following properties. A substitution σ can be represented as $\{x_i \to t_i, X_j \to C_j | i = 1, \ldots, n, j = 1, \ldots, m\}$, where $t_i, i = 1, \ldots, n$ is a ground term and $C_j, j = 1, \ldots, m$ is a ground context. σ operates on terms t by replacing all occurrences of variables x_i by $t_i, i = 1, \ldots, n$ and replacing all occurrences of context variables X_j by $C_j, j = 1, \ldots, m$. The replacement of X by $C[\cdot]$ means to replace all subterm occurrences of the form $X(s)$ by $C[s]$, and the replacement of X by Id is done by replacing all subterm occurrences of the form $X(s)$ by s. The ground substitution σ has as *domain* the set $\{x_i \mid i = 1, \ldots, n\} \cup \{X_j \mid j = 1, \ldots, m\}$ and as *codomain* the set $\{t_i \mid i = 1, \ldots, n\} \cup \{C_j \mid j = 1, \ldots, m\}$.

The tree addresses in terms and contexts are also called *positions*, which are words of positive integers. The expression $t|_p$ ($C|_p$) denotes the subterm (subcontext) of t (of C) at position p.

If C_1, C_2 are contexts, then we denote the context $C_1[C_2[\cdot]]$ also as C_1C_2. A *prefix* of a context C is a context C_1, such that $C_1C_2 \equiv C$ for some context C_2.

A *context unification problem (CUP)* is a set Γ of equations, denoted as $\{s_1 \doteq t_1, \ldots s_n \doteq t_n\}$. An equation of the form $X(s) \doteq Y(t)$ is called *flat* equation. The multiset $term(\Gamma)$ is defined to be $\{s, t \mid (s \doteq t) \in \Gamma\}$. The size of Γ is the sum of the sizes of the terms in $term(\Gamma)$, denoted as $size(\Gamma)$. With $Var(\Gamma)$ we denote the set of variables occurring in Γ, and with $Var_i(\Gamma)$ we denote the set $Var(\Gamma) \cap \mathcal{V}_i$ for $i = 1, 2$.
A *unifier* of Γ is a ground substitution σ with domain containing $Var(\Gamma)$, which solves all equations in Γ. I.e., $\sigma(s) \equiv \sigma(t)$ for all $(s \doteq t) \in \Gamma$. A CUP Γ is called *unifiable*, if there is a unifier of Γ.
A unifier σ of Γ is called *minimal* if there is no other unifier σ' of Γ with

$$\sum_{X \in Var_2(\Gamma)} (size(\sigma'(X))) < \sum_{X \in Var_2(\Gamma)} (size(\sigma(X))).$$

A ground substitution σ has *exponent of periodicity* n ([Mak77,SSS98]), iff n is the maximal number, such that there is some context variable X and ground contexts A, B, C, where B is not trivial, such that $\sigma(X) = A\underbrace{B\ldots B}_{n}C$.

Proposition 2.1. *([SSS98]) There is a constant c, such that for every unifiable context unification problem Γ and for every minimal unifier σ of Γ its exponent of periodicity is at most $2^{2.14*size(\Gamma)}$.*

3 An Overview of the Translated Algorithm

The idea of the translation is to represent the performance of the algorithm SCU in [SS99b] in a compressed data structure. The first order variables that are used for sharing in CSCU correspond to subterms in SCU. However, not every subterm in the problems in SCU is also represented in CSCU as a first order variable. Many subterms are implicitly represented in power expressions.

In the following we describe the translated rules. We have also to add a rigorous description of the sharing methods, but we omit arguments for soundness, correctness, and termination, since these are explained in [SS99b], and the algorithm in this paper mimics the algorithm SCU.

The algorithm "compressed stratified context unification decision algorithm" *CSCU* has an initial input Γ_I, which is without occurrences of power expressions. Let E_I be the upper bound on the exponent of periodicity given by the bound in Lemma 2.1 for Γ_I.

A preprocessing called flattening is required in order to exploit sharing on the term level. The decomposition rules are a bit more complicated, since some macro rules are added, which avoid the explicit representation of all subterms.

The rules for treating cycles and clusters have to be extended to show how they implement sharing of subcontexts.

4 The Data Structures for CSCU

The compressed representation of the data structures requires three different parts: term sharing using first order variables representing subterms, a representation of iterated context application and a sharing of contexts.

4.1 Description of the Compressed Data Structures

Let a *1-context* S be a context of the form $f(x_1, \ldots, x_{j-1}, [\cdot], x_{j+1}, \ldots, x_{ar(f)})$, where x_i are first order variables.

Let a *power expression* be an expression of the form $pow(X, n_1, n_2)$, where $0 \leq n_1 \leq n_2$ are nonnegative integers and X is a context variable. We denote them using the letter P. It is always assured that unifiers instantiate X by a nontrivial context. Given a substitution σ, power expressions $pow(X, n_1, n_2)$ denote the subcontext from main depth n_1 to n_2 of a sufficiently large power of X. Formally we define, given a ground substitution σ, $\sigma(pow(X, n_1, n_2)) := A$, where A is a ground context with $|A| = n_2 - n_1$, and there are ground contexts B, C with $|B| = n_1$, and a nonnegative integer n_3 with $\sigma(X^{n_3}) = BAC$.

As an example let $\sigma = \{X \to f(g([.]))\}$; then $\sigma(pow(X, 1, 7)) = g(f(g(f(g(f([.]))))))$.

The general form of power expressions enables a nice and short formulation of rules, and also guarantees nonincreasing space usage for rules extracting subcontexts of a power expression.
Let the extended *terms* be $t ::= x \mid f(t_1, \ldots, t_{ar(f)}) \mid X(t) \mid P(t)$, where x is a variable, f a function symbol, X a context variable, t, t_i are extended terms, and P is a power expression. In the following we will use only extended terms.

Definition 4.1. *A compressed context unification problem (CCUP) Γ is a set of equations of the following two kinds:*

- *equations $s \doteq t$, where s, t are (extended) terms. These are called* term equations.
- *equations $X \doteq S$, where S is a nonempty sequence $S_1 \ldots S_n$ and S_i are power expressions or 1-contexts. The sequence $S_1 \ldots S_n$ means $S_1(\ldots(S_n([.])))$. These equations are called* context sharing equations.

The subset of term equations in Γ is denoted as $T(\Gamma)$. The subset of context sharing equations in Γ is denoted as $D(\Gamma)$. We assume that $\doteq$ is symmetric for term equations, but not symmetric for sharing equations. The context variables occurring in left hand sides of context sharing equations are called defined context variables *(wrt. Γ). They are ranged over by U. The set of defined context variables is denoted as U_Γ. In the following "context variables" refers to context variables in $Var_2(\Gamma) \setminus U_\Gamma$, and "defined context variables" to the elements of U_Γ.*
Five conditions must hold:

- *We assume that there is no occurs-check situation in the context sharing equations. I.e., in Γ there is no chain $U_1 \doteq S_1, U_2 \doteq S_2, \ldots, U_n \doteq S_n$, such that U_{i+1} occurs in S_i for $i = 1, \ldots, n-1$ and U_1 occurs in S_n.*
- *If $U_1 \doteq S_1, U_2 \doteq S_2 \in D(\Gamma)$, then $U_1 \not\equiv U_2$.*
- *For every $U \doteq S \in D(\Gamma)$, it is $Var_2(S) \subseteq U_\Gamma$.*
- *Any defined context variable U can occur in $T(\Gamma)$ only in the form $pow(U, n_1, n_2)(x)$.*
- *For every occurrence of $pow(U, n_1, n_2)(x)$ in Γ, it is $U \in U_\Gamma$.*

Definition 4.2. *A ground substitution σ is a* unifier *of the CCUP Γ iff after applying σ, the left and right hand sides of all equations in Γ are syntactically equal.*

Note that the U in a power expression $pow(U, n_1, n_2)$ will be mapped to a nontrivial context by a unifier, and that application of a unifier to a power expressions results in a term without power expressions.

Definition 4.3. *Let Γ be a CCUP.* SO-prefixes *wrt. Γ are words in* $(\mathcal{V}_2 \setminus \mathcal{U}_\Gamma)^*$.

An SO-prefix of a position *p in a term t is the word consisting of the context variables from $\mathcal{V}_2 \setminus \mathcal{U}_\Gamma$ in head positions that are met going from the root to the position p. SO-prefixes of all positions in context sharing equations are defined to be empty.*
Let the following conditions hold: For all context variables $X \in \mathcal{V}_2 \setminus \mathcal{U}_\Gamma$ and all positions p_1, p_2 of the context variable X in equations, the SO-prefix of p_1 in Γ is the same as the SO-prefix of p_2 in Γ. For all first order variables x and all positions p_1, p_2 of the the first order variable x in equations, the SO-prefix of p_1 in Γ is the same as the SO-prefix of p_2 in Γ.
Then Γ is called stratified.
We abbreviate "stratified CCUP" as SCCUP.

In an SCCUP we can speak of the *SO-prefix of a context variable or of a first order variable, respectively.*

This definition is the translation of the definition in [SS99b] and consistent with [SS94,Lev96] for context unification problems without occurrences of power-expressions.

Example 4.4. The CUP $\{Y(X(x)) \doteq Z(y), Y(g(X(z))) = u\}$ is a stratified CCUP, where X has SO-prefix Y, and x, z have SO-prefix YX. The CUP $\{X(x) \doteq x\}$ is not stratified, since the SO-prefix of x is not unique.

As the syntactical size measure of terms and SCCUPs, we use the common size definitions. The only difference is that we define $size(pow(U, n_1, n_2)) := 2 + 2 * size(n_2)$. Note that $size(n_2) = O(log(n_2))$.

Example 4.5. In $\{X(f(g(a))) \doteq f(g(X(a)))\}$ it is possible to represent a unifier with 113 iterations of $f(g(\cdot))$ as $X \doteq pow(U, 0, 226), U \doteq f(\cdot)g(\cdot)$.

4.2 Expansions of Power Expressions

We assume a SCCUP Γ be given and explain the meaning and some auxiliary operations on the data structures.

In the SCCUP Γ we compute the main depth of contexts under any unifier: We denote this also using the $|\cdot|$-notation. The following holds:

1. $|pow(U, n_1, n_2)| = n_2 - n_1$
2. $|U| = |S|$, if $(U \doteq S) \in D(\Gamma)$.
3. $|S_1 \dots S_n| = |S_1| + \dots + |S_n|$ if $S_1 \dots S_n$ is a right hand side in $D(\Gamma)$.
4. $|K| = 1$ for a 1-context.

This defines in a unique way the main depth for all contexts U, S in sharing equations $U \doteq S$, and for all power expressions.

Definition 4.6. *Assume given a SCCUP Γ. Let $hdc(S, k)$, the top context of the subcontext of S at depth k, for contexts or sequences of contexts S with $0 \le k < |S|$ be defined as follows:*

$$\begin{array}{ll} hdc(K,0) & := K \textit{ for a 1-context } K \\ hdc(S_1 \cdot \ldots \cdot S_n, k) & := hdc(S_1, k) \quad \textit{if } k < |S_1| \\ hdc(S_1 \cdot \ldots \cdot S_n, k) & := hdc(S_2 \cdot \ldots \cdot S_n, k - |S_1|) \textit{ if } k \geq |S_1| \\ hdc(pow(U, n_1, n_2), k) & := hdc(S, n_1 + k \bmod |U|\ ), \quad \textit{if } U \doteq S \in D(\Gamma) \\ & \qquad \textit{and } n_1 + k < n_2 \\ hd(S) & := hdc(S, 0) \end{array}$$

Definition 4.7. *Assume given a SCCUP Γ. The function "expand" deshares the definitions of the defined context variables. Let $expand(pow(U, n_1, n_2))$ be defined as follows:*
$expand(pow(U, n_1, n_1)) := Id$
$expand(pow(U, n_1, n_2)) := hdc(pow(U, n_1, n_2), 0) \cdot expand(pow(U, n_1 + 1, n_2))$
This is also used for terms:
$expand(pow(U, n_1, n_2))(x) := expand(pow(U, n_1, n_2))[x]$.
If we say a variable occurs in $pow(U, n_1, n_2)$, *then we mean the occurrences in* $expand(pow(U, n_1, n_2))$.

Example 4.8. For $\{U \doteq f(x, [])g([]), y_1 \doteq pow(U, 3, 6)(y_2)\} \subseteq \Gamma$, we have $|U| = 2, |pow(U, 3, 6)| = 3$, $expand(pow(U, 3, 6)) = g(f(x, g([\cdot])))$, and $hd(pow(U, 3, 6)) = g([])$.

Let the top symbol of terms be defined as the top function symbol, i.e. the top symbol of $f(\ldots)$ is f, and the top symbol of a 1-context $f(\ldots)$ is f, and of a power expression it is the top symbol of its expansion.

5 Preprocessing of the Algorithm CSCU

Initially, and after a replacement of context variables, a flattening is mandatory. This intermediate flattening is only done for positions with empty SO-prefix.

Definition 5.1 (Flatten).

- $\dfrac{\{s \doteq t\} \cup \Gamma}{\{s \doteq x, x \doteq t\} \cup \Gamma}$ *if neither s nor t is a variable.*
- $\dfrac{\{f(s_1, \ldots, s_n) \doteq y\} \cup \Gamma}{\{f(x_1, \ldots, x_n) \doteq y, x_1 \doteq s_1, \ldots x_n \doteq s_n\} \cup \Gamma}$ *if some s_i is not a variable.*
- $\dfrac{\{pow(U, n_1, n_2)(s) \doteq y\} \cup \Gamma}{\{pow(U, n_1, n_2)(x) \doteq y, x \doteq s\} \cup \Gamma}$ *if s is not a variable.*

The introduced first order variables must be always fresh ones.

If the flattening rules are not applicable, then Γ is called *flattened.* In a flattened SCCUP, in term equations only terms of the form $x \mid f(x_1, \ldots, x_n) \mid X(x) \mid P(x)$ may occur, where x, x_i are first order variables.

The following definitions contain the required simple unification rules and the necessary decomposition rules. Decompositions for the situation $x \doteq pow(U, n_1, n_2), x \doteq pow(U', n_3, n_4)$ are done when needed in the treatments of clusters. An eager decomposition in this situation would potentially introduce an exponential number of (non-redundant) first order variables.

Definition 5.2 (decomposition rules).

1. *(variable replacement)* $\dfrac{\{x \doteq y\} \cup \Gamma}{\Gamma[y/x]}$.
2. *(decomposition)* $\dfrac{\{x \doteq f(x_1, \ldots, x_n), x \doteq f(y_1, \ldots, y_n)\} \cup \Gamma}{\{x \doteq f(x_1, \ldots, x_n), x_1 \doteq y_1, \ldots, x_n \doteq y_n\} \cup \Gamma}$
3. *(clash)* $\dfrac{\{x \doteq f(x_1, \ldots, x_n), x \doteq g(y_1, \ldots, y_m)\} \cup \Gamma}{Fail}$, *if* $f \neq g$.
4. *(decompose-fp) Let there be equations of the form* $x \doteq pow(U, n_1, n_2)(z)$, $x \doteq f(x_1, \ldots, x_n)$ *in* Γ *with* $n_1 < n_2$. *Let* $hd(pow(U, n_1, n_2)) = g(y_1, \ldots, y_{j-1}, [], y_{j+1}, \ldots, y_m)$.
   - *If* $g \neq f$, *then Fail.*
   - *If* $g = f$, *then remove the equation* $x \doteq pow(U, n_1, n_2)$ *and add the equations* $x_h \doteq y_h$ *for* $h \neq j$, *and* $x_j = pow(U, n_1 + 1, n_2)$.
5. *(decompose-pp) Let there be equations of the form* $x \doteq pow(U, n_1, n_2)(z_1)$, $x \doteq pow(U, n_1, n_3)(z_2)$ *in* Γ *with* $n_2 \leq n_3$ *Then remove the equation* $x \doteq pow(U, n_1, n_2)(z_1)$, *and add the equation* $z_1 \doteq pow(U, n_2, n_3)(z_2)$.
6. *(occurs-check) Fail, if there is a chain of equations* $x_1 \doteq t_1, \ldots, x_n \doteq t_n$, x_{i+1} *occurs in* t_i *for* $i = 1, \ldots, n-1$, x_1 *occurs in* t_n, *where at least one* t_i *is not a variable, and (remove-power) is not applicable.*
7. *(trivial) Remove equations* $t \doteq t$ *from* Γ.
8. *(remove-redundant) Remove* $x \doteq t$ *from* Γ *if* x *does not occur in* t *nor* Γ.
9. *(remove-redundant-U) Remove* $U \doteq S$ *from* Γ *if* U *does not occur in* Γ.
10. *(remove-power) Power expressions* $pow(U, n, n)$ *are replaced by* Id.
11. *(shift-power) A power expression of the form* $pow(U, n_1, n_2)$ *with* $|U| \leq n_1$ *is replaced by* $pow(U, n_1 - |U|, n_2 - |U|)$.

The decomposition rules are performed with high priority. If no decomposition rule is applicable, then we say Γ is *decomposed.*

6 Cycles and Clusters

We give the adapted definitions of SO-cycles and SO-clusters (cf. [SS99b]). We assume in this section that Γ is flattened and decomposed.

Definition 6.1. *A set of equations* $s_1 \doteq t_1, \ldots, s_n \doteq t_n$ *is called an* SO-cycle, *if the following holds:* s_i *is of the form* x_i *(or* $X_i(y_i)$*), and* X_{i+1} *(or* x_{i+1}*) occurs in* t_i *for* $i = 1, \ldots, n-1$, *and* X_1 *(or* x_1*) occurs in* t_n, *and at least one such occurrence is not at the top. The* length *of an SO-cycle is the number of context variables at the top positions in* s_i, t_i. *The variables* x_i *(*X_i*, respectively) are called the* cycle variables.

An SO-cycle is called ambiguous, *if for some* $i = 1, \ldots, n-1$, *the term* t_i *or* $expand(t_i)$ *contains more than one occurrence of* X_{i+1} *(or* x_{i+1} *respectively), or* t_n *or* $expand(t_n)$ *contains more than one occurrence of* X_1 *(or* x_1 *respectively). The* depth *of a nonambiguous SO-cycle is the sum of the depths of* X_{i+1} *(or* x_{i+1}*) in* t_i *for* $i = 1, \ldots, n-1$ *plus the depth if* X_1 *(or* x_1*) in* t_n *(or* $expand(t_n)$*).*

A SO-cycle is standardized, *iff the SO-cycle is of the form*

$$X_1(x_1) \doteq y_1, y_1 \doteq X_2(z_1), X_2(x_2) \doteq y_2, y_2 \doteq X_3(z_2), \ldots,$$
$$X_n(x_n) \doteq y_{n,1}, y_{n,1} \doteq C_{n,1}[y_{n,2}], \ldots, y_{n,k} \doteq C_{n,k}[y_{n,k+1}], y_{n,k+1} \doteq X_1(z_n)$$

A non-standardized SO-cycle L has j successive flat junctions iff there is a representation (perhaps revolving the SO-cycle) of L of the form

$$X_1(x_1) \doteq y_1, y_1 \doteq X_2(z_1), \ldots, X_j(x_j) \doteq y_j, y_j \doteq X_{j+1}(z_j), \ldots$$

The flat-length *of a non-standardized SO-cycle is the maximal j, such that L has j successive flat junctions.*

Definition 6.2. *Let Γ be an SCCUP. Let $\sim$ be the equivalence relation on variables in $Var(\Gamma) \setminus U_\Gamma$ which have empty SO-prefix generated by $X \sim x$ if there is an equation $X(y) \doteq x$ in Γ.*

Let $\succ$ be the relation $x \succ y$ if there is an equation $x \doteq t \in \Gamma$, $x \not\equiv y$ and y occurs in t.

Let $\gtrsim$ be the quasi-ordering on variables generated by the transitive and reflexive closure of $\succ \cup \sim$. If there are variables x, y with $x \succ y$ and $y \gtrsim x$, then we say Γ has a cycle conflict.

If Γ has no cycle conflict, then an equivalence class K of $\sim$ is called an SO-cluster. *An SO-cluster K is called a* top-cluster, *iff the variables in K are maximal w.r.t. $\gtrsim$. The set of equations in Γ, where the context variables from a top-cluster K occur, is denoted as $EQ(K)$, and the terms in the equations $EQ(K)$ are denoted as $EQT(K)$.*

A top-cluster K is called flat, *iff it is also $\gtrsim$-minimal.*

Remark 6.3. The set K consists exactly of the first order variables in $EQT(K)$ and the context variables X, such that $X(y)$ occurs in $EQT(K)$ for some y. If K is a top-cluster, then the only occurrences of variables from K are in $EQT(K)$. If K is a flat top-cluster, then all terms in $EQT(K)$ are of the form $X(y)$ or x.

In the following we write cycles and cluster instead of SO-cycles and SO-cluster.

7 The Termination Ordering

We give the termination measure adapted from [SS99b]. It follows from [SS99b] that every rule of CSCU strictly decreases this measure. The importance for the result in this paper is an upper estimation of the number of applications of cycle-rules.

Definition 7.1. *Let L be a cycle. The measure $\psi(L)$ is a lexicographic combination of the following components:*

1. *The length of L.*

2. The length of L minus the flat-length of L.

Definition 7.2. *The (well-founded) measure μ for termination, also written $\mu(\Gamma)$ is a lexicographic combination $(\mu_1, \ldots, \mu_5)$ of the following well-founded measures:*

1. μ_1: *The number of context variables in Γ.*
2. μ_2: *A measure for cycles: ∞, if there is no cycle, otherwise, the minimal $\psi(L)$ for all cycles in Γ. We use that $\infty > a$ for all $a \neq \infty$.*
3. μ_3: *If there is an cycle or if there is no flat top-cluster, then ∞. If there is a flat top-cluster, then the minimal number of context variables in a flat top-cluster.*
4. μ_4: *The number of occurrences of function symbols in $T(\Gamma)$.*
5. μ_5: *The number of equations in $T(\Gamma)$.*

8 The Main Rules of CSCU

8.1 Cycles

The algorithm SCU from [SS99b] has 3 rules for the treatment of cycles: We only mention the required actions, since otherwise we have to copy the paper [SS99b]. The rules have to be applied to a shortest cycle. Let $X_1, \ldots, X_h$ be the context variables in the cycle. We assume that the cycle is of the form

$$X_1(x_1) \doteq y_1, y_1 \doteq X_2(z_1), X_2(x_2) \doteq y_2, y_2 \doteq X_3(z_2), \ldots,$$
$$X_j(x_j) \doteq y_{j,1}, y_{j,1} \doteq C_{j,1}[y_{n,2}], \ldots, y_{j,k-1} \doteq C_{j,k}[y_{j,k}], y_{j,k} \doteq Y(z_j), \ldots$$

where $Y \equiv X_{j+1}$ or $j = h$ and $Y \equiv X_1$. In all the rules below, if not (CV-elimination) is selected, we have to add the sharing equation $U = C_{j,1} \ldots C_{j,k}$. The derailing has the obey the same conditions as in [SS99b].

Definition 8.1. *Rule (Standardize-cycle): This rule can only be applied to a non-standardized cycle.*

1. *(CV-elimination) Select some X_i and eliminate it.*
2. *(partial-prefix) Let $0 < m \leq |U|$. For all $i = 1, \ldots, j$, replace X_i by $pow(U, 0, m)(X_i')$ or by $pow(U, 0, m)$, where the replacement by $pow(U, 0, m)$ has to be selected at least once.*
3. *(full-prefix) For all $i = 1, \ldots, j$, replace X_i by $pow(U, 0, |U|)(X_i')$.*
4. *(derailing) Let $0 < m \leq |U|$. For all $i = 1, \ldots, j$, replace X_i by $pow(U, 0, m)$ $(f(x_{i,1}, \ldots, X_i', \ldots, x_{i,ar(f)}))$.*

Definition 8.2. *Rule (Solve-standardized-ambig-cycle): This rule can only be applied, if the cycle is standardized and ambiguous.*

1. *(CV-elimination) Select some X_i and eliminate it.*

2. *(partial-prefix) Let* $0 < m \leq |U|$. *For all* $i = 1, \ldots, j$, *replace* X_i *by* $pow(U, 0, m)(X_i')$ *or by* $pow(U, 0, m)$, *where the replacement by* $pow(U, 0, m)$ *has to be selected at least once.*
3. *(derailing) Let* $0 < m \leq |U|$. *For all* $i = 1, \ldots, h$, *replace* X_i *by* $pow(U, 0, m)(f(x_{i,1}, \ldots, X_i', \ldots, x_{i,ar(f)}))$.

Definition 8.3. *Rule (Solve-standardized-cycle): This rule can only be applied, if the cycle is standardized and not ambiguous.*

1. *(CV-elimination) Select some* X_i *and eliminate it.*
2. *(partial-prefix) Let* $0 < m \leq 2^{E_I} * |U|$. *For all* $i = 1, \ldots, j$, *replace* X_i *by* $pow(U, 0, m)(X_i')$ *or by* $pow(U, 0, m)$, *where the replacement by* $pow(U, 0, m)$ *has to be selected at least once.*
3. *(derailing) This selection requires that* $h > 1$. *Let* $0 < m \leq 2^{E_I} * |U|$. *For all* $i = 1, \ldots, h$, *replace* X_i *by* $pow(U, 0, m)(f(x_{i,1}, \ldots, X_i', \ldots, x_{i,ar(f)}))$.

After the application of the rules, decomposition has to be applied. The purpose of (decompose-pp), (remove-power), and (shift-power) is to mimic the decomposition effect of the algorithm SCU in [SS99b], however, using less space.

8.2 Treatment of Clusters

The main goal in this section is to describe the adapted rules for top-clusters, and to argue that they can be applied without increasing space usage.
We assume that Γ is flattened, decomposed, and that there are no cycles. If one application of the transformation rules introduces a cycle, then the rules for cycle-elimination will then be applied until a context variable will be eliminated.

Definition 8.4 (Non-flat top-cluster). *This rule is only applicable if there are no cycles, no flat top-clusters, but a non-flat top-cluster* K.

Let $K = \{x_1, \ldots, x_{h_1}, X_1, \ldots, X_{h_2}\}$ *be a non-flat top-cluster. Then select one of the following two possibilities:*

1. *(CV-elimination) This selection is only applicable, if* $K \cap \mathcal{V}_2 \neq \emptyset$. *Select some* $X_i, i \in [1..h_2]$ *and instantiate it by* Id.
2. *(rigid-flexible) The equations in* $EQ(K)$ *are of the form* $x \doteq f(x_1, \ldots, x_n)$, $x \doteq X(y)$, *or* $x \doteq pow(U, n_1, n_2)(y)$. *If there are different top function symbols of terms of the form* $f(\ldots)$ *or* $pow(\ldots)(\ldots)$ *in* $EQT(K)$, *then Fail. Otherwise, let* f *be the unique top symbol of these terms in* $EQT(K)$.
   For every first order variable $x_i \in K$, *let the instantiation be* $f(x_{i,1}, \ldots, x_{i,n})$, *and for a context variable* $X_i \in K$, *first select a position* k_i, *and let the instantiation be* $X_i \to f(x_{i,1}, \ldots, x_{i,k_i-1}, X_i', x_{i,k_i+1}, \ldots, x_{i,n})$. *The variables* $x_{i,j}$ *will be fresh ones.*
   Apply decomposition rules and flattening.

Definition 8.5 (flat top-clusters). *This rule is applicable if there are no cycles, but a flat top-cluster.*

Let $K = \{x_1, \dots, x_{h_1}, X_1, \dots, X_{h_2}\}$ be a flat top-cluster with a minimal number h_2 of context variables. The following selections are possible

1. *(CV-eliminate) Select some context variable in K and eliminate it.*
2. *(flexible-flexible branching) This case requires $|K| > 1$ and that the maximal arity of function symbols in the signature Σ is greater than 1.*
   Let F be a new function symbol with $2 \leq ar(F) \leq |K|$. For every context variable $X_i \in K$, select an index $1 \leq k_i \leq ar(F)$ and instantiate $X_i(\cdot)$ by $F(x_{i,1}, \dots, \underbrace{X_i'(\cdot)}_{k_i}, \dots, x_{i,ar(F)})$. There must be different selections of the index k_i. For every first order variable $x_i \in K$ replace x_i by $F(y_{i,1}, \dots, y_{i,ar(F)})$, where $x_{i,j}, y_{i,j}, X_i'$ are new.

Then decompose the equations that result from instantiating and flattening the equations in $EQ(K)$.

Proposition 8.6. *The application of the rules for top-clusters does not increase the size usage. The intermediate use of space is $O(n)$.*

Proof. 1. First consider the rule for flat top-clusters. Assume selection 2 is chosen. The variables from K do only occur in the equations from $EQT(K)$. Hence after the instantiation the equations can be decomposed. The only remaining equations are of the form $x = X'(y)$, where X' is a freshly introduced context variable. Thus the number of equations remains the same, even the sum of their sizes is the same.

2. The rule for non-flat top-cluster was applied.
   Assume, selection 2 is chosen. Then the equations in $EQT(K)$ are instantiated, and after the application of decompose, the number of equations is the same or smaller.
   - An equation $x \doteq X(y)$ leaves one equation of the form $x' \doteq X'(y')$
   - An equation $x \doteq f(x_1, \dots, x_n)$ leaves no equation at all, since the function symbol f (F, respectively) is removed by decomposition, as well as all first order variables in K.
   - An equation $x \doteq pow(U, n_1, n_2)(y)$ either leaves Id, or an equation $x' \doteq pow(U, n_1 + 1, n_2)(y)$.

   Since the size of power expressions is defined as if n_1 is represented using the same space as n_2, there is no size increase.

9 CSCU: Estimations

The (non-deterministic) algorithm CSCU has as input a stratified context unification problem Γ_I without any power expressions. First, Γ_I is flattened in order to exploit sharing. Then apply the following repeatedly until a Fail is signalled or success is signalled for empty Γ:

1. If there is a cycle, then apply cycle-elimination
2. If there is no cycle, then apply cluster-elimination.

Let D_I be the size of Γ_I, and let $\#CV := |Var_2(\Gamma_I)|$.

Lemma 9.1. *The rules for flattening make at most an $O(n)$ increase in size for the initial context unification problem.*

Unexpectedly, the numbers in power expressions will be responsible for a large contribution to the size requirements. Let Q be an upper bound for the size of power expressions.

Lemma 9.2. *The overall size increase by CSCU is $O(Q * D_I^7)$.*

Proof. The hard part are the cycle elimination rules: If a cycle S with h context variables is given, the cycle-elimination will be applied at most h^2 times until the number of context variables is strictly reduced by one. Thus the overall number of applications of cycle-elimination is at most $O(\#CV^3)$.
The following holds:

1. The number of occurrences of context variables is always $\leq D_I$.
2. The number of occurrences of function symbols in $T(\Gamma)$ is at most D_I.
3. The number of occurrences of power expressions in $T(\Gamma)$ is at most: (#occurrences of context variables)∗(#applications of cycle rules) + (#applications of (solve-standardized-cycle)), which is $< \#D_I * O(\#CV^3)$. i.e. $O(D_I^4)$.
4. The number of context sharing equations is at most the number of applications of cycle-rules, i.e., $O(D_I^3)$.
5. The number of components in the right hand side of a context sharing equation is bounded by the number of occurrences of function symbols and power expressions in $T(\Gamma)$. This is of order $O(D_I^4)$.

Now we can estimate the total size usage:

- $T(\Gamma)$: D_I^2 for terms of the form $f(x_1, \ldots, x_n)$, $O(Q * D_I^4)$ for power expressions, and D_I for context variables. The number of first order variables is at most the number of equations. The size is of order $O(Q * D_I^4)$.
- $D(\Gamma)$. There are at most D_I^3 right hand sides, and the size of the right hands is of order $O(Q * D_I^4)$. The size of $D(\Gamma)$ is of order $O(Q * D_I^7)$

In summary, the order of space usage is $O(Q * D_I^7)$.
We also have to check that there is no unexpected size explosion during the application of the rules. The only critical operation is computing the *hd* of power expressions, which can be done in the order of the size of right hand sides of context sharing equations, which is $O(Q * D_I^4)$. □

Lemma 9.3. *The size Q of power-expressions in CSCU is of order $O(D_I^4)$*

Proof. Enumerate the defined context variables as U_i in the sequence of their creation by the algorithm and define $d_i := D_I^4 * 2^{c*D_I} * d_{i-1}$ where $c = 2.14$ (see proposition 2.1). Using induction on i we see that $|U_i| \leq d_i$ for all i by estimating right hand sides of the context sharing equations. Thus the upper bound is $(D_I^4 * 2^{c*D_I})^{D_I^3}$, since the number of context sharing equations is $O(D_I^3)$. This is of order $O(2^{c*D_I^4})$. In summary, this means, the size of power expressions is of order $O(D_I^4)$. □

Now we can state an upper bound on the complexity of the algorithm:

Theorem 9.4. *CSCU can be performed in polynomial space.*

Proof. The space usage is $O(D_I^{11})$, which is polynomial. Furthermore, NPSPACE is contained in PSPACE. □

Corollary 9.5. *Stratified context unification is in PSPACE.*

Corollary 9.6. *Satisfiability of one-step rewrite constraints is in PSPACE.*

Note that the number of steps of the algorithm CSCU may be exponential. In retrospect, this result permits a worst case estimation of the size-complexity of the algorithm for stratified context unification in [SS99b]. First use replacement of first order variables. The result has an exponential number of power expressions. Since after exploding the power expressions the space requirement per power expression is doubly exponential, so the non optimized algorithm SCU in [SS99b] is likely to perform in doubly exponential space.

10 Conclusion

The paper shows that unifiability of stratified context unification problems can be decided in polynomial space, which is far better than the highly complex algorithm given in [SS99b]. The lower bound for stratified context unification is $\mathcal{NP}$-hard [SSS98], hence there remains a gap, which we leave for future research. Since context unification is equivalent to one-step rewrite constraints, the paper also shows PSPACE as an upper complexity bound for their decidability.

The techniques in this paper may enable to show improved upper bounds for bounded second order unification [SS99a] as well as for D-unification [SS98].

The author conjectures that stratified context unification is in $\mathcal{NP}$.

References

Com93. H. Comon. Completion of rewrite systems with membership constraints, part I: Deduction rules and part II: Constraint solving. Technical report, CNRS and LRI, Université de Paris Sud, 1993.

Com98a. H. Comon. Completion of rewrite systems with membership constraints. Part I: Deduction rules. *J. Symbolic Computation*, 25(4):397–419, 1998.

Com98b. H. Comon. Completion of rewrite systems with membership constraints. Part II: Constraint solving. *J. Symbolic Computation*, 25(4):421–453, 1998.

EN00. Katrin Erk and Joachim Niehren. Parallelism constraints. In *Proc. of the 11th RTA*, volume 1833 of *LNCS*, pages 110–126, 2000.

Far91. W.A. Farmer. Simple second-order languages for which unification is undecidable. *J. Theoretical Computer Science*, 87:173–214, 1991.

Gol81. W.D. Goldfarb. The undecidability of the second-order unification problem. *Theoretical Computer Science*, 13:225–230, 1981.

Gut98. C. Gutierrez. Satisfiability of word equations with constants is in exponential space. In *Proc. FOCS'98*, pages 112–119, Palo Alto, California, 1998. IEEE Computer Society Press.

Hue75. Gerard Huet. A unification algorithm for typed λ-calculus. *Theoretical Computer Science*, 1:27–57, 1975.

Lev96. Jordi Levy. Linear second order unification. In *Proc. of the 7th RTA*, volume 1103 of *LNCS*, pages 332–346, 1996.

LV00a. Jordi Levy and Margus Veanes. On the undecidability of second-order unification. *Information and Computation*, 159:125–150, 2000.

LV00b. Jordi Levy and Mateu Villaret. Linear second-order unification and context unification with tree-regular constraints. In *Proc. of the 11th RTA*, volume 1833 of *LNCS*, pages 156–171, 2000.

Mak77. G.S. Makanin. The problem of solvability of equations in a free semigroup. *Math. USSR Sbornik*, 32(2):129–198, 1977.

NPR97a. Joachim Niehren, Manfred Pinkal, and P. Ruhrberg. On equality up-to constraints over finite trees, context unification, and one-step rewriting. In *CADE '97*, volume 1249 of *LNCS*, pages 34–48, 1997.

NPR97b. Joachim Niehren, Manfred Pinkal, and Peter Ruhrberg. A uniform approach to underspecification and parallelism. In *Proc. of the 35th Annual Meeting of the Ass. of Computational Linguistics*, pages 410–417, Madrid, Spain, 1997.

NTT00. Joachim Niehren, Sophie Tison, and Ralf Treinen. On rewrite constraints and context unification. *Information Processing Letters*, 74:35–40, 2000.

Pie73. T. Pietrzykowski. A complete mechanization of second-order type theory. *J. ACM*, 20:333–364, 1973.

Pla99a. W. Plandowski. Satisfiability of word equations with constants is in NEXPTIME. In T. Leighton, editor, *Proc. STOC'99*, pages 721–725, Atlanta, Georgia, 1999. ACM Press.

Pla99b. W. Plandowski. Satisfiability of word equations with constants is in PSPACE. In *FOCS 99*, pages 495–500, 1999.

SG89. Wayne Snyder and Jean Gallier. Higher-order unification revisited: Complete sets of transformations. *J. Symbolic Computation*, 8:101–140, 1989.

SS94. Manfred Schmidt-Schauß. Unification of stratified second-order terms. Internal Report 12/94, Fachbereich Informatik, J.W. Goethe-Universität Frankfurt, Frankfurt, Germany, 1994.

SS98. Manfred Schmidt-Schauß. A decision algorithm for distributive unification. *TCS*, 208:111–148, 1998.

SS99a. Manfred Schmidt-Schauß. Decidability of bounded second order unification. Technical Report Frank-report-11, FB Informatik, J.W. Goethe-Universität Frankfurt am Main, 1999. submitted for publication.

SS99b. Manfred Schmidt-Schauß. A decision algorithm for stratified context unification. Frank-Report 12, Fachbereich Informatik, J.W. Goethe-Universität Frankfurt, Frankfurt, Germany, 1999. accepted for publication in Journal of Logic and Computation, preliminary version available at `http://www.ki.informatik.uni-frankfurt.de/papers/articles.html`.

SS99c. Manfred Schmidt-Schauß. An optimized decision algorithm for stratified context unification. Technical Report Frank-report-13, FB Informatik, J.W. Goethe-Universität Frankfurt am Main, 1999.

SSS98. Manfred Schmidt-Schauß and Klaus U. Schulz. On the exponent of periodicity of minimal solutions of context equations. In *Proc. of the 9th RTA*, volume 1379 of *LNCS*, pages 61–75, 1998.

SSS99. Manfred Schmidt-Schauß and Klaus U. Schulz. Solvability of context equations with two context variables is decidable. In *Proc. of the Int. Conf. on Automated Deduction*, volume 1632 of *LNCS*, pages 67–81, 1999.

Vor98. Sergei Vorobyov. $\forall\exists^*$-equational theory of context unification is Π_1^0-hard. In *MFCS 1998*, volume 1450 of *LNCS*, pages 597–606. Springer-Verlag, 1998.

Uniform Derivation of Decision Procedures by Superposition*

Alessandro Armando[1], Silvio Ranise[1,2], and Michaël Rusinowitch[2]

[1] DIST–Università degli Studi di Genova, via all'Opera Pia 13, 16145, Genova, Italy
{armando,silvio}@dist.unige.it
Phone: +39.010.353-2216 and Fax: +39.010.353-2948
[2] LORIA-INRIA-Lorraine, 615, rue du Jardin Botanique, BP 101,
54602 Villers les Nancy Cedex, France
{ranise,rusi}@loria.fr
Phone: (33) 03.83.59.30.20 and Fax: (33) 03.83.27.83.19

Abstract. We show how a well-known superposition-based inference system for first-order equational logic can be used almost directly as a decision procedure for various theories including lists, arrays, extensional arrays and combinations of them. We also give a superposition-based decision procedure for homomorphism.

Keywords: Automated Deduction, Equational Logic, Term Rewriting, Superposition, Decision Procedures, Lists, Arrays with Extensionality, Homomorphism

1 Introduction

In verification with proof assistants (such as PVS, COQ, HOL, and Nqthm), decision procedures are typically used for eliminating trivial subgoals represented for instance as sequents modulo a background theory. These theories axiomatize standard data-types such as arrays, lists, bit-vectors and have proved to be quite useful for, e.g., hardware verification. Elimination of trivial sequents often reduces to the **problem of proving the unsatisfiability of conjunctions of literals modulo a background theory** T, which is the problem we shall consider here.

The rewriting approach permits us the uniform design of decision procedures for eliminating these subgoals and also offers an efficient alternative to congruence closure techniques. This approach was inspired by Greg Nelson's thesis [Nel81] where it is suggested to apply Knuth-Bendix completion to derive decision procedures. Here, instead of the Knuth-Bendix completion procedure, we apply a standard complete superposition-based inference system for clausal equational logic (given for instance in [NR01]). This allows us not only to handle pure equality but also several interesting axiomatic theories that were not handled previously that way such as lists, arrays, and extensional arrays. The proof

* The authors would like to thank C. Ringeissen and L. Vigneron for their comments on a draft of this paper and the anonymous referees for helpful criticisms.

L. Fribourg (Ed.): CSL 2001, LNCS 2142, pp. 513–527, 2001.

that the decision procedures are correct is straightforward w.r.t. other correctness proofs given in the literature (compare for instance our decision procedure for arrays with extensionality of Section 6 with [SDBL01]). In our approach, combining theories is also immediate. As an illustration, we show how to decide a combination of lists and arrays.

A second contribution of the paper is in the same spirit of applying Knuth-Bendix completion to derive a decision procedure for the theory of homomorphism. This is the first decision procedure, to our knowledge, for this theory.

Related work. For lack of space we only discuss results that are closely related to ours. In previous work, the rewriting approach was mainly used for pure equality theories. For instance, [BT00] focus on abstracting the control of congruence closure algorithms, in order to give a uniform presentation of several known algorithms. A recent extension to deal with equality modulo AC is presented in [BRTV00].

In [NO80], Nelson and Oppen describe a decision procedure for the "quantifier-free theory of the LISP list structure". The procedure is obtained as an extension of a congruence closure algorithm with a mechanism which augments the graph by selected instances of the axioms of the theory. The proof of correctness is model theoretic and seems difficult to generalize. A discussion of the difficulties of deriving a general method to obtain decision procedures by extending congruence closure algorithms as well as a decision procedure for the theory of arrays (without extensionality) can be found in [Nel81]. This discussion has motivated our work.

In [SDBL01], the first decision procedure for an extensional theory of arrays is presented. The key ingredient is a modified congruence closure algorithm which is capable of handling (so called) partial equations. The correctness proof is rather complex and it takes the main part of the paper; it is model-theoretic and rather *ad-hoc*. In Section 6, we give a decision procedure for the same theory considered in [SDBL01]. Our procedure is simpler to understand since it amounts to applying (almost directly) standard equality reasoning in contrast to handling partial equalities and our proof of correctness relies on basic properties of skolemization. As a consequence, the decision procedure (as well as its correctness proof) for the theory of arrays with extensionality can be adapted to similar presentations for sets and multisets.

Finally, we notice that we can easily derive decision procedures for combinations of theories in a manner closely resembling the combination schema described in [NO78]. This is exemplified for a combination of the theory of lists and arrays in Section 7. Furthermore, the decision procedures derived in our framework can be extended so to provide the interface functionalities needed for them to be plugged into the Nelson and Oppen combination schema [NO78].

2 Preliminaries

We assume the usual (first-order) syntactic notions of *signature*, (*ground*) *term*, *position*, *substitution*, *replacement*, *rewrite relation* $\rightarrow$, as defined, e.g., in [DJ90].

If Σ is a signature and X is a set of variables, then $T(\Sigma, X)$ denotes the set of terms built out of the symbols in Σ and the variables in X. $T(\Sigma)$ abbreviates $T(\Sigma, \emptyset)$. 0-ary function symbols are called *individual constants*. Let l and r be elements of $T(\Sigma, X)$, then $l = r$ is a $T(\Sigma, X)$-*equality* and $\neg(l = r)$ (also written as $l \neq r$) is a $T(\Sigma, X)$-*disequality*. A $T(\Sigma, X)$-*literal* is either a $T(\Sigma, X)$-equality or a $T(\Sigma, X)$-disequality, i.e. an expression of the form $s \bowtie t$ where $\bowtie$ is either $=$ or $\neq$. A $T(\Sigma, X)$-*clause* is a disjunction of literals, i.e. an expression of the form $\neg A_1 \vee \cdots \vee \neg A_n \vee B_1 \vee \cdots \vee B_m$ (abbreviated with $A_1, \ldots, A_n \Rightarrow B_1, \ldots, B_m$) where $A_1, \ldots, A_n$, $B_1, \ldots, B_m$ are $T(\Sigma, X)$-equalities ($n \geq 0$ and $m \geq 0$). We simply use the terms equality, disequality, literals, and clauses when $T(\Sigma, X)$ is clear from the context. A *flat equality* is an equality of the form $f(t_1, \ldots, t_n) = t_0$ or $t_0 = f(t_1, \ldots, t_n)$ where f is an n-ary function symbol and t_i is either a variable or an individual constant for $i = 0, 1, \ldots, n$ with $n \geq 0$. A *distinction* is a disequality $t_1 \neq t_2$, where t_i is either a variable or an individual constant for $i = 1, 2$. A *flat literal* is either a flat equality or a distinction. A *flat clause* is a disjunction of flat literals.

We assume the usual (first-order) notions of interpretation, satisfiability, validity, logical consequence (in symbols, $\models$), and theory (see, e.g., [End72]). Let S be a set of ground literals, then we say that S is T-*satisfiable* (T-*unsatisfiable*) iff $T \cup S$ is satisfiable (unsatisfiable, resp.). All the theories we shall consider in this paper contain the quantifier-free theory of equality $\mathcal{E}$.

Example 1. Assume that the axiom of T is $h(f(x, y)) = f(h(x), h(y))$ (where x and y are implicitly universally quantified variables). We can show the T-unsatisfiability of $\{h(c) = c',\ h(c') = c,\ f(c, c') = h(h(a)),\ f(c', c) = a,\ h(h(h(a))) \neq a\}$.

The *satisfiability problem for a theory* T amounts to establishing whether any given finite set of ground literals is T-satisfiable or not. A *decision procedure for* T is any algorithm that solves the satisfiability problem for T.

3 Our Approach

In this paper, we propose a uniform approach based on superposition inference rules to build decision procedures for a variety of decidable theories. For all theories T, **the first step is to flatten all the input literals**. The soundness of this preprocessing step is ensured by the following fact.

Lemma 1. *Let T be a $T(\Sigma, X)$-theory and S be a finite set of $T(\Sigma)$-literals. Then there exists a finite set of flat $T(\Sigma')$-literals S' (where Σ' is obtained from Σ by adding a finite number of individual constants) such that S' is T-satisfiable iff S is.*

Notice that flattening augments the size of the input set S of literals to $O(n)$, where n is the number of subterms in S.

Example 2. The following set of flat literals can be derived from the previous example: $\{h(c) = c',\ h(c') = c,\ f(c, c') = c_2,\ f(c', c) = a, h(a) = c_1, h(c_1) = c_2,\ h(c_2) = c_3, c_3 \neq a\}$.

Table 1. Inference rules of $\mathcal{SP}$

Name	Rule	Applicability Conditions
Superposition	$\dfrac{\Gamma \Rightarrow \Delta, l[u'] = r \quad \Pi \Rightarrow \Sigma, u = v}{\sigma(\Gamma, \Pi \Rightarrow \Delta, \Sigma, l[v] = r)}$	$\sigma(u) \not\preceq \sigma(v)$, $\sigma(u = v) \not\preceq \sigma(\Pi \cup \Sigma)$, $\sigma(l[u']) \not\preceq \sigma(r)$, $\sigma(l[u'] = r) \not\preceq \sigma(\Gamma \cup \Delta)$
Paramodulation	$\dfrac{\Gamma, l[u'] = r \Rightarrow \Delta \quad \Pi \Rightarrow \Sigma, u = v}{\sigma(l[v] = r, \Gamma, \Pi \Rightarrow \Delta, \Sigma)}$	$\sigma(u) \not\preceq \sigma(v)$, $\sigma(u = v) \not\preceq \sigma(\Pi \cup \Sigma)$, $\sigma(l[u']) \not\preceq \sigma(r)$, $\sigma(l[u'] = r) \not\prec \sigma(\Gamma \cup \Delta)$
Reflection	$\dfrac{\Gamma, u' = u \Rightarrow \Delta}{\sigma(\Gamma \Rightarrow \Delta)}$	$\sigma(u' = u) \not\prec \sigma(\Gamma \cup \Delta)$
Factoring	$\dfrac{\Gamma \Rightarrow \Delta, u = t, u' = t'}{\sigma(\Gamma, t = t' \Rightarrow \Delta, u = t')}$	$\sigma(u) \not\preceq \sigma(t)$,$\sigma(u) \not\preceq \sigma(\Gamma)$, $\sigma(u = t) \not\prec \sigma(\{u' = t'\} \cup \Delta)$

Table 2. Simplification rules of $\mathcal{SP}$

Name	Rule	Applicability Conditions
Subsumption	$\dfrac{S \cup \{C, C'\}}{S \cup \{C\}}$	for some substitution $\theta(C) \subseteq C'$, and there is no substitution ρ such that $\rho(C') = C$
Simplification	$\dfrac{S \cup \{C[l'], l = r\}}{S \cup \{C[\theta(r)], l = r\}}$	$l' = \theta(l)$, $\theta(l) \succ \theta(r)$, and $C[\theta(l)] \succ (\theta(l) = \theta(r))$
Deletion	$\dfrac{S \cup \{\Gamma \Rightarrow \Delta, t = t\}}{S}$	

We will make use of a superposition calculus, $\mathcal{SP}$, comprising the inference rules of Table 1 and the simplification rules of Table 2. $\mathcal{SP}$ is taken from [NR01]. It extends the system from [Rus91] by the *equality factoring rule* [BG94], so that more ordering restrictions are possible (in the non-Horn case). The relation $\succ$ is a reduction ordering [DJ90], which is total on ground terms. $\succ$ is extended to literals in the following way: $(a \bowtie b) \succ (c \bowtie d)$ if $\{a, b\} \succ\!\!\succ \{c, d\}$, where $\succ\!\!\succ$ is the multiset extension of $\succ$. Multisets of literals are compared using the multiset extension of $\succ$ on literals.

An inference system including simplification rules is refutationally complete if *any fair application of the rules to an unsatisfiable set of clauses will derive the empty clause.* Fairness means that if some inference is possible it will be performed at some step unless one of the parent clauses gets simplified, subsumed, or deleted. The calculus $\mathcal{SP}$ is known to be refutationally complete for general first-order equational logic [BG94,NR01]. (Note that for Horn clauses *Equality Factoring* is useless [KR91].) In Table 1 the substitution σ is the most general unifier of u and u', and u' is not a variable in *Superposition* and *Paramodulation.* We shall write Factoring instead of Equality Factoring for conciseness. In this paper, a *saturation* of a set of clauses by $\mathcal{SP}$ is the final set of clauses generated by a fair derivation from S using rules in $\mathcal{SP}$ with higher priority given to the simplification rules. If the saturation terminates for the union of T and any set of ground flat literals then it is a decision procedure for T: if the final set of clauses contains the empty clause then the input set of literals is unsatisfiable; it is satisfiable, otherwise. This is a direct consequence of the refutational completeness of

$\mathcal{SP}$. From now on, we shall call $\mathcal{SP}$ any fair application of the inference system with priority given to the simplification rules.

3.1 A Decision Procedure for the Quantifier-Free Theory of Equality

The following result says that $\mathcal{SP}$ can be used as a decision procedure for the quantifier-free theory of equality $\mathcal{E}$.[1] In fact, the decision procedure we obtain is just a variant of the Knuth-Bendix completion procedure (similar to the rational reconstruction of Nelson and Oppen's congruence closure algorithm of [BT00]). We shall assume now and in the remainder of this paper that the ordering $\succ$ **is s.t.** $t \succ c$ **for each constant** c **and for each ground term** t **that contains a symbol of arity greater than** 0. Note that it is easy to satisfy this requirement with a suitable precedence ordering.

Lemma 2. *Let S be a finite set of flat $T(\Sigma)$-literals. All the saturations of S by $\mathcal{SP}$ are finite.*

Proof. Note that *Simplification* is applicable whenever *Superposition* is. Hence *Superposition* is useless since *Simplification* has higher priority. *Simplification* and *Paramodulation* generate ground flat literals. *Reflection* generates the empty clause (which subsumes all other clauses). Since the number of possible ground flat literals is finite, it readily follows that all saturations are finite. □

Theorem 1. *$\mathcal{SP}$ is a decision procedure for $\mathcal{E}$.*

Let n be the size of the input set of flattened literals. Each *Simplification* or *Paramodulation* replaces a subterm by a $\succ$-smaller constant (i.e. a term of type $f(c_1, \ldots, c_n)$ or c' by some c). Hence the maximal number of inference steps is equal to the number of subterms times the number of constants in Σ, i.e. $O(n^2)$. Since finding a *Simplification* or *Paramodulation* inference is polynomial, the whole saturation is polynomial.

4 A Decision Procedure for the Theory of Lists

Let $\Sigma_{\mathcal{L}}$ be a signature containing the function symbols car (unary), cdr (unary), and cons (binary), and let $\mathcal{L}$ be the theory obtained by adding the following two axioms, denoted with $Ax(\mathcal{L})$, to $\mathcal{E}$:

$$\mathsf{car}(\mathsf{cons}(x, y)) = x \tag{1}$$

$$\mathsf{cdr}(\mathsf{cons}(x, y)) = y. \tag{2}$$

For simplicity, $\mathcal{L}$ is only a sub-theory of the "LISP list structure" considered in [NO80]. However, a decision procedure for such a theory can be derived by preprocessing the set of ground literals using the technique of [NO80] to eliminate negative occurrences of the predicate recognizing atoms and by applying $\mathcal{SP}$.

[1] We do not claim this result to be new; it is stated here only to give the flavor of our approach in the simple case of the pure equational theory.

Lemma 3. *Let S be a finite set of flat $T(\Sigma_{\mathcal{L}})$-literals. The clauses occurring in the saturations of $S \cup Ax(\mathcal{L})$ by $\mathcal{SP}$ can only be the empty clause, ground flat literals, or the equalities in $Ax(\mathcal{L})$.*

Proof. The proof is by induction on the length of the derivations. No inference between axioms in $Ax(\mathcal{L})$ is possible. Thus, by inspection of the rules in $\mathcal{SP}$, there are four cases to consider: (a) a *Simplification* between a ground flat equality and a ground flat literal,[2] (b) application of *Reflection* to a ground distinction, (c) a *Superposition* between an equality in $Ax(\mathcal{L})$ and a ground flat equality of the form $\mathsf{cons}(c_1, c_2) = c_3$ (where c_i is an individual constant for $i = 1, 2, 3$), or (d) a *Paramodulation* from a ground flat equality into a ground distinction. It is straightforward to verify that in case (a) only ground flat literals are generated, in case (b) the empty clause is generated, in case (c) ground flat equalities are generated, and finally in case (d) ground distinctions are generated. □

Lemma 4. *Let S be a finite set of flat $T(\Sigma_{\mathcal{L}})$-literals. All the saturations of $S \cup Ax(\mathcal{L})$ by $\mathcal{SP}$ are finite.*

Proof. By Lemma 3, we know that the saturations of $S \cup Ax(\mathcal{L})$ by $\mathcal{SP}$ can only contain the empty clause or ground flat literals. It is trivial to see that only a finite number of flat literals can be built out of a finite set of symbols and variables. □

Theorem 2. *$\mathcal{SP}$ is a decision procedure for $\mathcal{L}$.*

Let n be the size of the input set of flattened literals. At most $O(n^2)$ flat literals can be created by *Superposition* during saturation. The size of the current set of literals in a derivation is always bounded by a constant k which is $O(n^2)$. Other inferences take polynomial time in k according to Section 3.1. Hence overall the decision procedure is polynomial.

5 A Decision Procedure for the Theory of Arrays

Let $\Sigma_{\mathcal{A}}$ be a signature containing the function symbols select (binary) and store (ternary), and let $\mathcal{A}$ be the theory obtained by adding the following two axioms, denoted by $Ax(\mathcal{A})$, to $\mathcal{E}$:

$$\mathsf{select}(\mathsf{store}(a, i, e), i) = e \tag{3}$$

$$i \neq j \Rightarrow \mathsf{select}(\mathsf{store}(a, i, e), j) = \mathsf{select}(a, j) \tag{4}$$

(where a, i, j, and e are variables and (4) denotes $i = j \vee \mathsf{select}(\mathsf{store}(a, i, e), j) = \mathsf{select}(a, j)$). We shall assume that the ordering $\succ$ **is s.t. any term that contains select or store is $\succ$-bigger than all ground terms not containing them; moreover, all non constant symbols are greater than the constant ones**. Using an LPO ordering [DJ90], this can easily be ensured by a suitable precedence relation.

[2] Notice that *Superposition* can never apply to ground flat literals since *Simplification* has higher priority.

Lemma 5. *Let S be a finite set of flat $T(\Sigma_{\mathcal{A}})$-literals. The clauses occurring in the saturations of $S \cup Ax(\mathcal{A})$ by $\mathcal{SP}$ can only be:*

i) the empty clause; *ii) the axioms in $Ax(\mathcal{A})$;* *iii) ground flat literals;*
iv) clauses of the form $t \bowtie t' \vee c_1 = c'_1 \vee \cdots \vee c_n = c'_n$ where $c_1, c'_1, \ldots, c_n, c'_n$ ($n \geq 0$) are individual constants and $t \bowtie t'$ is either a distinction between two individual constants or an equality between individual constants or terms of the form $\mathsf{select}(c_i, i)$ (for some individual constants c and c_i);
v) clauses of the form $\mathsf{select}(c, x) = \mathsf{select}(c', x) \vee c_1 = k_1 \vee \cdots \vee c_n = k_n$, where k_i (for $i = 1, ..., n$) is either the variable x or is one among the individual constants $c, c', c_1, c'_1, \ldots, c_n, c'_n$ ($n \geq 0$).

Proof. The proof is by induction on the length of the derivations. The base case is simple and therefore omitted. By the induction hypothesis there are five types of clauses produced after n inference steps: *i)–v)*. For inferences with *Reflexion* or *Factoring* on one clause the result is obvious. *Deletion* and *Subsumption* do not create new clauses. For the sake of brevity, let *replacement* be either a *Superposition* or *Paramodulation* step. Let us consider inference steps involving two clauses. There are several cases to consider according to the categories the clauses belong to:

- *ii)-ii)*: A *Superposition* can be applied to the axioms in $Ax(\mathcal{A})$ but it generates the trivial clause $i = i \vee \mathsf{select}(a, i) = e$ which is immediately eliminated by *Deletion*. No new clause can be produced this way.
- *ii)-iii)*: A *Superposition* from a flat equality into axiom (3) produces a ground flat equality, i.e. a clause of type *iii)*, whereas a *Superposition* into axiom (4) produces a clause of type *v)*.
- *iii)-iii)*: The only possible inference is *Simplification* or *Paramodulation* between a ground flat equality and a ground flat literal. It produces only ground flat literals, i.e. a clause of type *iii)*.
- *iii)-iv)*: A replacement produces a clause of type *iv)*.
- *iii)-v)*: A replacement produces a clause of type *iv)* or *v)*.
- *iv)-iv)*: A replacement produces a clause of type *iv)*.
- *iv)-v)*: A replacement produces a clause of type *iv)*.
- *v)-v)*: A replacement produces a clause of type *iv)* or *v)*.

There are no possible inference between axioms and clauses of type *iv)* or *v)*. □

Lemma 6. *Let S be a finite set of flat $T(\Sigma_{\mathcal{A}})$-literals. All the saturations of $S \cup Ax(\mathcal{A})$ by $\mathcal{SP}$ are finite.*

The proof of this Lemma is analogous to that of Lemma 4 and therefore it is omitted.

Theorem 3. *$\mathcal{SP}$ is a decision procedure for $\mathcal{A}$.*

Let n be the size of the input set of flattened literals. At most $O(2^{n^k})$ clauses can be generated by saturation for some k (in fact $k = 2$). Hence the decision procedure takes time $O(2^{n^k})$.

Finally, it is worth noticing that the above decision procedure is similar to the algorithm described in [Nel81].

6 A Decision Procedure for the Theory of Arrays with Extensionality

Let $\mathcal{A}^s$ be the many-sorted version of the theory $\mathcal{A}$ of Section 5, i.e. the many-sorted theory with sorts ELEM, INDEX, and ARRAY, with function symbols store and select of type $\mathrm{ARRAY}, \mathrm{INDEX}, \mathrm{ELEM} \longrightarrow \mathrm{ARRAY}$ and $\mathrm{ARRAY}, \mathrm{INDEX} \longrightarrow \mathrm{ELEM}$ respectively, and with the sorted version of (3) and (4) as axioms. (Notice that the use of sorts allows us to avoid problematic terms such as $\mathsf{store}(a, \mathsf{store}(a, i, e), \mathsf{select}(a, \mathsf{store}(a, i, e)))$.) Let $\mathcal{A}^s_e$ be the many-sorted theory of arrays with extensionality obtained from $\mathcal{A}^s$ by extending the set of axioms with

$$\forall i.(\mathsf{select}(a, i) = \mathsf{select}(b, i)) \Rightarrow a = b \qquad (5)$$

where a and b are variables of sort ARRAY and i is a variable of sort INDEX (by abuse of notation, (5) denotes its clausal form). $\Sigma_{\mathcal{A}^e_s}$ denotes a signature containing the function symbols select, store, and a finite set of function symbols s.t. **if f is a function symbol of type $s_0, \ldots, s_{n-1} \longrightarrow s_n$ distinct from select and store, then s_i is either INDEX or ELEM, for all $i = 0, 1, ..., n$ and** $n \geq 1$. Furthermore, we assume that $\Sigma_{\mathcal{A}^e_s}$ admits at least one ground term for each sort, i.e. it is a sensible signature. Finally, let $Ax(\mathcal{A}^s)$ and $Ax(\mathcal{A}^s_e)$ be the set of axioms of $\mathcal{A}^s$ and of $\mathcal{A}^s_e$, respectively.

Lemma 7. *Let S be a set of $T(\Sigma_{\mathcal{A}^e_s})$-literals and let S' be obtained from S by replacing all the inequalities of the form $t \neq t'$ with $\exists i.\mathsf{select}(t, i) \neq \mathsf{select}(t', i)$, where t and t' are terms of sort* ARRAY. *Then S is $\mathcal{A}^s_e$-satisfiable iff S' is $\mathcal{A}^s$-satisfiable.*

Proof. We must show that $S \cup \mathcal{A}^s_e$ is satisfiable iff $S' \cup \mathcal{A}^s$ is or, equivalently, that $S \cup Ax(\mathcal{A}^s_e)$ is satisfiable iff $S' \cup Ax(\mathcal{A}^s)$ is. The 'only if' case is easy. For the 'if' case, let I be a (many-sorted) model of $S' \cup Ax(\mathcal{A}^s)$. We define the binary relation $\sim$ over ARRAY^I to hold whenever $\mathsf{select}^I(a, i) = \mathsf{select}^I(b, i)$ for all $i \in \mathrm{INDEX}^I$, and we define $\sim$ over the INDEX^I and ELEM^I to be the identity relation. We now show that $\sim$ is a $\Sigma_{\mathcal{A}^s_e}$-congruence. It is clearly an equivalence. To prove that $\sim$ is a congruence it remains to show that if $a \sim b$, then $\mathsf{store}^I(a, i, e) \sim \mathsf{store}^I(b, i, e)$ for all $i \in \mathrm{INDEX}^I$ and $e \in \mathrm{ELEM}^I$.[3] Let us assume that $a \sim b$ but $\mathsf{store}^I(a, i, e) \not\sim \mathsf{store}^I(b, i, e)$ for some $i \in \mathrm{INDEX}^I$ and $e \in \mathrm{ELEM}^I$, i.e. that $\mathsf{select}^I(\mathsf{store}^I(a, i, e), k) \neq \mathsf{select}^I(\mathsf{store}^I(b, i, e), k)$ for some $i, k \in \mathrm{INDEX}^I$ and $e \in \mathrm{ELEM}^I$. There are two cases to consider. If $k = i$ then, since I is a model of (3), we can conclude that $e \neq e$, a contradiction. Otherwise (i.e. if $k \neq i$), since I is a model of (4), we can conclude that $\mathsf{select}^I(a, k) \neq \mathsf{select}^I(b, k)$. This is in contradiction with the assumption $a \sim b$. To conclude the proof, it is sufficient to check that $I' = I/\sim$ is a model of $S' \cup Ax(\mathcal{A}^s_e)$. □

[3] The case for select trivially follows from the definition of $\sim$. For a function symbol in $\Sigma_{\mathcal{A}^s_e}$ distinct from select and store, congruence immediately follows from the definition of $\sim$ and the properties of identity.

Lemma 8. *Let S be a conjunction of ground literals, then S is $\mathcal{A}^s$-satisfiable iff it is $\mathcal{A}$-satisfiable.*

The following theorem is the key of our reduction mechanism.

Theorem 4. *Let S be a set of $T(\Sigma_{\mathcal{A}^s_e})$-literals and let S' be obtained from S by replacing all the inequalities of the form $t \neq t'$ with $\mathsf{select}(t, sk(t,t')) \neq \mathsf{select}(t', sk(t,t'))$, where t and t' are terms of sort* ARRAY, *and sk is a Skolem function of type* ARRAY, ARRAY $\longrightarrow$ INDEX. *Then S is $\mathcal{A}^s_e$-satisfiable iff S' is $\mathcal{A}$-satisfiable.*

Proof. The Theorem readily follows from Lemma 7, Lemma 8, and basic properties of skolemization. □

A **decision procedure for the theory of arrays with extensionality** $\mathcal{A}^s_e$ is as follows. Given as input a finite set S of $T(\Sigma_{\mathcal{A}^s_e})$-literals, the procedure first replaces every occurrence of literals of the form $t \neq t'$ with $\mathsf{select}(t, sk(t,t')) \neq \mathsf{select}(t', sk(t,t'))$, where t and t' are terms of sort ARRAY, and sk is a Skolem function of type ARRAY, ARRAY $\longrightarrow$ INDEX. Then, it feeds the resulting set of literals to the decision procedure for $\mathcal{A}$ described in Section 5.

It is worth noticing that our decision procedure can be straightforwardly generalized to multi-dimensional arrays if we view them as arrays of arrays.

The worst-case time of the decision procedure for $\mathcal{A}^s_e$ is that of the procedure for $\mathcal{A}$, i.e. $O(2^{n^k})$ for a fixed natural number k, since the size of the set of input literals obtained by the pre-processing step described above is $O(n)$.

7 Combining Decision Procedures for Lists and Arrays

To emphasize the flexibility of our approach, we show how easy it is to combine the decision procedures for the theories of lists and arrays. Let $\Sigma_{\mathcal{U}}$ be a signature containing the function symbols select (binary), store (ternary), car (unary), cdr (unary), and cons (binary). Let $Ax(\mathcal{U})$ be the set of axioms obtained as the union of $Ax(\mathcal{A})$, $Ax(\mathcal{L})$, and $\mathcal{E}$. Furthermore, we shall assume that the simplification ordering $\succ$ (total on ground terms) satisfies the requirements of Section 5.

Lemma 9. *Let S be a finite set of ground flat $T(\Sigma_{\mathcal{U}})$-literals. The clauses occurring in the saturations of $S \cup Ax(\mathcal{U})$ by $\mathcal{SP}$ can only be of the type i), iii), iv), v) given in Lemma 5, of the types given in Lemma 3, or elements of $Ax(\mathcal{U})$.*

Proof. Every *Superposition* or *Paramodulation* between axioms in $Ax(\mathcal{U})$ generate a clause that can be deleted. Hence the proof is as that of Lemma 3 and Lemma 5. □

Lemma 10. *Let S be a finite set of ground flat $T(\Sigma_{\mathcal{U}})$-literals. All the saturations of $S \cup Ax(\mathcal{U})$ by $\mathcal{SP}$ are finite.*

The proof of this Lemma is analogous to that of Lemma 4.

Theorem 5. *$\mathcal{SP}$ is a decision procedure for $\mathcal{U}$.*

8 A Decision Procedure for the Theory of Homomorphism

In this Section, we present an adaptation of the Knuth-Bendix completion procedure [KB70] to work modulo the theory of homomorphism. The completion process always terminates for ground equations and gives a decision procedure for this theory.[4]

Let $\Sigma_{\mathcal{H}}$ be a signature containing the unary function symbol h and let $\mathcal{H}$ be the theory obtained by adding instances of the following axiom schema, denoted with $Ax(\mathcal{H})$, to $\mathcal{E}$:

$$\mathsf{h}(f(x_1,\ldots,x_n)) = f(\mathsf{h}(x_1),\ldots,\mathsf{h}(x_n)) \tag{6}$$

where f is any n-ary function symbol ($n > 0$) in a subset Σ' of $\Sigma_{\mathcal{H}} \setminus \{\mathsf{h}\}$. We want to decide the $\mathcal{H}$-unsatisfiability of the set of ground literals ψ.

Example 3. $\{\mathsf{h}(c) = c', \mathsf{h}(c') = c, f(c,c') = \mathsf{h}(\mathsf{h}(a)), \mathsf{h}(\mathsf{h}(\mathsf{h}(a))) \neq a,\ f(c',c) = a\}$ is $\mathcal{H}$-unsatisfiable.

By Lemma 1, we can assume that ψ is a set of flat literals. Our decision procedure consists of two steps. First, we complete the set of ground equalities in ψ modulo $\mathcal{H}$ in order to get a rewrite system R. Second, for each inequality $s \neq t$ in ψ, we compute the normal form $s\downarrow_R$ of s and the normal form $t\downarrow_R$ of t (w.r.t. R). Then, if there exists an inequality $s' \neq t'$ in ψ s.t. $s'\downarrow_R$ is identical to $t'\downarrow_R$, ψ is $\mathcal{H}$-unsatisfiable; otherwise, ψ is $\mathcal{H}$-satisfiable.

8.1 Orientation

We introduce an ordering over ground terms which allows us to orient equalities as rewrite rules in such a way that a superposition between a ground equality and an equality in $Ax(\mathcal{H})$ can only generate a ground equality.

We first define a weight function on the symbols in $\Sigma_{\mathcal{H}}$, denoted with $[e]$ where e is in $\Sigma_{\mathcal{H}}$: $[c] = 1$, for each constant symbol c in $\Sigma_{\mathcal{H}}$; $[\mathsf{h}] = 0$; and $[f] = 1$, for f in $\Sigma_{\mathcal{H}}$ s.t. f is not a constant and f is not h. The weight of a ground term t, denoted with $[t]$, is the sum of the weight of the symbols (of $\Sigma_{\mathcal{H}}$) occurring in it. Then, we consider a total precedence $\succ$ on symbols s.t. $\mathsf{h} \succ f \succ c$, for all constant symbol c and all non constant symbol f distinct from h of $\Sigma_{\mathcal{H}}$. In the following $f^0(t)$ stands for t and $f^n(t)$ abbreviates $f(f^{n-1}(t))$ for $n > 1$, where f is a unary function symbol and t is any term. The ordering on ground terms we shall use is defined as follows (similarly to the Knuth-Bendix ordering [KB70]): $s \succ t$ iff

[4] Note that the word problem for ground Associative-Commutative (AC) theories is decidable [NR91] but for ground AC+Distributivity is undecidable [Mar92]. A direct modification of the proof of this last result would show that ground AC+Homomorphism is undecidable too.

1. $[s] > [t]$ or
2. $[s] = [t]$, s is of the form $f(s_1, \ldots, s_m)$, t is of the form $g(t_1, \ldots, t_n)$, and one of the following condition holds:
  2.1. $f \succ g$
  2.2. $f = g$, $m = n$ and $(s_1, \ldots, s_m) \succ\!\!\succ_{lex} (t_1, \ldots, t_m)$ (where $\succ\!\!\succ_{lex}$ denotes the lexicographic extension of $\succ$).

Lemma 11. *The relation $\succ$ is transitive, irreflexive, and monotonic (i.e. $s \succ t$ implies $f(.., s, ...) \succ f(..., t, ...)$, where f is in $\Sigma_{\mathcal{H}}$). Furthermore, $\succ$ is well-founded and it satisfies:*

- *$f(c_1, ..., c_n) \succ \mathsf{h}^i(c_{n+1})$ for all $i \geq 0$, all f that are not constants and are different from h,*
- *$\mathsf{h}(f(x_1, ..., x_n)) \succ f(\mathsf{h}(x_1), ..., h(x_n))$ for all ground terms x_i $(i = 1, ..., n)$, and*
- *$\mathsf{h}^i(c) \succ \mathsf{h}^j(c')$ for all $i > j$ and for all constants c, c' in $\Sigma_{\mathcal{H}}$.*

Proof. The lemma is proved in exactly the same way as for the Knuth-Bendix ordering [KB70].

We denote by $l \to r$ the rule obtained by orienting an equality $l = r$ when $l \succ r$. Given a rewrite system R, We shall sometimes write $s \downarrow_R t$ to express that t is the normal form of s by R.

8.2 Computation of Critical Pairs

Now, we are in the position to orient the equalities in ψ by means of the ordering $\succ$ defined in Section 8.1 and to perform a completion on the resulting set of rewrite rules using superposition rules. Unfortunately, with a naive approach, the number of rules generated by completion would be infinite. For instance, from $\mathsf{h}(c) = c$, $f(c, c') = c$, and $Ax(\mathcal{H})$ we can generate $f(c, \mathsf{h}^n(c')) = c$ for $n \geq 0$. To cope with this problem, we will consider any rewrite rule r as a rule scheme (denoted $Gen(r, R)$ or $Gen(r)$ and defined below) and we compute all superpositions between instances of two rule schemes in one step by using a special purpose inference rule (cf. *Homomorphism* rule below).

Some preliminary definitions and lemmas are mandatory. We define an *f-term* as a term with f as root symbol and for which the only other non-constant function symbol is h, where f can be any symbol in $\Sigma_{\mathcal{H}}$ (in particular, f can possibly be h). We define an *f-rule* as a rewrite rule with an f-term as left-hand side and an h-term or a constant symbol as right-hand side. For instance $f(c, \mathsf{h}^(c'))$ is an f-term and $f(c, \mathsf{h}^2(c')) = \mathsf{h}^3(c)$ or $f(c, \mathsf{h}^2(c')) = c$ is an f-rules. Examples of h-rules are $\mathsf{h}^2(c') = c$ or $\mathsf{h}^2(c') = \mathsf{h}(c)$.

In the following, let R_{h} be a convergent set of h-rules. We recall that Σ' is the subset of $\Sigma_{\mathcal{H}} \setminus \{\mathsf{h}\}$ such that if f of arity n is in Σ', then $\mathsf{h}(f(x_1, \ldots, x_n)) = f(\mathsf{h}(x_1), \ldots, \mathsf{h}(x_n))$ is in $Ax(\mathcal{H})$.

Lemma 12. *The set* $R_{\mathsf{h}} \cup \{\mathsf{h}(f(x_1,\ldots,x_n)) = f(\mathsf{h}(x_1),\ldots,\mathsf{h}(x_n)) \mid f \in \Sigma'\}$ *is convergent (we shall denote it by* $R_{\mathsf{h}} \cup H$*).*

Lemma 13. *Given constants* c, c' *and two* h*-terms* $h^j(c), h^i(c')$*, the set* $\{n \mid n \in N$*, such that* $h^n(h^j(c)) \rightarrow^*_{R_{\mathsf{h}}} h^i(c')\}$ *is linear i.e. the union of a finite set of nonnegative integers and a finite set of arithmetic sequences. We denote it by* $P_{j,c,i,c'}$*.*

Proof. We may consider unary terms as words (for instance $\mathsf{h}^j(c)$ as $\mathsf{h}^j c$). Note that the set of ancestors $\{w | w \rightarrow^*_{R_{\mathsf{h}}} w'\}$ of a term w' by R_{h} can be effectively described by a context-free grammar. The set of h-terms with constant c is obviously regular. Hence the set of $\mathsf{h}^n \mathsf{h}^j c$ that reduces to $\mathsf{h}^i c'$ is the intersection of a regular language $\mathsf{h}^* \mathsf{h}^j c$ with a context-free language and therefore context-free. The set of lengths of words of a context-free language is linear.[5] □

Let J be the set of constants that do not occur in a left-hand side of R_{h}. If $c \notin J$ we say that c is *bounded* (in R_{h}).

Lemma 14. *Given an* h*-term* $\mathsf{h}^j(c)$ *and two constants* c, c' *s.t.* c' *is not bounded, the set* $\{i \mid \exists n \in N, \mathsf{h}^n(\mathsf{h}^j(c)) \rightarrow_{R_{\mathsf{h}}} \mathsf{h}^i(c')\}$ *is an interval* $[u, \infty]$ *denoted by* $P_{j,c,-,c'}$*.*

Proof. Note that $\mathsf{h}^i(c')$ is R_{h}-irreducible. If there exists u, v with $\mathsf{h}^v(\mathsf{h}^j(c)) \downarrow_{R_{\mathsf{h}}} \mathsf{h}^u(c')$ then for all $g \in N$ we have $\mathsf{h}^{v+g}(\mathsf{h}^j(c)) \rightarrow^*_{R_{\mathsf{h}}} \mathsf{h}^{u+g}(c')$. □

Given an f-rule $r: f(t_1,\ldots,t_n) \rightarrow t_{n+1}$, we define $\mathsf{h}^n(r) \downarrow_{R_{\mathsf{h}} \cup H}$ to be the rule $(\mathsf{h}^n(f(t_1,\ldots,t_n)) \downarrow_{R_{\mathsf{h}} \cup H}) \rightarrow (\mathsf{h}^n(t_{n+1}) \downarrow_{R_{\mathsf{h}} \cup H})$. By the convergence of $R_{\mathsf{h}} \cup H$ this is well defined.

Definition 1. *For* $f \in \Sigma'$*, we define* $Gen(r, R_{\mathsf{h}})$ *as the set* $\{\mathsf{h}^n(r) \downarrow_{R_{\mathsf{h}} \cup H} \mid n \in N\}$ *where* r *denotes any* f*-rule* $f(t_1,\ldots,t_n) \rightarrow t_{n+1}$*. For* $f \notin \Sigma'$ *we define* $Gen(r, R_{\mathsf{h}}) = \{r\}$*. We shall omit the argument* R_{h} *in Gen when it is clear from the context.*

Now, we derive a finite description for $Gen(r, R_{\mathsf{h}})$. We first classify the elements in $Gen(r)$ according to their bounded arguments. More specifically we introduce the equivalence relation $\sim$ on f-rules in $Gen(r)$:

Definition 2. *Given two normalized (by* R_{h}*) rules* $r_1: f(\mathsf{h}^{l_1}(c_1), ..., \mathsf{h}^{l_n}(c_n)) \rightarrow \mathsf{h}^{l_{n+1}}(c_{n+1})$ *and* $r_2: f(\mathsf{h}^{j_1}(d_1),\ldots,\mathsf{h}^{j_n}(d_n)) \rightarrow \mathsf{h}^{j_{n+1}}(d_{n+1})$ *and such that* $r_1, r_2 \in Gen(r)$*, we have* $r_1 \sim r_2$ *iff for all* k*,* $c_k = d_k$ *and for all* $c_k \notin J$*,* $l_k = j_k$*.*

For instance if $R_{\mathsf{h}} = \{\mathsf{h}(c) \rightarrow c\}$ then $(g(\mathsf{h}^3(c'), c) = \mathsf{h}^2(c')) \sim (g(\mathsf{h}^2(c'), c) = \mathsf{h}^3(c'))$. We have the following simple lemma:

Lemma 15. *The equivalence* $\sim$ *defined on* $Gen(r)$ *has finite index (i.e. the number of classes is finite).*

[5] For details, see ex. 6.8 at page 142 of [UAH74].

Proof. Simple and therefore omitted.

We are now in the position to give a finite representation for the equivalence class of a rule r' in $Gen(r)$

Definition 3. *Let r be an f-rule $r : f(\mathsf{h}^{l_1}(c_1), \ldots, \mathsf{h}^{l_n}(c_n)) = \mathsf{h}^{l_{n+1}}(c_{n+1})$ and $r' : f(\mathsf{h}^{j_1}(d_1), \ldots, \mathsf{h}^{j_n}(d_n)) = \mathsf{h}^{j_{n+1}}(d_{n+1})$. Then, we define $C_{r,r'} = \{r" \in Gen(r) \mid r' \sim r"\}$.*

Let us compute $C_{r,r'}$ more explicitly. We introduce

$$P_{r,r'} = \Big(\bigcap_{\substack{1 \leq m \leq n+1 \\ d_m \in J}} P_{l_m,c_m,j_m,d_m}\Big) \cap \Big(\bigcap_{\substack{1 \leq m \leq n+1 \\ d_m \notin J}} P_{l_m,c_m,-,d_m}\Big)$$

Let $p_{r,r'}$ be the minimal element of $P_{r,r'}$. Note that $p_{r,r'}$ is computable since it can be defined by a formula of Presburger arithmetic:

$$P_{r,r'}(x) \wedge (\forall\ y\ P_{r,r'}(y) \Rightarrow x \leq y)$$

We denote by $n(p, l, c, d)$ the natural number n (when it exists) such that $\mathsf{h}^p(\mathsf{h}^l(c)) \downarrow_{R_\mathsf{h}} \mathsf{h}^n(d)$. Then

$$\begin{aligned} C_{r,r'} = \{\ & f(\mathsf{h}^{t_1}(d_1), \ldots, \mathsf{h}^{t_n}(d_n)) = \mathsf{h}^{t_{n+1}}(d_{n+1}) \mid \text{for } 1 \leq m \leq n+1 \\ & t_m = j_m \text{ if } d_m \notin J \text{ and} \\ & t_m = p' - p_{r,r'} + n(p_{r,r'}, l_m, c_m, d_m) \text{ if } d_m \in J \text{ where } p' \in P_{r,r'}\} \end{aligned}$$

We define the size of an h-rule $\mathsf{h}^a(b) \to \mathsf{h}^c(d)$ to be $a + c$. By reduction to Presburger arithmetic, we can prove the following fact.

Lemma 16. *Given two f-rules r_1, r_2, the minimal non-trivial critical pairs between rules in $Gen(r_1)$ and $Gen(r_2)$, are computable.*

8.3 Completion Procedure

We now give the three inference rules defining the binary transition relation over sets of equalities (denoted with $\vdash$), which models our completion procedure (modulo $\mathcal{H}$). The first is the *Deletion* rule of Table 2. The second is the *Simplification* rule, obtained as an instance for unit clauses of the *Simplification* rule of Table 2 (i.e. $E \cup \{l[s] = r, s = t\} \vdash E \cup \{l[t] = r, s = t\}$, if $l[s] \succ r$ and $s \succ t$). The third is a special purpose inference which allows us to take into account finitely many selected instances of the axioms in $Ax(\mathcal{H})$ which suffices for correctness.

$$\mathit{Homomorphism} : E \cup \{r_1, r_2\} \vdash E \cup \{r_1, r_2, h_1, \ldots, h_k\}$$

where the r_i are f-rules and the h_j are the minimal critical pairs of $Gen(r_1, R_\mathsf{h})$ and $Gen(r_2, R_\mathsf{h})$. We recall that by Lemma 1, we assume that the initial set of rules is *flat* , which means by definition that the arguments of the non-constant symbols are constants.

Lemma 17. *When initially given a set of flat rules, inference rules Simplification and Homomorphism only generate equations of type* $f(\mathsf{h}^{i_1}(c_1), ..., \mathsf{h}^{i_n}(c_n)) = \mathsf{h}^{i_{n+1}}(c_{n+1})$ *or of type* $\mathsf{h}^i(c) = \mathsf{h}^{i'}(c')$.

Theorem 6. *Completion with priority given to rule Simplification always terminates.*

Proof. Note that any sequence of *Simplification* applications always terminates. Let $E_0, E_1, E_2 \dots$ be an infinite derivation such that E_i is the result of applying *Homomorphism* to E_{i-1} followed by a maximal sequence of *Simplification* applications. We assume that the set of constants is $\{c_1, \dots, c_k\}$. Let $M_j = (m_1^j, \dots, m_k^j)$ be the exponents of h in the h-rules of E_j. That is, if there is a rule in E_j with left-hand side $\mathsf{h}^m(c_i)$ then $m_i^j = m$. Note that there are no two rules of this type for the same constant c_i (otherwise one simplifies another) and therefore the vector M_j is well-defined. When no rule exists we put ∞ as a coordinate with $n < \infty$ for all integers.

The component-wise ordering on vectors M_j is well-founded and we always have $M_j \leq M_{j-1}$. Hence after some finite number of steps the left-hand sides of h-rules remain the same. Also the right-hand sides of rules may be simplified but only finitely many time (the reduction relation is well-founded too) Finally after some finite number of steps the set of h-rules is constant. Note also that this subset of rules is canonical. We shall denote it by R_h. In particular at most one rule applies to an h-term $\mathsf{h}^n(c)$.

Homomorphism generates only h-rules. Hence after a finite number of steps, say K, it will not produce any new rule. Note that the arguments of left-hand sides of f-rules are of type $\mathsf{h}^i(c_j)$ with $i < M_K(j)$ when c_j is bounded. □

Theorem 7. *Let E be the final finite set of rules obtained by the terminating completion procedure above. Let R_h be the final set of* h *rules in E. Then, $\overline{E} \cup H$ is convergent where $\overline{E}$ is the union of all sets $Gen(r, R_\mathsf{h})$ for all r in E.*

Corollary 1. *Given a set of ground equations E_0, and the set E derived from E_0 by completion then $E_0 \cup H \models a = b$ iff $a \downarrow_{\overline{E} \cup H} = b \downarrow_{\overline{E} \cup H}$.*

9 Conclusions and Future Work

We have shown how to apply a generic inference system to derive decision procedures for the theories of lists, arrays, arrays with extensionality, and combinations of them. A decision procedure (based on superposition) for the theory of homomorphism has been presented for the first time.

We envisage two main directions for future research. Firstly, our approach might be extended using different automated deduction techniques from e.g. [CP95,Lei90]. Secondly, we want to investigate possible cross-fertilizations with techniques used in heuristic theorem provers to effectively incorporating decision procedures, see e.g. [AR01].

References

AR01. A. Armando and S. Ranise. A Practical Extension Mechanism for Decision Procedures: the Case Study of Universal Presburger Arithmetic. *J. of Universal Computer Science*, 7(2):124–140, February 2001.

BG94. L. Bachmair and H. Ganzinger. Rewrite-based equational theorem proving with selection and simplification. *J. of Logic and Comp.*, 4(3):217–247, 1994.

BRTV00. L. Bachmair, I. V. Ramakrishnan, A. Tiwari, and L. Vigneron. Congruence closure modulo associativity and commutativity. In *Frontiers of Comb. Sys.'s (FroCos'2000)*, LNCS 1794, pages 245–259, 2000.

BT00. L. Bachmair and A. Tiwari. Abstract congruence closure and specializations. In D. A. McAllester, editor, *17th CADE*, LNAI 1831, pages 64–78, 2000.

CP95. R. Caferra and Peltier. Decision procedures using model building techniques. In *CSL: 9th Workshop on Computer Science Logic*. LNCS 1092, 1995.

DJ90. N. Dershowitz and J.-P. Jouannaud. Rewrite systems. In J. van Leeuwen, editor, *Hand. of Theoretical Comp. Science*, pages 243–320. 1990.

End72. H. B. Enderton. *A Mathematical Introduction to Logic*. Academic Pr., 1972.

KB70. D. E. Knuth and P. E. Bendix. Simple word problems in universal algebra. In J. Leech, editor, *Computational Problems in Abstract Algebra*, pages 263–297, Oxford, 1970. Pergamon Press.

KR91. E. Kounalis and M. Rusinowitch. On Word Problems in Horn Theories. *JSC*, 11(1&2):113–128, January/February 1991.

Lei90. A. Leitsch. Deciding horn classes by hyperresolution. In *CSL: 3rd Workshop on Computer Science Logic*. LNCS, 1990.

Mar92. C. Marché. The word problem of ACD-ground theories is undecidable. *International Journal of Foundations of Computer Science*, 3(1):81–92, 1992.

Nel81. G. Nelson. Techniques for Program Verification. Technical Report CSL-81-10, Xerox Palo Alto Research Center, June 1981.

NO78. G. Nelson and D.C. Oppen. Simplification by Cooperating Decision Procedures. Technical Report STAN-CS-78-652, Stanford CS Dept., April 1978.

NO80. Greg Nelson and Derek C. Oppen. Fast decision procedures based on congruence closure. *Journal of the ACM*, 27(2):356–364, 1980.

NR91. P. Narendran and M. Rusinowitch. Any ground associative-commutative theory has a finite canonical system. In *4th RTA Conf., Como (Italy)*, 1991.

NR01. R. Nieuwenhuis and A. Rubio. Paramodulation-based theorem proving. In A. Robinson and A. Voronkov, editors, *Hand. of Automated Reasoning*. 2001.

Rus91. M. Rusinowitch. Theorem-proving with Resolution and Superposition. *JSC*, 11(1&2):21–50, January/February 1991.

SDBL01. A. Stump, D. L. Dill, C. W. Barrett, and J. Levitt. A Decision Procedure for an Extensional Theory of Arrays. In *LICS'01*, 2001. To appear.

UAH74. J. D. Ullman, A. V. Aho, and J. E. Hopcroft. *The Design and Analysis of Computer Algorithms*. Addison-Wesley, Reading, 1974.

Complete Categorical Equational Deduction

Grigore Roşu

Research Institute for Advanced Computer Science
Automated Software Engineering Group
NASA Ames Research Center
Moffett Field, California, 94035-1000, USA
http://ase.arc.nasa.gov/grosu

Abstract. A categorical four-rule deduction system for equational logics is presented. We show that under reasonable finiteness requirements this system is complete with respect to equational satisfaction abstracted as injectivity. The generality of the presented framework allows one to derive conditional equations as well at no extra cost. In fact, our deduction system is also complete for conditional equations, a new result at the author's knowledge.

1 Introduction

Equational logic has advantages that give it a special place in computer science: is easily machanizable, is expressive, has simple semantic models, has complete deduction. It is supported by many known specification and verification systems, such as those in the OBJ family [13, 5, 8, 9]. Its expressivity is probably best reflected by the fact that any computable data type can be characterized by means of a finite equational specification [3]. Its models are just algebras, which are very simple and intuitive structures. We suggest [11, 22] for an introduction to many-sorted equational logics and its completeness.

There is a plethora of variants and generalizations of equational logics, ranging from unsorted [4] to partial [23], order sorted [12, 28], and hidden [10, 24] equational logics. Categorical generalizations allowed proving common results, such as variety and quasi-variety theorems, only once [2, 21, 25]. These categorical approaches abstract equational satisfaction by injectivity, which turn out to be equivalent concepts in concrete situations. Local equational logic [6] allows deduction to be done in any model, not only in the initial model. This was the basis of category-based equational logic [7] and of the present paper.

In this paper we take the categorical view of an equation as an epimorphism and show that there is a categorical deduction system at that abstract level, and that that deduction system is in fact complete under appropriate finiteness requirements. Since both unconditional and conditional equations can be viewed as epimorphisms, our deduction system can derive conditional equations as well and it is also complete for them. We are not aware of any similar result for any equational paradigm in the literature. Since our first class objects are the epimorphisms, there is a high chance that our results find applications in other fields as well, such as topology.

L. Fribourg (Ed.): CSL 2001, LNCS 2142, pp. 528–538, 2001.

The paper is structured as follows: Section 2 reminds the reader a few less frequently used categorical concepts and introduces our notations and conventions. Section 3 revise factorization systems and then Section 4 shows how equations, both unconditional and conditional, are equivalent to surjective morphisms when satisfaction is replaced by injectivity. Section 5 gives the categorical deduction rules and their completeness, and then the last section presents some challenges for further research.

Acknowledgments: This work started at the University of Bucharest under the supervision of Virgil Emil Căzănescu and Sergiu Rudeanu. Frequent discussions with Joseph Goguen were not only very productive, but they actually motivated writing this paper. Debates with Răzvan Diaconescu on relationships between the present approach and his category-based equational logic have led to the finiteness condition that is needed to prove the completeness. Thank you!

2 Preliminaries

The language of this paper is category theory and the reader is assumed familiar with basic concepts of both category theory and equational logics. The purpose of this section is to introduce our notations and conventions rather than to redefine known concepts, though some less frequent notions will be reminded. We suggest the books by MacLane [18] and by Herrlich and Strecker [15] for more detail on category theory. $|\mathcal{C}|$ is the class of objects of a category $\mathcal{C}$. By abuse, we often use set-theoretic notation, such as $P \in |\mathcal{C}|$. The composition of morphisms is written in diagrammatic order, that is, if $f\colon A \to B$ and $g\colon B \to C$ then $f;g\colon A \to C$. If the source or the target of a morphism is not important in a certain context, then we replace it by a bullet to avoid inventing new letters; for example, $f\colon A \to \bullet$. In situations where there are more than one such object, these objects may be different.

Given a class of morphisms $\mathcal{E}$ in a category $\mathcal{C}$, an object $P \in |\mathcal{C}|$ is called **$\mathcal{E}$-projective** if and only if for any $e\colon \bullet \to X$ in $\mathcal{E}$ and any $h\colon P \to X$, there is a g such that $g;e = h$. Dually, I is **$\mathcal{E}$-injective** if and only if for any $e\colon X \to \bullet$ and any $h\colon X \to I$, there is a g such that $e;g = h$. $\mathcal{C}$ is called **$\mathcal{E}$-co-well-powered** if and only if for any $X \in |\mathcal{C}|$ and any *class* $\mathcal{D}$ of morphisms in $\mathcal{E}$ of source X, there is a *set* $\mathcal{D}' \subseteq \mathcal{D}$ such that each morphism in $\mathcal{D}$ is isomorphic to some morphism in $\mathcal{D}'$. We often call $\mathcal{D}'$ a **representative set** of $\mathcal{D}$. There is a dual notion of well-poweredness but it is not needed in this paper. Unless otherwise specified, by colimit we mean *small* colimit, that is, colimit of a diagram whose nodes and arrows form a set.

If X is an object in a category $\mathcal{E}$, then $X \downarrow \mathcal{E}$ is the comma category containing morphisms $e, e', \ldots\colon X \to \bullet$ in $\mathcal{E}$ as objects and morphisms $h \in \mathcal{E}$ such that $e;h = e'$ as morphisms. Notice that if $\mathcal{E}$ contains only epimorphisms than there is at most one morphism between any two objects in $X \downarrow \mathcal{E}$.

A nonempty partially ordered set $(\mathcal{I}, \leq)$ is called **directed** provided that each pair of elements has an upper bound. A **directed colimit** in a category $\mathcal{K}$ is a colimit of a diagram $D\colon (\mathcal{I}, \leq) \to \mathcal{K}$, where $(\mathcal{I}, \leq)$ is a directed poset (regarded as a category). An object K of a category $\mathcal{K}$ is called **finitely presentable** provided that its hom-functor $Hom(K, \_)\colon \mathcal{K} \to \mathbf{Set}$ preserves directed colimits. It is easy to see that K is finitely presentable iff for each directed colimit $(\{\gamma_i\colon D(i) \to C\}_{i\in|\mathcal{I}|}, C)$ and each morphism $f\colon K \to C$, there is an $i \in |\mathcal{I}|$ and a unique morphism $f_i\colon K \to D(i)$ such that $f_i;\gamma_i = f$.

3 Factorization Systems

At the author's knowledge, the first formal definition of a factorization system of a category was given by Herrlich and Strecker[1] [15] in 1973, and a first comprehensive study of factorization systems containing different equivalent definitions was done by Németi [20] in 1982. However, the idea to form subobjects by factoring each morphism f as $e;m$, where e is an epimorphism and m is a monomorphism, seems to go back to Grothendieck [14] in 1957, and was intensively used by Isbell [16], Lambek [17], Mitchell [19], and many others. At our knowledge, Lambek was the first to explicitly state and prove a diagonal-fill-in lemma in 1966 [17].

Definition 1. *A* **factorization system** *of a category $\mathcal{C}$ is a pair $\langle\mathcal{E},\mathcal{M}\rangle$, s.t.:*

- *$\mathcal{E}$ and $\mathcal{M}$ are subcategories of epics and monics, respectively, in $\mathcal{C}$,*
- *all isomorphisms in $\mathcal{C}$ are both in $\mathcal{E}$ and $\mathcal{M}$, and*
- *every morphism f in $\mathcal{C}$ can be factored as $e;m$ with $e \in \mathcal{E}$ and $m \in \mathcal{M}$ "uniquely up to isomorphism", that is, if $f = e';m'$ is another factorization of f then there is a unique isomorphism α such that $e;\alpha = e'$ and $\alpha;m' = m$.*

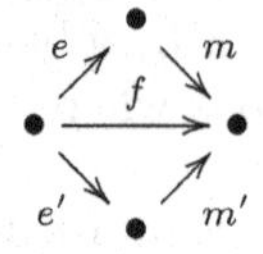

The following is one of the most important property of factorization systems, often used as equivalent definition:

Lemma 1. Diagonal-fill: *If $f;m = e;g$ then there is a "unique up to isomorphism" $h \in \mathcal{C}$ such that $e;h = f$ and $h;m = g$:*

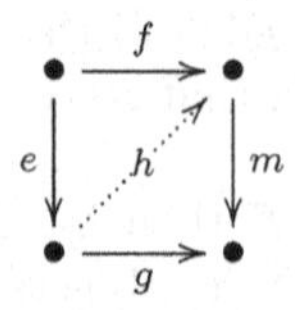

We have localized the use of factorization systems in this paper to only the following simple property which is crucial for the completeness result. Since we are not aware of any proof in the literature, we sketch one next:

[1] They called it *$\langle\mathcal{E},\mathcal{M}\rangle$-factorizable category.*

Proposition 1. *If $X \in |\mathcal{C}|$ and $\mathcal{C}$ has colimits then $X \downarrow \mathcal{E}$ has colimits.*

Proof. Let $\mathcal{D}$ be a diagram in $X \downarrow \mathcal{E}$ having as nodes the set $\{e_i : X \to \bullet\}_{i \in I}$, and let $\overline{\mathcal{D}}$ be the diagram in $\mathcal{C}$ obtained by "flattening" $\mathcal{D}$, that is, by merging both the objects and the arrows of $\mathcal{D}$. Let $(\{\gamma_i\}_{i \in I}, X_{\mathcal{D}})$ be a colimit of $\overline{\mathcal{D}}$ in $\mathcal{C}$. Then we claim that $(\{\gamma_i\}_{i \in I}, X_{\mathcal{D}})$ can be organized as a colimit of $\mathcal{D}$ in $X \downarrow \mathcal{E}$. The only interesting thing to show is that the morphism $h = e_i; \gamma_i$ (which is the same for all $i \in I$) is in $\mathcal{E}$, which follows by the diagonal-fill lemma: Factor h as $e_h; m_h$. Then for any $i \in I$ there is some β_i such that $e_i; \beta_i = e_h$ and $\beta_i; m_h = \gamma_i$, that is, $\{\beta_i\}_{i \in I}$ is a cocone for $\overline{\mathcal{D}}$, so there is a unique g such that $\gamma_i; g = \beta_i$ for all $i \in I$. Then $g; m_h = 1_{X_{\mathcal{D}}}$, so m_h is an isomorphism.

When $\mathcal{C}$ is additionally $\mathcal{E}$-co-well-powered, colimits in $X \downarrow \mathcal{E}$ exist even for large diagrams $\mathcal{D}$ whose nodes form a class: one takes the colimit of a *set* $\mathcal{D}' \subseteq \mathcal{D}$ with the property that each $e \in |\mathcal{D}|$ is isomorphic to some $e' \in |\mathcal{D}'|$.

Definition 2. $(\{\gamma_i\}_{i \in I}, e_{\mathcal{D}} : X \to X_{\mathcal{D}})$ *denotes the colimit of* $\mathcal{D} \subseteq X \downarrow \mathcal{E}$.

4 Equational Satisfaction as Injectivity

As advocated by Bannaschewski and Herrlich [2], and by Andréka, Németi and Sain [1, 21] among many others[2], satisfaction of equations is equivalent to injectivity. It is often more convenient to work with sets of equations than with individual equations. For example, Craig interpolation doesn't hold for individual equations but it holds for sets of equations [27] (see [26] for a categorical approach). In our present framework, it is most convenient to view the equations as finite or infinite sets of pairs of terms quantified over the same variables, for example $(\forall X)\ t_1 = t'_1, t_2 = t'_2, \ldots$. If the number of terms is finite then we informally call the equation finite.

Consider that $\mathcal{C}$ is the category of universal or many-sorted Σ-algebras over a (many-sorted) signature Σ. Each equation $(\forall X)\ t_1 = t'_1, t_2 = t'_2, \ldots$ generates a congruence relation on the term algebra $T_\Sigma(X)$ over variables in X, which implicitly gives a surjective morphism $e : T_\Sigma(X) \to \bullet$. It can be readily seen that an algebra A satisfies $(\forall X)\ t_1 = t'_1, t_2 = t'_2, \ldots$ if and only if it is $\{e\}$-injective. Conversely, each surjective morphism $e : T_\Sigma(X) \to \bullet$ of free algebra source generates a potentially infinite equation $(\forall X)\ \mathit{Ker}(e)$. It can also be readily seen that an algebra is $\{e\}$-injective if and only if it satisfies $(\forall X)\ \mathit{Ker}(e)$. Therefore, satisfaction of equations and Ω-injectivity where Ω contains morphisms with free sources, are equivalent concepts. It is often technically easier to abstract freeness by projectivity (any free algebra is projective; see [25] for conditions under which free objects are projective).

What is less known is that satisfaction of *conditional* equations is also equivalent to Ω-injectivity, but this time Ω can contain epimorphisms of non-free sources. To be more precise, let us consider the following conditional equation:

$$(\forall X)\ t_1 = t'_1, t_2 = t'_2, \ldots \texttt{ if } u_1 = u'_1, u_2 = u'_2, \ldots$$

[2] See also [25] for an approach based on inclusion systems.

Let Q be the quotient $T(X)/\{(u_1, u'_1), (u_2, u'_2), ...\}$ and let $e\colon\ Q \to \bullet$ be the canonical surjective morphism generated by the pairs $([t_1], [t'_1]), ([t_2], [t'_2]), ...$ on[3] Q. Then one can relatively easily see that an algebra A satisfies the conditional equation if and only if it is $\{e\}$-injective. Conversely, let $e\colon\ Q \to \bullet$ be any surjective morphism and let $e_Q\colon\ T_\Sigma(Q) \to Q$ be the unique extension of $1_Q\colon\ Q \to Q$ viewed as function to a morphism, where $T_\Sigma(Q)$ is the free algebra over Q regarded as a set of variables. In other words, e_Q is the co-unit of the free algebra adjunction. If one considers now the equation

$$(\forall Q)\ \mathit{Ker}(e_Q; e)\ \texttt{if}\ \mathit{Ker}(e_Q)$$

then one can verify that an algebra A satisfies it if and only if A is $\{e\}$-injective. $\mathit{Ker}(e_Q; e)$ and $\mathit{Ker}(e_Q)$ can be replaced by sets of generators for the kernels of the two morphisms. Therefore, satisfaction of conditional equations is also equivalent to injectivity.

In [25], it is shown that the difference wrt injectivity between epimorphisms of free or projective source and epimorphisms of any source is exactly as the difference wrt satisfaction between unconditional and conditional equations, that is, the first ones define varieties while the second define quasi-varieties.

The disadvantage to regard equations as epimorphisms $e\colon\ Q \to \bullet$ is that their kernel may not be finitely generated, so complete deduction systems do not seem to exist anymore. However, in the rest of the paper we give a categorical deduction system for epimorphisms $e\colon\ Q \to \bullet$ and show that it is complete for e under reasonable finiteness requirements on e. The following simple but important result gives intuition for further notions, where $\mathcal{E}$ is the category of surjective morphisms of $\mathcal{C}$:

Proposition 2. *e is finitely presentable in $Q \downarrow \mathcal{E}$ iff $\mathit{Ker}(e)$ is finitely generated.*

5 Complete Categorical Deduction

In this section we first give an inference system and then we show it sound for any epimorphisms, but complete only for finite epimorphisms. To make our results as general as possible we choose to work in an abstract, categorical framework:

Framework: A category $\mathcal{C}$ that
- admits a factorization system $(\mathcal{E}, \mathcal{M})$,
- is $\mathcal{E}$-co-well-powered,
- has colimits.

The category $\mathcal{C}$ can be thought of as the category of algebras over a given signature. We think that this framework is general enough to contain all equational approaches, but of course, it is not limited to only those; for example, one can consider $\mathcal{C}$ as the category of topological spaces. However, having in mind the previous section, we often abuse and call the epimorphisms in $\mathcal{E}$ *equations*. We next define satisfaction in this framework as injectivity:

[3] $[t]$ is the equivalence class of t in Q.

Definition 3. *Given an object A in $\mathcal{C}$ and $e\colon X \to \bullet$ in $\mathcal{E}$, then A* **satisfies** *e if and only if A is $\{e\}$-injective. As usual, we write $A \models e$ and extend it to $E \models e$ for any class of equations E.*

5.1 Inference Rules

The following introduces four rules by which one can derive epimorphisms in $\mathcal{E}$. If not explicitly stated otherwise, from now in the paper consider that $E \subseteq \mathcal{E}$ is a class of equations with $\mathcal{E}$-projective sources and that $e\colon X \to \bullet$ is any equation (its source is not required to be projective). We often use the same letter e for the equations in E, mentioning that there is no confusion because those have $\mathcal{E}$-projective sources denoted by P. The projectivity condition is not needed for the soundness, but for simplicity we prefer to add it here.

Definition 4. *$E \vdash e$ denotes the derivation relation generated by the rules:*

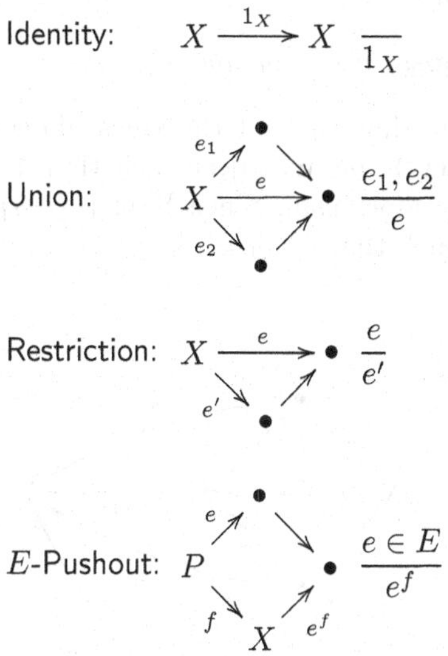

Identity: $X \xrightarrow{1_X} X \quad \dfrac{}{1_X}$

Union: $\begin{array}{ccc} & \bullet & \\ {}^{e_1}\nearrow & & \searrow \\ X & \xrightarrow{e} & \bullet \\ {}_{e_2}\searrow & & \nearrow \\ & \bullet & \end{array} \quad \dfrac{e_1, e_2}{e}$

Restriction: $\begin{array}{ccc} X & \xrightarrow{e} & \bullet \\ {}_{e'}\searrow & & \nearrow \\ & \bullet & \end{array} \quad \dfrac{e}{e'}$

E-Pushout: $\begin{array}{ccc} & \bullet & \\ {}^{e}\nearrow & & \searrow \\ P & & \bullet \\ {}_{f}\searrow & & \nearrow{}_{e^f} \\ & X & \end{array} \quad \dfrac{e \in E}{e^f}$

For a better intuition wrt the more traditional equational logics, one could imagine that X is the set of variables (seen as a free algebra) while the pair(s) of terms over those variables correspond to the kernel of the derived epimorphism. In this light, Identity corresponds to reflexivity, Union to symmetry, transitivity and congruence, while E-pushout corresponds to substitution. Notice that Union actually grows the set of "proved" facts; for example, if e_1 corresponds to $t = t'$ and e_2 corresponds to $t' = t''$, then their union corresponds to all $t = t', t' = t'', t = t''$, ... The role of Restriction is to select a subset of interest of those pairs, for example $t = t''$.

Despite the projectivity of sources of equations in E, notice that the derived equations may not have projective sources. In a standard equational terminology, that means that one can actually derive conditional equations as well.

Definition 5. *If $E \vdash e$ and e has source X then e is called an* X**-derivation of** E. *Let $\mathcal{D}_X(E)$ denote the full subcategory of $X \downarrow \mathcal{E}$ of X-derivations of E.*

Notice that $\mathcal{D}_X(E)$ can be a class in general because E can be a class. However, since $\mathcal{C}$ is $\mathcal{E}$-co-well-powered, $\mathcal{D}_X(E)$ still has colimits in $X \downarrow \mathcal{E}$; its colimit object is denoted by $e_{\mathcal{D}_X(E)} : X \to X_{\mathcal{D}_X(E)}$, as usual (see Definition 2).

Proposition 3. *$\mathcal{D}_X(E)$ is directed.*

Proof. This is because $\mathcal{D}_X(E)$ is closed under Union.

Again, if $\mathcal{D}_X(E)$ is not a set then it can be replaced by some representative set that it includes.

5.2 Soundness

In this subsection we show that the rules above are correct with respect to satisfaction as injectivity.

Theorem 1. **Soundness.** *$E \vdash e$ implies $E \models e$.*

Proof. It is easy to see that each of the rules above is sound. We only give the proof for union. Let Y be an object such that $Y \models e_1$ and $Y \models e_2$, and let $h\colon X \to Y$ be any morphism. Since Y is $\{e_1, e_2\}$-injective, there are two morphisms g_1 and g_2 such that $e_1; g_1 = e_2; g_2 = h$:

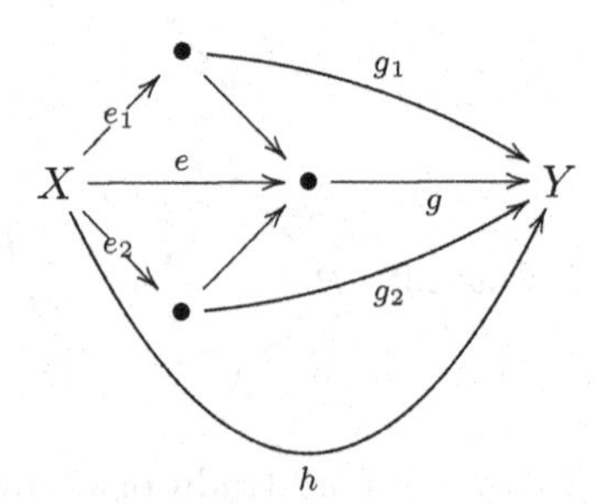

Then by the pushout property, there is a g such that $e; g = h$, i.e., $Y \models e$.

5.3 Closures under E-Pushouts

Closures under E-pushouts correspond to closures under substitutions in the usual equational setting. Formally,

Definition 6. *$\mathcal{D} \subseteq X \downarrow \mathcal{E}$ is* **closed under E-pushouts** *if and only if for any $e\colon P \to \bullet$ in E and any $f\colon P \to X$, it is the case that $e^f : X \to \bullet$ is in $\mathcal{D}$.*

With the notation in Definition 2,

Lemma 2. *If $\mathcal{D} \subseteq X \downarrow \mathcal{E}$ is closed under E-pushouts then $X_{\mathcal{D}} \models E$.*

Proof. Like in Definition 2, let us consider that γ_j is the coprojection associated to each $e_j \in \mathcal{D}$ (composed with an appropriate isomorphism if $\mathcal{D}$ is not a set), so that $e_j; \gamma_j = e_{\mathcal{D}}$.

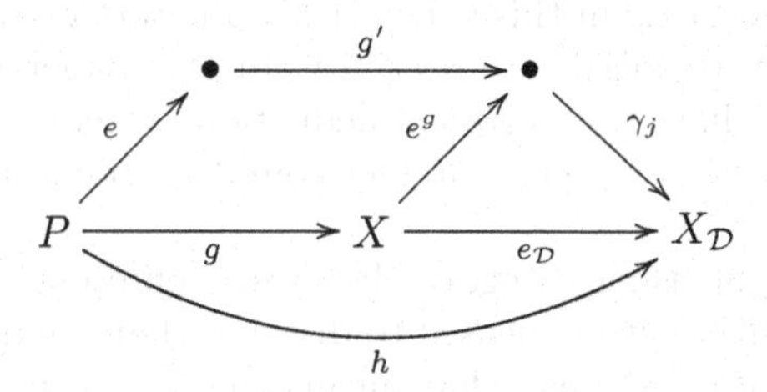

Let $e\colon P \to \bullet$ be any equation in E and $h\colon P \to X_{\mathcal{D}}$ any morphism. By the projectivity of P, there is a $g\colon P \to X$ such that $g; e_{\mathcal{D}} = h$. Since $\mathcal{D}$ is closed under E-pushouts, there is an $e_j \in \mathcal{D}$ such that $e^g = e_j$; let g' be the morphisms that completes the pushout diagram, that is, $g; e^g = e; g'$. It can be easily seen now that $e; (g'; \gamma_j) = h$, that is, $X_{\mathcal{D}} \models E$.

With the notation in Definition 5,

Proposition 4. $E \models e$ *iff* $X_{\mathcal{D}_X(E)} \models e$.

Proof. If $E \models e$ then one can replace $\mathcal{D}$ in the lemma above by $\mathcal{D}_X(E)$ and thus obtain that $X_{\mathcal{D}_X(E)} \models e$.

Conversely, if $X_{\mathcal{D}(E)} \models$ then there is an e' such that $e; e' = e_E$. Let $A \models E$ and let $h\colon X \to A$. Since $A \models \mathcal{D}_X(E)$, for each $e_j \in \mathcal{D}_X(E)$ there is a β_j such that $e_j; \beta_j = h$. Then A together with the morphisms β form a cocone in $\mathcal{C}$ for $\mathcal{D}_X(E)$, so there is a unique $g\colon X_{\mathcal{D}(E)} \to A$ such that $\gamma_j; g = \beta_j$ for all $e_j \in \mathcal{D}_X(E)$. It follows then that $e; (e'; g) = e_{\mathcal{D}_X(E)}; g = e_j; \gamma_j; g = e_j; \beta_j = h$, that is, $A \models e$.

5.4 Completeness

One cannot expect any deduction system to be complete for satisfaction as injectivity without some kind of finiteness. We find the following convenient enough:

Definition 7. *The equation $e\colon X \to \bullet$ is* **finite** *provided that it is finitely presentable in the comma category $X \downarrow \mathcal{E}$.*

According to Proposition 2, the intuition for the above is that $e\colon X \to \bullet$ is finite if and only if the (many-pair) equation it represents is finitely generated.

Theorem 2. <u>**Completeness.**</u> $E \models e$ *implies* $E \vdash e$ *whenever e is finite.*

Proof. Suppose that $E \models e$. Then by Proposition 4, $X_{\mathcal{D}_X(E)} \models e$, so there is an e' such that $e; e' = e_{\mathcal{D}_X(E)}$. Since e is finite, there is an e_j in $\mathcal{D}_X(E)$ and an e'_j such that $e; e'_j = e_j$. But $E \vdash e_j$, so by restriction, $E \vdash e$.

6 Conclusion and Future Work

A categorical deduction system was presented for a categorical abstraction of equational logics. This categorical abstraction together with generalizations of Birkhoff variety and quasi-variety results are known and considered folklore among category theorists; in this categorical framework, equational satisfaction can be equivalently replaced by injectivity. In this paper we showed that it is quite reasonable to have a categorical deduction system at that abstract level, and that that deduction system is in fact complete under appropriate finiteness requirements.

The advantage of using category theory to represent deduction systems is multiple. First of all, we were pleased to discover that the deduction is complete even for conditional equations. That means that one can deduce a conditional equation directly, as opposed to the common approach that uses the theorem of constants to add the hypothesis of the equation as a new axiom and then to deduce the conclusion. We are not aware of any other proof of this in the literature and we find it interesting that it comes for free in the categorical approach. Second, all equational frameworks with their complete deduction systems fall now under a common umbrella: any new results obtained at the abstract level, such as the derivation of conditional equations, can be pushed down to the level of each individual equational framework. Third, it is relatively straightforward to dualise the complete deduction system as well as all the results in the paper, and thus to get a complete deduction system for coequational logic.

We only considered epimorphisms of projective source as axioms in E. That corresponds to unconditional equations in concrete equational contexts. We admit that this can be viewed as a limitation at this stage, but we are confident that the results can be extended relatively easily to conditional axioms. We predict that in order to do this generalization, one would need to add to the framework at the beginning of section 5 the requirement that $\mathcal{C}$ has enough projectives.

An interesting subject of research is to implement the four-rule deduction system presented in the paper. In this way, one would have an arrow-based, maybe graphical, equational reasoning engine. Perhaps the most important direction of further research is to dualise the results presented here and to obtain a complete deduction system for coequations. It is not clear for us at this stage what its significance would be and if there is any relationship with modal logics.

References

1. Hajnal Andréka and István Németi. A general axiomatizability theorem formulated in terms of cone-injective subcategories. In B. Csakany, E. Fried, and E.T. Schmidt, editors, *Universal Algebra*, pages 13–35. North-Holland, 1981. Colloquia Mathematics Societas János Bolyai, 29.
2. Bernhard Banaschewski and Horst Herrlich. Subcategories defined by implications. *Houston Journal Mathematics*, 2:149–171, 1976.

3. Jan Bergstra and John Tucker. Characterization of computable data types by means of a finite equational specification method. In Jaco de Bakker and Jan van Leeuwen, editors, *Automata, Languages and Programming, Seventh Colloquium*, pages 76–90. Springer, 1980. Lecture Notes in Computer Science, Volume 81.
4. Garrett Birkhoff. On the structure of abstract algebras. *Proceedings of the Cambridge Philosophical Society*, 31:433–454, 1935.
5. Manuel Clavel, Steven Eker, Patrick Lincoln, and José Meseguer. Principles of Maude. In José Meseguer, editor, *Proceedings, First International Workshop on Rewriting Logic and its Applications*. Elsevier Science, 1996. Volume 4, *Electronic Notes in Theoretical Computer Science*.
6. Virgil Căzănescu. Local equational logic. In Zoltan Esik, editor, *Proceedings, 9th International Conference on Fundamentals of Computation Theory FCT'93*, pages 162–170. Springer-Verlag, 1993. Lecture Notes in Computer Science, Volume 710.
7. Răzvan Diaconescu. *Category-based Semantics for Equational and Constraint Logic Programming*. PhD thesis, University of Oxford, 1994.
8. Răzvan Diaconescu and Kokichi Futatsugi. *CafeOBJ Report: The Language, Proof Techniques, and Methodologies for Object-Oriented Algebraic Specification*. World Scientific, 1998. AMAST Series in Computing, volume 6.
9. Joseph Goguen, Kai Lin, and Grigore Roşu. Circular coinductive rewriting. In *Proceedings, Automated Software Engineering '00*, pages 123–131. IEEE, 2000. (Grenoble, France).
10. Joseph Goguen and Grant Malcolm. A hidden agenda. *Theoretical Computer Science*, 245(1):55–101, August 2000. Also UCSD Dept. Computer Science & Eng. Technical Report CS97–538, May 1997.
11. Joseph Goguen and José Meseguer. Completeness of many-sorted equational logic. *Houston Journal of Mathematics*, 11(3):307–334, 1985. Preliminary versions have appeared in: *SIGPLAN Notices*, July 1981, Volume 16, Number 7, pages 24–37; SRI Computer Science Lab, Report CSL-135, May 1982; and Report CSLI-84-15, Center for the Study of Language and Information, Stanford University, September 1984.
12. Joseph Goguen and José Meseguer. Order-sorted algebra I: Equational deduction for multiple inheritance, overloading, exceptions and partial operations. *Theoretical Computer Science*, 105(2):217–273, 1992. Drafts exist from as early as 1985.
13. Joseph Goguen, Timothy Winkler, José Meseguer, Kokichi Futatsugi, and Jean-Pierre Jouannaud. Introducing OBJ. In Joseph Goguen and Grant Malcolm, editors, *Software Engineering with OBJ: algebraic specification in action*, pages 3–167. Kluwer, 2000.
14. Alexandre Grothendieck. Sur quelques points d'algébre homologique. *Tôhoku Mathematical Journal*, 2:119–221, 1957.
15. Horst Herrlich and George Strecker. *Category Theory*. Allyn and Bacon, 1973.
16. J. R. Isbell. Subobjects, adequacy, completeness and categories of algebras. *Rozprawy Matematyczne*, 36:1–33, 1964.
17. Joachim Lambek. *Completions of Categories*. Springer-Verlag, 1966. Lecture Notes in Mathematics, Volume 24.
18. Saunders Mac Lane. *Categories for the Working Mathematician*. Springer, 1971.
19. B. Mitchell. *Theory of categories*. Academic Press, New York, 1965.
20. István Németi. On notions of factorization systems and their applications to cone-injective subcategories. *Periodica Mathematica Hungarica*, 13(3):229–335, 1982.

21. István Németi and Ildickó Sain. Cone-implicational subcategories and some Birkhoff-type theorems. In B. Csakany, E. Fried, and E.T. Schmidt, editors, *Universal Algebra*, pages 535–578. North-Holland, 1981. Colloquia Mathematics Societas János Bolyai, 29.
22. Peter Padawitz and Martin Wirsing. Completeness of many-sorted equational logic revisited. *Bulletin of the European Association for Theoretical Computer Science*, 24:88–94, October 1984.
23. H. Reichel. *Initial Computability, Algebraic Specifications, and Partial Algebras*. Oxford University Press, 1987.
24. Grigore Roşu. *Hidden Logic*. PhD thesis, University of California at San Diego, 2000. `http://ase.arc.nasa.gov/grosu/phd-thesis.ps`.
25. Grigore Roşu. Axiomatizability in inclusive equational logics. *Mathematical Structures in Computer Science*, to appear. `http://ase.arc.nasa.gov/grosu/iel.ps`.
26. Grigore Roşu and Joseph Goguen. On equational Craig interpolation. *Journal of Universal Computer Science*, 6(1):194–200, 2000.
27. Pieter Hendrik Rodenburg. A simple algebraic proof of the equational interpolation theorem. *Algebra Universalis*, 28:48–51, 1991.
28. Gert Smolka, Werner Nutt, Joseph Goguen, and José Meseguer. Order-sorted equational computation. In Maurice Nivat and Hassan Aït-Kaci, editors, *Resolution of Equations in Algebraic Structures, Volume 2: Rewriting Techniques*, pages 299–367. Academic, 1989.

Beyond Regularity: Equational Tree Automata for Associative and Commutative Theories

Hitoshi Ohsaki

National Institute of Advanced Industrial Science and Technology
Nakoji 3-11-46, Amagasaki 661-0974, Japan
hitoshi.ohsaki@aist.go.jp Phone: +81 6 6494 7823 Fax: +81 6 6491 5028

Abstract. A new tree automata framework, called *equational tree automata*, is presented. In the newly introduced setting, congruence closures of recognizable tree languages are recognizable. Furthermore, we prove that in certain useful cases, recognizable tree languages are closed under union and intersection. To compare with early related work, e.g. [7], we discuss the relationship between linear bounded automata and equational tree automata. As a consequence, we obtain some (un)decidability results. We further present a hierarchy of 4 classes of tree languages.

Keywords: Tree automata, equational theory, decidability

1 Introduction

Over the past decade tree automata theory have been extensively studied and many applications were developed in various areas, e.g. for verification of cryptographic protocols [10,12], subtyping in programming language [9] and reduction strategies in term rewriting [6]. The devised techniques are based on "regular" tree automata, which are the counterpart of classical finite automata. The tree automata framework is very useful in the sense that many decision problems are known to be decidable and recognizable tree languages are closed under boolean operations.

In contrast to the situations where regularity allows us to design terminating procedures easily, non-regular languages, such as term algebras modulo congruence, are considered to be troublesome in the framework. In fact, it is undecidable whether or not congruence closure of a regular tree language is regular [7]. Even, except for a few examples [11], AC-congruence closure of a regular tree language is not regular in general [5]. For instance, consider the signature $\mathcal{F} = \{ f, a, b \}$, where f is a binary function symbol, and a and b are constants. Let L be the set of (ground) terms t such that the number of occurrences of a in t is the same as the number of occurrences of b in t. The tree language L is *not* regular, because of PUMPING LEMMA [2], although L' defined below is regular and L is the AC-closure of L': $f(a, b) \in L'$ and $f(a, f(s, b)) \in L'$ for all $s \in L'$.

The aim of this paper is to introduce a new tree automata framework, called *equational tree automata* (ETA for short), in which congruence closures of recognizable languages are recognizable. Furthermore, we investigate the expressive

L. Fribourg (Ed.): CSL 2001, LNCS 2142, pp. 539–553, 2001.

power of the new tree language theory by comparing with other well-known classes. In the sense, we are concerned in the paper with questions about equational tree automata like in the following:

- $\forall L : \text{TL}.\ P(L) \Rightarrow \exists \mathcal{A}/\mathcal{E} : \text{ETA}.\ \mathcal{L}(\mathcal{A}/\mathcal{E}) = L$,
- $\forall L : \text{TL}.\ Q(L) \Leftarrow \exists \mathcal{A}/\mathcal{E} : \text{ETA}.\ \mathcal{L}(\mathcal{A}/\mathcal{E}) = L$.

In the above formulae, P and Q are predicates for tree languages (TL for short). In particular, we spend the most of spaces to explain the relationship between the standard (finite bottom-up regular) tree automata and our equational extension. For instance, we discuss sufficient conditions for equational systems $\mathcal{E}$ and tree languages L that satisfy

(Q1) $\exists \mathcal{A} : \text{TA}.\ \mathcal{E}(\mathcal{L}(\mathcal{A})) = L \Leftrightarrow \exists \mathcal{B}/\mathcal{E} : \text{ETA}.\ \mathcal{L}(\mathcal{B}/\mathcal{E}) = L$.

The question asks us: under which condition it holds or does not hold that a tree language is recognizable with a TA $\mathcal{A}$ if and only if the $\mathcal{E}$-congruence closure is recognizable with an ETA $\mathcal{B}/\mathcal{E}$. Another instance to be considered is whether or not it holds that for any tree language L,

(Q2) $\exists \mathcal{A} : \text{TA}.\ \mathcal{L}(\mathcal{A}) = L \Leftrightarrow \exists \mathcal{B}/\mathcal{E} : \text{ETA}.\ \mathcal{L}(\mathcal{B}/\mathcal{E}) = \mathcal{E}(L)$.

The paper is organized as follows. The basics of tree automata and related theory are introduced in the next section. We show several positive answers to (Q1) and (Q2) in Section 3. We also present some decidability results by studying the relationship between linear bounded automata and equational tree automata. In Section 4, we show closure properties of union and intersection. We conclude in Section 5 by showing a hierarchy of 4 classes of tree languages. Open questions related to equational tree automata are also mentioned.

2 Preliminaries

A signature is a finite set $\mathcal{F}$ of function symbols together with natural numbers n for every $f \in \mathcal{F}$. Here n is called the *arity* of f, denoted by $\mathtt{arity}(f) = n$. Function symbols of arity 0 are called *constants*. We assume the existence of countably infinite sets of variables $\mathcal{V}$. The set $\mathcal{T}(\mathcal{F}, \mathcal{V})$ of terms is inductively defined as follows: $\mathcal{V} \subseteq \mathcal{T}(\mathcal{F}, \mathcal{V})$ and $f(t_1, \ldots, t_n) \in \mathcal{T}(\mathcal{F}, \mathcal{V})$ if the arity of f is n and $t_i \in \mathcal{T}(\mathcal{F}, \mathcal{V})$ for all $1 \leqslant i \leqslant n$. The set $\mathcal{T}(\mathcal{F}, \varnothing)$ of *ground* terms is denoted by $\mathcal{T}(\mathcal{F})$. Let $\Box$ be a fresh constant, named *hole*. The set $\mathcal{T}(\mathcal{F} \cup \{\Box\}, \mathcal{V})$ of terms is denoted by $\mathcal{C}(\mathcal{F}, \mathcal{V})$. Elements of $\mathcal{C}(\mathcal{F}, \mathcal{V})$ are called *contexts*. The empty context is a hole. If C is a context with n holes and $t_1, \ldots, t_n$ are terms, $C[t_1, \ldots, t_n]$ denotes the term obtained from C by replacing the holes from left to right by $t_1, \ldots, t_n$. A *substitution* is a mapping σ from $\mathcal{V}$ to $\mathcal{T}(\mathcal{F}, \mathcal{V})$. We write $t\sigma$ for the result of applying σ to a term t, where σ is extended as $f(t_1, \ldots, t_n)\sigma = f(t_1\sigma, \ldots, t_n\sigma)$. The set $\mathsf{pos}(t)$ of *positions* in a term t is defined by

$$\mathsf{pos}(t) = \begin{cases} \{\varepsilon\} & \text{if } t \text{ is a variable,} \\ \{\varepsilon\} \cup \{i \cdot p \mid 1 \leqslant i \leqslant n \text{ and } p \in \mathsf{pos}(t_i)\} & \text{if } t = f(t_1, \ldots, t_n). \end{cases}$$

Here ε is the empty sequence and $p \cdot q$ denotes concatenation of sequences p and q of positive integers. The position ε in $\mathsf{pos}(t)$ is called the *root* of t and a symbol at ε is denoted by $\mathsf{root}(t)$. A *subterm* of t at a position p is denoted by $t_{|p}$ and that is inductively defined as follows:

$$t_{|p} = \begin{cases} t & \text{if } p = \varepsilon, \\ t_{i|q} & \text{if } t = f(t_1, \ldots, t_n) \text{ and } p = i \cdot q \text{ with } 1 \leqslant i \leqslant n. \end{cases}$$

The set $\mathsf{pos}(t)$ is divided into two sets $\mathsf{pos}_{\mathcal{V}}(t) = \{p \in \mathsf{pos}(t) \mid t_{|p} \in \mathcal{V}\}$ and $\mathsf{pos}_{\mathcal{F}}(t) = \mathsf{pos}(t) \setminus \mathsf{pos}_{\mathcal{V}}(t)$. Intuitively, $\mathsf{pos}_{\mathcal{V}}(t)$ is the set of variable positions in t and $\mathsf{pos}_{\mathcal{F}}(t)$ is the set of function symbols. The *length* of a term t, denoted by $|t|$, is the number of elements in $\mathsf{pos}(t)$. The number of occurrences of a function symbol f in a term t is denoted by $\|t\|_f$. We write $\|t\|$ for the number of elements in $\mathsf{pos}_{\mathcal{F}}(t)$. Note that $\|t\| = \sum_{f \in \mathcal{F}} \|t\|_f$. The set of variables appearing in t is denoted by $\mathsf{var}(t)$ and the set of function symbols in t is denoted by $\mathsf{fun}(t)$. Those multisets are denoted by $\mathsf{var}_{\mathrm{mul}}(t)$ and $\mathsf{fun}_{\mathrm{mul}}(t)$, respectively. The *height* of a term t, denoted by $\mathsf{height}(t)$, is defined by $\mathsf{height}(t) = 0$ if $t \in \mathcal{V}$; $\mathsf{height}(t) = 1 + \max\{\mathsf{height}(t_i) \mid 1 \leqslant i \leqslant n\}$ if $t = f(t_1, \ldots, t_n)$.

An *equation* over the signature $\mathcal{F}$ is a pair (s, t) of terms $s, t \in \mathcal{T}(\mathcal{F}, \mathcal{V})$. The equation (s, t) is denoted by $s \approx t$. An equation $l \approx r$ is called *linear* if neither l nor r contains multiple positions of the same variable. We say $l \approx r$ is *variable-preserving* if $\mathsf{var}_{\mathrm{mul}}(l) = \mathsf{var}_{\mathrm{mul}}(r)$. A variable-preserving equation $l \approx r$ is called *length-preserving* if $\|l\| = \|r\|$. An equation $l \approx r$ is called *ground* if $l, r \in \mathcal{T}(\mathcal{F})$, i.e. $\mathsf{var}(l) = \mathsf{var}(r) = \varnothing$. An *equational system* (ES for short) $\mathcal{E}$ is a set of equations. Given a set $\mathcal{F}'$ ($\subseteq \mathcal{F}$) of some binary function symbols. The set of associativity axioms $f(f(x, y), z) \approx f(x, f(y, z))$ for all $f \in \mathcal{F}'$ is denoted by $\mathrm{A}(\mathcal{F}')$, and the set of commutativity axioms $f(x, y) \approx f(y, x)$ for all $f \in \mathcal{F}'$ is $\mathrm{C}(\mathcal{F}')$. We write $\mathrm{AC}(\mathcal{F}')$ for the union of $\mathrm{A}(\mathcal{F}')$ and $\mathrm{C}(\mathcal{F}')$. If unnecessary to be explicit, we simply write A, C and AC. An ES $\mathcal{E}$ is called linear (variable-preserving, length-preserving, ground) if it consists of linear (variable-preserving, length-preserving, ground) equations. The binary relation $s \to_{\mathcal{E}} t$ is defined by letting $s = C[l\sigma]$ and $t = C[r\sigma]$ for some equation $l \approx r \in \mathcal{E}$, context $C \in \mathcal{C}(\mathcal{F}, \mathcal{V})$ and substitution σ over $\mathcal{T}(\mathcal{F}, \mathcal{V})$. In the paper, it is not guaranteed that $r \approx l \in \mathcal{E}$ even if $l \approx r \in \mathcal{E}$. The symmetric closure of $\to_{\mathcal{E}}$ is denoted by $\vdash\!\dashv_{\mathcal{E}}$ and the equivalence relation of $\to_{\mathcal{E}}$ (i.e., the reflexive-transitive closure of $\vdash\!\dashv_{\mathcal{E}}$) is denoted by $\sim_{\mathcal{E}}$.

A *tree automaton* (TA for short) $\mathcal{A} = (\mathcal{F}, \mathcal{Q}, \mathcal{Q}_f, \mathcal{R})$ consists of a signature $\mathcal{F}$, a finite set $\mathcal{Q}$ of states (special constants with $\mathcal{F} \cap \mathcal{Q} = \varnothing$), a set $\mathcal{Q}_f$ ($\subseteq \mathcal{Q}$) of final states and a finite set $\mathcal{R}$ of transition rules in one of the following forms:

$$f(p_1, \ldots, p_n) \to q$$

or

$$f(p_1, \ldots, p_n) \to f(q_1, \ldots, q_n)$$

for some $f \in \mathcal{F}$ and $p_1, \ldots, p_n, q_1, \ldots, q_n, q \in \mathcal{Q}$. In the latter form the root symbols in the left- and right-hand sides must be the same. An *equational tree* *automaton* $\mathcal{A}/\mathcal{E}$ is the combination of a TA $\mathcal{A}$ and an ES $\mathcal{E}$. We often denote $\mathcal{A}/\mathcal{E}$ by the 5-tuple $(\mathcal{F}, \mathcal{Q}, \mathcal{Q}_f, \mathcal{R}, \mathcal{E})$ for convenience. An ETA $\mathcal{A}/\mathcal{E}$ is called *regular* if $\mathcal{R}$ consists of transition rules in the shape of $f(p_1, \ldots, p_n) \to q$. We say $\mathcal{A}/\mathcal{E}$ is *quasi-regular* if for all $l \to r \in \mathcal{R}$ such that $\mathsf{root}(l) \notin \mathsf{fun}(\mathcal{E})$, $r \in \mathcal{Q}$. Here fun is extended to be $\mathsf{fun}(\mathcal{E}) = \bigcup_{l \approx r \in \mathcal{E}} (\mathsf{fun}(l) \cup \mathsf{fun}(r))$. Every TA is transformed to a regular TA with the same expressive power. The details are described in the next section. We say $\mathcal{A}/\mathcal{E}$ is a C-TA (commutative-tree automaton) if $\mathcal{E} = \mathrm{C}$. An ETA $\mathcal{R}/\mathcal{E}$ with $\mathcal{E} = \mathrm{A}$ is called an A-TA (associative-tree automaton). Likewise, if $\mathcal{E} = \mathrm{AC}$, it is called an AC-TA. We write $s \to_{\mathcal{A}/\mathcal{E}} t$ if there exist s', t' such that $s \sim_{\mathcal{E}} s'$, $s' = C[l]$, $t \sim_{\mathcal{E}} t'$ and $t' = C[r]$ for some transition rule $l \to r \in \mathcal{R}$ and context $C \in \mathcal{C}(\mathcal{F} \cup \mathcal{Q})$. The relation $\to_{\mathcal{A}/\mathcal{E}}$ on $\mathcal{T}(\mathcal{F} \cup \mathcal{Q})$ is called *move relation* of $\mathcal{A}/\mathcal{E}$. The transitive closure and reflexive-transitive closure of $\to_{\mathcal{A}/\mathcal{E}}$ are denoted by $\to^+_{\mathcal{A}/\mathcal{E}}$ and $\to^*_{\mathcal{A}/\mathcal{E}}$. For a TA $\mathcal{A}$, we simply write $\to_{\mathcal{A}}$, $\to^+_{\mathcal{A}}$ and $\to^*_{\mathcal{A}}$, instead. A term $t \in \mathcal{T}(\mathcal{F})$ is *accepted* by $\mathcal{A}/\mathcal{E}$ if $t \to^*_{\mathcal{A}/\mathcal{E}} q$ for some $q \in \mathcal{Q}_f$. Elements of $\mathcal{L}(\mathcal{A}/\mathcal{E})$ are ground terms accepted by $\mathcal{A}/\mathcal{E}$. A tree language L over $\mathcal{F}$ is some subset of $\mathcal{T}(\mathcal{F})$. We say a tree language L is *recognizable* with an ETA if there exists $\mathcal{A}/\mathcal{E}$ such that $L = \mathcal{L}(\mathcal{A}/\mathcal{E})$. A tree language L is called *regular* if $L = \mathcal{L}(\mathcal{A})$ for some regular TA $\mathcal{A}$. We write $\mathcal{E}(L)$ for $\{t \in \mathcal{T}(\mathcal{F}) \mid t \sim_{\mathcal{E}} s \text{ for some } s \in L\}$ and we say $\mathcal{E}(L)$ is $\mathcal{E}$-congruence closure of L. Note that $\mathcal{E}(\mathcal{E}(L)) = \mathcal{E}(L)$ for any tree language L, however, $\mathcal{E}_1(\mathcal{E}_2(L)) \neq (\mathcal{E}_1 \cup \mathcal{E}_2)(L)$. By definition, if a tree language L is recognizable with an $\mathcal{E}$-TA, so is $\mathcal{E}(L)$. In the questions (Q1) and (Q2), one direction '⇒' is trivial, because $\mathcal{E}(\mathcal{L}(\mathcal{A})) \subseteq \mathcal{L}(\mathcal{A}/\mathcal{E})$ in any case.

Finally we spend the remaining space for explaining some concepts on tree grammars [4]. A *tree grammar* $\mathcal{G}$ is the 4-tuple $(\mathcal{F}, \mathcal{Q}, \mathsf{q}_0, \mathcal{R})$, whose components are the signature $\mathcal{F}$, a finite set $\mathcal{Q}$ of state symbols with fixed arities, an initial state constant q_0 $(\in \mathcal{Q})$ and a finite set $\mathcal{R}$ of pairs (l, r) of terms $l, r \in \mathcal{T}(\mathcal{F} \cup \mathcal{Q}, \mathcal{V})$ such that $\mathsf{var}(r) \subseteq \mathsf{var}(l)$ and $\mathsf{fun}(l) \cap \mathcal{Q} \neq \varnothing$. We write $l \to r$ for a pair $(l, r) \in \mathcal{R}$ and we write $\to_{\mathcal{G}}$ for the induced binary relation. A tree language L is *generatable* if $L = \{t \in \mathcal{T}(\mathcal{F}) \mid \mathsf{q}_0 \to^*_{\mathcal{G}} t\}$ for some tree grammar $\mathcal{G}$. For instance, we consider the tree language $L_1 = \{f^n(g^n(h^n(a))) \mid n \geqslant 0\}$. The tree language L_1 is generatable. Actually, it is represented by the tree grammar $\mathcal{G}_1 = (\mathcal{F}, \{\mathsf{q}_0, \mathsf{q}_1, \mathsf{q}_2, \mathsf{q}_3\}, \mathsf{q}_0, \mathcal{R}_1)$, where $\mathcal{R}_1$:

$$\begin{array}{lll} \mathsf{q}_0 \to \mathsf{q}_1(a, a, a) & \mathsf{q}_2(x, g(y), z) \to \mathsf{q}_2(x, y, g(z)) & \mathsf{q}_3(a, z) \to z \\ \mathsf{q}_1(x, y, z) \to \mathsf{q}_1(f(x), g(y), h(z)) & \mathsf{q}_2(x, a, z) \to \mathsf{q}_3(x, z) & \\ \mathsf{q}_1(x, y, z) \to \mathsf{q}_2(x, y, z) & \mathsf{q}_3(f(x), z) \to \mathsf{q}_3(x, f(z)) & \end{array}$$

3 Recognizability and Some Decidability Results

We start this section by showing the previous tree language L_1 is *not* recognizable with an ETA. First we state the following property.

Lemma 1. *For every TA $\mathcal{A}$ there exists a regular TA $\mathcal{B}$ such that $\mathcal{L}(\mathcal{A}) = \mathcal{L}(\mathcal{B})$.*

Proof. Suppose $\mathcal{A} = (\mathcal{F}, \mathcal{Q}, \mathcal{Q}_f, \mathcal{R})$. We take $\mathcal{B} = (\mathcal{F}, \mathcal{Q}, \mathcal{Q}_f, \mathcal{R}')$ by letting $\mathcal{R}' = \{f(p_1, \ldots, p_n) \to q \mid f \in \mathcal{F}$ and $p_1, \ldots, p_n, q \in \mathcal{Q}$ such that $f(p_1, \ldots, p_n) \to^*_{\mathcal{A}}$

$q\}$. Then it is easy to prove that for all $t \in \mathcal{T}(\mathcal{F})$, $t \to_{\mathcal{A}}^* q \in \mathcal{Q}$ if and only if $t \to_{\mathcal{B}}^* q \in \mathcal{Q}$. □

We suppose to the contradiction that there is an ETA $\mathcal{A}/\mathcal{E} = (\mathcal{F}, \mathcal{Q}, \mathcal{Q}_f, \mathcal{R}, \mathcal{E})$ such that $\mathcal{L}(\mathcal{A}/\mathcal{E}) = L_1$. In this case $\mathcal{E}$ is non-empty; otherwise, L_1 is recognized by a regular TA (Lemma 1). We take a term $t = f^n(g^n(h^n(a)))$ such that $n > |\mathcal{Q}| + |l|$ and $n > |\mathcal{Q}| + |r|$ for every $l \approx r \in \mathcal{E}$. Since t is accepted by $\mathcal{A}/\mathcal{E}$, there exists a derivation $t \to_{\mathcal{A}/\mathcal{E}}^* \mathsf{q}$ for some $\mathsf{q} \in \mathcal{Q}_f$. Suppose $t \to_{\mathcal{A}}^* f^n(g^n(h^{n-m}(\mathsf{p}))) \to_{\mathcal{A}/\mathcal{E}}^* \mathsf{q}$ and $\mathsf{p} \in \mathcal{Q}$. In case $0 \leqslant m \leqslant |\mathcal{Q}|$, $f^n(g^n(h^{n-m}(\mathsf{p}))) \sim_{\mathcal{E}} s'$ if and only if $f^n(g^n(h^{n-m}(\mathsf{p}))) = s'$ (as there is no term $t' \notin L_1$ such that $t \sim_{\mathcal{E}} t'$). This admits a derivation $t \to_{\mathcal{A}}^* f^n(g^n(h^{n-m_1}(\mathsf{p}'))) \to_{\mathcal{A}}^+ f^n(g^n(h^{n-m_2}(\mathsf{p}'))) \to_{\mathcal{A}/\mathcal{E}}^* \mathsf{q}$ for some $\mathsf{p}' \in \mathcal{Q}$ and $m_1 < m_2 \leqslant |\mathcal{Q}|$. This implies $f^n(g^n(h^{n-m_1}(h^{(m_1-m_2)\times i}(h^{m_2}(a)))))$ is accepted for any $i \geqslant 0$, but it contradicts to the assumption. Therefore L_1 is not recognizable, and thus, every generatable tree language is not recognizable. One should notice that $\{f^n(g^n(a)) \mid n \geqslant 0\}$ is recognized by an ETA (but not by a TA) having an equation $f(g(x)) \approx f(f(g(g(x))))$.

On the other hand, every recognizable tree language $\mathcal{L}(\mathcal{A}/\mathcal{E})$ is generatable whenever $\mathcal{E}$ is linear. Let $\mathcal{A}/\mathcal{E} = (\mathcal{F}, \mathcal{Q}, \mathcal{Q}_f, \mathcal{R}, \mathcal{E})$. For all $f \in \mathcal{F}$, we take fresh state symbols q_f (for tree grammar) such that $\mathtt{arity}(\mathsf{q}_f) = \mathtt{arity}(f)$. Now we define the tree grammar $\mathcal{G} = (\mathcal{F}, \mathcal{Q}', \mathsf{q}_0, \mathcal{R}' \cup \mathcal{R}'')$ as follows: Let ϕ be the mapping defined by $\phi(t) = t$ if $t \in \mathcal{V}$ or $t \in \mathcal{Q}$; $\phi(t) = \mathsf{q}_f(\phi(t_1), \ldots, \phi(t_n))$ if $t = f(t_1, \ldots, t_n)$. Then $\mathcal{Q}' = \mathcal{Q} \cup \{\mathsf{q}_0, \mathsf{q}_*\} \cup \{\mathsf{q}_f \mid f \in \mathcal{F}\}$, $\mathcal{R}' = \{\phi(r) \to \phi(l) \mid l \to r \in \mathcal{R}\} \cup \{\mathsf{q}_f(x_1, \ldots, x_n) \to f(x_1, \ldots, x_n) \mid f \in \mathcal{F}\} \cup \{\mathsf{q}_0 \to q \mid q \in \mathcal{Q}_f\} \cup \{\mathsf{q}_* \to \mathsf{q}_f(\mathsf{q}_*, \ldots, \mathsf{q}_*) \mid f \in \mathcal{F}\} \cup \{\mathsf{q}_* \to q \mid q \in \mathcal{Q}\}$ and $\mathcal{R}''$ consists of rules $\phi(l)\sigma_1 \to \phi(r)\sigma_1\sigma_2$ for all $l \approx r \in \mathcal{E} \cup \mathcal{E}^{-1}$ and substitutions σ_1, σ_2. Here $\mathcal{E}^{-1} = \{s \approx t \mid t \approx s \in \mathcal{E}\}$, $\sigma_1 = \{x \mapsto q(x_1, \ldots, x_n)\}$ for some $q \in \mathcal{Q}' \setminus \{\mathsf{q}_0, \mathsf{q}_*\}$ and fresh variables $x_1, \ldots, x_n$ if $l = x (\in \mathcal{V})$; otherwise, $\sigma_1 = \varnothing$. Moreover, $\sigma_2 = \{y_1 \mapsto \mathsf{q}_*, \ldots, y_k \mapsto \mathsf{q}_* \mid y_i \in \mathsf{var}(r) \setminus \mathsf{var}(l)$ for all $1 \leqslant i \leqslant k\}$. The tree grammar $\mathcal{G}$ satisfies $\mathsf{q}_0 \to_{\mathcal{G}}^* t \in \mathcal{T}(\mathcal{F})$ if and only if $t \to_{\mathcal{A}/\mathcal{E}}^* q$ for some $q \in \mathcal{Q}_f$.

We now consider the initial questions (Q1) and (Q2). First we observe that $\mathcal{E}(\mathcal{L}(\mathcal{A})) \neq \mathcal{L}(\mathcal{A}/\mathcal{E})$. For instance, let $\mathcal{E}_1 = \{f(x,x) \approx g(x,x)\}$ and $\mathcal{A}_1 = (\{f, g, a, b\}, \{\mathsf{q}_1, \mathsf{q}_2\}, \{\mathsf{q}_2\}, \{a \to \mathsf{q}_1, b \to \mathsf{q}_1, f(\mathsf{q}_1, \mathsf{q}_1) \to \mathsf{q}_2\})$. It is trivial that $\mathcal{L}(\mathcal{A}_1) = \{f(a,a), f(a,b), f(b,a), f(b,b)\}$, and then, $\mathcal{E}_1(\mathcal{L}(\mathcal{A}_1)) = \mathcal{L}(\mathcal{A}_1) \cup \{g(a,a), g(b,b)\}$. On the other hand, $\mathcal{L}(\mathcal{A}_1/\mathcal{E}_1) = \{f(s,t), g(s,t) \mid s, t \in \{a, b\}\}$. Note that $g(a,b) \to_{\mathcal{A}_1}^+ g(\mathsf{q}_1, \mathsf{q}_1) \sim_{\mathcal{E}_1} f(\mathsf{q}_1, \mathsf{q}_1) \to_{\mathcal{A}_1} \mathsf{q}_2$, although $g(a,b) \notin \mathcal{E}_1(\mathcal{L}(\mathcal{A}_1))$.

Unfortunately, linearity of $\mathcal{E}$ is insufficient to guarantee $\mathcal{E}(\mathcal{L}(\mathcal{A})) = \mathcal{L}(\mathcal{A}/\mathcal{E})$. Consider $\mathcal{A}_2 = (\{f, a, b\}, \{\mathsf{q}_1, \mathsf{q}_2, \mathsf{q}_3\}, \{\mathsf{q}_2\}, \{a \to \mathsf{q}_1, b \to \mathsf{q}_2, f(\mathsf{q}_1, \mathsf{q}_2) \to f(\mathsf{q}_3, \mathsf{q}_3), f(\mathsf{q}_3, \mathsf{q}_2) \to \mathsf{q}_2\})$ and $\mathcal{E}_2 = \mathrm{A}(\{f\})$. Since only b is reduced to q_2 by $\mathcal{A}_2$, $\mathcal{L}(\mathcal{A}_2) = \{b\}$, and then $\mathcal{E}_2(\mathcal{L}(\mathcal{A}_2)) = \{b\}$. Let $t = f(f(a,b),b)$. The subterm $f(a,b)$ can be reduced to $f(\mathsf{q}_1, \mathsf{q}_2)$, then $t \to_{\mathcal{A}_2}^* f(f(\mathsf{q}_1, \mathsf{q}_2), \mathsf{q}_2) \to_{\mathcal{A}_2} f(f(\mathsf{q}_3, \mathsf{q}_3), \mathsf{q}_2)$. Due to associativity of f, $f(f(\mathsf{q}_3, \mathsf{q}_3), \mathsf{q}_2) \sim_{\mathcal{E}_2} f(\mathsf{q}_3, f(\mathsf{q}_3, \mathsf{q}_2))$, and thus, $f(\mathsf{q}_3, f(\mathsf{q}_3, \mathsf{q}_2)) \to_{\mathcal{A}_2}^* \mathsf{q}_2$. Hence t is accepted by $\mathcal{A}_2/\mathcal{E}_2$.

As a consequence, we obtain a partial solution to the initial questions.

Lemma 2. *Every regular ETA $\mathcal{A}/\mathcal{E}$ with $\mathcal{E}$ linear satisfies $\mathcal{E}(\mathcal{L}(\mathcal{A})) = \mathcal{L}(\mathcal{A}/\mathcal{E})$.*

Proof. Since $\mathcal{E}(\mathcal{L}(\mathcal{A})) \subseteq \mathcal{L}(\mathcal{A}/\mathcal{E})$ is trivial, we show the reverse. It suffices to prove $\rightarrow_{\mathcal{A}} \cdot \vdash\!\dashv_{\mathcal{E}} \subseteq \vdash\!\dashv_{\mathcal{E}} \cdot \rightarrow_{\mathcal{A}}^{=}$. Here $\rightarrow_{\mathcal{A}}^{=}$ denotes the reflexive closure of $\rightarrow_{\mathcal{A}}$. Let $\mathcal{A}/\mathcal{E} = (\mathcal{F}, \mathcal{Q}, \mathcal{Q}_f, \mathcal{R}, \mathcal{E})$ and suppose $s \rightarrow_{\mathcal{A}} t \vdash\!\dashv_{\mathcal{E}} u$ such that $s = C[l]$ and $t = C[r]$ for some $l \rightarrow r \in \mathcal{R}$, and moreover, $t = C'[l'\sigma]$ and $u = C'[r'\sigma]$ for some $l' \approx r' \in \mathcal{E} \cup \mathcal{E}^{-1}$. Since r is a state in $\mathcal{Q}$, there are the two cases as follows: If $l' \approx r'$ is applied above r, then r occurs below or at a variable position of l'. Suppose $\sigma = \{x_1 \mapsto t_1, \ldots, x_i \mapsto D[r], \ldots, x_n \mapsto t_n \mid x_i \in \mathsf{var}(l') \cup \mathsf{var}(r')$ for all $1 \leqslant i \leqslant n\}$. Then we take $\sigma' = (\sigma \backslash \{x_i \mapsto D[r]\}) \cup \{x_i \mapsto D[l]\}$, and we obtain $s = C'[l'\sigma']$. Since $l' \approx r'$ is linear, we also obtain $C'[l'\sigma'] \vdash\!\dashv_{\mathcal{E}} C'[r'\sigma']$ and $C'[r'\sigma'] \rightarrow_{\mathcal{A}}^{=} u$. Otherwise (i.e., if $l' \approx r'$ is applied at a parallel position of r), $s \rightarrow_{\mathcal{A}} t \vdash\!\dashv_{\mathcal{E}} u$ obviously implies $s \vdash\!\dashv_{\mathcal{E}} t' \rightarrow_{\mathcal{A}} u$ for some t'. □

In this case the emptiness problem (i.e., a question if $\mathcal{L}(\mathcal{A}/\mathcal{E}) = \varnothing$) is decidable, because $\mathcal{L}(\mathcal{A}) = \varnothing$ if and only if $\mathcal{E}(\mathcal{L}(\mathcal{A})) = \varnothing$. In case $\mathcal{E}$ is also length-preserving, membership and finiteness problems are decidable.

Along the same lines of the proof of Lemma 1, we obtain another statement.

Lemma 3. *Every ETA $\mathcal{A}/\mathrm{C}$ has a TA $\mathcal{B}$ that satisfies $\mathcal{L}(\mathcal{A}/\mathrm{C}) = \mathcal{L}(\mathcal{B})$.*

Proof. We use the similar construction of the proof of Lemma 1. Let $\mathcal{A}' = (\mathcal{F}, \mathcal{Q}, \mathcal{Q}_f, \mathcal{R}')$, where $\mathcal{R}' = \{f(p_1, \ldots, p_n) \rightarrow q \mid f \in \mathcal{F}$ and $p_1, \ldots, p_n, q \in \mathcal{Q}$ such that $f(p_1, \ldots, p_n) \rightarrow_{\mathcal{A}/\mathrm{C}}^{*} q\}$. It is easy to show $\mathcal{L}(\mathcal{A}/\mathrm{C}) = \mathcal{L}(\mathcal{A}'/\mathrm{C})$. By Lemma 2 we have $\mathcal{L}(\mathcal{A}'/\mathrm{C}) = \mathrm{C}(\mathcal{L}(\mathcal{A}'))$. Moreover, every C-congruence closure of a regular tree language is recognizable with a regular TA (e.g. Exercise 12(3) in [2]). Hence there exists a TA $\mathcal{B}$ such that $\mathrm{C}(\mathcal{L}(\mathcal{A}')) = \mathcal{L}(\mathcal{B})$. □

Accordingly, we have the positive partial solutions: if $\mathcal{E}$ is a linear ES,

$$\forall L : \mathrm{TL},\ \exists \mathcal{A} : \text{regular TA}.\ \mathcal{E}(\mathcal{L}(\mathcal{A})) = L \ \Leftrightarrow\ \exists \mathcal{B}/\mathcal{E} : \text{regular ETA}.\ \mathcal{L}(\mathcal{B}/\mathcal{E}) = L.$$

As a special case, if $\mathcal{E} = \mathrm{C}$, the regularity condition for $\mathcal{A}$ and $\mathcal{B}$ is unnecessary (Lemma 3). Moreover, we showed C-TA's have the same expressive power as regular TA's have. Another particular known case is $\mathcal{E}$ ground. Dauchet and Tison [3] showed $\mathcal{E}(\mathcal{L}(\mathcal{A}))$ is regular whenever $\mathcal{E}$ is ground, so there exists a regular ETA $\mathcal{B}/\mathcal{E}$ such that $\mathcal{L}(\mathcal{B}/\mathcal{E}) = \mathcal{E}(\mathcal{L}(\mathcal{A}))$ (as there exists a regular TA $\mathcal{B}$ such that $\mathcal{L}(\mathcal{B}) = \mathcal{E}(\mathcal{L}(\mathcal{B})) = \mathcal{L}(\mathcal{B}/\mathcal{E})$). We can also prove $\mathcal{L}(\mathcal{B}/\mathcal{E})$ is regular for any ETA $\mathcal{B}/\mathcal{E}$ with $\mathcal{E}$ ground, because $\rightarrow_{\mathcal{B}} \cdot \vdash\!\dashv_{\mathcal{E}} \subseteq \vdash\!\dashv_{\mathcal{E}} \cdot \rightarrow_{\mathcal{B}}$ in this case.

In contrast to the situation in Lemma 2, $\mathcal{A}$ and $\mathcal{B}$ in the above formula are not necessarily the same. This leads us to a new question; that is, whether or not the regularity of $\mathcal{A}$ and $\mathcal{B}$ is really essential in the formula. We are concerned with this question in the following part.

Ground term rewriting and regular tree automata are closely related each other. In fact, the "word problem" for ground theory is solvable, by reducing to TA's intersection-emptiness problem which is decidable. The same result holds for ground C-theory. In contrast, it is known that word problem for ground A-theory is undecidable, which was proved by Post [13]. In term rewriting, Deruyver and Gilleron [5] showed reachability of ground A-term rewriting is undecidable.

In equational tree automata framework a similar phenomenon can be found. The problem is regularizability of transition rules of A-symbols in A-TA. One should notice that every A-TA can be transformed to a *quasi*-regular A-TA.

Lemma 4. *Every $\mathcal{A}/\mathrm{A}$ has a quasi-regular $\mathcal{B}/\mathrm{A}$ that satisfies $\mathcal{L}(\mathcal{A}/\mathrm{A})=\mathcal{L}(\mathcal{B}/\mathrm{A})$.*

Proof. Let $\mathcal{A}/\mathrm{A} = (\mathcal{F}, \mathcal{Q}, \mathcal{Q}_f, \mathcal{R}, \mathrm{A})$. We take $\mathcal{A}'/\mathrm{A} = (\mathcal{F}, \mathcal{Q}, \mathcal{Q}_f, \mathcal{R}', \mathrm{A})$, where $\mathcal{R}' = \{f(p_1,\dots,p_n) \to q \mid f \in \mathcal{F} \setminus \mathsf{fun}(\mathrm{A})$ and $p_1,\dots,p_n, q \in \mathcal{Q}$ such that $f(p_1,\dots,p_n) \to^*_{\mathcal{A}} q\}$. We take $\mathcal{B}/\mathrm{A} = (\mathcal{F}, \mathcal{Q}, \mathcal{Q}_f, \mathcal{S}, \mathrm{A})$ by letting $\mathcal{S}$ be the union of $\mathcal{R}'$ and $\{l \to r \in \mathcal{R} \mid \mathsf{root}(l) \in \mathsf{fun}(\mathrm{A})\}$. Suppose $t \in \mathcal{T}(\mathcal{Q} \cup \mathcal{F})$ such that $t \to^*_{\mathcal{A}/\mathrm{A}} q$ for some $q \in \mathcal{Q}$. Using the induction on the size of terms we show $t \to^*_{\mathcal{B}/\mathrm{A}} q$ below. If t is a constant $c\,(\notin \mathcal{Q})$, there exists a rule $c \to q \in \mathcal{R}$, and thus $c \to q \in \mathcal{R}'$. If t is a state then $t = q$. Otherwise, there exist a term $t' \in \mathcal{T}(\mathcal{Q} \cup \mathcal{F})$ and a derivation $t \to^*_{\mathcal{A}/\mathrm{A}} t' \to^*_{\mathcal{A}/\mathrm{A}} q$ such that $t = C[t_1,\dots,t_n]$, $t' = C'[q_1,\dots,q_n]$, $C \sim_{\mathrm{A}} C'$, $t_i \to^*_{\mathcal{A}/\mathrm{A}} q_i\,(\in \mathcal{Q})$ and $t_i \in \mathcal{T}(\mathcal{Q} \cup (\mathcal{F} \setminus \mathsf{fun}(\mathrm{A})))$ for all $1 \leqslant i \leqslant n$. Here we assume $t_1,\dots,t_n$ are maximal subterms. Let $\mathcal{R}_1 = \{l \to r \in \mathcal{R} \mid \mathsf{root}(l) \in \mathsf{fun}(\mathrm{A})\}$ and $\mathcal{R}_2 = \{l \to r \in \mathcal{R} \mid \mathsf{root}(l) \notin \mathsf{fun}(\mathrm{A})\}$. Since $t_i \to^*_{\mathcal{A}/\mathrm{A}} q_i$ is performed by $\mathcal{R}_2$, we obtain $t_i \to^*_{\mathcal{A}'} q_i$. This implies $t \to^*_{\mathcal{A}'/\mathrm{A}} t' \to^*_{\mathcal{A}/\mathrm{A}} q$. Again we observe that there exist a term t'' and a derivation $t' \to^*_{\mathcal{A}/\mathrm{A}} t'' \to^*_{\mathcal{A}/\mathrm{A}} q$ such that $t' = D[t'_1,\dots,t'_m]$, $t'' = D'[p_1,\dots,p_m]$, $D \sim_{\mathrm{A}} D'$, $t'_i \to^*_{\mathcal{A}/\mathrm{A}} p_i\,(\in \mathcal{Q})$ and $t'_i \in \mathcal{T}(\mathsf{fun}(\mathrm{A}))$ for all $1 \leqslant i \leqslant m$. Here we assume $t'_1,\dots,t'_m$ are maximal. In this case $t'_i \to^*_{\mathcal{B}/\mathrm{A}} p_i$ (actually $t'_i \to^*_{\mathcal{R}_1/\mathrm{A}} p_i$), and thus, $t' \to^*_{\mathcal{B}/\mathrm{A}} t''$. Since $|t''| < |t|$, $t'' \to^*_{\mathcal{B}/\mathrm{A}} q$ by induction hypothesis. Moreover, from the fact that $\to_{\mathcal{A}'} \subseteq \to_{\mathcal{B}}$, we obtain $t \to^*_{\mathcal{B}/\mathrm{A}} q$.

The inverse is an easy consequence of the property that $\to_{\mathcal{B}/\mathrm{A}} \subseteq \to^+_{\mathcal{A}/\mathrm{A}}$. □

However, every quasi-regular A-TA is not always simplified to a regular A-TA with the same expressive power. In the remaining (but major) part of this section we explain the reason, by introducing *linear bounded automata* found in Hopcroft and Ullman [8]. A linear bounded automaton (LBA for short) M is the 7-tuple $(\Sigma, \mathcal{Q}, \mathcal{Q}_f, \mathsf{q}_0, \#, \$, \mathcal{S})$. Each of the components denotes:

- Σ: a finite set of tape symbols,
- $\mathcal{Q}$: a finite set of state symbols such that $\Sigma \cap \mathcal{Q} = \varnothing$,
- $\mathcal{Q}_f(\subseteq \mathcal{Q})$: a set of final states,
- $\mathsf{q}_0(\in \mathcal{Q})$: an initial state symbol,
- $\#\,(\notin \Sigma)$: left-endmark,
- $\$\,(\notin \Sigma)$: right-endmark,
- $\mathcal{S}$: a finite set of string rewrite rules in the form of either $a\,q\,b \to q'\,a\,b'$ or $q\,b \to b'\,q'$ for some $a, a', b, b' \in \Sigma$ and $q, q' \in \mathcal{Q}$. (Basic notions of string rewriting are explained, e.g. in [1].) As a special case of the former rule, it is allowed to be $b = \$$ whenever $b' = \$$. Similarly, in the latter case, $b = \#$ is allowed whenever $b' = \#$. If there exists a rule $a\,q\,b \to q'\,a\,b'$ for some $a \in \Sigma$, we assume $\mathcal{S}$ contains $c\,q\,b \to q'\,c\,b'$ for all $c \in \Sigma$.

LBA is a Turing machine whose tape length is finitely bounded. As we showed in the example in Section 2, Turing machine (or tree grammar) is too general to discuss the expressive power of ETA. In the following part we show equivalence of LBA and (a special case of) ETA, and so we use such a resource bounded Turing machine.

A word w is a finite (possibly empty) sequence of alphabets over Σ. The empty word is denoted by ϵ and the set of all words over Σ is Σ^*. A language is a subset of Σ^*. A move relation on $(\Sigma \cup \mathcal{Q} \cup \{\#, \$\})^*$ with respect to M, denoted by $\to_M$, is defined as follows: $u \to_M v$ if there exists a rule $l \to r \in \mathcal{S}$ such that $u = u_1\, l\, u_2$ and $v = u_1\, r\, u_2$. A word w is called accepted by M if $\mathsf{q}_0\, \#\, w\, \$ \to_M^* u\, \mathsf{q}\, v$ for some $\mathsf{q} \in \mathcal{Q}_f$ and $u, v \in (\Sigma \cup \{\#, \$\})^*$. The set $\mathcal{L}(M)$ consists of words accepted by M. By definition, it is allowed to be $\epsilon \in \mathcal{L}(M)$. We say a language L is recognizable with an LBA if there exists an LBA M such that $L = \mathcal{L}(M)$.

It is known that emptiness problem is undecidable for LBA; that means, there is no algorithm deciding whether a language recognized by an arbitrary LBA is empty. This implies that for an arbitrary LBA M, if there exists an ETA $\mathcal{A}/\mathrm{A}$ simulating M, we may not find a regular ETA $\mathcal{B}/\mathrm{A}$ such that $\mathcal{L}(\mathcal{A}/\mathrm{A}) = \mathcal{L}(\mathcal{B}/\mathrm{A})$. Otherwise, we can determine whether $\mathcal{L}(M) = \varnothing$ by examining $\mathcal{L}(\mathcal{B}) = \varnothing$, because $\mathcal{L}(\mathcal{B}/\mathrm{A}) = \varnothing$ if and only if $\mathcal{L}(\mathcal{B}) = \varnothing$ due to Lemma 2.

Thus, all we have to do in the remaining part is to show that for an arbitrary LBA M, there exists an associated ETA $\mathcal{A}_M/\mathrm{A}$ such that $\mathcal{L}(M) = \varnothing$ if and only if $\mathcal{L}(\mathcal{A}_M/\mathrm{A}) = \varnothing$.

Given an LBA $M = (\Sigma, \mathcal{Q}, \mathcal{Q}_f, \mathsf{q}_0, \#, \$, \mathcal{S})$. Let us take $\mathcal{F}_M = \Sigma \cup \{\mathsf{q}_0, \#, \$\} \cup \{\mathsf{f}\}$ such that f is a fresh binary function symbol assumed to be associative. The set $\mathcal{Q}_M$ of state symbols is the union of $\mathcal{Q}_1 = \{\, \alpha_q, \overline{\alpha}_q \mid q \in \mathcal{Q} \,\}$ and $\mathcal{Q}_2 = \{\, \beta_a \mid a \in \Sigma \cup \{\#, \$\} \,\}$ together with fresh state symbols, $*_{\mathsf{q}_0}$, $*_\#$, $\circ_1$, $\circ_2$, $\diamond$ such that $\mathcal{Q}_{Mf} = \{\diamond\}$. The set $\mathcal{R}_M$ consists of the following transition rules:

1. $\mathsf{q}_0 \to *_{\mathsf{q}_0}$,
2. $\# \to *_\#$,
3. $\mathsf{f}(*_{\mathsf{q}_0}, *_\#) \to \mathsf{f}(\alpha_{\mathsf{q}_0}, \beta_\#)$,
4. $a \to \beta_a$ for all $a \in \Sigma \cup \{\$\}$,
5. $\mathsf{f}(\alpha_p, \beta_a) \to \mathsf{f}(\beta_b, \alpha_q)$ if $p\,a \to b\,q \in \mathcal{S}$ for some $p, q \in \mathcal{Q}$ and $a, b \in \Sigma$,
6. $\mathsf{f}(\alpha_p, \beta_a) \to \mathsf{f}(\overline{\alpha}_q, \beta_b)$ if $c\,p\,a \to q\,c\,b \in \mathcal{S}$ for some $p, q \in \mathcal{Q}$ and $a, b, c \in \Sigma$,
7. $\mathsf{f}(\beta_a, \overline{\alpha}_q) \to \mathsf{f}(\alpha_q, \beta_a)$ for all $a \in \Sigma \cup \{\#, \$\}$ and $q \in \mathcal{Q}$,
8. $\mathsf{f}(\alpha_q, \beta_a) \to \mathsf{f}(\circ_1, \beta_a)$ and $\mathsf{f}(\alpha_q, \beta_\#) \to \circ_2$ for all $q \in \mathcal{Q}_f$ and $a \in \Sigma \cup \{\$\}$,
9. $\mathsf{f}(\beta_a, \circ_1) \to \circ_1$ and $\mathsf{f}(\circ_2, \beta_a) \to \circ_2$ for all $a \in \Sigma$,
10. $\mathsf{f}(\beta_\#, \circ_1) \to \circ_2$
11. $\mathsf{f}(\circ_2, \beta_\$) \to \diamond$.

Henceforth, we write $t = C_f[\![t_1, \ldots, t_n]\!]$ if $t = C[t_1, \ldots, t_n]$ such that C is a non-empty and maximal context consisting of a function symbol f. If unnecessary f to be explicit, we simply write $t = C[\![t_1, \ldots, t_n]\!]$.

The idea of the previous construction is described below. We take an ETA $\mathcal{A}_M/\mathrm{A} = (\mathcal{F}_M, \mathcal{Q}_M, \mathcal{Q}_{Mf}, \mathcal{R}_M, \mathrm{A})$ with $\mathrm{A} = \{\mathsf{f}(\mathsf{f}(x,y),z) \approx \mathsf{f}(x,\mathsf{f}(y,z))\}$. In the setting, a word $w = a_1\, a_2\, a_3 \,\dots\, a_n$ is represented by a term $C[\![a_1, a_2, a_3, \dots, a_n]\!]$ and an *initial instantaneous description* for w, i.e. $\mathsf{q}_0 \,\#\, a_1\, a_2 \,\dots\, a_n\,\$$, is represented as $C[\![\mathsf{q}_0, \#, a_1, a_2, \dots, a_n, \$]\!]$.

The first three rules 1–3 examine whether q_0 and # are located in the order of $\mathsf{q}_0\,\#$ at the initial stage. Using the transition rules 4, each tape symbol together with right-endmark of a term is replaced by a corresponding state symbol in $\mathcal{Q}_M$. This step is not necessarily performed at once.

In case M admits the move relation $\#\,\mathsf{p}\, a_1\, a_2 \,\dots\, a_n\,\$ \to_M \#\, b_1\, \mathsf{q}\, a_2 \,\dots\, a_n\,\$$, there exists the corresponding derivation

$$\begin{aligned} C[\![\beta_\#, \alpha_\mathsf{p}, \beta_{a_1}, \beta_{a_2}, \beta_{a_n}, \beta_\$]\!] \sim_\mathrm{A}\ & \mathsf{f}(\mathsf{f}(\dots\mathsf{f}(\mathsf{f}(\beta_\#, \mathsf{f}(\alpha_\mathsf{p}, \beta_{a_1})), \beta_{a_2})\dots, \beta_{a_n}), \beta_\$) \\ \to_{\mathcal{A}_M}\ & \mathsf{f}(\mathsf{f}(\dots\mathsf{f}(\mathsf{f}(\beta_\#, \mathsf{f}(\beta_{b_1}, \alpha_\mathsf{q})), \beta_{a_2})\dots, \beta_{a_n}), \beta_\$). \end{aligned}$$

If there is a rule $b_1\, \mathsf{q}\, a_2 \to \mathsf{r}\, b_1\, b_2 \in \mathcal{S}$ and it is applied at the next step, then

$$\begin{aligned} C[\![\beta_\#, \beta_{b_1}, \alpha_\mathsf{q}, \beta_{a_2}, \dots, \beta_{a_n}, \beta_\$]\!] \sim_\mathrm{A}\ & \mathsf{f}(\mathsf{f}(\dots\mathsf{f}(\beta_\#, \mathsf{f}(\beta_{b_1}, \mathsf{f}(\alpha_\mathsf{q}, \beta_{a_2})))\dots, \beta_{a_n}), \beta_\$) \\ \to_{\mathcal{A}_M}\ & \mathsf{f}(\mathsf{f}(\dots\mathsf{f}(\beta_\#, \mathsf{f}(\beta_{b_1}, \mathsf{f}(\overline{\alpha}_\mathsf{r}, \beta_{b_2})))\dots, \beta_{a_n}), \beta_\$) \\ \sim_\mathrm{A}\ & \mathsf{f}(\mathsf{f}(\dots\mathsf{f}(\beta_\#, \mathsf{f}(\mathsf{f}(\beta_{b_1}, \overline{\alpha}_\mathsf{r}), \beta_{b_2}))\dots, \beta_{a_n}), \beta_\$) \\ \to_{\mathcal{A}_M}\ & \mathsf{f}(\mathsf{f}(\dots\mathsf{f}(\beta_\#, \mathsf{f}(\mathsf{f}(\alpha_\mathsf{r}, \beta_{b_1}), \beta_{b_2}))\dots, \beta_{a_n}), \beta_\$). \end{aligned}$$

Lemma 5. *If an LBA M admits a move relation*

$$\mathsf{q}_0\,\#\,a_1 \,\dots\, a_n\,\$ \to_M^* b_0\, b_1\, b_2 \,\dots\, b_{i-1}\,\mathsf{p}\, b_i \,\dots\, b_n\, b_{n+1},$$

the associated ETA $\mathcal{A}_\mathcal{M}/\mathrm{A}$ simulates the computation sequence by resulting in the derivation

$$C[\![\mathsf{q}_0, \#, a_1, \dots, a_n, \$]\!] \to_{\mathcal{A}_\mathcal{M}/\mathrm{A}}^* C[\![\beta_{b_0}, \beta_{b_1}, \dots, \beta_{b_{i-1}}, \alpha_\mathsf{p}, \beta_{b_i}, \dots, \beta_{b_n}, \beta_{b_{n+1}}]\!].$$

Proof. Use induction on the length of M-move relation. □

Hence $\mathcal{A}_M/\mathrm{A}$ results in a term $C[\![\beta_{b_0}, \beta_{b_1}, \dots, \beta_{b_{i-1}}, \alpha_\mathsf{p}, \beta_{b_i}, \dots, \beta_{b_n}, \beta_{b_{n+1}}]\!]$, provided $\mathsf{q}_0\,\#\,a_1 \,\dots\, a_n\,\$ \to_M^* b_0\, b_1 \,\dots\, b_{i-1}\,\mathsf{p}\, b_i \,\dots\, b_n\, b_{n+1}$. In case $\mathsf{p} \in \mathcal{Q}_f$, a subterm $\mathsf{f}(\alpha_\mathsf{p}, \beta_{b_i})$ is replaced by $\mathsf{f}(\circ_1, \beta_{b_i})$ using the rules 8. Moreover, the whole term is simplified to $\mathsf{f}(\circ_2, \beta_\$)$ by the rules 9–10. Finally $\diamond$ is obtained by applying the rule 11.

This obviously implies soundness of the construction, with respect to acceptability. To be formalized, it is represented as follows. Let $\Gamma = \Sigma \cup \mathcal{Q} \cup \{\#, \$\}$ and define the mapping

$$\langle w \rangle = \begin{cases} \mathsf{f}(a, \langle u \rangle) & \text{if } w = a\,u \text{ for some } a \in \Gamma \text{ and } u \in \Gamma^+, \\ w & \text{if } w \in \Gamma. \end{cases}$$

Due to Lemma 5 together with the preceding observation, the soundness property is established.

Lemma 6. *Every $\mathcal{A}_M/\mathrm{A}$ associated with an LBA $M = (\Sigma, \mathcal{Q}, \mathcal{Q}_f, \mathsf{q}_0, \#, \$, \mathcal{S})$ satisfies $\langle \mathsf{q}_0 \# w \$\rangle \to^*_{\mathcal{A}_M/\mathrm{A}} \diamond$ for all $w \in \mathcal{L}(M)$.* □

Next we show the reverse also holds. Looking at the transition rules of M, we can observe that: if $t \to^*_{\mathcal{A}_M/\mathrm{A}} \diamond$, there exists a derivation represented as follows. Let $p \in \mathcal{Q}_f$.

$$\begin{aligned} t \sim_{\mathrm{A}} & \; C[\![\mathsf{q}_0, \#, a_1, \ldots, a_n, \$]\!] \\ \to^+_{\mathcal{A}_M/\mathrm{A}} & \; C[\![\alpha_{\mathsf{q}_0}, \beta_\#, \beta_{a_1}, \ldots, \beta_{a_n}, \beta_\$]\!] & \cdots (1) \\ \to^*_{\mathcal{A}_M/\mathrm{A}} & \; C[\![\beta_{b_0}, \beta_{b_1}, \ldots, \beta_{b_{i-1}}, \alpha_{\mathsf{p}}, \beta_{b_i}, \ldots, \beta_{b_n}, \beta_{b_{n+1}}]\!] & \cdots (2) \\ \to^+_{\mathcal{A}_M/\mathrm{A}} & \; \mathsf{f}(\circ_2, \beta_\$) \\ \to_{\mathcal{A}_M/\mathrm{A}} & \; \diamond. \end{aligned}$$

More precisely, we have a derivation $t \to^*_{\mathcal{A}_M/\mathrm{A}} \diamond$ only in the case $t \to^*_{\mathcal{A}_M/\mathrm{A}} \mathsf{f}(\circ_2, \beta_\$)$. On the other hand, t has to contain the initial state symbol q_0 and the endmarks # and $ such that q_0 is located left-next to #. And, for any $s, s' \in \mathcal{T}(\mathcal{F}_M \cup \mathcal{Q}_M)$, it holds that if $s \to^*_{\mathcal{A}_M/\mathrm{A}} s'$ then $\|s\|_\$ + \|s\|_{\beta_\$} + \|s\|_\diamond = \|s'\|_\$ + \|s'\|_{\beta_\$} + \|s'\|_\diamond$ and

$$\begin{aligned} & \textstyle\sum_{q \in \mathcal{Q}} (\|s\|_q + \|s\|_{\alpha_q} + \|s\|_{\overline{\alpha}_q}) + \|s\|_{*_{\mathsf{q}_0}} + \|s\|_{\circ_1} + \|s\|_{\circ_2} + \|s\|_\diamond \\ = & \textstyle\sum_{q \in \mathcal{Q}} (\|s'\|_q + \|s'\|_{\alpha_q} + \|s'\|_{\overline{\alpha}_q}) + \|s'\|_{*_{\mathsf{q}_0}} + \|s'\|_{\circ_1} + \|s'\|_{\circ_2} + \|s'\|_\diamond. \end{aligned}$$

Moreover, $\|t\|_\# = \|t\|_{\mathsf{q}_0}$ by the transition rule 3. Then,

$$t = C[\![c_1, \ldots, c_{j-1}, \mathsf{q}_0, \#, c_j, \ldots, c_{j+k}, \$, c_{j+k+1}, \ldots, c_m]\!]$$

or

$$t = C[\![c_1, \ldots, c_{j-1}, \$, c_j, \ldots, c_{j+k}, \mathsf{q}_0, \#, c_{j+k+1}, \ldots, c_m]\!]$$

for some context C and $c_1, \ldots, c_m \in \Sigma$. Since $t \to^*_{\mathcal{A}_M/\mathrm{A}} \diamond$, we can assume without loss of generality that

$$t \to^*_{\mathcal{A}_M/\mathrm{A}} C[\![\beta_{c_1}, \ldots, \beta_{c_{j-1}}, \alpha_{\mathsf{q}_0}, \beta_\#, \beta_{c_j}, \ldots, \beta_{c_{j+k}}, \beta_\$, \beta_{c_{j+k+1}}, \ldots, \beta_{c_m}]\!] \to^*_{\mathcal{A}_M/\mathrm{A}} \diamond$$

or

$$t \to^*_{\mathcal{A}_M/\mathrm{A}} C[\![\beta_{c_1}, \ldots, \beta_{c_{j-1}}, \beta_\$, \beta_{c_j}, \ldots, \beta_{c_{j+k}}, \alpha_{\mathsf{q}_0}, \beta_\#, \beta_{c_{j+k+1}}, \ldots, \beta_{c_m}]\!] \to^*_{\mathcal{A}_M/\mathrm{A}} \diamond.$$

Since $t \to^*_{\mathcal{A}_M/\mathrm{A}} \mathsf{f}(\circ_2, \beta_\$)$, the former derivation is the case, which can be proved by induction on the length of a derivation $s \to^*_{\mathcal{A}_M/\mathrm{A}} \mathsf{f}(\circ_2, \beta_\$)$. Moreover, we obtain $j = 1$ and $k = n - 1$ such that $c_i = a_i$ for all $1 \leqslant i \leqslant n$. Before we apply a transition rule 8, there is no applicable rules other than rules 5–7.

We let u be a term appearing in between (1) and (2), and we define the mapping `str` as follows.

$$\mathtt{str}(s) = \begin{cases} \mathtt{str}(s_1)\ \mathtt{str}(s_2) & \text{if } s = \mathsf{f}(s_1, s_2), \\ q & \text{if } s = \alpha_a \text{ for some } q \in \mathcal{Q}, \\ a & \text{if } s = \beta_a \text{ for some } a \in \Sigma \cup \{\#, \$\}. \end{cases}$$

Then we obtain the following property.

Lemma 7. *Let v be a term in (2). If $u \to^*_{\mathcal{A}_M/\mathrm{A}} v$ and $\mathtt{str}(u) \in (\Sigma \cup \{\#, \$\} \cup \mathcal{Q})^*$, then $\mathtt{str}(u) \to^*_M \mathtt{str}(v)$.*

Proof. We use induction on the length of $u \to^*_{\mathcal{A}_M/\mathrm{A}} v$. The base case is trivial, because $u = v$. For the induction step we suppose $u \to^+_{\mathcal{A}_M/\mathrm{A}} v$. By assumption, u does not contain $\overline{\alpha}_{\mathsf{q}}$ for any $\mathsf{q} \in \mathcal{Q}$. Let $u \to_{\mathcal{A}_M/\mathrm{A}} u' \to^*_{\mathcal{A}_M/\mathrm{A}} v$. If there exists a transition rule $\mathsf{f}(\alpha_{\mathsf{q}}, \beta_a) \to \mathsf{f}(\beta_b, \alpha_{\mathsf{r}}) \in \mathcal{R}_M$ and it is applied to u, then $\mathtt{str}(u') = \# d_1 \ldots d_{i-2}\, b\, \mathsf{r}\, d_i \ldots d_n \,\$$ for some $d_1, \ldots, d_{i-2}, b, d_i, \ldots, d_n \in \Sigma$. So, $\mathtt{str}(u) = \# d_1 \ldots d_{i-2}\, \mathsf{q}\, a\, d_i \ldots d_n \,\$$, and thus, $\mathtt{str}(u) \to_M \mathtt{str}(u')$. Otherwise, there is a rule $\mathsf{f}(\alpha_{\mathsf{q}}, \beta_a) \to \mathsf{f}(\overline{\alpha}_{\mathsf{r}}, \beta_b) \in \mathcal{R}_M$ and it is applied to u. In this case there is also a (and only) transition rule $\mathsf{f}(\beta_c, \overline{\alpha}_{\mathsf{r}}) \to \mathsf{f}(\alpha_{\mathsf{r}}, \beta_c)$ which is applicable to u'. So, $u' \to_{\mathcal{A}_M/\mathrm{A}} u''$ such that $\mathtt{str}(u'') = \# d_1 \ldots d_{i-2}\, c\, \mathsf{r}\, b\, d_{i+1} \ldots d_n \,\$$ for some $d_1, \ldots, d_{i-2}, c, b, d_{i+1}, \ldots, d_n \in \Sigma$. This implies $\mathtt{str}(u) = \# d_1 \ldots d_{i-2}\, c\, a\, \mathsf{q}$ $d_{i+1} \ldots d_n \,\$$ and $\mathtt{str}(u) \to_M \mathtt{str}(u'')$. □

As a consequence, completeness (with respect to acceptability) is established.

Lemma 8. *Every $\mathcal{A}_M/\mathrm{A}$ associated with an LBA $M = (\Sigma, \mathcal{Q}, \mathcal{Q}_f, \mathsf{q}_0, \#, \$, \mathcal{S})$ satisfies: for all $t \in \mathcal{T}(\mathcal{F}_M)$, if $t \to^*_{\mathcal{A}_M/\mathrm{A}} \diamond$ then $t \sim_{\mathrm{A}} \langle \mathsf{q}_0 \# w \,\$ \rangle$ and $w \in \mathcal{L}(M)$.* □

We know $\mathcal{L}(M)$ is empty if and only if $\mathcal{L}(\mathcal{A}_M/\mathrm{A})$ is empty. Moreover, the former property ($\mathcal{L}(M) = \varnothing$) is known to be undecidable, and so is the latter.

Corollary 1. *For an arbitrary* A-*TA it is undecidable whether a tree language recognized by the* A-*TA is empty.* □

Hence A-TA is not always regularized, although it can be quasi-regularized.

Theorem 1. *There exists an ETA $\mathcal{A}/\mathrm{A}$ such that $\mathcal{L}(\mathcal{A}/\mathrm{A}) \neq \mathrm{A}(\mathcal{L}(\mathcal{B}))$ for any TA $\mathcal{B}$.* □

In fact, the language $P = \{w \in \{a\}^* \mid |w| = 2^n \text{ and } n \geqslant 0\}$ is recognizable with LBA, and thus, the tree language $T = \{t \mid t \sim_{\mathrm{A}} \langle \mathsf{q}_0 \# w \,\$ \rangle \text{ and } w \in W\}$ is recognizable with an A-TA. However, T is not recognizable with a regular A-TA, and then $\{\langle \mathsf{q}_0 \# w \,\$ \rangle \mid w \in P\}$ is not recognizable with a regular TA. In other words, even if there exists a tree language L such that $\mathrm{A}(L) = \mathcal{L}(\mathcal{A}/\mathrm{A})$, L is not recognizable with a TA in general.

Undecidability of finiteness is also obtained, because finiteness problem of LBA is undecidable. Note that $\{\langle \mathsf{q}_0 \# w \,\$ \rangle \mid w \in \mathcal{L}(M)\}$ is finite if and only if $\mathcal{L}(\mathcal{A}_M/\mathrm{A})$ is finite.

Corollary 2. *For an arbitrary* A-*TA it is undecidable whether a tree language recognized by the* A-*TA is finite.* □

Furthermore, a question if $\mathcal{L}(M) = \Sigma^*$ is known to be undecidable for an arbitrary LBA M. This yields the following undecidability results.

Corollary 3. *Let $\mathcal{A}/\mathrm{A}$ and $\mathcal{B}/\mathrm{A}$ be ETA. It is undecidable to test the subset relation $\mathcal{L}(\mathcal{A}/\mathrm{A}) \subseteq \mathcal{L}(\mathcal{B}/\mathrm{A})$. Equivalence test $\mathcal{L}(\mathcal{A}/\mathrm{A}) = \mathcal{L}(\mathcal{B}/\mathrm{A})$ is also undecidable.*

Proof. Let M_1 be an arbitrary LBA and M_2 be an LBA such that $\mathcal{L}(M_2) = \Sigma^*$, e.g $M_2 = (\Sigma, \{\mathsf{q}_0\}, \{\mathsf{q}_0\}, \mathsf{q}_0, \#, \$, \varnothing)$. Then we take $\mathcal{A} = \mathcal{A}_{M_2}$ and $\mathcal{B} = \mathcal{A}_{M_1}$ together with $\mathrm{A} = \{\, \mathsf{f}(\mathsf{f}(x,y),z) \approx \mathsf{f}(x,\mathsf{f}(y,z))\,\}$. As we can see, $\mathcal{L}(\mathcal{A}/\mathrm{A}) \subseteq \mathcal{L}(\mathcal{B}/\mathrm{A})$ (and $\mathcal{L}(\mathcal{A}/\mathrm{A}) = \mathcal{L}(\mathcal{B}/\mathrm{A})$) if and only if $\Sigma^* = \mathcal{L}(M_1)$. □

4 Closure Properties

As we discussed in the previous section, equational tree automata are sometimes too powerful. Nevertheless, the recognizable tree languages are still useful in a certain situation, as they are closed under two operations: union and intersection. In this section we discuss the closure properties of $\mathcal{E}$-tree languages.

Theorem 2. *If $\mathcal{E}$ is a variable-preserving ES, the union of tree languages L_1, L_2 recognized by ETA's $\mathcal{A}/\mathcal{E}$ and $\mathcal{B}/\mathcal{E}$ is recognizable with an ETA $\mathcal{C}/\mathcal{E}$.*

Proof. Let $\mathcal{A}/\mathcal{E} = (\mathcal{Q}_\mathcal{A}, \mathcal{F}, \mathcal{Q}_{\mathcal{A}f}, \mathcal{R}_\mathcal{A}, \mathcal{E})$ and $\mathcal{B}/\mathcal{E} = (\mathcal{Q}_\mathcal{B}, \mathcal{F}, \mathcal{Q}_{\mathcal{B}f}, \mathcal{R}_\mathcal{B}, \mathcal{E})$. We assume without loss of generality that $\mathcal{Q}_\mathcal{A} \cap \mathcal{Q}_\mathcal{B} = \varnothing$. We take the TA $\mathcal{C} = (\mathcal{Q}, \mathcal{F}, \mathcal{Q}_f, \mathcal{R}, \mathcal{E})$ as follows. $\mathcal{Q} = \mathcal{Q}_\mathcal{A} \cup \mathcal{Q}_\mathcal{B}$, $\mathcal{Q}_f = \mathcal{Q}_{\mathcal{A}f} \cup \mathcal{Q}_{\mathcal{B}f}$ and $\mathcal{R} = \mathcal{R}_\mathcal{A} \cup \mathcal{R}_\mathcal{B}$. Below we show the two properties: (1) $s \to^*_{\mathcal{C}/\mathcal{E}} p \in \mathcal{Q}_\mathcal{A}$ if and only if $s \to^*_{\mathcal{A}/\mathcal{E}} p$ and (2) $s \to^*_{\mathcal{C}/\mathcal{E}} q \in \mathcal{Q}_\mathcal{B}$ if and only if $s \to^*_{\mathcal{B}/\mathcal{E}} q$. Since the "if" parts of both properties are trivial, it suffices to show the "only if". We observe that if $t \to_{\mathcal{C}/\mathcal{E}} t'$ and $\mathsf{fun}(t) \cap \mathcal{Q}_\mathcal{A} \neq \varnothing$ then $\mathsf{fun}(t') \cap \mathcal{Q}_\mathcal{A} \neq \varnothing$, because of $\mathcal{E}$ variable-preserving. Moreover, if $t \to_{\mathcal{A}/\mathcal{E}} t'$ then $\mathsf{fun}(t') \cap \mathcal{Q}_\mathcal{A} \neq \varnothing$. The same property holds also for $\mathcal{Q}_\mathcal{B}$ and $\to_{\mathcal{B}/\mathcal{E}}$. This implies that if $t \to^*_{\mathcal{C}/\mathcal{E}} q \notin \mathcal{Q}_\mathcal{B}$ then $t \to^*_{\mathcal{A}/\mathcal{E}} q$. Hence the property (1) holds. Similarly, the property (2) can be proved. □

Let P be a set of ill-formed tape statuses. For instance, P contains $\#\, w\, \$$ (missing a state symbol in a tape) and $\#\, \mathsf{p}\, \#\, w\, \$$ (extra right-endmark). The set $\{\langle p\rangle \mid p \in P\}$ is regular, and then $L = \{t \mid t \sim_\mathrm{A} \langle p\rangle \text{ and } p \in P\}$ is recognizable with a (regular) A-TA. We take the union of the tree languages L and $\mathcal{L}(\mathcal{A}_M/\mathrm{A})$, which is recognizable with an A-TA due to the above theorem. Since universality problem of LBA, i.e. a question if $\{\mathsf{q}_0\, \#\, w\, \$ \mid w \in \Sigma^*\} = \mathcal{L}(M)$, is undecidable, so is to test $L \cup \mathcal{L}(\mathcal{A}_M/\mathrm{A}) = \mathcal{T}(\mathcal{F}_M)$.

Corollary 4. *For an arbitrary ETA $\mathcal{A}/\mathrm{A}$ over the signature $\mathcal{F}$ it is undecidable whether $\mathcal{L}(\mathcal{A}/\mathrm{A}) = \mathcal{T}(\mathcal{F})$.* □

Next we discuss the intersection. Regular tree languages are closed under intersection [2]. Tree languages recognizable with C-TA are also closed under intersection (Corollary 3). The remaining questions to be considered as useful cases are closedness of A- and AC-TA's.

Theorem 3. *If $\mathcal{E} = \mathrm{A}$ or $\mathcal{E} = \mathrm{AC}$, the intersection of tree languages L_1, L_2 recognized by ETA's $\mathcal{A}/\mathcal{E}$ and $\mathcal{B}/\mathcal{E}$ is recognizable with an ETA $\mathcal{C}/\mathcal{E}$.*

Proof. We show the proof sketch for the A-case. Note that the same proof construction can be applied to the AC-case. Let $\mathcal{G}\ (\subseteq \mathcal{F})$ be the set of binary symbols of A-axioms and let $\mathcal{A}/\mathrm{A} = (\mathcal{F}, \mathcal{Q}_{\mathcal{A}}, \mathcal{Q}_{\mathcal{A}f}, \mathcal{R}_{\mathcal{A}}, \mathrm{A})$ and $\mathcal{B}/\mathrm{A} = (\mathcal{F}, \mathcal{Q}_{\mathcal{B}}, \mathcal{Q}_{\mathcal{B}f}, \mathcal{R}_{\mathcal{B}}, \mathrm{A})$ such that $\mathcal{Q}_{\mathcal{A}} \cap \mathcal{Q}_{\mathcal{B}} = \varnothing$. Due to Corollary 4, we assume without loss of generality that $\mathcal{A}/\mathrm{A}$ and $\mathcal{B}/\mathrm{A}$ are quasi-regular. Define the ETA $\mathcal{C}/\mathrm{A} = (\mathcal{F}, \mathcal{Q}, \mathcal{Q}_f, \mathcal{R}, \mathrm{A})$ as follows. $\mathcal{Q} = (\mathcal{Q}_{\mathcal{A}} \times \mathcal{Q}_{\mathcal{B}}) \cup \mathcal{Q}_{\mathcal{A}} \cup \mathcal{Q}_{\mathcal{B}}$ and $\mathcal{Q}_f = \mathcal{Q}_{\mathcal{A}f} \times \mathcal{Q}_{\mathcal{B}f}$. The set $\mathcal{R}$ of transition rules are the union of the 4 sets $\mathcal{R}_\times$, $\mathcal{R}_{\overline{\mathcal{A}}}$, $\mathcal{R}_{\overline{\mathcal{B}}}$ and $\mathcal{R}_{\mathcal{G}}$ defined below.

$$\mathcal{R}_\times : f((p_1, q_1), \ldots, (p_n, q_n)) \to (p, q) \qquad \begin{array}{l} \forall f \in \mathcal{F} \setminus \mathcal{G} \\ \forall f(p_1, \ldots, p_n) \to p \in \mathcal{R}_{\mathcal{A}} \\ \forall f(q_1, \ldots, q_n) \to q \in \mathcal{R}_{\mathcal{B}} \end{array}$$

$$\mathcal{R}_{\overline{\mathcal{A}}} : \begin{array}{rl} g((p_1, q_1), (p_2, q_2)) \to g((p, q_1), q_2) & \forall g \in \mathcal{G} \\ g(p_1, (p_2, q_2)) \to (p, q_2) & \forall q_1, q_2 \in \mathcal{Q}_{\mathcal{B}} \\ & \forall g(p_1, p_2) \to p \in \mathcal{R}_{\mathcal{A}} \\ \text{and} & \\ g((p_1, q_1), (p_2, q_2)) \to g((r_1, q_1), (r_2, q_2)) & \forall g(p_1, p_2) \to g(r_1, r_2) \in \mathcal{R}_{\mathcal{A}} \\ g(p_1, (p_2, q_2)) \to g(r_1, (r_2, q_2)) & \end{array}$$

$$\mathcal{R}_{\overline{\mathcal{B}}} : \begin{array}{rl} g((p_1, q_1), (p_2, q_2)) \to g((p_1, q), p_2) & \forall g \in \mathcal{G} \\ g(q_1, (p_2, q_2)) \to (p_2, q) & \forall p_1, p_2 \in \mathcal{Q}_{\mathcal{A}} \\ & \forall g(q_1, q_2) \to q \in \mathcal{R}_{\mathcal{B}} \\ \text{and} & \\ g((p_1, q_1), (p_2, q_2)) \to g((p_1, r_1), (p_2, r_2)) & \forall g(q_1, q_2) \to g(r_1, r_2) \in \mathcal{R}_{\mathcal{B}} \\ g(q_1, (p_2, q_2)) \to g(r_1, (p_2, r_2)) & \end{array}$$

$$\mathcal{R}_{\mathcal{G}} : \begin{array}{rl} g((p, q_1), q_2) \to g(q_1, (p, q_2)) & \forall g \in \mathcal{G} \\ g((p_1, q), p_2) \to g(p_1, (p_2, q)) & \forall p_1, p_2, p \in \mathcal{Q}_{\mathcal{A}} \\ g(q, p) \to (p, q) & \forall q_1, q_2, q \in \mathcal{Q}_{\mathcal{B}} \end{array}$$

The ETA $\mathcal{C}/\mathrm{A}$ satisfies that for any term $t \in \mathcal{T}(\mathcal{F})$, $t \to^*_{\mathcal{C}/\mathrm{A}} (p, q) \in \mathcal{Q}_f$ if and only if $t \to^*_{\mathcal{A}/\mathrm{A}} p \in \mathcal{Q}_{\mathcal{A}f}$ and $t \to^*_{\mathcal{B}/\mathrm{A}} q \in \mathcal{Q}_{\mathcal{B}f}$. □

This theorem holds also for $\mathcal{E}$-TA whose ES $\mathcal{E}$ consists of equations in the shape of $\mathsf{f}(x, \mathsf{f}(y, z)) \approx \mathsf{f}(y, \mathsf{f}(x, z))$. Kaji *et al.* [10] pointed out that in order to express some key-exchange protocols using term rewriting, those axioms are required.

5 Concluding Remarks

In the paper we introduced equational tree automata together with the undecidability results. We also showed the closure properties of union and intersection for equational tree automata. The newly introduced tree automata framework is almost optimal from the beneficial reason and it obtains our goal: to propose a class of tree languages in which congruence closures of recognized languages are recognizable. Furthermore, we presented the relationship between the standard

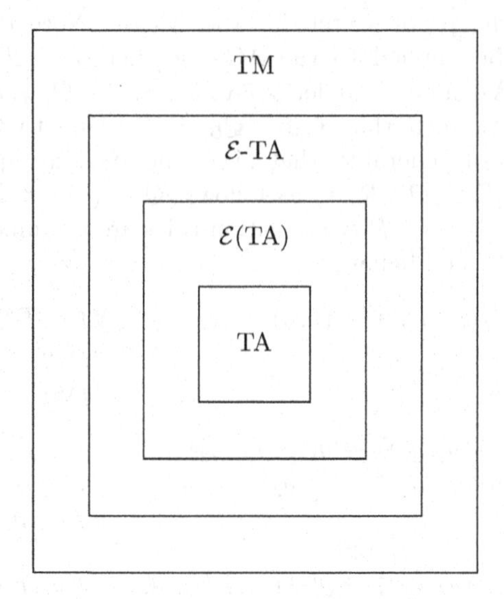

Fig. 1. Hierarchy of tree languages

TA and our equational extension. Fig. 1 illustrates the result on a hierarchy of 4 classes of tree languages (in case $\mathcal{E}$ is linear). In the above figure the smallest area TA denotes the class of tree languages recognizable with a regular TA. The second smallest area $\mathcal{E}(\text{TA})$ is $\mathcal{E}$-congruence closure of TA, and $\mathcal{E}$-TA is the class of ETA. The largest area TM denotes the set of generatable tree languages. These inclusion relations are strict each other. It is unclear so far whether the strict inclusion holds between TM and $\mathcal{E}$-TA also for $\mathcal{E}$ non-linear.

In order to discuss (minimal) equational extension of tree automata it would be important to consider whether the following question is positive:

– If $\mathcal{E} = \text{A}$ (or $\mathcal{E} = \text{AC}$),

$$\forall \mathcal{A}, \mathcal{B} : \text{regular TA}, \ \exists \mathcal{C} : \text{regular TA}. \ \mathcal{E}(\mathcal{L}(\mathcal{A})) \cap \mathcal{E}(\mathcal{L}(\mathcal{B})) = \mathcal{E}(\mathcal{L}(\mathcal{C}))?$$

There are two more interesting questions about equational tree automata.

– Regularizability of AC-TA;

$$\forall \mathcal{A}/\text{AC} : \text{ETA}, \ \exists \mathcal{B}/\text{AC} : \text{regular ETA}. \ \mathcal{L}(\mathcal{A}/\text{AC}) = \mathcal{L}(\mathcal{B}/\text{AC})?$$

– Closure under complement of A- and AC-TA; if $\mathcal{E} = \text{A}$ (or $\mathcal{E} = \text{AC}$),

$$\forall \mathcal{A}/\mathcal{E} : \text{ETA}, \ \exists \mathcal{B}/\mathcal{E} : \text{ETA}. \ \overline{\mathcal{L}(\mathcal{A}/\mathcal{E})} = \mathcal{L}(\mathcal{B}/\mathcal{E})?$$

We observe that equational tree automata are closely related to *context-sensitive grammar* (Section 9.3, [8]), so it is conjectured that the second question is positively solved.

Acknowledgements

I would like to thank Aart Middeldorp for his continuous help and encouragement.

References

1. R.V. Book and F. Otto: *String-Rewriting Systems*, Texts and Monographs in Computer Science, Springer-Verlag, 1993.
2. H. Comon, M. Dauchet, R. Gilleron, F. Jacquemard, D. Lugiez, S. Tison and M. Tommasi: *Tree Automata Techniques and Applications*, draft, 1997. Available on `http://www.grappa.univ-lille3.fr/tata/`.
3. M. Dauchet and S. Tison: *The Theory of Ground Rewrite Systems is Decidable*, Proc. 5th LICS, Philadelphia (Pennsylvania), pp. 242–248, 1990.
4. M. Dauchet and S. Tison: Structual Complexity of Classes of Tree Languages, In *Tree Automata and Languages*, Studies in Computer Science and Artificial Intelligence 10, pp. 327–353, Elsevier Science Publishers B.V., 1992.
5. A. Deruyver and R. Gilleron: *The Reachability Problem for Ground TRS and Some Extensions*, Proc. CAAP '89, Barcelona (Spain), LNCS 351, pp. 227–243, 1989.
6. I. Durand and A. Middeldorp: *Decidable Call by Need Computations in Term Rewriting (Extended Abstract)*, Proc. 14th CADE, Townsville (Australia), LNAI 1249, pp. 4–18, 1997.
7. R. Gilleron and S. Tison: *Regular Tree Languages and Rewriting Systems*, Fundamenta Informaticae 24, pp. 157–175, 1995.
8. J.E. Hopcroft and J.D. Ullman: *Introduction to Automata Theory, Languages, and Computation*, Addison-Wesley Publishing Company, 1979.
9. H. Hosoya, J. Vouillon and B.C. Pierce: *Regular Expression Types for XML*, Proc. 5th ICFP, Montreal (Canada), SIGPLAN Notices 35(9), pp. 11–22, 2000.
10. Y. Kaji, T. Fujiwara and T. Kasami: *Solving a Unification Problem under Constrained Substitutions Using Tree Automata*, Journal of Symbolic Computation 23, pp. 79–117, 1997.
11. D. Lugiez and J.L. Moysset: *Tree Automata Help One to Solve Equational Formulae in AC-Theories*, Journal of Symbolic Computation 18(4), pp. 297–318, 1994.
12. D. Monniaux: *Abstracting Cryptographic Protocols with Tree Automata*, Proc. 6th SAS, Venice (Italy), LNCS 1694, pp. 149–163, 1999.
13. E.L. Post: *Recursive Unsolvability of a Problem of Thue*, Journal of Symbolic Logic 13, pp. 1–11, 1947.

Normalized Types

Pierre Courtieu

Université Paris-Sud, Laboratoire de recherche en Informatique (LRI),
Bâtiment 490 – 91405 Orsay Cedex, France
`http://www.lri.Fr/`

Abstract. We present a new method to specify a certain class of quotient in intentional type theory, and in the calculus of inductive constructions in particular. We define the notion of "normalized types". The main idea is to associate a *normalization function* to a type, instead of the usual relation. This function allows to compute on a particular element for each equivalence class, avoiding the difficult task of computing on equivalence classes themselves. We restrict ourselves to quotients that allow the construction of such a function, i.e. quotient having a canonical member for each equivalence class. This method is described as an extension of the calculus of constructions allowing normalized types. We prove that this calculus has the properties of strong normalization, subject reduction, decidability of typing. In order to show the example of the definition of $\mathbb{Z}$ by a normalized type, we finally present a pseudo Coq session.

1 Introduction

1.1 The Calculus of Inductive Constructions

Type theory is a fruitful formalism for automated proof systems like Coq, Lego, Agda/alfa etc. The expressive power of this framework is comparable to set theory. However it appeared that the definition of complex structures is easier and computationally more powerful if the type theory is enriched with some constructions. In particular *the calculus of inductive constructions* ($\mathcal{CIC}$) [14] is an extension of the calculus of construction [6] where it is possible to define new types by giving the list of its typed constructors, in a sort of ML-style. For example:

Inductive Nat :=	*0:Nat — S:Nat → Nat.*
Inductive AList :=	*nil: (A:Set) (AList A)*
	— cons:(A:Set) A → (AList A) → (AList A).
Inductive BListn :=	*nil: (BListn 0)*
	— cons: (n:Nat) Bool → (BListn n) → (BListn (S n)).

By definition the terms of an *inductive* type T is *the least set of terms* recursively built from its constructors. This is stated by the induction schemes, also called *elimination principles*, like for example the one on natural: $\forall P\ \{(P\ 0) \wedge (\forall x(P\ x) \Rightarrow (P\ (S\ x)))\} \Rightarrow \forall x(P\ x)$. It is possible to define functions on an

L. Fribourg (Ed.): CSL 2001, LNCS 2142, pp. 554–569, 2001.

inductive type by pattern matching in ML-style, that can be evaluated in the usual way. For example:

$$parity = \lambda x{:}Nat.\ \ Case\ x\ of\ 0 \Rightarrow true\ —\ (S\ n) \Rightarrow (not\ (parity\ n))\ end$$

In the calculus of inductive constructions, these notions are stated by typed terms, in the *Curry-Howard* spirit. The elimination principle above is for example expressed by the term:

$$Nat\_rec{:}(P{:}Nat{\rightarrow}\mathbf{Set})(P\ 0){\rightarrow}((m{:}Nat)(P\ m){\rightarrow}(P\ (S\ m))){\rightarrow}(m{:}Nat)(P\ m)$$

where an expression of the form $(x{:}A)T$ (also written $\Pi x : A.T$) is the notation for the dependent product, abbreviated in $A \rightarrow T$ when T does not depend on x.

Using *Nat_rec*, we can define (possibly recursive) functions on *Nat* by pattern matching on constructors. We can also make proofs by cases or by induction on *Nat*. We can deduce from *Nat_rec* a non dependent principle, easier to use when the type P is not dependent:

$$Nat\_rec'{:}\ (P{:}Set)\ P{\rightarrow}(Nat{\rightarrow}P{\rightarrow}P){\rightarrow}Nat{\rightarrow}P$$

The function parity is written in the $\mathcal{CIC}$ using *Nat_rec'*:

$$parity = (Nat\_rec'\ bool\ true\ ([y{:}nat][b{:}bool](not\ b)))$$

where $[x{:}A]t$ stands for $\lambda x : A.t$. In order to evaluate such functions in the $\mathcal{CIC}$, new rewriting rules are added for the recursors:

$$(Nat\_rec'\ T\ t_0\ t_1\ 0) \quad \rightarrow_\iota t_0$$
$$(Nat\_rec'\ T\ t_0\ t_1\ (S\ x)) \rightarrow_\iota (t_1\ x\ (Nat\_rec'\ T\ t_0\ t_1\ x))$$

where t_0 and t_1 are two terms of type T and $nat{\rightarrow}T{\rightarrow}T$ respectively. Notice how structural recursion is simulated and how the head constructor of the last argument (0 or S) is used to choose which *branch* of the recursor must be used.

In the version of the $\mathcal{CIC}$ that we consider, this new reduction, called ι-reduction is part of the internal reduction of the system, as for β (evaluation of applications). We do not consider η (evaluation of dummy abstractions) in our work. More precisely, the internal reduction $\longrightarrow_{\mathcal{CIC}}$ which defines the evaluation mechanism on terms, is:

$$\longrightarrow_{\mathcal{CIC}} \ = \longrightarrow_\iota \cup \longrightarrow_\beta$$

From the theorem prover perspective, it is very important that the evaluation mechanism terminates. So in $\mathcal{CIC}$ restrictions on the definition of functions are made to allow only terminating functions to be defined (see appendix B, and [14]). This evaluation mechanism is closely related to the internal notion of equality of $\mathcal{CIC}$, as we explain in the next section.

1.2 The Notion of Equality

The notion of equality in type theory is a delicate problem. The undecidability in general of the usual notion of equality in mathematics makes impossible any implementation of a general decision procedure. However, a lot of work has been done to find weaker decidable equivalence relations between terms of the type theory, in order to:

- Make decision procedures to minimize the number of proofs of equality to be made by hand.
- Make typing decidable in presence of a *conversion rule* of the form (needed in a dependently typed framework):

$$\textbf{Conv}: \frac{\Gamma \vdash t : T_1 \quad T_1 \equiv T_2}{\Gamma \vdash t : T_2}$$

where $\equiv$ is an equivalence relation between terms, Γ is a typing environment, and t, T_1 and T_2 are terms. We see here that in order to keep a decidable typing procedure, $\equiv$ must remain decidable.

A decidable typing is not an absolute requirement, systems like PVS do not have a decidable typing. However, there are evident practical advantages in decidability of typing and internal equality and we will consider it necessary in our paper according to the point of view that prevailed in the conception of Coq.

Since the classical (undecidable) notion of equality is still necessary, the $\mathcal{CIC}$, as other systems of type theory, uses several notions of equality. In $\mathcal{CIC}$ we first have a notion of *internal equality*, also called definitional equality[1]. This equality is a congruence on terms (and types because we are in a type dependent framework) and is the equality used in the *conversion rule* of the type system. More precisely, the internal equality $\equiv_{\mathcal{CIC}}$ of $\mathcal{CIC}$ is the reflexive, symmetric and transitive closure of $\longrightarrow_{\mathcal{CIC}}$.

We have a second notion of equality, called *propositional equality* ($\stackrel{p}{\equiv}_{\mathcal{CIC}}$) also called *Leibniz equality*. It is defined in the calculus and can be used and extended by the user. It will not be considered in the typing rules and is not necessarily decidable. It is defined such that $\equiv_{\mathcal{CIC}} \subseteq \stackrel{p}{\equiv}_{\mathcal{CIC}}$, thus we will call *user equality*, noted $\stackrel{u}{\equiv}_{\mathcal{CIC}}$, the minimal relation such that: $\stackrel{p}{\equiv}_{\mathcal{CIC}} = \equiv_{\mathcal{CIC}} \cup \stackrel{u}{\equiv}_{\mathcal{CIC}}$ that is the part of $\stackrel{p}{\equiv}_{\mathcal{CIC}}$ that is not in $\equiv_{\mathcal{CIC}}$.

Notice here that nothing prevents the user from generating an inconsistency, for example by defining P such that $P(True, False)$. We see in the next section that the fact that the propositional equality is not restricted to $\equiv_{\mathcal{CIC}}$ creates more subtle problems.

To sum up this section, we can say that $\stackrel{p}{\equiv}_{\mathcal{CIC}}$ is the mathematical equality and $\equiv_{\mathcal{CIC}}$ is the sub-relation of $\stackrel{p}{\equiv}_{\mathcal{CIC}}$ that is considered (and decided) during typing. Of course extending $\equiv_{\mathcal{CIC}}$ is interesting as more terms will be identified internally.

1.3 The Problem of Non-free Structures

As we saw in section 1.1 the ι rule only checks the head constructor of the argument of a function to decide which reduction rule to apply. This mechanism works well when the structures defined by inductive types are *free*, which means that two terms starting with two different constructors cannot be equal (even for $\stackrel{u}{\equiv}_{\mathcal{CIC}}$). But it fails in presence of equations between head constructor terms[2]

[1] In fact, Martin-Löf in its type theory gives four notions of equality: intentional (definitional) equality, judgment equality, type equality, and propositional equality. In the $\mathcal{CIC}$ these categories are not exactly relevant as intentional equality is much more powerful than in Martin-löf's theory, and judgment and type equalities are somehow replaced by internal or propositional equality.

[2] Terms whose head symbol is a constructor.

(in $\stackrel{u}{\equiv}_{CIC}$). Indeed, in this case, two equal terms (by $\stackrel{p}{\equiv}_{CIC}$)$t_1$ and t_2, starting by two different constructors, can generate an incoherence by $\longrightarrow_{CIC}$. Suppose for example that we want to define the natural numbers modulo 2, we can naively state the axiom *Eq0mod2:* $0 \stackrel{u}{\equiv}_{CIC} (S\ (S\ 0))$ but then consistency of the system is compromised, as shown by the simple function f[3]:

$$f = \lambda x{:}Nat.\ \ Case\ x\ of\ \ \ \ 0 \Rightarrow True \\ \qquad\qquad\qquad\qquad -\ (S\ n) \Rightarrow False\ end$$

generating two equalities $(f\ 0) \equiv_{CIC} True$ and $(f\ 0) \stackrel{u}{\equiv}_{CIC} (f\ (S\ S\ 0)) \equiv_{CIC} False$ leading trivially to $True \stackrel{p}{\equiv}_{CIC} False$.

For this reason, inductive types allow only to specify free structures. This means that it is not possible to define quotients on inductive types in a simple way in systems like Coq.

However, mathematical structures like integers ($\mathbb{Z}$), integers modulo ($\mathbb{Z}/n\mathbb{Z}$), rationals ($\mathbb{Q}$), or Sets are intrinsically quotients, that we want to define from an underlying type and an equivalence on this type.

Our contribution, the *normalized types* is a method to specify structures where $\stackrel{in}{\equiv}$ is safely extended by the user, allowing to define a certain class of quotient. It has been inspired by two existing methods, the *quotient types* [9], [11], [5], and the *congruence types* [2], that we both briefly present in the following.

1.4 Related Work

The definition of non free structures in intentional type theory has been studied by Backhouse, Hofmann, Barthes, Geuvers, Jacobs and recently by S. Boutin. Roughly two methods were proposed: quotient types and congruence types, both inspired by previous works on extensional type theory.

Quotient types as presented in [5] are an axiomatization of quotients from the set theory, that has been implemented in the Coq system. A quotient type T/R is built from a type T (that we shall call the *underlying type*), and a relation R on this type. For all $x : T$ we define the term $(In\ x) : T/R$. A set of axioms defines the properties of T/R according with the classical notion of quotient, for example: $\forall x, y : T.(R\ x\ y) \rightarrow (In\ x) \equiv_{user} (In\ y)$ that defines the equality among the elements of the quotient. This axiomatization, by taking a set theoretical approach, is very general, but because of its axiomatic nature, does not catch the computational aspect of quotients (In particular because of the use of $\equiv_{user}$ above instead of $\equiv_{CIC}$).

Martin Hofmann, in [10] and [9] already defines quotients this way. By extending the Martin-Löf type theory with quotient types and by giving an interpretation of it in the pure Calculus of Constructions, he gives good properties to quotient types. Our present work follows a similar method.

A common aspect of these works is that the elimination principles for quotient is split in two principles. The first is weak and allows to define functions on the

[3] Notice that strictly speaking we use here a strong version of `Nat_rec'`.

quotient from functions on the underlying type. This *lifting* operation is allowed when the function is *compatible* with the relation of the quotient, according to the classical set theory mechanism. The second is strong and allows to make inductions on quotient types, again by lifting proofs made on the underlying type. There is no compatibility condition for this principle. But it is clear that in some cases it is possible to define better induction schemes.

Congruence types are a generalization of inductive types which could be called "inductive types with relation" ([13] [5]), or "inductive types with rewriting" [2]. This last point of view is the closest of our approach. It consists in the association of an inductive type $\underline{T}$ and a canonical (i.e. confluent and terminating) term rewriting system(TRS) ρ, which will be added to the internal reduction and equality of the system. This defines a new type T. Despite not being an inductive type, T has a good computational behavior, because we can use ρ to link any closed term of $\underline{T}$ to a unique term in T: its normal form by ρ. This method allows a satisfying representation of quotients when the relation R can be oriented in a canonical TRS. In particular better induction schemes can be proved by hand using a notion of *fundamental constructor*. However, adding rewriting systems to internal reduction dynamically leads to the difficult problems of termination criteria and interaction between rewriting systems [3].

1.5 Plan

We give a general description of normalized types in the next section. Then we define formally our extension of the calculus of constructions in section 2. We prove properties of this system in section 3. Finally we conclude with some considerations on our system and with some ideas of further work in 4. In appendix A the example of the definition of $\mathbb{Z}$ in a pseudo Coq session is developed. We give in the appendix B a short definition of the calculus of construction as defined in [14].

2 The Calculus of Inductive Constructions with Normalized Types

Let A be a type and nf a function from A to A, we define a new type $\mathsf{Norm}(A, nf)$ called "type A normalized by nf". Its elements are, and are only, of the form: $\mathsf{Class}(A, nf, t)$, where t is a term of type A. This will expressed by the elimination principle. The main idea of this work is to make $\mathsf{Class}(A, nf, t)$ and $\mathsf{Class}(A, nf, u)$ equivalent for internal equality (convertible) if $(nf\ t)$ and $(nf\ u)$ are equivalent. Reduction is defined to avoid the coherency problems cited in section 1.3.

Our system is a variant of quotient types in the spirit of congruence types of Barthe and Geuvers. We take advantages of both methods:

- We use the idea of the association of a type with a computational object. Barthe uses a TRS, we will use a normalization function nf, i.e. a term of the original system, avoiding the problem of mixing reductions cited previously.

– We will define a new calculus extending the calculus of inductive constructions. We use an interpretation from our calculus into $\mathcal{CIC}$ to define our notion of internal equality.

By a slight addition to the reduction rules we can add the normalized types to the Calculus of Inductive Constructions. The modification mainly consists in the enrichment of the conversion and reduction rules in order to make $\mathsf{Class}(A, nf, t)$ and $\mathsf{Class}(A, nf, u)$ convertible provided that t and u have the same canonical form. The subject of the end of this paper is the formal definition and the study of the main properties of this extension.

We use a method similar to [1] but in the context of inductive types, we first extend the syntax, typing and reduction rules of $\mathcal{CIC}$ and then we prove the properties of the new calculus ($\mathcal{CIC}^{nf}$) using a translation from $\mathcal{CIC}^{nf}$ to $\mathcal{CIC}$ that has some strong properties.

2.1 Syntax

The syntax of $\mathcal{CIC}^{nf}$ is based on the notations of B. Werner's thesis [14] where a precise description of the $\mathcal{CIC}$ can be found (a short one can be found in appendix B.

We take here the following hierarchy of sorts: Set:Type:Extern.

Variables: $V ::= x, y, z \ldots$

Sorts: $S ::= \mathsf{Set} \mid \mathsf{Type} \mid \mathsf{Extern}$

Terms (all terms of $\mathcal{CIC}$ belong also to $\mathcal{CIC}^{nf}$):

$$\begin{aligned} T ::= {} & V \mid S \mid [V:T]T \mid (V:T)T \mid TT \\ & \mid \mathsf{Ind}(V:T)(\boldsymbol{T}) \mid \mathsf{Constr}(n,T)\ n \in \mathbb{N} \mid \mathsf{Elim}(T,T,\boldsymbol{T},T)\{\boldsymbol{T}\} \\ & \mid \mathsf{Norm}(T,T) \mid \mathsf{Class}(T,T,T) \mid \mathsf{Elimnorm}(T,T,T,T) \end{aligned}$$

$\boldsymbol{T}$ denotes a sequence of terms. Ind, Constr and Elim are the usual constructions for inductive types. Norm, Class and Elimnorm are respectively the type constructor, term constructor and destructor for normalized types.

We will use the usual notation $A \to B$ in place of $(x:A)B$ when B does not depend on x. To increase readability, we will often use Coq-like notations for pattern matching expressions:

$$\textit{Cases } t \textit{ of } \text{<pattern1>} \Rightarrow u_1 \mid \textit{<pattern2>} \Rightarrow u_2 \ldots \textit{end.}$$

in place of the corresponding $\mathsf{Elim}(t_1, t_2, t_3, t)\ \{\boldsymbol{u_i}\}$, hiding arguments t_1, t_2 and t_3 that can be deduced from context.

We will as usual note $t[x \leftarrow u]$ the term t where all free occurrences of x have been replaced by u. The notion of free variable extends well to our new constructions, and all usual properties of substitutions are preserved.

Finally, we define typing environments as sets of pairs of the form $(x:T)$. As usual [] will denote the empty environment, and $\Gamma :: x : T$ the environment $\Gamma \cup \{(x:T)\}$. We will omit parenthesis when it is not ambiguous.

2.2 Computation

Definition 1. *Let $\longrightarrow_{nf}$ be the following rewriting rule:*

$$\mathsf{Elimnorm}(A, nf, f, \mathsf{Class}(A, nf, t)) \longrightarrow_{nf} (f\ (nf\ t))$$

the reduction $\longrightarrow_{\mathcal{CIC}^{nf}}$ of $\mathcal{CIC}^{nf}$ is the congruent closure of $\longrightarrow_{nf} \cup \longrightarrow_{\iota} \cup \longrightarrow_{\beta}$

Remark 1. Let t_1 and t_2 be two terms of $\mathcal{CIC}$ such that $t_1 \longrightarrow_{\iota\beta} t_2$ in $\mathcal{CIC}$, then we have also $t_1 \longrightarrow_{\iota\beta} t_2$ in $\mathcal{CIC}^{nf}$. $\longrightarrow_{\mathcal{CIC}^{nf}}$ reduction preserves typing.

2.3 Conversion and Internal Equality

Conversion is defined using a combination of an interpretation φ of terms of $\mathcal{CIC}^{nf}$ into other terms of $\mathcal{CIC}^{nf}$, and a new reduction $\longrightarrow_{\iota\beta+nf'}$ applied to the interpreted terms. This unusual definition is necessary since we want two terms to be convertible when there canonical forms (calculated by the interpretation) are equivalent for a certain relation (the new notion of reduction).

Since φ is not necessarily idempotent, it is impossible to define the conversion as the closure of a reduction. We see here that deciding equality ($\equiv_{nf+\mathcal{CIC}}$) and computing ($\longrightarrow_{\mathcal{CIC}^{nf}}$) are not anymore the same issues, but the two notions have to be compatible, as it is stated in property 1.

Definition 2. *We define φ on terms and environments recursively as follows:*

- $\varphi(\mathsf{Class}(A, nf, t)) = \mathsf{Class}(\varphi(A), \varphi(nf), (\varphi(nf)\ \varphi(t)))$,
- $C[t] = C[\varphi(t)]$ *where C is a context different than* $\mathsf{Class}(\_)$.
- $\varphi([]) = []$ *and* $\varphi(\Gamma :: x : T) = \varphi(\Gamma) :: x : \varphi(T)$

Definition 3. *We define $\longrightarrow_{nf'}$, $\equiv_{\iota\beta+nf'}$, $\equiv_{\mathcal{CIC}^{nf}}$ as follows:*

- $\mathsf{Elimnorm}(A, nf, f, \mathsf{Class}(A, nf, t)) \longrightarrow_{nf'} (f\ t)$
- $\equiv_{\iota\beta+nf'}$ *is the congruent closure of $\longrightarrow_{\iota}$, $\longrightarrow_{\beta}$ and $\longrightarrow_{nf'}$.*
- *t_1 and t_2 are convertible ($t_1 \equiv_{\mathcal{CIC}^{nf}} t_2$) iff $\varphi(t_1) \equiv_{\iota\beta+nf'} \varphi(t_2)$.*

Property 1. If $t \longrightarrow_{\mathcal{CIC}^{nf}} u$ then $t \equiv_{\mathcal{CIC}^{nf}} u$.

2.4 Typing

The typing system contains the rules of $\mathcal{CIC}$ (given in appendix B), except the conversion rule, plus the following rules:

$$\textbf{FormN}: \frac{\Gamma \vdash A : \mathsf{Set} \quad \Gamma \vdash nf : A \to A}{\Gamma \vdash \mathsf{Norm}(A, nf) : \mathsf{Set}} \qquad \textbf{IntroN}: \frac{\Gamma \vdash t : A \quad \Gamma \vdash \mathsf{Norm}(A, nf) : \mathsf{Set}}{\Gamma \vdash \mathsf{Class}(A, nf, t) : \mathsf{Norm}(A, nf)}$$

$$\textbf{Conv}: \frac{\Gamma \vdash t : T_1 \quad \Gamma \vdash T_1, T_2 : s \quad T_1 \equiv_{\mathcal{CIC}^{nf}} T_2}{\Gamma \vdash t : T_2}$$

$$\textbf{ElimN}_{\textbf{nodep}}: \frac{\Gamma \vdash P : Sort \quad \Gamma \vdash t : \mathsf{Norm}(A, nf) \quad \Gamma \vdash H : A \to P}{\Gamma \vdash \mathsf{Elimnorm}(A, nf, H, t) : P}$$

$$\textbf{ElimN}: \frac{\Gamma \vdash P : \mathsf{Norm}(A, nf) \to Sort \quad \Gamma \vdash t : \mathsf{Norm}(A, nf) \quad \Gamma \vdash H : (s : A)(P\ \mathsf{Class}(A, nf, s))}{\Gamma \vdash \mathsf{Elimnorm}(A, nf, H, t) : (P\ (IdNorm(A, nf, t)))}$$

with $IdNorm(\mathrm{A}, nf, \mathrm{t}) =_{def} \mathsf{Elimnorm}(\mathrm{A}, nf, ([\mathrm{x}:\mathrm{A}]\ \mathsf{Class}(\mathrm{A}, nf, \mathrm{x})), \mathrm{t})$

In the rule **ElimN** we use the term *IdNorm*, which maps $\mathsf{Class}(\ldots, x)$ to $\mathsf{Class}(\ldots, (nf\ x))$. It is a consequence of the implicit normalization done with the reduction $\longrightarrow_{nf}$ (and φ) defined previously. It is necessary to ensure that the reduction $\longrightarrow_{nf}$ preserves typing. We see here that $\mathsf{Elimnorm}(A, nf, H, t)$ is not the proof that P is verified by t but by the canonical form of t.

Now it is easy to replace the property $(nf\ s\ \stackrel{p}{\equiv}_{\mathcal{CIC}}\ s)$ by any equivalent inductive predicate to have a powerful principle to define function or make inductive proofs on normalized types. See appendix A for an example.

3 Properties of $\mathcal{CIC}^{nf}$

Important results about this system are:

1. *subject reduction*, which proof is classical, similar to what can be found in [14] or [8].
2. *strong normalization* $\longrightarrow_{\mathcal{CIC}^{nf}}$ on well typed terms, that is proved in the following sections as a consequence of the analog property of $\mathcal{CIC}$.
3. *decidability of typing*, that is a consequence of the decidability of $\equiv_{\mathcal{CIC}^{nf}}$, that is a consequence of 4 and 5
4. *strong normalization* of $\longrightarrow_{\iota\beta+nf'}$ on well typed terms, which proof is completely similar to 2.
5. *confluence (Church-Rosser)* of $\longrightarrow_{\iota\beta+nf'}$, proved using a classical the notion of *parallel reduction*.

3.1 A Translation from $\mathcal{CIC}^{nf}$ to $\mathcal{CIC}$

To prove strong normalization of our reduction $\longrightarrow_{\iota\beta+nf}$ we use the translation $\langle\rangle$ from $\mathcal{CIC}^{nf}$ to $\mathcal{CIC}$, such that if $t \longrightarrow_{\mathcal{CIC}^{nf}} t'$, then $\langle t\rangle \longrightarrow_{\mathcal{CIC}} \langle t'\rangle$. Thus, since $\langle\rangle$ preserves also typing, $\mathcal{SN}(\mathcal{CIC}^{nf})$ is a consequence of $\mathcal{SN}(\mathcal{CIC})$, which is well known.

Definition 4. *$\langle\rangle$ is defined by induction as follows:*

$$\begin{array}{lcr}
\langle \mathsf{Norm}(A, nf)\rangle & = & (Indnorm\ \langle A\rangle\ \langle nf\rangle) \\
\langle \mathsf{Class}(A, nf, t)\rangle & = & (indclass\ \langle A\rangle\ \langle nf\rangle\ \langle (nf\ t)\rangle) \\
\langle \mathsf{Elimnorm}(A, nf, f, t)\rangle & = & (\langle f\rangle\ (rep\ \langle A\rangle\ \langle nf\rangle\ \langle t\rangle)) \\
\langle C[\mathbf{t}]\rangle & = & C[\langle \mathbf{t}\rangle] \\
 & & \text{For any other construction } C
\end{array}$$

We extend trivially $\langle\rangle$ to environments:

$$\begin{array}{lcr}
\langle []\rangle & = & [] \\
\langle \Gamma :: x : T\rangle & = & \langle \Gamma\rangle :: x : \langle T\rangle
\end{array}$$

Notice the interpretation of the new terms of $\mathcal{CIC}^{nf}$ into an inductive type (*Indnorm* defined below) of $\mathcal{CIC}$. Terms of the form Class(...) are normalized by $\langle\rangle$ via nf. The type and terms of $\mathcal{CIC}$used in the translation are defined as follows:

- *Indnorm* is a parameterized inductive type, it corresponds to the type Norm: *Indnorm := [A:Set] [nf:A → A] Ind (X:Set) {A → X}*
- the unique constructor of *Indnorm* corresponds to the construction Class: *indclass:= [A:Set] [nf:A → A] Constr(1,(Indnorm A nf))* [4]
- the destructor *rep* of *Indnorm* corresponds to Elimnorm: *rep:=[A:Set][nf:A→A][t:(Indnorm A nf)]*
  Cases t of (indclass A nf x)⇒x end.

  The type of *rep* is the following: *(A:Set) (nf:A → A) (Indnorm A nf) → A.*

3.2 Properties of the Translation $\langle\rangle$

In this section we prove properties of $\langle\rangle$ that will allow us to state in the following section the strong normalization, subject reduction and confluence modulo conversion of $\mathcal{CIC}^{nf}$. This is the technical part. The two main properties of $\langle\rangle$ are that it preserves the typing relation and reduction.

We will note $t \overset{\langle\rangle}{=} t'$ when t is equal to t' by definition of $\langle\rangle$.

In order to prove strong normalization of $\mathcal{CIC}^{nf}$ from strong normalization of $\mathcal{CIC}$, $\langle\rangle$ needs to preserve the typing relation. Before proving this property we state a small lemma:

Lemma 1. *For all terms X, Y and T, and for all environment Γ, if $\Gamma \vdash_{\mathcal{CIC}} (Indnorm\ X\ Y) : T$ or $\Gamma \vdash_{\mathcal{CIC}} T : (Indnorm\ X\ Y)$, then $\Gamma \vdash_{\mathcal{CIC}} X : \mathsf{Set}$ and $\Gamma \vdash_{\mathcal{CIC}} Y : X \to X$.*

Proof. This is deduced from the typing rules and the type of *Indnorm* which is *(A:Set) (nf:A → A) Set.*

Lemma 2. *If $\Gamma \vdash_{\mathcal{CIC}^{nf}} u : T$ then $\langle\Gamma\rangle \vdash_{\mathcal{CIC}} \langle u\rangle : \langle T\rangle$.*

Proof. By induction on the proof that $\Gamma \vdash_{\mathcal{CIC}^{nf}} u : T$. All cases from $\mathcal{CIC}$ are immediate since the typing rules of $\mathcal{CIC}$ are also typing rules of $\mathcal{CIC}^{nf}$. Finally the only difficulty is the case of the rule **ElimN**:

- Last rule used is **ElimN**, we know that:

$$u = \mathsf{Elimnorm}(A, nf, f, t), \text{ and thus } \langle u\rangle \overset{\langle\rangle}{=} (\langle f\rangle\ (rep\ \langle A\rangle\ \langle nf\rangle\ \langle t\rangle))$$
$$\Gamma \vdash_{\mathcal{CIC}^{nf}} \mathsf{Elimnorm}(A, nf, f, t) : (P\ (IdNorm(A, nf, t)))$$

From the rule **ElimN** we know that the following assertions hold:

[4] To make things clear, we can say that this is what is defined in Coq when we write the following definition:

Inductive Indnorm [A:Set, nf:A → A]:= indclass : A → (Indnorm A nf).

Where A and nf are *parameters* of the inductive type *Indnorm*.

(i) $\Gamma \vdash_{\mathcal{CIC}^{nf}} f : (s : A)(P\ \mathsf{Class}(A, nf, s))$
So by ind. hyp.: $\langle \Gamma \rangle \vdash_{\mathcal{CIC}} \langle f \rangle : (s : \langle A \rangle)((\langle P \rangle\ (indclass\ \langle A \rangle\ \langle nf \rangle\ ((\langle nf \rangle\ s)))$
(ii) $\Gamma \vdash_{\mathcal{CIC}^{nf}} t : \mathsf{Norm}(A, nf)$
So by ind. hyp.: $\langle \Gamma \rangle \vdash_{\mathcal{CIC}} \langle t \rangle : (Indnorm\ \langle A \rangle\ \langle nf \rangle)$
(iii) $\Gamma \vdash_{\mathcal{CIC}^{nf}} P : \mathsf{Norm}(A, nf) \to Sort.$

By lemma 1 and (ii), we have:
$\langle \Gamma \rangle \vdash_{\mathcal{CIC}} \langle A \rangle : \mathsf{Set}$, and $\langle \Gamma \rangle \vdash_{\mathcal{CIC}} \langle nf \rangle : \langle A \rangle \to \langle A \rangle$
and so, from the type of *rep* we have:
$\langle \Gamma \rangle \vdash_{\mathcal{CIC}} (rep\ \langle A \rangle\ \langle nf \rangle\ \langle t \rangle) : \langle A \rangle$
let us denote the term $(rep\ \langle A \rangle\ \langle nf \rangle\ \langle t \rangle)$ by v. By applying the typing rule for application, we can conclude that:
$\langle \Gamma \rangle \vdash_{\mathcal{CIC}} (\langle f \rangle\ v) : ((\langle P \rangle\ (indclass\ \langle A \rangle\ \langle nf \rangle\ ((\langle nf \rangle\ s)))[s \leftarrow v].$
Therefore:

$$\langle \Gamma \rangle \vdash_{\mathcal{CIC}} \langle u \rangle : ((\langle P \rangle\ (indclass\ \langle A \rangle\ \langle nf \rangle\ ((\langle nf \rangle\ (rep\ \langle A \rangle\ \langle nf \rangle\ \langle t \rangle)))) \quad (1)$$

On the other hand, we can verify that:
$\langle P\ (IdNorm(A, nf, t)) \rangle \equiv_{\mathcal{CIC}^{nf}} ((\langle P \rangle\ (indclass\ \langle A \rangle\ \langle nf \rangle\ (rep\ \langle A \rangle\ \langle nf \rangle\ \langle t \rangle)))$
which is the type of $\langle u \rangle$ in $\langle \Gamma \rangle$(Cf. (1)). So by the conversion rule of $\mathcal{CIC}$ we conclude that:
$\langle \Gamma \rangle \vdash_{\mathcal{CIC}} \langle u \rangle : \langle P\ (IdNorm(A, nf, t)) \rangle.$

The second property that we want in order to prove that $\mathcal{CIC}^{nf}$ is normalizing is that $\langle\rangle$ preserves reductions (lemma 4). We need to prove first that $\langle\rangle$ is coherent with substitutions, which is rather immediate.

Lemma 3. *If* $t = u[x \leftarrow v]$ *is a term of* $\mathcal{CIC}^{nf}$, *then* $\langle t \rangle = \langle u \rangle[x \leftarrow \langle v \rangle]$.

Proof. By induction on u and then by cases. We suppose without loss of generality that all occurrences of x in u are free.

Now we can state the preservation of reductions:

Lemma 4. *For all well typed terms* t_1 *and* t_2 *of* $\mathcal{CIC}^{nf}$, *if* $t_1 \to_{\iota\beta+nf} t_2$ *then* $\langle t_1 \rangle \to^+_{\iota\beta} \langle t_2 \rangle$.

Proof. By induction on t_1.

- Base case: $t_1 = x$, Extern, Set or Type, then t_1 is not reducible by $\to_{\iota\beta+nf}$.
- $t_1 = (x : u)v$, then $\langle t_1 \rangle = (x : \langle u \rangle)\langle v \rangle$ the reduction is necessarily done on a strict sub-term of t_1. There are two cases:
  1. $t_2 = (x : u')v$ with $u \to_{\iota\beta+nf} u'$, then $\langle t_2 \rangle = (x : \langle u' \rangle)\langle v \rangle$. By induction hypothesis, $\langle u \rangle \to^+_{\iota\beta} \langle u' \rangle$, thus $\langle t_1 \rangle = (x : \langle u \rangle)\langle v \rangle \to^+_{\iota\beta} (x : \langle u' \rangle)\langle v \rangle = \langle t_2 \rangle$. OK.
  2. or $t_2 = (x : u)v'$ with $v \to_{\iota\beta+nf} v'$, same argument. OK.
- The following case are similar to the previous case: $t_1 = [x : u]v$, $t_1 = \mathsf{Ind}(x : u)\boldsymbol{v}$, $t_1 = \mathsf{Constr}(i, u)$, $t_1 = \mathsf{Norm}(u, v, w)$, $t_1 = \mathsf{Class}(t, u, v, w)$.
- $t_1 = (u\ v)$, then $\langle t_1 \rangle = (\langle u \rangle\ \langle v \rangle)$, we distinguish two cases:

1. if the reduction is in a strict sub-term of t_1, then the same argument as in previous cases holds again.
2. if the reduction is on the head of t_1, then it is a β-reduction, $t_1 = ([x : T]u'\ v)$, and $t_2 = u'[x \leftarrow v]$. So $\langle t_1 \rangle = ([x : \langle T \rangle]\langle u' \rangle\ \langle v \rangle)$. By lemma 3 $\langle t_2 \rangle = \langle u' \rangle [x \leftarrow \langle v \rangle]$ and therefore $\langle t_1 \rangle \rightarrow_\beta \langle t_2 \rangle$. OK.

- $t_1 = \mathsf{Elim}(u_1, u_2, \boldsymbol{v_i}, u_3)\{\boldsymbol{f_j}\}$, this case is similar to the previous one, giving details here would involve to define precisely ι-reduction, which is rather long.
- $t_1 = \mathsf{Elimnorm}(A, nf, f, t)$, then $\langle t_1 \rangle = (\langle f \rangle\ (rep\ \langle A \rangle\ \langle nf \rangle\ \langle t \rangle))$, there are two cases:
  1. if the reduction is in a sub-term of t_1, then the above arguments holds.
  2. if the reduction is done on the head of t_1, then it is a nf-reduction, and we know that:
     - $t = \mathsf{Class}(A, nf, u)$ and therefore
       $\langle t_1 \rangle = (\langle f \rangle\ (rep\ \langle A \rangle\ \langle nf \rangle\ \langle \mathsf{Class}(A, nf, u) \rangle))$
       $\stackrel{\langle\rangle}{=} (\langle f \rangle\ (rep\ \langle A \rangle\ \langle nf \rangle\ (indclass\ \langle A \rangle\ \langle nf \rangle\ (\langle nf \rangle\ \langle u \rangle))))$.
       Which can be reduced by β and ι to
       $\rightarrow^+_{\beta i} (\langle f \rangle\ (\langle nf \rangle\ \langle u \rangle))$
     - $t_2 = (f\ (nf\ u))$. and therefore $\langle t_2 \rangle$ is equal to $(\langle f \rangle\ (\langle nf \rangle\ \langle u \rangle))$.

     We have proved that $\langle t_1 \rangle \rightarrow_\beta \langle t_2 \rangle$. OK.

3.3 Strong Normalization

Theorem 1. *If there exists an infinite reduction Δ starting from a well typed term t of $\mathcal{CIC}^{nf}$ by $\rightarrow_{\iota\beta+nf}$, then there exists an infinite reduction Δ' starting from a well typed term of $\mathcal{CIC}$ ($\langle t \rangle$) by $\rightarrow_{\iota\beta}$.*

Thus $\mathcal{CIC}^{nf}$ is strongly normalizing on well typed terms.

Proof. The reduction exists by iteration of lemma 4, and $\langle t \rangle$ is well typed by lemma 2.

4 Conclusion

We presented a method to specify a certain class of quotient. Our choice of a function instead of a term rewriting system as in [2] is motivated by the fact that functions are the computational object of the type theory. This allows to use a rather simple extension of $\mathcal{CIC}$ and its reduction. However, defining *nf* by a rewriting system remains a good idea for several reasons, in particular because it is possible to reduce in a term at any position, which is not possible in general with a recursive function and third it is more efficient.

The class of quotients that we can represent this way is the same as for [2], i.e. quotient whose relation can be "oriented" into a computation.

For Rationals we can define *nf* as the function that reduces fractions to irreducible fractions, but one needs more work to have a nice definition of $\mathbb{Q}$.

As we said in the beginning of the paper, focusing on a particular member of an equivalence class at the moment of computation can be seen as a weakness of our approach. But the fact that we use this artifice only when *computing* allows to stay at the level of equivalence classes when *reasoning*. Indeed, the terms `Class(...,OI)` and `Class(..., (SI (PI OI)))` are identified only at conversion level (of course they are propositionally equal) but are not reduced one to the other.

Anyway it is clear that an implementation of normalized types should propose several principles as we said previously. It should by the way be interesting to see how automated induction methods as in [4] could be used to generate them automatically. Indeed, such methods are for example able to generate `NormInt` of previous section.

A Example: The Integers

We shall now describe by an example the use of normalized types. We define the integers ($\mathbb{Z}$), using a type `Int` that has 3 constructors `O`, `S` and `P`, and a function `nf` that eliminates the useless combinations: `(S (P _))` and `(P (S _))`. To improve readability, we present this example as a pseudo Coq session. Coq is a tool that allows to interactively define, and type, terms of $\mathcal{CIC}$. One way to define terms is to build them with a *proof engine* interactively by means of a set of *tactics*. We will suppose here that it has been extended to $\mathcal{CIC}^{nf}$. Actually, normalized types have only been axiomatized in the real Coq. `Int`, `nf` and `I` have been defined, in particular `I` has been proved. Of course when we use normalized types, we pretend that normalized types have been implemented in the system Coq, which is not the case (though it will be implemented in a non official version of the system in a short future).

A.1 Basic Definitions

We assume that the propositional equality is defined like it is presently in Coq, i.e. as the least reflexive relation:

```
Inductive eq [A:Set; x:A] : A->Prop :=  refl_equal : (eq A x x)
```

We first define the underlying type `Int`:

```
Inductive Int : Set :=  O: Int | S: Int -> Int | P: Int -> Int.
```

Then we define the function `nf`:

```
Fixpoint nf [n:Int]: Int :=
  Cases n of
    O    => O
  | (S a) => Cases a of
               O  => (S O)
             | (S y) =>  Cases (nf a) of O  => (S O)
                                        | (S x) => (S (S x))
                                        | (P x) => x  end
```

```
             |(P y) => (nf y) end
  | (P a) =>  ...  end.
```

Notice that it is written for a *head-first* strategy. That way, a term of the form (nf (P (S (P x)))) will reduce to (nf (P x)), which is not the case with the more obvious function (the two terms are only provably equal). This is a weakness of the normalized types that we briefly discuss in the conclusion.

A.2 Definition and Use of the Normalized Type

From now we still present the example in Coq syntax but it is supposed that the system has been modified to deal with normalized types as it is described in this paper. We define the normalized type representing $\mathbb{Z}$:

```
Definition Z: Set := Norm (Int,nf).
```

Assuming that our notion of convertibility has been implemented in the system, Class(Z,nf,0) and Class(Z,nf,(S (P 0))) are convertible, therefore we can make the following proof of $S\ P\ 0 = 0$:

```
Lemma (eq Z Class(Int,nf,0) Class(Int,nf,(S (P 0)))).
Exact (refl_equal Z Class(Int,nf,0)). Save.
```

Let us show now the definition of a function. We first define a function IdInt on the underlying type and then we define IdZ on Z using Elimnorm and IdInt:

```
Definition IdInt : Int -> Z := [x:Z] Class(Int,nf,x).
Definition IdZ : Z -> Z := [x:Z] Elimnorm(Int,nf,IdInt,x).
```

Now we can apply IdZ to several terms and see how it is computed.

```
Eval Compute in (IdZ Class(Int,nf,0))  = Class(Int,nf,0)
Eval Compute in (IdZ Class(Int,nf,(S (P 0))))  = Class(Int,nf,0)
```

Let us follow step by step the second reduction:

(IdZ Class(Int,nf,(S (P 0))))

$\rightarrow_\beta$ Elimnorm(Int,nf,IdInt,Class(Int,nf,(S (P 0))))

$\rightarrow_{nf}$ (IdInt (nf (S (P 0)))) $\rightarrow^*$ (IdInt 0) $\rightarrow^*$ Class(Int,nf,0)

Notice that because of the reduction $\rightarrow_{nf}$, (S (P 0)) is reduced to 0 before being applied to IdInt, this is why the two terms above are reduced to the same. We see here an example of the way incoherence by ι (explained in section 1) is avoided in our system: *nf is applied before any ι-reduction on a normalized term can occur.*

A.3 Better Induction Scheme

The following principle is a stronger elimination scheme in case that *nf* is idempotent:

```
Elim_A: (A:Set) (nf: A -> A) ((x:Int) (nf (nf x))= (nf x))
          -> (P: A -> Prop) (H:(s:A) (nf s = s) (P Class(A,nf,s)))
          -> (t:(Norm (A,nf))) (P t)
```

It is easily provable in $\mathcal{CIC}^{nf}$, and we can deduce from it a stronger elimination principle for Z. We have to prove is that nf is Idempotent, and then instantiate Elim_A:

```
Lemma I:(x:Int) (nf (nf x))= (nf x). ...¡tactics¿Save.
Lemma ElZ:
      (P:Z->Prop)(H:(x:Int)(nf x=x)(P Class(Int,nf,x)))->(t:Z)(P t).
```

which is already a better induction principle. We can build an even more useful one by defining inductive predicate `NormInt` equivalent to the proposition `(nf x = x)`:

```
Inductive pos:Int->Prop:=p0:(pos (S 0)) | pS:(x:Int)(pos x)->(pos (S x))
Inductive neg:Int->Prop:=n0:(neg (P 0)) | pS:(x:Int)(neg x)->(neg (S x))
Inductive NormInt: Int -> Prop :=  norm0   : (NormInt I0)
                                  | normpos: (x:Int) (pos x) (NormInt x)
                                  | normneg: (x:Int) (neg x) (NormInt x).
Lemma Normint_correct : (x:Int) (NormInt x) <-> (nf x = x). ... Save.
```

Finally we can replace one by the other and obtain the well known principle on $\mathbb{Z}$:

```
Lemma IZ:
(P: Z -> Prop) (H:(x:Int) (NormInt x) (P Class(Int,nf,x))) -> (t:Z) (P t)
```

B The Calculus of Inductive Constructions

Here is a short description of the calculus of inductive constructions first defined in [7], following notations of [14] and [12]. We first give the syntax, then the typing and reduction rules.

The Syntax is the following:

Variables: $V ::= x, y, z \ldots$

Sorts: $S ::= \mathsf{Set} \mid \mathsf{Type} \mid \mathsf{Extern}^5$.

Terms:

$$T ::= V \mid S \mid [V : T]T \mid (V : T)T \mid TT$$
$$\mid \mathsf{Ind}(V : T)(\boldsymbol{T}) \mid \mathsf{Constr}(n, T)\ n \in \mathbb{N} \mid \mathsf{Elim}(T, T, \boldsymbol{T}, T)\{\boldsymbol{T}\}$$

Ind, Constr and Elim are respectively the type constructor, term constructor and destructor for inductive types.

Reduction of $\mathcal{CIC}$ is the congruent closure of β and ι-reductions. See [14] for precisions on the term $\Delta[t_1, t_3, t_2]$. It roughly applies the good arguments ($\vec{m}$) to the *branch* (f_k) of the recursive definition.

$$([x : t]t_1\ t_2) \to_\beta t_1[x \backslash t_2]$$
$$\mathsf{Elim}(I, Q, \boldsymbol{a}, \mathsf{Constr}(k, I')\boldsymbol{m})\{\boldsymbol{f_i}\} \to_\iota (\Delta[C_k(I), f_k, \mathit{Fun\_Elim}(I, Q, \boldsymbol{f_i})]\boldsymbol{m})$$
$$\text{where} \quad I = \mathsf{Ind}(X : A)\{\boldsymbol{C_i(X)}\}$$

The type system

$$(\mathrm{Ax}_1) [] \vdash \mathsf{Set} : \mathsf{Type} \qquad (\mathrm{Ax}_2) [] \vdash \mathsf{Type} : \mathsf{Extern}$$

$$(\mathrm{Prod\text{-}s})\frac{\Gamma :: (x : t_1) \vdash t_2 : s}{\Gamma \vdash (x : t_1)t_2 : s} \qquad (\mathrm{Lam\text{-}s})\frac{\Gamma \vdash (x : t_1)t_2 : s \qquad \Gamma :: (x : t_1) \vdash t : t_2}{\Gamma \vdash [x : t_1]t : (x : t_1)t_2}$$

[5] it is possible to replace Extern by a universe hierarchy, actually it is the case for the system implemented in Coq and Lego, it seems not difficult to extend our work to a universe hierarchy.

$$(\text{W-SET})\frac{\Gamma \vdash t : \mathsf{Set} \quad \Gamma \vdash A : B \quad a \notin \Gamma}{\Gamma :: (a : t) \vdash A : B} \qquad (\text{W-TYPE})\frac{\Gamma \vdash t : \mathsf{Type} \quad \Gamma \vdash A : B \quad \alpha \notin \Gamma}{\Gamma :: (\alpha : t) \vdash A : B}$$

$$(\text{VAR})\frac{\Gamma \vdash t_1 : t_2 \qquad (x : t_1) \in \Gamma}{\Gamma \vdash x : t_1} \qquad (\text{APP})\frac{\Gamma \vdash t_2 : (x : T_1)T_2 \qquad \Gamma \vdash t_1 : T_1}{\Gamma \vdash (t_2\ t_1) : T_2[x \leftarrow t_1]}$$

$$(\text{CONV})\frac{\Gamma \vdash t : T_1 \quad \Gamma \vdash T_1 : s \quad \Gamma \vdash T_2 : s \quad T_1 \equiv_{\beta\iota} T_2}{\Gamma \vdash t : T_2}$$

$$(\text{IND})\frac{Ar(A, \mathsf{Set}) \quad \Gamma \vdash A : \mathsf{Type} \quad \forall i.(\Gamma :: (X : A) \vdash C_i(X) : \mathsf{Set}) \quad \forall i.constr(C_i(X))}{\Gamma \vdash \mathsf{Ind}(X : A)\{\boldsymbol{C_i(X)}\} : A}$$

$$(\text{INTRO})\frac{\Gamma \vdash \mathsf{Ind}(X : A)\{\boldsymbol{C_i(X)}\} : T \quad 1 \leq n \leq |\ \boldsymbol{C_i(X)}\ |}{(\Gamma \vdash \mathsf{Constr}(n, \mathsf{Ind}(X : A)\{\boldsymbol{C_i(X)}\}) : C_n(\mathsf{Ind}(X : A)\{\boldsymbol{C_i(X)}\}))}$$

$$(\text{W-ELIM})\frac{\begin{array}{c} A \equiv (\boldsymbol{x} : \boldsymbol{A})Set \quad I = \mathsf{Ind}(X : A)\{\boldsymbol{C_i(X)}\} \quad \Gamma \vdash \boldsymbol{u} : \boldsymbol{A} \quad \Gamma \vdash t : (I\ \boldsymbol{u}) \\ \Gamma \vdash Q : (\boldsymbol{x} : \boldsymbol{A})(I\ \boldsymbol{x}) \rightarrow \mathsf{Set} \quad \forall i.(\Gamma \vdash f_i : \Delta\{C_i(I), Q, \mathsf{Constr}(i, I)\}) \end{array}}{\Gamma \vdash \mathsf{Elim}(I, Q, \boldsymbol{u}, t)\{\boldsymbol{f_i}\} : (Q\ \boldsymbol{u}\ t)}$$

$$(\text{S-ELIM})\frac{\begin{array}{c} A \equiv (\boldsymbol{x} : \boldsymbol{A})Set \quad I = \mathsf{Ind}(X : A)\{\boldsymbol{C_i(X)}\} \quad \Gamma \vdash \boldsymbol{u} : \boldsymbol{A} \quad \Gamma \vdash t : (I\ \boldsymbol{u}) \\ \Gamma \vdash Q : (\boldsymbol{x} : \boldsymbol{A})(I\ \boldsymbol{x}) \rightarrow \mathsf{Type} \\ \forall i.Small(C_i(X)) \quad \forall i.(\Gamma \vdash f_i : \Delta\{C_i(I), Q, \mathsf{Constr}(i, I)\}) \end{array}}{\Gamma \vdash \mathsf{Elim}(I, Q, \boldsymbol{u}, t)\{\boldsymbol{f_i}\} : (Q\ \boldsymbol{u}\ t)}$$

The system is composed of the usual set of typing rules for a pure type system (PTS) and a specific set of rules for inductive types (Ind and below).

The ELIM rules make use of the term $\Delta\{C_i(I), Q, \mathsf{Constr}(i, I)\}$, which is again defined in[14], that builds the type that each branch of a recursive function defined with Elim should have.

The constructions $constr(C_i(X))$, $Ar(A, \mathsf{Set})$ and $Small(C_i(X))$ are very important syntactic conditions that must be satisfied to ensure normalization and coherency of the system. There exact formulation is not very important for us, the important point is that the resulting calculus has nice properties like strong normalization, confluence and subject-reduction, and that normalized types can therefore be define from it.

References

1. G. Barthe. Extensions of pure type systems. In *proc TLCA'95*, volume 902 of *lncs*. Springer-Verlag, 1995.
2. G. Barthe and H. Geuvers. *Congruence types.* In H. Kleine Buening, editor, *Proc. Conf. Computer Science Logic*, volume 1092 of *Lecture Notes in Computer Science*, pages 36–51, 1995.
3. F. Blanqui, J.-P. Jouannaud, and M. Okada. The Calculus of Algebraic Constructions. In Paliath Narendran and Michael Rusinowitch, editors, *10th International Conference on Rewriting Techniques and Applications*, volume 1631 of *Lecture Notes in Computer Science*, Trento, Italy, July 1999. Springer-Verlag.
4. Adel Bouhoula and Jean-Pierre Jouannaud. Automata-driven automated induction. In *Twelfth Annual IEEE Symposium on Logic in Computer Science*, pages 14–25, Warsaw,Poland, June 1997. IEEE Comp. Soc. Press.
5. S. Boutin. *Réflexions sur les quotients.* thèse d'université, Paris 7, April 1997.
6. Thierry Coquand and Gérard Huet. The calculus of constructions. *Information and Computation*, 76:95–120, February 1988.

7. Thierry Coquand and Christine Paulin-Mohring. Inductively defined types. In P. Martin-Löf and G. Mints, editors, *Proceedings of Colog'88*, volume 417 of *Lecture Notes in Computer Science*. Springer-Verlag, 1990.
8. H. Geuvers. The church rosser property for $\beta\eta$-reduction in typed λ-calculi. In *Proc. 7th IEEE Symp. Logic in Computer Science, Santa Cruz*, pages 453–460, 1992.
9. M. Hofmann. *Extensional concepts in intensional type theory*. Phd thesis, Edinburgh university, 1995.
10. M. Hofmann. A simple model for quotient types. In *proc TLCA'95*, volume 902 of *lncs*. Springer-Verlag, 1995.
11. B. Jacobs. Quotients in simple type theory. Drafts and Notes. http://www.cwi.nl/ bjacobs/, 1991.
12. Christine Paulin-Mohring. Inductive definitions in the system COQ. In *Typed Lambda Calculi and Applications*, pages 328–345. Springer-Verlag, 1993. LNCS 664.
13. G. Malcolm R. Backhouse, P. Chisholm and E. Saaman. Do-it-yourself type theory. *Formal Aspects of Computing*, 1989.
14. Benjamin Werner. *Méta-théorie du Calcul des Constructions Inductives*. Thèse de doctorat, Univ. Paris VII, 1994.

Markov's Principle for Propositional Type Theory*

Alexei Kopylov and Aleksey Nogin

Department of Computer Science
Cornell University
Ithaca, NY 14853
{kopylov,nogin}@cs.cornell.edu

Abstract. In this paper we show how to extend a constructive type theory with a principle that captures the spirit of Markov's principle from constructive recursive mathematics. Markov's principle is especially useful for proving termination of specific computations. Allowing a limited form of classical reasoning we get more powerful resulting system which remains constructive and valid in the standard constructive semantics of a type theory. We also show that this principle can be formulated and used in a propositional fragment of a type theory.

1 Introduction

1.1 Overview

The main goal of this paper is to support limited classical reasoning in a generally intuitionistic framework. We use a `squash` operator for this purpose. This operator can creates a proposition stating that a certain type is non-empty without providing an inhabitant, i.e. `squash` "forgets" proofs. It was first introduced in [6] and also used in [15] and MetaPRL system [9,10]. Using `squash` it is possible to define the notion of squash–stability, which is similar to self–realizability.

The `squash` operator can be considered as a modality. The propositional logic equipped with this modality can express a principle that allows turning classical proofs of squash–stable propositions into constructive ones. This principle is valid in a standard type theory semantics if we consider it in the classical metatheory. Therefore this principle does not destroy the constructive nature of type theory in the sense that we can always extract a witness term from a derivation.

It turns out that this principle implies Markov's principle providing us a propositional analog of Markov's principle. It is rather surprising that such analog exists because normally one needs quantifiers in order to formulate Markov's principle.

We also show an equivalent way of defining the same principle using a membership type instead of the `squash` operator.

* This work was partially supported by AFRL grant F49620-00-1-0209

L. Fribourg (Ed.): CSL 2001, LNCS 2142, pp. 570–584, 2001.

1.2 Markov's Constructivism

Constructive mathematics is interesting in Computer Science because of program correctness issues. There are several approaches to constructivism (see [4,5,19] for an overview). We are especially interested in the constructive recursive mathematics (CRM) approach developed by Markov [12,13] and in constructive type theories (especially those that are based on Martin-Löf type theory [14]) since we believe them to be highly relevant to Computer Science. In this paper we demonstrate how to apply the ideas of CRM to a constructive type theory thus creating a more powerful type theory that combines the strengths of both approaches to constructive mathematics.

According to Markov's CRM approach, all objects are algorithms, where algorithms are understood as finite strings in a finite alphabet. All logical connectives are understood in a constructive way. That is, a statement is true if and only if there exists an algorithm that produces a *witness* of this statement. For example, the witness for $\forall x.A(x) \vee \neg A(x)$ is an algorithm that for a given x tells us either that $A(x)$ is true (and provides a witness for $A(x)$) or that $\neg A(x)$ is true (and provides a witness for $\neg A(x)$). That means that $\forall x.A(x) \vee \neg A(x)$ is true only for decidable predicates A. Since not all predicates are decidable, Markov's school has to reject the rule of excluded middle.

Note that a witness of a proposition does not "prove" that proposition. For example, $\forall x.A(x) \vee \neg A(x)$ is *true* when there is a decision algorithm for A, but it does not mean that there is a *proof* [1] that this algorithm works properly (i.e. always terminates and gives the correct answer). In this respect the constructive recursive mathematics differs from the Brouwer–Heyting–Kolmogorov's intuitionism. We will return to the topic of differences between "proof witnesses" and "algorithm witnesses" in Section 1.4.

The question arisen in the CRM is *which means is one allowed to use in order to establish that a particular algorithm is indeed a witness for the given proposition?* This is not an obvious question since the termination problem is undecidable. Even if algorithm terminates for every input, we can not test it explicitly, because there are infinitely many possible inputs. But to establish that an algorithm is applicable to an object a, the algorithm does not have to be executed explicitly from the beginning to the end. According to Markov [13] we can prove this by contradiction. That is, we are allowed use some classical reasoning to prove that a particular algorithm has some particular properties.

1.3 Markov's Principle

The Markov school uses the intuitionistic predicate arithmetic with an additional principle (known as Markov's principle):

$$\forall x.(A(x) \vee \neg A(x)) \rightarrow \neg\neg\exists x.A(x) \rightarrow \exists x.A(x) \tag{1.1}$$

[1] In this paper we use terms *proof* and *derivation* interchangeably.

where variables range over natural numbers. Note that this principle does not hold for Brouwer–Heyting–Kolmogorov's intuitionism, thus Markov's principle distinguishes these two schools of constructivism.

Here is the justification of this principle in the CRM framework. Assume $\forall x.(A(x) \vee \neg A(x))$. Then there exists an effective procedure that for every x decides whether $A(x)$ or $\neg A(x)$. To establish (1.1), we need to write a program that produces the witness of $\exists x.A(x)$. We can achieve that by writing a program that will try every natural number x and check $A(x)$ until it finds such an x that $A(x)$ is true. We know that it is impossible that such x would not exist (because of $\neg\neg\exists x.A(x)$). Therefore it is impossible that this algorithm does not terminate. Hence according to recursive constructivism it eventually stops.

Markov's principle is an important technical tool for proving termination of computations. Adding Markov's principle to a traditional constructive type theory would considerably extend the power of the latter in a pivotal class of verification problems.

1.4 Type Theory

We assume the type theory under consideration adheres to the *propositions–as–types principle.* This principle means that a proposition is identified with the type of all its witnesses. A proposition is considered true if the corresponding type is inhabited and is considered false otherwise. This makes terms *an element of a type* and *a witness of a proposition* synonyms. The elements of a type are actually λ–terms, i.e. programs that evaluates to a "canonical" element of this type.

We also assume that the type theory is extensional. That is, to prove that a term f is a function from A to B, it should be sufficient to show that for any $a \in A$ the application fa eventually evaluates to an element of the type B. This allows us to deal with recursive functions that we can prove will always terminate. We will use the `fix` operator to define recursive functions, where $\mathtt{fix}(f.p[f])$ is defined as $(\lambda x.p[xx])(\lambda x.p[xx])$, i.e. `fix` is the operator with the following property $\mathtt{fix}(f.p[f]) \mapsto p[\mathtt{fix}(f.p[f])]$. Although the general typing rule for `fix`

$$\frac{f : A \to A \vdash p[f] \in A \to A}{\vdash \mathtt{fix} f.p[f] \in A \to A}$$

is unsound, but for some particular p we can prove that $\mathtt{fix}(f.p[x])$ is a well-typed function.

Note that in an extensional type theory *a witness of a proposition* T is not the same as *a derivation of a proposition* T. In general, a witness (i.e. an element) of the type T may potentially range from full encoding of some derivation of T to a trivial constant. For example, if V is an empty type, then *every* function has type $V \to W$, e.g. a function $\lambda x.foo$ is an element of $(A \wedge \neg A) \to \bot$ (although $\lambda x.foo$ does not encode any derivations of proposition $(A \wedge \neg A) \to \bot$). Note that the question of whether a particular term is a witness of a particular proposition is in general undecidable.

We assume that the type theory has a membership type – "$t \in T$" which stands for the proposition "t is an element of type T". The only witness of a membership proposition is a special constant $\bullet$[2].

We assume that $t \in A$ implies A. The inverse should also be true:

Property 1.1. If we can prove $\Gamma \vdash A$ then there is a term t such that $\Gamma \vdash t \in A$.

Remark 1.2. The reason we want judgments of the form $\Gamma \vdash T$ and not just $\Gamma \vdash t \in T$ is that we are interested in type theories that can be used as a foundation for theorem provers. In a theorem prover situation we want user to be able to state and prove a judgment of the for $\Gamma \vdash T$ and have the system "extract" t from the resulting derivation instead of being required to figure out and provide t upfront.

In this paper we present several formal derivations. These derivations were machine–checked in the MetaPRL system [10]. The rules used in those derivations are summarized in Appendix A. The results of Sections 2, 3 and 4 are valid for any type theory containing this set of rules and satisfying Property 1.1. The results of the sections Sections 5 and 6 also require an intentional semantics. NuPRL is an example of a type theory that satisfies all these constraints.

However, most of our ideas can be easily applied to an even wider class of type theories. For example, typing rules are not really essential. Additionally, we do not need a membership type to express Markov's principle, we can use the `squash` operator instead (Section 4). And as Section 7 shows, our form of Markov's principle can be used even in a purely propositional fragment of type theory without arithmetic and quantifiers.

2 Constructive Recursive Mathematics in a Type Theory with a Membership Type

Suppose one has proved that A implies $t \in T$ and $\neg A$ also implies $t \in T$. Then *classically* we can conclude that T is inhabited. Moreover the philosophy of recursive constructivism allows us to conclude that T is true *constructively*, because we can explicitly provide a constructive object t as an element of T. In other words, since the witness of T (which is just t) does not depend on the proof of $t \in T$, then T has a uniform witness regardless whether A is true or not (although the *proof* that t is a witness of T may depend on A).

This argument establishes that the following type theory rule is valid according to recursive constructivism:

$$\frac{\Gamma; x : A \vdash t \in T \quad \Gamma; y : \neg A \vdash t \in T \quad \Gamma \vdash A\,\mathrm{Type}}{\Gamma \vdash t \in T} \tag{2.1}$$

This rule formalizes exactly the philosophy of the recursive constructivism.

[2] MetaPRL system [9,10] uses the unit element () or "it" as a $\bullet$, NuPRL uses **Ax** and [18] uses *Triv*.

Remark 2.1. Note that in NuPRL–like type theories $t \in T$ is well-formed only when t is in fact an element of T. Therefore the rule stating that $\neg\neg(t \in T)$ implies $t \in T$ would be useless. On the other hand, in NuPRL type theory (2.1) is equivalent to

$$\frac{\Gamma \vdash (s \in T) \quad \Gamma \vdash (t \in T) \quad \Gamma \vdash \neg\neg(s = t \in T)}{\Gamma \vdash s = t \in T} \tag{2.2}$$

but the proof of (2.2) $\Rightarrow$ (2.1) is very NuPRL–specific.

3 Squashed Types and Squash–Stability

Below we will assume that our type theory contains a "**squash**" operator. For each type A we define a type $[A]$ ("squashed A") which is empty *if and only if* A is empty and contains a single element $\bullet$ when A is inhabited. Informally one can think of $[A]$ as a proposition that says that A is a *non-empty type.*

We define the **squash** operator as a primitive type constructor with the following rules (cf. [15])[3]:

$$\frac{\Gamma \vdash A}{\Gamma \vdash [A]}(I_{sq}) \qquad \frac{\Gamma \vdash A}{\Gamma \vdash \bullet \in [A]}(M_{sq})$$

$$\frac{\Gamma; x : A; \Delta \vdash [B]}{\Gamma; v : [A]; \Delta \vdash [B]}(E_{sq}) \ (v,\, x \text{ are not free in } \Delta,\, B)$$

$$\frac{\Gamma \vdash [t \in T]}{\Gamma \vdash t \in T}(MembershipSqstable)$$

$$\frac{\Gamma \vdash A\,\text{Type}}{\Gamma \vdash [A]\,\text{Type}}(W_{sq})$$

Note that (E_{sq}) and $(MembershipSqstable)$ can be replaced by one rule

$$\frac{\Gamma; x : A; \Delta[\bullet] \vdash t \in T}{\Gamma; v : [A]; \Delta[v] \vdash t \in T}(E^2_{sq}) \ (v,\, x \text{ are not free in } \Delta,\, t,\, T)$$

The (I_{sq}) rule allows us to prove that $A \vdash [A]$. Note that $[A]$ does not imply A, because $[A]$ does not provide a witness for A. We can only derive that $[A] \vdash \neg\neg A$.

Squash operator allows us to formulate an important notion of squash stability. Although $[A]$ does not provide a witness for A in general, in some cases

[3] In a type theory that has a set type (also sometimes called "subset type") constructor, the squashed type $[A]$ can also be defined as $\{x : \mathsf{Unit} \mid A\}$ where Unit is a singleton type which contains only $\bullet$ and x is a variable that does not occur in A.

we know what witness would be in the case when A is non-empty. For example we know that if $t \in T$ is true, then $\bullet$ is the witness for the type $t \in T$. We will refer to such types as *squash–stable* types. For such types we can conclude that $[A] \vdash A$.

Definition 3.1. *A type T is squash–stable (in context Γ) when $\Gamma; x : T \vdash t \in T$ is provable for some t that does not have free occurrences of x.*

Using `squash` operator we can formulate the squash stability as a proposition in a type theory.

Lemma 3.2. *T is squash–stable in a context Γ if one can derive*

$$\Gamma; v : [T] \vdash T$$

Proof. Suppose T is squash–stable. Then we have the following derivation of $\Gamma; v : [T] \vdash T$:

$$\dfrac{\dfrac{\Gamma; x : T \vdash t \in T}{\Gamma; v : [T] \vdash t \in T}(E_{sq}^2)}{\Gamma; v : [T] \vdash T}$$

Now assume that $\Gamma; v : [T] \vdash T$. Then for some term t we have $\Gamma; v : [T] \vdash t \in T$. Term t may depend on v, i.e. v may be a free variable of t. Let $t = t[v]$. Now let x be an arbitrary variable not occurring freely in t. Now we can derive the following:

$$\dfrac{\dfrac{\overline{\Gamma; x : T \vdash T}}{\Gamma; x : T \vdash \bullet \in [T]}(M_{sq}) \qquad \Gamma; v : [T] \vdash t[v] \in T}{\Gamma; x : T \vdash t[\bullet] \in T}(Let)$$

Therefore T is squash–stable.

We have already seen that $t \in T$ is a squash–stable type. Squash type itself is also squash–stable because we know that $\bullet \in [A]$ whenever $[A]$ is true. Other examples of squash–stable types include empty type ($\bot$), negations of arbitrary types ($\neg A$). Note that conjunction of two squash–stable types is also squash–stable. But a disjunction of two squash–stable types is not necessarily squash–stable since there is no way to figure out which of the disjuncts is true when we only know that at least one of them must be true.

4 Classical Reasoning on Squashed Types

Squash operator gives us an alternative way of formulating constructive recursive mathematics in a type theory. Let us consider a problem similar to the one we have considered in Section 2.

Suppose we have constructively proved that $A \to B$ and $(\neg A) \to B$. It means that there is an algorithm that produces an element of B when A is true, and another algorithm that produces an element of B when A is false. Classically we know that B is true, because in each case we can produce an element of B. But in the case when A is undecidable, B is not necessary known to be constructively true, since we do not have a uniform algorithm for producing an element of B. In intuitionistic mathematics we can only prove $\neg\neg B$ in this case.

Now suppose B is squash–stable. Then there exists an element b, such that $b \in B$ whenever B is non-empty. We know that B is not an empty type regardless of whether A is true. The constant algorithm that returns b does not depend on the truth of A. Therefore in constructive recursive mathematics we can conclude that B is *constructively* true, because we have an element b such that $b \in B$.

This reasoning establishes the following rule:

$$\frac{\Gamma; x : A \vdash B \quad \Gamma; y : \neg A \vdash B \quad \Gamma; v : \left[B\right] \vdash B \quad \Gamma \vdash A\,\mathrm{Type}}{\Gamma \vdash B} \tag{4.1}$$

This rule allows us to turn classical proofs of *squash–stable* statements into constructive ones. It is clear that rule (2.1) from Section 2 is a particular instance of (4.1). We will show that these two rules are in fact equivalent. We can also write a simpler version of the same rule:

$$\frac{\Gamma \vdash \neg\neg A \quad \Gamma \vdash A\,\mathrm{Type}}{\Gamma \vdash \left[A\right]} \tag{4.2}$$

This rules states that $\left[A\right] \Leftrightarrow \neg\neg A$ or informally, A is *a non-empty type* if and only if it is *not an empty type*.

Another way to formulate the same principle is to allow classical reasoning inside `squash` operator:

$$\frac{\Gamma \vdash A\,\mathrm{Type}}{\Gamma \vdash \left[A \vee \neg A\right]} \tag{4.3}$$

The following two theorems state that all the above rules are equivalent and that they imply Markov's principle. This shows that we can formulate Markov's principle in a very simple language – we only need propositional language with the modal operator "`squash`".

Theorem 4.1. *The rules* (2.1), (4.1), (4.2) *and* (4.3) *are equivalent.*

Proof. (4.1) $\Rightarrow$ (2.1). Take $B = (t \in T)$. We know that $t \in T$ is squash stable, therefore we can apply rule (4.1) to derive (2.1).

(2.1) ⇒ (4.2).

$$\dfrac{\dfrac{\dfrac{}{\Gamma; x : A \vdash A}}{\Gamma; x : A \vdash \bullet \in [A]} \qquad \dfrac{\dfrac{\Gamma \vdash \neg\neg A \quad \dfrac{}{y : \neg A; z : \neg\neg A \vdash \bot}}{\Gamma; y : \neg A \vdash \bot}(Cut)}{\Gamma; y : \neg A \vdash \bullet \in [A]}(E_{\bot}, Cut) \qquad \Gamma \vdash A\,\mathrm{Type}}{\dfrac{\Gamma \vdash \bullet \in [A]}{\Gamma \vdash [A]}(E_{\in})}(2.1)$$

(4.2) ⇒ (4.3).

It is easy to establish that $\neg\neg(A \vee \neg A)$ is an intuitionistic tautology. Therefore we have the following derivation:

$$\dfrac{\dfrac{\Gamma \vdash A\,\mathrm{Type}}{\Gamma \vdash \neg\neg(A \vee \neg A)} \quad \Gamma \vdash A\,\mathrm{Type}}{\Gamma \vdash [A \vee \neg A]}(4.2)$$

(4.3) ⇒ (4.1).

$$\dfrac{\dfrac{\Gamma \vdash A\,\mathrm{Type}}{\Gamma \vdash [A \vee \neg A]}(4.3) \qquad \dfrac{\dfrac{\dfrac{\Gamma; x : A \vdash B \quad \Gamma; y : \neg A \vdash B}{\Gamma; z : A \vee \neg A \vdash B}(E_{\vee})}{\Gamma; z : A \vee \neg A \vdash [B]}(I_{sq})}{\Gamma; v : [A \vee \neg A] \vdash [B]}(E_{sq}) \qquad \Gamma; [B] \vdash B}{\Gamma \vdash B}(Cut)$$

Theorem 4.2. *Using one of these rules one can prove Markov's principle in a type theory:*

$$\forall x : \mathbb{N}.(A(x) \vee \neg A(x)) \to \neg\neg\exists x : \mathbb{N}.A(x) \to \exists x : \mathbb{N}.A(x)$$

Proof. We need to show that the following sequent is derivable:

$$d : \forall x : \mathbb{N}.(A(x) \vee \neg A(x)); \, v : \neg\neg\exists x : \mathbb{N}.A(x) \vdash \exists x : \mathbb{N}.A(x)$$

The proof is just a formalization of Markov's reasoning [13]. We are given the element d of the type $\forall x : \mathbb{N}.(A(x) \vee \neg A(x))$. That means that d is an algorithm that given a natural number x decides whether $A(x)$ holds. Now we construct a function f_d that would find an x such that $A(x)$. Let f_d be the following function

$$\mathtt{fix}\Big(f.\lambda x.decide\big(d(x); a.\,\langle x, a\rangle\,; b.f(x+1)\big)\Big)$$

that is a function such that

$$f_d(x) = \begin{cases} \langle x, a\rangle\,, & \text{if } A(x) \text{ is true and } a \in A(x) \\ f_d(x+1), & \text{if } A(x) \text{ is false} \end{cases}$$

If we are given natural n such that $A(n)$ is true, then we have a bound for computation of $f(n-k)$. One can prove that $\forall k \leq n. f_d(n-k) \in \exists x : \mathbb{N}.A(x)$ by induction on k. Therefore $f_d(0) \in \exists x : \mathbb{N}.A(x)$. Then we have the following derivation:

$$\dfrac{\dfrac{d : \forall x : \mathbb{N}.(A(x) \vee \neg A(x));\ n : \mathbb{N};\ u : A(n)\ \vdash\ f_d(0) \in \exists x : \mathbb{N}.A(x)}{d : \forall x : \mathbb{N}.(A(x) \vee \neg A(x));\ \exists x : \mathbb{N}.A(x)\ \vdash\ f_d(0) \in \exists x : \mathbb{N}.A(x)}(E_\exists)}{d : \forall x : \mathbb{N}.(A(x) \vee \neg A(x));\ \left[\exists x : \mathbb{N}.A(x)\right]\ \vdash\ f_d(0) \in \exists x : \mathbb{N}.A(x)}(E^2_{sq})$$

Now we are left to show that $\neg\neg\exists x : \mathbb{N}.A(x)$ implies $\left[\exists x : \mathbb{N}.A(x)\right]$. This is true because of the rule (4.2).

5 Semantical Consistency of Markov's Principle

Theorem 5.1. *The rule* (4.3) *(as well as its equivalents –* (2.1), (4.1) *and* (4.2)*) is valid in S. Allen's semantics [1,2] if we consider it in a classical meta-theory.*

Proof. We need to show that $\Gamma \vdash \left[A \vee \neg A\right]$ is true when A is a type. It is clear that $\left[A \vee \neg A\right]$ is a well–formed type. To prove that it is a true proposition we have to find a term in this type. Let us prove that $\bullet$ is the witness of $\left[A \vee \neg A\right]$. Since we are in a classical meta-theory, for every instantiation of variables introduced by Γ, A is either empty or not. If A is non-empty, then $A \vee \neg A$ is non-empty and so $\bullet \in \left[A \vee \neg A\right]$. If A is an empty type, then $\neg A$ is non-empty type and so, $\bullet$ is again in $\left[A \vee \neg A\right]$. Therefore $\bullet \in \left[A \vee \neg A\right]$ always holds.

Note that we can not prove that $\Gamma \vdash A \vee \neg A$ is valid even using a classical meta-theory, because there is no uniform witness for for $A \vee \neg A$.

Corollary 5.2. *The rule* (4.3) *(and its equivalents) is consistent with the NuPRL type theory containing the theory of partial functions [7].*

Note however that the rule of excluded middle $\Gamma \vdash A \vee \neg A$ is known to be inconsistent with the theory of [7]. In particular, in that theory we can prove that there exists an undecidable proposition. That is, for some P the following is provable:

$$\neg(\forall n : \mathbb{N}.P(n) \vee \neg P(n)) \tag{5.1}$$

Therefore even using rule (4.3) we can not prove that

$$\left[\forall n : \mathbb{N}.P(n) \vee \neg P(n)\right]$$

(which would contradict (5.1)). But we can prove a weaker statement

$$\forall n : \mathbb{N}.\left[P(n) \vee \neg P(n)\right]$$

that does not contradict (5.1).

6 Squash Operator as Modality

The squash operator can be considered as an intuitionistic modality. It turns out that it behaves like the lax modality (denoted by $\bigcirc$) in the Propositional Lax Logic (PLL) [8]. This logic was developed independently for several different purposes (see [8] for an overview).

PLL is the extension of intuitionistic logic with the following rules (in Gentzen style):

$$\frac{\Gamma \vdash A}{\Gamma \vdash [A]} \qquad \frac{\Gamma; A \vdash [B]}{\Gamma; [A] \vdash [B]}$$

PLL$^+$ is PLL+$(\neg[\bot])$, i.e. PLL$^+$ has an additional rule:

$$\frac{\Gamma; A \vdash \bot}{\Gamma; [A] \vdash \bot}$$

PLL* is PLL$^+$+$([A] \leftrightarrow \neg\neg A)$. We can write this axiom as the rule in Gentzen style:

$$\frac{\Gamma; \neg A \vdash \bot}{\Gamma \vdash [A]}$$

PLL$^+$ and PLL* are decidable and have natural Kripke models [8]. They meet cut elimination property. PLL$^+$ has the subformula property. PLL* also has the subformula property if we define $\neg A$ to be a subformula of $[A]$.

Theorem 6.1. *Let A be a propositional formula with the* squash *modality. Let Γ be a set of hypothesis of the form $x : (p\,\text{Type})$ for all propositional variables p in A. Then*

(i) PLL$^+$ $\vdash A$ iff $\Gamma \vdash A$ is derivable in the type theory without 4.3

(ii) PLL* $\vdash A$ iff $\Gamma \vdash A$ is derivable in the type theory with 4.3

Proof. From left to right this theorem can be proved by induction on derivation in PLL$^+$ (PLL*). The right to left direction needs a semantical reasoning. We will only outline the proof for PLL*.

Let A' be the formula A where all subformulas of the form $[B]$ are replaced by $\neg\neg B$. If $\Gamma \vdash A$ is derivable in the type theory with 4.3 then this sequent is valid in the standard semantics in classical meta-theory (Theorem 5.1). Since $[B] \leftrightarrow \neg\neg B$ is true in this semantics then $\Gamma \vdash A'$ is also true. A' is a modal-free formula. Therefore A' is a valid intuitionistic formula. Hence A' is derivable in the intuitionistic propositional logic. Since we have $[B] \leftrightarrow \neg\neg B$ in PLL*, we can derive A in PLL*.

Remark 6.2. It is possible to consider the lax modality in PLL$^+$ as the diamond modality in the natural intuitionistic analog of S4 (in the style of [20]) with an additional rule $\Box A \leftrightarrow A$. Note that since in intuitionistic logics $\Box$ and $\Diamond$ are not interdefinable, $\Box A \leftrightarrow A$ does not imply $\Diamond A \leftrightarrow A$.

Example 6.3. We can prove some basic properties of **squash** in PLL$^+$:
$[A] \to \neg\neg A$
$[A] \leftrightarrow [[A]]$
$[A \wedge B] \leftrightarrow ([A] \wedge [B])$
$[A \to B] \to ([A] \to [B])$, but $([A] \to [B]) \to [A \to B]$ is true only in PLL*
$[\neg A] \leftrightarrow \neg[A]$
$([A] \vee [B]) \to [A \vee B]$, however $[A \vee B] \not\to [A] \vee [B]$ even in PLL*

We can express the notion of squash–stability in this logic as $sqst(A) = [A] \to A$.

Example 6.4. The following properties of squash-stability are derivable in PLL$^+$:
$sqst(\bot)$
$sqst(\neg A)$
$sqst([A])$
$sqst(A) \wedge sqst(B) \to sqst(A \wedge B)$, but $sqst(A) \vee sqst(B) \not\to sqst(A \vee B)$
$sqst(B) \to sqst(A \to B)$
In PLL* we can also prove
$sqst(A) \to (\neg\neg A \to A)$

7 Example: What Logic Do We Use in the Court?

It is clear that we do not use pure classical logic in the court. Let us consider the following cases.

Case 7.1. One night a jewelry shop was robbed. The same night a barber shop was robbed in another town. It was clear that these two robberies were committed by different people. There were two suspects, X and Y for both cases. It was determined that no one else could have committed these crimes. Is this information enough to sentence someone?

Let us formulate this problem in logic. Let

- J stand for "the jewelry shop was robbed",
- J_t stand for "t is guilty in the robbing of the jewelry shop",
- B stand for "the barber shop was robbed",
- B_t stand for "t is guilty in the robbing of the barber shop",
- G_t stand for "t is guilty".
  where t is X or Y.

We know the following:

1. $J \wedge B$ (the jewelry and barber shops were robbed)
2. $(\neg J_X \wedge \neg J_Y) \to \neg J$ (no one but X or Y robbed the jeweler)
3. $(\neg B_X \wedge \neg B_Y) \to \neg B$ (no one but X or Y robbed the barber)
4. $\neg(J_t \wedge B_t)$, where $t = X, Y$ (no one could rob both shops)
5. $J_t \to G_t$, where $t = X, Y$ (Criminal Law)
6. $B_t \to G_t$, where $t = X, Y$ (Criminal Law)

Using this assumptions one can *classically* prove G_X and G_Y. Here is the proof that X is guilty in the prosecutor's words.

"X is guilty! Only X and Y could rob the jeweler. If X robbed the jewelry shop, then he is guilty. Suppose for a moment that X is innocent in this robbery. Then Y robbed the jewelry shop. Therefore he could not have robbed the barber. Hence the barber was robbed by X, because no one else did this. Again X is guilty."

Should the jury accept this *classical* reasoning? Even if they were convinced by the prosecutor, they would be unable to bring in a verdict of "guilty" on X since such verdict must specify a particular crime of which X is found guilty. This is especially clear if the punishment for robbing jewelry shops is different from the punishment for robbing barber shops. Judge would not be able to "extract a constructive element" (i.e. determine the sentence) from the prosecutor's proof.

In terms of Section 3, G_t is not a squash–stable proposition.

Case 7.2. The jewelry shop was robbed again. Now there were three suspects, two twin brothers X and Y, and their friend Z. At the trial, it was determined that only X, Y and Z were able to commit the robbery. It was also determined that the shop was robbed by at least two robbers. One of the twins X or Y was seen in another town at the night of the crime, but unfortunately, since they are twins, there was no way to determine exactly who it was. Can jury find someone guilty?

It is easy to see that we can prove that Z is guilty using classical reasoning. But as we have learned from the previous case classical proofs are not sufficient in court. However this case differs from the previous one! We can prove (classically, of course) not only that Z is guilty, but also that Z is guilty of a particular crime. Therefore the judge has enough information to pass sentence on Z.

In the terms of Section 3 we can said that while the proposition "Z is guilty" is not squash–stable, the proposition "Z robbed a particular jewelry store on a certain date" is squash–stable (since we know how to pass a sentence when this proposition is true). This assumption together with the rule (4.1) allows us to bring Z to justice.

Here is a formal *constructive* proof (again J_t stands for "t robbed the jewelry shop" and G_t stands for "t is guilty"):

$$\dfrac{\dfrac{\dfrac{\dfrac{\text{X or Y has an alibi}}{J_X \vdash \neg J_Y} \quad \dfrac{\text{two of suspects are robbers}}{\neg J_Y \vdash J_X \wedge J_Z}}{J_X \vdash J_Z}(Cut) \quad \dfrac{\dfrac{\text{two of suspects are robbers}}{\neg J_X \vdash J_Y \wedge J_Z}}{\neg J_X \vdash J_Z} \quad \dfrac{\text{J_Z is squash–stable}}{[J_Z] \vdash J_Z}}{\vdash J_Z}(4.1)}{\vdash G_Z}\text{(Criminal Law)}$$

8 Related Work

Our notion of *squash–stability* is very similar to the squash–stability defined in [9, Section 14.2] and to the notion of *computational redundancy* [3, Section 3.4].

The `squash` operator we use is similar to the notion of proof irrelevance [11,17]. Each object in a proof irrelevance type is considered to be equal to any other object of this type. In [17] proof irrelevance was expressed in terms of a certain modality $\triangle$. If A is a type then $\triangle A$ is a type containing all elements of A considered equal. Using NuPRL notation we can write $\triangle A = A//True$, where $A//P$ is a quotient of a type A over relation P. We can prove the following chain:

$$A \to \triangle A \to [A] \to \neg\neg A$$

The main difference between $[A]$ and $\triangle A$ is that there is no uniform element for $\triangle A$. Therefore $\triangle A$ is not squash–stable and $[A]$ does not imply $\triangle A$. However it seems that modal logic of $\triangle$ modality is the same as logic of `squash` (i.e. PLL$^+$).

As far as we know Markov's principle in type theory was considered only by Erik Palmgren in [16]. He proved that a fragment of *intentional* Martin-Löf type theory is closed under Markov's rule:

$$\frac{\Gamma \vdash \neg\neg\exists x : A.P[x]}{\Gamma \vdash \exists x : A.P[x]}$$

where $P[x]$ is an equality type (i.e. $P[x]$ is $t[x] = s[x] \in T$. It is easy to see that this formulation of Markov's rule is not valid for type theories with undecidable equality and, in particular, in extensional type theories.

A The "Minimal" Set of Rules

The judgments of the type theory are the sequents of the following form

$$x_1 : A_1; x_2 : A_2[x_1]; \ldots; x_n : A_n[x_1, \ldots, x_{n-1}] \vdash C[x_1, \ldots, x_n]$$

This sequent is true if we have a uniform witness $t[x_1, \ldots, x_n]$ such that for every $x_1, \ldots, x_n$ if $x_i \in A_i[x_1, \ldots, x_{i-1}]$ then $t[x_1, \ldots, x_n]$ is a member of $C[x_1, \ldots, x_n]$.

The inference rules are presented below[4]. For every type constructor we have a well-formedness rule (W), an introduction rule (I), an elimination rule (E) and a membership introduction rule (M).

Syntax rules:

$$\frac{}{\Gamma; x{:}A; \Delta[x] \vdash A}(ax) \qquad \frac{\Gamma;\Delta \vdash A \quad \Gamma; x{:}A; \Delta \vdash C}{\Gamma;\Delta \vdash C}(Cut) \qquad \frac{\Gamma \vdash a \in A \quad \Gamma; x{:}A \vdash C[x]}{\Gamma \vdash C[a]}(Let)$$

Membership:

$$\frac{\Gamma \vdash t \in A}{\Gamma \vdash (t \in A)\ \mathrm{Type}}(W_\in) \qquad \frac{\Gamma \vdash t \in A}{\Gamma \vdash \bullet \in (t \in A)}(M_\in) \qquad \frac{\Gamma \vdash A\ \mathrm{Type}}{\Gamma \vdash \bullet \in (A\ \mathrm{Type})}(M_{\mathrm{Type}})$$

$$\frac{}{\Gamma; x{:}A; \Delta[x] \vdash x \in A}(I_\in) \qquad \frac{\Gamma \vdash t \in A}{\Gamma \vdash A}(E_\in)$$

---

[4] Some of this rules are redundant. For example most of introduction rules are derivable from their membership introduction counterparts. The (*Let*) rule is derivable from the (*Cut*) rule using function type.

Disjunction:

$$\frac{\Gamma\vdash A\,\text{Type} \quad \Gamma\vdash B\,\text{Type}}{\Gamma\vdash A\vee B\,\text{Type}}(W_\vee) \qquad \frac{\Gamma\vdash a\in A}{\Gamma\vdash \mathtt{inl}\;a\in A\vee B}(M^1_\vee) \qquad \frac{\Gamma\vdash b\in B}{\Gamma\vdash \mathtt{inr}\;b\in A\vee B}(M^2_\vee)$$

$$\frac{\Gamma\vdash A}{\Gamma\vdash A\vee B}(I^1_\vee) \quad \frac{\Gamma\vdash B}{\Gamma\vdash A_1\vee A_2}(I^2_\vee) \quad \frac{\Gamma;x{:}A;\Delta[\mathtt{inl}\;x]\vdash C[\mathtt{inl}\;x] \quad \Gamma;y{:}B;\Delta[\mathtt{inr}\;y]\vdash C[\mathtt{inr}\;y]}{\Gamma;z{:}A;\Delta[z]\vee B\vdash C[z]}(E_\vee)$$

Universal quantifier:

$$\frac{\Gamma;x{:}A\vdash B[x]\,\text{Type}}{\Gamma\vdash \forall x{:}A.B[x]\,\text{Type}}(W_\forall) \qquad \frac{\Gamma;x{:}A\vdash f x\in B[x]}{\Gamma\vdash f\in\forall x{:}A.B[x]}(M_\forall)$$

$$\frac{\Gamma;x{:}A\vdash B[x]}{\Gamma\vdash\forall x{:}A.B[x]}(I_\forall) \qquad \frac{\Gamma;f{:}\forall x{:}A.B[x];\Delta[f]\vdash a\in A}{\Gamma;f{:}\forall x{:}A.B[x];\Delta[f]\vdash f a\in B[a]}(E_\forall)$$

Existential quantifier:

$$\frac{\Gamma;x{:}A\vdash B[x]\,\text{Type}}{\Gamma\vdash\exists x{:}A.B[x]\,\text{Type}}(W_\exists) \qquad \frac{\Gamma;x{:}A\vdash a\in A \quad \Gamma;x{:}A\vdash b\in B[a]}{\Gamma\vdash\langle a,b\rangle\in\exists x{:}A.B[x]}(M_\exists)$$

$$\frac{\Gamma;x{:}A\vdash a\in A \quad \Gamma;x{:}A\vdash B[a]}{\Gamma\vdash\exists x{:}A.B[x]}(I_\exists) \qquad \frac{\Gamma;x{:}A;y{:}B[x];\Delta[\langle x,y\rangle]\vdash C[\langle x,y\rangle]}{\Gamma;z{:}\exists x{:}A.B[x];\Delta[z]\vdash C[z]}(E_\exists)$$

False:

$$\frac{}{\Gamma\vdash\bot\,\text{Type}}(W_\bot) \qquad \frac{}{\Gamma;x{:}\bot;\Delta[x]\vdash C}(E_\bot)$$

Computation:

$\frac{\Gamma\vdash b\in T}{\Gamma\vdash a\in T}$ where $a\mapsto b$ Usual reduction rules: $\lambda x.a[x]\;b\longrightarrow a[b]$, *etc*

Arithmetic:
Induction, *etc*

We assume the following definitions:

$$A\to B=\forall x:A.B \quad A\wedge B=\exists x:A.B,\text{ where } x \text{ is not free in } B$$
$$\neg A = A\to\bot \qquad \mathtt{fix}(f.p[f]) = (\lambda x.p[xx])(\lambda x.p[xx])$$

We can establish the property 1.1 in this fragment by a straightforward induction on the derivation.

Acknowledgments

The authors would like to thank Bob Constable for his general guidance in this research and for extremely productive discussions of various constructive theories. We would also like to thank Sergei Artemov for his very useful comments on the early drafts of this paper. We would also like to thank the anonymous reviewers for their valuable comments and pointers.

References

1. Stuart F. Allen. *A Non-Type-Theoretic Semantics for Type-Theoretic Language.* PhD thesis, Cornell University, 1987.
2. Stuart F. Allen. A Non-type-theoretic Definition of Martin-Löf's Types. In *Proceedings of the Second Symposium on Logic in Computer Science*, pages 215–224. IEEE, June 1987.
3. Roland C. Backhouse, Paul Chisholm, Grant Malcolm, and Erik Saaman. Do-it-yourself type theory. *Formal Aspects of Computing*, 1:19–84, 1989.
4. M. J. Beeson. *Foundations of Constructive Mathematics.* Springer-Verlag, 1985.

5. Douglas Bridges and Fred Richman. *Varieties of Constructive Mathematics.* Cambridge University Press, Cambridge, 1988.
6. Robert L. Constable, Stuart F. Allen, H. M. Bromley, W. R. Cleaveland, J. F. Cremer, R. W. Harper, Douglas J. Howe, T. B. Knoblock, N. P. Mendler, P. Panangaden, James T. Sasaki, and Scott F. Smith. *Implementing Mathematics with the NuPRL Development System.* Prentice-Hall, NJ, 1986.
7. Robert L. Constable and Karl Crary. Computational complexity and induction for partial computable functions in type theory. In *Preprint*, 1998.
8. Matt Fairtlough and Michael Mendler. Propositional lax logic. *Information and Computation*, 137(1):1–33, 1997.
9. Jason J. Hickey. *The MetaPRL Logical Programming Environment.* PhD thesis, Cornell University, January 2001.
10. Jason J. Hickey, Aleksey Nogin, et al. MetaPRL home page. `http://metaprl.org/`.
11. Martin Hofmann. *Extensional concepts in intensional Type theory.* PhD thesis, University of Edinburgh, Laboratory for Foundations of Computer Science, 1995.
12. A.A. Markov. On the continuity of constructive function. *Uspekhi Matematicheskikh Nauk*, 9/3(61):226–230, 1954. In Russian.
13. A.A. Markov. On constructive mathematics. *Trudy Matematicheskogo Instituta imeni V.A. Steklova*, 67:8–14, 1962. In Russian. English Translation: A.M.S. Translations, series 2, vol.98, pp. 1-9. MR 27#3528.
14. Per Martin-Löf. Constructive mathematics and computer programming. In *Proceedings of the Sixth International Congress for Logic, Methodology, and Philosophy of Science*, pages 153–175, Amsterdam, 1982. North Holland.
15. Aleksey Nogin. Quotient types – a modular approach. Technical report, Cornell University, 2001. To appear.
16. Erik Palmgren. The Friedman translation for Martin-Löf's type theory. *Mathematical Logic Quarterly*, 41:314–326, 1995.
17. Frank Pfenning. Intensionality, extensionality, and proof irrelevance in modal type theory. In *Proceedings of the 16th Annual Symposium on Logic in Computer Science*, Boston, Massachusetts, June 2001. LICS. To appear.
18. Simon Thompson. *Type Theory and Functional Programming.* Addison-Wesley, 1991.
19. A.S. Troelstra and D. van Dalen. *Constructivism in Mathematics, An Introduction, Vol. I,II.* North-Holland, Amsterdam, 1988.
20. Duminda Wijesekera. Constructive modal logics I. *Annals of Pure and Applied Logic*, 50:271–301, 1990.

Recursion for Higher-Order Encodings

Carsten Schürmann*

Yale University
Department of Computer Science
New Haven, CT 06511
carsten@cs.yale.edu

Abstract. This paper describes a calculus of partial recursive functions that range over arbitrary and possibly higher-order objects in LF [HHP93]. Its most novel features include recursion under λ-binders and matching against dynamically introduced parameters.

1 Introduction

Logical frameworks are meta-languages that are designed *to represent* deductive systems [Pfe99], but not all provide the functionality necessary to manipulate those representations. For the class of logical frameworks based on inductive definitions such as in Isabelle/HOL [Pau94] and Coq [DFH$^+$93,PM93] the connection to programming has been thoroughly studied and is well understood. However it is not at all well understood for logical frameworks that support true higher-order encodings such as LF [HHP93]. In this paper we study this particular connection and present a calculus of recursive functions that manipulate higher-order, dependently typed encodings.

LF allows concise and elegant encodings of many inference systems including their side conditions, such as natural deduction, sequent calculi, type systems, operational semantics, compilers, etc. It draws its expressive power from dependent types together with higher-order representation techniques both of which directly support common concepts in deductive systems such as variable binding, capture-avoiding substitutions, parametric and hypothetical judgments and substitution properties. The fact that these notions are an integral part of the logical framework would seem to make it an ideal candidate to not only reason within but also to express functions about various inference systems.

Unfortunately, those higher-order representation techniques generally clash with common programming techniques. The problem arises as a side effect of the higher-order nature of encodings because recursive calls may have to traverse λ-binders. This stands in direct conflict with the central requirement of standard inductive types. The positivity condition [PM93] rules out most datatype definitions that seem natural in LF. Carelessly mixing higher-order representation

* This work was sponsored by NSF Grant CCR-9619584 and by the Advanced Research Projects Agency CSTO under the title "The Fox Project: Advanced Languages for Systems Software", ARPA Order No. C533, issued by ESC/ENS under Contract No. F19628-95-C-0050.

L. Fribourg (Ed.): CSL 2001, LNCS 2142, pp. 585–599, 2001.

techniques with function definition by cases and recursion triggers the problem of exotic terms which undermines the adequacy of encodings.

In functional programming languages, arguments to functions are typically closed and therefore functions need not be defined on free parameters. We call this property the *closed world assumption.* In fact, the positivity condition implies the closed world assumption, and is built into all logical frameworks based on inductive definitions such as Coq and Isabelle/HOL. However, when programming with higher-order encodings, the closed world assumption is too restrictive.

Higher-order encodings make frequent use of abstraction as a means to represent variable binding. For example, in a higher-order logical framework such as LF or the simply typed λ-calculus, the identity function '$\texttt{fn}\ x \Rightarrow x$' may be represented as 'lam $(\lambda x.\, x)$' for some constant lam. Under the closed world assumption, however, there is no reasonable scheme for recursion and function definition by cases because recursive calls may have to traverse λ-binders.

In this paper, we propose a solution to the tension between closed worlds and higher-order encodings. We propose to weaken the closed world assumption and allow arguments to be open and depend on parameters. Worlds must not be arbitrary; on the contrary they must be regularly formed in order to allow function extensions by those new cases that match parameters. We call this property the *regular world assumption* and it guides our design of a calculus of partial recursive functions for higher-order encodings which we call $\mathcal{T}_\omega^+$. That $\mathcal{T}_\omega^+$ can be restricted to a meta-logic for LF via a realizability interpretation of proofs is an entirely orthogonal issue and is studied in another forthcoming paper.

This paper is organized as follows. In Section 2 we discuss the technique of higher-order representation, demonstrate its advantages, and compare it to alternative representation techniques. In Section 3 we motivate and analyze the regular world assumption by the means of a function that embeds natural deduction derivations into the sequent calculus. We then describe the type system $\mathcal{T}_\omega^+$ in Section 4, whose elements are partial functions that range over possibly open LF objects. $\mathcal{T}_\omega^+$'s operational semantics is specified in Section 5, and its meta-theory is summarized in Section 6. A sample application is given in Section 7.

2 Higher-Order Encodings

As a motivating example, consider the standard formulation of the calculus of natural deductions [Gen35] for the implicational fragment of propositional logic depicted in Figure 1. The rules define the judgment "*G is true*" ($\Gamma \vdash G$). We take the freedom to assign names to hypotheses $\Gamma ::= \cdot \mid \Gamma, u :: G$. There are many possible solutions of how to represent natural deduction derivations formally and we discuss some of them below. We argue that the representation in a dependently typed, higher-order type theory, such as LF [HHP93] is the most natural. We write $\ulcorner \cdot \urcorner$ for the representation function and begin with the straightforward representation of formulas.

$$\frac{}{\Gamma, u :: G \vdash G}\, u \qquad \frac{\Gamma, u :: G_1 \vdash G_2}{\Gamma \vdash G_1 \supset G_2} \supset \mathrm{I}^u \qquad \frac{\Gamma \vdash G_1 \supset G_2 \qquad \Gamma \vdash G_1}{\Gamma \vdash G_2} \supset \mathrm{E}$$

Fig. 1. Natural deduction calculus

$$\ulcorner G_1 \supset G_2 \urcorner = \ulcorner G_1 \urcorner \operatorname{imp} \ulcorner G_2 \urcorner \qquad \begin{array}{ll} \mathrm{o} & : \mathrm{type} \\ \mathrm{imp} & : \mathrm{o} \to \mathrm{o} \to \mathrm{o} \end{array}$$

'imp' is used as an infix operator throughout this paper. In LF, judgments are represented as types, and derivations as objects. Following standard practice [Pfe91] we omit all implicit Π-abstractions from types and we take $\beta\eta$-conversion as the notion of definitional equality [Coq91].

Example 1 (Representation of Figure 1 in LF).

$$\begin{array}{ll} \mathrm{nd} & : \mathrm{o} \to \mathrm{type} \\ \mathrm{impi} & : (\mathrm{nd}\ G_1 \to \mathrm{nd}\ G_2) \to \mathrm{nd}\ (G_1 \operatorname{imp} G_2) \\ \mathrm{impe} & : \mathrm{nd}\ (G_1 \operatorname{imp} G_2) \to \mathrm{nd}\ G_1 \to \mathrm{nd}\ G_2 \end{array}$$

$\Gamma \vdash A$ is a hypothetical judgment since the premiss to '$\supset$ I' in Figure 1 discharges the hypothesis u. A good choice for the representation of Γ is the LF context itself, because then all (admissible) structural rules including 'u', weakening, contraction, and exchange, are inherited from LF and need not to be encoded individually. Consequently, hypothetical judgments are represented as functions, and thus the premiss of $\supset$ I corresponds to an object of type (nd $G_1 \to$ nd G_2). The substitution operation for hypotheses may be encoded as a simple LF β-redex. The representation of natural deduction derivations in LF by 'impi' and 'impe' is adequate, i.e. they are in one-to-one correspondence with their canonical forms [HHP93].

Despite its elegance, the encoding of the natural deduction calculus in LF has one fundamental drawback. It is not inductive in the sense of [PM93]. 'nd G_1' in the type of 'impi' violates the positivity condition, and thus recursive functions cannot be defined by cases on an argument of type 'nd G'. It is the main contribution of this paper to resolve this seeming contradiction.

There are many other logical frameworks that represent natural deductions while supporting function definition by cases, however many of these encodings are less immediate. For example, the encoding from Example 1 can be turned into an inductive type by a technique proposed by Despeyroux et al. [DFH95]. This technique enforces the positivity condition by replacing the negative occurrence 'nd G_1' in 'impi' by a new parameter type 'var G_1' at the expense of the advantages that come with higher-order abstract syntax. Structural rules and substitution are no longer directly supported.

The modal λ-calculus [DPS97,DL99,Hof99] supports true higher-order encodings and function definition by iteration. Its version of iteration however is quite rigid and sometimes poses problems when applying weakening and substitution lemmas or using dependent types.

The programming language FreshML [GP99,PG00] imports higher-order abstract syntax into a functional programming setting and takes care of α-renaming internally. However, neither dependent types nor substitutions are directly supported by FreshML, and must therefore be programmed explicitly.

The meta-logic $\mathcal{M}_2$ is a direct predecessor of this work [SP98]. It supports higher-order encodings, but it does not allow recursion to traverse λ-binders. Consequently, $\mathcal{M}_2$'s expressiveness is limited and those functions that motivate this work are simply not expressible.

The calculi of partial inductive definitions [Hal87] and definitional reflection [SH93] are purely logical systems. Even when interpreted functionally, their proofs require all arguments to be closed and recursion must not traverse λ-binders. Figure 1 can also be encoded in a logical framework implemented in the meta-logic $FO\lambda^{\Delta I\!N}$ [MM97]. $FO\lambda^{\Delta I\!N}$, however, is a sequent calculus, and does not support functions to range over those encodings.

Miller [Mil90] suggests to extend ML datatypes by a parameter mechanism to support higher-order abstract syntax. New parameters can be dynamically generated and (recursive) function dynamically extended by new cases, however dependent types are not supported.

The representation of natural derivations in Example 1 is concise and elegant and it leverages off properties inherent to LF. In this paper we design a type system $\mathcal{T}_\omega^+$ of partial recursive functions that range over higher-order encodings. The functions are definable by case analysis and recursion.

3 The Regular World Assumption

As running example we use a mapping of natural deduction derivations into derivations of the sequent calculus. Although we present only one connective, the example scales to full first-order logic. The judgment for sequent derivations is $\Gamma \Longrightarrow G$, its rules and its adequate encoding are depicted in Figure 2 [Pfe95]. Hypotheses to the left of the sequent symbol '$\Longrightarrow$' are encoded using 'hyp' in order to distinguish them from the conclusion to the right which is encoded using 'conc'. Only by representing them as two separate type families we can guarantee the adequacy of the encoding which also relies on the fact that the hard-wired structural properties of hypotheses such as weakening and contraction are implicit in the encoding in LF.

Gentzen has shown in [Gen35] that such a mapping from natural deductions $\mathcal{D}$ into sequent derivations $\mathcal{E}$ exists. This algorithm is depicted in the left column of Figure 3, and we refer to it as Gentzen's algorithm. Because natural deductions and sequent derivations are represented differently, we let $\Gamma_u = u_1 :: G_1, \ldots, u_n :: G_n$ be the natural deduction context, and $\Gamma_h = h_1 :: G_1, \ldots, h_n :: G_n$ be the sequent context.

Intuitively, Gentzen's algorithm should be expressible as a recursive function that maps natural deduction derivations $\ulcorner \mathcal{D} \urcorner$ to sequent derivations $\ulcorner \mathcal{E} \urcorner$ where $\ulcorner \mathcal{D} \urcorner$ and $\ulcorner \mathcal{E} \urcorner$ satisfy the invariant:

$$\frac{}{\Gamma, h :: G \Longrightarrow G}\,\text{init} \qquad \frac{\Gamma \Longrightarrow G_1 \qquad \Gamma, h :: G_1 \Longrightarrow G_2}{\Gamma \Longrightarrow G_2}\,\text{cut}^h \qquad \frac{\Gamma, h :: G_1 \Longrightarrow G_2}{\Gamma \Longrightarrow G_1 \supset G_2}\,{\supset}\text{R}^h$$

$$\frac{\Gamma, h :: G_1 \supset G_2 \Longrightarrow G_1 \qquad \Gamma, h :: G_1 \supset G_2, h_2 :: G_2 \Longrightarrow G_3}{\Gamma, h :: G_1 \supset G_2 \Longrightarrow G_3}\,{\supset}\text{L}^{h_2}$$

$$\begin{array}{ll}
\text{hyp} & : \text{o} \rightarrow \text{type} \\
\text{conc} & : \text{o} \rightarrow \text{type} \\
\text{init} & : \text{hyp}\ G \rightarrow \text{conc}\ G \\
\text{cut} & : \text{conc}\ G_1 \rightarrow (\text{hyp}\ G_1 \rightarrow \text{conc}\ G_2) \rightarrow \text{conc}\ G_2 \\
\text{impR} & : (\text{hyp}\ G_1 \rightarrow \text{conc}\ G_2) \rightarrow \text{conc}\ (G_1\ \text{imp}\ G_2) \\
\text{impL} & : \text{conc}\ G_1 \rightarrow (\text{hyp}\ G_2 \rightarrow \text{conc}\ G_3) \rightarrow (\text{hyp}\ (G_1\ \text{imp}\ G_2) \rightarrow \text{conc}\ G_3)
\end{array}$$

Fig. 2. Sequent Calculus and its Representation in LF

Input: A derivation $\mathcal{D}$ of $\Gamma \vdash G$

Case 1: $\mathcal{D} = \dfrac{}{\Gamma_u, u :: G \vdash G}\,u$

Compute $\mathcal{E}$ of $\Gamma_h, h :: G \Longrightarrow G$ by init
Return $\mathcal{E}$.

Case 2: $\mathcal{D} = \dfrac{\begin{matrix}\mathcal{D}' \\ \Gamma_u, u :: G_1 \vdash G_2\end{matrix}}{\Gamma_u \vdash G_1 \supset G_2}\,{\supset}\text{I}^u$

Compute $\mathcal{E}'$ of $\Gamma_h, h :: G_1 \Longrightarrow G_2$ by recursion on $\mathcal{D}'$
Compute $\mathcal{E}$ of $\Gamma_h \Longrightarrow G_1 \supset G_2$ by rule $\supset$R on $\mathcal{E}'$
Return $\mathcal{E}$

Case 3: $\mathcal{D} = \dfrac{\begin{matrix}\mathcal{D}_1 \\ \Gamma_u \vdash G_1 \supset G_2\end{matrix} \qquad \begin{matrix}\mathcal{D}_2 \\ \Gamma_u \vdash G_1\end{matrix}}{\Gamma_u \vdash G_2}\,{\supset}\text{E}$

Compute $\mathcal{E}_1$ of $\Gamma_h \Longrightarrow G_1 \supset G_2$ by recursion on $\mathcal{D}_1$
Compute $\mathcal{E}_2$ of $\Gamma_h \Longrightarrow G_1$ by recursion on $\mathcal{D}_2$
Compute $\mathcal{E}_2'$ of $\Gamma_h, h :: G_1 \supset G_2 \Longrightarrow G_1$ by weakening on $\mathcal{E}_2$
Compute $\mathcal{E}_3$ of $\Gamma_h, h :: G_1 \supset G_2, h_2 :: G_2 \Longrightarrow G_2$ by init
Compute $\mathcal{E}_4$ of $\Gamma_h, h :: G_1 \supset G_2 \Longrightarrow G_2$ by $\supset$L on $\mathcal{E}_2'$ and $\mathcal{E}_3$
Compute $\mathcal{E}$ of $\Gamma_h \Longrightarrow G_2$ by cut on $\mathcal{E}_1$ and $\mathcal{E}_4$
Return $\mathcal{E}$.

fun ndseq $\underline{u}$ = (init $\underline{h}$)
  | **ndseq** (impi D') =
    let
      new u : nd G_1, h : hyp G_1
      val $(E\ h)$ = **ndseq** $(D'\ u)$
    in
      (impR $(\lambda h.\, E\ h)$)
    end
  | **ndseq** (impe D_1 D_2) =
    let
      val (E_1) = **ndseq** D_1
      val (E_2) = **ndseq** D_2
    in
      (cut E_1 $(\lambda h_1.\,$impL E_2
        $(\lambda h_2.\,$init $h_2)\ h_1))$
    end

Fig. 3. Gentzen's algorithm

$$u_1 : \text{nd}\ulcorner G_1 \urcorner, \ldots, u_n : \text{nd}\ulcorner G_n \urcorner \overset{\Pi}{\vdash} \ulcorner \mathcal{D} \urcorner : \text{nd}\ulcorner G \urcorner$$
$$h_1 : \text{hyp}\ulcorner G_1 \urcorner, \ldots, h_n : \text{hyp}\ulcorner G_n \urcorner \overset{\Pi}{\vdash} \ulcorner \mathcal{E} \urcorner : \text{conc}\ulcorner G \urcorner$$

Here, the function must accommodate the fact that $\ulcorner \mathcal{D} \urcorner$ and $\ulcorner \mathcal{E} \urcorner$ are open objects since the LF contexts to the left of the $\overset{\Pi}{\vdash}$ symbols are not empty, and therefore violate the so called *closed world assumption.* Non-standard is also the recursive call in **Case 2** which extends the natural deduction context to $\Gamma_u, u :: G_1$. This extension corresponds to the traversal of the λ-binder '$u : \ulcorner G_1 \urcorner$' in LF.

The main contribution of this paper is a characterization of partial recursive functions such as the one described above. To address this situation we propose to change the status quo of the closed world assumption and generalize it to the *regular world assumption.* If we accept this new assumption, arguments to recursive functions do not need to be closed. They may be open in a parameter context which conforms to an a priori specified *regular* grammar, and it is this restriction which makes the construction work.

The form of the regular world for Gentzen's algorithm is best explained by interpreting the invariant associated with the algorithm. It says, that Γ_u and Γ_h may only differ in names, but not in length. Thus, the grammar that defines the regular world for our example results from interleaving Γ_u and Γ_h as follows:

$$u_1 : \text{nd} \ulcorner G_1 \urcorner, h_1 : \text{hyp} \ulcorner G_1 \urcorner, \ldots, u_n : \text{nd} \ulcorner G_n \urcorner, h_n : \text{hyp} \ulcorner G_n \urcorner$$

Now the u_i's and the h_i's always come in pairs and therefore, any valid world can be expressed by the following grammar that defines the *regular world* Φ.

$$\Phi ::= \cdot \mid \Phi, u : \text{nd } G, h : \text{hyp } G$$

A world Φ is a finite LF context that results from repeatedly unfolding the definition of Φ. With each unfolding operation, a new *parameter block* is added to the world. This particular definition of Φ is custom-designed to fit Gentzen's algorithm, recursive functions from other problem domains however will require different Φ's. The results presented in this paper are not linked to one particular definition of Φ, they are valid for any Φ. Another example is given in Section 7.

In this world Φ, there exists a recursive function **ndseq** which is depicted (in informal syntax) in the right column of Figure 3. Parallel to the informal case analysis performed on $\mathcal{D}$ in Gentzen's algorithm, **ndseq** is defined by cases.

The first case of **ndseq** formalizes **Case 1**. It applies in a situation where $\ulcorner \mathcal{D} \urcorner$ is bound to some parameter u in Φ. Automatically, by invariant, there will also be a parameter h (the sibling of u). We use $\underline{u}$ and $\underline{h}$ as variables that range exclusively over parameters from the regular world Φ.

The second case of **ndseq** formalizes **Case 2**. The argument $D' = \ulcorner \mathcal{D}' \urcorner$ is of functional type 'nd $G_1 \to$ nd G_2', and therefore **ndseq** must traverse the λ-binder before it recurses. It does so by extending the regular world by a new parameter block (indicated by the **new**-notation) u : nd G_1, h : hyp G_1. The result of the recursive call on $(D'\ u)$ yields a sequent derivation $(E'\ h)$ which is parametric in h. One more application of 'impR' forms the return object.

The third case formalizes **Case 3** and simply returns an appropriate LF object after recursing on D_1 and D_2. Note, that in constructing this result object all appeals to weakening lemmas are automatic since they are inherited from LF. In other examples, appeals to substitution lemmas are formalized equally elegantly.

The regular world assumption provides many new concepts and advantages. It generalizes recursive functions to higher-order encodings *without* an explicit positivity condition [PM93]. It permits arguments passed to recursive functions to be open. It gives recursive calls the freedom to traverse λ-binders. It allows recursive function to define cases for parameters. It is elegant in that recursive functions can be formulated concisely and directly which are not so easy to write in modal λ-calculi [DPS97,DL99,Hof99]. And finally, it is compatible with dependent types.

4 Type System $\mathcal{T}_\omega^+$

We begin now with the formal presentation of the type system $\mathcal{T}_\omega^+$ which supports function definition by recursion as used above for the definition of Gentzen's algorithm **ndseq**. A detailed presentation of $\mathcal{T}_\omega^+$ and its meta-theory can be found in the author's thesis [Sch00]. $\mathcal{T}_\omega^+$'s recursive functions range over higher-order LF objects and can therefore take advantage of all properties which are associated with hypothetical judgments, such as, substitutions, weakening, contraction, and exchange.

$\mathcal{T}_\omega^+$'s type system is simple. It provides a dependent function type $\forall$, a dependent product type $\exists$, and a unit type $\top$. $\mathcal{T}_\omega^+$'s function types can contain ω many alternations of $\forall$ and $\exists$ types. The type constructors of $\mathcal{T}_\omega^+$ are inspired by logic, however, for the purpose of this paper we view them solely as type constructors. **ndseq**, for example, has type $\forall G : \mathrm{o}. \forall D : \mathrm{nd}\ G. \exists E : \mathrm{conc}\ G. \top$.

That **ndseq** in Figure 3 expects only one argument is due to the fact that the first argument can be inferred from the second, and is therefore omitted when using informal syntax. **ndseq**'s type states that in any regular world Φ, for every valid formula $G : \mathrm{o}$ and for every valid natural deduction derivation D ($\Phi \overset{\Pi}{\vdash} D : \mathrm{nd}\ G$), the execution of **ndseq** applied to G and D might or might not terminate normally. However, if it terminates normally, then **ndseq** has computed an instance of type $\exists E : \mathrm{conc}\ G. \top$ and the left projection yields a valid LF object representing a sequent derivation in Φ ($\Phi \overset{\Pi}{\vdash} E : \mathrm{conc}\ G$). For the purpose of this paper, we assume the specification of Φ to be fixed, as we do, for example, with the specification of the LF signature. $\forall$ and $\exists$ types are designed to bind variables whose structure can be analyzed by case analysis.

$$\mathcal{T}_\omega^+ \textit{ types: } F ::= \forall x : A. F \mid \exists x : A. F \mid \top$$

Here, A ranges over LF types [HHP93]. The $\mathcal{T}_\omega^+$ type system can be easily extended by product types $F_1 \wedge F_2$ which we omit because of space constraints.

4.1 Subordination

Recursive functions in $\mathcal{T}_\omega^+$ are executed in regular worlds. During execution, they may extend the current world by new parameters. Thus, the world grows with each traversal of a λ-binder before a recursive call and it shrinks upon return when parameters are discharged. In general, a returned object may need not depend on all but some of the parameters to be discharged. In the second case of **ndseq**, for example, executing $(D'\ u)$ yields an E which may only depend on h but not on u.

That E does not depend on u is justified by the way sequent derivations are constructed. They are simply not defined in terms of natural deduction derivations. Virga [Vir99] proposed a technique that decides if a parameter can possibly occur in an object of a given type or not. His procedure inspects the LF signature and derives the so called "subordination relation". A type A is said to be subordinate to A' ($A \prec A'$), if objects of type A may occur within objects of type A'.

Lemma 1 (Strengthen). *If* $\Gamma, x : A \overset{\Pi}{\vdash} M : A'$ *and* $A \not\prec A'$ *then* $\Gamma \overset{\Pi}{\vdash} M : A'$.

Indeed, because 'hyp' $\prec$ 'conc', and 'nd' $\not\prec$ 'conc' it follows from Lemma 1 that E cannot depend on u : nd G_1, and thus E : hyp $G_1 \to$ conc G_2. In general, if a return object has type A_2 before discharging parameters, then it has type 'abs $\Gamma.\, A_2$' after the discharge takes place.

Definition 1 (Abstraction).

$$\text{abs}\ \Gamma.\, A_2 = \begin{cases} A_2 & \textit{if}\ \Gamma = \cdot \\ \text{abs}\ \Gamma'.\, A_2 & \textit{if}\ \Gamma = \Gamma', x : A_1\ \textit{and}\ A_1 \not\prec A_2 \\ \Pi x : A_1.\, (\text{abs}\ \Gamma'.\, A_2) & \textit{if}\ \Gamma = \Gamma', x : A_1\ \textit{and if}\ A_1 \prec A_2 \end{cases}$$

In the example 'hyp $G_1 \to$ conc G_2 = abs (u : nd G_1, h : hyp G_1). (conc G_2)' is the type of E after parameter discharge. Lemma 1 guarantees that no parameters escape their scope. Similarly, on the object level, abstraction must bind each parameter it discharges if there is any chance that the parameter occurs free in the object.

Definition 2 (Raising). *Let* M *be of type* A_2.

$$\text{raise}\ \Gamma.\, M = \begin{cases} M & \textit{if}\ \Gamma = \cdot \\ \text{raise}\ \Gamma'.\, M & \textit{if}\ \Gamma = \Gamma', x : A_1\ \textit{and}\ A_1 \not\prec A_2 \\ \lambda x : A_1.\, (\text{raise}\ \Gamma'.\, M) & \textit{if}\ \Gamma = \Gamma', x : A_1\ \textit{and}\ A_1 \prec A_2 \end{cases}$$

By a simple inductive argument we can show that this design is sound.

Lemma 2 (Soundness).

1. *If* $\Gamma, \Gamma' \overset{\Pi}{\vdash} A : \textit{type}$ *then* $\Gamma \overset{\Pi}{\vdash}$ abs $\Gamma'.\, A : \textit{type}$.
2. *If* $\Gamma, \Gamma' \overset{\Pi}{\vdash} M : A$ *then* $\Gamma \overset{\Pi}{\vdash}$ raise $\Gamma'.\, M$: abs $\Gamma'.\, A$.

4.2 The Core of $\mathcal{T}_\omega^+$

$\mathcal{T}_\omega^+$'s functions use two kinds of variable binding, and their use is demonstrated by **ndseq** in Figure 3. First, there are variables x that bind arbitrary LF objects M valid in any well-formed regular world Φ. These variables are used, for example, in the patterns of the second and third case. In general, they are bound by the $\forall$ and $\exists$ quantifiers and any instantiation may be analyzed by cases.

Second, there are *parameter variables* $\underline{u}$ and $\underline{h}$ as used in the first case of **ndseq**. They are designed exclusively to range over parameter blocks. In this sense, they behave more like constructors than variables, and yet, they are variables because they bind concrete parameter occurrences declared in Φ at run-time. The fact that parameters are always organized in blocks is also reflected in the formal system of $\mathcal{T}_\omega^+$. Parameter variables never occur alone, they only occur as compound *variable blocks* $\rho = (\underline{u}$: nd $G, \underline{h}$: hyp $G)$. Variable blocks can only range over parameter blocks declared in Φ, and binding is defined componentwise. Although our informal syntax of $\mathcal{T}_\omega^+$ functions used in Figure 3 hides variable blocks, they are important in the formal exposition.

In our example, Φ is defined in terms of one block, but in the general case it can be defined in terms of several [Sch00]. We distinguish each of them by labels, and assign the label L to the one block in our example.

$$\Phi ::= \cdot \mid \Phi, (u : \text{nd } G, h : \text{hyp } G)^L$$

Variable blocks ρ^L are indexed by the appropriate label L. Ψ is a *context* that denotes not only the set of free LF-variables, but also what is know about the regular world.

$$\textit{Object context: } \Psi ::= \cdot \mid \Psi, x : A \mid \Psi, \rho^L$$

The context $D : \text{nd } G_1, (\underline{u} : \text{nd } G_2, \underline{h} : \text{hyp } G_2)^L$, for example, expresses that D may be instantiated with an object of type 'nd G_1', valid in a regular world Φ and that $(\underline{u} : \text{nd } G_2, \underline{h} : \text{hyp } G_2)^L$ is bound to one parameter block in this world. The validity judgments for contexts extend the standard definition of valid LF contexts in a straightforward way.

$\mathcal{T}_\omega^+$ is a two level system, where x and $\underline{x}$ are variables ranging over LF objects and parameters, respectively. Variables on the meta-level are separate, they are called *meta-variables* $\mathbf{x}$ and range over $\mathcal{T}_\omega^+$-objects of type F. The $\mathbf{x}$'s form a *meta-context* Δ on the meta-level.

$$\textit{Meta-context: } \Delta ::= \cdot \mid \Delta, \mathbf{x} \in F$$

The syntax we have used to present **ndseq** in Figure 3 contains syntactic sugar in order to make this material accessible. Its desugared version (given in the right column of Figure 6) however is less intuitive and requires several mutually dependent syntactic categories including programs and declarations. These programs and declarations resemble the proof terms of an intuitionistic sequent calculus with the important addition of the $\nu\rho^L.\, D$ declaration, which – operationally speaking – extends the world by new parameters before evaluating D.

$$\begin{array}{ll} \textit{Programs:} & P ::= \mathbf{x} \mid \Lambda x : A.\, P \mid \langle M, P\rangle \mid \langle\rangle \mid \textbf{let } D \textbf{ in } P \\ \textit{Declarations:} & D ::= \cdot \mid \nu\, \rho^L.\, D \mid \mathbf{x} \in F = P\ M, D \mid \langle x : A, \mathbf{y} \in F\rangle = P, D \end{array}$$

The syntactic category of programs is extended below to accommodate case analysis and recursion. For the sake of brevity we omit the straightforward desugaring algorithm that maps our informal syntax into $\mathcal{T}_\omega^+$.

Programs and declarations give rise to the definition of two typing judgments. We write $\Psi; \Delta \vdash P \in F$ if program P is of type F and for declarations we write $\Psi; \Delta \vdash D \in \Psi'$. From an operational point of view, all variable declarations in Ψ' will be instantiated and returned after D is computed. We use '$\in$' as typing symbol for $\mathcal{T}_\omega^+$ in order to clearly distinguish it from the ':' for LF.

$$\begin{array}{ll} \textit{Typing programs:} & \Psi; \Delta \vdash P \in F \\ \textit{Typing declarations:} & \Psi; \Delta \vdash D \in \Psi' \end{array}$$

The typing rules for these judgments are mutually dependent and are given in Figure 4. We omit the validity judgment for Ψ from the rules axvar, $\mathsf{P}\top$, and

$$\frac{(\mathbf{x} \in F) \text{ in } \Delta}{\Psi; \Delta \vdash \mathbf{x} \in F}\ \mathsf{axvar} \qquad \frac{}{\Psi; \Delta \vdash \langle\rangle \in \top}\ \mathsf{P}\top \qquad \frac{\Psi; \Delta \vdash D \in \Psi' \quad \Psi, \Psi'; \Delta \vdash P \in F}{\Psi; \Delta \vdash \mathbf{let}\ D\ \mathbf{in}\ P \in F}\ \mathsf{sel}$$

$$\frac{\Psi, x : A; \Delta \vdash P \in F}{\Psi; \Delta \vdash \Lambda x : A.\, P \in \forall x : A.\, F}\ \mathsf{P}\forall \qquad \frac{\Psi \vdash M : A \quad \Psi; \Delta \vdash P \in F[M/x]}{\Psi; \Delta \vdash \langle M, P\rangle \in \exists x : A.\, F}\ \mathsf{P}\exists$$

$$\frac{}{\Psi; \Delta \vdash \cdot \in \cdot}\ \mathsf{D}\cdot \qquad \frac{\Psi, \rho^L; \Delta \vdash D \in \Psi' \quad \text{abs } \rho.\, \Psi' = \Psi''}{\Psi; \Delta \vdash \nu\, \rho^L.\, D \in \Psi''}\ \mathsf{D}\nu$$

$$\frac{\Psi; \Delta \vdash P \in \forall x : A.\, F \quad \Psi \vdash M : A \quad \Psi; \Delta, \mathbf{y} \in F[M/x] \vdash D \in \Psi'}{\Psi; \Delta \vdash (\mathbf{y} \in F[M/x] = P\ M, D) \in \Psi'}\ \mathsf{D}\forall$$

$$\frac{\Psi; \Delta \vdash P \in \exists x : A.\, F \quad \Psi, x : A; \Delta, \mathbf{y} \in F \vdash D \in \Psi'}{\Psi; \Delta \vdash (\langle x : A, \mathbf{y} \in F\rangle = P, D) \in (x : A, \Psi')}\ \mathsf{D}\exists$$

Fig. 4. Typing Rules of the Core of $\mathcal{T}_\omega^+$

D·. The premiss $\Psi \vdash M : A$ of the rules P∃ and D∀ is the standard LF typing judgment. It allows M to depend on free variables and parameters declared in Ψ. The Dν rule specifies how new parameter blocks are inserted into the regular world and how they are discharged thereafter. The premiss 'abs $\rho.\, \Psi' = \Psi''$' of the Dν rule shifts all types in Ψ' componentwise (see Definition 1).

4.3 Case Analysis and Recursion

The design of $\mathcal{T}_\omega^+$'s case analysis mechanism is non-standard. Because of dependencies, variables can occur non-linearly within the types of other variables. Thus case analysis is not a local operation that effects only one variable but can simultaneously effect others as well. In Gentzen's algorithm (Figure 3), for example, matching against the declaration D determines the form of the formula G in the 'impi'-case. Therefore, case analysis must simultaneously distinguish cases on *all* variables declared in Ψ.

To this effect, our design employs a substitution (or environment) η as case subject. It enables us to consider cases over all objects stored in η simultaneously. Similarly, patterns ψ that trigger each case are also substitutions, and thus the matching problem "if ψ matches η" reduces to the question "if ψ is a *more general* substitution than η". We define that ψ *matches* η iff there exists an η', such that $\eta = \psi \circ \eta'$.

Environments η and patterns ψ are substitutions with domain Ψ. They substitute objects M for variables x and variable blocks ρ' for variable blocks ρ. We write $\Psi' \vdash \psi \in \Psi$ if ψ is valid.

$$\textit{Object-substitutions: } \psi ::= \cdot \mid \psi, M/x \mid \psi, \rho'/\rho$$

The choice of a substitution η as case subject has another advantage namely that each branch remains invariant under substitution application. If substitution σ is applied to a case program, it is absorbed by simple substitution

$$\frac{\Psi; \Delta, \mathbf{x} \in F \vdash P \in F}{\Psi; \Delta \vdash \mu \mathbf{x} \in F.\, P \in F}\,\mathsf{fix} \qquad \frac{\Psi \vdash \eta \in \Psi' \qquad \Psi'; \Delta' \vdash \Omega \in F}{\Psi; \Delta'[\eta] \vdash \mathsf{case}\ \eta\ \mathsf{of}\ \Omega \in F[\eta]}\,\mathsf{case}$$

$$\frac{}{\Psi; \Delta \vdash \cdot \in F}\,\mathsf{base} \qquad \frac{\Psi' \vdash \psi \in \Psi \qquad \Psi; \Delta \vdash \Omega \in F \qquad \Psi'; \Delta[\psi] \vdash P \in F[\psi]}{\Psi; \Delta \vdash \Omega, (\Psi' \triangleright \psi \mapsto P) \in F}\,\mathsf{alt}$$

Fig. 5. Case Analysis and Recursion in $\mathcal{T}_\omega^+$

```
fun ndseq u = (init h)
  | ndseq (impi D') =
    let
      new u : nd G1, h : hyp G1
      val  (E h) = ndseq (D' u)
    in
      (impR (λh. E h))
    end
  | ndseq (impe D1 D2) =
    let
      val (E1) = ndseq D1
      val (E2) = ndseq D2
    in
      (cut E1 (λh1. impL E2
           (λh2. init h2) h1))
    end
```

```
μndseq. ΛG : o. ΛD : nd G.
    case (G/G, D/D)
     of  (G : o, (u : nd G, h : hyp G)^L)
              ▷ (G/G, u/D) ↦ (init h)
      |  (G1 : o, G2 : o, D' : nd G1 → nd G2)
              ▷(G1 imp G2/G,
                  impi (λu : nd G1. D'u)/D)
              ↦ let
                  ν (u : nd G1, h : hyp G1)^L.
                  x0 = ndseq G2,
                  x1 = x0 (D' u),
                  ⟨E, _⟩ = x1, ·
                in (impR (λh. E h))
      |  (G1 : o, G2 : o, D1 : nd (G1 imp G2),
           D2 : nd G1)
              ▷ (G2/G, impe D1 D2/D)
              ↦ let
                  x0 = ndseq (G1 imp G2),
                  x1 = x0 D1,
                  ⟨E1, _⟩ = x1,
                  x2 = ndseq (G1),
                  x3 = x2 D2,
                  ⟨E2, _⟩ = x3, ·
                in (cut E1 (λh1. impL E2
                            (λh2. init h2) h1)))
 ∈ ∀G : o. ∀D : nd G. ∃E : conc G. ⊤
```

Fig. 6. Gentzen's algorithm in $\mathcal{T}_\omega^+$

composition. The new case subject is $\eta \circ \sigma$. Each individual case is therefore closed and defined in terms of the pattern ψ, its co-domain Ψ, and a body P. $\mathcal{T}_\omega^+$ also provides a recursion operator μ with the standard static semantics.

$$\begin{array}{ll} \textit{Programs:} & P ::= \ldots \mid \mathsf{case}\ \eta\ \mathsf{of}\ \Omega \mid \mu \mathbf{x} \in F.\, P \\ \textit{Cases:} & \Omega ::= \cdot \mid (\Omega \mid (\Psi \triangleright \psi \mapsto P)) \end{array}$$

The rules defining case analysis and recursion can be found in Figure 5. The premisses of the case rule ensure that the subject of the case program is well-typed as well as Ω. The latter requires a new judgment: $\Psi; \Delta \vdash \Omega \in F$. For the sake of brevity, we have omitted a strictness side condition on alt, that guarantees that matching against ψ instantiates all variables declared in Ψ'.

Figure 6 shows the definition of **ndseq** (from Figure 3) on the left in informal syntax and on the right in $\mathcal{T}_\omega^+$ syntax. For improved readability we omit the type annotations from declarations.

5 Operational Semantics of $\mathcal{T}_\omega^+$

The rules defining the operational semantics are constructed in such a way, that once evaluation is invoked in a regular world Φ its result is well-defined in

$$\frac{}{\Phi \vdash \Lambda x : A.\, P \hookrightarrow \Lambda x : A.\, P}\,\textsf{ev\_lam} \qquad \frac{\Phi \vdash D \hookrightarrow \eta \qquad \Phi \vdash P[\eta] \hookrightarrow V}{\Phi \vdash \mathbf{let}\ D\ \mathbf{in}\ P \hookrightarrow V}\,\textsf{ev\_let}$$

$$\frac{\Phi \vdash P \hookrightarrow V}{\Phi \vdash \langle M, P\rangle \hookrightarrow \langle M, V\rangle}\,\textsf{ev\_inx} \qquad \frac{}{\Phi \vdash \langle\rangle \hookrightarrow \langle\rangle}\,\textsf{ev\_unit}$$

$$\frac{\Phi \vdash P[\mu \mathbf{x} \in F.\, P/\mathbf{x}] \hookrightarrow V}{\Phi \vdash \mu \mathbf{x} \in F.\, P \hookrightarrow V}\,\textsf{ev\_rec} \qquad \frac{\Phi \vdash \eta \sim \Omega \hookrightarrow V}{\Phi \vdash \mathsf{case}\ \eta\ \mathsf{of}\ \Omega \hookrightarrow V}\,\textsf{ev\_case}$$

$$\frac{}{\Phi \vdash \cdot \hookrightarrow \cdot}\,\textsf{ev\_empty} \qquad \frac{\Phi, \rho^L \vdash D \hookrightarrow \eta' \qquad \mathsf{raise}\ \rho.\, \eta' = \eta''}{\Phi \vdash \nu\, \rho^L.\, D \hookrightarrow \eta''}\,\textsf{ev\_new}$$

$$\frac{\Phi \vdash P \hookrightarrow \langle M, V\rangle \qquad \Phi \vdash D[M/x; V/\mathbf{y}] \hookrightarrow \eta'}{\Phi \vdash \langle x : A, \mathbf{y} \in F\rangle = P, D \hookrightarrow M/x, \eta'}\,\textsf{ev\_split}$$

$$\frac{\Phi \vdash P \hookrightarrow \Lambda x : A.\, P' \qquad \Phi \vdash P'[M/x] \hookrightarrow V \qquad \Phi \vdash D[V/\mathbf{y}] \hookrightarrow \eta'}{\Phi \vdash \mathbf{y} \in F = P\ M, D \hookrightarrow \eta'}\,\textsf{ev\_app}$$

$$\frac{\Phi \vdash P[\eta'] \hookrightarrow V \qquad \psi \circ \eta' = \eta}{\Phi \vdash \eta \sim (\Omega, (\Psi \triangleright \psi \mapsto P)) \hookrightarrow V}\,\textsf{ev\_yes} \qquad \frac{\Phi \vdash \eta \sim \Omega \hookrightarrow V}{\Phi \vdash \eta \sim (\Omega, (\Psi \triangleright \psi \mapsto P)) \hookrightarrow V}\,\textsf{ev\_no}$$

Fig. 7. Operational semantics of $\mathcal{T}_\omega^+$

exactly the same world. During evaluation, new parameters can be dynamically introduced and discharged. The operational semantics is call-by-value and it relates programs to be executed with the result of their computation. Altogether, there are three evaluations judgments for programs, declarations, and cases, respectively, each indexed by the world Φ.

Program evaluation:	$\Phi \vdash P \hookrightarrow V$
Declarations evaluation:	$\Phi \vdash D \hookrightarrow \eta$
Selection:	$\Phi \vdash \eta \sim \Omega \hookrightarrow V$

The first judgment relates P with the result of its evaluation. The second returns a list of values in the form of an environment instantiating all variables in Ψ' (the type of D). And the third judgment selects a case from Ω with pattern ψ where η matches ψ and evaluates the respective body to V.

$$\textit{Values: } V ::= \langle\rangle \mid \langle M, V\rangle \mid \Lambda x : A.\, P$$

The rules defining the operational semantics are given in Figure 7. We comment only on the non-standard rule ev_new since it requires all objects summarized in η' to be abstracted to their new abstract types. This is established by a straightforward generalization of abstraction to substitutions 'raise $\rho.\, \eta'$'.

6 Meta Theory

$\mathcal{T}_\omega^+$ is a type system of partial recursive functions. It is defined with respect to the regular world assumption. If we assume that only well-typed programs, declarations, or cases are executed, we can show that the operational semantics is type preserving. See [Sch00] for a proof.

Theorem 1 (Type-preservation).

1. *If* $\Phi \vdash P \hookrightarrow V$ *and* $\Phi; \cdot \vdash P \in F$ *then* $\Phi; \cdot \vdash V \in F$
2. *If* $\Phi \vdash D \hookrightarrow \eta$; *and* $\Phi; \cdot \vdash D \in \Psi$ *then* $\Phi \vdash \cdot, \eta : \Phi, \Psi$
3. *If* $\Phi \vdash \eta \sim \Omega \hookrightarrow V$ *and* $\mathcal{F} :: \Phi \vdash \eta : \Psi$ *and* $\Psi; \cdot \vdash \Omega \in F$ *then* $\Phi; \cdot \vdash V \in F[\eta]$

Recursive functions in $\mathcal{T}_\omega^+$ are partial. Their execution does not always terminate nor does it always make progress. For example, the execution of '$\mu \mathbf{x} \in F. \mathbf{x}$' is non-terminating, the execution of 'case η of $\cdot$' gets always stuck for any environment η.

7 Applications

In our experience many algorithms related to programming languages and logics can be directly represented in $\mathcal{T}_\omega^+$. In the setting of first-order intuitionistic logic, for example, we have encoded an algorithm that transforms sequent derivations with cut into sequent derivations without cut [Pfe95]. In the interest of space, we can only give the definition of the regular world and the types of the two recursive $\mathcal{T}_\omega^+$ functions. The sequent calculus for full first-order intuitionistic logic extends Figure 2 with additional left and right rules for the other connectives. Let 'i' be the LF type for terms, and Φ a regular world of the form

$$\Phi ::= \cdot \mid \Phi, (u : \text{hyp } G)^{L_1} \mid \Phi, (a : \text{i})^{L_2} \mid \Phi, (p : \text{o})^{L_3}.$$

The block labeled L_1 covers extensions of the world by new hypotheses, L_2 by new term parameters, and L_3 by new formulas. We write 'conc$^-$' for the cut-free sequent calculus. **ca** and **ce**, which implement the admissibility and the elimination of cut

$$\mathbf{ca} \in \forall G : \text{o}. \forall H : \text{o}. \forall D_1 : \text{conc}^- \ H. \forall D_2 : \text{hyp } H \to \text{conc}^- \ G. \exists D' : \text{conc}^- \ G. \top$$
$$\mathbf{ce} \in \forall G : \text{o}. \forall D : \text{conc } G. \exists D' : \text{conc}^- \ G. \top$$

have direct and elegant implementations in $\mathcal{T}_\omega^+$.

8 Conclusion

We have presented a type system $\mathcal{T}_\omega^+$ whose partial recursive functions range over (possibly higher-order) objects and are defined by cases and recursion. Most importantly, $\mathcal{T}_\omega^+$ allows recursive calls to traverse λ-binders, and function definitions to match parameters. In addition $\mathcal{T}_\omega^+$ supports dependent types and its functions can take direct advantage of all properties that are associated with the representation in LF, e.g. weakening, contraction, and substitution. These features make $\mathcal{T}_\omega^+$ an excellent candidate for the formalization of algorithms over deductive systems that are encoded in the logical framework LF [HHP93] via higher-order abstract syntax and hypothetical judgments.

$\mathcal{T}_\omega^+$'s concept of partial recursive functions is justified on the basis of the regular world assumption. It successfully resolves the tension between higher-order encodings and function definition by cases and recursion. Once specified, the form of regular worlds limits the ways that parameters are introduced during execution, and accounts for the exact form of a parameter block. Conversion procedures from one logic to another, bracket abstraction, and normalization via parallel reduction are only few of the examples that can be directly and cleanly expressed as recursive functions in $\mathcal{T}_\omega^+$.

Acknowledgment

I want to thank my advisor Frank Pfenning for all his contributions to this work, and comments on earlier versions of this draft.

References

Coq91. Thierry Coquand. An algorithm for testing conversion in type theory. In Gérard Huet and Gordon Plotkin, editors, *Logical Frameworks*, pages 255–279. Cambridge University Press, 1991.

DFH+93. Gilles Dowek, Amy Felty, Hugo Herbelin, Gérard Huet, Chet Murthy, Catherine Parent, Christine Paulin-Mohring, and Benjamin Werner. The Coq proof assistant user's guide. Rapport Techniques 154, INRIA, Rocquencourt, France, 1993. Version 5.8.

DFH95. Joëlle Despeyroux, Amy Felty, and André Hirschowitz. Higher-order abstract syntax in Coq. In M. Dezani-Ciancaglini and G. Plotkin, editors, *Proceedings of the International Conference on Typed Lambda Calculi and Applications*, pages 124–138, Edinburgh, Scotland, April 1995. Springer-Verlag LNCS 902.

DL99. Joelle Despeyroux and Pierre Leleu. Primitive recursion for higher-order abstract syntax with dependent types. In *Proceedings of the Workshop on Intuitionistic Modal Logics and Applications (IMPL'99)*, Trento, Italy, July 1999.

DPS97. Joëlle Despeyroux, Frank Pfenning, and Carsten Schürmann. Primitive recursion for higher-order abstract syntax. In R. Hindley, editor, *Proceedings of the Third International Conference on Typed Lambda Calculus and Applications (TLCA'97)*, pages 147–163, Nancy, France, April 1997. Springer-Verlag LNCS. An extended version is available as Technical Report CMU-CS-96-172, Carnegie Mellon University.

Gen35. Gerhard Gentzen. Untersuchungen über das logische Schließen. *Mathematische Zeitschrift*, 39:176–210, 405–431, 1935.

GP99. Murdoch Gabbay and Andrew Pitts. A new approach to abstract syntax involving binders. In G. Longo, editor, *Proceedings of the 14th Annual Symposium on Logic in Computer Science (LICS'99)*, pages 214–224, Trento, Italy, July 1999. IEEE Computer Society Press.

Hal87. Lars Hallnäs. A note on the logic of a logic program. In *Proceedings of the Workshop on Programming Logic*. University of Göteborg and Chalmers University of Technology, Report PMG-R37, 1987.

HHP93. Robert Harper, Furio Honsell, and Gordon Plotkin. A framework for defining logics. *Journal of the Association for Computing Machinery*, 40(1):143–184, January 1993.

Hof99. Martin Hofmann. Semantical analysis for higher-order abstract syntax. In G. Longo, editor, *Proceedings of the 14th Annual Symposium on Logic in Computer Science (LICS'99)*, pages 204–213, Trento, Italy, July 1999. IEEE Computer Society Press.

Mil90. Dale Miller. An extension to ML to handle bound variables in data structures: Preliminary report. In *Proceedings of the Logical Frameworks BRA Workshop*, Nice, France, May 1990.

MM97. Raymond McDowell and Dale Miller. A logic for reasoning with higher-order abstract syntax: An extended abstract. In Glynn Winskel, editor, *Proceedings of the Twelfth Annual Symposium on Logic in Computer Science*, pages 434–445, Warsaw, Poland, June 1997.

Pau94. Lawrence C. Paulson. *Isabelle: A Generic Theorem Prover*. Springer-Verlag LNCS 828, 1994.

Pfe91. Frank Pfenning. Logic programming in the LF logical framework. In Gérard Huet and Gordon Plotkin, editors, *Logical Frameworks*, pages 149–181. Cambridge University Press, 1991.

Pfe95. Frank Pfenning. Structural cut elimination. In D. Kozen, editor, *Proceedings of the Tenth Annual Symposium on Logic in Computer Science*, pages 156–166, San Diego, California, June 1995. IEEE Computer Society Press.

Pfe99. Frank Pfenning. Logical frameworks. In Alan Robinson and Andrei Voronkov, editors, *Handbook of Automated Reasoning*. Elsevier Science Publishers, 1999. In preparation.

PG00. A. M. Pitts and M. J. Gabbay. A metalanguage for programming with bound names modulo renaming. In R. Backhouse and J. N. Oliveira, editors, *Mathematics of Program Construction, MPC2000, Proceedings, Ponte de Lima, Portugal, July 2000*, volume 1837 of *Lecture Notes in Computer Science*, pages 230–255. Springer-Verlag, Heidelberg, 2000.

PM93. Christine Paulin-Mohring. Inductive definitions in the system Coq: Rules and properties. In M. Bezem and J.F. Groote, editors, *Proceedings of the International Conference on Typed Lambda Calculi and Applications, TLCA'93*, pages 328–345, Utrecht, The Netherlands, March 1993. Springer-Verlag LNCS 664.

Sch00. Carsten Schürmann. *Automating the Meta-Theory of Deductive Systems*. PhD thesis, Carnegie Mellon University, 2000. CMU-CS-00-146.

SH93. Peter Schroeder-Heister. Rules of definitional reflection. In M. Vardi, editor, *Proceedings of the Eighth Annual IEEE Symposium on Logic in Computer Science*, pages 222–232, Montreal, Canada, June 1993.

SP98. Carsten Schürmann and Frank Pfenning. Automated theorem proving in a simple meta-logic for LF. In Claude Kirchner and Hélène Kirchner, editors, *Proceedings of the 15th International Conference on Automated Deduction (CADE-15)*, pages 286–300, Lindau, Germany, July 1998. Springer-Verlag LNCS 1421.

Vir99. Roberto Virga. *Higher-Order Rewriting with Dependent Types*. PhD thesis, Department of Mathematical Sciences, Carnegie Mellon University, 1999. Forthcoming.

Monotone Inductive and Coinductive Constructors of Rank 2

Ralph Matthes

Institut für Informatik der Ludwig-Maximilians-Universität München
Oettingenstraße 67, D-80538 München, Germany
matthes@informatik.uni-muenchen.de

Abstract. A generalization of positive inductive and coinductive types to monotone inductive and coinductive constructors of rank 1 and rank 2 is described. The motivation is taken from initial algebras and final coalgebras in a functor category and the Curry-Howard-correspondence. The definition of the system as a λ-calculus requires an appropriate definition of monotonicity to overcome subtle problems, most notably to ensure that the (co-)inductive constructors introduced via monotonicity of the underlying constructor of rank 2 are also monotone as constructors of rank 1. The problem is solved, strong normalization shown, and the notion proven to be wide enough to cover even highly complex datatypes.

1 Introduction

Only in the last couple of years, the functional programming community has become interested in heterogeneous datatypes, i. e., polymorphic datatypes whose constructors relate different instances of the datatype – unlike the lists where cons takes a list of some type and produces a list of the same type. One might wonder whether there are interesting examples. It is by now well-known that de Bruijn notation indeed may be represented as a nested datatype [3, 2]. However, the question arose whether there are also examples taken from "the outside world". In [12], a nested datatype is shown which represents arbitrary square matrices over some type with elements accessible in logarithmic time. It uses type constructors of rank 2. On the theoretical side, [1] studies even coinductive type constructors of rank 2 with considerable nesting.

The present paper intends to give a thoroughly justified general framework for the description of terminating algorithms involving inductive and coinductive constructors of rank 2. It tries to be as general as possible while not departing from higher-order parametric polymorphism which – as a logical system – is higher-order intuitionistic propositional logic, expressed in natural deduction style. This (so-called Curry-Howard-)correspondence allows to insert a logical understanding into the system design: Inductive constructors $\mu F\mathcal{F}$ and $\nu F\mathcal{F}$ are not restricted to some kind of positivity, i. e., to requirements that free occurrences of F in $\mathcal{F}$ are only allowed an even number to the left of $\to$. Positivity is replaced by proven monotonicity: a term inhabiting the type expressing that $\lambda F\mathcal{F}$ is monotone. For rank 1 (inductive types) this has been studied at length

L. Fribourg (Ed.): CSL 2001, LNCS 2142, pp. 600–614, 2001.

in [10] – with the obvious notion of monotonicity expressible in system F. The main technical contribution is the definition of rank 2 monotonicity which will be justified at many places in this article.

The next section shortly reviews system F^ω, the long section 3 proposes and discusses the extension of F^ω to the system MICC^2 of monotonone inductive and coinductive constructors of rank 1 and rank 2. One of the most crucial properties is the subject of section 4: $\mu F\mathcal{F}$ and $\nu F\mathcal{F}$ are monotone if $\lambda F\mathcal{F}$ is. This result is not needed, however, for the reduction-preserving embedding into F^ω, shown in section 5. MICC^2's strong normalization follows from that of F^ω. Section 6 gives the idea of an extension even to primitive recursion of rank 2 while many closure properties of monotone constructors are shown in section 7. A few examples are considered in section 8.

2 System F^ω

Girard's system of higher-order parametric polymorphism [7] is an extension of system F where the set of types is extended to a simply-"typed" λ-calculus of constructors. The "types" of the constructors are called kinds and are built from the base kind $*$ (the universe of types) and the binary $\Rightarrow$ intended to form function kinds. We denote kinds by κ (possibly with indices). The constructors all have a kind and are built from kinded constructor variables as shown below. We use the following conventions: Constructor variables are denoted by $X, Y, Z, \ldots$ and constructors by $\mathcal{X}, \mathcal{Y}, \mathcal{Z}, \ldots$, constructors of kind $*$ are called types, and constructor variables of kind $*$ are (consequently) called type variables. They are denoted by $\alpha, \beta, \gamma, \ldots$ and types by $\rho, \sigma, \tau, \ldots$ The phrase "$\mathcal{X}$ is a constructor of kind κ" will be shortened to $\mathcal{X} : \kappa$, later also with superscript, i. e., $\mathcal{X}^\kappa$. The constructors are given inductively by:

- $X^\kappa : \kappa$.
- If $\mathcal{X} : \kappa_2$ then $\lambda X^{\kappa_1}\mathcal{X} : \kappa_1 \Rightarrow \kappa_2$.
- If $\mathcal{X} : \kappa_1 \Rightarrow \kappa_2$ and $\mathcal{Y} : \kappa_1$ then $\mathcal{X}\mathcal{Y} : \kappa_2$.
- $\rho \to \sigma : *$, $\rho \times \sigma : *$, $\rho + \sigma : *$, $\forall X^\kappa \rho : *$.

Let $=_\beta$ be the (decidable) congruence closure of $(\lambda X^{\kappa_1}\mathcal{X}^{\kappa_2})\mathcal{Y}^{\kappa_1} =_\beta \mathcal{X}[X := \mathcal{Y}]$. We identify constructors which are equal w. r. t. $=_\beta$.

Terms are built from typed variables and have a type (term variables are denoted x, y, z, u, v, f and terms $r, s, t, \ldots$ The phrase "r is a term of type ρ" will be shortened to $r : \rho$, later also with superscript, i. e., r^ρ):

- $x^\rho : \rho$.
- If $r : \rho$ then $\Lambda X^\kappa r : \forall X^\kappa \rho$, provided X does not occur free in the types of the free term variables of r.
- If $r : \sigma$ then $\lambda x^\rho r : \rho \to \sigma$.
- If $r : \rho$ and $s : \sigma$ then $\langle r, s\rangle : \rho \times \sigma$.
- If $r : \forall X^\kappa \rho$ then $r\mathcal{X}^\kappa : \rho[X := \mathcal{X}]$.
- If $r : \rho \to \sigma$ and $s : \rho$ then $rs : \sigma$.
- If $r : \rho \times \sigma$ then $r\mathsf{L} : \rho$ and $r\mathsf{R} : \sigma$.

Finally, β-reduction $\rhd$ on terms is defined as the term closure of the following rules: $(\Lambda X^\kappa r)\mathcal{X} \rhd r[X := \mathcal{X}]$, $(\lambda x^\rho r)s \rhd r[x := s]$, $\langle r, s\rangle\mathsf{L} \rhd r$ and $\langle r, s\rangle\mathsf{R} \rhd s$.

Without being specific, we also assume to have term rules and β-reduction rules for the sum types $\rho + \sigma$ (those rules will not be used in the sequel since proofs pertaining to them had to be omitted).

$\Rightarrow$ for kinds and $\rightarrow$ for types shall associate to the right, while constructor application and term application are considered to be parenthesized to the left.

The rank of a kind is defined as expected:

$$\begin{aligned} \mathrm{rk}(*) &:= 0 \\ \mathrm{rk}(\kappa_1 \Rightarrow \kappa_2) &:= \max(\mathrm{rk}(\kappa_1) + 1, \mathrm{rk}(\kappa_2)) \end{aligned}$$

The rank of a constructor is defined to be the rank of its kind.

3 Definition of System MICC^2

System F^ω is extended by monotone inductive and coinductive constructors of rank 1 and rank 2, hence the name MICC^2.

Constructor variables of kind $* \Rightarrow *$ are denoted by $F, G, H, \ldots$ and constructors of kind $* \Rightarrow *$ by $\mathcal{F}, \mathcal{G}, \mathcal{H}, \ldots$

Define the abbreviation $\mathcal{F} \leq \mathcal{G} := \forall\alpha.\mathcal{F}\alpha \rightarrow \mathcal{G}\alpha$. We will only need monotonicity for the "pure" kinds $* \Rightarrow *$ and $(* \Rightarrow *) \Rightarrow * \Rightarrow *$ of rank 1 and 2, respectively. Monotonicity of constructors can and will simply be expressed by the definition of the following *types* (hence $\mathcal{F}\,\mathrm{mon}$ and $\mathcal{X}\,\mathrm{mon}$ are nothing but abbreviations for certain types):

$$\mathcal{F}^{* \Rightarrow *}\,\mathrm{mon} := \forall\alpha\forall\beta.(\alpha \rightarrow \beta) \rightarrow \mathcal{F}\alpha \rightarrow \mathcal{F}\beta$$

$$\boxed{\begin{aligned} &\mathcal{X}^{(* \Rightarrow *) \Rightarrow * \Rightarrow *}\,\mathrm{mon} := \Big(\forall F.F\,\mathrm{mon} \rightarrow \mathcal{X}F\,\mathrm{mon}\Big) \times \\ &\qquad \Big(\forall F \forall G.F \leq G \rightarrow (F\,\mathrm{mon} \rightarrow \mathcal{X}F \leq \mathcal{X}G) \times (G\,\mathrm{mon} \rightarrow \mathcal{X}F \leq \mathcal{X}G)\Big) \end{aligned}}$$

There is no doubt that for rank 1 this is the expected and only reasonable definition. In contrast to that, there is a lot of freedom in giving a monotonicity definition for rank 2. Our definition is designed to meet the following criteria:

1. If F occurs only "positively" in some $\mathcal{F}$ then $\lambda F\mathcal{F}\,\mathrm{mon}$ should be inhabited by one and only one "canonical" term.
2. Iteration and coiteration should be driven by the same monotonicity witnesses.
3. If $\lambda F\mathcal{F}\,\mathrm{mon}$ is inhabited then also $(\mu F\mathcal{F})^{* \Rightarrow *}\,\mathrm{mon}$ and $(\nu F\mathcal{F})^{* \Rightarrow *}\,\mathrm{mon}$ should be inhabited in a "canonical" way.

However, the definition is slightly blown up in comparison with the following – intuitively isomorphic – possibility:[1]

$$\Big(\forall F.F\,\mathrm{mon} \rightarrow \mathcal{X}F\,\mathrm{mon}\Big) \times \Big(\forall F \forall G.F \leq G \rightarrow (F\,\mathrm{mon} + G\,\mathrm{mon}) \rightarrow \mathcal{X}F \leq \mathcal{X}G\Big)$$

[1] Clearly, the naive definition of monotonicity would have asserted $F\,\mathrm{mon} \times G\,\mathrm{mon}$ instead of $F\,\mathrm{mon} + G\,\mathrm{mon}$.

Both definitions are intuitively isomorphic since, in general, $\rho_1 + \rho_2 \to \sigma$ and $(\rho_1 \to \sigma) \times (\rho_2 \to \sigma)$ are. However, in applications we would need to refer to full extensional equality for sum types (even beyond permutative conversions) which is still decidable [6] but difficult to treat since it leaves the realm of rewriting. Since our focus is on intensional equality as expressed by term rewriting, the slightly less elegant definition had to be made.

First, we extend system F^ω by monotone inductive constructors of rank 1, i.e., inductive types $\mu\alpha\rho$ depending on the monotonicity of the constructor $\lambda\alpha\rho$ of rank 1. This is exactly done as for the iterative part of the elimination-based system EMIT of monotone inductive types in [10]: For every type ρ we assume the type $\mu\alpha\rho$ in the system of constructors (and that $\mu\alpha$ binds every free occurrence of α in ρ).[2] The type $\mu\alpha\rho$ is called a monotone inductive constructor of rank 1 (or monotone inductive type) although its *formation* does not at all rest on a monotonicity assumption. This notion is justified since it is the usage of those constructors as specified in the *term* formation rules which constitutes a constructor concept. In our case, this is made clear in the introduction rule (corresponding to fold in the functional programming literature) since we introduce the type $\mu\alpha\rho$ by

$$\frac{m : \lambda\alpha\rho\,\mathsf{mon} \qquad t : \rho[\alpha := \mu\alpha\rho]}{\mathsf{C}_{\mu\alpha\rho} m t : \mu\alpha\rho}\,(\mu\text{-}I)$$

(This rule is – following the standard practice in type theory – to be read as: If m is a term of type $\lambda\alpha\rho\,\mathsf{mon}$ and t is a term of type $\rho[\alpha := \mu\alpha\rho]$ then $\mathsf{C}_{\mu\alpha\rho} m t$ is a term of type $\mu\alpha\rho$. Thus, $\mathsf{C}_{\mu\alpha\rho}$ is considered as a binary function symbol.)

If there is a closed $m : \lambda\alpha\rho\,\mathsf{mon}$, we call $\mu\alpha\rho$ a *proven* monotone inductive constructor of rank 1 (or proven monotone inductive type) and m a monotonicity *witness* of $\lambda\alpha\rho$. The type $\mu\alpha\rho$ is eliminated by an iterative concept:

$$\frac{r : \mu\alpha\rho \qquad s : \rho[\alpha := \sigma] \to \sigma}{r\mathsf{E}_\mu s : \sigma}\,(\mu\text{-}E)$$

For $s : \rho[\alpha := \sigma] \to \sigma$ define $\mathcal{E}_\mu s := \lambda x^{\mu\alpha\rho}.x\mathsf{E}_\mu s : \mu\alpha\rho \to \sigma$.[3]

The β-rule of *iteration* is

$$(\mathsf{C}_{\mu\alpha\rho} m t)\mathsf{E}_\mu s \rhd s\Big(m(\mu\alpha\rho)\sigma(\mathcal{E}_\mu s)t\Big)$$

These rules obviously generalize the system of positive inductive types, e. g., system $2\lambda\mathsf{J}$ of Leivant [9] in which $\mu\alpha\rho$ is only allowed if α only occurs "positively" in ρ. In our language, all those $\mu\alpha\rho$ are proven monotone inductive types.[4] Therefore, the iteration rule for $2\lambda\mathsf{J}$ simply makes use of those predetermined monotonicity witnesses. In [10, 4.2.2 and 5.1.1], the embedding of (a variant of)

[2] Alternatively, one could have assumed a constructor constant $\mu : (* \Rightarrow *) \Rightarrow *$ of rank 2 and set $\mu\alpha\rho := \mu(\lambda\alpha\rho)$.

[3] Why don't we introduce $\mathcal{E}_\mu s$ as official syntax for the μ-elimination rule? Because we favour natural deduction style as opposed to Hilbert style. For studies on fixed-points in λ-calculi this seems to be a reasonable choice, see also [15].

[4] This is an easy consequence of theorem 3 below.

2λJ into the present system is carried out in all details (even with primitive recursion instead of iteration only).

A more categorical motivation (still with fixed monotonicity witnesses) may be found in [5]: $\mu\alpha\rho$ may be conceived as a weakly initial $\lambda\alpha\rho$-algebra (for rank 2, this is carried out below).

Now we also extend the system by monotone inductive constructors of rank 2, i. e., inductive constructors $\mu F\mathcal{F}$ depending on the monotonicity of the constructor $\lambda F\mathcal{F}$ of rank 2. The constructor $\mu F\mathcal{F}$ has kind $* \Rightarrow *$. Hence, as a constructor, $\mu F\mathcal{F}$ is of rank 1, but it is called a monotone inductive constructor of rank 2. More precisely, we extend the system of constructors by another constructor formation rule: If $\mathcal{F}$ is a constructor of kind $* \Rightarrow *$ then $\mu F\mathcal{F}$ is a constructor of kind $* \Rightarrow *$, and free occurrences of F in $\mathcal{F}$ are bound by μF.[5] The constructor $\mu F\mathcal{F}$ is called a monotone inductive constructor of rank 2 although again its formation does not at all rest on a monotonicity assumption. Since $\mu F\mathcal{F}$ is not a type, the introduction and eliminination rules describe how to introduce and how to eliminate the types $(\mu F\mathcal{F})\rho$ for arbitrary ρ. We start with the introduction rule:

$$\frac{m : \lambda F\mathcal{F}\,\mathrm{mon} \qquad t : \mathcal{F}[F := \mu F\mathcal{F}]\rho}{\mathsf{C}_{\mu F\mathcal{F}}mt : (\mu F\mathcal{F})\rho}\,(\mu{-}I)$$

For $m : \lambda F\mathcal{F}\,\mathrm{mon}$ define

$$\mathcal{C}_{\mu F\mathcal{F}}m := \Lambda\alpha\lambda x^{\mathcal{F}[F:=\mu F\mathcal{F}]\alpha}.\mathsf{C}_{\mu F\mathcal{F}}mx : \mathcal{F}[F := \mu F\mathcal{F}] \leq \mu F\mathcal{F}.$$

If there is a closed $m : \lambda F\mathcal{F}\,\mathrm{mon}$, we call $\mu F\mathcal{F}$ a *proven* monotone inductive constructor of rank 2 and m a monotonicity *witness* of $\lambda F\mathcal{F}$.[6]

The elimination rule (corresponding to an iterator) is:

$$\frac{r : (\mu F\mathcal{F})\rho \qquad n : \mathcal{G}\,\mathrm{mon} \qquad s : \mathcal{F}[F := \mathcal{G}] \leq \mathcal{G}}{r\mathsf{E}_\mu ns : \mathcal{G}\rho}\,(\mu{-}E)$$

For $n : \mathcal{G}\,\mathrm{mon}$ and $s : \mathcal{F}[F := \mathcal{G}] \leq \mathcal{G}$ define

$$\mathcal{E}_\mu ns := \Lambda\alpha\lambda x^{(\mu F\mathcal{F})\alpha}.x\mathsf{E}_\mu ns : \mu F\mathcal{F} \leq \mathcal{G}.$$

The β-rule of *iteration* for rank 2 is

$$(\mathsf{C}_{\mu F\mathcal{F}}mt)\mathsf{E}_\mu ns \rhd s\rho\Big(m\mathsf{R}(\mu F\mathcal{F})\mathcal{G}(\mathcal{E}_\mu ns)\mathsf{R}n\rho t\Big)$$

Some comments are in order. First we collect the information contained in the term rules while ignoring the terms themselves. We get the following inference rules:

[5] Analogous to rank 1, we could have assumed a constant $\mu : ((* \Rightarrow *) \Rightarrow * \Rightarrow *) \Rightarrow * \Rightarrow *$ of rank 3 and set $\mu F\mathcal{F} := \mu(\lambda F\mathcal{F})$.

[6] In Theorem 1 below it is proved that if $\mu F\mathcal{F}$ is a proven monotone constructor of rank 2, then $\mu F\mathcal{F}$ – as a constructor of rank 1 – is monotone (of rank 1), and hence $\mu\alpha.(\mu F\mathcal{F})\alpha$ becomes a proven monotone inductive constructor of rank 1 which, as a constructor, is simply a type, hence of rank 0.

$$\frac{\lambda F \mathcal{F}\,\mathsf{mon}}{\mathcal{F}[F := \mu F\mathcal{F}] \leq \mu F\mathcal{F}}(\mu\text{-}I) \quad \text{and} \quad \frac{\mathcal{G}\,\mathsf{mon} \qquad \mathcal{F}[F := \mathcal{G}] \leq \mathcal{G}}{\mu F\mathcal{F} \leq \mathcal{G}}(\mu\text{-}E)$$

This means that $\mu F\mathcal{F}$ is a pre-fixed-point of $\lambda F\mathcal{F}$ (provided $\lambda F\mathcal{F}$ is monotone) and that it is less than or equal to any monotone pre-fixed-point $\mathcal{G}$ of $\lambda F\mathcal{F}$ (whose monotonicity is not required this time). In case $\lambda F\mathcal{F}$ is monotone, we will prove below (see footnote 6) that $\mu F\mathcal{F}$ is monotone, too, whence $\mu F\mathcal{F}$ becomes a minimal monotone pre-fixed-point of $\lambda F\mathcal{F}$. Thus, $\mu F\mathcal{F}$ follows the lattice-theoretic understanding of inductive definitions. But there are also the terms and the term rewrite rule of iteration. Viewing category theory as constructive lattice theory, we can justify the iteration rule by reference to the concept of a weakly initial algebra as has been done for inductive *types* in [5], mentioned earlier. We will do this only rather sloppily. The "category" consists of the monotone constructors of kind $* \Rightarrow *$, the "morphisms" from $\mathcal{F}$ to $\mathcal{G}$ being the terms of type $\mathcal{F} \leq \mathcal{G}$, and the "composition" being the functional composition of terms. An "endofunctor" is a monotone $\lambda F\mathcal{F}$, in this situation a "$\lambda F\mathcal{F}$-algebra" is a monotone $\mathcal{G}$ together with a term $s : \mathcal{F}[F := \mathcal{G}] \leq \mathcal{G}$. The constructor $\mu F\mathcal{F}$ now turns out to be a *weakly*[7] initial $\lambda F\mathcal{F}$-algebra, i. e., $\mu F\mathcal{F}$ together with $\mathcal{C}_{\mu F\mathcal{F}}m$ is a $\lambda F\mathcal{F}$-algebra – as expressed by the introduction rule (concerning monotonicity of $\mu F\mathcal{F}$, see footnote 6 once more). And, whenever $\mathcal{G}$ and s form a $\lambda F\mathcal{F}$-algebra, which amounts to monotonicity of $\mathcal{G}$ – witnessed by a term $n : \mathcal{G}\,\mathsf{mon}$ – and $s : \mathcal{F}[F := \mathcal{G}] \leq \mathcal{G}$, then there is the term $\mathcal{E}_\mu ns$ which is a morphism from $\mu F\mathcal{F}$ to $\mathcal{G}$, such that the following diagram "commutes":

$$\begin{array}{ccc}
\mathcal{F}[F := \mu F\mathcal{F}] & \xrightarrow{\mathcal{C}_{\mu F\mathcal{F}}m} & \mu F\mathcal{F} \\
{\scriptstyle \mathcal{F}[F:=\mathcal{E}_\mu ns]}\downarrow & & \downarrow{\scriptstyle \mathcal{E}_\mu ns} \\
\mathcal{F}[F := \mathcal{G}] & \xrightarrow[s]{} & \mathcal{G}
\end{array}$$

The expression $\mathcal{F}[F := \mathcal{E}_\mu ns]$ shall denote the action of the functor $\lambda F\mathcal{F}$ on the morphism $\mathcal{E}_\mu ns$. The most natural understanding is given by the following definition: $\mathcal{F}[F := \mathcal{E}_\mu ns] := m\mathsf{R}(\mu F\mathcal{F})\mathcal{G}(\mathcal{E}_\mu ns)\mathsf{R}n$.[8] Therefore, commutation of the diagram means that for any type ρ and any term $t : \mathcal{F}[F := \mu F\mathcal{F}]\rho$, we have

$$\mathcal{E}_\mu ns\rho(\mathcal{C}_{\mu F\mathcal{F}}m\rho t) = s\rho\Big(m\mathsf{R}(\mu F\mathcal{F})\mathcal{G}(\mathcal{E}_\mu ns)\mathsf{R}n\rho t\Big).$$

Obviously, the left-hand side is β-equivalent (using only rules already present in system F) with the left-hand side of the iteration rule, and the right-hand side is that of the iteration rule. Subject reduction will now also be clear, i. e., the right-hand side of the iteration rule receives the same type as the left-hand side.

By dualization, we also introduce monotone coinductive constructors of rank 1 and rank 2: Let also $\nu\alpha\rho$ be a type and $\nu F\mathcal{F}$ be a constructor of kind $* \Rightarrow *$ (with

[7] Full initiality includes a uniqueness statement and would lead to an extensional notion of equality.

[8] Notice how our definition of monotonicity for rank 2 allows to avoid the reference to the monotonicity of $\mu F\mathcal{F}$.

binding as for $\mu\alpha\rho$ and $\mu F\mathcal{F}$). The type $\nu\alpha\rho$ is called a monotone coinductive constructor of rank 1 (a monotone coinductive type), and $\nu F\mathcal{F}$ is called a monotone coinductive constructor of rank 2. The term formation rules pertaining to $\nu\alpha\rho$ are an elimination rule (corresponding to unfold in the functional programming literature) and a mechanism for coiteration. We start with the elimination rule:

$$\frac{r : \nu\alpha\rho \qquad m : \lambda\alpha\rho\,\mathrm{mon}}{r\mathsf{E}_\nu m : \rho[\alpha := \nu\alpha\rho]}(\nu\text{-}E)$$

If there is a monotonicity witness of $\lambda\alpha\rho$, we call $\nu\alpha\rho$ a *proven* monotone coinductive constructor of rank 1 (or proven monotone coinductive type). Terms of type $\nu\alpha\rho$ are introduced by a coiterative concept:

$$\frac{s : \sigma \to \rho[\alpha := \sigma] \qquad t : \sigma}{\mathsf{C}_{\nu\alpha\rho} st : \nu\alpha\rho}(\nu\text{-}I)$$

For $s : \sigma \to \rho[\alpha := \sigma]$ define $\mathcal{C}_{\nu\alpha\rho} s := \lambda x^\sigma.\mathsf{C}_{\nu\alpha\rho} sx : \sigma \to \nu\alpha\rho$.

The β-rule of *coiteration* is

$$(\mathsf{C}_{\nu\alpha\rho} st)\mathsf{E}_\nu m \rhd m\sigma(\nu\alpha\rho)(\mathcal{C}_{\nu\alpha\rho} s)(st)$$

Rank 2 is treated analogously. The elimination rule is

$$\frac{r : (\nu F\mathcal{F})\rho \qquad m : \lambda F\mathcal{F}\,\mathrm{mon}}{r\mathsf{E}_\nu m : \mathcal{F}[F := \nu F\mathcal{F}]\rho}(\nu\text{-}E)$$

For $m : \lambda F\mathcal{F}\,\mathrm{mon}$ define

$$\mathcal{E}_\nu m := \Lambda\alpha\lambda x^{(\nu F\mathcal{F})\alpha}.x\mathsf{E}_\nu m : \nu F\mathcal{F} \le \mathcal{F}[F := \nu F\mathcal{F}].$$

If there is a monotonicity witness of $\lambda F\mathcal{F}$, we call $\nu F\mathcal{F}$ a *proven* monotone coinductive constructor of rank 2.

The types $(\nu F\mathcal{F})\rho$ are introduced via:

$$\frac{n : \mathcal{G}\,\mathrm{mon} \qquad s : \mathcal{G} \le \mathcal{F}[F := \mathcal{G}] \qquad t : \mathcal{G}\rho}{\mathsf{C}_{\nu F\mathcal{F}} nst : (\nu F\mathcal{F})\rho}(\nu\text{-}I)$$

For $n : \mathcal{G}\,\mathrm{mon}$ and $s : \mathcal{G} \le \mathcal{F}[F := \mathcal{G}]$ define

$$\mathcal{C}_{\nu F\mathcal{F}} ns := \Lambda\alpha\lambda x^{\mathcal{G}\alpha}.\mathsf{C}_{\nu F\mathcal{F}} nsx : \mathcal{G} \le \nu F\mathcal{F}.$$

Finally, the β-rule of *coiteration* for rank 2 becomes:

$$(\mathsf{C}_{\nu F\mathcal{F}} nst)\mathsf{E}_\nu m \rhd m\mathsf{R}\mathcal{G}(\nu F\mathcal{F})(\mathcal{C}_{\nu F\mathcal{F}} ns)\mathsf{L}n\rho(s\rho t)$$

It is again worth studying these notions in the light of category theory. Firstly, the system without terms has the following rules:

$$\frac{\lambda F\mathcal{F}\,\mathrm{mon}}{\nu F\mathcal{F} \le \mathcal{F}[F := \nu F\mathcal{F}]}(\nu\text{-}E) \qquad \text{and} \qquad \frac{\mathcal{G}\,\mathrm{mon} \qquad \mathcal{G} \le \mathcal{F}[F := \mathcal{G}]}{\mathcal{G} \le \nu F\mathcal{F}}(\nu\text{-}I)$$

Hence, $\nu F\mathcal{F}$ is a post-fixed-point of $\lambda F\mathcal{F}$ if $\lambda F\mathcal{F}$ is monotone and that it is greater or equal to any monotone post-fixed-point of $\lambda F\mathcal{F}$ (even without the assumption of $\lambda F\mathcal{F}$'s monotonicity). Since $\nu F\mathcal{F}$ is also monotone (see theorem 1), $\nu F\mathcal{F}$ is a maximal monotone post-fixed-point of $\lambda F\mathcal{F}$ provided $\lambda F\mathcal{F}$ is monotone. Therefore, it is appropriate to call $\nu F\mathcal{F}$ a coinductive constructor. Concerning terms and term rewrite rules, if we again have an endofunctor, i. e., a monotone $\lambda F\mathcal{F}$, a "$\lambda F\mathcal{F}$-coalgebra" is a monotone constructor $\mathcal{G}$ (which constructively means that we have a term $n : \mathcal{G}\,\mathrm{mon}$) together with a term $s : \mathcal{G} \leq \mathcal{F}[F := \mathcal{G}]$. The constructor $\nu F\mathcal{F}$ is a weakly final $\lambda F\mathcal{F}$-coalgebra: It is a $\lambda F\mathcal{F}$-coalgebra and for any $\lambda F\mathcal{F}$-coalgebra, given by $\mathcal{G}$, $n : \mathcal{G}\,\mathrm{mon}$ and s, the term $\mathcal{C}_{\nu F\mathcal{F}}ns$ is a morphism from $\mathcal{G}$ to $\nu F\mathcal{F}$ such that the following diagram commutes:

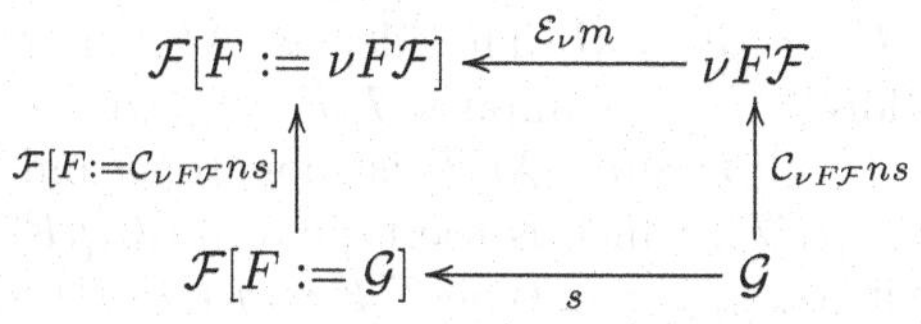

$$\begin{array}{ccc}
\mathcal{F}[F := \nu F\mathcal{F}] & \xleftarrow{\;\mathcal{E}_\nu m\;} & \nu F\mathcal{F} \\
{\scriptstyle \mathcal{F}[F:=\mathcal{C}_{\nu F\mathcal{F}}ns]}\uparrow & & \uparrow{\scriptstyle \mathcal{C}_{\nu F\mathcal{F}}ns} \\
\mathcal{F}[F := \mathcal{G}] & \xleftarrow[\;s\;]{} & \mathcal{G}
\end{array}$$

This time, the easiest definition of the action of the functor on such a morphism is given by $\mathcal{F}[F := \mathcal{C}_{\nu F\mathcal{F}}ns] := m\mathsf{R}\mathcal{G}(\nu F\mathcal{F})(\mathcal{C}_{\nu F\mathcal{F}}ns)\mathsf{L}n$ since it avoids using $\nu F\mathcal{F}$'s monotonicity. Hence, the possibility to choose whether to use the monotonicity of the first or second argument to $m\mathsf{R}$ is vital to fulfill criterion 2 in the list on page 602.

Observe that the comparison of the figures displays the duality between $\mu F\mathcal{F}$ and $\nu F\mathcal{F}$ much better than that of the rules of iteration and coiteration.

This completes the definition of the extension of system F^ω by monotone inductive and coinductive constructors of rank 1 and rank 2, henceforth called system MICC^2. (Although the different extensions have been described separately, it is understood that they are meant to be done simultaneously which implies, e. g., the presence of the type $\mu\alpha.(\nu F\mathcal{F})\alpha$ in our system for arbitrary constructors $\mathcal{F}$.)

4 Monotone (Co-)Inductive Constructors are Monotone

Monotone inductive and monotone coinductive constructors of rank 2 are constructors $\mu F\mathcal{F}$ and $\nu F\mathcal{F}$ having rank 1. Therefore, we may ask whether they are in turn monotone (of rank 1) if $\lambda F\mathcal{F}$ were monotone (of rank 2). This is answered in the affirmative, and consequently, e. g., $\nu\alpha.(\mu F\mathcal{F})\alpha$ is a proven monotone coinductive type if $\mu F\mathcal{F}$ is a proven monotone inductive constructor.

Theorem 1. *There are closed terms* $M_{\mu F\mathcal{F}} : \lambda F\mathcal{F}\,\mathrm{mon} \to \mu F\mathcal{F}\,\mathrm{mon}$ *and* $M_{\nu F\mathcal{F}} : \lambda F\mathcal{F}\,\mathrm{mon} \to \nu F\mathcal{F}\,\mathrm{mon}$.

This shows that criterion 3 in the list on page 602 is met (the construction is "canonical" since the monotonicity witness is not analyzed at all – it is merely fed in as an argument[9]).

Proof. We argue slightly informally by assuming a term $m : \lambda F\mathcal{F}$ mon and showing $\mu F\mathcal{F}$ mon and $\nu F\mathcal{F}$ mon using m. We begin with the inductive constructor. Introduce $\mu F\mathcal{F}$'s "positivization" $\mathcal{G} := \lambda\alpha\forall\beta.(\alpha \to \beta) \to (\mu F\mathcal{F})\beta$. In order to prove monotonicity of $\mu F\mathcal{F}$, it suffices to show that $\mu F\mathcal{F} \leq \mathcal{G}$. Of course, this is done by proving that $\mathcal{G}$ gives rise to a $\lambda F\mathcal{F}$-algebra. We have the canonical monotonicity witness $n := \Lambda\alpha\Lambda\gamma\lambda f^{\alpha\to\gamma}\lambda x^{\mathcal{G}\alpha}\Lambda\beta\lambda y^{\gamma\to\beta}.x\beta(\lambda v^\alpha.y(fv)) : \mathcal{G}$ mon (α occurs only non-strictly positively in $\forall\beta.(\alpha \to \beta) \to (\mu F\mathcal{F})\beta$). Abbreviate $\ell := \Lambda\alpha\lambda x^{\mathcal{G}\alpha}.x\alpha(\lambda y^\alpha y) : \mathcal{G} \leq \mu F\mathcal{F}$ and

$$s := \Lambda\alpha\lambda x^{\mathcal{F}[F:=\mathcal{G}]\alpha}\Lambda\beta\lambda f^{\alpha\to\beta}.\mathsf{C}_{\mu F\mathcal{F}}m(m\mathsf{R}\mathcal{G}(\mu F\mathcal{F})\ell\mathsf{L}n\beta(m\mathsf{L}\mathcal{G}n\alpha\beta fx))$$

which has type $\mathcal{F}[F := \mathcal{G}] \leq \mathcal{G}$. The idea is: $\lambda F\mathcal{F}$ is monotone, hence also $\mathcal{F}[F := \mathcal{G}]$. Therefore, $\alpha \to \beta$ implies $\mathcal{F}[F := \mathcal{G}]\alpha \to \mathcal{F}[F := \mathcal{G}]\beta$. Thus, $m\mathsf{L}\mathcal{G}n\alpha\beta fx : \mathcal{F}[F := \mathcal{G}]\beta$. Since $\lambda F\mathcal{F}$ is monotone and $\mathcal{G} \leq \mu F\mathcal{F}$, we have $\mathcal{F}[F := \mathcal{G}] \leq \mathcal{F}[F := \mu F\mathcal{F}]$, which is the type of $m\mathsf{R}\mathcal{G}(\mu F\mathcal{F})\ell\mathsf{L}n$. Consequently, the second argument to $\mathsf{C}_{\mu F\mathcal{F}}$ gets type $\mathcal{F}[F := \mu F\mathcal{F}]\beta$, hence the term starting with $\mathsf{C}_{\mu F\mathcal{F}}$ has type $(\mu F\mathcal{F})\beta$. Since $x\mathsf{E}_\mu ns$ has type $\mathcal{G}\alpha$, the proof is completed by $\Lambda\alpha\Lambda\beta\lambda f^{\alpha\to\beta}\lambda x^{(\mu F\mathcal{F})\alpha}.x\mathsf{E}_\mu ns\beta f : \mu F\mathcal{F}$ mon.

We turn to the coinductive constructor. $\mathcal{G}$ is a positivization of $\mu F\mathcal{F}$ which is less than or equal to $\mu F\mathcal{F}$. Dually, we seek for a positivization of $\nu F\mathcal{F}$ which is greater than or equal to $\nu F\mathcal{F}$.[10] Instead of taking $\lambda\beta\exists\alpha.(\alpha \to \beta) \times (\nu F\mathcal{F})\alpha$, we avoid the $\exists$ type and set $\mathcal{H} := \lambda\beta\forall\gamma.\Big(\forall\alpha.(\alpha \to \beta) \to (\nu F\mathcal{F})\alpha \to \gamma\Big) \to \gamma$. We will recycle the names n, ℓ, s. Since $\mathcal{H}$ is also non-strictly positive, it has a canonical mononicity witness

$$n := \Lambda\beta\Lambda\delta\lambda f^{\beta\to\delta}\lambda x^{\mathcal{H}\beta}\Lambda\gamma\lambda y^{\forall\alpha.(\alpha\to\delta)\to(\nu F\mathcal{F})\alpha\to\gamma}.x\gamma(\Lambda\alpha\lambda u^{\alpha\to\beta}.y\alpha(\lambda v^\alpha.f(uv)))$$

Set $\ell := \Lambda\beta\lambda x^{(\nu F\mathcal{F})\beta}\Lambda\gamma\lambda z^{\forall\alpha.(\alpha\to\beta)\to(\nu F\mathcal{F})\alpha\to\gamma}.z\beta(\lambda y^\beta y)x : \nu F\mathcal{F} \leq \mathcal{H}$ and show that we have got a $\lambda F\mathcal{F}$-coalgebra by finding a term of type $\mathcal{H} \leq \mathcal{F}[F := \mathcal{H}]$:

$$s := \Lambda\beta\lambda x^{\mathcal{H}\beta}.x(\mathcal{F}[F := \mathcal{H}]\beta)(\Lambda\alpha\lambda f^{\alpha\to\beta}\lambda z^{(\nu F\mathcal{F})\alpha}t)$$

where we used $t := m\mathsf{L}\mathcal{H}n\alpha\beta f\Big(m\mathsf{R}(\nu F\mathcal{F})\mathcal{H}\ell\mathsf{R}n\alpha(z\mathsf{E}_\nu m)\Big) : \mathcal{F}[F := \mathcal{H}]\beta$. (Note that $z\mathsf{E}_\nu m$ has type $\mathcal{F}[F := \nu F\mathcal{F}]\alpha$ and that the idea for finding t is essentially the same as in the proof of $\mu F\mathcal{F}$'s monotonicity.) The proof is finished with

$$\Lambda\alpha\Lambda\beta\lambda f^{\alpha\to\beta}\lambda x^{(\nu F\mathcal{F})\alpha}.\mathsf{C}_{\nu F\mathcal{F}}ns(\Lambda\gamma\lambda y^{\forall\alpha.(\alpha\to\beta)\to(\nu F\mathcal{F})\alpha\to\gamma}.y\alpha fx) : \nu F\mathcal{F}\text{ mon}$$

[9] Even $\lambda F\mathcal{F}$ is only a parameter not to be analyzed: We could even provide an inhabitant of $\forall X^{(*\Rightarrow*)\Rightarrow*\Rightarrow*}.X\text{ mon} \to (\mu F.XF)\text{ mon} \times (\nu F.XF)\text{ mon}$.

[10] The lattice-theoretic version of $\mathcal{G}$ is called the lower monotonization of an operator in [10, p.83]. Its dual (the "upper monotonization" of an operator), when written in type theory, gives $\mathcal{H}$. Both play a central role in understanding inductive and coinductive types à la Mendler, see [11] for the definition, [15] for the explanation and [10] for applications to monotone inductive types.

which is again a bit weird due to the impredicative encoding of the existential quantifier. □

Notice that for the proofs it has been essential to be free to decide whether to provide monotonicity for the first or the second constructor argument when using monotonicity of $\lambda F\mathcal{F}$. In the proof, the other argument has always been that for which we tried to prove monotonicity! Obviously, one cannot require both arguments to be monotone and get the proof through. Neither can we dispense with monotonicity of the arguments altogether since we would not get, e. g., a witness for $\lambda F\lambda\alpha.F(F\alpha)$ mon and hence arrive at too narrow a concept. If we required monotonicity only of the first argument, then the rule of iteration would have to refer to the monotonicity of $\mu F\mathcal{F}$. Although our monotonicity proof would still go through, this considerably complex term would enormously blow up the terms during iteration (not to speak about the deviation from the wanted behaviour). On the contrary, if we required monotonicity only of the second argument, iteration would not have to be changed, but there would be no way to recover the proof of monotonicity for $\mu F\mathcal{F}$. Moreover, the situation for coiteration would always be the other way round, and it would hardly be acceptable if different notions of monotonicity were required for the treatment of inductive and coinductive constructors.

The categorial perspective did not at all suggest our definition of monotonicity for rank 2. Categorially, one would consider the category of endofunctors (see [1] for a recent treatment of rank 2 in this framework) and would never speak about other entities whereas we also make requirements for non-monotone arguments. And, of course, monotonicity is only a part of functoriality, since "coherence" is not dealt with: functor laws cannot be expressed in our type system.

5 MICC^2 Embeds into System F^ω

It is well-known that iteration and coiteration may be encoded impredicatively, and that this can be done such that strong normalization of the system with iteration and coiteration can be inferred from that of the impredicative system, see e. g. [5, 15]. This also worked easily for monotone inductive types [13, 10]. Hence, it does not come as a surprise that MICC^2 embeds into F^ω.

Theorem 2. *There is an embedding* $-'$ *of* MICC^2 *into* F^ω*, i. e., for every constructor* $\mathcal{X}$ *of kind* κ *in* MICC^2*, there is a constructor* $\mathcal{X}'$ *of kind* κ *in* F^ω*, and for every term* r *of type* ρ *in* MICC^2*, there is a term* r' *of type* ρ' *in* F^ω *such that whenever* $r \rhd s$ *in* MICC^2*, then* $r' \rhd^+ s'$ *in* F^ω *(every step in* MICC^2 *is translated into at least one step in* F^ω*).*

Proof. The definition is by iteration on constructors and by iteration on terms, respectively. We only consider the most interesting non-homomorphic clauses. Note that only (co-)inductive constructors are affected and that we therefore have $(\mathcal{F} \leq \mathcal{G})' = \mathcal{F}' \leq \mathcal{G}'$ and $(\mathcal{X}\,\mathrm{mon})' = \mathcal{X}'\,\mathrm{mon}$. It will also be clear that $(\mathcal{X}[X := \mathcal{Y}])' = \mathcal{X}'[X := \mathcal{Y}']$, $(r[X := \mathcal{Y}])' = r'[X := \mathcal{Y}']$ and $(r[x^\rho := s])' = r'[x^{\rho'} := s']$. Define

$$
\begin{aligned}
(\mu\alpha\rho)' &:= \forall\alpha.(\rho' \to \alpha) \to \alpha\\
(r\mathsf{E}_\mu s)' &:= r'\sigma' s'\\
(\mathsf{C}_{\mu\alpha\rho} m t)' &:= \Lambda\alpha\lambda x^{\rho'\to\alpha}.x\Big(m'(\mu\alpha\rho)'\alpha(\lambda z^{(\mu\alpha\rho)'}.z\alpha x)t'\Big)\\
(\mu F\mathcal{F})' &:= \lambda\alpha\forall F.F\,\mathrm{mon} \to \mathcal{F}' \le F \to F\alpha\\
(r\mathsf{E}_\mu n s)' &:= r'\mathcal{G}'n's'\\
(\mathsf{C}_{\mu F\mathcal{F}} m t)' &:= \Lambda F\lambda y^{F\,\mathrm{mon}}\lambda x^{\mathcal{F}'\le F}.x\rho'\Big(m'\mathsf{R}(\mu F\mathcal{F})'F(\Lambda\alpha\lambda z^{(\mu F\mathcal{F})'\alpha}.zFyx)\mathsf{R}y\rho' t'\Big)
\end{aligned}
$$

We do not set $(\nu\alpha\rho)'$ to $\exists\alpha.(\alpha \to \rho') \times \alpha$ but avoid the existential type by the following definitions:

$$
\begin{aligned}
(\nu\alpha\rho)' &:= \forall\beta.(\forall\alpha.(\alpha \to \rho') \to \alpha \to \beta) \to \beta\\
(\mathsf{C}_{\nu\alpha\rho} s t)' &:= \Lambda\beta\lambda z^{\forall\alpha.(\alpha\to\rho')\to\alpha\to\beta}.z\sigma' s' t'\\
(r\mathsf{E}_\nu m)' &:= r'(\rho[\alpha := \nu\alpha\rho])'\Big(\Lambda\alpha\lambda z_1^{\alpha\to\rho'}\lambda z_2^{\alpha}.m'\alpha(\nu\alpha\rho)'\\
&\qquad (\lambda x^{\alpha}\Lambda\beta\lambda z^{\forall\alpha.(\alpha\to\rho')\to\alpha\to\beta}.z\alpha z_1 x)(z_1 z_2)\Big)\\
(\nu F\mathcal{F})' &:= \lambda\alpha\forall\beta.(\forall F.F\,\mathrm{mon} \to F \le \mathcal{F}' \to F\alpha \to \beta) \to \beta\\
(\mathsf{C}_{\nu F\mathcal{F}} n s t)' &:= \Lambda\beta\lambda z^{\forall F.F\,\mathrm{mon}\to F\le\mathcal{F}'\to F\rho'\to\beta}.z\mathcal{G}'n's't'\\
(r\mathsf{E}_\nu m)' &:= r'(\mathcal{F}[F := \nu F\mathcal{F}]\rho)'(\Lambda F\lambda y^{F\,\mathrm{mon}}\lambda z_1^{F\le\mathcal{F}'}\lambda z_2^{F\rho'}.m'\mathsf{R}F(\nu F\mathcal{F})'\\
&\qquad \Big(\Lambda\alpha\lambda x^{F\alpha}\Lambda\beta\lambda z^{\forall F.F\,\mathrm{mon}\to F\le\mathcal{F}'\to F\alpha\to\beta}.zFyz_1x)\mathsf{L}y\rho'(z_1\rho' z_2)\Big)
\end{aligned}
$$

It is not too hard to check the translation of steps of MICC^2 into those of F^ω. □

Corollary 1. *System* MICC^2 *is strongly normalizing, i. e.,* $\rhd$ *is well-founded.*

Proof. It is well-known that F^ω is strongly normalizing ([7] only considered weak normalization although his proof method copes with strong normalization as well) and, certainly, strong normalization is inherited via embeddings. □

In this impredicative encoding, the previous section's concern becomes irrelevant since $(\mu F\mathcal{F})'$ and $(\nu F\mathcal{F})'$ are always monotone regardless of monotonicity of $\lambda F\mathcal{F}$. In some sense, α occurs only strictly positively in the kernel of $(\mu F\mathcal{F})'$ and only non-strictly positively in the kernel of $(\nu F\mathcal{F})'$. In fact, it easily follows from theorem 3 below that there are closed terms of type $(\mu F\mathcal{F})'\,\mathrm{mon}$ and of type $(\nu F\mathcal{F})'\,\mathrm{mon}$. In the general lattice-theoretic situation (that of Tarski's fixed-point theorem), this phenomenon also occurs: The pointwise infimum of all the pre-fixed-points of an operator on a lattice of monotone set-theoretic functions is automatically monotone.

6 Extension by Primitive (Co-)Recursion

System MICC^2 may now easily be extended by the concept of primitive recursion which is extremely unlikely to have a reasonable embedding into system F^ω [14]. We just have to add another elimination rule for $\mu F\mathcal{F}$ and a new β-reduction rule of primitive recursion. We abbreviate $\mathcal{F} \times \mathcal{G} := \lambda\alpha.\mathcal{F}\alpha \times \mathcal{G}\alpha$.

$$
\frac{r : (\mu F\mathcal{F})\rho \qquad s : \mathcal{F}[F := \mu F\mathcal{F} \times \mathcal{G}] \le \mathcal{G}}{r\mathsf{E}_\mu^+ s : \mathcal{G}\rho}\,(\mu\text{-}E^+)
$$

For $s : \mathcal{F}[F := \mu F\mathcal{F} \times \mathcal{G}] \leq \mathcal{G}$ define $\mathcal{E}^+_\mu s := \Lambda\alpha\lambda x^{(\mu F\mathcal{F})\alpha}.x\mathsf{E}^+_\mu s : \mu F\mathcal{F} \leq \mathcal{G}$.

The β-rule of *primitive recursion* for rank 2 is

$$(\mathsf{C}_{\mu F\mathcal{F}}mt)\mathsf{E}^+_\mu s$$
$$\rhd s\rho\Big(m\mathsf{R}(\mu F\mathcal{F})(\mu F\mathcal{F} \times \mathcal{G})(\Lambda\alpha\lambda x^{(\mu F\mathcal{F})\alpha}.\langle x, (\mathcal{E}^+_\mu s)\alpha x\rangle)\mathsf{L}(M_{\mu F\mathcal{F}}m)\rho t\Big)$$

Notice that we no longer require $\mathcal{G}$ mon since we anyway have to refer to $\mu F\mathcal{F}$'s monotonicity as expressed by $M_{\mu F\mathcal{F}}$.

The respective treatment for rank 1 in [10] is now obvious:

$$\frac{r : \mu\alpha\rho \qquad s : \rho[\alpha := \mu\alpha\rho \times \sigma] \to \sigma}{r\mathsf{E}^+_\mu s : \sigma}\,(\mu\text{–}E^+)$$

For $s : \rho[\alpha := \sigma] \to \sigma$ define $\mathcal{E}^+_\mu s := \lambda x^{\mu\alpha\rho}.x\mathsf{E}^+_\mu s : \mu\alpha\rho \to \sigma$. The β-rule of primitive recursion of rank 1 is

$$(\mathsf{C}_{\mu\alpha\rho}mt)\mathsf{E}^+_\mu s \rhd s\Big(m(\mu\alpha\rho)(\mu\alpha\rho \times \sigma)(\lambda x^{\mu\alpha\rho}.\langle x, (\mathcal{E}^+_\mu s)x\rangle)t\Big)$$

In [10] it is shown that this schema may be embedded into that of primitive recursion only for non-strictly positive inductive types. Also, this primitive recursion captures the intuitive notion of primitive recursion. In the rank 2 case, the author has no knowlegde of interesting examples. On the contrary, its dualization to corecursion would be useful in defining substitution for non-wellfounded terms involving binding. For lack of space, the definition of corecursion of rank 2 is left out.

7 Proven Monotone (Co-)Inductive Constructors

There is a wealth of constructors having monotonicity witnesses. Below a list of closure properties will be given. It turns out, however, that this wealth cannot be grasped without the companion notion of anti-monotonicity. Somewhat surprising, we only need it for rank 1. Set $\mathcal{F}\,\mathrm{mon}^- := \forall\alpha\forall\beta.(\alpha \to \beta) \to \mathcal{F}\beta \to \mathcal{F}\alpha$. Aiming at a concise description, let p always range over the set of *polarities* $\{+, -\}$, set $-- := +$ and $-+ := -$ and $\mathcal{F}\,\mathrm{mon}^+ := \mathcal{F}\,\mathrm{mon}$. Hence $\mathcal{F}\,\mathrm{mon}^{\mathrm{p}}$ and $\mathcal{F}\,\mathrm{mon}^{-\mathrm{p}}$ are always types.

Theorem 3. *There are closed terms of the following types:*

1. $\lambda\alpha\rho\,\mathrm{mon}^{\mathrm{p}}$ *if* α *does not occur free in* ρ
2. $\lambda\alpha\alpha\,\mathrm{mon}^+$
3. $\lambda\alpha\rho\,\mathrm{mon}^{\mathrm{p}} \to \lambda\alpha\sigma\,\mathrm{mon}^{\mathrm{p}} \to \lambda\alpha.\rho \times \sigma\,\mathrm{mon}^{\mathrm{p}}$
4. $\lambda\alpha\rho\,\mathrm{mon}^{\mathrm{p}} \to \lambda\alpha\sigma\,\mathrm{mon}^{\mathrm{p}} \to \lambda\alpha.\rho + \sigma\,\mathrm{mon}^{\mathrm{p}}$
5. $\lambda\alpha\rho\,\mathrm{mon}^{-\mathrm{p}} \to \lambda\alpha\sigma\,\mathrm{mon}^{\mathrm{p}} \to \lambda\alpha.\rho \to \sigma\,\mathrm{mon}^{\mathrm{p}}$
6. $(\rho \to \lambda\alpha\sigma\,\mathrm{mon}^{\mathrm{p}}) \to \lambda\alpha.\rho \to \sigma\,\mathrm{mon}^{\mathrm{p}}$ *if* α *does not occur free in* ρ
7. $(\forall X^\kappa.\lambda\alpha\rho\,\mathrm{mon}^{\mathrm{p}}) \to \lambda\alpha\forall X^\kappa\rho\,\mathrm{mon}^{\mathrm{p}}$
8. $(\forall\alpha.\lambda\gamma\rho\,\mathrm{mon}^+) \to (\forall\gamma.\lambda\alpha\rho\,\mathrm{mon}^{\mathrm{p}}) \to \lambda\alpha\mu\gamma\rho\,\mathrm{mon}^{\mathrm{p}}$

9. $(\forall\alpha.\lambda\gamma\rho\,\mathrm{mon}^+) \to (\forall\gamma.\lambda\alpha\rho\,\mathrm{mon}^{\mathsf{p}}) \to \lambda\alpha\nu\gamma\rho\,\mathrm{mon}^{\mathsf{p}}$
10. $\lambda F\lambda\alpha\rho\,\mathrm{mon} \to \lambda\alpha\rho\,\mathrm{mon}^+$ *if* F *does not occur free in* $\lambda\alpha\rho$
11. $\lambda F\mathcal{F}\,\mathrm{mon} \to (\forall F.F\,\mathrm{mon} \to \forall\gamma(\lambda\alpha.\mathcal{F}\gamma\,\mathrm{mon})) \to \lambda\alpha.(\mu F\mathcal{F})\alpha\,\mathrm{mon}^+$
12. $\lambda F\mathcal{F}\,\mathrm{mon} \to (\forall F.F\,\mathrm{mon} \to \forall\gamma(\lambda\alpha.\mathcal{F}\gamma\,\mathrm{mon})) \to \lambda\alpha.(\nu F\mathcal{F})\alpha\,\mathrm{mon}^+$
13. $\lambda\alpha\rho\,\mathrm{mon}^+ \to \lambda F\lambda\alpha\rho\,\mathrm{mon}$ *if* F *does not occur free in* $\lambda\alpha\rho$
14. $\lambda FF\,\mathrm{mon}$
15. $\lambda F\mathcal{F}\,\mathrm{mon} \to \lambda F\mathcal{G}\,\mathrm{mon} \to \lambda F.\mathcal{F}\times\mathcal{G}\,\mathrm{mon}$
16. $\lambda F\mathcal{F}\,\mathrm{mon} \to \lambda F\mathcal{G}\,\mathrm{mon} \to \lambda F.\mathcal{F}+\mathcal{G}\,\mathrm{mon}$ *with* $\mathcal{F}+\mathcal{G} := \lambda\alpha.\mathcal{F}\alpha+\mathcal{G}\alpha$
17. $(\forall F.\lambda G\mathcal{F}\,\mathrm{mon}) \to (\forall G.\lambda F\mathcal{F}\,\mathrm{mon}) \to \lambda F\mu G\mathcal{F}\,\mathrm{mon}$
18. $(\forall F.\lambda G\mathcal{F}\,\mathrm{mon}) \to (\forall G.\lambda F\mathcal{F}\,\mathrm{mon}) \to \lambda F\nu G\mathcal{F}\,\mathrm{mon}$
19. $\lambda F\mathcal{F}\,\mathrm{mon} \to \lambda F\mathcal{G}\,\mathrm{mon} \to \lambda F\lambda\alpha.\mathcal{F}(\mathcal{G}\alpha)\,\mathrm{mon}$

An informal claim is the "canonicity" of all those proofs as long as their goal does not degenerate to be covered by 1 or 13.[11]

Proof. The first 9 properties are quite common since they (except for number 6) form the basis of the definition of canonical monotonicity witnesses for positive inductive and coinductive types (see e. g. [9, 8, 5, 10]).

Rules 11 and 12 are proved by first establishing $\forall\gamma(\lambda\alpha.(\mu F\mathcal{F})\gamma\,\mathrm{mon})$ and $\forall\gamma(\lambda\alpha.(\nu F\mathcal{F})\gamma\,\mathrm{mon})$, using the assumptions. For this to work, one has to make use of theorem 1.

The other rules are more or less adaptations of those for rank 1 to rank 2.

The interesting new case is 19. We assume $m_1 : \lambda F\mathcal{F}\,\mathrm{mon}$, $m_2 : \lambda F\mathcal{G}\,\mathrm{mon}$ and show $\lambda F\lambda\alpha.\mathcal{F}(\mathcal{G}\alpha)\,\mathrm{mon}$. Clearly, the proof will be a pair. Its first component having type $\forall F.F\,\mathrm{mon} \to \lambda\alpha.\mathcal{F}(\mathcal{G}\alpha)\,\mathrm{mon}$ is

$$\Lambda F\lambda x^{F\,\mathrm{mon}}\Lambda\alpha\Lambda\beta\lambda f^{\alpha\to\beta}.m_1\mathsf{L}Fx(\mathcal{G}\alpha)(\mathcal{G}\beta)(m_2\mathsf{L}Fx\alpha\beta f)$$

We abbreviate $\mathcal{F}' := \mathcal{F}[F := G]$ and $\mathcal{G}' := \mathcal{G}[F := G]$. Then the second component is given by $\Lambda F\Lambda G\lambda x^{F\leq G}.\langle\lambda y^{F\,\mathrm{mon}}\Lambda\alpha.t_1, \lambda y^{G\,\mathrm{mon}}\Lambda\alpha.t_2\rangle$ with t_1 and t_2 of type $\mathcal{F}(\mathcal{G}\alpha) \to \mathcal{F}'(\mathcal{G}'\alpha)$, defined by

$$t_1 := \lambda z^{\mathcal{F}(\mathcal{G}\alpha)}.m_1\mathsf{R}FGx\mathsf{L}y(\mathcal{G}'\alpha)\Big(m_1\mathsf{L}Fy(\mathcal{G}\alpha)(\mathcal{G}'\alpha)(m_2\mathsf{R}FGx\mathsf{L}y\alpha)z\Big)$$

$$t_2 := \lambda z^{\mathcal{F}(\mathcal{G}\alpha)}.m_1\mathsf{L}Gy(\mathcal{G}\alpha)(\mathcal{G}'\alpha)(m_2\mathsf{R}FGx\mathsf{R}y\alpha)(m_1\mathsf{R}FGx\mathsf{R}y(\mathcal{G}\alpha)z)$$

Their ideas are as follows: In t_1 we first show $\mathcal{G}\alpha \to \mathcal{G}'\alpha$, then lift this to $\mathcal{F}(\mathcal{G}\alpha) \to \mathcal{F}(\mathcal{G}'\alpha)$. We compose this with $\mathcal{F}(\mathcal{G}'\alpha) \to \mathcal{F}'(\mathcal{G}'\alpha)$. In this case, $\mathcal{F}'\,\mathrm{mon}$ is not available. In t_2, we show $\mathcal{F}(\mathcal{G}\alpha) \to \mathcal{F}'(\mathcal{G}\alpha)$, then we lift $\mathcal{G}\alpha \to \mathcal{G}'\alpha$ to $\mathcal{F}'(\mathcal{G}\alpha) \to \mathcal{F}'(\mathcal{G}'\alpha)$ and finally compose. Here, $\mathcal{F}\,\mathrm{mon}$ is not available. This shows that although two different strategies for proving $\mathcal{F}(\mathcal{G}\alpha) \to \mathcal{F}'(\mathcal{G}'\alpha)$ exist, we have no choice when producing t_1 and t_2. □

[11] This could be formulated slightly better by referring to universally quantified types such as $\forall X^{(*\Rightarrow*)\Rightarrow*\Rightarrow*}\forall Y^{(*\Rightarrow*)\Rightarrow*\Rightarrow*}.X\,\mathrm{mon} \to Y\,\mathrm{mon} \to \lambda F\lambda\alpha.XF(YF\alpha)\,\mathrm{mon}$ instead of 19.

8 Examples and Future Work

We start with an example not to be found from theorem 3 alone: There is a witness for $\lambda F \lambda \beta. \Big(\lambda \alpha.(((\alpha \to \gamma) \to \alpha) \to \alpha) \to \alpha\Big)(F(F\beta))$ mon. This follows from theorem 3 as soon as we give a monotonicity witness for the constructor $\lambda \alpha.(((\alpha \to \gamma) \to \alpha) \to \alpha) \to \alpha$, proposed by U. Berger as the first example of a non-positive and monotone constructor of rank 1: The last but one occurrence of α is at a negative position. Nevertheless, it is not hard to find a monotonicity witness. For the examples to follow, set $0 := \mu\alpha\alpha$ and $1 := \nu\alpha\alpha$ which are proven monotone.

Untyped λ-calculus may be represented with de Bruijn indices, and this can be reflected in the constructor $\mathcal{F} := \mu F \lambda \alpha.\alpha + F(1 + \alpha) + (F\alpha \times F\alpha)$. $\mathcal{F}\rho$ would be the type of untyped λ-terms having free variables taken from ρ. Clearly, theorem 3 implies that $\mathcal{F}$ is a proven monotone inductive constructor. By iteration on $\mathcal{F}$, substitution can be defined. The result is the same as that of the structurally inductive approach in [2]. The representation is also studied in [3] but with the concept of generalized folds. [3] also exhibits a variation, called an "extension of de Bruijn's notation", which in our notation reads $\mu F \lambda \alpha.\alpha + F(1 + F\alpha) + (F\alpha \times F\alpha)$. No doubt, theorem 3 shows that this is a proven monotone inductive constructor. Note the nesting of F which clearly requires the relativization to monotone F in the definition of rank 2 monotonicity. Again, substitution can be easily defined by rank 2 iteration. Another proven monotone variation would be $\mu F \lambda \alpha.\alpha + F(1+\alpha) + (F\alpha \times F\alpha) + ((F\alpha \to 0) \to 0)$ which would add a kind of continuation terms to the untyped λ-calculus.

As an example of the main result in [1], functional programs for the computation of the isomorphisms between $(\mu\alpha.1 + \alpha \times \alpha) \to \sigma$ and $(\nu F \lambda \alpha.\alpha \times F(F\alpha))\sigma$ are given. Obviously, both $\mu\alpha.1 + \alpha \times \alpha$ and $\nu F \lambda \alpha.\alpha \times F(F\alpha)$ are proven monotone by theorem 3. By iteration and coiteration, the "isomorphisms" can be established and have as reduction behaviour that described in the program in [1] which is based on pattern-matching. While [1] establishes that those are in fact isomorphisms w. r. t. semantic equality we cannot expect such a result for our intensional setting. However, since MICC^2 is strongly normalizing, those functional programs are guaranteed to terminate.

Notice that these examples did not use the possibility of interleaved μ and ν. However, the author still does not know useful examples for rank 2 using interleaving.

Open questions: We are still in need of a good notion of positivity for rank 2. Can we expect monotonicity witnesses for positive constructors to be "unique"? It would be interesting to study course-of-value iteration for rank 2 or the extension of other concepts in [15] to rank 2. The relation to [3]'s generalized folds should be studied carefully. Can their behaviour be simulated in MICC^2? This is easy to see in the situation of [3] but not for the general approach to generalized folds [4]. Can one find data structures which need deeper nested inductive constructors than the example of square matrices in [12]? Is there a chance to get a similar clean view on constructors of higher rank than 2?

References

1. Thorsten Altenkirch. Representations of first order function types as terminal coalgebras. In Samson Abramsky, editor, *Proceedings of TLCA 2001*, volume 2044 of *Lecture Notes in Computer Science*, pages 8–21. Springer Verlag, 2001.
2. Thorsten Altenkirch and Bernhard Reus. Monadic presentations of lambda terms using generalized inductive types. In Jörg Flum and Mario Rodríguez-Artalejo, editors, *Computer Science Logic, 13th International Workshop, CSL '99, 8th Annual Conference of the EACSL, Madrid, Spain, September 20-25, 1999, Proceedings*, volume 1683 of *Lecture Notes in Computer Science*, pages 453–468. Springer Verlag, 1999.
3. Richard S. Bird and Ross Paterson. De Bruijn notation as a nested datatype. *Journal of Functional Programming*, 9(1):77–91, 1999.
4. Richard Bird and Ross Paterson. Generalised folds for nested datatypes. *Formal Aspects of Computing*, 11(2):200–222, 1999.
5. Herman Geuvers. Inductive and coinductive types with iteration and recursion. In Bengt Nordström, Kent Pettersson, and Gordon Plotkin, editors, *Proceedings of the 1992 Workshop on Types for Proofs and Programs, Båstad, Sweden, June 1992*, pages 193–217, 1992. Only published electronically. Available at `ftp://ftp.cs.chalmers.se/pub/cs-reports/baastad.92/proc.dvi.Z`.
6. Neil Ghani. $\beta\eta$-equality for coproducts. In Mariangiola Dezani-Ciancaglini and Gordon Plotkin, editors, *Proceedings of the Second International Conference on Typed Lambda Calculi and Applications (TLCA '95), Edinburgh, United Kingdom, April 1995*, volume 902 of *Lecture Notes in Computer Science*, pages 171–185. Springer Verlag, 1995.
7. Jean-Yves Girard. *Interprétation fonctionnelle et élimination des coupures dans l'arithmétique d'ordre supérieur.* Thèse de Doctorat d'État, Université de Paris VII, 1972.
8. Brian Howard. *Fixed Points and Extensionality in Typed Functional Programming Languages.* PhD thesis, Stanford University, 1992.
9. Daniel Leivant. Contracting proofs to programs. In Piergiorgio Odifreddi, editor, *Logic and Computer Science*, volume 31 of *APIC Studies in Data Processing*, pages 279–327. Academic Press, 1990.
10. Ralph Matthes. *Extensions of System F by Iteration and Primitive Recursion on Monotone Inductive Types.* Doktorarbeit (PhD thesis), University of Munich, 1998.
11. Nax P. Mendler. Recursive types and type constraints in second-order lambda calculus. In *Proceedings of the Second Annual IEEE Symposium on Logic in Computer Science, Ithaca, N.Y.*, pages 30–36. IEEE Computer Society Press, 1987.
12. Chris Okasaki. From fast exponentiation to square matrices: An adventure in types. In *Proceedings of the fourth ACM SIGPLAN International Conference on Functional Programming (ICFP '99), Paris, France, September 27-29, 1999*, volume 34 of *SIGPLAN Notices*, pages 28–35. ACM, 1999.
13. Christine Paulin-Mohring. *Définitions Inductives en Théorie des Types d'Ordre Supérieur.* Habilitation à diriger les recherches, ENS Lyon, 1996.
14. Zdzisław Spławski and Paweł Urzyczyn. Type Fixpoints: Iteration vs. Recursion. In *Proceedings of the fourth ACM SIGPLAN International Conference on Functional Programming (ICFP '99), Paris, France, September 27-29, 1999*, volume 34 of *SIGPLAN Notices*, pages 102–113. ACM, 1999.
15. Tarmo Uustalu and Varmo Vene. Least and greatest fixed points in intuitionistic natural deduction. *Theoretical Computer Science.* To appear.

Author Index

Lecture Notes in Computer Science

For information about Vols. 1–2065
please contact your bookseller or Springer-Verlag

Vol. 1982: S. Näher, D. Wagner (Eds.), Algorithm Engineering. Proceedings, 2000. VIII, 243 pages. 2001.

Vol. 1996: P.L. Lanzi, W. Stolzmann, S.W. Wilson (Eds.), Advances in Learning Classifier Systems. Proceedings, 2000. VIII, 273 pages. 2001. (Subseries LNAI).

Vol. 2066: O. Gascuel, M.-F. Sagot (Eds.), Computational Biology. Proceedings, 2000. X, 165 pages. 2001.

Vol. 2068: K.R. Dittrich, A. Geppert, M.C. Norrie (Eds.), Advanced Information Systems Engineering. Proceedings, 2001. XII, 484 pages. 2001.

Vol. 2069: C. Peters (Ed.), Cross-Language Information Retrieval and Evaluation. Proceedings, 2000. IX, 389 pages. 2001.

Vol. 2070: L. Monostori, J. Váncza, M. Ali (Eds.), Engineering of Intelligent Systems. Proceedings, 2001. XVIII, 951 pages. 2001. (Subseries LNAI).

Vol. 2071: R. Harper (Ed.), Types in Compilation. Proceedings, 2000. IX, 207 pages. 2001.

Vol. 2072: J. Lindskov Knudsen (Ed.), ECOOP 2001 – Object-Oriented Programming. Proceedings, 2001. XIII, 429 pages. 2001.

Vol. 2073: V.N. Alexandrov, J.J. Dongarra, B.A. Juliano, R.S. Renner, C.J.K. Tan (Eds.), Computational Science – ICCS 2001. Part I. Proceedings, 2001. XXVIII, 1306 pages. 2001.

Vol. 2074: V.N. Alexandrov, J.J. Dongarra, B.A. Juliano, R.S. Renner, C.J.K. Tan (Eds.), Computational Science – ICCS 2001. Part II. Proceedings, 2001. XXVIII, 1076 pages. 2001.

Vol. 2075: J.-M. Colom, M. Koutny (Eds.), Applications and Theory of Petri Nets 2001. Proceedings, 2001. XII, 403 pages. 2001.

Vol. 2076: F. Orejas, P.G. Spirakis, J. van Leeuwen (Eds.), Automata, Languages and Programming. Proceedings, 2001. XIV, 1083 pages. 2001.

Vol. 2077: V. Ambriola (Ed.), Software Process Technology. Proceedings, 2001. VIII, 247 pages. 2001.

Vol. 2078: R. Reed, J. Reed (Eds.), SDL 2001: Meeting UML. Proceedings, 2001. XI, 439 pages. 2001.

Vol. 2079: E. Burke, W. Erben (Eds.), Practice and Theory of Automated Timetabling III. Proceedings, 2001. XII, 359 pages. 2001.

Vol. 2080: D.W. Aha, I. Watson (Eds.), Case-Based Reasoning Research and Development. Proceedings, 2001. XII, 758 pages. 2001. (Subseries LNAI).

Vol. 2081: K. Aardal, B. Gerards (Eds.), Integer Programming and Combinatorial Optimization. Proceedings, 2001. XI, 423 pages. 2001.

Vol. 2082: M.F. Insana, R.M. Leahy (Eds.), Information Processing in Medical Imaging. Proceedings, 2001. XVI, 537 pages. 2001.

Vol. 2083: R. Goré, A. Leitsch, T. Nipkow (Eds.), Automated Reasoning. Proceedings, 2001. XV, 708 pages. 2001. (Subseries LNAI).

Vol. 2084: J. Mira, A. Prieto (Eds.), Connectionist Models of Neurons, Learning Processes, and Artificial Intelligence. Proceedings, 2001. Part I. XXVII, 836 pages. 2001.

Vol. 2085: J. Mira, A. Prieto (Eds.), Bio-Inspired Applications of Connectionism. Proceedings, 2001. Part II. XXVII, 848 pages. 2001.

Vol. 2086: M. Luck, V. Mařík, O. Štěpánková, R. Trappl (Eds.), Multi-Agent Systems and Applications. Proceedings, 2001. X, 437 pages. 2001. (Subseries LNAI).

Vol. 2087: G. Kern-Isberner, Conditionals in Nonmonotonic Reasoning and Belief Revision. X, 190 pages. 2001. (Subseries LNAI).

Vol. 2088: S. Yu, A. Păun (Eds.), Implementation and Application of Automata. Proceedings, 2000. XI, 343 pages. 2001.

Vol. 2089: A. Amir, G.M. Landau (Eds.), Combinatorial Pattern Matching. Proceedings, 2001. VIII, 273 pages. 2001.

Vol. 2090: E. Brinksma, H. Hermanns, J.-P. Katoen (Eds.), Lectures on Formal Methods and Performance Analysis. Proceedings, 2000. VII, 431 pages. 2001.

Vol. 2091: J. Bigun, F. Smeraldi (Eds.), Audio- and Video-Based Biometric Person Authentication. Proceedings, 2001. XIII, 374 pages. 2001.

Vol. 2092: L. Wolf, D. Hutchison, R. Steinmetz (Eds.), Quality of Service – IWQoS 2001. Proceedings, 2001. XII, 435 pages. 2001.

Vol. 2093: P. Lorenz (Ed.), Networking – ICN 2001. Proceedings, 2001. Part I. XXV, 843 pages. 2001.

Vol. 2094: P. Lorenz (Ed.), Networking – ICN 2001. Proceedings, 2001. Part II. XXV, 899 pages. 2001.

Vol. 2095: B. Schiele, G. Sagerer (Eds.), Computer Vision Systems. Proceedings, 2001. X, 313 pages. 2001.

Vol. 2096: J. Kittler, F. Roli (Eds.), Multiple Classifier Systems. Proceedings, 2001. XII, 456 pages. 2001.

Vol. 2097: B. Read (Ed.), Advances in Databases. Proceedings, 2001. X, 219 pages. 2001.

Vol. 2098: J. Akiyama, M. Kano, M. Urabe (Eds.), Discrete and Computational Geometry. Proceedings, 2000. XI, 381 pages. 2001.

Vol. 2099: P. de Groote, G. Morrill, C. Retoré (Eds.), Logical Aspects of Computational Linguistics. Proceedings, 2001. VIII, 311 pages. 2001. (Subseries LNAI).

Vol. 2100: R. Küsters, Non-Standard Inferences in Description Logocs. X, 250 pages. 2001. (Subseries LNAI).

Vol. 2101: S. Quaglini, P. Barahona, S. Andreassen (Eds.), Artificial Intelligence in Medicine. Proceedings, 2001. XIV, 469 pages. 2001. (Subseries LNAI).

Vol. 2102: G. Berry, H. Comon, A. Finkel (Eds.), Computer-Aided Verification. Proceedings, 2001. XIII, 520 pages. 2001.

Vol. 2103: M. Hannebauer, J. Wendler, E. Pagello (Eds.), Balancing Reactivity and Social Deliberation in Multi-Agent Systems. VIII, 237 pages. 2001. (Subseries LNAI).

Vol. 2104: R. Eigenmann, M.J. Voss (Eds.), OpenMP Shared Memory Parallel Programming. Proceedings, 2001. X, 185 pages. 2001.

Vol. 2105: W. Kim, T.-W. Ling, Y-J. Lee, S.-S. Park (Eds.), The Human Society and the Internet. Proceedings, 2001. XVI, 470 pages. 2001.

Vol. 2106: M. Kerckhove (Ed.), Scale-Space and Morphology in Computer Vision. Proceedings, 2001. XI, 435 pages. 2001.

Vol. 2107: F.T. Chong, C. Kozyrakis, M. Oskin (Eds.), Intelligent Memory Systems. Proceedings, 2000. VIII, 193 pages. 2001.

Vol. 2108: J. Wang (Ed.), Computing and Combinatorics. Proceedings, 2001. XIII, 602 pages. 2001.

Vol. 2109: M. Bauer, P.J. Gymtrasiewicz, J. Vassileva (Eds.), User Modelind 2001. Proceedings, 2001. XIII, 318 pages. 2001. (Subseries LNAI).

Vol. 2110: B. Hertzberger, A. Hoekstra, R. Williams (Eds.), High-Performance Computing and Networking. Proceedings, 2001. XVII, 733 pages. 2001.

Vol. 2111: D. Helmbold, B. Williamson (Eds.), Computational Learning Theory. Proceedings, 2001. IX, 631 pages. 2001. (Subseries LNAI).

Vol. 2116: V. Akman, P. Bouquet, R. Thomason, R.A. Young (Eds.), Modeling and Using Context. Proceedings, 2001. XII, 472 pages. 2001. (Subseries LNAI).

Vol. 2117: M. Beynon, C.L. Nehaniv, K. Dautenhahn (Eds.), Cognitive Technology: Instruments of Mind. Proceedings, 2001. XV, 522 pages. 2001. (Subseries LNAI).

Vol. 2118: X.S. Wang, G. Yu, H. Lu (Eds.), Advances in Web-Age Information Management. Proceedings, 2001. XV, 418 pages. 2001.

Vol. 2119: V. Varadharajan, Y. Mu (Eds.), Information Security and Privacy. Proceedings, 2001. XI, 522 pages. 2001.

Vol. 2120: H.S. Delugach, G. Stumme (Eds.), Conceptual Structures: Broadening the Base. Proceedings, 2001. X, 377 pages. 2001. (Subseries LNAI).

Vol. 2121: C.S. Jensen, M. Schneider, B. Seeger, V.J. Tsotras (Eds.), Advances in Spatial and Temporal Databases. Proceedings, 2001. XI, 543 pages. 2001.

Vol. 2123: P. Perner (Ed.), Machine Learning and Data Mining in Pattern Recognition. Proceedings, 2001. XI, 363 pages. 2001. (Subseries LNAI).

Vol. 2124: W. Skarbek (Ed.), Computer Analysis of Images and Patterns. Proceedings, 2001. XV, 743 pages. 2001.

Vol. 2125: F. Dehne, J.-R. Sack, R. Tamassia (Eds.), Algorithms and Data Structures. Proceedings, 2001. XII, 484 pages. 2001.

Vol. 2126: P. Cousot (Ed.), Static Analysis. Proceedings, 2001. XI, 439 pages. 2001.

Vol. 2129: M. Goemans, K. Jansen, J.D.P. Rolim, L. Trevisan (Eds.), Approximation, Randomization, and Combinatorial Optimization. Proceedings, 2001. IX, 297 pages. 2001.

Vol. 2130: G. Dorffner, H. Bischof, K. Hornik (Eds.), Artificial Neural Networks – ICANN 2001. Proceedings, 2001. XXII, 1259 pages. 2001.

Vol. 2132: S.-T. Yuan, M. Yokoo (Eds.), Intelligent Agents. Specification. Modeling, and Application. Proceedings, 2001. X, 237 pages. 2001. (Subseries LNAI).

Vol. 2136: J. Sgall, A. Pultr, P. Kolman (Eds.), Mathematical Foundations of Computer Science 2001. Proceedings, 2001. XII, 716 pages. 2001.

Vol. 2138: R. Freivalds (Ed.), Fundamentals of Computation Theory. Proceedings, 2001. XIII, 542 pages. 2001.

Vol. 2139: J. Kilian (Ed.), Advances in Cryptology – CRYPTO 2001. Proceedings, 2001. XI, 599 pages. 2001.

Vol. 2141: G.S. Brodal, D. Frigioni, A. Marchetti-Spaccamela (Eds.), Algorithm Engineering. Proceedings, 2001. X, 199 pages. 2001.

Vol. 2142: L. Fribourg (Ed.), Computer Science Logic. Proceedings, 2001. XII, 615 pages. 2001.

Vol. 2143: S. Benferhat, P. Besnard (Eds.), Symbolic and Quantitative Approaches to Reasoning with Uncertainty. Proceedings, 2001. XIV, 818 pages. 2001. (Subseries LNAI).

Vol. 2146: J.H. Silverman (Eds.), Cryptography and Lattices. Proceedings, 2001. VII, 219 pages. 2001.

Vol. 2147: G. Brebner, R. Woods (Eds.), Field-Programmable Logic and Applications. Proceedings, 2001. XV, 665 pages. 2001.

Vol. 2149: O. Gascuel, B.M.E. Moret (Eds.), Algorithms in Bioinformatics. Proceedings, 2001. X, 307 pages. 2001.

Vol. 2150: R. Sakellariou, J. Keane, J. Gurd, L. Freeman (Eds.), Euro-Par 2001 Parallel Processing. Proceedings, 2001. XXX, 943 pages. 2001.

Vol. 2152: R.J. Boulton, P.B. Jackson (Eds.), Theorem Proving in Higher Order Logics. Proceedings, 2001. X, 395 pages. 2001.

Vol. 2154: K.G. Larsen, M. Nielsen (Eds.), CONCUR 2001 – Concurrency Theory. Proceedings, 2001. XI, 583 pages. 2001.

Vol. 2157: C. Rouveirol, M. Sebag (Eds.), Inductive Logic Programming. Proceedings, 2001. X, 261 pages. 2001. (Subseries LNAI).

Vol. 2161: F. Meyer auf der Heide (Ed.), Algorithms – ESA 2001. Proceedings, 2001. XII, 538 pages. 2001.

Vol. 2162: Ç. K. Koç, D. Naccache, C. Paar (Eds.), Cryptographic Hardware and Embedded Systems – CHES 2001. Proceedings, 2001. XIV, 411 pages. 2001.

Vol. 2164: S. Pierre, R. Glitho (Eds.), Mobile Agents for Telecommunication Applications. Proceedings, 2001. XI, 292 pages. 2001.

Vol. 2166: V. Matoušek, P. Mautner, R. Mouček, K. Taušer (Eds.), Text, Speech and Dialogue. Proceedings, 2001. XIII, 452 pages. 2001. (Subseries LNAI).